CRYOCOOLERS 19

A publication of the International Cryocooler Conference

CRYOCOOLERS 19

Edited by

Saul D. Miller

and

Ronald G. Ross, Jr.

ICC Press

Boulder, Colorado

Proceedings of the 19th International Cryocooler Conference, held June 20-23, 2016
in San Diego, California

Library of Congress Control Number: 2016950156

ISBN 978-1-934021-05-7

ISSN: 1549-1757

Printed in the United States of America

www.cryocooler.org

Preface

The objective of *Cryocoolers 19* is to archive the latest developments and performance measurements in the field of cryocoolers by drawing upon the work of leading international experts. In particular, this book is based on the 77 peer reviewed manuscripts that were prepared for the 19th International Cryocooler Conference held in San Diego, California on June 20-23, 2016. Although this is the nineteenth meeting of the conference, which has met every two years since 1980, the authors' works have only been made available to the public in hardcover book form since *Cryocoolers 8* in 1994.

Starting with *Cryocoolers 14*, we began publishing the series of *Cryocoolers X* books in-house at ICC Press using all-electronic manuscripts and digital printing. This has allowed us to also include a CD of the book's contents, in color, within the back cover of each book. Also, consistent with the trend toward instant electronic access to important technical works, color PDFs of each contribution in the book are available over the internet via the "past proceedings" link on the ICC web site http://www.cryocooler.org.

Because the book's content is designed for users of cryocoolers as much as for developers of cryocoolers, extra effort has been made to provide a full-text search capability on the CD and web site. Within the book itself, a thorough Subject Index covers the referenced cryocoolers by type and manufacturer's name, as well as by the scientific or engineering subject matter. Contributing organizations are also listed in the Subject Index to assist in finding the work of a known institution, laboratory, or manufacturer. To aide those attempting to locate a particular contributor's work, a separate Author Index is provided, listing all authors and coauthors.

Prior to 1994, proceedings of the International Cryocooler Conference were published as informal reports by the particular government organization sponsoring the conference — typically a different organization for each conference. A listing of previous conference proceedings is presented in the Proceedings Index, at the rear of this book. Most of the previous proceedings were printed in limited quantity and are out of print at this time. However, digital PDF copies of these past proceedings are available via the "past proceedings" link on the ICC web site http://www.cryocooler.org.

The contents of *Cryocoolers 19* is organized into 13 chapters, covering the various types of cryocoolers and their applications. At the beginning is a chapter on multistage cryocoolers to provide low-temperature cooling in the 2-10 K temperature range for space and military applications. This is followed by a chapter on two-stage Stirling, pulse tube and hybrid Stirling/PT coolers designed to provide cooling in the 10 to 80 K range. Single-stage Stirling and Pulse Tube coolers for the common 50-150 K temperature range are covered next in two chapters covering both very small and medium to large coolers. Following them are chapters on regenerative-cooler modeling and performance investigations, and on associated regenerator, compressor, and drive electronics research.

Following the chapters on regenerative cryocooler technologies, are two chapters covering recuperative cryocoolers, including Joule-Thomson, Sorption, and Brayton cryocoolers. These are followed by a chapter on unique sub-Kelvin, magnetic, and novel refrigerators. Applications requiring sub-Kelvin temperatures include space bolometers and x-ray sensors, and ground-based sensors in materials research, nuclear research, quantum materials research, quantum information technology, metrology, astronomy, and scanning tunneling microscopy (STM).

The last two chapters of the book deal with integration technologies such as use in liquefaction and zero-boil-off systems, lessons learned in representative applications, and product reliability studies. These articles contain particularly useful information for the potential user of cryocoolers as well as for the developer.

In reviewing the contributions contained in *Cryocoolers 19*, we note the continued strong interest in the development of pulse tube cryocoolers for a growing variety of long-life, high-reliability cryocooler applications. Pulse tube coolers can be driven by several competing compressor technologies. One class of pulse tube coolers is referred to as "Stirling type" because they are based on the linear Oxford Stirling-cooler type compressor; these generally provide cooling in the 10 to 100 K temperature range and operate at frequencies from 30 to 100 Hz. A second type of pulse tube cooler is the so-called "Gifford-McMahon or GM-type." Pulse tube coolers of this type use a GM-type compressor and lower frequency operation (~1 Hz) to achieve temperatures in the 2 to 10 K temperature range.

Also of note, is the significant number of very small (micro coolers) and very large (multi-kilowatt coolers) entering the marketplace...these expand the available operating capacity range to over three orders of magnitude!

Example applications of cryocoolers include sensors for tiny CubeSat satellites, space infrared sensors for large space instruments, precooling for cryogen-free sub-Kelvin applications, cooling of HTS and LTS superconducting magnets and electronics, and helium and hydrogen liquefaction and control of cryogen boil-off.

In summary, it is hoped that this book will serve as a valuable source of reference to all those faced with the challenges of taking advantage of the enabling physics of cryogenics temperatures. The expanding availability of low-cost, reliable cryocoolers continues to enable major advances in a number of fields.

The Editors

Acknowledgments

The International Cryocooler Conference Board wishes to express its appreciation to the Conference Organizing Committee, whose members dedicated many hours to organizing and managing the conduct of the 19th ICC Conference. In particular, we wish to express our deepest appreciation to the NASA Jet Propulsion Laboratory for their extensive contributions managing the conference's day-to-day operation and preparations. Members of the conference Organizing Committee and Board included:

CONFERENCE CHAIR
- Dean Johnson, *CHAIR, Jet Propulsion Lab*

CONFERENCE CO-CHAIRS
- Jose Rodriguez, *CO-CHAIR, Jet Propulsion Lab*
- Sidney Yuan, *CO-CHAIR, Aerospace Corp.*

TREASURER
- Ray Radebaugh, *NIST*

CO-EDITORS
- Saul Miller, *Retired*
- Ron Ross, *Jet Propulsion Lab*

PROGRAM (CHAIR & DEPUTY)
- Carl Kirkconnell, *West Coast Solutions*
- Mark Zagarola, *Creare*

ICC BOARD
- Paul Bailey, *Oxford Univ., UK*
- Peter Bradley, *NIST*
- Martin Crook, *STFC, UK*
- Rich Dausman, *Cryomech, Inc.*
- Lionel Duband, *CEA/SBT, France*
- Zhihua Gan, *Zhejiang Univ., China*
- Dean Johnson, *Jet Propulsion Lab*
- Ali Kashani, *Atlas Scientific*
- Takenor Numazawa, *NIMS, Japan*
- Jeff Olson, *Lockheed Martin*
- Limin Qui, *Zhejiang Univ., China*
- Jeff Raab, *NGAS, Retired*
- Ray Radebaugh, *NIST*
- Ron Ross, *Jet Propulsion Lab*
- Thierry Trollier, *Absolut Sys., France*
- Mark Zagarola, *Creare*

In addition to the above Organizing Committee members, a number of people deserve special mention for their extensive efforts in support of the conference. In particular, the Board is grateful to Nga Vu-Lintag, Yvonne Chen, Denise Gutierrez and Wing-Sze Liu of JPL for their planning and successful handling of the on-site registration and publications duties throughout the conference, and for providing the on-site signage. We are indebted to Kirsten Maynard (JPL) for preparing the Program and Abstract Book; to Ian McKinley (JPL) for developing the speaker timer to keep us all on time; and to Steve Staso who has maintained our Cryocooler.org website for the past three conferences. We also sincerely thank Richard Bachman and the many other sales and catering staff at the Town & Country Resort and Convention Center for coordinating all the conference activities and refreshments; and a special thank you to both Corinne Youngholm and Jamie Klages of Arrangements Unlimited for the coordination and execution of a most enjoyable banquet event held at the San Diego Natural History Museum. Finally, thanks go to Jeff Raab and Mark Michaelian, who as interns to the proceedings editing activity, edited a number of submitted manuscripts in support of Saul Miller and Ron Ross.

Contents

Micro & Miniature 50-200K Single-Stage Coolers 65

Medium to High Capacity 50-150 K 1-Stage Coolers 141

Sub-Kelvin, Magnetic, and Novel Refrigerators 479

Cryocooler Integration and Application Lessons 505

Cryocooler Durability Investigations 613

Proceedings Index 647

Mid InfraRed Instrument (MIRI) Cooler Compressor Assembly Characterization

M. Petach, M. Michaelian, T. Nguyen, R. Colbert and J. Mullin

Northrop Grumman Aerospace Systems
Redondo Beach, CA 90278 USA

ABSTRACT

The Cooler Subsystem for the Mid InfraRed Instrument (MIRI) of the James Webb Space Telescope (JWST) utilizes a remotely mounted subassembly containing a compressor and gas precooler to provide pressurized and precooled gas to the Cold Head Assembly as the final cooling stage of the precooled Joule-Thomson cooler. The flight model subassembly, called the Cooler Compressor Assembly, has been built and has completed its in-process performance testing prior to delivery to acceptance testing. This paper describes the Cooler Compressor Assembly and summarizes its cryogenic refrigeration performance results.

INTRODUCTION

The Mid InfraRed Instrument Cooler Subsystem (MIRI Cooler) is a closed cycle helium Joule-Thomson (JT) cryocooler precooled by a three-stage pulse tube cryocooler. The basic architecture and design parameters of this cooler have been described previously [1,2,3,4,5,6,7]. The MIRI Cooler consists of four major subassemblies that are integrated into the spacecraft separately and then interconnected. These four subassemblies (Figure 1) are:

- The Cryocooler Electronics Assembly (CCEA)
- The Cryocooler Compressor Assembly (CCA)
- The Cryocooler Tower Assembly (CTA)
- The Cold Head Assembly (CHA)

This paper focuses on the Cryocooler Compressor Assembly (CCA), which is the cooler subassembly that compresses and precools the helium working fluid. The CCA (Figure 1) resides within the JWST observatory's spacecraft bus. The major assemblies of the CCA are the three-stage Pulse Tube Precooler, the PT Precooler Helium compressor, and a series of recuperators that provide the thermal isolation between the room temperature helium from the JT compressor and as well as between the precooler's three stages. In the first box at the left margin of Figure 1, "CCEA" stands for Cryocooler Control Electronics Assembly, "RSA" stands for Relay Switch Assembly, and "JT P, JT R, PT P, and PT R" stand for primary and redundant JT and PT cryocooler control electronics. In the highlighted box "CCA", "PT" indicates the various Pulse Tube cold heads, "P_{high}" and "P_{low}" indicate pressure transducers and "HX" stands for various heat exchangers between the precooler and the 1st, 2nd, and 3rd stage recuperators. The next box to the right, "CTA" stands for the Cooler

Cryocoolers 19, edited by S.D. Miller and R.G. Ross, Jr.

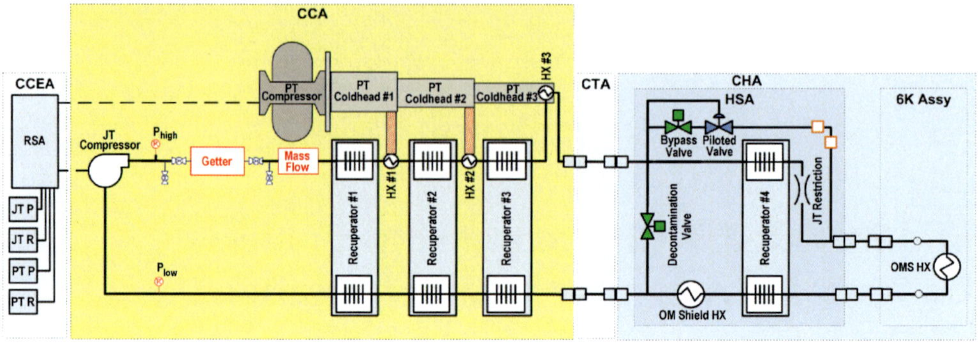

Figure 1. Functional block diagram of the entire MIRI Cryocooler system, highlighting the Cryocooler Compressor Assembly (CCA).

Tower Assembly. In the large box to the right, "CHA" stands for the remote Cold Head Assembly; it is divided into the "HSA" (Heat Sink Assembly) and the 6 K Assembly. Within the CHA, "OM_Shield HX" stands for the optical module shield heat exchanger, and "OMS HX" is the low temperature (6 K) heat exchanger that cools the MIRI optical bench and its focal plane modules.

The two major subassemblies of the CCA (JT compressor and PT precooler) were tested independently before being integrated together on the CCA's thermal interface plate. The JT compressor contains a single stage which provides a nominal helium gas flow of 25mg/s at approximately 12 bar on the high side and 4 bar on the low side. The PT precooler combines the PT compressor and the three-stage pulse tube cold heads that provide cooling at the nominally 150K, 50K and 18K helium precooling stages. The three recuperators associated with the PT precooler provide thermal isolation between the aforementioned precooling stages.

COOLER SUBSYSTEM PERFORMANCE TESTING RESULTS

JT Compressor

The Flight Model and Flight Spare JT Compressors were tested as stand-alone components for performance and exported forces. The performance of each compressor was tested by pumping helium through a load valve with a pneumatic impedance that was varied to mimic the range of flow impedances experienced with the flight system. The performance of each compressor is shown in Table 1. The summary includes key pertinent performance characteristics, including reject temperature, frequency, pressures, mass flows and compressor powers. The two flight compressors met subsystem requirements.

The JT compressor exported force was measured on a 6-axis force dynamometer, with data shown in Figure 2. The JT compressor was operated at acceptance level power with a representative pneumatic restriction used in place of the nominal system JT restriction.

Table 1. JT Compressor Performance Summary

	Date/ Time	T_{reject}, K	Freq, Hz	P_{hi}, bar abs.	P_{lo}, bar abs.	Mass flow, mg/s	Comp. Power, W
Flight Model CCA1 ZF610	10-27-14, 13:07	305	90.06	11.53	4.10	24.5	67.6
Flight Spare CCA2 ZF607	7-24-15, 18:51	305	90.06	11.48	4.01	24.4	78.6

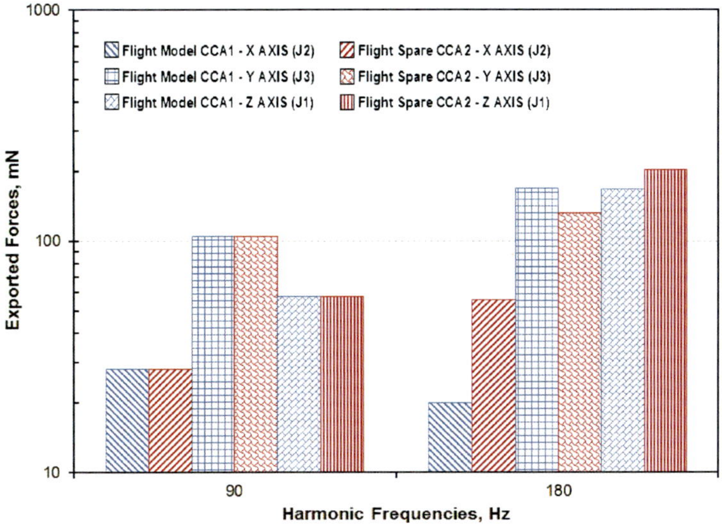

Figure 2. Exported force results at the fundamental drive frequency (90 Hz) and 2nd harmonic (180 Hz) for the JT Flight Model and Flight Spare compressors.

The amplitudes of the various exported forces are plotted for each compressor axis in units of mN peak, with Flight Model and Flight Spare results adjacent to one another for each axis. The key result from this figure is that the exported forces in the JT compressor drive axis were on the order of 50 mN and less for both the drive frequency and 2nd harmonic, whereas the off-axis exported forces were all less than 200 mN.

Launch vibration testing of the JT compressor was not conducted at the component level and was instead deferred to CCA level vibe tests. The basis for this was that the JT compressor is the same as the TRL9 NGAS High Efficiency Cryocooler (HEC) compressor (Figure 3), with over a dozen units in Flight Operation, and the only major change from these units was the reed valve design that was validated with valve-module-level vibration tests [2]. Vibration testing of the JT compressor at CCA assembly level was successful and will be described in a future paper.

PT Precooler

The PT Precooler design was derived from the NGAS 10K Pulse Tube Cooler [2,7]. The precooler contains a three-stage pulse tube with stages operating at 150K, 50K and 18K. The PT

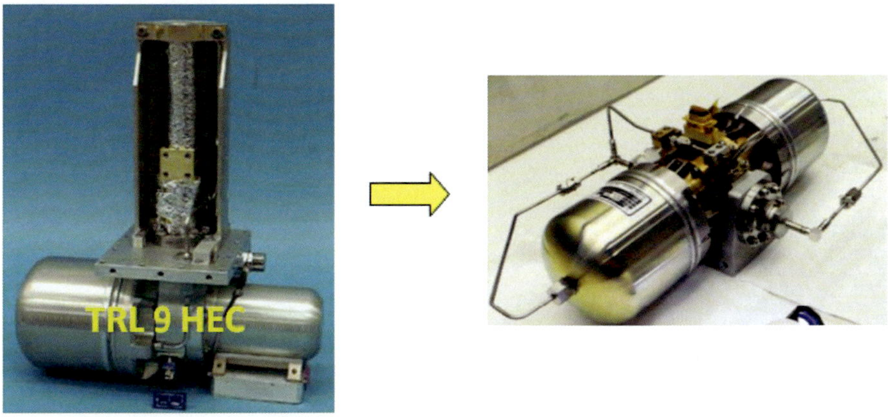

Figure 3. TRL 9 HEC Heritage of the MIRI JT compressor

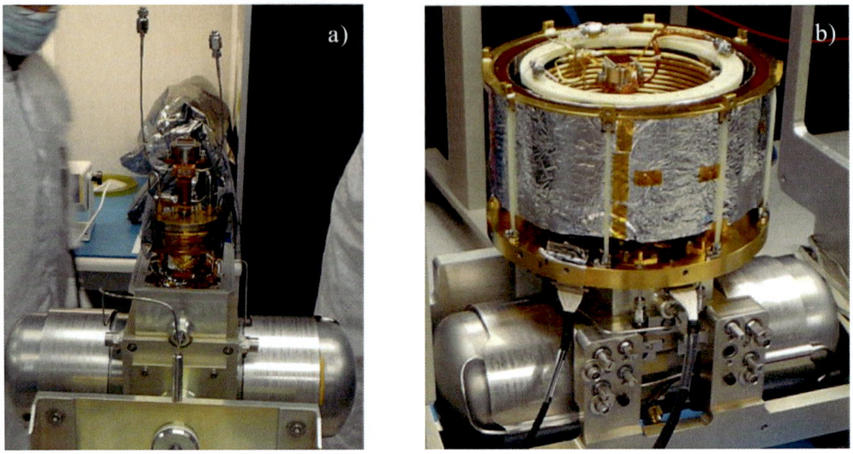

Figure 4. PT Precooler configurations: a) Cooler-Only (without recuperators and flight shields) (left figure) and b) Cooler plus flight recuperators and flight shields (right figure)

cooler was first tested at the Cooler-Only level (Part# 460026), which contains only the compressor and the pulse tube cold heads, as shown in Figure 4a. It was tested again at a higher level of assembly (Part# 460120) – containing the cooler along with flight recuperators and the two flight thermal shields that its upper stages cool, as shown in Figure 4b. The second test was completed in order to verify minimal impact to the system performance as a result of the added flight hardware. In the case of the Cooler plus flight recuperators/thermal shields, the refrigeration performance was initially measured *without helium gas flow* to confirm the minimal impact of the parasitic thermal loads of the recuperative and flight shield hardware. Later, using the same cooler configuration, precooler performance was measured *with helium gas flow* through the JT circuit using a representative pneumatic restriction to reproduce approximate flight mass flow rates and pressures, with electrical heat added to measure lift. Figure 5 shows pinch-point-condition [6] load lines for the Flight Model and Flight Spare PT precoolers for the third stage at the Cooler-only level (Part# 460026). The

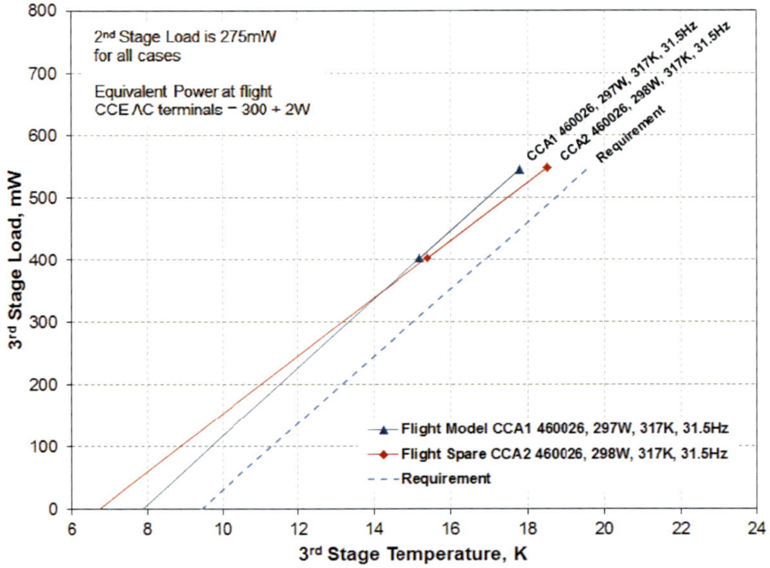

Figure 5. Three-stage Pulse Tube Precooler Refrigeration performance at Cooler-Only level (460026).

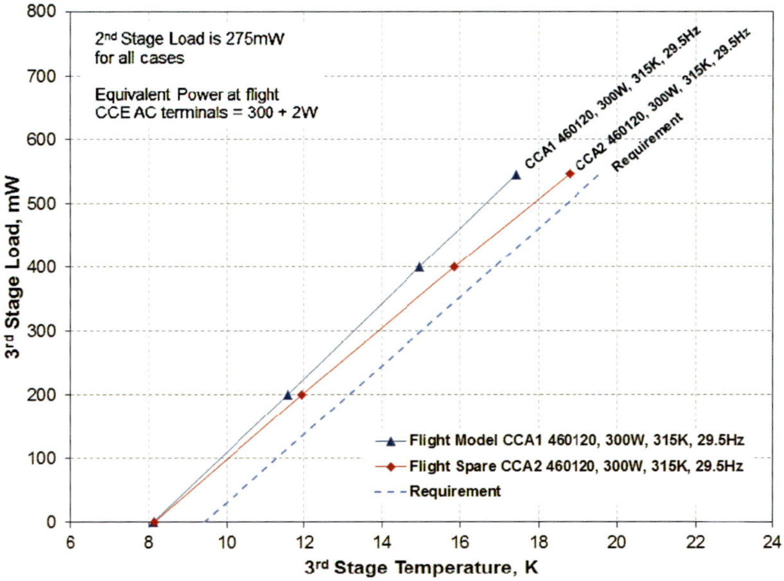

Figure 6. Three-stage Pulse Tube Precooler Refrigeration performance at Cooler plus flight recuperators/ thermal shields level (460120).

environmental condition for all data shown has a constant 2nd Stage load of 275 mW applied at approximately 50 K to simulate the expected flight loads on the 2nd Stage shield. The power applied to the PT compressor in all cases was an equivalent compressor power at Flight CCEA AC terminals of 300 W. Figure 6 shows the pinch-point condition load line performance at the next higher level of assembly with cooler plus flight recuperators/thermal shields (Part# 460120). Figure 7 shows the combined plot of pinch point load line data from the two aforementioned assembly levels. The Precooler refrigeration performance data at both assembly levels were better than the requirement and were consistent with previous end-to-end cooler tests using the DM precooler and FM Cold Head Assembly [6].

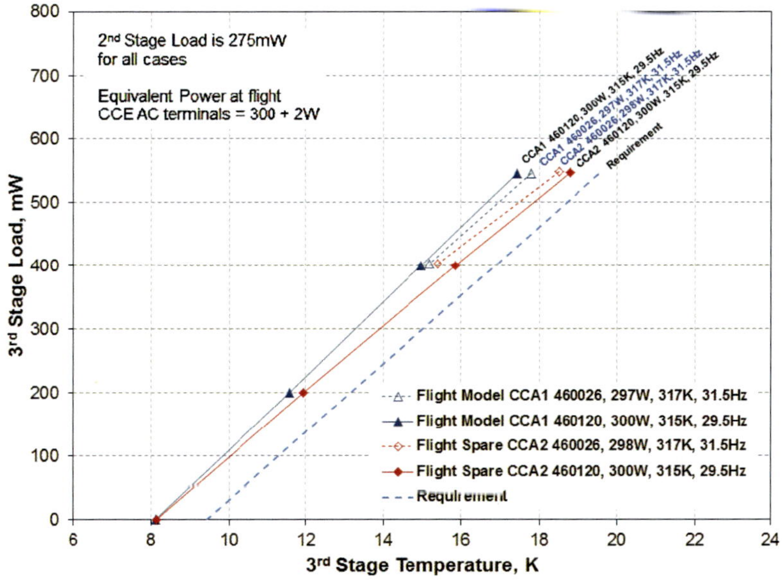

Figure 7. Three-stage Pulse Tube Precooler Refrigeration performance at both assembly levels.

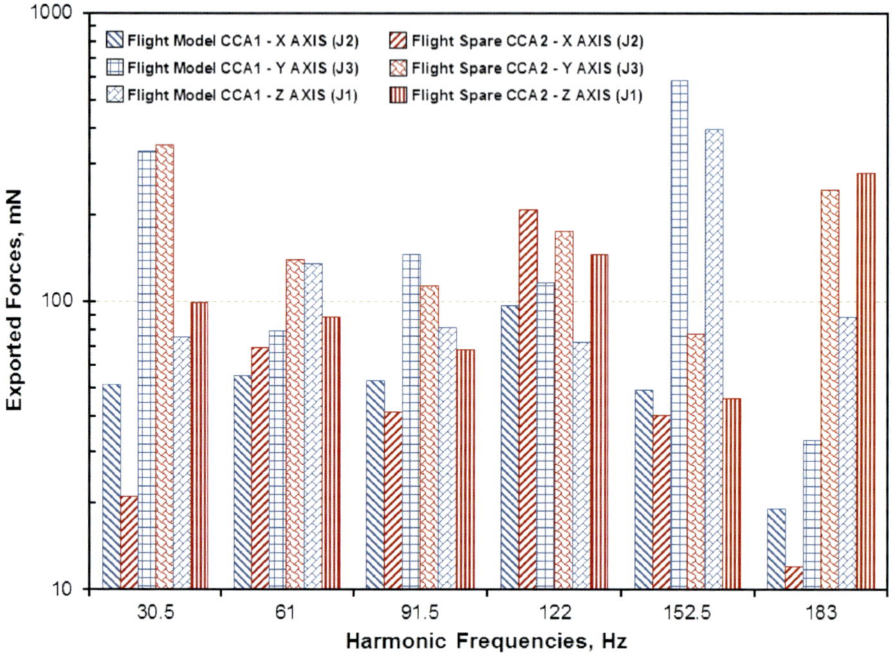

Figure 8. Exported force results for PT Flight Model and Flight Spare compressors.

The three-stage PT Precooler was also tested to quantify its exported forces. The three-stage PT Precoolers were tested on the same 6-axis force dynamometer as the JT compressors and were tested at the cooler plus flight recuperator/thermal shields level (Part# 460120), with the recuperative heat exchangers and flight thermal shields attached. Like the refrigeration performance testing, the external loads on the precooler were kept constant at 275mW at 50K and greater than 500mW at 18K. The results from the three-stage Precoolers are shown in Figure 8. All six axes of the Flight Model and Flight Spare assemblies were measured, with X, Y and Z forces reported here. The abscissa of Figure 8 shows a nominal drive frequency of 30.5, with the 2nd through 5th harmonics shown with increasing frequency. The 1st and 2nd bars in each frequency correspond to the drive, or X-axis exported forces, the 3rd and 4th bars delineating the Y-axis exported forces and the 5th and 6th bars corresponding to the Z-axis exported forces, as measured on the 6-axis dynamometer. The key result from this figure is that the exported forces due to the PT compressor drive axis were less than 100 mN throughout the drive frequency through the first six harmonics, whereas the off-axis exported forces were all less than 600 mN, with many around 100 mN or less.

The three-stage pulse-tube Precooler was additionally exposed to random vibration environmental testing. The Flight Model and Flight Spare were tested at acceptance levels of approximately 0.1 g2/Hz. All post-vibe checkouts indicated nominal results. The post-vibe refrigeration performance was later measured at the CCA level and confirmed the operational integrity of the CCA. CCA testing results will be reported in a future paper.

The PT and JT compressors are the only shock sensitive components in the CCA, thus they needed to be flight qualified for resistance to shock. As previously mentioned, the HEC Compressor design used in the JT Compressor for the JT loop had been qualified for this shock environment on several previous flight programs, so additional flight qualification for that compressor was not needed.

In contrast, the HCC Compressor used for the three-stage PT precooler needed to be qualified for the shock levels of the MIRI program. This was accomplished via a ringing plate test of an NGAS-owned compressor that is essentially identical to the MIRI compressor. Figure 9 shows the shock level profile established for the MIRI cooler mounting environment and the results of the

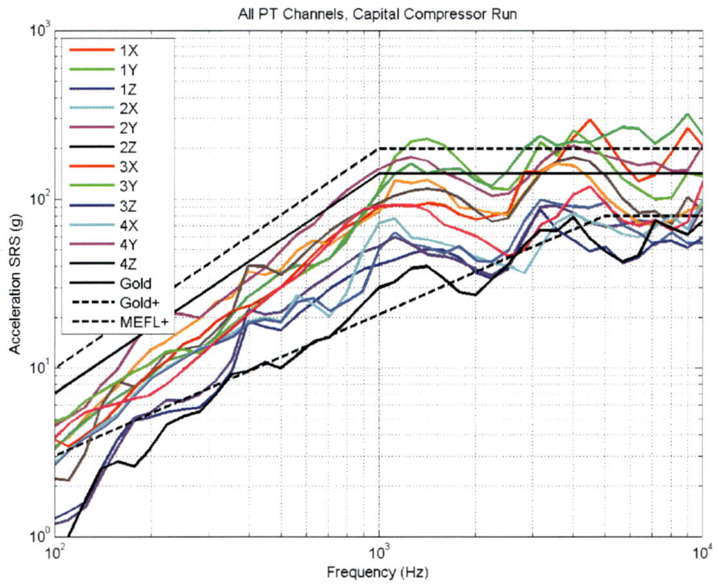

Figure 9. Shock acceleration profile established for the MIRI PT cooler mounting environment compared with measured levels achieved during the PT-cooler shock qualification test.

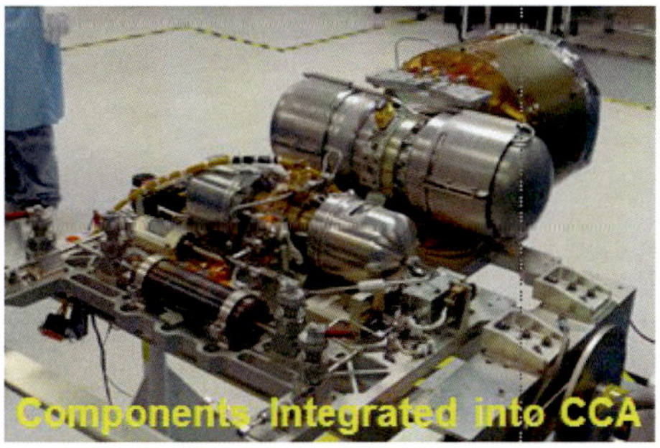

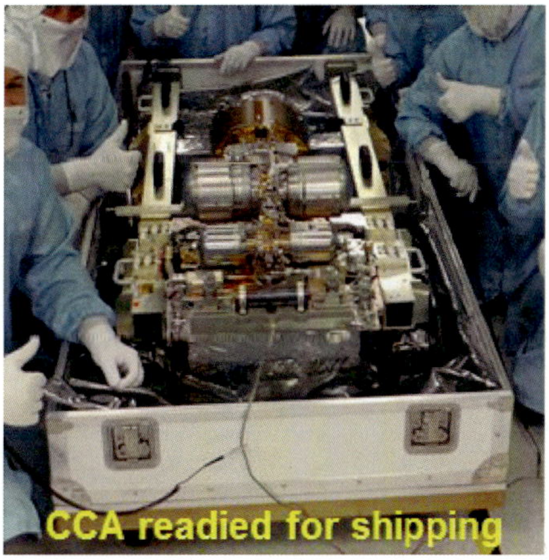

Figure 10. MIRI Cryocooler Compressor Assembly integrated and ready for shipping.

actual test profiles achieved in the test of the NGAS-owned compressor. Post-test measurements confirmed that the compressor was not affected by the shock exposure.

CONCLUSION

The MIRI Cooler Compressor Assembly components have all been characterized by in-process tests before integration onto the CCA. The Flight Model and Flight Spare CCA subassemblies have been successfully delivered to Acceptance testing at the CCA level (Figure 10).

The Component level characterization data acquired as a result of the MIRI Cooler CCA Sub Assembly test campaign are applicable to future space missions which might use a different configuration of these components.

ACKNOWLEDGMENT

This work was performed for the Jet Propulsion Laboratory, California Institute of Technology, sponsored by the United States Government under a Prime Contract between the California Institute of Technology and NASA.

REFERENCES

1. D. Durand, J. Raab, R. Colbert, M. Michaelian, T. Nguyen, M. Petach, and E. Tward, "NGST Advanced Cryocooler Technology Development Program (ACTDP) Cooler System," *Advances in Cryogenic Engineering: Transactions of the Cryogenic Engineering Conference – CEC*, vol. 51, edited by J. G. Weisend II (2006), pp. 615-622.
2. D. Durand, R. Colbert, C. Jaco M. Michaelian, T. Nguyen, M. Petach, and E. Tward, "NGST Advanced Cryocooler Technology Development Program (ACTDP) Cooler System," *Cryocoolers 14,* ICC Press, Boulder, Colorado (2007), pp. 21-29.
3. D. Durand, R. Colbert, C. Jaco M. Michaelian, T. Nguyen, M. Petach, and J. Raab, "Mid InfraRed Instrument (MIRI) Cooler Subsystem Prototype Demonstration," *Advances in Cryogenic Engineering: Transactions of the Cryogenic Engineering Conference – CEC*, vol. 53, edited by J. G. Weisend II (2008), pp. 807-814.
4. D. Durand, D. Adachi, D. Harvey C. Jaco, M. Michaelian, T. Nguyen, M. Petach, and J. Raab, "Mid InfraRed Instrument (MIRI) Cooler Subsystem Design," *Cryocoolers 15,* ICC Press, Boulder, Colorado (2009), pp. 7-12.
5. M. Petach, M. Michaelian "MIRI Cooler System Design Update," *Cryocoolers 16*, ICC Press, Boulder, Colorado (2011), pp. 9-12.
6. M. Petach, M. Michaelian "Mid InfraRed Instrument (MIRI) Cooler Cold Head Assembly Acceptance Testing and Characterization," *Cryocoolers 18*, ICC Press, Boulder, Colorado (2014), pp. 11-17.
7. T. Nguyen, R. Colbert, D. Durand, C. Jaco M. Michaelian, and E. Tward, "10K Pulse Tube Cooler," *Cryocoolers 14,* ICC Press, Boulder, Colorado (2007), pp. 27-32.

Development of a 2K Joule-Thomson Closed-Cycle Cryocooler

M. Crook, T. Bradshaw, G. Gilley, M. Hills, S. Watson, B. Green, C. Pulker, T. Rawlings

STFC Rutherford Appleton Laboratory, Harwell, Oxford OX11 0QX

ABSTRACT

The Rutherford Appleton Laboratory (RAL) is currently developing a *2K Joule-Thomson cooler* targeted at future space science missions requiring low temperatures, for example as part of the cryogenic chain for the ESA Athena X-ray telescope. The cooler builds on previous closed cycle cooler developments at RAL, in particular the very successful 4K-JT cooler that was flown on the ESA Planck mission. However the working fluid for the 2K-JT cooler will be ^{3}He, and two extra stages of compression have been added to provide a much lower effluent pressure and a greater overall compression ratio. The design of the 2K-JT cooler is described, with reference to lessons learnt from Planck in-orbit data, and results of the test program to date are presented.

INTRODUCTION

Cryocooler development for space applications has been ongoing for over 30 years at RAL[1], and has resulted in a range of closed cycle mechanical cryocoolers covering temperatures from 2 K to 80 K and above. These coolers have been licensed to both UK and US industry and have subsequently underpinned a large variety of high profile scientific and operational missions; ranging from space exploration and the origins of the universe, to earth observation, climate change, and weather forecasting.

One of those developments was a 4K Joule-Thomson (4K-JT) cooler[2] which was flown on the ESA Planck mission[3] to observe the fluctuations of the Cosmic Microwave Background. As part of the cryogenic chain[4] for the High Frequency Instrument (HFI), the 4K-JT cooler was precooled passively at 54 K by the spacecraft radiators and actively at 18 K by a hydrogen JT cooler supplied by JPL. The cooler was able to produce 20 mW of cooling at 4.5 K with a pre-cooler temperature of 17 K. Some in-orbit results are presented below and discussed within the context of the current RAL JT cooler developments.

Following on from the Planck 4K-JT cooler, RAL are currently developing a 2K-JT cooler targeted toward the Athena X-ray telescope[5], which has been selected as the second L-class mission in ESA's Cosmic Vision plan, with launch foreseen in 2028. As part of the cryogenic chain[6] for the X-ray Integral Field Unit instrument (X-IFU)[7] the present baseline is that the 2KJT cooler will be pre-cooled passively at 120 K-150 K by the spacecraft radiators and actively at 15 K by a pulse tube cooler. The cooler is required to provide 20 mW of pre-cooling at 2 K to service a hybrid ^{3}He sorption-ADR cooler operating at 50 mK. The 2K-JT cooler will retain much of the Planck 4K-JT cooler heritage, in particular the heat exchanger and thermal interface architecture. However, the requirement of a 2 K interface drives the use of ^{3}He as the working fluid, operating at a vapour pressure of $\leq$200 mbar in the effluent line.

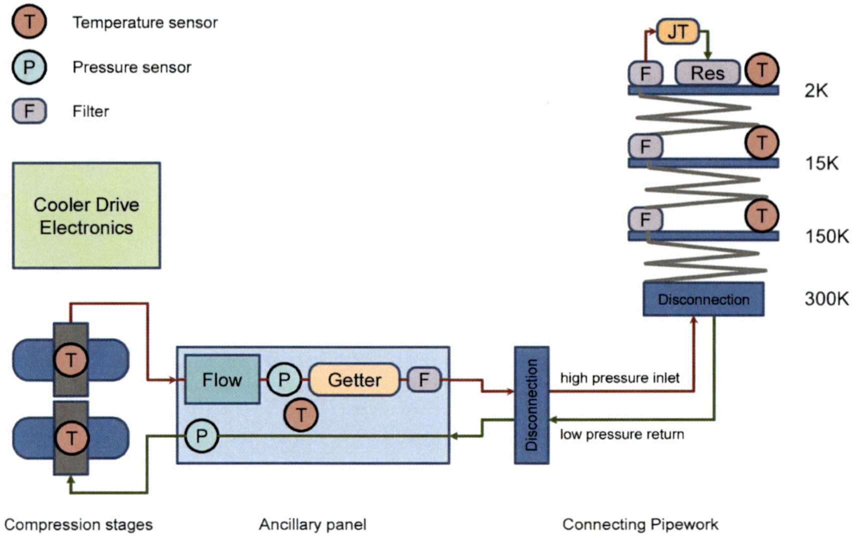

Figure 1. Schematic of the 2K-JT cooler system.

COOLER ARCHITECTURE AND INTEGRATION WITH SPACECRAFT

The 2K-JT cooler schematic is shown in Figure 1. The cooler comprises four reciprocating linear motor compression stages to perform an expansion of the gas across a JT restriction. The compressors are equipped with an arrangement of reed valves and buffer volumes in order to produce a steady flow of gas. The gas must be pre-cooled prior to the expansion taking place, and, to reduce the heat rejected at the pre-cooling interfaces, counter-flow heat exchangers are used between the stages. An ancillary panel carries gas handling and measuring equipment, as well as filters and a reactive getter to ensure gas cleanliness, which is critical to the long term operation of the cooler. The cooler is controlled by a set of drive electronics which perform all commanding and controlling functions as well as providing the electrical input power for the compressors and returning the cooler housekeeping data.

Also shown is a set of disconnection plates and connecting pipework that allow the system to be broken into several pieces to aid integration. An advantage of the JT cooler architecture is that the cooling is delivered remote from the compressor mechanisms which means that most of the cooler parts, i.e. almost all the mass and all those requiring significant electrical power, may reside in the Service Module part of the spacecraft at ambient temperatures. This greatly reduces the possibility of unwanted mechanical and electrical perturbations from those items being manifest at the detectors. In addition the heat exchanger pipework lends itself to easy manipulation which gives great flexibility when selecting a route through the Payload Module to deliver cooling remotely. A disadvantage of the distributed nature is that it is undesirable to integrate as a complete unit. The connecting pipework, which can be designed to suit the specific spacecraft architecture, allows these two parts to be delivered and integrated to the spacecraft separately, with a final purge and fill procedure being carried out afterwards.

HEAT EXCHANGER ASSEMBLY

The heat exchangers are a tube-in-tube counter-flow design that makes use of the effluent gas to remove enthalpy from the high pressure inlet gas in order to reduce the heat rejected to each of the pre-cooling stages. At the interfaces to each of these stages there are filters to ensure good thermal contact to the stage and also to trap any impurities in the gas. On the 2K-JT thermal interface there is a filter prior to the JT orifice and a sintered reservoir afterwards. The reservoir retains any liquid produced during the expansion and prevents sloshing and flash evaporation affecting the temperature stability. Figure 2 shows the masses and size envelopes of the pre-cooling and 2K-JT interfaces, and shows an annotated photograph of the 150 K interface under test configuration.

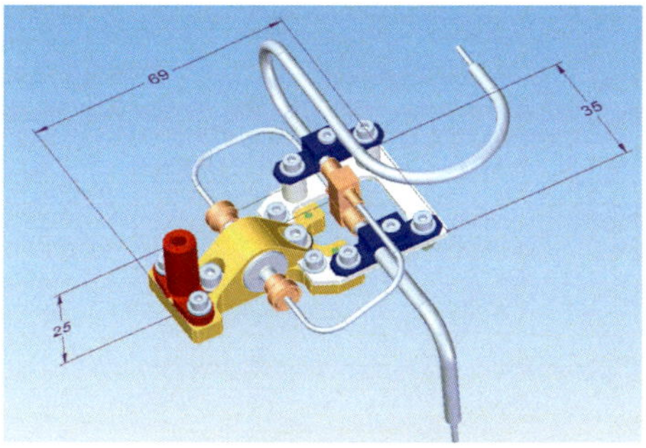

15K/150K interface:
100g, 69mm x 35mm x 21mm

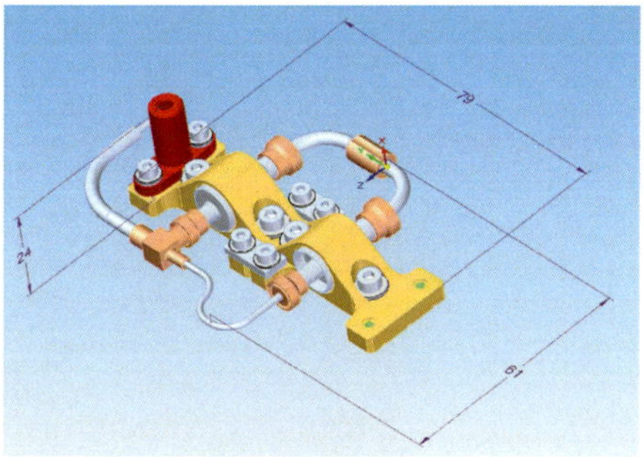

2K interface:
125g, 79mm x 61mm x 24mm

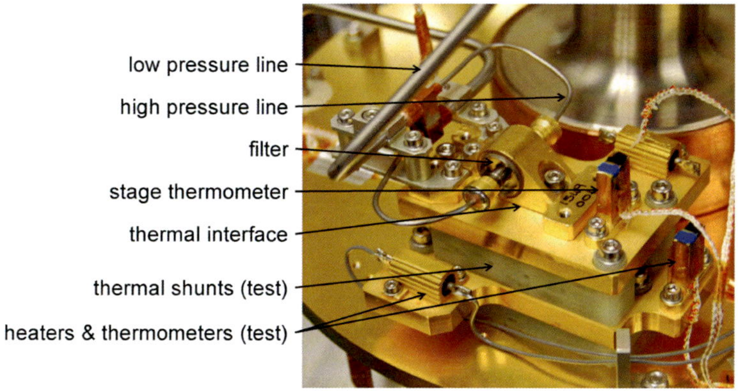

Figure 2. 2K cooler heat exchanger interfaces (top) and the 150K interface under test (bottom).

Heat Exchanger Modelling

The design of the heat exchangers dictates the heat load on the pre-cooling stages to a great extent. A model of the heat exchanger system originally developed for the 4K-JT cooler was extended to cover the operating and fluid parameters of the 2K-JT cooler. A summary of the required performance is given in Table 1 and the model has been used to optimise the design of the heat exchangers whilst respecting those criteria. To ease the requirements on the low pressure compressor, a maximum pressure drop has been allowed in the return line. The selected heat exchanger configuration is shown in Table 2.

Table 1. Cooling power requirements.

Requirement	Value
Total heat exchanger system pressure drop	< 50 mbar
Load on the first stage	< 300 mW
Load on the second stage	< 150 mW
JT cooling power	≥ 20 mW @ T≤2 K

Table 2. Selected heat exchanger configuration.

HX	Tube	Gauge	ID [mm]	OD [mm]	Wall [mm]	Length [m]
1	HP	17	1.10	1.47	.185	1
	LP	8	3.28	4.04	.380	1
2	HP	17	1.10	1.47	.185	2.5
	LP	8	3.28	4.04	.380	2.5
3	HP	17	1.10	1.47	.185	2
	LP	10	2.50	3.30	.400	2

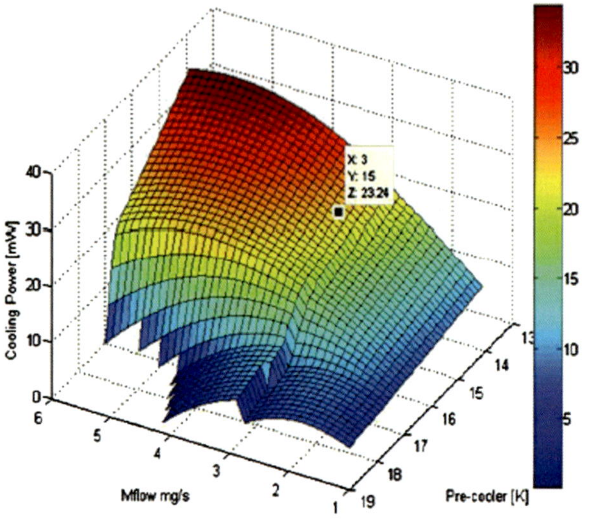

Figure 3. Modelled parameter space for the 2K cooler heat exchanger system.

Table 3. The cooling power parameter space of the selected heat exchanger configuration. Data in the highlighted region meet the design requirements.

2K stage Cooling power (W)	Mass Flow (mg.s-1)								
High pressure line (Bar_a)	2.00	2.25	2.50	2.75	3.00	3.25	3.50	3.75	4.00
6.50	0.0137	0.0153	0.0171	0.0187	0.0201	0.0214	0.0226	0.0237	0.0246
7.00	0.0146	0.0162	0.0177	0.0198	0.0213	0.0226	0.0239	0.0249	0.0258
7.50	0.0154	0.0170	0.0185	0.0208	0.0223	0.0237	0.0249	0.0260	0.0269
8.00	0.0161	0.0178	0.0194	0.0217	0.0232	0.0246	0.0259	0.0269	0.0278
8.50	0.0168	0.0185	0.0201	0.0225	0.0241	0.0254	0.0267	0.0277	0.0285
9.00	0.0174	0.0192	0.0207	0.0232	0.0248	0.0262	0.0273	0.0283	0.0291
9.50	0.0179	0.0197	0.0212	0.0225	0.0254	0.0267	0.0279	0.0288	0.0295
10.00	0.0184	0.0202	0.0217	0.0230	0.0259	0.0272	0.0283	0.0292	0.0299
10.50	0.0189	0.0207	0.0222	0.0234	0.0264	0.0277	0.0288	0.0296	0.0303
11.00	0.0193	0.0211	0.0226	0.0238	0.0269	0.0281	0.0292	0.0300	0.0306

With a nominal high pressure of 8 bar, a plot of the cooling power as a function of mass flow and pre-cooler temperature is shown in Figure 3. Calculations indicate a diminishing return in cooling power for an increasing inlet pressure beyond 8-10 bar. For a given heat exchanger geometry the changes in the operating parameters do not always satisfy the operating criteria. To illustrate this, the cooling power of the selected heat exchanger, for a fixed pre-cooling temperature of 15 K, is shown in Table 3 as a function of inlet pressure and mass flow.

The requirements for the compressor assembly derive directly from the outputs of the heat exchanger model; in summary the low pressure return should be less than 150 mbar in order to meet the temperature requirement, and the high pressure supply should be around 8 bar with a mass flow in the region 2-4 mg/s in order to meet the cooling power requirement.

COMPRESSOR ASSEMBLY

The compressors are arranged in back-to-back pairs such that the exported vibrations are minimised. This can be done most effectively using identical mechanisms operating at equal stroke amplitudes and with their individual piston diameters tuned to meet resonance conditions for each stage of compression. The compressors have a capacitive position sensor incorporated and are operated under closed loop drive control.

Table 4. Selected compressor configuration and operating parameters.

Compression stages	4
Total mass	8.5 kg including baseplate
size envelope	300 mm x 400 mm x 100 mm
operating frequency	45-55 Hz
Input power	<100 W

Each pair is supported directly from a central mount, providing an alignment interface and acting as a heat spreader. The reed valves and buffer volumes are arranged within the mount. This configuration is more compact than the Planck 4K-JT cooler, allows a better axial alignment, and allows more effective heat transfer to the spacecraft panel.

Compressor Modelling

A model has been developed for combined operation of the compressor and reed valve assembly, including compression chamber heat transfer and clearance seal leakage, valve conduction and valve leakage. A wide range of compressor architectures and configurations were modelled, the most efficient configuration was found to be four stages of compression with each stage being a single ended piston acting into an external buffer volume. The model was used to investigate the sensitivities to design parameters. At low pressures valve operation is shown as being critical; in particular the reed thickness should be minimised whilst respecting flatness and reliability considerations, as well as the response time with respect to the operating frequency. In addition, pressure drop and dead volumes associated with the valve housing entrance and exit channels must be carefully minimised. At high pressures, the model shows that piston clearance seal leakage is the major concern.

The model is used in conjunction with a kinematic model that incorporates the motor electrical characteristics to determine piston sizes that minimise the total input power. In general, operation at longer strokes and higher frequencies offer the potential for smaller motor force requirements, but this places conflicting engineering constraints on the flexure spring suspension and motor designs, respectively, and ultimately a trade-off between mass and efficiency considering system level requirements must be made. With that considered, the selected compressor configuration and operating parameters are shown in Table 4.

ANCILLARY PANEL ASSEMBLY AND GAS CLEANLINESS CONSIDERATIONS

The ancillary panel houses gas handling and measuring elements comprising: disconnection valves, particulate filters, a reactive getter, a mass flowmeter, and pressure sensors.

JT coolers are susceptible to clogging from impurity condensates at the JT restriction, a rigorous bake, purge and fill procedure is required to guarantee working fluid purity and a decontamination heater may be provided as back-up. The reactive getter provides additional gas cleaning, particularly for those impurities that may evolve from the internal surfaces of the cooler during its lifetime. The Planck 4K-JT cooler employed a hot reactive getter operating at 320 °C, which was large, heavy and consumed 13 W of electrical power. Much smaller ambient temperature getters are now available and are used on the 2K-JT cooler.

The mass flowmeter, in conjunction with the pressure sensors, is provided to give a measure of the cooling power in operation and to diagnose possible clogging. A space qualified commercial version was flown on Planck but this was bulky and heavy. A much smaller bespoke orifice flowmeter, employing a sensitive differential pressure transducer (Kistler 42464A) in the high pressure line, has been developed for the 2K-JT cooler. After calibration with ^{4}He the flow meter has been found to be accurate to a few percent at the flows of interest.

PLANCK 4K-JT COOLER IN-ORBIT RESULTS AND LESSONS LEARNT

Temperature Stability

The 4K-JT cooler on the Planck spacecraft operated for 4.5 years with no interruptions, blockages or issues from just after launch until the spacecraft was retired. The temperature stability over

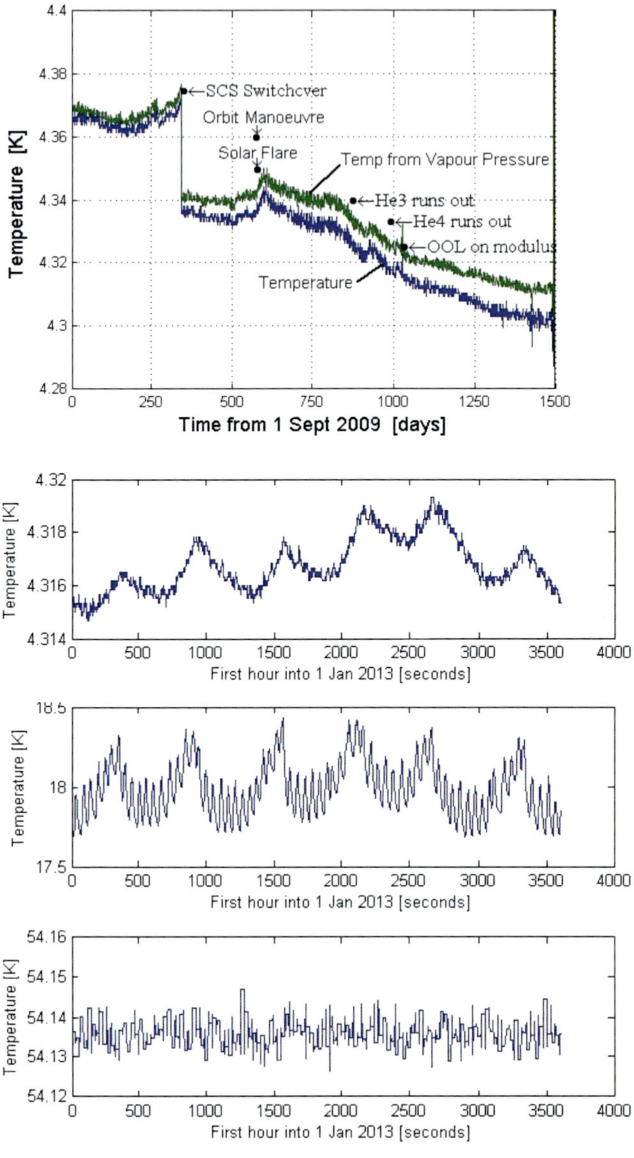

Figure 4. In orbit temperature stability of the Planck 4K-JT cooler. Top – 4 K temperature vs time from vapour pressure and thermometry. Bottom – short term fluctuations at 4 K and at the 18 K and 54 K pre-cooling stages.

this period was excellent. Figure 4 shows the temperature measured close to the 4 K interface and also that inferred from the pressure as measured on the ancillary panel as a cross check against sensor drift. Major life events are also shown; the "SCS Switchover" shows the switch from the nominal to redundant 18 K hydrogen sorption pre-cooler, and the two He events refer to the open cycle dilution refrigerator.

Short term fluctuations are well correlated (~2.5 mK/K) with perturbations (~0.6 K, 700 s) of the 18 K pre-cooling stage which are due to sorption bed switching during normal operation.

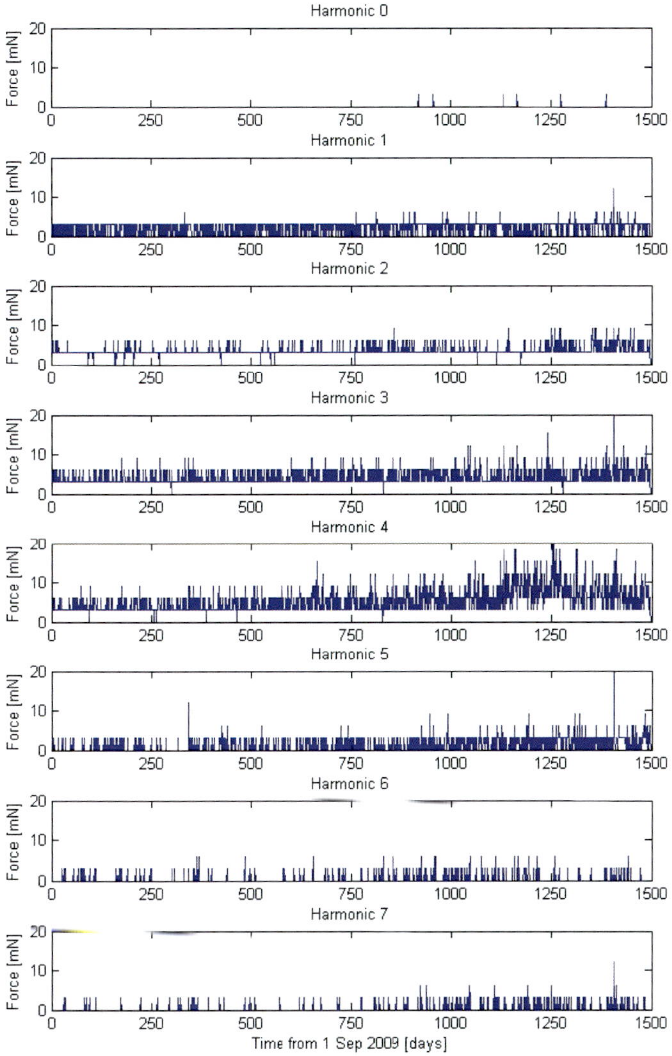

Figure 5. In orbit exported vibrations of the Planck 4K-JT cooler.

Exported Vibrations

The compressors on the Planck 4K-JT cooler were mounted to the spacecraft structure through force transducers. These were part of a closed loop vibration control system and also provided a measurement of the residual vibrations transmitted to the spacecraft structure. The requirement was to reduce the exported forces to less than 40 mN; note that a single uncompensated mechanism generates ~50 N. The harmonic content of the exported force is shown in Figure 5 over a one month period. The measurement is resolution limited at the fundamental frequency, only the occasional switching of a single bit can be seen.

As part of the control algorithm, harmonic corrections are applied to the compressor drive waveforms. The amplitude and phase of these corrections were monitored over the 4.5 year mission duration, it was anticipated that these would evolve as the mechanisms aged. However, this was not found to be the case; instead, only minor changes in the harmonic correction table were observed indicating that there was no age-induced pathology appearing in the mechanisms. This is an interesting and encouraging result for this type of flexure bearing compressor and imparts additional detail to laboratory life testing results.

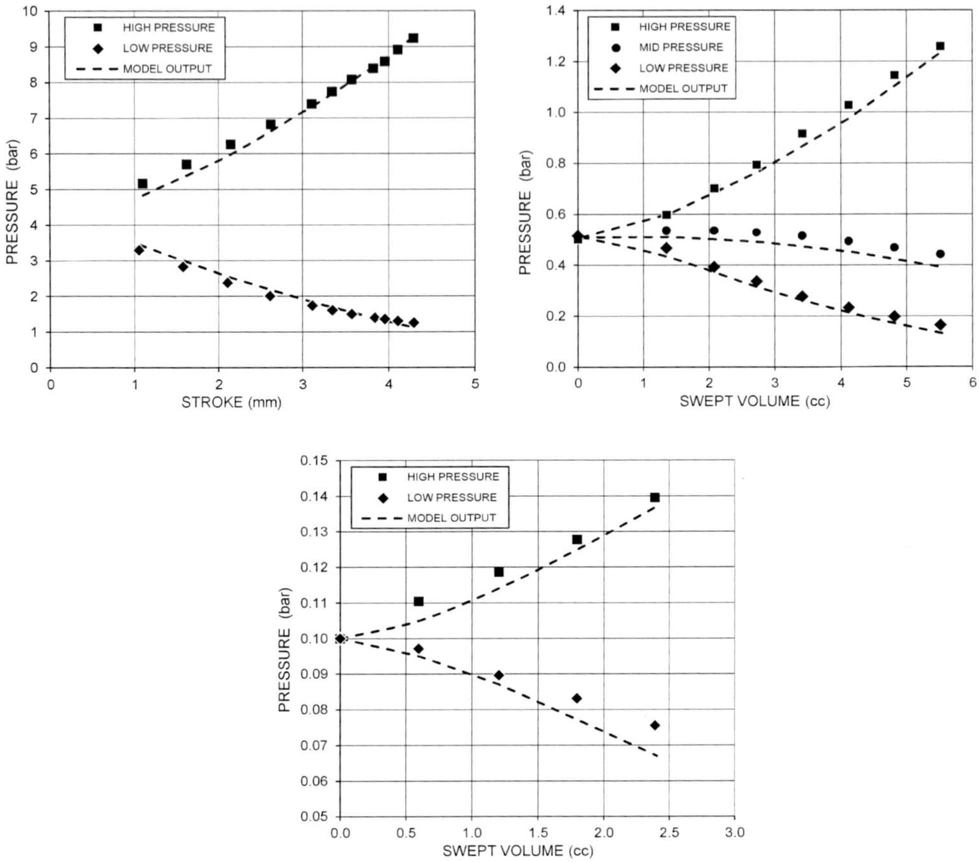

Figure 6. Verification of the 2K-JT compressor and valve model.

Heritage and Lessons Learnt

Many lessons were learnt and heritage was acquired during the development of the Planck 4K-JT cooler that will be taken forward in the development of the 2K-JT cooler. In the context of in-orbit operational experience it is worth highlighting that:

- Exceptional temperature stability over long periods can be obtained provided the compressor amplitude is well controlled.
- Long periods of operation without any evidence of clogging are achievable, despite the use of plastic materials as part of the internal construction and the small orifice size (~12 μm), provided good contamination control procedures are in place.[8] A decontamination heater was supplied on Planck, but was never used.
- No significant changes to the mechanism drive waveforms were required to maintain active vibration control, suggesting no age pathology and that continuous monitoring and vibration control may not always be necessary, depending on mission sensitivity.

2K-JT COOLER BREADBOARD TEST RESULTS

Compressor Breadboard Tests

The 2K-JT cooler compressor and valve model described above was verified by a breadboard test campaign using ^4He in a JT-representative loop and with compressor pairs of several different capacities. Measurements were made over a wide range of operating conditions, results are shown in Figure 6 for operation at different fill pressures most pertinent to the 2K-JT cooler. A needle

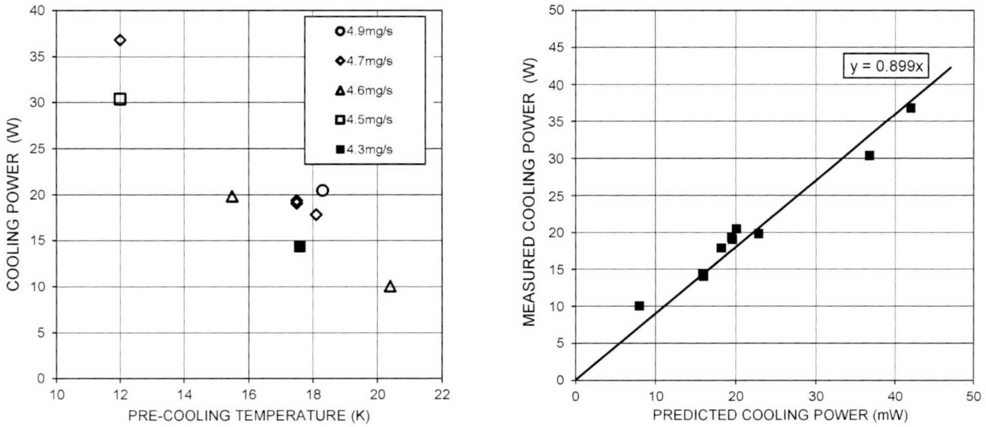

Figure 7. Verification of the 2K-JT heat exchanger assembly.

valve was used as an adjustable orifice. The compressor, reed valve assemblies and the pipework volumes were carefully characterized and those measured parameters were used as inputs to the model. No further adjustments are made as a function of fill pressure, stroke amplitude, or operating frequencies. Measurements were made over a wide range of operating conditions, and results are shown in Figure 6 for operation at the different fill pressures most pertinent to the 2K-JT cooler.

In general, the agreement between the model and the measurements is reasonably good. The most notable discrepancies are at the inlet to the low pressure stage at the lowest operating pressure where the model tends to over-predict the compressor performance. An additional 50 mbar compressor performance margin, in addition to the 50 mbar allowance for the heat exchanger pipework, has therefore been taken when selecting the compressor configuration and operating parameters shown earlier in Table 4.

Heat Exchanger Open Loop Tests

The performance of the selected 2K-JT cooler heat exchanger configuration (see Table 2) has been verified by representative cryogenic tests under thermal vacuum conditions. These early tests were carried out during the compressor set manufacture and were performed in an open loop configuration, using ^4He, with a pressurised cylinder and a commercial scroll pump providing the required pressures at the inlet and outlet, respectively. Pre-cooling was provided by a GM cooler with heaters used to regulate the stage temperatures. Thermal shunts (see Fig. 2) were used to measure the heat rejected at the interfaces.

A JT orifice diameter of 20 µm was selected to provide the nominal flowrate with ^3He. Using ^4He for these tests, flow rates of 4-5 mg/s were typical, dependent upon the high pressure. A temperature of ~3.6 K was achieved at the JT interface, which was limited by the scroll pump performance and its positioning outside the vacuum chamber.

The results of the tests are summarized in Figure 7, which shows the cooling power at ~3.6 K as a function of mass flowrate and second stage pre-cooling temperature. A maximum cooling power of 36.8 mW was achieved with a pre-cooling temperature of 12 K and a mass flow rate of 4.7 mg/s. Also shown in the figure, for the same data, is the measured cooling power against that predicted by the heat exchanger model described above, with ^4He as the working fluid. The agreement is reasonable over the full data range, with the linear fit showing that the model generally under-predicts the cooling power by around 10%. The results give confidence that the heat exchangers will provide the desired performance when under the nominal 2K-JT cooler operating conditions and with ^3He.

CURRENT STATUS AND NEXT STEPS

A Demonstrator Model 2K-JT cooler is under development and is targeted for use on ESA Athena mission. Much of the heritage from the Planck 4K-JT cooler is being carried forward in this development. In particular, the heat exchanger and thermal interface architecture is directly retained. Improvements in technology are being implemented in the compressor and ancillary panel assemblies that will significantly reduce their mass, their size envelope, and their input power requirements.

To date, the performance of the heat exchanger assembly has been measured in a standalone configuration under thermal vacuum conditions. The results are very encouraging and in line with expectations. The ancillary panel was also part of those tests, during which the correct operation of the bespoke mass flowmeter was verified.

At present, the compressor motor modules are undergoing characterisation prior to the compressor final assembly procedure. Once completed the compressor assembly will be integrated with the ancillary panel for performance verification with ^4He in a loop configuration incorporating an adjustable needle valve as a JT-orifice. Full system testing with the heat exchanger configuration under thermal vacuum conditions will then be carried out with ^3He. Afterward it is planned for the compressor and ancillary panel assembly to undergo thermal cycling and high level sine and random vibration testing.

Subsequent to the Demonstrator Model 2K-JT cooler test campaign, it will be delivered to ESA to be incorporated into a 50 mK cryogenic chain demonstrator as part of the X-IFU instrument consortium development activities.

ACKNOWLEDGMENTS

The 2K-JT cooler development is funded by ESA as part of the Basic Technology Research Programme. We would like to acknowledge Karin Dassis and Ludivigne Vignon for their help in obtaining the Planck 4K-JT cooler in-orbit data.

REFERENCES

1. Ross, R.G., "Aerospace Coolers: a 50-Year Quest for Long-life Cryogenic Cooling in Space," *Cryogenic Engineering*, Springer, New York, (2007), pp. 225–284.

2. Bradshaw, T.W., Orlowska, A.H., "Technology Developments on the 4K Cooling System for Planck and FIRST," *Sixth European Symposium on Space Environmental Control Systems*, (1997), pp. 465–470.

3. Planck Collaboration, "Planck 2013 results. I. Overview of products and scientific results," *Astronomy & Astrophysics 571*, (2014), A1.

4. Planck Collaboration, "Planck early results. II. The thermal performance of Planck," *Astronomy & Astrophysics 536*, (2011), A2.

5. Barret, D., Nandra, K., Barcons, X. et al, "ATHENA+ : The first Deep Universe X-ray Observatory," *Société Française d'Astronomie et d'Astrophysique*, (2013).

6. Branco, M.B.C., Charles, I., Butterworth, J., "ATHENA X-IFU Detector Cooling Chain" *Proc. SPIE 9144, Space Telescopes and Instrumentation 2014: Ultraviolet to Gamma Ray*, (2014), 91445V.

7. Ravera, L. et al, "The X-ray integral field unit (X-IFU) for Athena," *Proc. SPIE 9144, Space Telescopes and Instrumentation 2014: Ultraviolet to Gamma Ray*, (2014), 91442L.

8. Bradshaw, T.W., Orlowska, A.H., Jewell, C., "Life Test and Performance Testing of a 4K Cooler for Space Applications," *Cryocoolers 10*, Springer, (2002), pp 521-528.

Demonstration of Two-Stage Temperature Control for Raytheon Hybrid Cryocoolers

T. Conrad, B. Schaefer, D. Kuo, D. Bruckman, M. Kieffer, R. Yates

Raytheon Space and Airborne Systems
El Segundo, CA 90025

ABSTRACT

A two stage temperature control algorithm has been demonstrated for the Raytheon Stirling/pulse tube hybrid (RSP2) cryocooler. This algorithm adjusts the input power and Stirling displacer phase command in order to independently control the temperatures of both cryocooler stages without the addition of any trim heat. The ability to independently control the temperatures of the two stages is a unique capability of Stirling/pulse tube hybrid cryocoolers to shift cooling capacity between the stages through the adjustment of the Stirling displacer phase. The Raytheon Low Temperature RSP2 (LT-RSP2) cryocooler was used to exercise the algorithm and demonstrate its performance in several scenarios, including during cool-down and under changing load conditions at constant 1st and 2nd stage temperatures.

INTRODUCTION

A temperature control algorithm has been developed and demonstrated for the Raytheon Stirling/pulse tube hybrid (RSP2) two stage cryocooler. The algorithm is capable of independently controlling temperature at both stages and adjusting to changing loads over the capacity range of the RSP2 cryocoolers without the use of any trim heaters. This capability improves the overall efficiency of the cryocooler in an integrated configuration, as the use of trim heat for temperature control is directly translated to an increase in input power required. It also provides flexibility to both instrument designers and users and is unique to the RSP2 architecture. The benefits of efficiency and flexibility pay dividends in several different ways over the mission life cycle.

The typical design cycle for space-based instruments using active cryogenic cooling often requires cryocooler selection during the preliminary design phase, at which point the exact instrument load requirements are still being determined. For instruments utilizing a single stage cooler, this uncertainty is mitigated by requiring additional capacity margin from the cryocooler. When a two stage cryocooler is needed, e.g. to simultaneously cool both a detector and optical bench, this uncertainty becomes more significant. Requiring additional cooling capacity at both stages may cause the cryocooler to be oversized for the final application. Also, most two stage coolers have relatively constant ratios of first to second stage capacity as input power is varied; thus, any changes to the ratio of first and second stage load requirements will force the use of a heater on one of the stages if both temperatures must be controlled precisely. Stirling/pulse tube hybrid cryocoolers

Cryocoolers 19, edited by S.D. Miller and R.G. Ross, Jr.

have a unique capability to shift load capacity between stages which, combined with the ability to perform independent two stage temperature control. This capability allows the RSP2 coolers to share margin between their two stages and efficiently adjust to the final load requirement ratio.

An additional consideration in the instrument design phase is that program cost and schedule constraints often preclude the design and qualification of a new cryocooler for each instrument. It then becomes desirable to re-use an existing cryocooler for temperatures and loads other than those that it was initially optimized for. For a single stage cooler, this may be done by simply modulating the input power, but for a two stage cryocooler this adaptation is generally more complex and may require hardware modifications or the use of trim heaters if both stage temperatures must be controlled precisely. The load shifting capability of the Stirling/pulse tube hybrid cooler significantly expands the operating range of temperatures and loads which can be achieved without making hardware changes. In combination with the ability to independently control each stage temperature, which eliminates the need for trim heat, this allows the RSP2 cryocoolers to operate more efficiently at off-nominal loads and temperatures.

During the integration and test phase of a mission, thermal model predictions of instrument load and temperature requirements are replaced with experimentally determined values. The actual cooling requirement which the cryocooler must meet often changes substantially at this point, particularly if changes in the operating temperatures are needed to improve the overall instrument performance. At this point, the combination of the load shifting and two stage independent temperature control capabilities of the RSP2 Stirling/pulse tube hybrid cryocoolers provides unique flexibility to optimize overall instrument performance while simultaneously minimizing the required cryocooler input power through the elimination of trim heat.

Finally, during operation on orbit the capabilities of load shifting and independently controlling stage temperatures deliver flexibility and convenience to the instrument operator. The RSP2 temperature control algorithm is capable of maintaining temperature set points at both stages without additional user input as heat rejection temperatures, parasitic loads, and instrument power dissipation vary. Also, as an instrument progresses towards its end of life it may experience performance degradation, increased parasitic loads, or a reduction in available power. In these situations the ability to shift load and independently control cryocooler stage temperatures may help to extend the mission lifetime by allowing flexibility in prioritizing available cooling capacity.

BACKGROUND

Raytheon's Stirling / Pulse Tube Hybrid 2-Stage (RSP2) cryocoolers have been matured through an active development effort now spanning approximately 15 years [1,2]. They combine an actively driven Stirling first stage with a passive pulse tube second stage. The architecture of each RSP2 expander module is identical, the main differences being the size of the compressor module and the size of the Stirling piston and the pulse tube on the expander. In this way, the RSP2 architecture can be adapted to accommodate a wide variety of 1st and 2nd stage temperatures and load requirements. Nominal design temperatures (1st stage / 2nd stage) for the RSP2 coolers have included 55 K / 10 K for the LT-RSP2, 85 K / 35 K for the High Capacity (HC) RSP2, and 110 K / 58 K for the Mid-Capacity (MC) RSP2. Even with the compressor and expander sizes fixed, however, there remains a great deal of flexibility in operating temperature and load capacity for each of the RSP2 coolers; for example, the initial 60 K / 12 K testing at the beginning of the LT-RSP2 development effort was performed using the HC-RSP2, nominally designed for 85 K / 35 K operation [3].

For a given set of operating temperatures, the RSP2 cryocoolers are able to produce a range of cooling capacities at each stage. An example of this is shown in Figure 1, a load map for a production LT-RSP2 cryocooler operating at 55 K and 10 K for the first and second stage temperatures, respectively. In this figure it can be seen that for a given input power, the ratio of 55 K and 10 K capacities varies significantly with the Stirling phase angle. For example, at 350 W of input power the capacity can be varied from 4.4 W at 55 K and 0.2 W at 10 K to 2.2 W at 55 K and 0.36 W at 10 K by adjusting the Stirling phase angle from 85 degrees to 65 degrees. This was the tested range of Stirling phase angles for this cooler and these temperatures, however both higher and lower phases are possible and thus additional range in stage cooling capacities is possible as well. This

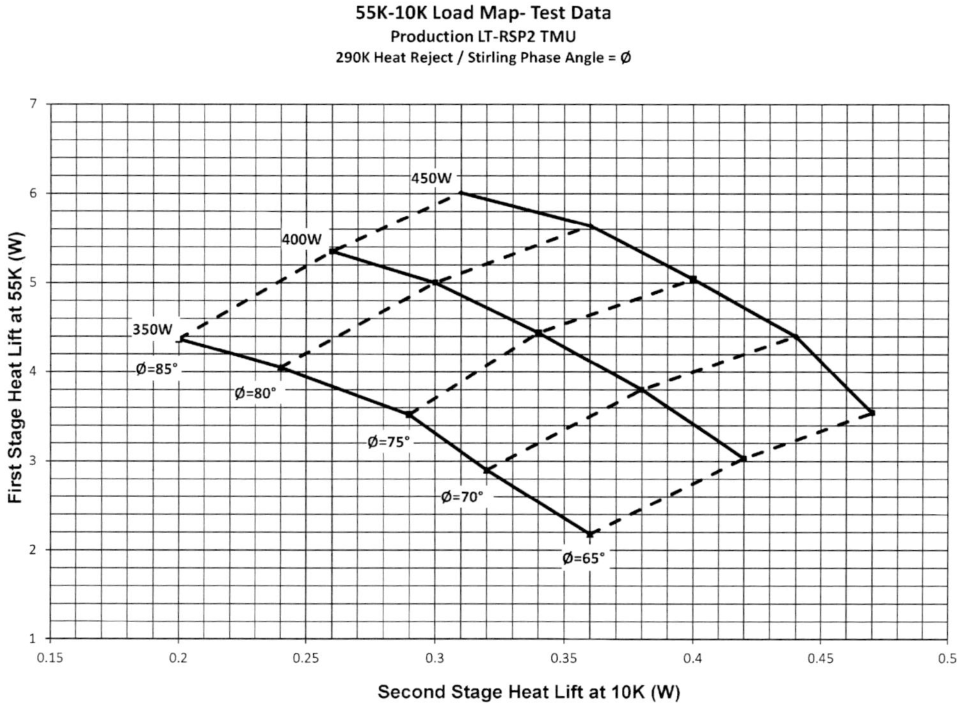

Figure 1. Load map showing LT-RSP2 cooling capacity at 55 K and 10 K as a function of Stirling phase.

unique capability of Stirling / pulse tube hybrid cryocoolers to trade capacity between their stages through variation of the Stirling phase angle is referred to as load shifting. For the RSP2 cryocoolers, it is implemented very easily through modulation of the displacer drive waveform produced by the cryocooler electronics. This load shifting capability also provides the basic mechanism for the independent two stage temperature control algorithm.

TWO STAGE TEMPERATURE CONTROL

From Figure 1, it is apparent that each combination of first and second stage loads is met by a unique input power and Stirling phase angle. In its most basic form the two stage temperature control algorithm can be thought of as automatically seeking this combination of input power and phase, thereby finding the minimum input power which can hold the combination of loads. The actual implementation is a little more complex, as any combination of loads and stage temperatures within the cryocooler's capacity envelope can be commanded and a unique combination of input power and Stirling phase will be determined by the control algorithm. In a typical application the loads are not actually commanded; rather, they are the sum of the instrument parasitic and dissipative loads. The desired stage temperatures are simply commanded to the temperature control algorithm and the required input power and Stirling phase are determined.

Figure 2 shows a demonstration of the control algorithm's response to step changes in applied loads on each stage and the resulting cryocooler performance. The top chart in the figure shows a history of the first and second stage loads, which were varied during the test. The middle chart shows the Stirling phase and compressor input power adjustments made by the algorithm in response to the load changes.

At the bottom, the resulting first and second stage temperature histories are shown. The temperature set points were 55 K and 10 K for the first and second stages, respectively. During the test, a step change in the first stage applied load was made, followed by a step in the second stage

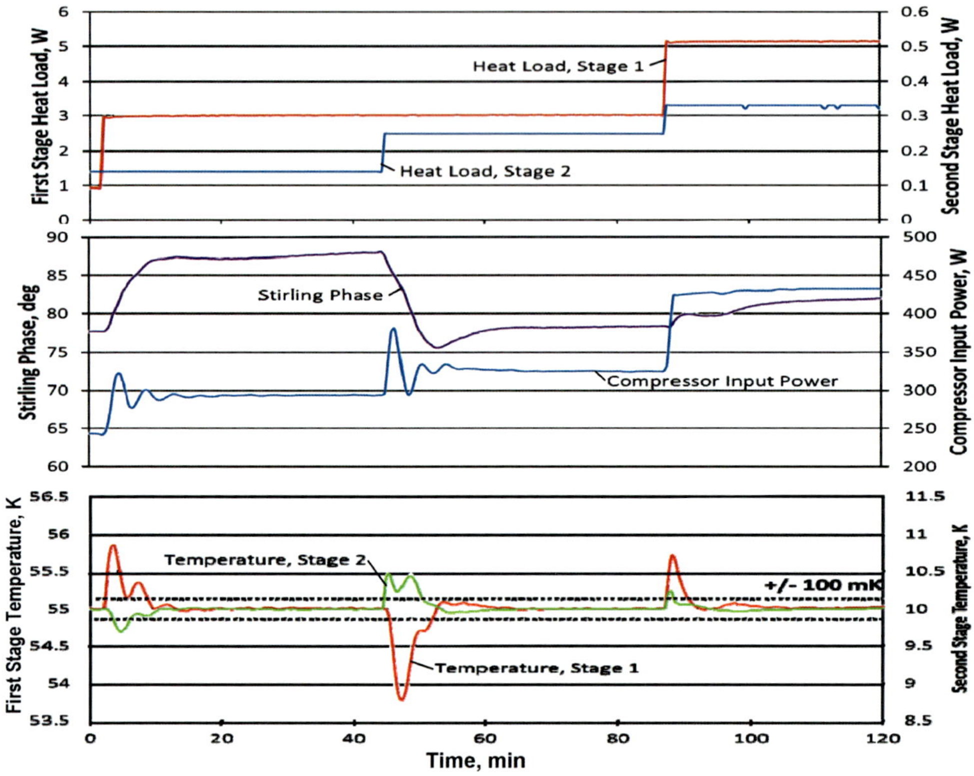

Figure 2. Demonstration of dual stage temperature control response to step functions in applied loads

load and finally a simultaneous step at both stages. The control algorithm responded to the first step by increasing both the input power and Stirling phase, consistent with the load map of Figure 1 which showed that an increase in phase results in an increase in first stage capacity and a decrease in second stage capacity. Likewise, in response to the second load step which was on the second stage, the control algorithm again increased the input power but decreased the phase angle. Finally, in response to the simultaneous load step the input power increased and a smaller adjustment was made to the Stirling phase angle. In each case the temperatures at both stages exhibited a brief transient response before converging again to the commanded temperatures of 55 K and 10 K. Temperature stability outside of the transient responses was well within the range of +/- 100 mK, a typical cryocooler temperature stability requirement. The transient responses are exaggerated in this test compared to what would be expected in an integrated configuration due to both a lack of thermal mass attached to either stage, which would greatly slow the temperature responses, and the generally unrealistic step changes in load.

A similar demonstration of controller performance in response to ramp changes in load is shown in Figure 3. First the first stage load is ramped from 1 W to 3 W over a 15 minute period, then the second stage load is ramped from 0.20 W to 0.42 W over a 10 minute interval. Again, temperature setpoints were 55 K and 10 K. As was the case for the step inputs, the controller increased the input power and Stirling phase in response to the increased first stage load, then increased the power and decreased the phase in response to the second stage load ramp. Temperature stability during controller adjustments was better for the load ramp inputs than it was for the step inputs, particularly for the first load increases on the first stage. Again, the temperature fluctuations seen in these controller demonstration tests are much larger than what would be expected for an integrated cryocooler and instrument due to the additional thermal mass of the instrument. The applied load changes are also greater in magnitude and occurring over a much shorter period of time than is typical for a cryocooler and instrument in an operational environment.

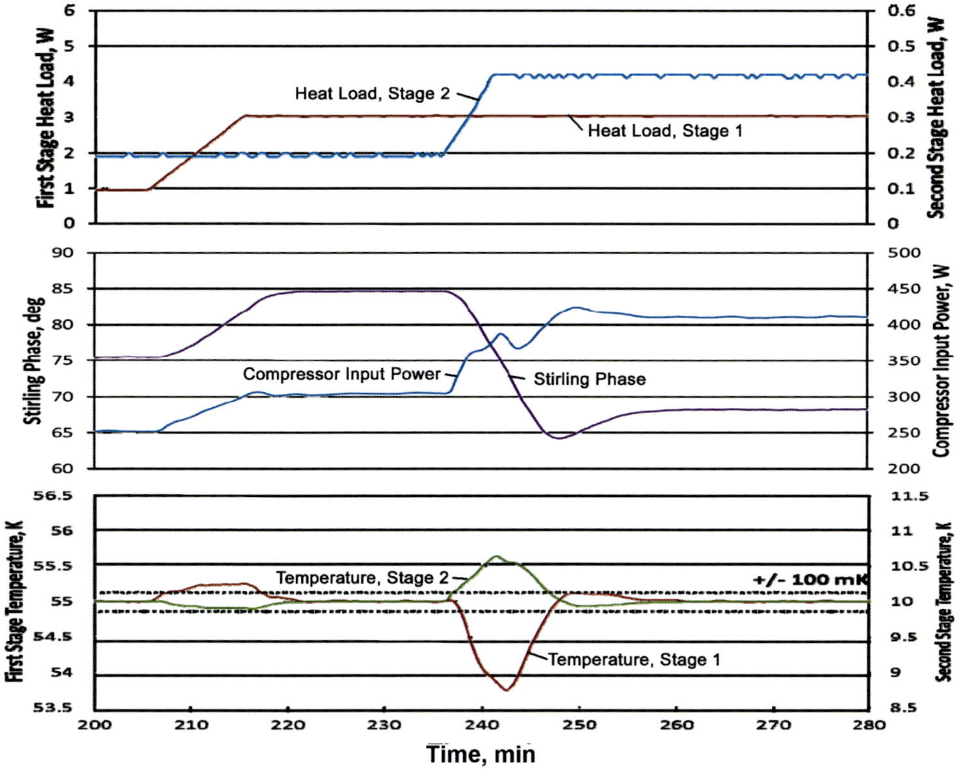

Figure 3. Demonstration of dual stage temperature control response to ramp functions in applied loads

The developed dual stage temperature control algorithm is capable of driving the RSP2 cryocoolers to a commanded set of first and second stage temperatures from any starting condition; the cryocooler does not need to be operating at or near the commanded temperatures before the controller can be utilized. A demonstration of this capability is given in Figure 4, which shows cooldown data for the LT-RSP2 cryocooler from an ambient (295 K) power off condition to 55 K and 8 K set points on the first and second stages. The controller begins by increasing the input power until it reaches its stroke limit while simultaneously shifting the Stirling phase from its starting point of 70 degrees to its nominal upper limit of 85 degrees.

The preference for higher phase angles and thus increased first stage cooling during cooldown stems from the larger thermal mass of the first stage, which typically drives the cooldown time of the RSP2 cryocoolers. From the figure, it can be seen that as the 1st and 2nd stage temperatures approach their set points, the input power is reduced along with the Stirling phase until the input power and phase required to maintain the desired stage temperatures are determined. No trim heat was added during this test and no user inputs were made except for the initial input of the desired temperature set points.

CONCLUSIONS

A dual stage temperature control algorithm has been demonstrated for the Raytheon Stirling pulse tube hybrid two stage (RSP2) cryocoolers which is capable of independently controlling the temperature at each stage without the addition of any trim heat. The control algorithm utilizes the unique capability of Stirling / pulse tube cryocoolers to alter their ratio of first to second stage cooling capacities by variation of the Stirling phase angle. The controller seeks the unique combination of phase angle and input power which exactly holds the applied loads at a commanded set of temperatures; by doing so, it also minimizes the input power required to maintain the temperature

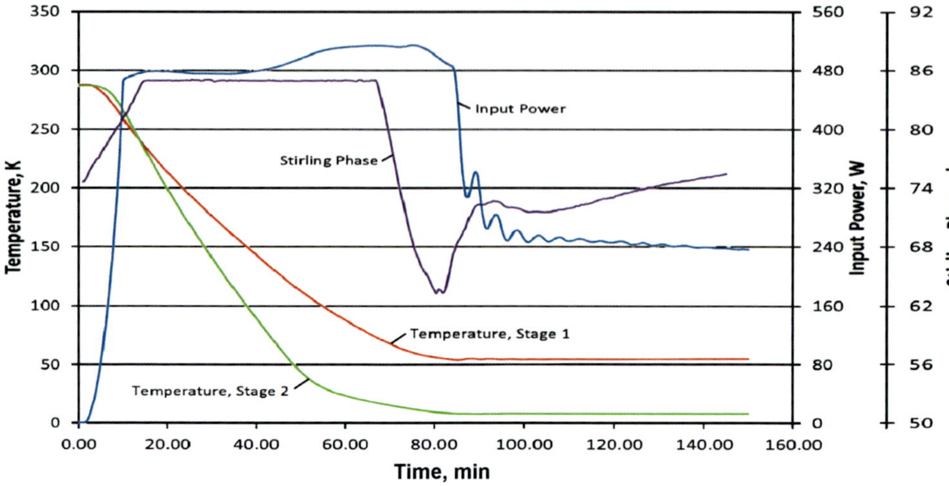

Figure 4. Demonstration of dual stage temperature control during cooldown from 295 K to 55 K / 8 K first/second stage temperatures.

setpoints. The capability to independently adjust the cooling capacity at both stages provided by the RSP2 cryocoolers and the dual stage temperature control algorithm provides valuable flexibility at every stage of a typical instrument design and mission cycle, from the preliminary design phase to end of life operation.

ACKNOWLEDGEMENTS

Raytheon would like to thank the Air Force Research Laboratory (AFRL) at Kirtland Air Force Base for their broad support of the RSP2 cryocooler development program.

REFERENCES

1. Kirkconnell, C .S., Hon, R. C. and Roberts, T., "Raytheon Stirling/Pulse Tube Cryocooler Maturation Programs," *Cryocoolers 15*, ICC Press, Boulder, CO (2009), pp. 31-37.

2. Schaefer, B.R., Bellis, L., Conrad, T., Bruckman, D., Yates, R., Pillar, M., and Barr, M., "Raytheon Low Temperature RSP2 Production Program," *Cryocoolers 18*, ICC Press, Boulder, CO (2015), pp. 19-25.

3. Hon, R.C., Kirkconnell, C.S., and Schrago, J.A., "Raytheon RSP2 Cryocooler Low Temperature Testing and Design Enhancements," *Adv. in Cryogenic Engineering*, Vol. 55, Amer. Institute of Physics, Melville, NY (2010), pp. 371 – 380.

Low Temperature RSP2 Production Cryocooler and Electronics Performance

T. Conrad, B. Schaefer, R. Yates, D. Bruckman, M. Barr, M. Kieffer

Raytheon Space and Airborne Systems
El Segundo, CA 90245

ABSTRACT

The Low-Temperature Raytheon Stirling / Pulse Tube 2-stage (LT-RSP2) hybrid cryocooler is a long-life, robust machine designed to operate efficiently at a first stage temperature of 55K with a capacity of 5W and a second stage temperature of 10 K with a capacity of 300mW. The LT-RSP2 design was finalized in mid-2009, with fabrication of the prototype unit taking place in late 2009 and early 2010 and execution of the production program in 2011 – 2016. It has recently been mated to a set of production electronics and subsequently undergone characterization and dynamometer testing. The results of this testing are presented along with aspects of the production cryocooler and electronics designs.

INTRODUCTION

The development of practical low temperature (sub-12K) closed-cycle cryogenic coolers is being actively pursued by Raytheon Space and Airborne Systems due to the beneficial detector chemistries that are enabled by such a system. Arsenic doped silicon (Si:As) focal planes offer exceptional performance for Long Wave Infrared (LWIR) astronomy and earth sensing applications; however, cryogenic cooling to temperatures below 12 K is required for operation. Existing state of the art space and airborne closed cycle cryocooler systems are generally incapable of simultaneously holding the required loads below 12 K / 55 K, and stored-cryogen systems have typically been employed instead. The amount of cryogen required is considerable and can easily exceed the mass and volume of the instrument. Launch mass and volume limitations thus provide severe constraints on mission lifetime. A closed loop cryogenic solution would therefore offers not only smaller mass and volume, but longer mission lifetime and reduced logistics costs. To date Raytheon has designed, built and tested three different Thermo-Mechanical Units (TMUs) that meet the requirements of Si:As and other systems: the AFRL funded High Capacity-RSP2 (HC-RSP2), an IRAD funded LT-RSP2, and a production LT-RSP2.

BACKGROUND

Stirling / Pulse Tube Hybrid Cryocoolers

Raytheon's Stirling / Pulse Tube Hybrid 2-Stage (RSP2) cryocoolers have been matured through an active development effort spanning approximately 15 years [1, 2]. They combine an actively driven Stirling first stage with a passive pulse tube second stage. The architecture of each RSP2

Cryocoolers 19, edited by S.D. Miller and R.G. Ross, Jr.

expander module is identical, the main differences being the sizing of the compressor module and the size of the Stirling piston and the pulse tube on the expander. In this way, the RSP2 architecture can be adapted to accommodate a wide variety of 1st and 2nd stage temperatures and load requirements. Nominal design temperatures (1st stage / 2nd stage) for the RSP2 coolers have included 55 K / 10 K for the LT-RSP2, 85 K / 35 K for the HC-RSP2, and 110 K / 58 K for the Mid-Capacity (MC-RSP2). Even with the compressor and expander sizes fixed, however, there remains a great deal of flexibility in operating temperature and load capacity for each of the RSP2 coolers. For example, the initial 60 K / 12 K testing at the beginning of the LT-RSP2 development effort was performed using the HC-RSP2, nominally designed for 85 K / 35 K operation [3].

The flexibility in operating temperatures and capacity of the RSP2 cryocoolers is due in large part to their ability to load shift between their first and second stages. This capability is unique to the Stirling / pulse tube hybrid architecture and provides a means for temperature control without the need for trim heaters [4]. Load shifting allows for the ratio of refrigeration capacities at the two stages to be altered via a simple software command of the Stirling phase angle (ø), without a significant loss of overall efficiency.

The capability to load shift is extremely valuable to systems utilizing two-stage cryocoolers, providing benefits during the system design, integration, and operational phases of a typical mission. During the design and integration phases, the operating temperature and capacity demands placed on the cryocooler are often revised, with the final requirement being determined during an integrated system test. The load shifting capability allows the Stirling / pulse tube hybrid coolers to efficiently adapt to the changing requirements without adding trim heat to one or both stages. Similarly, using operation load shifting and the related capability to perform independent temperature control at each stage, the cryocooler can maintain its operating temperatures even as its heat rejection temperatures, parasitic loads, and the instrument power dissipation vary over time.

LT-RSP2 Development History

The development of the LT-RSP2 cryocooler will only be briefly summarized here as more detailed descriptions have previously been presented [1-3]. Initial testing at second stage temperatures of 12 K and below was performed starting in 2008 using the HC-RSP2 cryocooler, initially developed with support from the Air Force Research Laboratory (AFRL) for cooling at 85 K and 35 K. Upon completion of the testing internal investment was secured to develop an RSP2 cryocooler optimized for operation at 55 K and 10 K, capable of efficiently producing relatively large amounts of cooling at 2nd stage temperatures <10 K. The design of the IRAD LT-RSP2 cooler was started in 2009 and initial testing was performed in 2011. The IRAD unit successfully achieved its performance goals, leading to the design and fabrication of the first production unit. The evolution from the IRAD to production unit involved mostly minor changes aimed at improving robustness, incorporating standard connectors, and optimizing heat rejection interfaces. The design work on the production cooler began in late 2011 and fabrication was completed in 2013.

PRODUCTION LT-RSP2 PERFORMANCE

Thermodynamic Performance

Initial characterization testing of the production LT-RSP2 cryocooler began in July of 2013 with no-load temperature test showing 35.5 K on the 1st stage and 5.46 K on the 2nd stage for an input power of 466 W, as shown in Figure 1. After optimizing the fill pressure in the production cooler, thermodynamic performance characterization was performed at its nominal operating temperatures of 55 K and 10 K. This is summarized in the load map shown in Figure 2.

Figure 2 shows that the LT- RSP2 cryocooler is able to produce a range of cooling capacities at each of its stages depending upon the input power and Stirling phase angle. Variation in the Stirling phase angle acts to modify the ratio of first and second stage capacities for a given input power. For example, at 350 W of input power the capacity can be varied from 4.4 W at 55 K and 0.2 W at 10 K to 2.2 W at 55 K and 0.36 W at 10 K by adjusting the Stirling phase angle from 85 degrees to 65 degrees. This was the tested range of Stirling phase angles for this cooler and these temperatures,

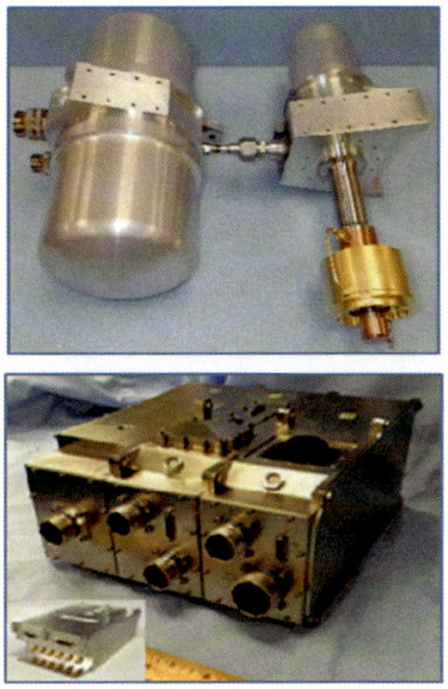

Parameter	Test Data
No Load Temperature and Power	35.5 K and 5.46 K at 466 W of TMU Power
Nominal Operating Point	4.3 W at 55K + 0.18 W at 10K 367 W TMU Power / 434 W System Power
Power Limit	600 W TMU / 800 W Electronics
Exported Forces and Torques at Nominal	< ± 50 mN or N-m in all Axis up to 243 Hz
Temperature Control	< ± 50 mK on Both Stages No Trim Heaters Required

Figure 1. Production LT-RSP2 Cryocooler and Control Electronics (left) and typical operating parameters (right).

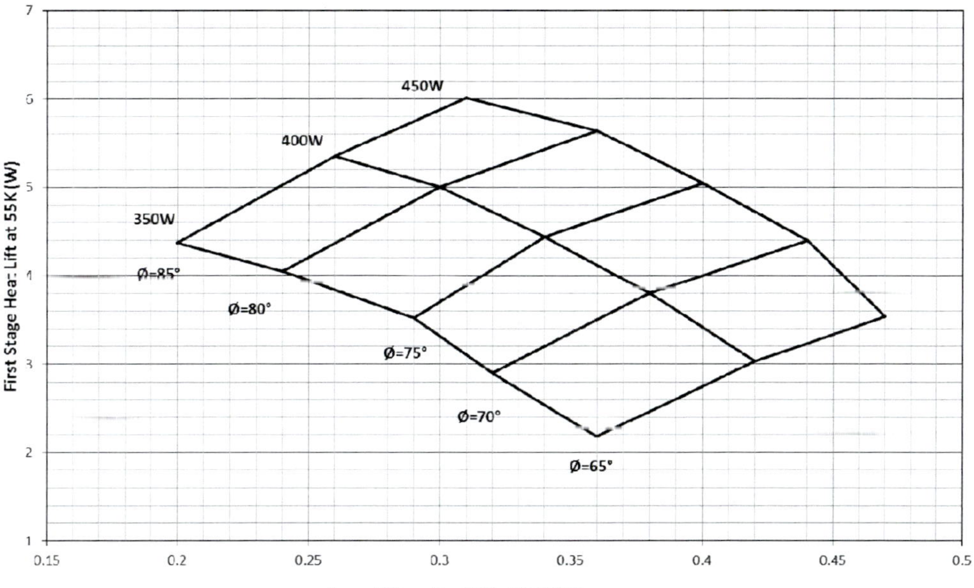

Figure 2. Production LT-RSP2 heat load capacity at 55 K and 10 K for input powers of 350 W to 450 W and Stirling phase angles of 65° to 85°.

however both higher and lower phases are possible, and thus additional range in stage cooling capacities is possible as well. This unique capability of Stirling / pulse tube hybrid cryocoolers to trade capacity between their stages through variation of the Stirling phase angle is referred to as load shifting. For the LT-RSP2 cryocooler, it is implemented very easily through modulation of the displacer drive waveform produced by the cryocooler electronics.

Cryocooler Thermal Control Electronics (CTCE)

During 2014-2016 the Raytheon Cryocooler Product Line designed, assembled and tested a modern set of common control electronics that is capable of powering and controlling all Raytheon cryocoolers. The Cryocooler Thermal Control Electronics (CTCE) combines modular, high-power

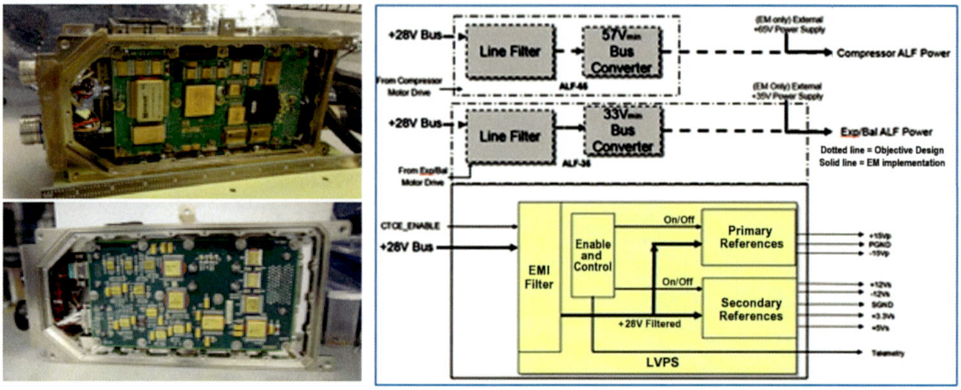

Figure 3. CTCE Individual Slices (left) and Cryocooler Drive Electronics Block Diagram (right).

motor drives with precision telemetry capabilities in a design that utilizes radiation hardened components and is capable of driving cryocoolers with maximum power requirements up to 800W. The CTCE is comprised of the following slices: Expander-Balancer-Control (EBC), Compressor Drive (CD), Low Voltage Power Supply (LVPS)/ Active Line Filter-35V (ALF-35), Cryocooler Mounted Electronics (CME), and the Active Line Filter-65V. Currently, these electronics slices are at TRL 5 having been tested with the LT-RSP2 cryocooler at power levels ~ 400W. An Engineering Model (EM) version populated with space prototype-grade components has been successfully tested with the RSP2 cooler with two stage temperature control with stability < 50mK on both stages and vibration control resulting in less than 350mN over the first 10 harmonics in all axes (X,Y,Z).

The Expander/Balancer Control processor hosts the piston position, temperature and vibration control algorithms. It generates all the power amp commands, stores vibration and Adaptive Feed Forward (AFF) drive signal data in memory, and provides piston position control. The driver amp circuits convert commands into motor drive Pulse Frequency Modulated (PFM) signals, compensates motor drive power with varying input voltages, and provides telemetry feedback. There are three drive circuits of 800W (Compressor) or 150W (Expander/Balancer) maximum drives supporting a range of cryocoolers. The LVPS and Active Line Filter block diagram is shown in Figure 3 and shows that the input power required is +28V from one or two feeds (dependent on cryocooler selection).The low-rate digital computations (AFF, Temperature Control and CTCE Control) are performed within the space-qualified microprocessor. The high-rate digital computations (Piston Position and Motor Current servos, Telemetry Data Acquisition) are performed in the FPGAs. The Low Voltage Power Supply (LVPS) provides the low voltage power to the CTCE system via high efficiency switching regulators. The active line filters (ALF) provides 35V and 65V cryo motor drive power while actively controlling conducted emissions back onto bus power, which is a significantly smaller and lighter solution than traditional passive EMI filters used on legacy designs.

The overall reliability of our third generation cryocooler electronics is excellent. Figure 4 shows the overall electronics reliability plot using dual string electronics and GSFC level 2 grade parts which achieve over 97% reliability at 10 years. The reliability is high, as our third generation electronics use fewer total parts than legacy systems. Our high density packaging design allows excellent thermal management keeping most components less than 20 degrees above our baseplate temperature and our hottest components junction temperatures at just under 35 degrees above our base plate temperature. Raytheon's third generation electronics are built from a tightly managed bill of materials as part of the Core Payload Architecture efforts, and the parts were carefully selected to enable procurement to support class A through class D missions. This provides a range of price points and reliability options just from changing the component quality. The architecture also allows single or dual string configurations, which further control costs on short life lower reliability missions. Testing in 2016 without the ALFs demonstrated efficiency in the rage of 87% - 89% without AFF vibration control active and 84% - 87% with AFF active controlling the vibration to the levels

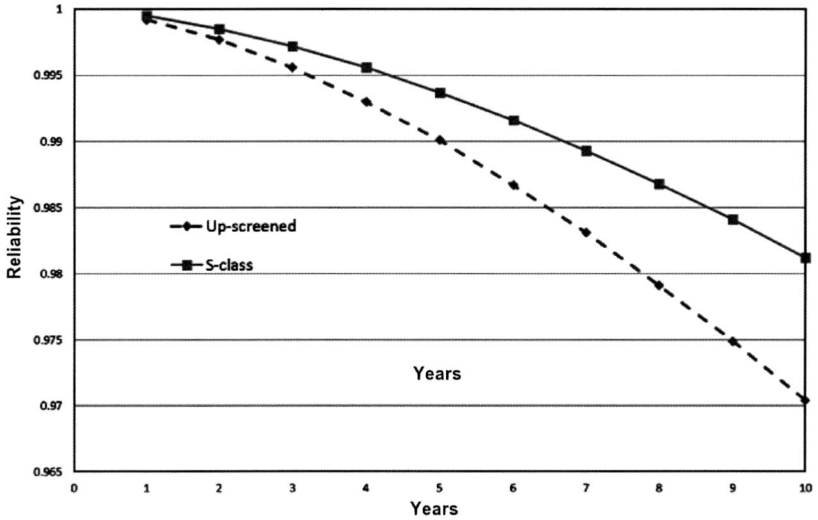

Figure 4. Calculated CTCE reliability depending on part types.

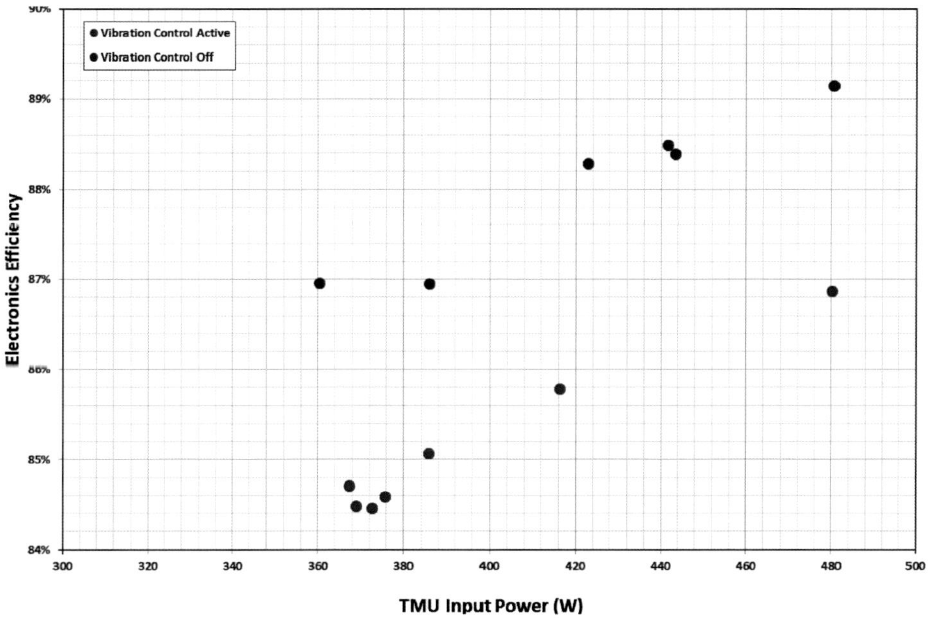

Figure 5. LT-RSP2 CTCE measured efficiency with production LTRSP2 TMU at 55K and 10K with vibration control on and off.

shown in Figure 5. Once the ALFs are integrated into the system the efficiency is expected to be close to 75%.

Exported Disturbance

Prior to the integration of a cryocooler with an instrument, the exported disturbance of the cryocooler must generally be characterized to ensure that it will not degrade the instrument's performance. Such characterization testing has been performed for the production LT-RSP2, which was operated on a dynamometer to determine its exported disturbance levels in the X, Y and Z axes. Prior

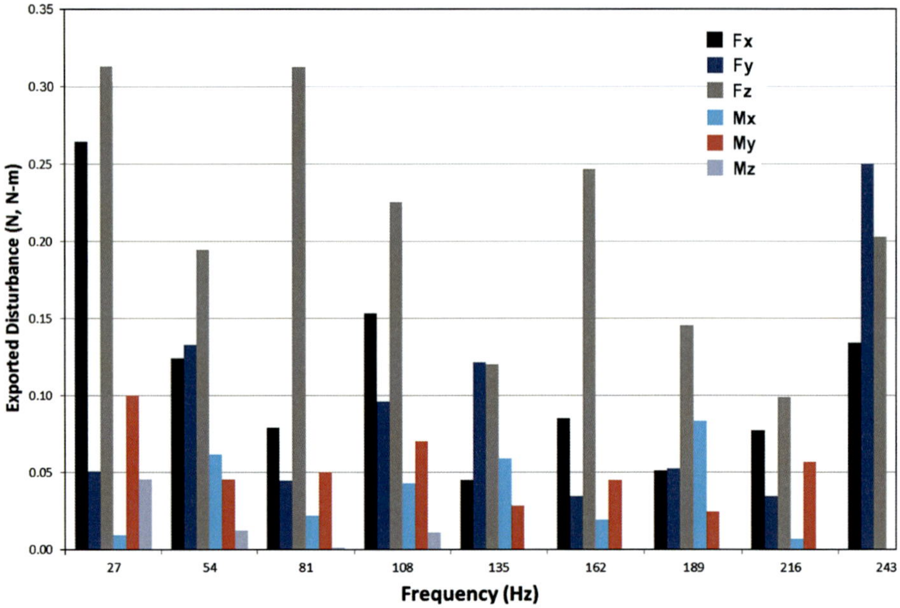

Figure 6. LT-RSP2 cryocooler system production unit exported disturbance levels. Note: levels not shown were not able to be determined due to STE interactions with dynamometer.

to this testing, the dynamometer fixture itself was characterized in order to determine its frequency-dependent transfer functions. Figure 6 shows measured exported disturbance levels for the production LT-RSP2 utilizing Raytheon's Adaptive Feed Forward (AFF) vibration cancellation algorithm implemented while operating at a compressor input power of 367 W. What is interesting is that these exported disturbance levels are similar to what was seen on a Raytheon single stage Stirling cryocooler at 115 W of input power. The results of the exported disturbance testing demonstrate that the production LT-RSP2 produces relatively low levels of exported disturbance, particularly for a cryocooler of its capacity, and that Raytheon's patented AFF algorithms are capable of handling a wide variety of input power levels.

Independent Dual Stage Temperature Control

The capability of the LT-RSP2 cryocooler to shift loads between its first and second stages also provides the basic mechanism for performing independent temperature control at each stage. From Figure 2, it is apparent that each combination of first and second stage loads is accomplished by a unique input power and Stirling phase angle. In its most basic form the two stage temperature control algorithm can be described as automatically seeking this combination of input power and phase, thereby finding the minimum input power which can hold the combination of loads. The actual implementation is a little more complex, as any combination of loads and stage temperatures within the cryocooler's capacity envelope can be commanded and a unique combination of input power and Stirling phase will be determined by the control algorithm. In a typical application the loads are not actually commanded; rather, they are the sum of the instrument parasitic and dissipative loads. The desired stage temperatures are simply commanded to the temperature control algorithm and the required input power and Stirling phase are generated by the controller.

Figure 7 shows a demonstration of the control algorithm's response to step changes in applied loads on each stage and the resulting cryocooler performance. The top chart in the figure shows a history of the first and second stage loads, which were varied during the test. The middle chart shows the Stirling phase and compressor input power adjustments made by the algorithm in response to the load changes. At the bottom, the resulting first and second stage temperature histories are shown. The temperature set points were 55 K and 10 K for the first and second stages, respectively.

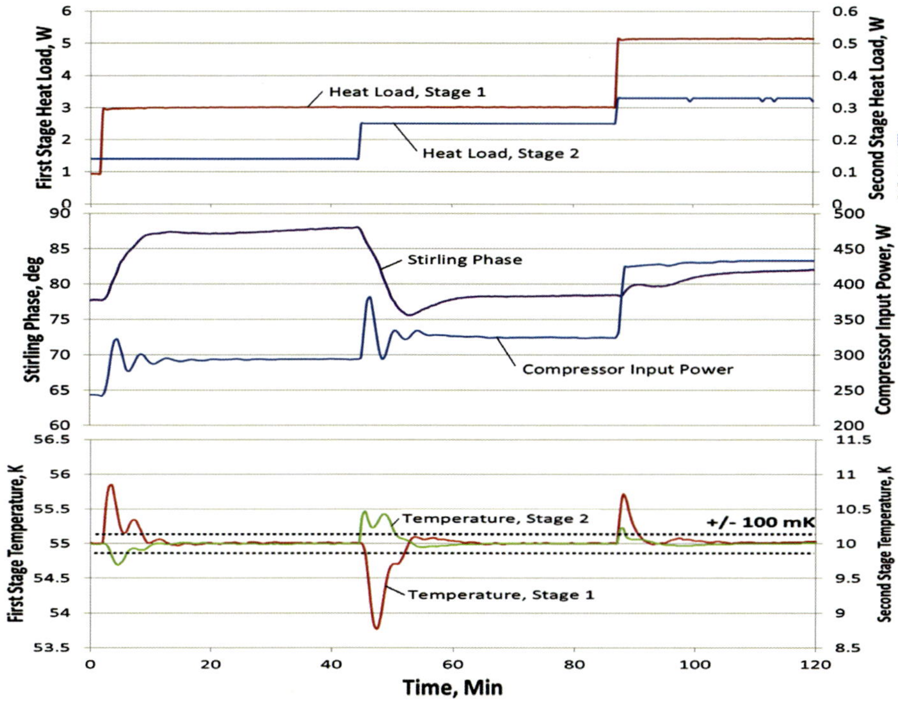

Figure 7. Demonstration of dual stage temperature control response to step functions in applied loads.

During the test, a step change in the first stage applied load was made, followed by a step in the second stage load and finally a simultaneous step at both stages. The control algorithm responded to the first step by increasing both the input power and Stirling phase, consistent with the load map of Figure 1 which showed that an increase in phase results in an increase in first stage capacity and a decrease in second stage capacity. Likewise, in response to the second load step which was on the second stage, the control algorithm again increased the input power but decreased the phase angle. Finally, in response to the simultaneous load step the input power increased and a smaller adjustment was made to the Stirling phase angle. In each case the temperatures at both stages exhibited a brief transient response before converging again to the commanded temperatures of 55 K and 10 K. Temperature stability outside of the transient responses was well within the range of +/- 100 mK, a typical cryocooler temperature stability requirement. The transient responses are exaggerated in this test compared to what would be expected in an integrated configuration due to both a lack of thermal mass attached to either stage, which would greatly slow the temperature responses, and the generally unrealistic step changes in load.

Integrated Performance

The production LT-RSP2 cryocooler has been integrated and tested with an instrument utilizing its cooling capacity at both stages. During this testing, the independent dual stage temperature control and adaptive feed forward vibration cancellation algorithms were successfully demonstrated along with the cryocooler's ability to meet the instrument's cooling requirements. In the integrated configuration, the adaptive feed forward algorithm was able to greatly reduce the magnitude of its feedback signals, provided by load washers, into or near their noise floor and significantly reduce the qualitatively observed level of exported force.

The dual stage temperature control algorithm was also demonstrated during the integrated testing. As shown in the top plot of Figure 8, the additional thermal mass of the instrument improves the temperature stability attained by the control algorithm. In this figure, a temperature set point change

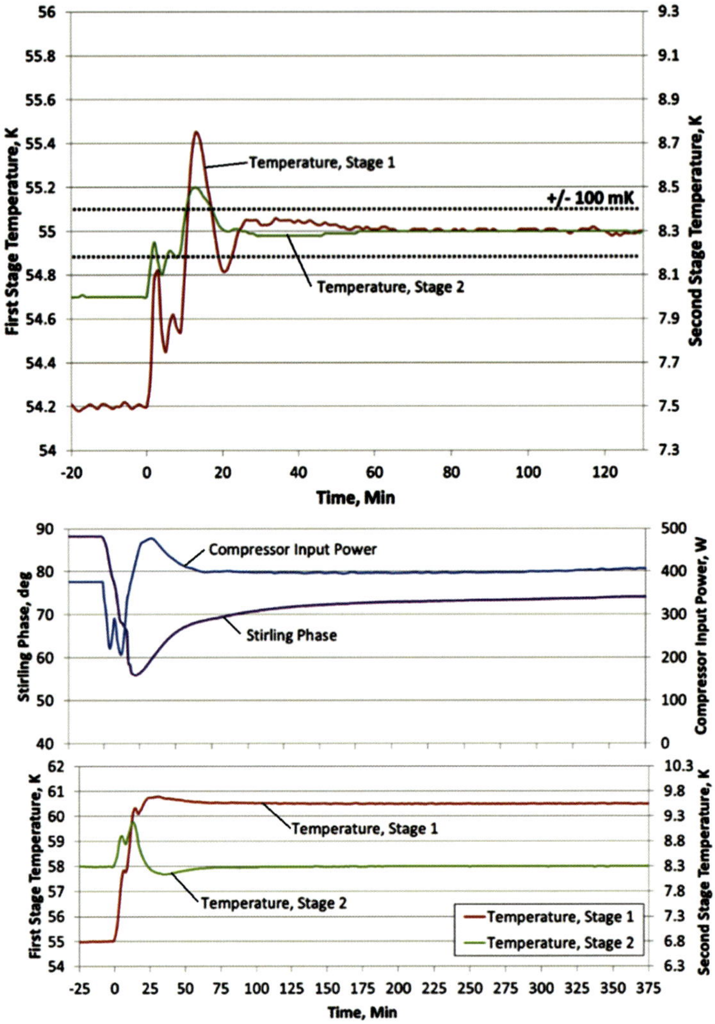

Figure 8. Integrated performance (top) and cryocooler temperature control response (bottom).

is commanded from 54.2 K and 8.0 K to 55 K and 8.3 K on the first and second stages, respectively. Both first and second stage temperatures converged within 100 mK of their set points within 20 minutes of the command being given at 0 minutes.

Similarly, the bottom plot in Figure 8 shows a set point change from 55 K to 60 K on the first stage, while the second stage is maintained at 8.3 K. In the top curves of the bottom chart, histories of input power and Stirling displacer phase are shown while the bottom curves of this chart show the two stage temperature histories. The transition begins at 0 minutes, but this time requires approximately 60 minutes to complete due to the thermal mass attached to the first stage. The control algorithm reduces the Stirling phase in order to change the ratio of first to second stage capacity and shift capacity to the second stage.

CONCLUSION

Raytheon has demonstrated that a relatively simple two-stage linear cryocooler can be used for 10 K refrigeration. The LT-RSP2 provides an attractive option for space and airborne sensors in this temperature range because it avoids the complexity of an auxiliary J-T cryocooler while providing the operational flexibility inherent to the RSP2 architecture. This flexibility results from the unique load shifting ability of the Stirling / pulse tube hybrid cryocooler, which has been harnessed to produce a dual stage independent temperature control capability for the RSP2 machines. The LT-RSP2 has been successively matured over three design iterations, resulting in the current production unit which has now undergone integrated testing with an instrument utilizing both stages of its cooling capacity and a set or control electronics. During this testing, the independent dual stage temperature control and adaptive feed forward vibration cancellation algorithms were successfully demonstrated along with the cryocooler's ability to meet the instrument's cooling requirements.

ACKNOWLEDGMENT

Raytheon would like to thank the Air Force Research Laboratory at Kirtland Air Force Base for their broad support of the RSP2 development and the use of the HC-RSP2 cryocooler in performing initial low temperature testing. The authors would like to further acknowledge the outstanding contributions of the LT-RSP2 design team.

REFERENCES

1. Kirkconnell, C .S., Hon, R. C. and Roberts, T., "Raytheon Stirling/Pulse Tube Cryocooler Maturation Programs," *Cryocoolers 15*, edited by S. D. Miller and R. G. Ross, Jr., ICC Press, Boulder, CO, 2009, pp. 31-37.

2. Schaefer, B.R., Bellis, L., Conrad, T., Bruckman, D., Yates, R., Pillar, M., and Barr, M., "Raytheon Low Temperature RSP2 Production Program," *Cryocoolers 18*, edited by S. D. Miller and R. G. Ross, Jr., ICC Press, Boulder, CO (2015), pp. 19-25.

3. Hon, R.C., Kirkconnell, C.S., and Schrago, J.A., "Raytheon RSP2 Cryocooler Low Temperature Testing and Design Enhancements," *Adv. in Cryogenic Engineering*, vol 55 (2010), pp. 371-380.

4. Conrad, T., Schaefer, B., Bruckman, D., Bellis, L., and Kieffer, M., "Demonstration of Two Stage Temperature Control for Raytheon Hybrid Cryocoolers," *Cryocoolers 19*, edited by S. D. Miller and R. G. Ross, Jr., ICC Press, Boulder, CO (2016), (this proceedings).

5. Kirkconnell, C. S. and Price, K.D., "Thermodynamic Optimization of Multi-Stage Cryocoolers," *Cryocoolers 11*, edited by R. G. Ross, Jr., Kluwer Academic/Plenum Publishers, New York (2001), pp. 69-78.

Design on a Two-Stage Stirling/Pulse Tube Hybrid Cryocooler

C.H Yin, Z.H Gan, B. Wang, Y.X. Guo

Institute of Refrigeration and Cryogenics, Zhejiang University
Hangzhou 310027, P.R. China
Key Laboratory of Refrigeration and Cryogenic Technology of Zhejiang Province
Hangzhou 310027, P.R. China

ABSTRACT

In the field of military and aerospace, there are strict requirements for cryocoolers, such as high efficiency, compactness, light weight, high reliability, long life and low vibration limits. At present, most mechanical cryocoolers which can be applied in space are Stirling cryocoolers and Stirling type pulse tube cryocoolers. Two-stage cryocoolers are sought-after for aerospace applications, since the performance of single-stage cryocoolers is limited below 35 K, especially at 20 K. Compared with two-stage Stirling cryocoolers, Stirling/pulse tube hybrid cryocoolers eliminate the moving parts of the second stage, which leads to longer life. Moreover, hybrid cryocoolers can achieve higher pressure ratios and efficiency in their regenerator due to a better match in charging pressure between stages. In this study, we designed a two-stage Stirling/pulse tube hybrid cryocooler at 20 K with low temperature inertance tubes and a surge volume to adjust the phase difference. We simulated the performance of the designed cryocooler using the Sage code, and proved that the as-designed cryocooler can provide 0.54 W cooling power at 20 K with 180 W electrical power input. In addition, the effects of the phase difference, mass flow, and pressure ratio in the displacer are also discussed.

INTRODUCTION

Regenerative cryocoolers have been widely applied in aerospace and other fields due to their promising features, such as high efficiency, compactness, light weight, high reliability, long life and low vibration. At 80 K, single-stage Stirling cryocoolers and Stirling type pulse tube cryocoolers predominate. However, below 35K, especially near 20K, the performance of single-stage cryocooler is limited and multi stage cryocoolers are required.

Compared with two-stage Stirling cryocoolers, Stirling/pulse tube hybrid cryocoolers eliminate the moving parts of the second stage, which leads to longer life. However, Stirling type pulse tube cryocoolers require a larger gas volume that must be cyclically pressurized and depressurized. It is known that the larger gas volume increases the gas mass flow and reduces the efficiency of the regenerator[1]; moreover, Stirling type PTCs need higher charge pressure to approach their optimal phase condition, and this increases the pressure drop and reduces the efficiency of the system.

Stirling/pulse tube hybrid cryocoolers evade the wear and tear of Stirling cryocoolers and the low efficiency of Stirling type pulse tube cryocoolers, taking advantage of the best of these two kinds of cryocoolers to achieve high efficiency and long life.

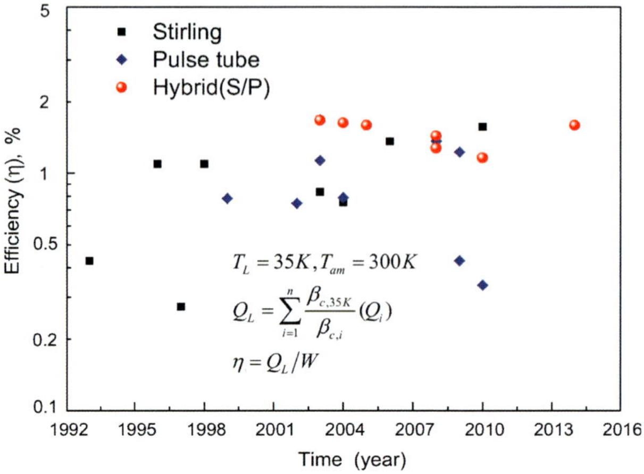

Figure 1. Comparison of the efficiency of Stirling cryocoolers at 35 K region.

We collected efficiency data for Stirling/pulse tube hybrid cryocoolers, Stirling cryocoolers and Stirling type pulse tube cryocoolers between 1992-2014, as shown in Figure 1. From the plot, it is apparent that the efficiency of hybrid cryocoolers is higher than those of others.

Besides the high efficiency and long life, Stirling/pulse tube hybrid cryocoolers also exhibit a better pressure-matching and higher pressure ratio at the cold end. For high frequency Stirling type pulse tube cryocoolers, the different temperature ranges correspond to different best charge pressure. For example, the best charging pressure for Stirling type PTC in the 80 K region is 2-2.5 MPa, while in the 20 K region it is 1.0 to 1.5 MPa. Therefore, for multistage gas coupling cryocoolers, it is challenging to optimize the charging pressure at different temperature regions simultaneously. In order to guarantee the performance in the low temperature region, cryocoolers must sacrifice efficiency in the 80 K region. For multistage cryocoolers with thermal coupling, they can use an independent compressor system to achieve the maximum efficiency of each stage. While the system becomes complex, the charging pressure of the Stirling cryocooler should be between 1 to 2 MPa, so that it can match well with the secondary Stirling type pulse tube cryocooler's charging pressure preference. A good pressure match on different stages can reduce the pressure drop in the regenerator and raise the pressure ratio at the cold end. This means that Stirling/pulse tube hybrid cryocoolers have the potential to achieve better performance than multistage Stirling type pulse tube cryocoolers.

The Raytheon Company first proposed the concept of a Stirling/pulse tube hybrid cryocooler and designed a prototype in June 2000. The prototype adopted a U structure for the second stage and achieved 2.2 W at 80 K and 0.5 W at 35 K.[2]

The Raytheon Company improved the cryocooler in 2002 and formed the RSP2 model which increased performance up to 1W at 35K plus 7W at 100 K for a drive motor whose input power is less than 170 W. The total mass is less than 7 kg. [3]

In 2005, the Raytheon Company redesigned the gas transmission route and further improved the performance to 1.1 W at 60 K and 4.3 W at 110 K for a PV power consumption of 72 W.[4]

In 2008, the Raytheon Company developed a Dual-Use Cryocooler (DUC) with a coaxial structure and compact components. At the same time, Raytheon continuously improved the HC-RSP2 models, and developed MC-RSP2 models. The performance of the MC-RSP2 is 2.7 W at 58 K and 6.7 W at 110 K with a PV power consumption of 210 W.[5]

In 2009, the Raytheon Company further improved the HC - RSP2 and MC - RSP2 coolers to make their performance better. Test results were: 2.6 W at 35 K and 16.2 W at 85 K with an electric power consumption of 513 W for HC-RSP2, and 2.4 W at 58 K and 6.1 W at 110 K the for MC-RSP2 design with an electric power consumption of 166 W. The initial design requirements were basically met. What's more, it was determined that the HC-RSP2 could provide 0.38 W of cooling capacity at 12 K, and the lowest refrigerating temperature was below 10 K.[5]

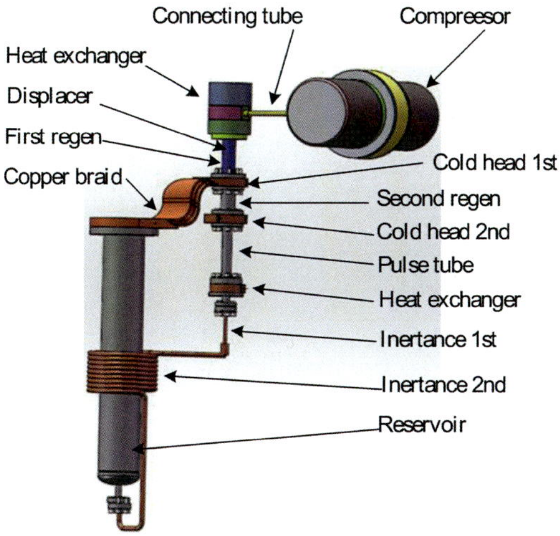

Figure 2. Design sketch of the Stirling/pulse tube hybrid cryocooler.

In 2010, in order to explore the application of 4-10 K region, Raytheon Company designed a cryocooler LT-RSP2, based on the HC-RSP2 design, which was focused on low-temperature cryogenic applications. The design goal for LT-RSP2 was to achieve 1 W at 12 K and 5W at 60K with an input power of 500 W; the lowest cooling temperature was close to 9 K.[6]

In 2013, the Raytheon Company reworked the DUC system designed in 2008. Using an advanced regenerator they streamlined the assembly, and the efficiency of the thermodynamics was improved as a result; the resulting Carnot efficiency for the DUC system was increased by 2%.[7]

DESIGN FOR THE STIRLING/PULSE TUBE HYBRID CRYOCOOLER

In light of the above-mentioned advantages, a two-stage hybrid cryocooler was designed. Figure 2 depicts the cryocooler's structure. The first stage is the POLAR SC-7 COM Stirling cryocooler, and the second stage is a pulse tube hybrid cryocooler. The structure is gas coupled. A low temperature inertance tube and surge volume were adopted to adjust the phase difference. The copper surge volume is heat sunk to the cold head of the first stage through a copper braid.

We simulated the performance of the as-designed PTC using Sage. The performance of the POLAR SC-7 COM Stirling cryocooler is 6 W at 77 K with an input power of 180 W. Based on the experiments and simulation, the cooler can provide a cooling capacity of 9.2 W at 104 K when the input electric power is 180 W, frequency is 37 Hz, and pressure is 1.7 MPa.

The temperatures at different positions in the cryocooler are shown in Figure 3. For the first stage, the temperature of hot end is about 310 K, while the temperature of cold end is about 104 K. It provides 9.2 W cooling capacity. For the second stage, the temperature of the cold end of regenerator is about 20 K. The hot end of pulse tube surge volume and inertance tubes are connected together, thus their temperatures are the same 104 K.

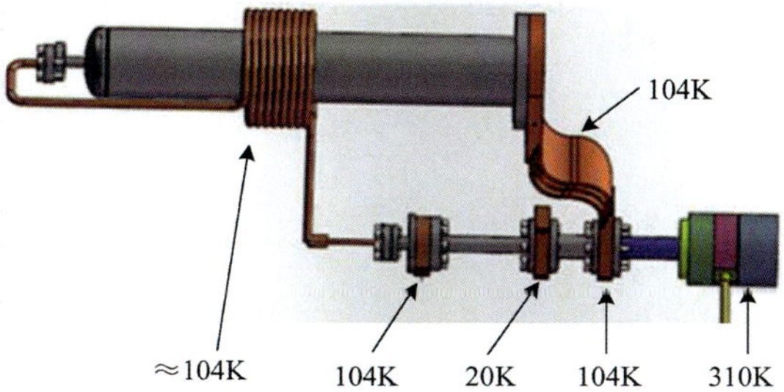

Figure 3. The temperature distribution in the cryocooler

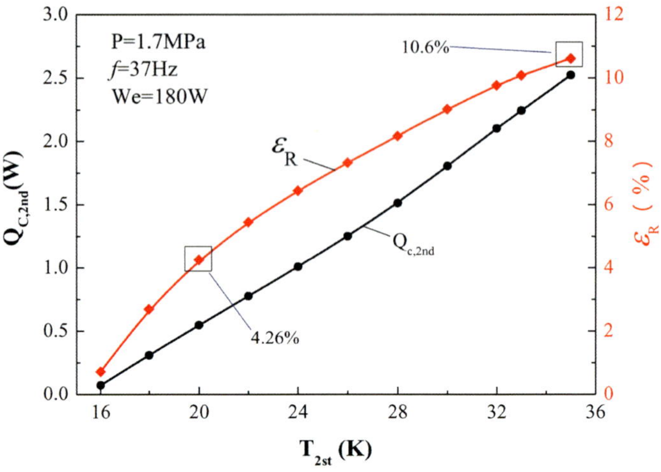

Figure 4. The efficiency of the designed hybrid cryocooler.

We simulated the refrigerating performance of the proposed hybrid cryocooler, and plotted the second stage cooling power and efficiency versus cold head temperature, as shown in Figure 4. The cryocooler can reach 0.54 W cooling capacity with a relative Carnot efficiency of 4.26% at 20 K. When the refrigerating temperature increases to 35 K, the cooling capacity is 2.52 W, and the relative Carnot efficiency is 10.6%.

THE KEY PARAMETERSANALYSIS

The key operating parameters of the Stirling/pulse tube hybrid cryocooler are analyzed as follows:

Dependence of the First-stage Temperature on Pressure and Frequency

Operating parameters such as frequency, pressure and input power have significant effects on the cryocooler performance. As seen from the Figure 5, when the input power is 180 W and the frequency is 37 Hz, the cooling capacity of the second stage (20K) increases first and then decreases as pressure increases. The best pressure is 1.7 MPa, which corresponds to the cooling power of 0.54 W. In addition, with the increasing pressure, the acoustic power of expansion chamber entrance increases. That is to say, increasing of cooling capacity of the first stage leads to reduced temperature at its cold head, as shown in Figure 6.

Similarly, when the input power is 180 W and the pressure is 1.7 MPa, the secondary cooling capacity increases first and then reduces with increasing frequency. The optimal frequency is 37-38 Hz,

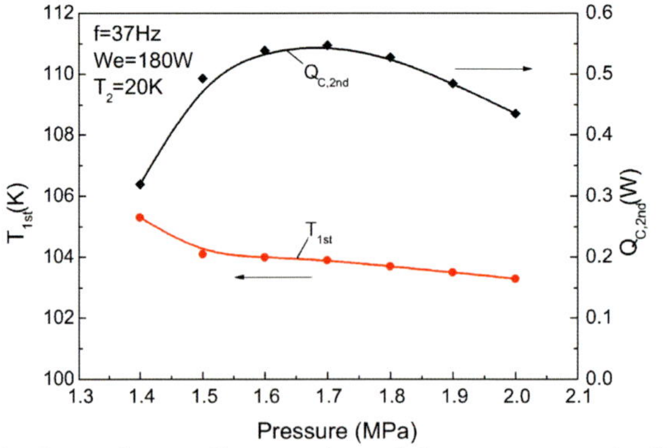

Figure 5. Graph of secondary cooling capacity and the temperature of cold head for the first stage when pressure changes.

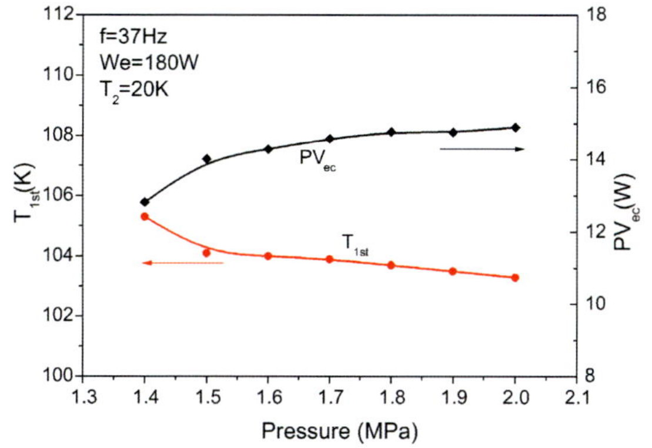

Figure 6. Graph of the temperature of cold head for the first stage and the acoustic power of expansion chamber entrance.

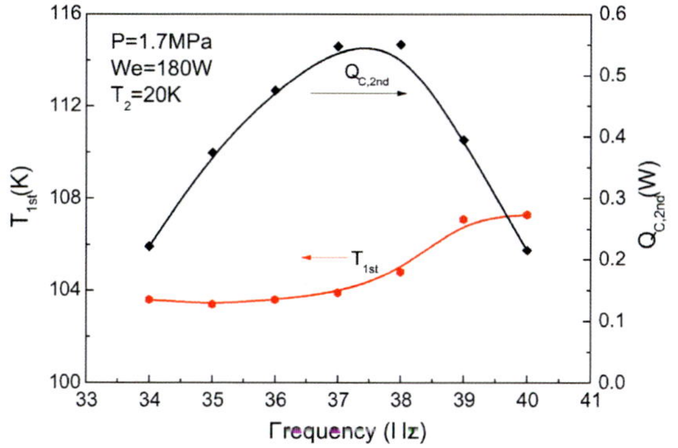

Figure 7. Graph of secondary cooling capacity and the temperature of cold head for the first stage when frequency changes.

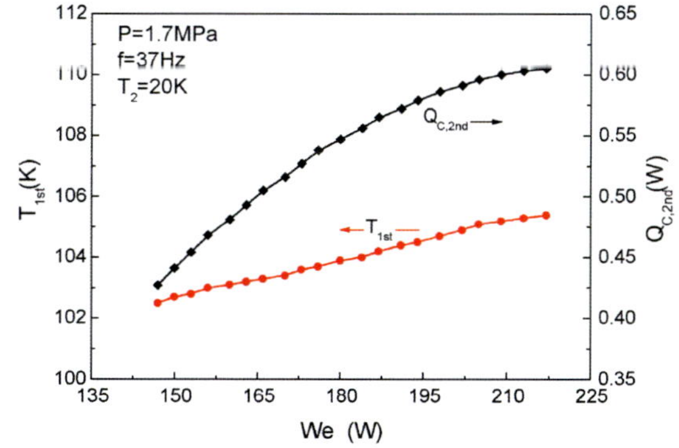

Figure 8. Graph of secondary cooling capacity and the temperature of cold head for the first stage when input power changes.

and correspondingly, the cooling capacity is 0.54 W. In addition, with the increasing frequency, the first stage's refrigerating temperature increases as shown in Figure 7. This is because the effect of frequency on the thermal efficiency of the regenerator is significant. A frequency which is too high or too low will cause the efficiency drops, leading to a drop of pressure.

Figure 8 shows the dependence of the secondary cooling capacity and the temperature of the cold head for the first stage on input power. Increasing the input power, the first stage's refrigerating

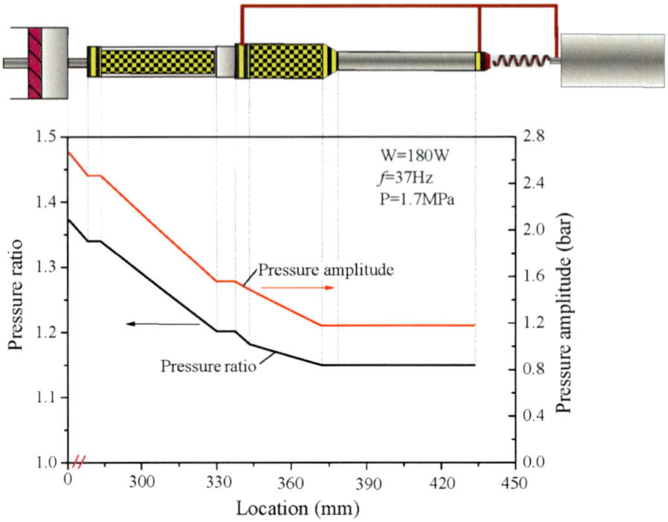

Figure 9. Changes of pressure ratio and pressure amplitude in the cryocooler.

temperature and secondary cooling capacity also increase. What's more, the first stage's refrigerating temperature increases almost linearly with the input power, but the secondary cooling capacity increases more and more slowly. When the input power increases to a certain extent, the secondary cooling capacity tends to be stable.

The Variance of Pressure Ratio and Acoustic Power in the As-designed PTC

In order to analyze the changes of parameters, such as pressure ratio and mass flow, the structure of this cryocooler is simplified. As shown in Figure 9, from left to right, the components are compressor, connecting tube, heat exchanger for the hot end, the first regenerator, expansion chamber, the intermediate heat exchanger, secondary regenerator, heat exchanger for the cold end, pulse tube, heat exchanger for the hot end of pulse tube, inertance tubes, surge volume and thermal bridge connecting secondary cold end and inertance tubes.

In Figure 9, the changes of pressure ratio and pressure amplitude are synchronized. Pressure drop is mainly caused by the connecting tube, the first regenerator and the secondary regenerator. This is because the regenerators are filled with wire mesh, which increases the flow resistance. Although there is no wire mesh compared with the heat exchanger and regenerators, the connecting tube is long and its diameter is small, which causes a significant pressure drop.

As mentioned above, the pressure drop happens in the connecting tube and regenerators, which can attenuate the acoustic power correspondingly. Figure 10 shows the variance of acoustic power within each component in the cryocooler. Along the connecting tube, heat exchanger for the hot end and the first regenerator, acoustic power decreases from 98 W to around 15 W until the expansion chamber, and continues to decrease to 4.3 W in the second regenerator again. In the pulse tube, acoustic power stays stable.

The Changes of Phase Difference and Mass Flow

As shown in Figure 11, the mass flow decreases significantly in the hot end heat exchanger, increases between the first and secondary state regenerators, and decreases in the pulse tube again. This is because the temperature in the hot end heat exchanger rises, which decreases the density of helium. Correspondingly, in the first and secondary regenerator, the temperature reduces, which makes the density of helium increase. In the pulse tube, temperature rises from around 20 K to 104 K, which makes the density of helium decrease again. In the cryocooler, the phase difference changes greatly at the hot end heat exchanger: from 28 ° to nearly 1°. In the first regenerator, it changes from 1° to -27 °.

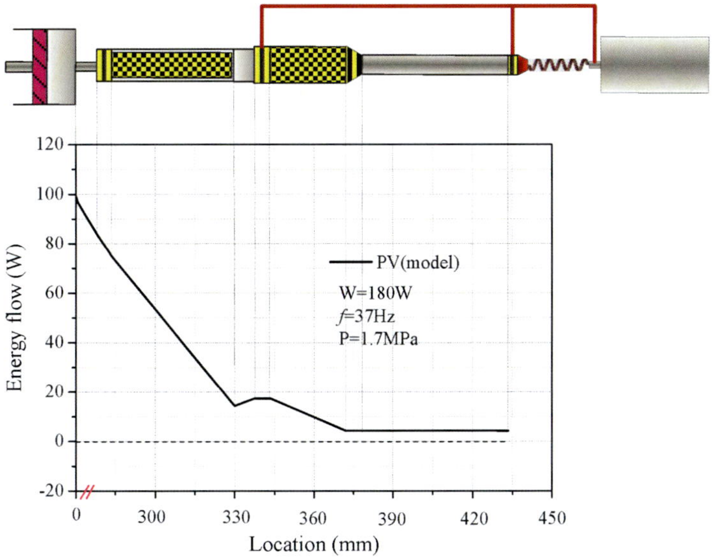

Figure 10. Changes in acoustic power within each component in the cryocooler.

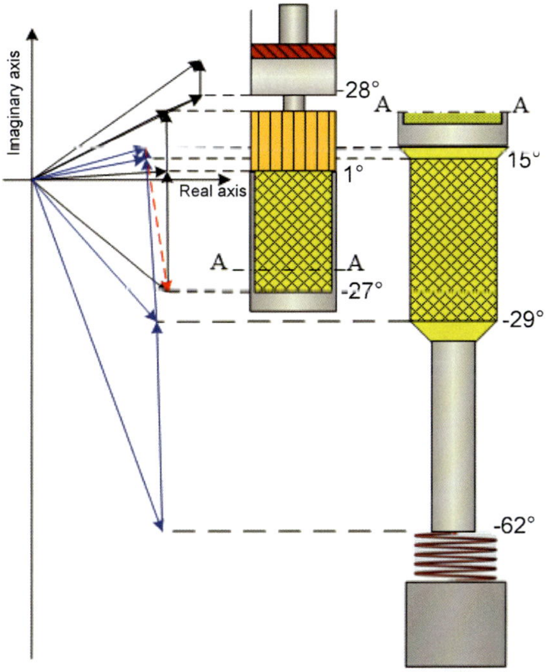

Figure 11. Changes of mass flow and phasor in the cryocooler.

Helium is shunted between the first and the secondary stage. As a result, the phase difference at the hot end of secondary regenerator is about 15° and at the cold end is - 29°. The structure of low temperature inertance tubes and surge volume lead to the phasor at the hot end of pulse tube reaching -62°.

CONCLUSION

In this paper, we compared the performance of a two-stage Stirling cryocooler, Stirling PTC and Stirling/pulse tube hybrid cryocooler. We also proposed a model for a two-stage gas coupled Stirling/pulse tube hybrid cryocooler. We simulated the performance of the as-designed cryocooler by studying the variance of operating parameters, such as the pressure ratio, acoustic power, mass flow and phase difference. The proposed two-stage Stirling/pulse tube hybrid cryocooler can provide 0.54 W at 20 K with 180 W input power. The experiments will be carried out soon.

ACKNOWLEDGMENT

This work is financially supported by the National Natural Science Foundation of China (No. 51376157).

REFERENCES

1. Price, K.D., Kirkconnell, C.S. "Two Stage Hybrid Cryocooler Development," *Cryocoolers 12*, Kluwer Academic/Plenum Publishers, New York (2003), pp. 233-239.
2. Price, K.D., Urbancek, V. "Raytheon 95K High Efficiency Cryocooler Program," IEEE. 2000.
3. Kirkconnell, C., Price, K. Ciccarelli, K., Harvey, J., "Second Generation Raytheon Stirling/Pulse Tube Hybrid Cold Head Design and Performance," *Cryocoolers 13*, Springer Science+Business Media, New York (2005), pp. 127-132.
4. Hon, R.C., Kirkconnell, C.S. and Roberts, T. "Raytheon Dual-use Cryocooler System Development," *Adv. in Cryogenic Engineering*, Vol. 53, Amer. Institute of Physics, Melville, NY (2008), pp. 538-548.
5. Kirkconnell, C.S., Hon, R.C. "Raytheon Stirling / PulseTube Cryocooler Maturation Programs," *Cryocoolers 15*, ICC Press, Boulder, CO (2009), pp. 31-37.
6. Hon, R.C., Kirkconnell, C.S. and Shrago, J.A. "Raytheon RSP2 Cryocooler Low Temperature Testing and Design Enhancement," *Adv. in Cryogenic Engineering*, Vol. 55, Amer. Institute of Physics, Melville, NY (2010), p. 371.
7. Schaefer, B.R. et al, "Raytheon Low Temperature RSP2 Cryocooler Airborne Testing," *Adv. in Cryogenic Engineering*, Vol. 59, Amer. Institute of Physics, Melville, NY (2014), pp. 1318-1324.

Engineering Model of a High Power Low Temperature Pulse Tube Cryocooler for Space Application

Y. Pennec[1], J. Butterworth[1], G. Coleiro[1], P. Barbier[1], S. Martin[1,2], P. Crespi[1], I. Charles[2], J-M Duval[2], J. Mullié[3], G. de Jonge[3], T. Benschop[3], M. Linder [4]

[1]Air Liquide Advanced Technologies,
38360 Sassenage, France
[2]Univ. Grenoble Alpes, INAC-SBT, F-38000 Grenoble, France
[3]Thales Cryogenics,
5626 DC Eindhoven, The Netherlands
[4]European Space Research and Technology Center,
Noordwijk, The Netherlands

ABSTRACT

Air Liquide Advanced Technologies (ALAT) has designed, built and commissioned a novel low temperature high power pulse tube cryocooler for space applications. This work was funded by the European Space Agency (ESA) under a Core Technology Program (CTP) in collaboration with Commissariat à l'Energie Atomique (CEA) and Thales TCBV. The cryocooler builds on ALAT commercial LPTC heritage and boosts its performance using a two cold finger configuration with an active phase shifter. The cryocooler provides more than 400 mW at 15K with low level of vibrations and electro-magnetic (EM) emission. The cryocooler passed initial qualification stress tests (temperature, vibrations, vacuum) and is currently at Technological Readiness Level 5 (TRL5).

INTRODUCTION

Previous developments provided ESA programs with single stage pulse tube coolers. They offer reliable, powerful and efficient cooling down to 45K. An example of such a cooler is the Air Liquide Advanced Technology (ALAT) Large Pulse Tube Cooler (LPTC) offering up to 3 W at 50K. Looking forward, Future Earth Observation missions will require cooling well below the 20K mark, to either extend the wavelength range of MCT Infrared detectors or others THz detectors. Ambitious scientific missions aimed at unraveling the mystery of the birth of the universe require cooling sensitive detectors down to the 50mK range or below. In this context our high power low temperature pulse tube cooler will provide the necessary foundation for effective cryochains. It can fulfill multiple roles such as supplying primary cooling to cryogenic shields or providing pre-cooling for helium Joule Thomson (JT) coolers. In particular 2K JT performances are extremely sensitive to heat reject temperature as the Joule Thomson coefficient improves significantly from to 20K down to 15K. Last but not least, achieving cooling capacity in the 20K and below will enable the use of standard superconducting materials. Direct applications can be seen for future electric propulsion systems, plasma confinement magnets, or loss free transmission antennas.

Cryocoolers 19, edited by S.D. Miller and R.G. Ross, Jr.

CRYOCOOLER ARCHITECTURE

Pulse tube cryocooler design and implementation becomes increasingly challenging as the temperature decreases. When applied to the space industry environment further constraints apply with tremendous focus not only on base performance but also on risk management and reliability. Hereafter, it was ALAT's concern to devise the simplest cryocooler achieving the highest thermodynamic performance using as much scientific, technical and manufacturing heritage as possible.

As the temperature decreases, it is a common solution to increase the number of stages and/or cold fingers to improve the thermodynamic efficiency. To mitigate the complexity level we opted for a configuration based on a first single stage cold finger cooling the intermediate heat exchanger of a secondary single stage cold finger, both driven by a common compressor. This arrangement (see Figure 1) allowed us to leverage extensively on our heritage. The first cryofinger is based on the LPTC [1], the second cryofinger is inspired by the previously developed 20K-50K model [2] and the compressor draws on heritage from the LPTC and LPT9710 compressors manufactured by Thales Cryogenics.

The cooler incorporates two breakthrough technologies in space coolers: the use of a novel low temperature regenerator and an active phase shifter for the second stage. The regenerator employs specifically developed materials. An extensive experimental campaign has been carried out to fine tune the regenerator performance at the 15K operating point [3]. Achieving optimal phase angle between the pressure wave and the mass flow has often been a limiting factor in pulse tube efficiency. Not to mention that at low temperature, the phase lag generated by a capillary inertance may be insufficient. Following [4], we have implemented a secondary small compressor as an active phase shifter. The significant performance and reliability gain outweighs the small weight, power and complexity penalties. Not only the phase shift can be arbitrarily set and fine tuned but the amplitude of the expansion stroke can also be chosen. It allows operating the low temperature cold finger as close as possible to its thermodynamics limits.

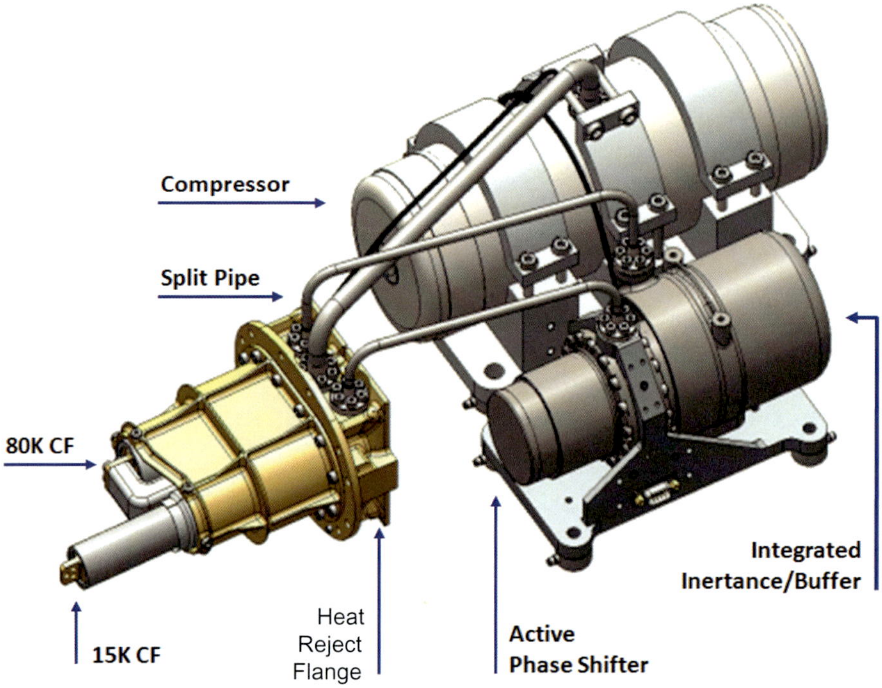

Figure 1. Cryocooler architecture: A common high power compressor drives the assembly of one intermediate single stage cold finger and one low temperature cold finger. The inertance of the intermediate stage is integrated around the active phase shifting miniature compressor for the low temperature finger.

The cooler is compact with a compressor envelope less than 100mm dia. x 300mm and a cold finger unit of less than 100 mm x 150mm x 225mm. The cooler was manufactured using space grade materials and fabrication methods. The compressor and support weigh 12 kg, the active phase shifter 3 kg and the cold finger assembly weighs 2.5 kg. The active phase shift compressor was used for our Miniature Pulse Tube Cooler. It has been commissioned previously and is also at TRL 5.

THERMAL PERFORMANCE

The target core performance of the cooler was to generate maximum power at 15K for efficient pre-cooling of a JT cooler. The unit provides above 400 mW cooling with 300 W electrical input at the compressor. If only limited cooling power is required in an application such as passive detector cooling a base temperature below 10K is accessible. The complete cooling power curve is shown in Figure 2A.

Thanks to the dual cold finger configuration, an additional cooling power of 3.5 W at 90K is also available at the intermediate stage. This feature is particularly interesting for thermalizing primary cryogenics shields for example. The base temperature of the intermediate stage can be as low as 70K. The complete curve is shown in Figure 2B. Lowering the intermediate stage does improve significantly the cooling capacity of the low temperature stage. To this point, we note that the low temperature stage performance is isolated from the warm flange temperature fluctuations as the additional heat load is intercepted by the intermediate stage.

The cooler is naturally stable in temperature as can be seen in Figure 2 D/E showing temperature drift in the course of $\sim$3 h without any active temperature control. < 10 mK/h at the low temperature stage and < 30 mK/h at the intermediate stage is achieved.

If the end application is stringent on weight, the cooler can be operated with a passive phase shifter for the low temperature stage combining an inertance and a double inlet. This configuration was found to be able to reproduce cooling power only slightly lower than with the active phase shifter but cooling capacity reproducibility and temperature and power stability do become a concern. Several technological pathways are foreseen to resolve this limitation. For programmatic reasons it has been decided to use the active phase shifter as the base line.

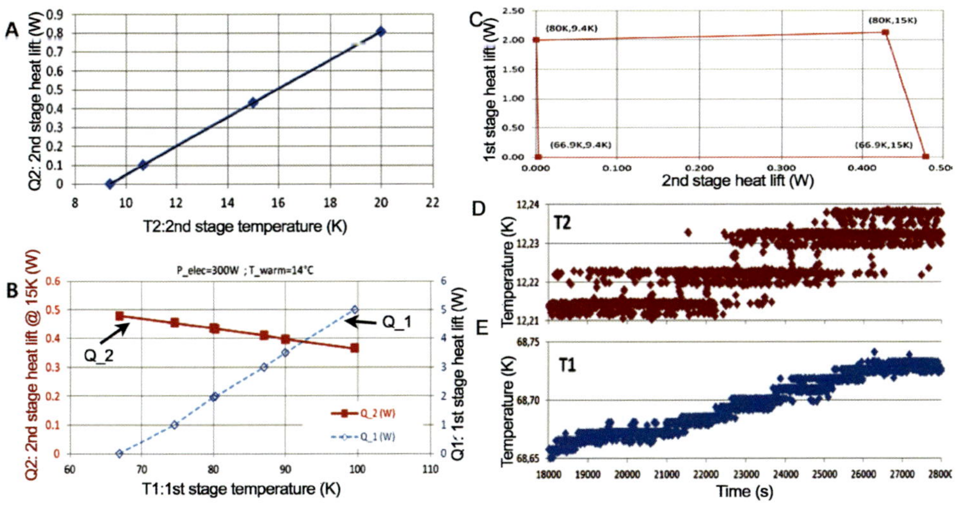

Figure 2. Thermal performance charts: A) Low temperature stage cooling power vs. temperature. B) Cooling power at both stages vs. intermediate stage temperature. C) Cooling power performances map. D/E) Temperature stability of the low and intermediate stage vs time.

EXPORTED ENVIRONMENT INTERFERENCE

Satellite platform, components and sensors are increasingly sensitive to environmental interferences. In particular admissible vibration and electromagnetic levels are reaching ever lower thresholds. Vibrations have direct impact on optical imaging quality. They can affect the satellite pointing accuracy and can also be a source of heating for ultra low temperature scientific missions. EMI level can affect neighboring electronics or add parasitic noise to high sensitivity sensors. Early focus on these environmental performance indicators are essential for the cooler maturation to ensure that potential trade-off against core thermal performances can be implemented if necessary.

The compressor design follows the standard dual opposed pistons design. The symmetric configuration naturally leads to a low vibration levels. The bare compressor is balanced generating less than 5N along the pistons and less than 1.5N off-axis at the driving frequency as seen in Figure 3A/B. Separate control over of the two pistons stroke with an active electronics allows further reduction of the exported vibrations down to the 100mN threshold. The peak at ~25 Hz is generated by the chiller water circulation connected to the warm flange. Note that the active control does not pertain to the off-axis vibrations.

The compressor exports both static and dynamic magnetic field. Leaks in the permanent magnet circuit of the two motors generate a complex 3-D vector field. Figure 4A shows the worst case measurement where a DC field up to 100 μT is found at the compressor face. It decays rapidly after 30 cm. The signal changes sign since it includes both the compressor and terrestrial field which here are of opposite sign. Figure 4C shows the spectrum of the AC field. A series of peaks are found at the driving frequency and its harmonics. The levels are on par with the DC values, ranging from 100μT at the compressor level and decaying below 10uT at 30cm.

ENVIRONMENTAL STRESS TESTING

A spatial cooler must endure intense mechanical, thermal and vacuum stress during the launch phase. The cold finger assembly and the compressor have been submitted to Ariane-like levels:

- 25g sine wave along X, Y and Z from 5Hz up to 100Hz with 2 octaves/minute sweep rate.
- Random excitation 20-100Hz +3dB/Oct, 100-300Hz 0.3 g^2/Hz, 300-2KHz -5 dB/Oct, 12.1g RMS.

The temperature was swept from -50 °C up to +90 °C across two days while the leak rate was monitored with a He leak detector.

The cold finger assembly has a first resonance mode above 180 Hz. The compressor passed also the random test. A passive dissipative resonant launch lock was implemented to reduce pistons motion during launch.

The compressor fully welded construction yields the expected leak free behavior even after the vibration stress test. The compressor was able to operate at -30 °C and +60°C. The cooler did not show any loss of performances after the stress tests.

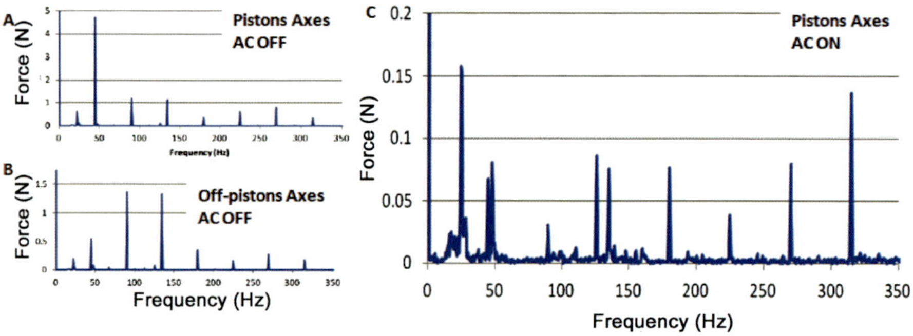

Figure 3. Exported forces from the compressor. Dataset presented on Fig A/B were acquired without active control electronics and show vibrations level in the Newton range. Application of active pistons amplitude and phase control allows reaching <150 mN level along the pistons axis.

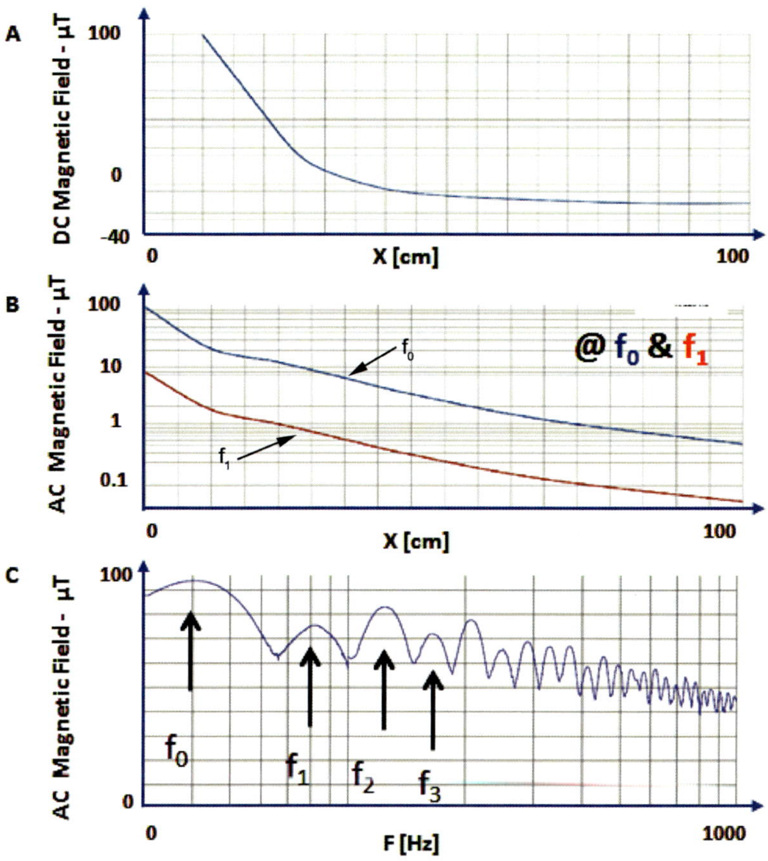

Figure 4. Compressor Exported Electromagnetic Interferences Static A) and dynamic B) vs. distance. C) EMI spectrum tracks the fundamental and harmonics of the motor driving frequency.

PROGRAM STATUS

At this time the cryocooler Engineering Model development is completed: base thermal performances are within the initial scope of work and the preliminary qualification tests have not exposed any major technological limitations.

The cooler is now entering a maturation phase. Several performances improvement pathways are available. Current R&D development in novel regenerator materials are promising. They could allow reducing the base temperature to below 8K or increase the available cooling power up to 25%. The compressor resonant frequency is currently several Hz above the cold finger optimum operating frequency. Improvement in frequency matching between the cold finger and the Compressor will raise the Power Factor and reduce accordingly the Joule losses. Up to 20% improvement in system efficiency may be in reach. Although the cold finger and the compressor are now at TRL 5, standard laboratory hardware was used to commission the unit. The core of the development effort will focus on a dedicated electronics. The new electronics will leverage on the existing standard from the LPTC while improving the modularity of the architecture to match each customer's specific technical and geographical needs while minimizing the budget. This phase will also include a lifetime study. The cooler is designed to sustain more than 10 years of continuous operation as is its LPTC predecessor.

The cooler is on the baseline of the Cryochain for the X-IFU/Athena mission scheduled for launch in 2028. Other candidate missions includes the Cosmic Microwave Background polarization explorer Core +. A flight model could be available as early as 2020.

CONCLUSION

ALAT and its partners designed, manufactured and commissioned a novel high power low temperature Pulse Tube Cryocooler. This campaign has successfully addressed the needs raised in the original scope of work. The cooler offers above 400mW at 15K of cooling power with an additional 5W at 100K in a compact sub 18kg package and 300W of electrical input at compressor level. The cooler generates less than 1.5N forces (sub 0.15N along pistons axis). Static and dynamic EMI levels are below 100μT at the cooler and below 10μT at 30cm. The cooler has undergone preliminary launch vibration and thermal stress tests and is currently rated at Technological Readiness Level 5.

REFERENCES

1. J. Butterworth, G. Aigouy , C. Chassaing, B. Debray and A. Huguet, "Pulse tube coolers for Meteosat Third Generation," *Adv. in Cryogenic Engineering*, Vol. 59, Amer. Institute of Physics, Melville, NY (2014), p. 520.
2. J. Tanchon, , T. Trollier, S. Triqueneaux, A. Ravex, "20–50 K and 40–80 K pulse tube coolers: Two candidates for a low temperature cooling chain", *Cryogenics*, Vol. 50, Issue 1 (2010), pp. 55–60.
3. Duval, J. M., I. Charles, J. Butterworth, J. Mullié, and M. Linder. "7K-15K Pulse Tube Cooler for Space," *Cryocoolers 17*, ICC Press, Boulder, CO (2013), pp. 17–23.
4. Charles, I., L. Golanski, A. Gauthier, A. Coynel, and J.M. Duval, "20 K coaxial pulse tube using passive precooling," *Adv. in Cryogenic Engineering*, Vol. 53, Amer. Institute of Physics, Melville, NY (2008), pp. 887–94.

Progress on 30K-50K Two-Stage EM PT Cold Finger for Space Applications

T. Prouvé[1], I. Charles[1], H. Leenders[2], J. Mullié[2], J. Tanchon[3], T. Trollier[3], T. Tirolien[4]

[1]Univ. Grenoble, Alpes, CEA INAC-SBT, Grenoble, France
[2]Thales Cryogenics BV, Eindhoven, The Netherlands
[3]Absolut System SAS, Seyssinet Pariset, France
[4]ESA-ESTEC, Noordwijk, The Netherlands

ABSTRACT

Within the framework of an ESA Technology and Research Program lead by TCBV, and in collaboration with Absolut System, CEA/SBT has improved the performance and the TRL level of the two-stage coaxial pulse tube developed at CEA/SBT in 2012 under a R&T program co-funded by CNES.

The developments presented in this paper are focused on the thermal performance of the second stage and involve breadboarding of regenerator materials and examining alternateve phase shifting inertance designs and a double inlet (V2) implementation. The EM mechanical developments have been focused on the pulse tube warm flange with electrical feedthrough, an optional double-inlet V2 implementation, and the flange sealing assembly. The TCBV compressor (180 W electrical power) is integrating design improvements from the LPTC cryocooler for which TCBV is currently manufacturing MTG compressor flight models.

Without the double-inlet V2 design, the no-load temperatures of the two-stage pulse tube are expected to be 86 K and 22 K, respectively, at first and second stage. With a load of 3W-1W, temperatures will rise to 111 K and 36.7 K. With the double-inlet V2, the no-load temperature will drop to 81 K and 21 K at first and second stage. With a load of 3W-1W, these expected temperatures will rise to 107 K and 36 K.

After environmental qualification, the cryocooler will be mounted in a breadboard cryostat devoted to infrared Earth observation and developed by Absolut System in the framework of this TRP program. The cryocooler and the cryostat design are made in close collaboration in order to demonstrate interest in such a 2-stage configuration for demanding Earth observation missions.

INTRODUCTION

There has been a trend towards increasing heat loads for cryogenically-cooled Earth Observation instruments in recent years. This is the case at both the current operational temperature levels (~50K), as well as at lower operational temperature levels (30-50K). One solution to meet this trend is to use existing pulse tube technology in a double stage configuration. By doing so, it is possible to intercept parasitic loads at higher temperatures,

Cryocoolers 19, edited by S.D. Miller and R.G. Ross, Jr.

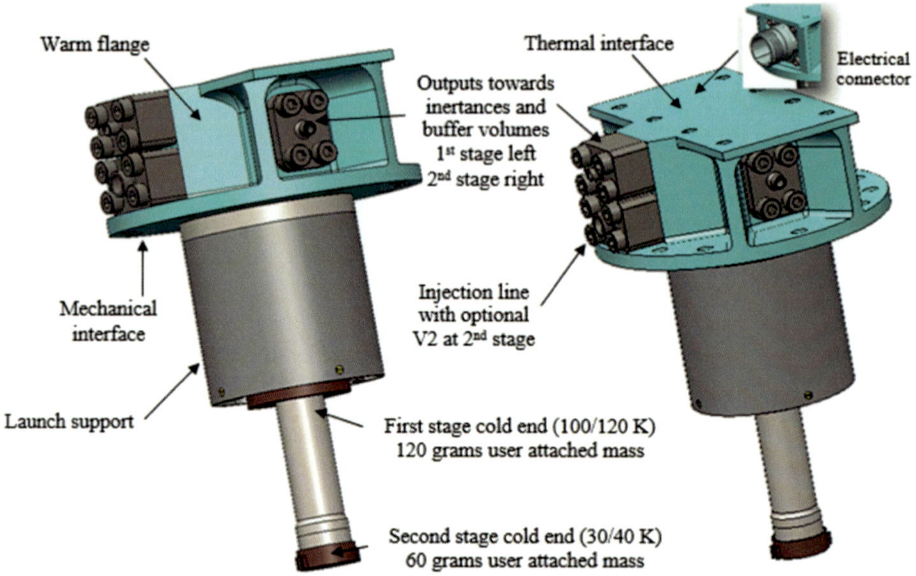

Figure 1. General view of the ESA EM CFA

thereby reducing the thermal load at the lower temperature stage, and making cooling power available at lower temperatures.

The TRP requirements of the 30K-50K pulse tube cooler are given as user loads and include the cold redundant parasitic heat losses. There are two specifications, and both have been updated with CEA parasitics heat losses values. The second specification has been updated to match the thermal budget of the test cryostat built by Absolut System [4]. The two specifications are:

1. The 2-Stage 30-50 K PT cooler shall provide 636 mW at 33 K and 2740 mW at 120 K simultaneously with max 180 W rms input power to the compressor motors and at a rejection temperature of 20°C.

2. The 2-Stage 30-50 K PT cooler shall provide 1080 mW at 42 K and 2730 mW at 126 K simultaneously with max 160 W rms input power to the compressor motors and at a rejection temperature of 20°C.

COLD FINGER ASSEMBLY (CFA) GENERAL OVERVIEW

Mechanical Design

A general view of the coaxial two stage EM cold finger is represented in Figure 1. This design is based on the CNES EM cold finger [1,3]. The warm flange has increased in height to implement an electrical connector, metallic seals, and the optional V2 assembly. The cold tips have been equipped with sensor location areas. The second stage thermo-mechanical interface has been redesigned. The CNES model has mechanical and thermal interfaces made by a disc surface and 4xM2.5 threaded holes. The ESA EM interface is the outer annular surface of the second stage cold tip. This surface is larger than the CNES model, and it enables the use of constriction clamping devices for a strap connection which has been developed by Absolut System [4]. The second stage cold tip still has 2xM2.5 threaded holes which can be used to attach heaters during ground tests. Thermal and mechanical analyses have been done to validate the design, and PA/QA at the ESA EM level has been carried out. The Cold Finger Assembly (CFA) can support 120 grams at the first stage and 60 grams at the second stage under vibration environments of SIN 25g and random 13.1g rms (20-2000 Hz with 0.3 g^2/Hz 100-300 Hz).

Figure 2. Overview of the complete ESA 30K-50K 2 stage EM pulse tube cryocooler mounted on testing support structure - TCBV design.

2nd Stage Regenerator and Phase Shifting

The second stage regenerator packing material has been optimized during the breadboard phase for operating temperatures of 30K-40K. The phase shifting at the first and second stage is "initially" achieved by inertance and buffer volumes. Along with TCBV, the operating frequency, the filling pressure, and inertance winding shape have been studied and tuned for meeting the best cryogenic performance and EM integration. Figure 2 gives an overview of the complete cryocooler. Inertances are wound into a square shape and placed around the compressor assembly and buffer volumes. An integrated double inlet (V2 assembly) has been designed and will be potentially implemented during the characterization phase in between the split-pipe and second stage inertance. The V2 assembly will be placed on the CFA in the split-pipe / inertance 2nd stage flanges.

Compressor and Supporting Structure

The TCBV compressor assembly (CPA) (180W electrical power) integrates design improvements in the framework of this TRP and has heritage from the LPTC cryocooler, from which TCBV is currently manufacturing the MTG compressor flight models. CEA/TCBV have decided to use a support structure, as shown in Figure 2, to assemble, characterize the cryocooler (to TRL 5) and to make the coupling with the dedicated Absolut System cryostat at a later time. This support structure has been designed by TCBV. The structure has openings to give direct access to the CFA and CPA thermal interfaces for heat rejection. The CFA mechanical supports connecting to the round circular CFA mechanical interface are not shown in Figure 2.

BREADBOARD PHASE ON THE CFA

The improvement of the CFA is a direct result of findings of tests performed on a U-shaped DM pulse tube, a dismountable coaxial pulse tube, and on the CNES EM CFA. The DM coaxial CFA has a dismountable heat exchanger at the second stage cold end. This CFA is identical to the CNES EM except for the dismountable second stage cold heat exchanger, and the fact that the copper heat exchangers were not gold plated. Since MLI is being used on the CFA in place of the gold plating, the measured parasitic heat losses should be equivalent, and we expect reliable results from the ESA EM CFA. To test the two stage cold fingers we measure the temperatures at four thermal-equilibrium points with first/second stage applied powers of 0 W / 0 W, 0 W / 1 W, 3 W / 1 W, 3 W / 0 W. Intermediate performance values are then interpolated in between these points to yield what we call "square measurements."

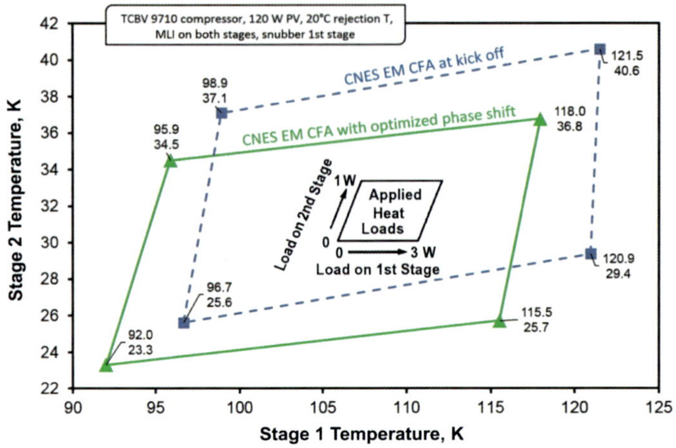

Figure 3. CFA phase shifting optimization (Loading pressure, operating frequency, inertance winding shape).

Phase Shifting

Figure 3 shows the performance obtained with the CNES EM CFA during the TRP program. A significant improvement is seen in the ultimate temperatures and load conditions. The optimization is done by running the cooler at 59.5 Hz and 34 bar (compared to 57.5 Hz and 30 bar for the CNES EM) and also by changing the winding diameter of the inertances.

The phase shifting adjustment on the CFA was optimized further by the use of a double inlet on the second stage pulse tube. This was investigated by using a V2 orifice between the split-pipe inlet and the second stage inertance inlet on the warm flange. The CFA design was made such that the two flanges are on top of each other's. Figure 4 indicates the best results obtained so far with an "EM-like" V2 orifice and operating conditions / inertance shape depicted in the previous paragraph. It is clear from Figure 4 that the second stage V2 orifice improves the first stage significantly by about 5 K at each measurement point. The gain at the second stage is less significant with a decrease of 0.7 K in the cold temperatures, but the inertance length was not optimized for this case. Nevertheless, the measurements were stable over many hours, and the thermal performance using a V2 orifice was reproducible. This clear improvement is why the second stage V2 orifice has been selected as an option on the ESA EM CFA.

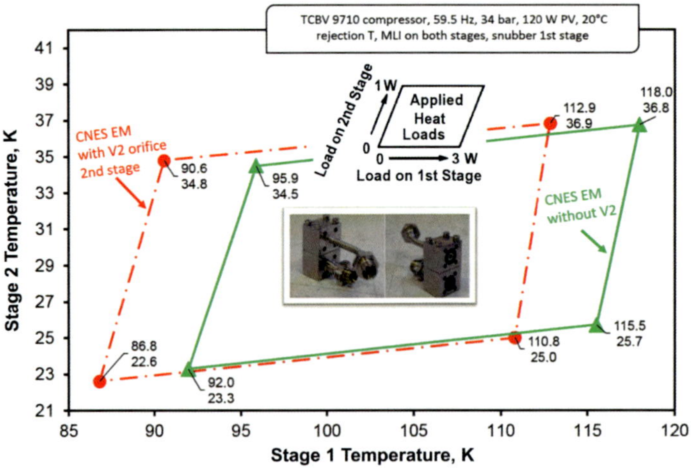

Figure 4. Measured performance gain on CNES EM using a V2 phase shifter.

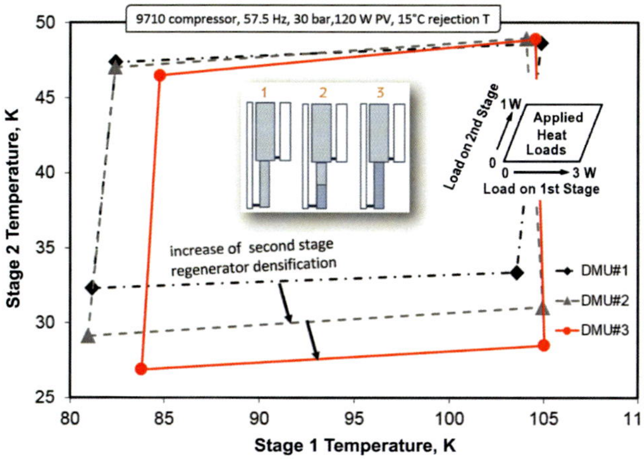

Figure 5. Effect of second stage regenerator densification on a 2 stages coaxial pulse tube

Regenerator Grading

Among the initial requirements of the TRP that have been reworked by Absolut System to fulfill the needs of the associated cryostat under development, the second stage cooling power was identified to be the most critical factor to achieving these updated requirements. Consequently, the regenerator packing study focused on the second stage. Tests were performed on the U-shape CFA first, and the results are presented in Figure 5. Three configurations were tested. Starting (DMU#1) with the same 2nd stage regenerator packing as the CNES EM CFA, the density of the regenerator was increased in its coldest part. It is clear that the U-shape CFA second stage (32 K ultimate) did not perform as well as the CNES EM CFA (25.6 K ultimate) at no load [1]. With that in mind, it was observed that increasing the density of the second stage regenerator lowered the no-load temperature of the second stage by about 5 K. However, it did not increase the cooling capacity at 1 W. Similar tests were then performed with the DM coax CFA and the gain was reduced by 2.5 K. The lower gain was attributed to the lower starting operating temperature of 25 K for this configuration. The impact of increasing the regenerator density on the DM coax CFA was nevertheless the same as on the U-shape CFA.

Assembly Process

The CNES EM regenerator configuration was tested on the DM coax and a drastic difference in the first stage thermal performance was measured. Figure 6 gives the results of the

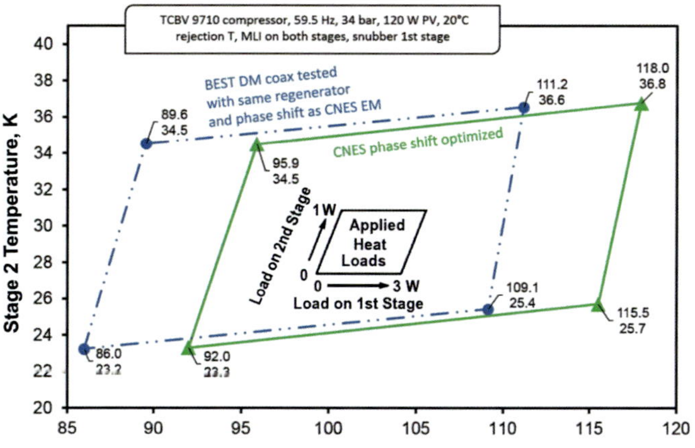

Figure 6. Performance discrepancies between CNES EM and DM coax probably to assembly process.

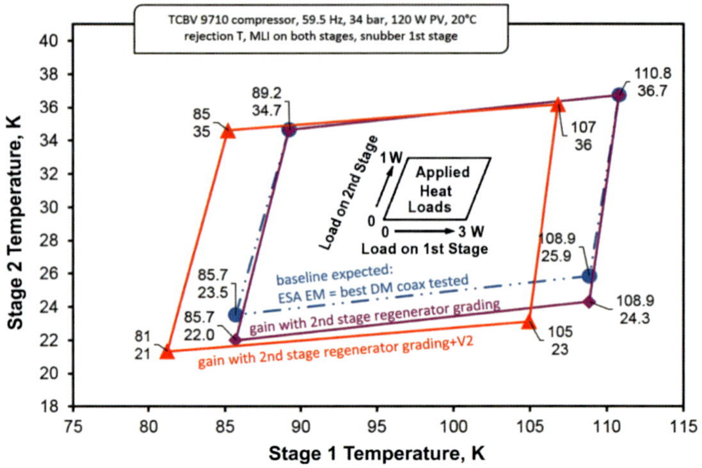

Figure 7. Expected ESA EM CFA performances with optimized phase shift, V2, 2nd stage regenerator grading and DM coax assembly process.

measurements. At the four set points the gain in first stage temperature is about 7 K, and the second stage has an equivalent temperature to the CNES EM. The assembly process of the DM coax was slightly different from the CNES EM. The ESA EM CFA design will have the same assembly process as the DM coax.

EXPECTED PERFORMANCES FOR THE ESA EM CFA

As a result of the improvements made during the breadboard phase activity, some predictions can be made regarding the expected performance of the ESA CFA that is currently being manufactured. Figure 7 gives the detail of the expected cooling capacity of the CFA with the different options that will be implemented in it. The CFA being built is expected to be at least as good as the DM coax version made during the breadboard phase. Then a grading in the regenerator packing at the second stage will be implemented which is expected to lower the ultimate temperature of the second stage as much as was seen on the DM coax, since the CFA has the same range of cold temperature at the second stage. Finally, a double inlet assembly at the second stage will be implemented that is expected to improve the first stage cooling capacity and will continue to lower the ultimate temperature of the second stage close to 20 K.

Parasitic Heat Losses

Parasitic heat losses have been studied in a cold redundant method with the CNES EM CFA and the coax DM CFA. Figure 8 shows the test bed using the coax DM cooler to cool down the OFF CNES EM cooler. The coolers are placed into the test bed cryostat that is used to characterize the cooler. This cryostat has an aluminum warm flange thermalized at 20°C and the coolers are surrounded by a 20°C copper shield thermalized on the vacuum side of the warm flange. The parasitic heat loss values are given for 15 layers of MLI at the first and second stage. The coolers are operating in a vertical (cold tips down) position. In this test configuration the OFF thermal loads of the CFA at 120K/40K are:

- 1.275 W at the first stage heat exchanger
- 0.26 W at the second stage heat exchanger

CONCLUSION

Progress on improving the CFA performance within the framework of this TRP is summarized in Figure 9. Phase shift has been a large focus of this study, since it improves the

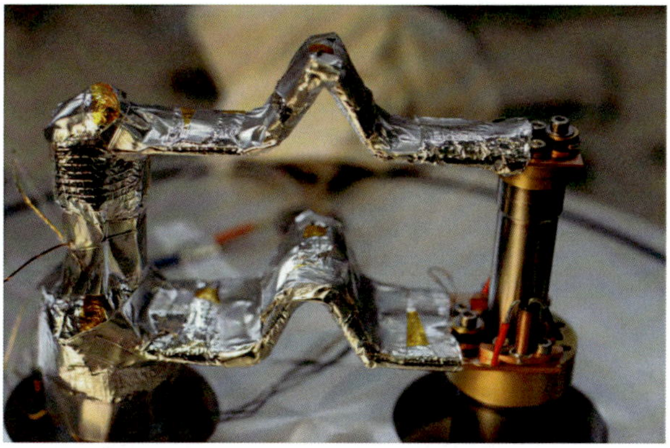

Figure 8. Cold redundant parasitic heat losses test bench. Left: coax DM cooler ON , Right: CNES EM cooler OFF.

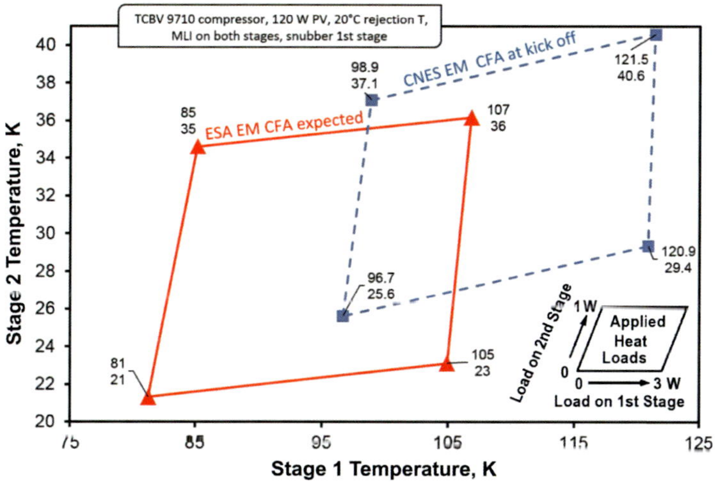

Figure 9. Comparisons between CNES EM and future expected ESA EM CFA.

cooling performances of the CFA at all operating conditions. The double inlet has given good results versus thermal stability and reproducibility. Coupled to that, regenerator optimization has shown the possibility for this CFA to achieve lower temperatures. The regenerator was not optimized further, since the TRP specifications are above 600 mW at the second stage, and the focus was on increased cooling capacity rather than targeting lower temperatures. The 15 K-20 K temperature range should be achievable at the second stage with a passive phase shift and using a specific regenerator packing.

Due to expected TCBV compressor efficiency improvements, we are confident in achieving the TRP requirements presented in the introduction of this paper and are looking forward to demonstrating the feasability of such a cooling solution into the Absolut System flight-like cryostat in the near future.

ACKNOWLEDGMENT

The project is supported by the Atomic and new Energies committee (CEA) and the European Space Agency (ESA) under TRP project (TRP-4000109933/14/NL/RA).

The authors acknowledge Alexandre COYNEL (CEA/SBT) for his significant contribution during design, manufacturing, assembly and thermal testing phases. We would like also to thanks Florian BANCEL (CEA/SBT) for his numerical calculation efforts during CDR activities.

REFERENCES

1. Charles, et al., "Thermal Testing of an EM Two-Stage Coaxial Pulse Tube Cold Finger for Earth Observation Missions", *Cryocoolers 18*, ICC Press, Boulder, CO (2014), pp. 141-149.
2. Bruins, et al., "A 30-50 K Dual-Stage Pulse Tube Space Cooler", *Cryocoolers 18*, ICC Press, Boulder, CO (2014), pp. 33-37.
3. Charles, et al., "Development of a Two Stage High Temperature Pulse Tube Cooler for Space Applications", *Cryocoolers 15*, ICC Press, Boulder, CO (2009), pp. 63-70.
4. Tanchon, et al., "Design of a Flight Like Cryostat for 30-50K Two-Stage Pulse Tube Cooler Integration", *Cryocoolers 19*, ICC Press, Boulder, CO (2016), (this proceedings).

Development of a Two-Stage Stirling Type Pulse Tube Cooler with Precooling inside the Pulse Tube

X. Pang[1,2], X. Wang[1], W. Dai[1], E. Luo[1], Y. Zhao[1,2]

[1]Key Laboratory of Cryogenics, Technical Institute of Physics and Chemistry of Chinese Academy of Sciences, Beijing 100190, China
[2]Graduate University of Chinese Academy of Sciences, Beijing 100049, China

ABSTRACT

This paper presents the performance of a thermally-coupled two-stage pulse tube cooler. In the cryocooler, an independent 50 Hz Stirling type pulse tube cooler is used to precool the middle of the regenerator as well as the pulse tube of another Stirling type pulse tube cooler with a lower operating frequency. The other cooler forms a configuration referred as a thermally-coupled two-stage configuration. The phase shifter of the first stage is an inertance tube plus reservoir. In order to enhance the performance, a double-inlet orifice is added in the second stage. Theoretical studies and optimization based on Sage® software is first carried out and then experimental apparatus is set up. Experimental results show that with precooling inside the 2nd stage pulse tube, a cooling capacities of 0.73 W at 20 K is obtained, which is more than 2 times as high as the original cooler without precooling (0.27 W). Total electric power input is 540 W (first stage is about 350 W and the second stage is about 190 W). The influence of the pressure ratio, precooling temperature, mean pressure and frequency are also studied.

INTRODUCTION

Refrigerators working at liquid hydrogen temperature are needed for superconductor cooling applications, mid-infrared sensors and space applications. There is an urgent need for efficient coolers with low maintenance and high reliability. Among the different coolers, pulse tube coolers have the advantages of a simple structure, low vibration over traditional regenerative coolers such as Gifford Mcmahon (G-M) and Stirling coolers due to no moving parts in the low temperature region [1]. Stirling-type pulse tube coolers with oil-free compressor are attractive due to their light weight, high reliability and high efficiency [2]. Stirling-type single stage pulse tube coolers at liquid nitrogen temperature [3-4] have been successfully developed and commercialized. For temperature around 20 K, both two stage and single stage configuration have been reported.

L. Yang et al., in 2008 developed a thermally-coupled two stage pulse tube cooler with a double inlet orifice as the phase shifter [5]. A no load temperature of 12.8 K was obtained with 400 W of electric power input. The cooling power at 20 K was about 0.3 W. M. Dietrich et al., developed a gas-coupled two stage Stirling-type pulse tube cooler in 2010 [6], a no load temperature of 13.7 K with a cooling power of 12.9 W at 25 K was achieved at an electrical input power of 4.7 kW. A

Cryocoolers 19, edited by S.D. Miller and R.G. Ross, Jr.

two stage pulse tube cooler with a self-precooled pulse tube was tested by L. Qiu et al., in 2012 [7]. After precooling at near the pulse tube wall, the no load temperature decreased from 26.6 K to 18.02 K. By further optimization, a no load temperature of 15.87 K was reached with 740 W of electrical power input and the cooling power at 20 K was less than 0.2 W. Besides the two stage configuration, there are also some studies of a single stage configuration. L. Chen et al., developed a single stage multi-bypass pulse tube cooler in 2014 [8]. A no-load temperature of 14.7 K was obtained and the maximum cooling power at 20 K was 0.39 W with 250 W electric power input. Q. Zhou et al., further optimized it [9] with a no load temperature of 13.9 K, which is the lowest temperature of a single stage pulse tube cooler ever reported. A further improvement in the efficiency of the Stirling-type pulse tube coolers working within this temperature region is one of the research focuses in this field.

This paper introduces the development of a two-stage Stirling pulse tube cooler which has a thermally-coupled configuration. The configuration provides the ability to study the influence by easily changing components. This serves as the first stage of a project aimed at developing a compact pulse tube cooler system at 20 K. The configuration details and the simulation tool are briefed first, the experimental setup is introduced next, and followed by the presentation of experimental results and discussions last. Finally, some conclusions are drawn.

GEOMETRICAL CONFIGURATION AND NUMERIC MODEL

System configuration

Figure 1 illustrates the system configuration with a coaxial configuration for both stages. Each stage is driven by a compressor respectively. The phase shifter for the first stage is an inertance tube plus reservoir. The double-inlet value is added in the second stage to obtain better performance. Normally, the thermally-coupled two stage pulse tube cooler only has the precooling thermal link at the middle of the 2nd stage regenerator. As shown in Figure 1, the 2nd stage pulse tube is also precooled to improve the cooling performance and is investigated in this paper.

Numeric model and simulation results

For the first stage, we directly use an existing pulse tube cooler in the laboratory and no simulation is required. The parameters of the pulse tube cooler are shown in Table 1.

The second stage cooler is simulated through Sage®[10] software developed by Gedeon Associates, which can simulate and optimize a variety of thermal systems based on an oscillating flow. The geometric parameters are selected according to the numeric calculation results which are shown in Table 2. For Regenerator 21 (RG 21), 300 mesh stainless steel screens with diameter of

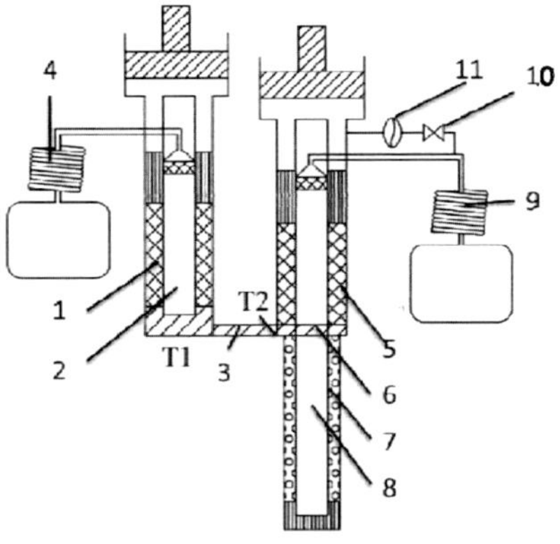

Figure 1. Schematic of the two stage pulse tube cooler: 1 Regenerator 1 (RG 1), 2 Pulse tube 1 (PT 1), 3 Thermal bridge , 4 1st inertance tube, 5 Regenerator 21 (RG 21), 6 Middle heat exchanger, 7 Regenerator 22 (RG 22), 8 Pulse tube 2 (PT 2), 9 2nd inertance tube, 10 double-inlet value, 11 membrane.

Table 1. Geometric parameters of the first stage pulse tube cooler

Component	Parameters (mm) (Diameter*Length)
RG 1	Φ 25*45
PT 1	Φ 12.8*64
1st inertance tube	Φ 3*3.4

30 μm are used as the matrix material. HoCu2 spheres with diameter of around 0.2 mm are used in Regenerator 22 (RG 22).

Figure 2 shows the influence of a double-inlet orifice on the system performance with a constant acoustic power input of 130 W. With only a inertance tube plus reservoir as the phase shifter, the system cannot obtain the desired performance and the no load temperature only reaches 31 K. The double-inlet orifice improves the phase difference between the pressure wave and the volume flow rate at the hot end of the pulse tube which is helpful for obtaining better cooling performance. However, the simulation results show that decreasing the no-load temperature is not a sole function of the phase difference. With increasing the openings of the double-inlet, the phase difference also increases, but there is a minimum no load temperature corresponding to the optimum double-inlet opening.

EXPERIMENTAL SETUP

Based on the numerical simulations, a two-stage pulse tube cooler has been developed with parameters shown in Table 1 and Table 2. Figure 3 shows the photo of the cold head. A middle heat exchanger is set inside the 2nd stage pulse tube, which is also precooled by the cold tip of the first stage (see Figure 1). The thermal bridge, which connects the first stage cooler and the second stage cooler, is composed of 0.1 mm thick copper plates with bends to provide flexibility. The corresponding overall length, width and height is about 155 mm × 70 mm × 10 mm. A copper radiation shield, which is thermally attached to the thermal bridge, surrounds the low temperature part of the second stage. In order to further reduce the radiation losses, aluminized foil is wrapped around the two stages and the copper shield.

Table 2. Geometric parameters of the second stage pulse tube cooler

Component	Parameters (mm) (Diameter*Length)
RG 21	Φ 25*35
RG 22	Φ 18.8*65
PT 2	Φ 10*154.5

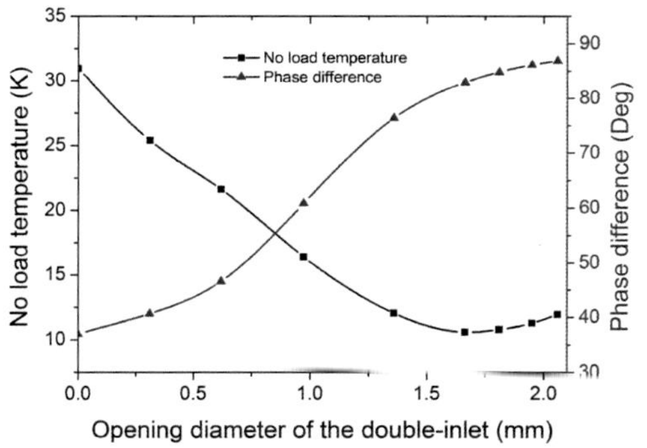

Figure 2. Influence of the double-inlet on system performance, 130 W acoustic power input

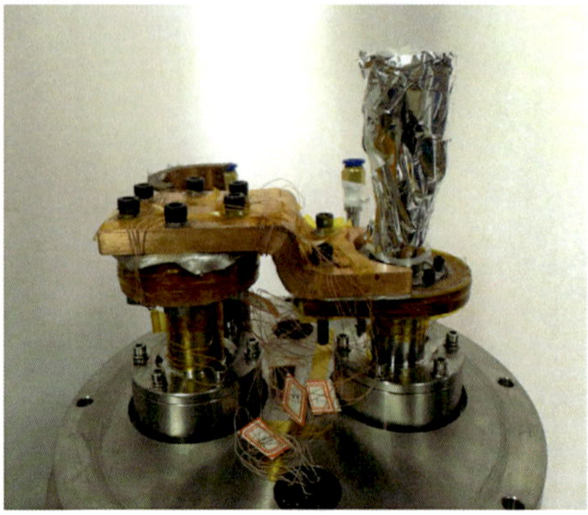

Figure 3. Photo of the two stage cold head

The cold head of each stage is driven by a linear compressor respectively of the moving-magnet type. A dual-opposed motor configuration is used to ensure then easy cancellation of vibration. The pressure waves are measured by dynamic pressure transducers (113B21, PCB piezotronics Inc.) which are placed at the compression chamber and the backside chamber of the compressors. Two platinum resistance thermometers (PT 100) with an accuracy of ± 0.1 K are attached to the heat exchanger to measure the temperature of T1 and T2 (show in Figure 1). A calibrated resistance thermometer (LakeShore) with 0.01 K accuracy is used to measure the temperature of T3 (2nd stage cold end). Cooling power is measured through a heating wire which is mounted on the cold end heat exchanger and powered by a DC voltage source.

EXPERIMENTAL RESULTS AND DISCUSSIONS

Cooling performance

During the experiments, the first stage is working at 50 Hz and 3.5 MPa. The second stage is working at 30 Hz and 2.5 MPa, if not specially mentioned. The cool down curves are given in Figure 4. Compared with the cooler with non-precooled pulse tube, the system with precooling inside the middle of pulse tube obtains a better cooling performance. No load temperature decreases from 18.86 K to 15.35 K. The cooling power at 20 K is more than 2 times higher than with the non-precooled pulse tube (0.73 W vs. 0.27 W), which shows the effectiveness of the middle heat exchanger inside the 2nd stage pulse tube. The electric power input is about 540 W (first stage about

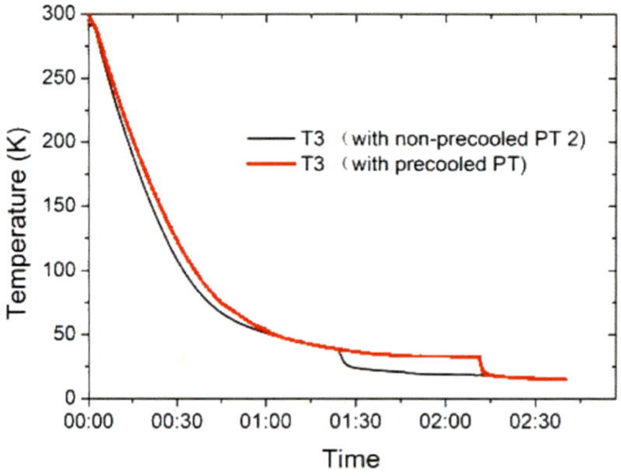

Figure 4. Cooling down process of the system

350 W and the second stage about 190 W). The following studies focus on the pulse tube cooler with precooling inside the middle of the 2nd stage pulse tube.

Influence of important parameters

The inertance tube affects the cooling performance. Figure 5 shows the no load temperature with different lengths of the inertance tube (optimized opening of the double-inlet). The acoustic power is approximately the same during the experiments, about 130 W. There is an optimum length (around 3.3 m) for the inertance tube to obtain a minimum no load temperature.

In addition to the phase shifter, the influence of the pressure ratio, the precooling temperature and the mean pressure are also investigated. Figure 6 shows the effect of the pressure ratio. The maximum value is about 1.22 due to the limitations of the piston displacement. The optimized no load temperature is about 15.35 K. From the trend of the curve, if the pressure ratio can be improved, the performance will be better.

As shown in Figure 7, the no load temperature changes almost linearly with the precooling temperature with approximately 130 W acoustic power input. Due to the limited cooling capacity of the first stage, the minimum precooling temperature is about 76 K. With a precooling temperature increase of 5.5 K, the no load temperature increases by about 1 K.

The operating frequency is also an important parameters. The effect of the frequency is shown in Figure 8. With the frequency changing from 28 Hz to 32 Hz, the change in no load temperature

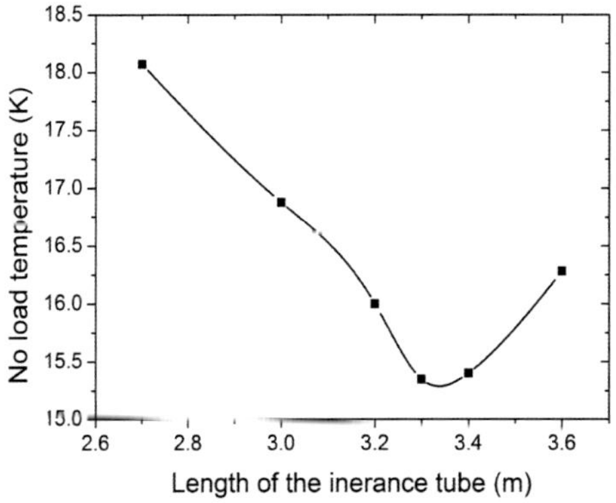

Figure 5. No load temperature vs. length of the inertance tube

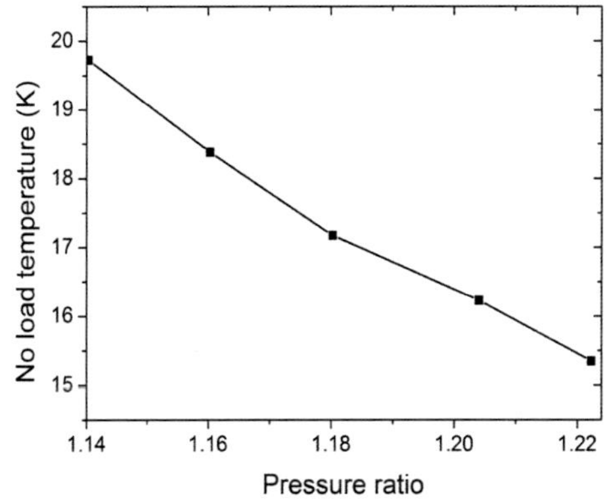

Figure 6. No load temperature vs. pressure ratio

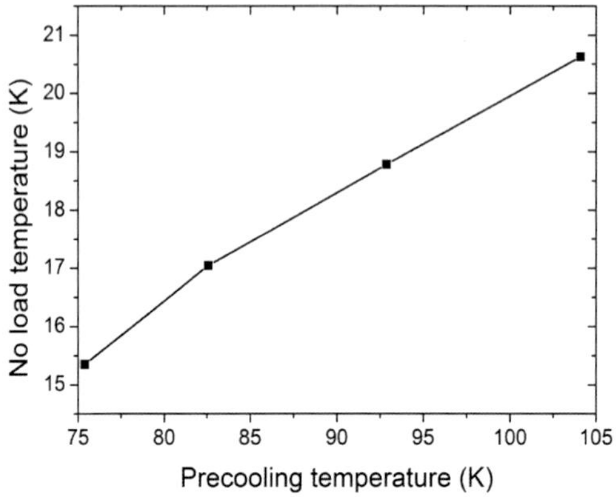

Figure 7. No load temperature vs. precooling temperature

is less than 1 K. The optimized operating frequency is about 30.5 Hz, which is very close to the designed frequency (30 Hz). Figure 8 also indicates that when the operating frequency is close to the optimized value, the effect on the no load temperature is small.

The relationship between the filling pressure and cooling performance is given in Figure 9. In the experiments, the maximum acoustic power input is different with different filling pressures. Considering the limitations of the compressor piston displacement, the swept volume of the compressor is kept the same for the experiments with different filling pressure. As shown in Figure 9, the no load temperature is nearly the same with four different filling pressures. However, with a higher filling pressure, the cooling power at 20 K is higher. A cooling power of 0.94 W at 20 K is obtained with 3.3 MPa mean pressure.

CONCLUSION

A thermally-coupled two stage pulse tube cooler with precooling inside the second stage pulse tube has been developed and tested. A no-load temperature of 15.35 K and cooling capacities of 0.73 W at 20 K are obtained with 540 W electric power input (first stage about 350 W and the second stage about 190 W). The cooling performance is much better than that with non-precooled pulse

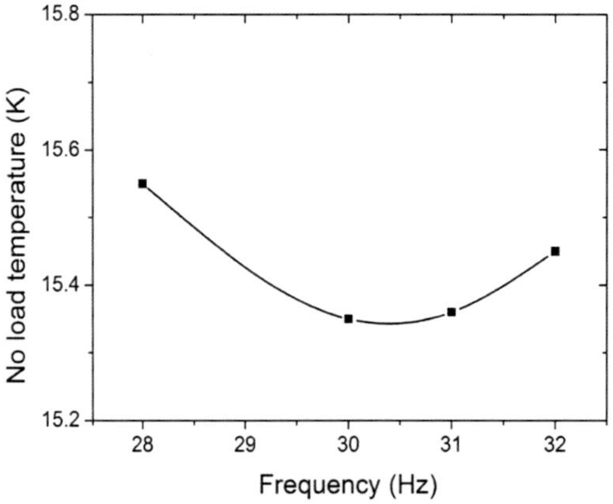

Figure 8. No load temperature vs. frequency

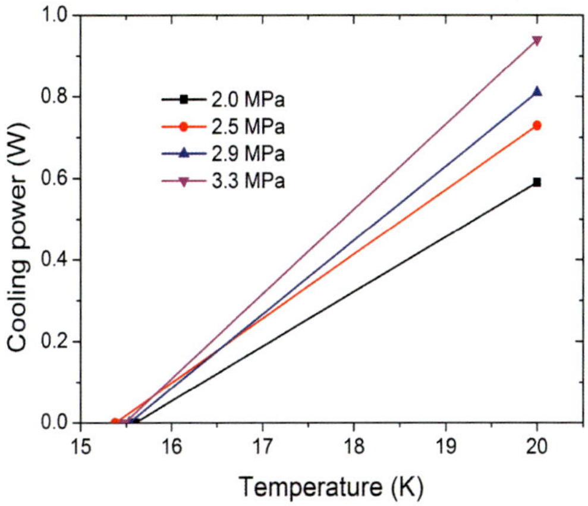

Figure 9. Cooling power vs. temperature

tube. The study reveals that the system can obtain a better performance with higher pressure ratio and lower precooling temperature. Through increasing the filling pressure, a cooling power of 0.94 W has been obtained under 3.3 MPa.

ACKNOWLEDGMENT

This work is financially supported by the National Natural Science Foundation of China (Contract number 51376187, 51576205) .

REFERENCES

1. Radebaugh, R., "The development and application of cryocoolers since 1985," *Proceedings of ICCR* 2003, International Academic Publishers/Beijing World Publishing Corporation (2003), pp. 858-70.
2. Radebaugh, R., "Development of the pulse tube refrigerator as an efficient and reliable cooler," *Proc. Institute of Refrigeration*, London, vol. 96 . 1999-2000; 11-31.
3. Hu, J.Y., Zhang, L.M., Zhu, J., Chen, S., Luo, E.C., Dai, W. and Li, H.B., "A high-efficiency coaxial pulse tube cryocooler with 500W cooling capacity at 80K," *Cryogenics*, Vol. 62(2014), pp. 7-10
4. Wang, X., Dai, W., Hu, J., Luo, E. and Zhou, Y., "Performance of a Stirling-Type Pulse Tube Cooler for High Efficiency Operation at 100Hz," *Cryocoolers 16*, ICC Press, Boulder, CO (2011).
5. Yang, L., "Investigation on a thermal-coupled two-stage Stirling-type pulse tube cryocooler," *Cryogenics*, vol. 48 (2008), pp. 492-496.
6. Dietrich, M. and Thummes, G., "Two-stage high frequency pulse tube cooler for refrigeration at 25 K," *Cryogenics*, vol. 50 (2010), pp. 281-286
7. Qiu, L.M., Zhi, X.Q., Han, L., Cao, Q. and Gan, Z.H., "Performance improvement of multi-stage pulse tube cryocoolers with a self-precooled pulse tube," *Cryogenics*, vol. 52 (2012), pp. 575-579.
8. Chen L, Zhou Q, Zhu X, et al., "Lightweight Stirling-type pulse tube cryocooler with temperature below 20 K," *Journal of Engineering Thermophysics*, vol. 35(5) (2014), pp. 858-861.
9. Zhou, Q., Chen, L., Zhu, X., Zhu, W., Zhou, Y. and Wang, J., "Development of a high-frequency coaxial multi-bypass pulse tube refrigerator below 14 K," *Cryogenics*, vol. 67 (2015), pp.28-30.
10. Gedeon, D., "Sage User's Guide," Gedeon Associates, Athens, OH, (2009).

Development of Microcryocoolers for Space and Avionics Applications

T. Nast, J. Olson, P. Champagne, E. Roth,
E. Saito[1], V. Loung[1], B. McCay[1], A. Kenton[2], C. Dobbins[3]

Advanced Technology Center, Lockheed Martin Space Systems Co., Palo Alto, CA
[1]Santa Barbara Focalplane, Lockheed Martin Missiles & Fire Control, Goleta, CA
[2]DCS Corporation, Niceville, FL
[3]U.S. Army Aviation and Missile Research Development and Engineering Center (AMRDEC), Redstone Arsenal, AL

ABSTRACT

Lockheed Martin's Advanced Technology Center has developed a series of Microcryocoolers for a range of applications including avionics and space sensors and instruments. We report the development and initial testing of the recently developed first-article, single-stage, compact, coaxial, Fast Cooldown Pulse Tube Microcryocooler (FC-PTM). The new cryocooler supports cooling requirements for emerging large, high operating temperature (105-150K) infrared focal plane array sensors with nominal cooling loads of ~300 mW at105K at 293K ambient. This is a sequel development that builds on our inline and coaxial pulse tube Microcryocoolers reported at CEC 2013[7], ICC 18[8,9], and CEC 2015.[10] The new FC-PTM and the prior units all share our long life space technology attributes, which typically have 10 year life requirements.[1]

The new prototype Microcryocooler builds on the previous development by incorporating cold head design improvements in two key areas: 1) reduced cool-down time, and 2) novel repackaging that greatly reduces envelope. The new coldhead and dewar were significantly redesigned from the earlier versions in order to achieve a cooldown time of 2-3 minutes— a projected requirement for tactical applications. A design approach was devised to reduce the cold head length from 115mm to 55mm, while at the same time reducing cooldown time. We present new FC-PTM performance test measurements with comparisons to our previous pulse-tube Microcryocooler measurements and design predictions. The FC-PTM exhibits attractive small size, volume, weight, power and cost (SWaP-C) features with sufficient cooling capacity over required ambient conditions that apply to an increasing variety of space and tactical applications.

This paper presents performance characteristics and qualification status of the various Microcryocoolers.

INTRODUCTION

Background and Motivation

A primary driver for smaller size, weight, and power (SWaP) cryocoolers is the application area of infrared (IR) cameras and other sensors. IR cameras are integral to portable ground tactical reconnaissance systems that perform many functions useful to the warfighter. Reducing size, weight, and power requirements while maintaining or improving imaging performance is perpetually desired. IR cameras are

Cryocoolers 19, edited by S.D. Miller and R.G. Ross, Jr.

increasingly demanded for installation within very small payloads, such as inside a 6-12 inch diameter ball turrets for tactical unmanned airborne systems (UAS). A need for lightweight, on-gimbal cryocooling also exists. Scientific instruments on small CubeSat class space payloads and outer planetary missions often require cryogenic cooling. Examples include imaging spectrometers, interferometers and IR sensors. Reduced size, weight and power, coupled with high reliability and long life are critical needs for space applications. Reduced power requirements can lead to smaller, lower weight power systems and batteries. The examples cited are of direct current interest and are difficult to meet with available existing capabilities. For some systems, such as the one described here, fast cool down is a critical requirement in order to provide quality imagery as quickly as possible.

Reduced SWaP-C includes reduced cost. A cryocooler is a large part of the cost of a cryogenic sensor subsystem. Reducing the cost in any subsystem component is an important overarching need today. For tactical systems, the unit cost can be reduced over the analogous space applications by increased demand and production. Modularizing subsystems and reusing designs and components also lead to cost benefits. Life-cycle costs include cooler replacements due to failure or precaution as projected lifetimes are approached. In many cases, cooler replacement may be impossible (e.g., deep space) or very expensive (e.g., aircraft refurbishment, on-orbit satellites, etc.). Overall costs for both science applications and military systems are under increasing scrutiny, and costs must decrease where possible so that the experiment or system is affordable. A key goal of this effort is to reduce the cool down time of the system to several minutes while retaining the low weight, excellent reliability and low cost potential. This is achieved by utilization of our present compressor design and a completely redesigned cold head to provide fast cooling and efficient power usage.

The Lockheed Martin Microcryocooler development exploits a long history of excellent space cooler development and products by the Advanced Technology Center (ATC) spanning 40 years.[1] Over the last 20 years, the development of closed-cycle mechanical pulse tube coolers has been very successfully accomplished and has mostly supplanted earlier stored cryogen space coolers.[6] A hallmark of a successful space cooler development then and today is superior lifetime and reliability. The Microcryocooler development program leveraged this expertise to design, develop, and manufacture an advanced pulse tube Microcryocooler (PTM) suitable for both tactical and space applications. The expected PTM lifetime is comparable to high cost, low volume space coolers. This is projected to result in an order of magnitude improvement in tactical cryocooler lifetime. Tactical cryocoolers typically have mean time to failure (MTTF) of less than 25,000 hours. The projected *tactical* PTM MTTF goal is currently 100,000 hours, and this value is predicted to increase as the PTM development and test program proceeds. Another feature of the PTM is reduced costs because of improved assembly, manufacturing, and test methods. The labor associated with component alignment and assembly has been reduced by leveraging ATC's experience, leading to an efficient mechanical design with creative assembly methods that minimize integration time. This translates to direct and potential cost savings per manufactured unit.

Advanced Infrared Camera Applications

High performance Infrared (IR) camera focal plane arrays (FPAs) demand cooling to cryogenic temperatures to overcome background noise effects.[2] Typical tactical IR cameras using proven materials for long-wave and mid-wave IR applications (LWIR: 8-14 μm and MWIR: 3-5 μm) are cooled to ~70-80 K for terrestrial tactical applications. Advances during the last five years include the commercial advent of new photonic semiconductor materials using advanced barrier detector device architectures (e.g., nBn detectors using new alloys and/or strained layer superlattice (SLS) structures that reduce dark current, and thus reduce noise).[3,4] Emerging MWIR FPAs using nBn material, such as Lockheed Martin Santa Barbara Focalplane's WhispIR® technology, permit high operating temperature (HOT) FPAs approaching 150 K while retaining high camera performance operation comparable to classical InSb FPAs.[5] Furthermore, typical LWIR materials like Mercury Cadmium Telluride (MCT) can be used at temperatures up to 120 K when operating at MWIR. Thus, cooler requirements for IR cameras now span the 70-150 K temperature range, potentially with lower net cooling requirements. The required cooling power depends heavily on the cold-side electronics power, but operating at higher temperatures clearly translates to higher cooler efficiency and less parasitic heat load. Typical net cooling lift/power requirements are 100–1000 mW depending on image size and data/framing rates. A net cooling lift of 300 mW is sufficient for many advanced cameras with a well-designed Integrated Detector Dewar MicroCryocooler Assembly (IDDMCA).

The initial Lockheed Martin PTM development was for cooling HOT MWIR materials, particularly the Lockheed Martin Santa Barbara Focalplane HOT nBn MWIR FPAs with the associated integral digital read-out-integrated circuits (ROICs) and new collaborative IDDMCA development.[7,9] The initial operating goal was 125-150 K with a corresponding net cooling power of 400-600 mW and a nominal 10 W electrical drive power. These do not necessarily target the smallest camera configurations, but provide a baseline system that is configurable or adaptable to a wide range of diverse existing and future applications. This paradigm is one way to reduce subsystem costs in accordance with a number of emerging commercial and DoD initiatives designed to exploit the modular use of subcomponents for future systems.

Advanced Microcompressor Development

Key to the Microcryocooler development, current configuration, and future configurations is the PTM's Microcompressor. The design and development of the miniaturized mechanical system is a hallmark of the effort and has been the priority interest to date. The early test data used a prototype in-line pulse tube coldhead optimized for cooling at 150 K, followed by a "standard" coaxial pulse tube Microcooler optimized at 125K.[7,8,9] Recent testing (reported herein) uses a new *fast cool-down* compact coaxial pulse tube Microcooler also optimized at 125 K. The *fast cool* coaxial pulse tube and Microcompressor are in the initial stages of microcryocooler integration, followed by integration with the ruggedized IDDMCA also under development. A new first-article fast cool-down compact pulse tube microcryocooler with a high definition (HD) 1280x1024 12μm pitch HOT nBn MWIR FPA and the SBF207 ROIC integrated MWIR camera is planned for completion by the end of 2016.

Collaboration on the Microcryocooler and HOT MWIR Camera Development

This work began as collaboration of Lockheed Martin Santa Barbara Focalplane (SBF) – a developer and manufacturer of advanced IR FPAs (primarily for tactical camera systems) and Lockheed Martin Advanced Technology Center (ATC) – a producer of space qualified cryogenic coolers. The emergence of HOT IR materials, increasing interest in improved SWaP cameras, and a desire for longer-life tactical cryocoolers motivated the development of this Microcryocooler subsystem. It was recognized that by leveraging ATC's space-related experience, much smaller and more efficient cryocoolers are possible for tactical applications with significant improvement in device lifetime. Lower cost is also a predicted benefit. The U.S. Army Aviation and Missile Research, Development, and Engineering Center (AMRDEC) has requirements for lower SWaP-C IR cameras as an integral part of ground tactical systems, and has partnered since 2010 with Lockheed Martin SBF and ATC to enhance and test devices leading to this goal under its Advanced Technology for Next Generation IR Sensors and Seekers R&D project. Internal R&D at SBF and ATC, coupled with AMRDEC development, integration and test support, were critical for funding this development effort. The Pulse Tube Microcryocooler developments, their integration into high performance HOT MWIR demonstration cameras, and Microcooler performance comparisons form the overall goals for completion of the AMRDEC project.

MICROCRYOCOOLER OVERALL REQUIRMENTS

In order to be broadly useful for a variety of current and future mission needs, many different cryocooler characteristics had to be considered when designing the Microcryocooler. The cryocooler must have sufficient cooling power and good power efficiency. It must be able to meet mission requirements at a variety of heat rejection temperatures, and must be able to effectively reject its waste heat. The cryocooler mass, size, and ease of packaging and integration with the sensor are also critically important. Very long cryocooler lifetime is a major motivation for this development program, but reducing unit, quantity and life-cycle costs is also important. It was desirable to have a low-vibration compressor which can also survive operation in a high-vibration environment such as on a military aircraft.

The desired characteristics of the Microcryocooler often conflict with each other. For example, making parts as lightweight as possible often leads to additional machining steps, with higher cost. Conse-

Table 1. Initial design goals for the Lockheed Martin Pulse Tube Microcryocooler.[7,9]

Cryocooler Parameter	*Design Goal*
Net Cooling Power	300 mW net cooling at 105 - 150 K
Specific Power	< 15 W/W at 150 K
Compressor Mass	< 200 grams
Compressor Length	< 100 mm length
Compressor Configuration	Back-to-back vibrationally-balanced pistons
Cryocooler Lifetime	MTTF > 250,000 hours (typical for space)
Coldfinger Configuration	Compact coaxial coldfinger
Cryocooler Configuration	Split cryocooler configuration
Heat Rejection	Air, water or heat pipe cooling
Production and Lifecycle Costs	Low

quently, Lockheed Martin ATC and SBF technical personnel worked closely with AMRDEC to develop goals for the Microcryocooler, which derive from broader goals for the integrated IR camera system. These design goals are summarized in Table 1.

MICROCRYOCOOLER SPECIFICATIONS

The pulse tube Microcryocooler, as designed, is shown schematically in Figure 1.[7,9] The first Microcompressor was built in 2012, and is shown in Figure 2. The compressor consists of two compressor modules, each with a piston and motor, which are mounted together in a central hub. The compressor motors utilize ATC's standard moving magnet configuration that is favored because it places the motor coil outside the gas space. This eliminates the need for moving electrical leads and places the coil potting organics outside the working gas, thus reducing the risk of gas contamination. The motor modules utilize the long-life "Oxford-style" clearance seal and flexure bearing design, which has proven to have lifetimes on orbit in excess of 10 years.

Visible at the top of the compressor hub is the port for the transfer line to the pulse tube coldhead. This port uses a metal c-seal to prevent gas leakage. Also visible on the hub is one of the mechanical mounting screws. One of the Microcompressor motor modules is shown in Figure 3. The external Hall Effect position sensor is shown at the end of the module. Also visible are the electrical leads coming from the motor coil. The electrical connector was not yet installed when the photo was taken.

The compressor length is 90 mm, somewhat smaller than the design goal shown in Table 1. This length includes the position sensors, which sit flush with the compressor dome and do not affect the overall length. The compressor diameter is 32 mm. The measured mass of the compressor with no electrical connector is 190 g, slightly lower than the goal shown in Table 1. The compressor was designed to survive 1500 PSI

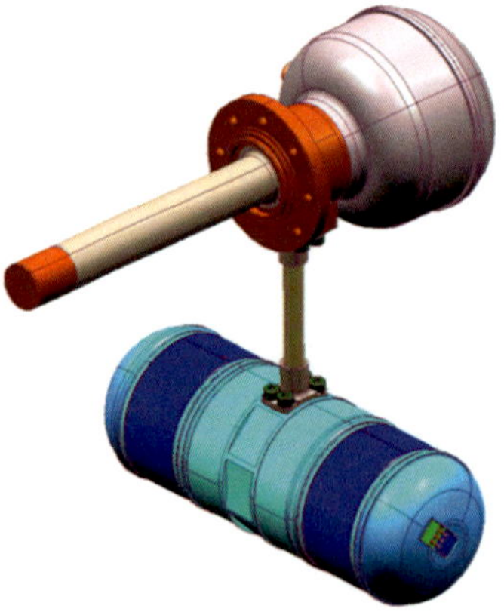

Figure 1. Coaxial pulse tube microcryocooler layout (split configuration).

Figure 2. Prototype microcompressor, with a mass of 190 grams and a length of 90 mm.

Figure 3. Microcompressor motor module containing piston and linear moving magnet motor.

Figure 4. Pulse tube coaxial coldhead. The coldhead mass as shown is 120 grams.

and operate at 1000 psi pressure, but the maximum proof pressure tested to date was 1200 psi for operation at 750 psi. The compressor operating frequency is 90-140 Hz, depending on model.

The first pulse tube Microcryocooler built and tested was an in-line coldhead configuration having dimensions which could be readily converted to a coaxial configuration (i.e., with the same regenerator and pulse tube lengths). An in-line prototype coldhead was preferred for the initial test because of its lower cost to build and ease in exchanging individual components. However, the thermodynamically critical features are similar for both the in-line and coaxial designs.

The completed cold head is shown in Figure 4. This coldhead has an overall length of 115 mm including the gas reservoir volume at the end. The reservoir diameter is 42 mm. The coldfinger length from the mounting flat to the end of the cold tip is 66 mm. The screw holes for the mechanical mounting flange are also shown. The fill port for the Microcryocooler is a copper pinch-off tube which extends from the reservoir volume (not shown in Figure 4).

The initial total coldhead mass is 120 grams. Structural analysis of the reservoir volume has shown that parts of it can be significantly thinned to reduce mass. Furthermore detailed thermal modeling has shown that the copper material in the warm flange can be replaced with aluminum to further reduce the coldhead mass. The predicted mass of this lighter-weight coldfinger is only 65 grams.

This coldhead was designed with a thin layer of titanium on the end of the copper coldfinger so that the FPA will bond directly to the end of the coldfinger with less thermal contraction mismatch. Based on mission needs, other thermal interfaces are possible. The warm flange was designed for a bolted heat sink. This heat sink can be finned for air cooling.

OVERVIEW OF PULSE TUBE MICROCRYOCOOLERS AT ATC

The three varieties of Microcryocoolers which have been developed at ATC are shown in Figures 5-8.

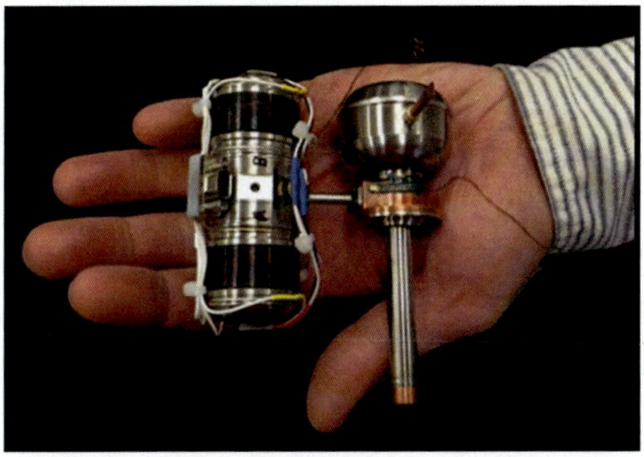

Figure 5. Standard Micro.

Figure 6. High Power (HP) Micro.

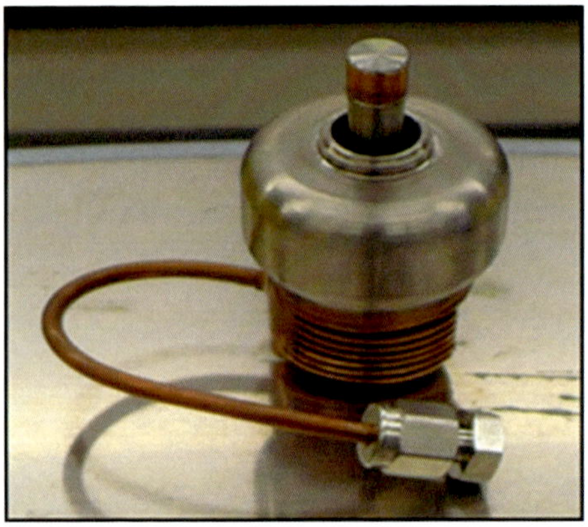

Figure 7. Fast cool-down micro cold head.

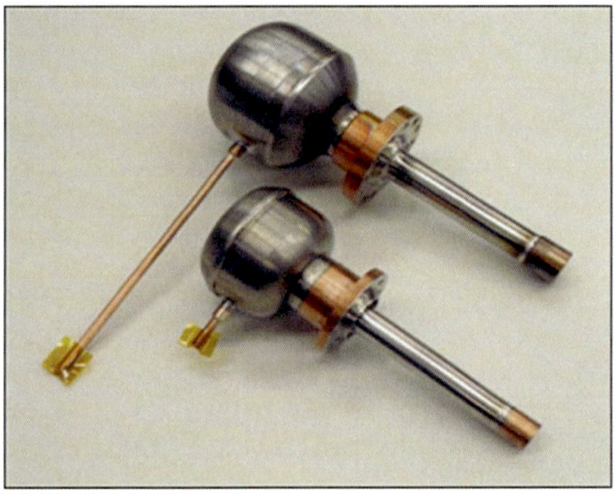

Figure 8. Comparison of standard and high power cold heads.

The primary design requirements of these units were:

- Standard PTM; design 1 W cooling at 150 K, Cooldown <8 minutes to 150 K,<11 minutes to 125 K (20 W)
- High Power PTM; 2 W at 105 K
- Fast Cool-Down PTM; 0.28 W at 125 K, <10 W Steady State Cooldown < 3 Minutes to 125 K (<40 W)

MICROCRYOCOOLER PERFORMANCE

Standard and High Power Coaxial Pulse Tube Microcryocooler Performance

The High Power PTM is somewhat heavier and larger than the standard unit and has a power capability of 60 W, while the standard unit is typically utilized with power inputs in the range of 10-25 W. The cooling capability for these two units is shown in Figure 9 and 10.[10]

Both systems have been qualified to TRL6 through temperature cycling and launch vibration.[8,9,10] In addition, the microcompressors have operated at heatsink temperatures as low as 130 K without problems. This capability allows operation of the systems for deep space and planetary missions, where power is limited. Operation at these heatsink temperatures greatly improves the specific power input (watts of power/watts of cooling). In addition, the high power system has undergone continuing life testing and has accumulated 7200 hours (as of May 2016) with no problems. Also a single motor module has been operating continuously for over two years, with no problems and continues to accumulate hours.

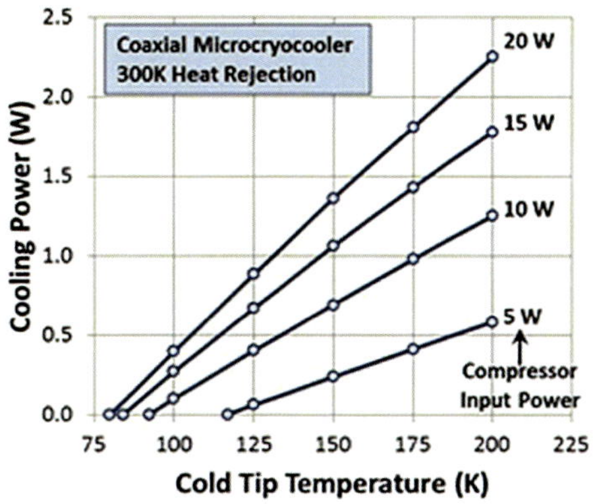

Figure 9. Load lines for standard microcooler.[9]

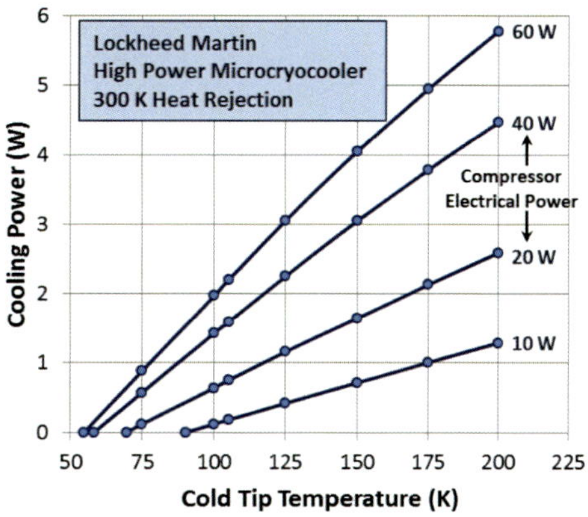

Figure 10. Load lines for high power microcryocooler.[10]

Induced Vibration Considerations

Although our cryocoolers are well balanced by the opposing dual piston motion, and operate at 100 Hz (standard micro) and 140 Hz (high power micro) there is some residual vibration that may require consideration for some sensitive instruments. These instruments are typically staring instruments, which require limited amounts of jitter. We have conducted several investigations of this vibration and means to reduce the levels, and the Jet Propulsion Laboratory has also measured the levels. Some electronic drives incorporate vibration cancellation to reduce levels to milliNewton (mN) levels.

The unregulated induced vibration is partly a result of very small differences in moving mass, motor characteristics and potentially very small misalignment of the drive axes of the two motors. The vibration from this source can be reduced by matching the motor modules resonant frequencies individually and then selecting the best matching pair.

When the units are operating at 100-140 Hz, this reduces the amplitude of the vibration and limits the motions, which reduces the vibration effects.

The Jet Propulsion laboratory has recently operated our cooler with an electronic controller which incorporates vibration cancellation and reduced levels to below 10 mN, and ~0 peak levels for the first four harmonics on the compressor axis.

We have achieved levels as low as700 mN by matching of stroke and phase between the two motors.[10] More work is required to further minimize and control the induced vibration levels; however we believe the values demonstrated to date are adequate for the majority of instruments and applications.

Fast Cool-Down Compact Coaxial Pulse Tube Microcryocooler Performance

The primary requirement for the fast cool down system (FC-PTM) was to achieve a coldtip temperature of 125 K in under 3 minutes. The allowable power input during this time was up to 40 W, and steady state performance required cooling of 0.28 W at 125 K. The design of this system was a compromise between fast cool down time and the steady state performance. It was evident from the start that the cold head should be optimized to achieve the cool down time. Initially our model was validated against cool down data from our standard Microcryocooler as shown in Figure 11.[9]

Excellent correlation between the model and the test data was confirmed with 20 W compressor power, and the optimization process was started. The effective mass of the FPA assembly was represented in this test case with 11.5 grams of added copper to the cold tip. Trades were conducted which compared the cool down time with the power required for steady operation, and a shortened cold finger assembly was selected. In addition, it was desired to modify the design of the cold head to make it more compact. The design which evolved from these optimizations is shown in Figure 12. The resulting design, which packages the plenum volume around the cold finger reduces the overall length by a factor of two compared with the regular cold head. This greatly reduces the package volume and also allows the utilization of this cooler in a 1U CubeSat.

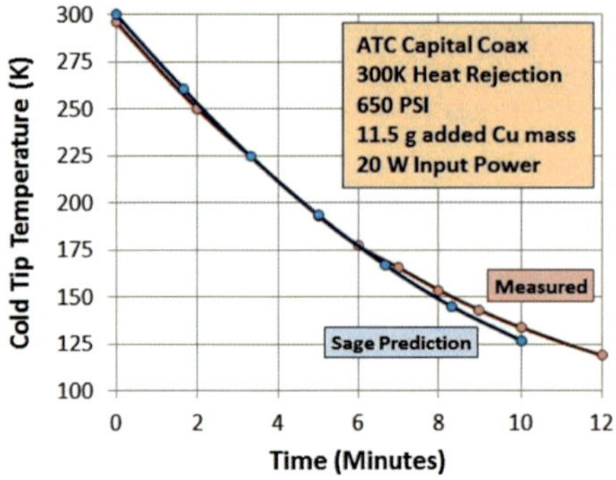

Figure 11. Comparison of prediction and test data for standard pulse tube microcryocooler cool down.

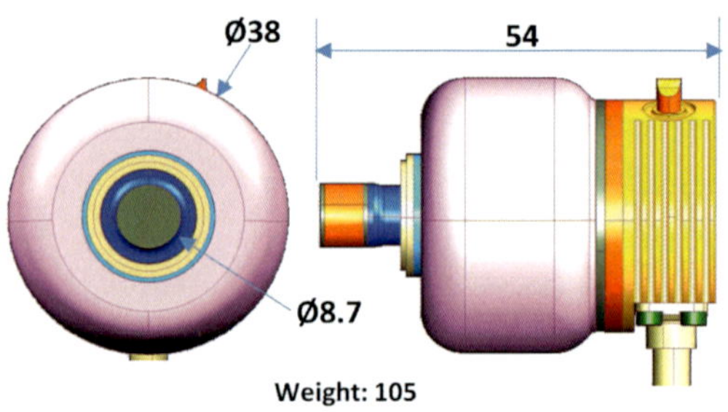

Figure 12. Configuration of the fast cool-down cold head (dimensions in mm).

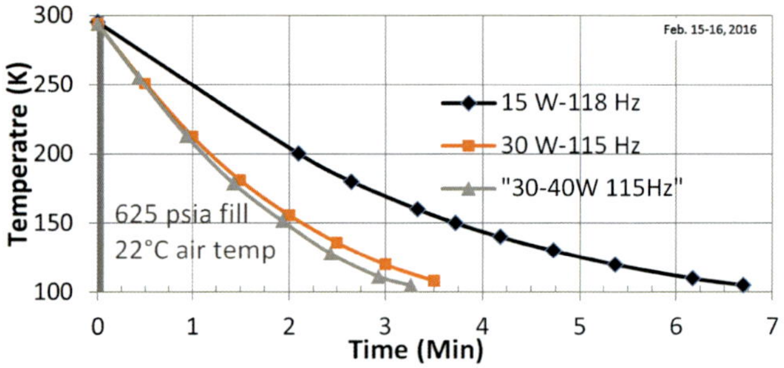

Figure 13. Cooldown test data for the Fast Cooldown FC-PTM system (4.1 grams copper of simulated focal plane).

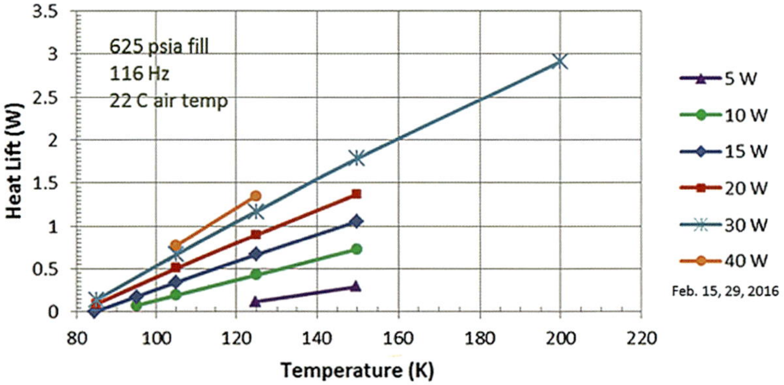

Figure 14. Steady state load lnes for the Fast Cooldown FC-PTM system.

The results of the initial FC-PTM cool down tests agreed very well with the modeling and are shown in Figure 13. An mass of copper of 4.1 grams was added to the cold tip to represent the effective enthalpy of the FPA. The results are excellent and agree closely with predicted values. Cooling to 150 K was achieved in under 2 minutes, and 2.5 minutes were required to reach 125 K — thus, meeting project goals.

The recently measured FC-PTM load lines under steady state conditions are presented in Figures 14 and 15. Figure 14 presents performance with ambient air temperatures of 293 K, while Figure 15 exhibits load lines for a span of three ambient air temperatures at a constant 15 W microcompressor input power. All tests were conducted in an environmental chamber with air circulation. It is interesting to note that the

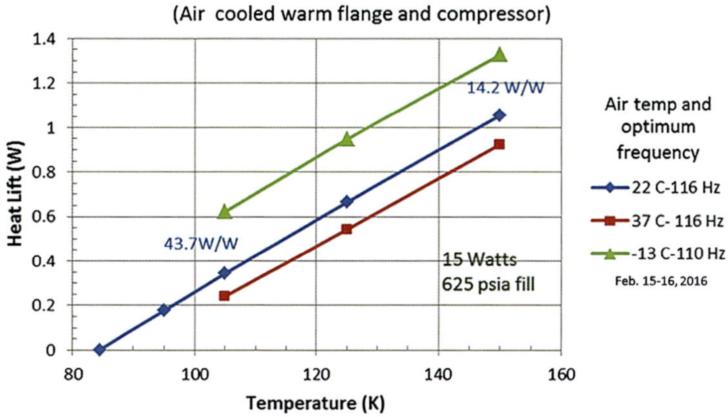

Figure 15. Effect of ambient air temperature on cooling performance for the Fast Cooldown FC-PTM System.

specific power (W/W) of the fast cool down system with the shortened cold finger is very similar to other units developed without restriction on cool down time or cold finger length.

FUTURE WORK

Work was started on the first space version of the pulse tube Microcryocooler in the summer of 2013.[7] Since then, LM-ATC has built and delivered several ground prototype Microcryocoolers. This includes two systems for LM-SBF IR cameras under AMRDEC funding.[9] Also, a unit was delivered to JHU-JPL for MatISSE, which is a spectrometer for planetary missions requiring cooling to 150-170 K.[10,11] We are currently assembling a high power Microcryocooler for a JHU-APL gamma ray instrument. We are also investigating a Microcryocooler solution designed to operate at 77 K for a different JPL planetary science instrument.

The small size and input power requirement of the Microcryocooler make it an excellent candidate for CubeSat and other micro and small satellite missions, and we hope to identify potential cryogenic CubeSat missions in the near future. In particular, the small length envelope of the fast cool-down cold head is compatible with packaging on a 1U CubeSat. Candidate space missions include surveillance and Earth science missions, with possible deployment of constellations of CubeSats for covering the entire globe.

Additional investigations ongoing include the utilization of two-stage cold heads for lower temperatures, and operation with Microcooler temperatures in the 100-150 K range.

There are numerous evolving applications for avionics and vetronics systems which require extended life times. However, transition to manufacturing and moderate-to-high volume production runs will be required to bring cost down to predictable and comparable levels to current commercial minicoolers.

ACKNOWLEDGMENTS

The development of the fast cool down system is supported by the AMRDEC 6.2 Missile Systems R&D Program—Next Generation Tactical Cooled/Uncooled IR Sensors and Seekers Project. Earlier work on the high power and standard Microcryocoolers has been supported by Lockheed Martin Space Systems Advanced Technology Center and Lockheed Martin Missiles & Fire Control Santa Barbara Focal plane collaborative Internal Research & Development (IR&D) projects, as well as AMRDEC. Appreciation is expressed to AMRDEC's Weapons Development and Integration Directorate for their support of Microcryocooler technology development and applied infrared technology RDT&E for the ground tactical U.S. Army warfighter.

REFERENCES

1. T. Nast, J. Olson, P. Champagne, B. Etimov, D. Frank, E. Roth, T. Renna, "Overview of Lockheed Martin Cryocoolers" *Cryogenics* **46** (2006), pp. 164-168.

2. Rogalski, Antoni., *Infrared Detectors 2nd Edition.* Boca Raton: CRC Press, 2011.

3. S. Maimon and G.W. Wicks, "nBn detector, an infrared detector with reduced dark current and higher operating temperature," Applied Physics Letters **89**, 151109 (2006).

4. D. Z. Ting, S. A. Keo, J. K. Liu, J. M. Mumolo, A. Khoshakhlagh, A. Soibel, J. Nguyen, L. Höglund, S.B. Rafol, C. J. Hill, S. D. Gunapala, "Barrier infrared detector research at the Jet Propulsion Laboratory," Proc. SPIE, Infrared Remote Sensing and Instrumentation XX, **8511**, 851102 (2012) and references therein.

5. A. L. Adams, E. Chang, J.W. Scott, A. E. Smith, M. A. Goodnough, "Infrared focal plane arrays for spectroscopic applications," Proc. SPIE, **7319**, Next-Generation Spectroscopic Techniques II, 731903 (2009).

6. J.R. Olson, E. Roth, P. Champagne, B. Evtimov, T. Nast, "High Performance Pulse Tube Cryocoolers," *Adv. in Cryogenic Engineering*, Vol. 53, Amer. Institute of Physics, Melville, NY (2008), pp. 514-521.

7. J.R. Olson, P. Champagne, E. Roth, T. Nast, E. Saito, V. Loung, A.C. Kenton, and C.L. Dobbins, "Microcryocooler for Tactical and Space Applications", *Adv. in Cryogenic Engineering*, Vol. 59, Amer. Institute of Physics, Melville, NY (2014), pp. 357-364.

8. T.C. Nast, E. Roth, J.R. Olson, P. Champagne, D. Frank, "Qualification of Lockheed Martin Micro Pulse Tube Cryocooler to TRL6," *Cryocoolers 18*, ICC Press, Boulder, CO (2014), pp. 45-50.

9. J.R. Olson, P. Champagne, E.W. Roth, G.B. Kaldas, T. Nast, E. Saito, V. Loung, B.S. McCay, A.C. Kenton, and C.L. Dobbins, "Coaxial Pulse Tube Microcryocooler", *Cryocoolers 18*, ICC Press, Boulder, CO (2014), pp. 51-57.

10. J.R. Olson, G. Kaldas, P. Champagne, E. Roth, T. Nast, "MatISSE Microcryocooler," *Adv. in Cryogenic Engineering: Proceedings of the Cryogenic Engin. Conf. (CEC) 2015*, Vol. 101, IOP Publishing, Bristol, UK (2015), Article 012025.

11. I.M. McKinley, D.L. Johnson and J.I.Rodriguez, "Characterization Testing of Lockheed Martin Standard Micro Pulse Tube Cryocooler," *Cryocoolers 19*, ICC Press, Boulder, CO (2016), (this proceedings).

Characterization Testing of Lockheed Martin Standard Micro Pulse Tube Cryocooler

I.M. McKinley, D.L. Johnson, and J.I. Rodriguez

Jet Propulsion Laboratory, California Institute of Technology
Pasadena, CA 91109

ABSTRACT

This paper describes the thermal performance, exported vibration, and magnetics testing and results for a Lockheed Martin standard micro pulse tube cryocooler. The thermal performance of the microcooler was measured in vacuum with heat reject temperatures between 150 and 300 K. The cooler was driven with Thales XPCDE4865 drive electronics for input powers ranging from 4 to 20 W and drive frequency between 84 and 98 Hz. The optimal drive frequency was dependent upon both input power and heat reject temperature. In addition, the exported forces and torques of the cooler were measured with the cooler driven by Thales CDE7232 drive electronics for input powers ranging from 4 to 20 W and drive frequency between 88 and 96 Hz. Moreover, the automatic vibration reduction function of the drive electronics was able to decrease the force along the compressor axis to below 10 mN 0-peak. Finally, the DC and AC magnetic fields around the cooler were measured at various locations.

INTRODUCTION

The Jet Propulsion Laboratory (JPL) has identified the Lockheed Martin standard micro pulse tube cryocooler as a low-cost candidate to provide active cooling on future cost-capped scientific missions. This tactical cooler offers a long lifetime due to the flexure bearings in its compressor and a pulse tube coldhead containing no moving parts. Lockheed Martin has published a number of papers focusing on this cooler in recent years [1, 2, 3, 4]. This cooler has been optimized for a coldtip temperature of 125 K to 150 K and operates with a drive frequency near 100 Hz [2]. In addition, it is capable of operating with a heat rejection temperature below 150 K [3]. This capability is important for deep space missions because of the low survival temperature required for the power savings in survival mode. The compressor weighs 210 grams and the entire cooler weighs 328 grams [3]. Moreover, it has been qualified to a technology readiness level (TRL) 6 [3]. Finally, the cooler tested in this work consisted of physically the same coldhead and compressor as that described in [4] except with a different transfer line. This work seeks to advance the Lockheed Martin standard microcooler towards use in space flight missions by characterizing its thermal performance, exported forces, and electromagnetic interference over a wide range of operating conditions.

Cryocoolers 19, edited by S.D. Miller and R.G. Ross, Jr.

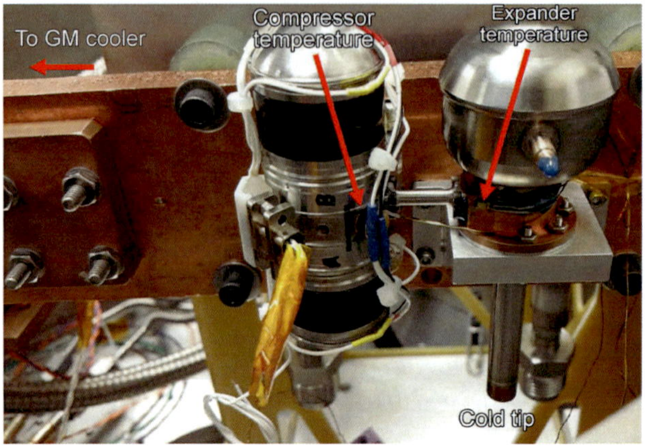

Figure 1. LM cooler in the TVAC test setup. Both the coldblock and MLI are not shown.

THERMAL PERFORMANCE TESTS

Test Setup and Procedure

Figure 1 shows a photograph of the test setup used for thermal performance measurements. The Lockheed Martin (LM) cooler's thermal performance was measured inside a thermal vacuum (TVAC) chamber at JPL's cryocooler characterization facility. Rejection of the LM cooler's heat was achieved by cooling a copper mounting plate with a Gifford-McMahon (GM) laboratory cooler. The cooler heat rejection temperature was defined as the temperature measured on the expander surface. It was maintained between 150 and 300 K with a Lake Shore 340 temperature controller powering Dale resistors. In addition, the compressor temperature was measured and was approximately 5°C warmer than the expander for the duration of the tests. The LM cooler was powered with Thales XPCDE4865 drive electronics supplying between 3.5 and 16 W to the compressor at drive frequencies between 83 and 98 Hz. A coldblock consisting of a copper ring equipped with resistors and a Lake Shore DT670 diode maintained the cold tip temperature between 55 and 225 K by supplying heat loads of between 0 and 2.5 W. Note that all of the cold surfaces inside the vacuum chamber were covered with multi-layer insulation (MLI). In addition, Lockheed Martin provided a recommended maximum drive voltage based on the motor characteristics as a function of frequency, compressor temperature, and current.

Effect of Drive Frequency

Figure 2 shows the specific power of the LM cooler and the motor efficiency vs. drive frequency. The specific power was defined as the compressor input power divided by the cooling power (or coldblock load). The motor efficiency was defined as the quantity of the i^2R losses in the coils subtracted from the compressor input power divided by the compressor input power [5]. In this case, the coil resistance was defined as 5.3 ohms and corresponded to the value at room temperature obtained with a 4-wire measurement. It is evident that the minimum specific power and the maximum motor efficiency both depended on both heat reject temperature and compressor input power. However, for a given heat reject temperature and cooler input power, the respective minimum and maximum fall nearly on the same drive frequency. This indicates that the peak thermodynamic frequency of the coldhead and the compressor resonant frequency are well matched [4].

Effect of Heat Rejection Temperature

Figure 3 shows compressor input power vs. coldblock load (Ross plots) for various expander temperatures with the cooler driven at 96 Hz. It also shows a Ross plot for the 150 K expander tem-

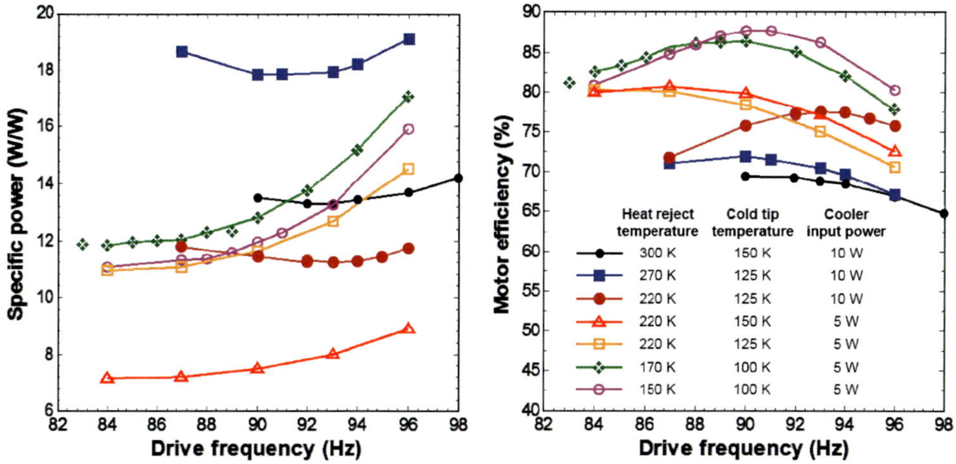

Figure 2. Specific power and motor efficiency vs. drive frequency for various heat rejection (expander) temperatures, cold tip temperatures, and cooler input powers.

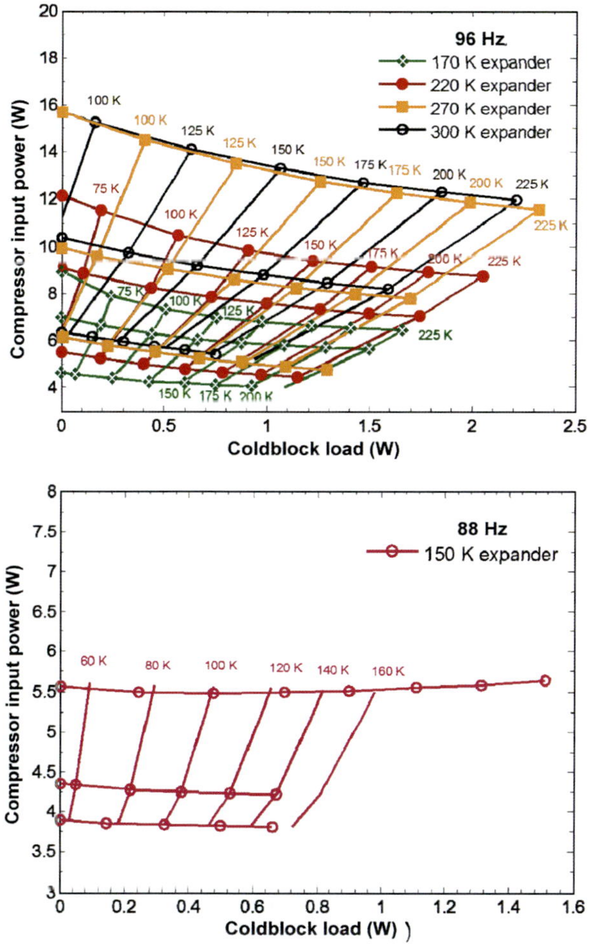

Figure 3. Ross plots for various expander temperatures with the cooler driven at 96 Hz and a Ross plot for 150 K expander temperature with the cooler driven at 88 Hz.

perature with the cooler driven at 88 Hz. Load lines were measured with a fixed compressor input voltage. For a given input voltage, the input power decreased as the cold tip temperature increased. The cold tip temperature for each load line was fit as a function of both compressor input power and coldblock load. These fits were used to obtain the isotherms shown in Figure 3. For a given cold tip temperature and input power, the cooling capability of the cooler increased with decreasing expander temperature. Note that the compressor input power was limited to 16 W for the duration of these tests. The maximum input power was not necessarily reached during these tests. In fact, Lockheed Martin has input up to 20 W to this model cooler at 300 K expander temperature for cold tip temperatures up to 200 K corresponding to 2.25 W of cooling [2].

EXPORTED FORCES AND TORQUES

Test setup and Procedure

Figure 4 shows a photograph of the LM cooler mounted on a Kistler dynamometer that measured exported forces from the cooler. It also shows the coldblock previously described. The complete test setup has been described in the literature previously [6, 7]. The cooler was operated with the Thales CDE7232 drive electronics for compressor input powers ranging from 5 to 18 W and drive frequency ranging from 88 to 96 Hz. In addition, the automatic vibration reduction (AVR) feature of this drive electronics was tested. Similar to previous works, the force signal in the compressor axis was used as feedback for this feature [6, 7]. The expander temperature of the cooler was not actively controlled but did not exceed 30°C for the duration of these tests. In addition, the cold tip was not contained in a vacuum housing during these measurements. Also note that the compressor mount was not bonded to the compressor when it was delivered to JPL and for the entire duration of the thermal performance and electromagnetic interference tests. However, the mount was bonded to the compressor with Nusil CV-2946 for the duration of the exported force results presented here.

Effect of Drive Frequency

Figure 5 shows the 0-peak force vs. harmonic in all three axes for 10 and 15 W compressor input power for drive frequency between 88 and 96 Hz. Note that harmonic 0 corresponded to the drive frequency. The compressor axis force at the drive frequency decreased with decreasing frequency. In most cases, the force for a given harmonic, direction, and input power decreased with decreasing drive frequency. In addition, the fourth harmonic of the force in the transfer line axis was largest among the harmonics. The fourth harmonic in the vertical axis at 15 W input power was also very large. This possibly could be attributed to non-uniform magnetic materials in the compressor or small motor misalignment. Indeed, magnetic analysis performed by Lockheed Martin indicated that non-uniform magnetic side loads could lead to lateral forces at the higher harmonics. However, it should be noted that the jitter introduced by large forces at higher harmonics is significantly less than that due to smaller harmonics. For example, 2 N at 400 Hz leads to the same displacement as 0.02 N at 40 Hz.

Figure 4. Photograph of the LM cooler mounted on the Kistler dynamometer.

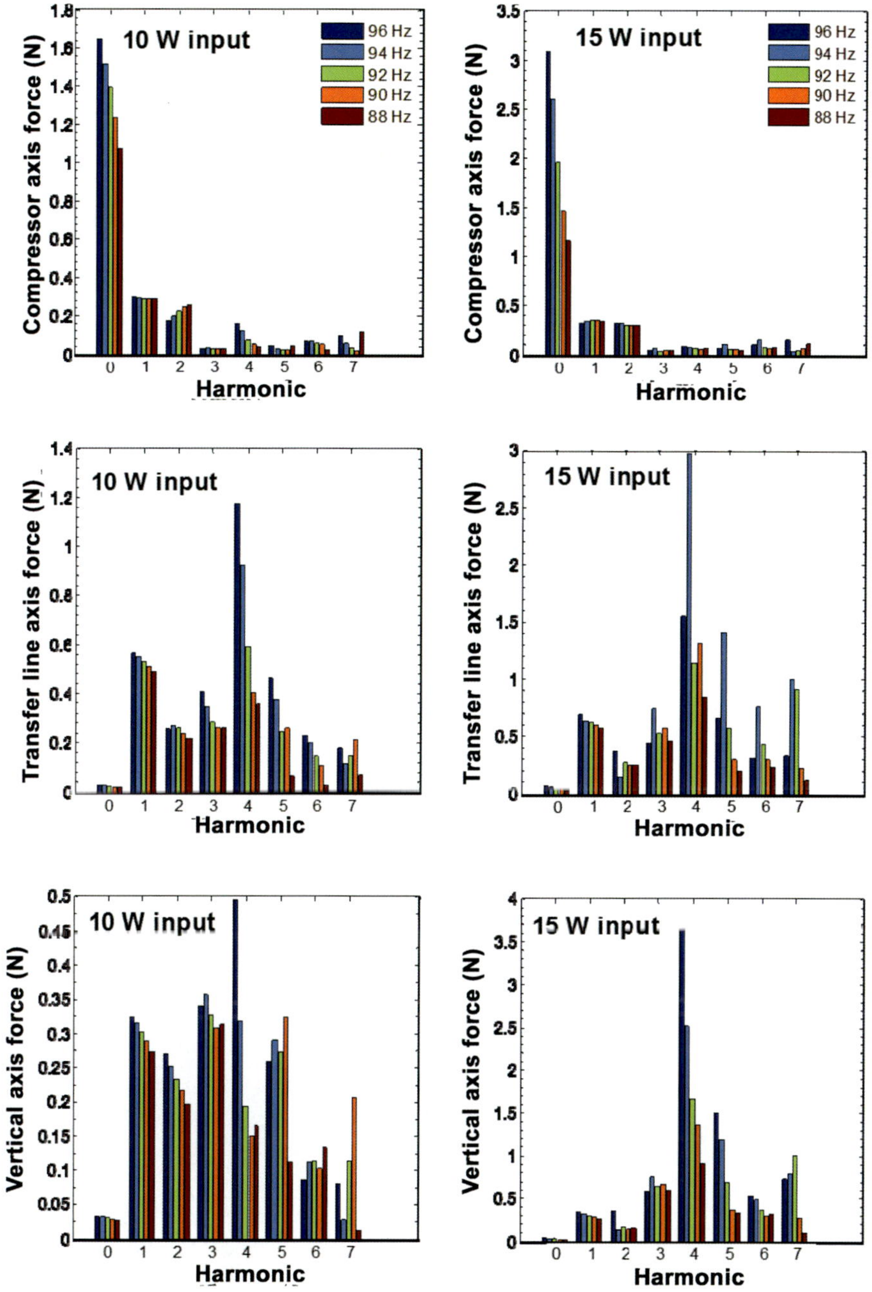

Figure 5. Force vs. harmonic in all three axes for 10 and 15 W compressor input power for drive frequency between 88 and 96 Hz. Note the vertical upper limit is different on each plot.

Effect of Input Power

Figure 6 shows the force vs. harmonic in all three axes for various input voltages as well as the same data as force vs. compressor input power for various harmonics with the cooler driven at 96 Hz. It is evident that for a given harmonic and axis, the force decreased with decreasing input power/voltage. In addition, the majority of the compressor axis force was contained in the drive frequency. However, the fourth harmonic of the transfer line and vertical axis forces was the largest. The transfer line axis force was the smallest reaching a peak of 1.6 N 0-peak. Moreover, the vertical axis force exceeded that of the compressor axis for input voltages above 11 V_{rms}. In fact, there was a sharp increase in vertical axis force for input voltages above 10 V_{rms} and the cooler compressor

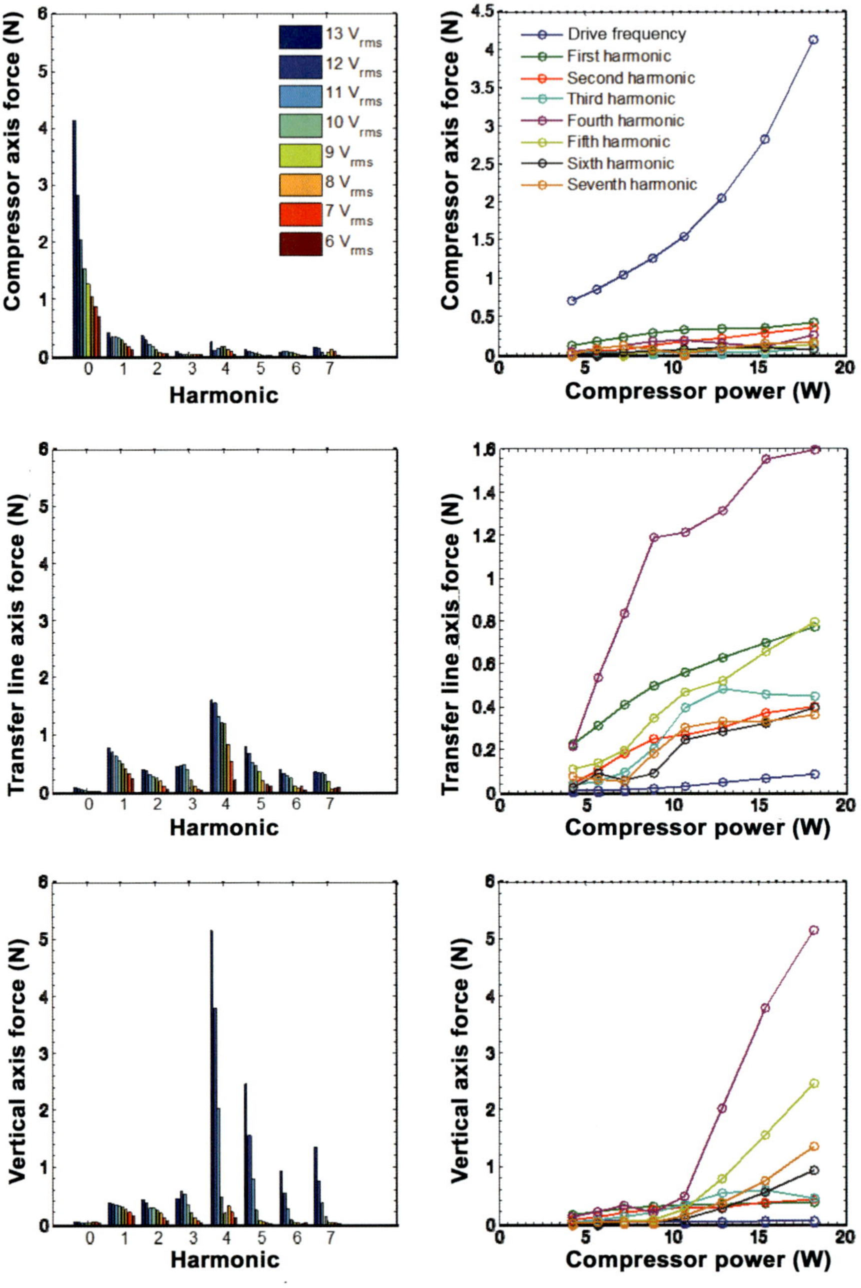

Figure 6. Force vs. harmonic in all three axes for various input voltages as well as force vs. compressor input power in all three axes for various harmonics. The cooler was driven at 96 Hz in all cases without AVR. The data in the left and right columns is the same. Note the vertical upper limit is different for each plot in the right column.

made a louder noise at a different tone. Again, the large forces in the vertical axis could possibly be attributed to non-uniform magnetic side loads in the compressor.

Effect of Automatic Vibration Reduction

Figure 7 shows the force in the compressor axis as a function of harmonic for various input voltages at 96 Hz with the automatic vibration reduction function of the drive electronics off and on. The AVR function was able to reduce the first four harmonics of the compressor axis force to below 10 mN 0-peak for input voltages up to 11 V_{rms}. For voltages above 11 V_{rms}, the function was not able to complete its calibration period and continuously changed the input wave forms to the compressor motors.

ELECTROMAGETIC INTERFERENCE

DC measurements

Figure 8 (left) shows the test setup used to measure the DC magnetic fields surrounding the LM cooler. The cooler was without any mechanical supports/mounting structures. A Lake Shore 475 DSP Gaussmeter with a Lake Shore HMMA-1808-VF axial probe was used to measure the DC field in the axial and radial directions. An axial position of 50 cm corresponded to the center of the compressor. Figure 9 (left) shows the magnitude of the magnetic field as a function of axial position at a radial distance of 7 cm from the edge of the compressor. The magnitude of the field was determined by taking the square root of the sum of the squares of the radial and axial measurements at

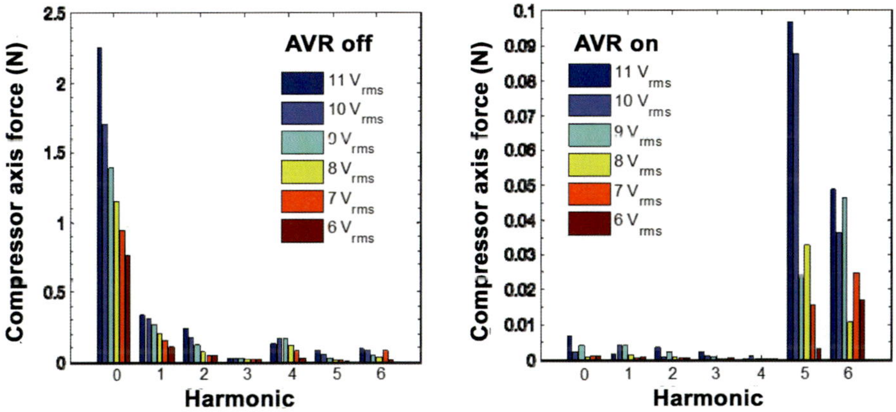

Figure 7. Force in the compressor axis vs. harmonic for various input voltages at 96 Hz with AVR off and on. Note the vertical upper limit is different on each plot.

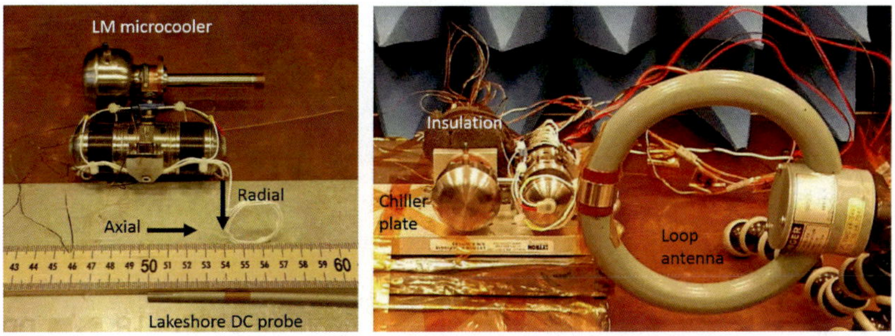

Figure 8. Photographs of the test setup for DC (left) and AC (right) magnetic field measurements.

a given location. Note that the magnetic field of the Earth was subtracted from the measured field. Figure 9 (right) shows the axial magnetic field as a function of distance from the compressor end at the radial center of the compressor. In this case, the radial component of the field was equal to zero. Figure 9 (right) indicates the expected 1/distance^3 relationship between DC field of a permanent magnet and distance from it [8]. In addition, magnetic mapping was performed around one rotational axis and additional measurements were taken about multiple rotational axes in order to establish the magnetic dipole moment of the cooler. The magnetic mapping yielded a quadrupole magnetic field of 14.41 nT pk-pk at a distance of 1 m. This was a result of the compressor containing more than one magnet. The overall magnetic dipole moment was calculated from the measurements about multiple rotational axes and was 9.18 mA-m^2. Finally, the magnetic dipole field was calculated from the dipole moment and was 3.67 nT pk-pk at 1 m.

AC measurements

Figure 8 (right) shows the test setup used to measure the AC magnetic fields surrounding the LM cooler. The cooler was mounted on a chiller plate that was used to maintain the expander at room temperature while the cooler operated. Insulation was placed around the cold tip to reduce the buildup of ice. The cooler was operated at input powers between 10 and 20.5 W at 96 Hz. The AC magnetic field was measured as a function of frequency in the radial and axial directions at various locations using a Singer 94605-1 loop antenna connected to a Rohde & Schwarz ESIB40 spectrum analyzer. Peaks in the AC field vs. frequency occurred at the drive frequency and its higher harmonics. The magnitude of the magnetic field at the drive frequency and first harmonic (96 and 192 Hz) were calculated by taking the square root of the sum of the squares of the axial and radial components. For a given radial distance, the magnitude of the field was nearly independent of axial location along the compressor. In addition, for a given location, the magnitude of the field was weakly dependent on cooler input power and varied by less than 5% for cooler input power ranging from 10 W to 20.5 W. Figure 10 shows the magnitude of the AC magnetic field at 96 Hz and 192 Hz as a function of axial distance at from the center of the compressor end (left) and of radial distance from the center of the compressor edge (right). The magnitude of the AC field decreased with increasing distance from the compressor. The Military Standard RE101 requirement [9] was met at 7 cm from the compressor end for both Army and Navy applications. However, it was not met 7 cm radially from the side of the compressor for Navy applications. These results indicate that the cooler would need magnetic shielding to meet the RE101 requirement for Navy applications.

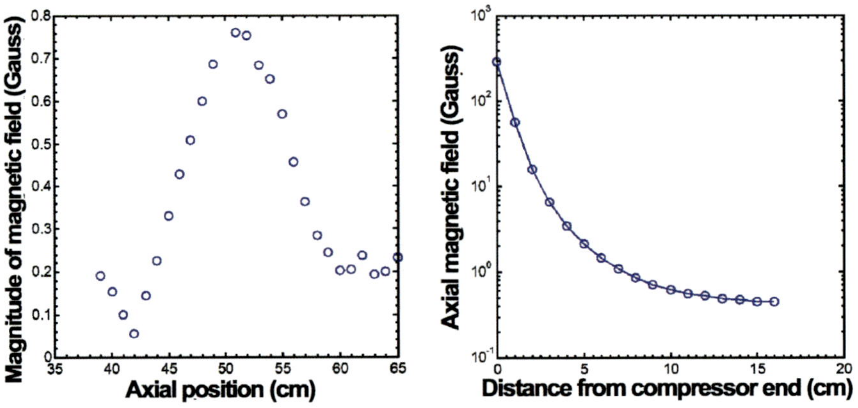

Figure 9. The magnitude of the DC magnetic field as a function of axial position at a radial distance of 7 cm from the compressor edge (left) and axial magnetic field on axis as a function of the distance from the

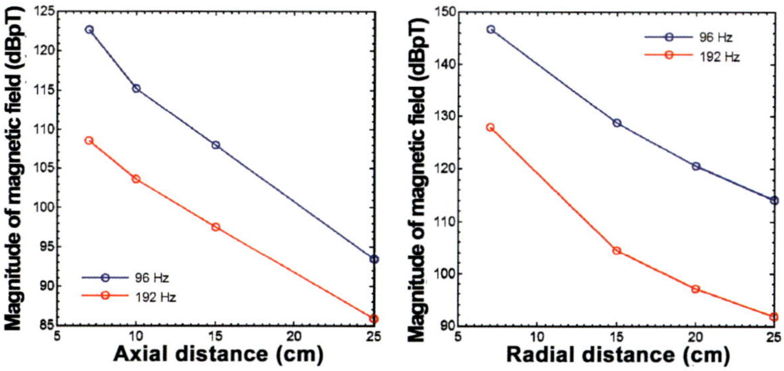

Figure 10. The magnitude of the AC magnetic field for the operating LM cooler as a function of axial distance from the compressor end (left) and as a function of radial distance from the compressor edge at the

CONCLUSION

This paper described the thermal performance, exported vibration, and magnetics testing and results of a Lockheed Martin standard micro pulse tube cryocooler. The thermal performance of the microcooler driven with Thales XPCDE4865 drive electronics was reported for heat reject temperatures between 150 and 300 K, input powers ranging from 4 to 20 W, and drive frequency between 84 and 98 Hz. The optimal drive frequency was dependent on both input power and heat reject temperature. In addition, the exported forces and torques of the cooler were measured with the cooler driven by Thales CDE7232 drive electronics for input powers ranging from 4 to 20 W and drive frequency between 88 and 96 Hz. The exported forces were dependent on both input power and drive frequency. Furthermore, the automatic vibration reduction function of the drive electronics was able to decrease the force in the compressor axis to below 10 mN 0-peak for input power up to approximately 12 W. Finally, the DC and AC magnetic fields around the cooler were measured at various locations. The DC measurements revealed the cooler to have a magnetic dipole field of 3.67 nT pk-pk at a distance of 1 m. The AC measurements revealed that the cooler did not meet the RE101 requirement at a distance of 7 cm for Navy applications. Overall, this cooler remains an excellent candidate for future space missions.

ACKNOWLEDGMENT

The work described in this paper was carried out at the Jet Propulsion Laboratory, California Institute of Technology, and was sponsored by the NASA Science Mission Directorate, Maturation of Instruments for Solar System Exploration (MatISSE) program. In addition, the authors would like to acknowledge Jeff Olson, Evan Droz, Andy Lamborn, Jerry Gutierrez, Katherine Dang, Pablo Narvaez, and Diana Blayney for their contributions to this work.

REFERENCES

1. J. Olson, P. Champagne, E. Roth, T. Nast, E. Saito, V. Loung, A. Kenton and C. Dobbins, "Microcryocooler for tactical and space applications," *Adv. in Cryogenic Engineering*, Vol. 59, Amer. Institute of Physics, Melville, NY (2014).

2. J. Olson, P. Champagne, E. Roth, G. Kaldas, T. Nast, E. Saito, V. Loung, B. McCay, A. Kenton and C. Dobbins, "Coaxial Pulse Tube Microcryocooler," *Cryocoolers 18*, ICC Press, Boulder, CO (2015).

3. T. Nast, E. Roth, J. Olson, P. Champagne and D. Frank, "Qualification of Lockheed Martin micro pulse tube cryocooler to TRL6," *Cryocoolers 18*, ICC Press, Boulder, CO (2015).

4. J. Olson, G. Kaldas, P. Champagne, E. Roth and T. Nast, "MatISSE Microcryocooler," in *IOP Conf. Series: Materials Science and Engineering 101*, 2015.

5. A. Orlowska and G. Davey, "Measurement of losses in a Stirling cycle cooler," *Cryogenics,* vol. 27, no. 11, pp. 645-651, 1987.

6. D. Johnson, I. McKinley, J. Rodriguez, B. Carroll and H. Tseng, "Characterization testing of the Thales LPT9310 pulse tube cryocooler," *Cryocoolers 18*, ICC Press, Boulder, CO (2015).

7. D. Johnson, I. McKinley and J. Rodriguez, "Flight Qualification Testing of the Thales LPT9510 Pulse Tube Cooler," *Cryocoolers 18*, ICC Press, Boulder, CO (2015).

8. R. Weidner and R. Sells, *Elementary Classical Physics Volume 2: Electromagnetism and Wave Motion*, Allyn and Bacon, Inc., Boston, MA 1965.

9. Department of Defense, "Requirement for the Control of Electromagnetic Interference Characteristics of Subsystems and Equipment," 2015.

Miniature Stirling Cryocoolers at Thales Cryogenics: Qualification Results and Integration Solutions

D. Willems, R. Arts, G. de Jonge, J. Mullié, T. Benschop

Thales Cryogenics BV, Eindhoven, The Netherlands

ABSTRACT

In 2015, Thales Cryogenics presented new miniature cryocoolers and cooler drive electronics for high operating temperatures (HOT). In this paper, an update is given regarding the qualification program performed on these new products. Integration aspects are discussed, including an examination of the influence of the dewar cold finger on the sizing and performance of the cryocooler.

The UP8197 and UP8497 and the corresponding cooler drive electronics will be placed in the reference frame of the Thales product range of high reliability linear cryocoolers. Compatibility of the cryocoolers design with new and existing ¼" dewar designs is examined, and potential future developments are presented.

INTRODUCTION

During the 2015 SPIE Defense, Security and Sensing exhibition, Thales Cryogenics presented an extension of their product range into the small form-factor domain [1], aimed towards detector cooling at high operating temperatures (HOT), typically 120 K to 150 K.

At that time, the cooler designs presented had not undergone qualification testing. Now, a year later, an update can be provided on these products and their qualification status.

Both the miniature linear coolers (UP8197 and UP8497) have completed their full tactical qualification programs. Following this successful qualification, both products are available on the market. In addition to the linear coolers, the rotary Size, Weight and Power (SWAP) cooler introduced at the same time has been redesigned for use with a new, generic, cold finger interface [1-3].

In this paper, the updated qualification results of the linear coolers and the matching cooler drive electronics will be presented, together with integration options for different available cold finger interfaces.

POSITION IN PORTFOLIO

The different miniature coolers in the current Thales portfolio should be regarded in the context of detector operating regimes. Different competing medium wave infrared technologies have matured to a point where true HOT operation – with focal plane temperatures exceeding 140 K - becomes possible in an actual application next to laboratory demonstrators.

The availability of successful and cost effective HOT detectors and the adoption of these true HOT detectors by infrared camera manufacturers occurs slowly and guardedly, but surely.

Cryocoolers 19, edited by S.D. Miller and R.G. Ross, Jr.

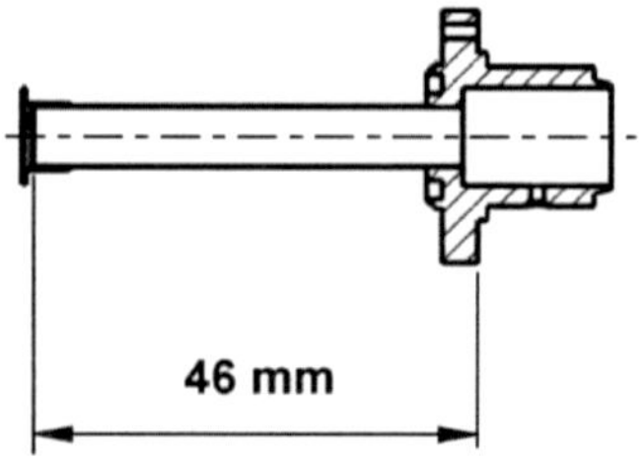

Figure 1. Thales CFA6 interface

Miniature coolers are required for many applications. For the foreseeable future, this means that both miniature coolers for the conventional temperature range around 77K are expected to be necessary in many applications. But in addition, new coolers for the true HOT operating domain will be required as well.

For the range of miniature coolers for detector temperatures around 77 K, a range of mature products is currently available. Many of these cryocoolers interface with the same cold finger sleeve, in order to provide maximum commonality for Integrated Detector/Cooler Assembly (IDCA) manufacturers. The standard ¼" cold finger sleeve (Thales CFA6 or C0012 definition) is seen in Figure 1.

The CFA6 cold finger is used in many cryocoolers, including larger coolers such as the RM4, designed for Long Wave Infrared (LWIR) detectors with operating temperatures around 60 K, or the LSF9597, designed for long-life operation for 24/7 surveillance cameras. Three miniature coolers are available within this range. These coolers are shown in Table 1.

For true HOT applications, smaller coolers should be used, with the selection depending on the SWAP constraints of the application, as well as the operating temperature. The UP8197 linear cooler (Figure 2) was designed for this purpose. Intended operating temperatures are 140 K and higher, but it has a potential application range down to 120 K.

This cooler is presented in the next section. First, the results of the qualification program are presented. Later, the available interfaces and the required future work are presented.

Table 1. Thales miniature coolers for low temperatures.

RM2 Rotary Monobloc	**UP8497**	**LSF9997**
	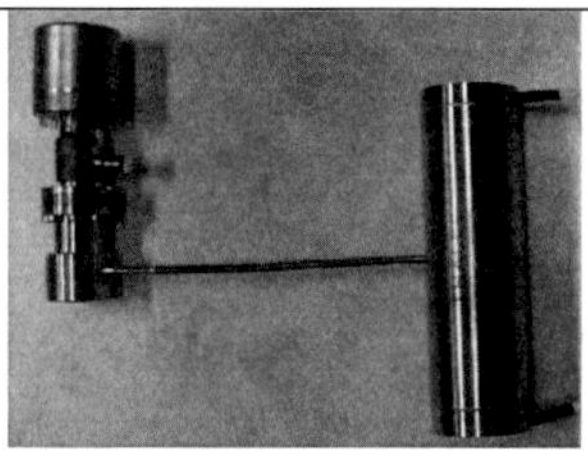	
Compressor 82x51.5x46 mm	Compressor 119 x 35mm	Compressor 119 x 44 mm
High power density Suitable for 77 K operation Balanced design for low vibrations and acoustic noise Produced in large series	High power density Designed for 110 K Suitable for 77 K operation Redesign (2015) qualified [1, 5], undergoing industrialization.	Smallest flexure-bearing compressor [5] Designed for 110 K Suitable for 77 K operation Moving-magnet design Designed for reliability In series production

Figure 2. Photograph of the UP8197 cooler.

UP8197 QUALIFICATION RESULTS

Performance qualification results

Based on the initial characterization of the cooler [1], the nominal parameters of the cooler were changed slightly compared to the original specification, in order to increase the available heat lift (allowable AC input power to the cooler) at +23 C and +71 C. Updated performance measurements are shown in Figure 3.

These performances were measured with a laboratory test dewar not optimized for heat load– the estimated parasitic heat load of the dewar itself was 250 mW at 150 K, 23 C. The characterization of this heat load at 150 K was done with the warm-up method [9], as measurements with the boil-off method are not practical at these tip temperatures. The result of the multi-slope warm-up is shown in Figure 4.

Compared to the previously reported performance numbers [1], the maximum input power to the cooler has increased by 33% from 15 W to 20 W. The advantage of the increased input power is that both the absolute available cooling power has increased, and the cool down time will decrease.

Cool down time has been tested in the same laboratory test dewar. The thermal characteristics of this lab dewar are different from a typical tactical dewar. Therefore, the test results are not fully relevant for cool down times in common tactical dewars. The expected cool down time in tactical dewars is scaled from the measured CDT in the lab dewar. Results are shown in Figure 5. It can be concluded that typical cool down times in tactical dewars are between 120 and 130 seconds

The method to determine the expected cool down time in actual dewars is as follows.

- The cooler is fully characterized in the lab dewar. The characterization consists of the cool down time, and the performance mapping for relevant operating conditions (input power, ambient temperature, and tip temperature).

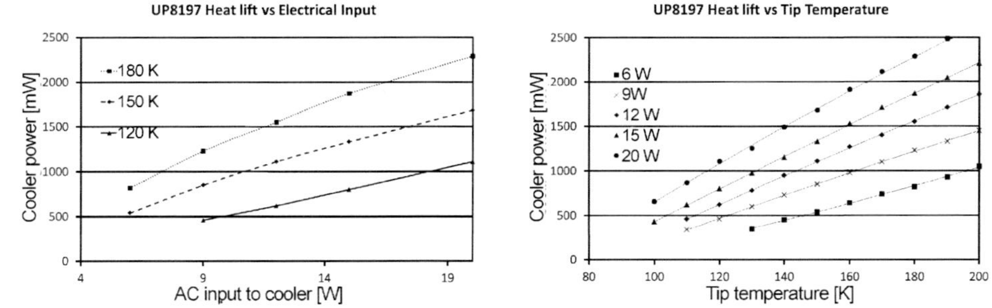

Figure 3. Typical performance of the UP8197 cooler. Left graphs show the cooling power versus input power at different tip temperatures. The right graph shows cooling power versus tip temperature at different input powers.

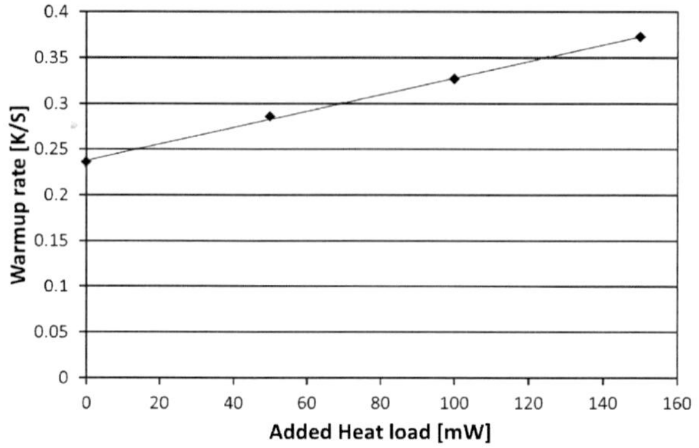

Figure 4. Determination of dewar heat load at 150 K using the warm-up method [9]. The resulting heat load is 250 mW.

- A transient thermal model of the cooler with the test dewar is built in the ThermXL thermal modelling software [6]. This model contains the full integration of the cooler in the application, including the impact of thermal resistances, heat sinking, etc.
- The transient model will predict tip temperature as a function of time. This prediction is compared to the measured cool down curve on the cooler to validate the model.
- The model is then used to calculate the transient behavior of the cooler in a dewar of typical characteristics. In this case, a detector is used with a dewar heat load of 220 mW between 23°C and 150 K, and a thermal mass of 180 J between 23°C and 150K.

The results are shown in Figure 5. It can be seen that the model results on the lab dewar are close to the measurement results. This validates the accuracy of the model. The other two results show typical values for two scenarios. The different scenarios use the effect that the maximum drive voltage for linear coolers at startup is lower than when the cooler is cold. In the first scenario, the safe 'warm' startup voltage is used for the entire cool-down. In the second scenario, the voltage ramps up from the low value at startup to the value at the cold tip temperature. This reduces the cool down time of the detector. This feature, called 'slow start', is a standard feature in all Thales linear cooler drive electronics .

Exported vibrations

One important advantage of a dual opposed-piston linear cryocooler is that the vibrational forces exported by the cooler are already at a very low level, without the need for a tuned balancer

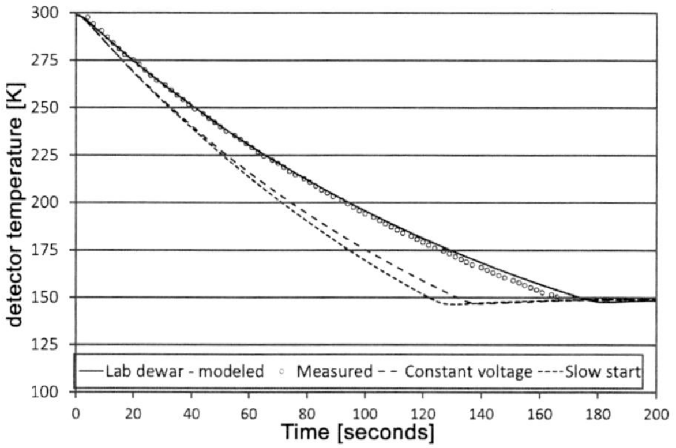

Figure 5. Cool down curves for the UP8197. Markers show the measured value in the lab dewar. Solid line is the simulated performance in the lab dewar. The two dashed lines are the simulated cool down curves in a representative tactical dewar.

or flexible mounting. This also has the advantage of resulting in low acoustic noise exported by the camera.

Two qualification models of the UP8197 were tested for exported vibrations. This was done using a three-axis accelerometers with either the compressor or the cold finger free-standing. No test dewar was mounted during exported vibrations testing. A measured vibration spectrum of the cold finger can be seen in Figure 6. This spectrum is measured at full input power. In regulation, the induced vibrations will scale with input power. As expected, the largest vibration exported by the cold finger occurs at the drive frequency, in the direction of motion of the displacer.

As the UP8197 is equipped with separate connections for each linear drive motor, it is possible to apply Thales AVR to the cooler [7, 8], enabling even lower vibration levels. In case an accelerometer with sufficient sensitivity is used, a similar reduction can be expected as with prior tests – which is at least a factor 50 reduction along the displacer axis.

Environmental conditions

The UP8197 has undergone a full environmental qualification program. The tests conducted are shown in Table 2. All environmental qualification tests were performed successfully. An example test can be seen in Figure 7.

UP8197 COLD FINGER INTERFACES

Cold finger interface definitions

An aspect of miniaturization is the reduction of the overall system length along the optical axis. The total length from the rear of the cooler to the front of the dewar window becomes an important parameter. HOT operation enables shorter overall lengths without much impact on performance. Therefore, interfaces for the UP8197 in HOT applications are no longer based on standard CFA6 interface as shown in Figure 2.

The 2015 product definition of the UP8197 [1] is based on a Thales proprietary short cold finger definition. Based on market feedback since then, two more variants of the UP8197 cold

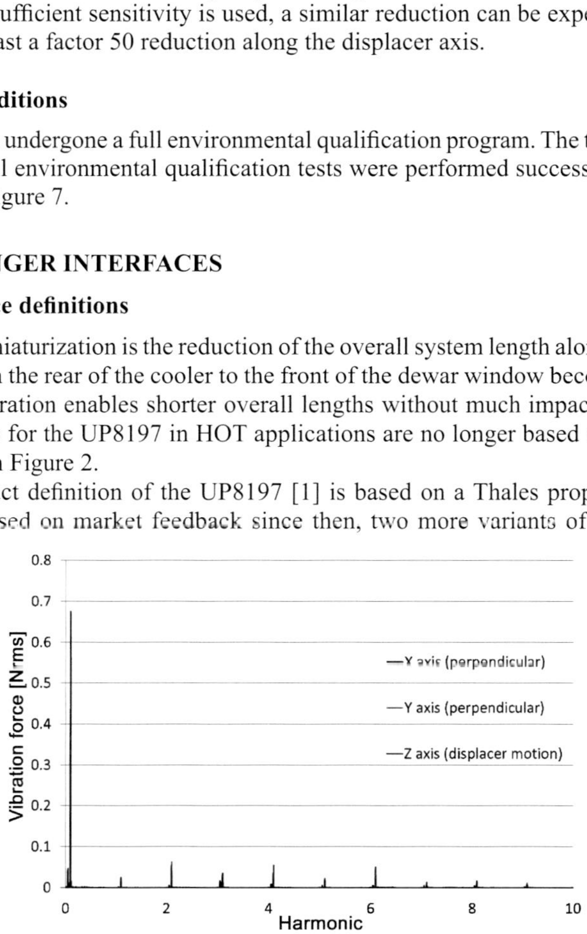

Figure 6. Induced vibration levels of the cold finger in three axes, measured with an accelerometer.

Table 2. Environmental tests

Test type	Test
Thermal	Dry heat, operating and storage
	Cold, operating and storage
	Change of temperature
Mechanical	Repetitive shock (bump)
	Shock, operating
	Random vibration, operating
	Sine vibration
	Acceleration, operating

Figure 7. UP8197 mounted on the shaker table undergoing operating random vibration testing.

finger have been proposed to meet various requirements. One variant that has been under discussion since 2015 is a harmonized cold finger sleeve, designed to interface with small coolers from several manufacturers.

The final configuration of this interface has been discussed with several manufacturers. A copy of this drawing is available on request. The other variant interfaces with the existing CFA6 cold finger. The advantage is that the cooler can be combined with existing detectors already available. The three variants are summarized in Table 3.

Both alternative versions are still under development. For the version with the common interface, all dimensions critical to the coolers performance remain unchanged, so no impact on cooler performance is expected. Only limited delta qualification is expected to be necessary once test equipment with this interface becomes available.

The other foreseen modification is a cold finger with the standard ¼" CFA6 interface (Figure 2). This variant is also currently under development. This variant is expected to have a different performance than the other two variants.

Because the cold finger is longer, a longer regenerator is possible. This will lead to higher efficiency because of reduced regenerator losses. On the other hand, because of the larger overall gas volume in the system the maximum input power will decrease. The same absolute cooling power is available but at a lower input power.

An indication of the expected efficiency difference (regenerator influence only) can be seen in Figure 8. This estimation does not yet take into account the lower losses due to the increased conduction length of the cold finger, which can be another benefit of a longer cold finger.

COOLER DRIVE ELECTRONICS FOR MINIATURE LINEAR COOLERS

The LPCDE1220 cooler drive electronics is designed for low power coolers. In particular, the UP8197 and UP8497 coolers can both be driven by it (see Figure 9). The LPCDE has undergone a full qualification program similar to that performed for the UP8197. A full feature set is available [1]. Typical performance characteristics are the programming capability over the serial port combined with dedicated I/O lines for specific functions such as 'Cooler Ready' and 'Remote On/Off'. Furthermore, the electronics allows the voltage ramp up feature during cool down and dual set points for a Standby function.

Typical performance characteristics of the cooler drive electronics are high efficiency and regulation stability. In Figure 10 the temperature drift measurements are shown. In the left graph, the temperature drift with varying ambient temperature is shown. This is measured with a dual diode test dewar. One diode is used as the input device for the cooler drive electronics. The other is used as an independent temperature measurement. Over the temperature range of -50 °C to 90 °C the error in diode voltage due to temperature drift is less than 0.2 mV, corresponding to 0.1 K.

The right graph of Figure 10 shows the regulation stability over a 15 minute interval. Within this interval, the diode voltage remains constant within a 0.05 mV bandwidth, corresponding roughly to 30 mK.

Table 3. Interface options for the UP8197

Thales Short Sleeveless CFA6/S	**Common Cold Finger**	**Thales Standard ¼" CFA6**
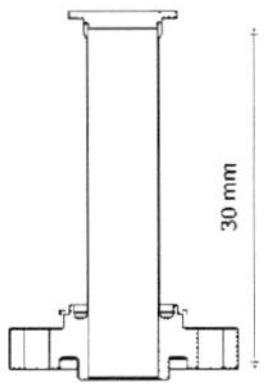	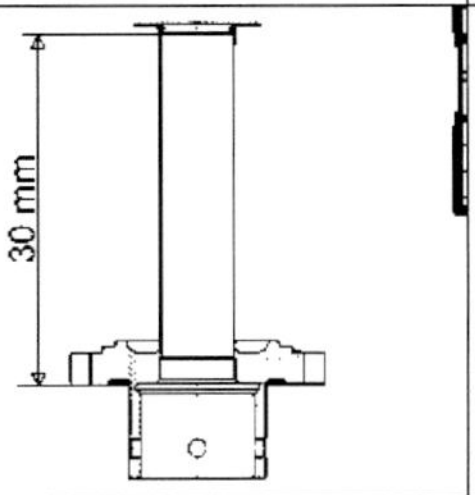	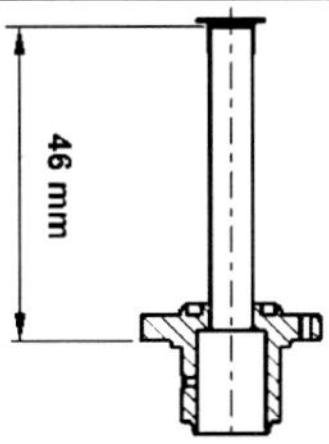
Novel sleeveless design Potential for further miniaturization Lower performance spread expected Cost advantage 30mm inner length	Sleeved design with heritage Single Cold Finger design for multiple cryocooler manufacturers Flexibility for customers; cryocooler choice can be made per application without changing dewar design 30mm inner length	Sleeved design with heritage Compatible with large range of existing coolers. Flexibility for customers as existing dewars can be used. 46mm inner length
Fully qualified	**Delta qualification required**	**Under development**

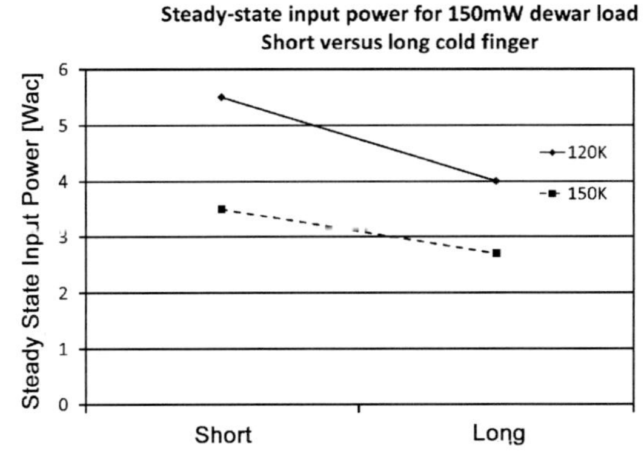

Figure 8. Estimated influence of regenerator length on cooler efficiency. Shown is the necessary input power for a dewar load of 150 mW. The longer regenerator has a higher efficiency and a lower required input power.

Figure 9. Photograph of the LPCDE1220 Cooler drive electronics.

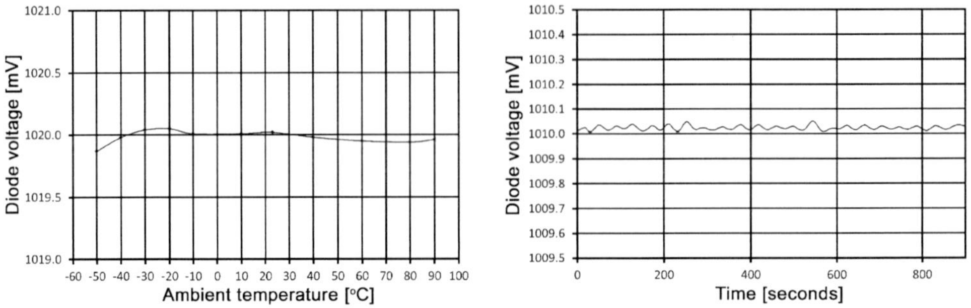

Figure 10. Temperature stability measurements with the LPCDE. The left graph shows the temperature stability (drift) with different ambient temperatures. The right graph shows the regulation stability over a 15 minute period.

The efficiency of the drive electronics is measured when driving the UP8197. The measured efficiency depends on the output power and the DC voltage. For an output power larger than 4 W, the efficiency is better than 90 %.

CONCLUSIONS

While HOT technologies are maturing, and system integrators consider how to benefit from this new technology, more accurate requirements are expressed for the necessary coolers. Thales Cryogenics works with thermal imager manufacturers to provide a full range of products suitable for the entire gamma of current HOT SWaP requirements.

Two new products, the UP8197 and UP8497 linear cryocoolers are now available to complement the Thales product portfolio for these miniature applications. The UP8197 for true HOT application of 120 K and higher, and the UP8497 for the gap between this and conventional 77 K applications. Both coolers and the matching cooler drive electronics LPCDE1220 are presently qualified and available.

New variants of the UP8197 with different cold finger options, such as a common interface for interchangeability with other manufacturer's coolers and existing ¼" interfaces for higher efficiency and interchangeability with a large range of existing dewar detector combinations, are foreseen for the near future.

REFERENCES

1. Arts, R., Martin, J-Y., Willems, D., Seguineau, C., Van Acker, S., Mullié, J.C., Göbel, A., Tops, M., Le Bordays, J., Etchanchu, T., Benschop, A.A.J., "Miniature cryocooler developments for high operating temperatures at Thales Cryogenics," *Proceedings of SPIE 9451* (2015).
2. Seguineau, C., Arts, R., Martin, J-Y., Willems, D., Van Acker, S., Mullié, J.C., de Jonge, G., Tops, M., Le Bordays, J., Etchanchu, T., "SWAP Cryocooler developments for high operating temperatures: a complementary approach based on linear and rotary Stirling coolers," *Proceedings of SPIE OPTRONICS* (2016).
3. Martin, J-Y., Cauquil, J-M., Benschop, A., Freche, S., "Thales Cryogenics rotary cryocoolers for HOT applications," *Proceedings of SPIE 8353* (2012).
4. Arts, R. "Adaptive vibration reduction on dual-opposed piston free displacer Stirling cooler,"*Proceedings of SPIE 9070* (2014).
5. Van den Groep, W., Mullié, J., Willems, D., Van Wordragen, F., Benschop, A., "The development of a new generation of miniature long-life linear coolers," *Cryocoolers 16*, ICC Press, Boulder, CO (2011), pp. 111-119.
6. ThermXL thermal modelling software, ESATAN-TMS, ITP Engines UK Ltd.
7. Willems, D., Tops, M., Bots, F., Arts, R., De Jonge, G., Benschop, A., Etchanchu, T., "Developments in advanced cooler drive electronics," *Cryocoolers 18*, ICC Press, Boulder, CO (2015), pp. 359-366.

8. Arts, R., De Bruin, A., Willems, D., De Jonge, G., Benschop, A., "Adaptive vibration reduction on dual-opposed piston free displacer Stirling cooler", *Proceedings of SPIE 9070* (2014).

9. Veprik, A., Shlomovich, B., Tuito, A., "Multi-slope warm-up calorimetry of Integrated Dewar-Detector Assemblies," *Proceedings of SPIE 9451* (2015).

LPT9310 COTS Cooler for ECOSTRESS

R. Arts[1], J. Mullié[1], D. Johnson[2], I. McKinley[2], J. Rodriguez[2], T. Benschop[1]

[1]Thales Cryogenics B.V., Eindhoven, The Netherlands
[2]Jet Propulsion Laboratory, California Institute of Technology, Pasadena, CA, USA

ABSTRACT

For the past several years, a significant effort has been performed at the Jet Propulsion Laboratory to characterize commercial-off-the-shelf (COTS) pulse tube cryocoolers for use in cost-effective spaceflight applications. This has resulted in the selection of the Thales LPT9310 cryocooler for the ECOSTRESS instrument that will fly on-board the Japanese Experiment Module – Exposed Facility (JEM-EF) of the International Space Station.

The Thales LPT9310 cryocooler nominally provides over 4 W of cooling capacity at 80 K, and has been produced in large quantities with a proven capability for multi-year continuous operation without an instance of cooler failure.

However, this capability has only been proven in terrestrial (commercial) applications. In order to provide sufficient justification for using an off-the-shelf cooler for a flight application, additional tests have been performed on the delivered flight coolers, to attain a sufficiently controlled level of quality while leveraging the heritage of the COTS cooler. The test program philosophy will be explained, and results will be discussed.

Restrictions in both the available electrical power and the heat exchanger fluid inlet temperature to the instruments onboard the JEM-EF eroded the performance margin maintained for the cryocoolers, prompting the need to find more efficient cryocoolers. Thales Cryogenics responded with an upgrade to the LPT9310, replacing the stainless steel pulse tube with titanium alloy, and a re-optimization of the regenerator matrix for 60K while leaving the baseline, qualified design parameters intact. The resulting performance enhancement from the additional cryocooler efficiency at lower temperatures will be presented.

INTRODUCTION

Figure 1 shows a Thales Cryogenics LPT9310 pulse-tube cooler. It is composed of a dual opposed-piston flexure-bearing moving magnet compressor coupled to a pulse-tube cold finger. The design of the cooler has strong ties to past and present space cryocoolers. The pulse-tube cold finger can be regarded as the commercial off-the-shelf sibling for the LPTC pulse tube, developed by CEA under ESA funding [1]. The 93xx compressor has served as the starting point for designed-for-space cooler developments, such as the LSF9330 flexure-bearing Stirling cooler [2] developed under ESA funding.

However, contrary to the LPTC and LSF9330 coolers, the LPT9310 was optimized for cost as well as minimum exported vibration and reliability, where some performance was sacrificed for a lower cost. This has resulted in a baseline LPT9310 design that is based on relatively inexpensive

Cryocoolers 19, edited by S.D. Miller and R.G. Ross, Jr.

Figure 1. LPT9310 cooler

parts and processes suitable for series production while incorporating all the design principles required for an extremely reliable cooler. As a result, it has been produced and deployed in large numbers for cooling Germanium detectors, an application where low vibrations and high reliability for 24/7 operation are required. It has accumulated a large amount of failure-free running hours [3].

PRIOR WORK: TESTING OF THE LPT9310 AT JPL

The LPT9310's proven reliability has resulted in interest from JPL in using this cooler for cost-sensitive space applications. Initial work performed by Paine [4] resulted in the successful use of this cooler in the PHyTIR technology demonstrator. Further work by Johnson et al., [5] was done to characterize the cooler and to perform preliminary robustness testing.

This prior work resulted in the LPT9310 being selected for the ECOSTRESS instrument.

ECOSTRESS INSTRUMENT DESIGN AND PROJECT

The ECOsystem Spaceborne Thermal Radiometer Experiment on Space Station (ECOSTRESS) will measure the temperature of plants. This temperature data will be used to better understand how much water plants need and respond to heat stress. Figure 2 and Figure 3 show photographs of the instrument. ECOSTRESS will address three overarching science questions: 1) how is the terrestrial biosphere responds to changes in water availability? 2) how do changes in diurnal vegetation water stress impact the global carbon cycle? 3) can agricultural vulnerability be reduced through advanced monitoring of agricultural water consumptive use and improved drought estimation? The ECOSTRESS mission will answer these questions by accurately measuring the temperature of plants. Plants regulate their temperature by releasing water for evaporative cooling through tiny pores on their leaves. If they have sufficient water they can maintain their temperature but if there

Figure 2. Radiometer buildup with two LPT9310 coolers shown.

Figure 3. Close up of coolers with compressor and expander high performance heat exchangers.

is insufficient water, their temperatures rise and this temperature rise can be measured accurately with a space-based sensor. ECOSTRESS will use a multispectral thermal infrared radiometer to measure plant surface temperatures and will acquire the most detailed temperature images on the surface ever acquired from space. ECOSTRESS is manifested for launch in 2018 onboard the trunk of the SpaceX Dragon vehicle.

ECOSTRESS will be implemented by placing the existing Prototype HyspIRI Thermal Infrared Radiometer (PHyTIR) with design updates needed for flight qualification on the International Space Station (ISS) and using it to gather the measurements needed to address the science goals and objectives. PHyTIR was developed under the NASA Earth Science Technology Office (ESTO) Instrument Incubator Program (IIP). From the ISS, ECOSTRESS will provide data with 38-m in-track by 69-m cross-track spatial resolution and a predicted temperature sensitivity of ≤0.3 K. The ISS orbit allows excellent coverage of the selected targets including diurnal coverage. The existing hardware was developed to reduce the cost and risk for the thermal infrared radiometer on the future Hyperspectral Infrared Imager (HyspIRI) mission, A double-sided scan mirror, rotating at a constant 23.3 rpm, allows the telescope to view a 51°-wide nadir cross-track swath as well as two internal blackbody calibration targets every 1.29 seconds.

The optical signal is focused by a telescope onto a cryocooler cooled 65K focal plane containing a custom 13.2-µm-cutoff mercury-cadmium-telluride (MCT) infrared detector array. Spectral filters on the focal plane define 5 spectral bands in the 8-12.5 µm range and an additional band at 1.6 µm for geolocation and cloud detection (six bands total). The focal plane is cooled by two LPT9310 Thales cryocoolers while a third LPT9310 cools a cold shield to ~120K to reduce parasitic heat loads on the focal plane. The Thales COTS cooler drive electronics XPCDE4865 are used to power and control the coolers. An electronic switch with two XPCDE4865 units will drive each of the coolers independently providing sufficient redundancy for this class of mission. The LPT9310 cooler along with the XPCDE4865 were qualified to TRL 6 at JPL.

Heat rejection for the ECOSTRESS cryocoolers and electronics is provided by the cooling fluid loop on the ISS JEM-EF. The heat rejection system (HRS) was designed to remove ~950W of instrument power dissipation with 540W alone from the three cryocoolers. In October 2015 (near instrument PDR), JAXA along with the ISS recognized that the JEM-EF power and cooling resources were not sufficient to meet the new demands from current and future payloads. ECOSTRESS along with the other payloads in development were asked to reduced electrical power requirements and re-design their HRS with reduced coolant flow rates to a minimum of 155 kg/hr and maximum fluid inlet temperature of 24°C. The consequence of this was the erosion of all design margins for the cryogenic system and the estimated power saving was small with the 65K focal plane requirement. The focal plane performance was measured as a function of temperature and results indicated that 67K would yield adequate science but a lower temperature would be preferred. Further design trade-offs led to the final implementation of high performance fluid heat exchangers from MaxQ Technology (Tempe AZ, USA) for the cryocoolers to allow operation at lower skin temperatures and led to the choice for high efficiency cryocoolers. Thermal analysis of the re-designed LPT9310 cryocooler and the MaxQ heat rejection systems predicted sufficient cryogenic margin to proceed forward with a 25% reduction in cooler input power while maintaining the focal plane at 65K.

With uncertainty on the available instrument power and on the inlet temperature and flow rate of the heat transfer coolant that would be provided by JAXA, it was unclear if the three coolers; one cooler to cool the cold shield, and the two detector coolers could provide <65K cooling to the detectors. The limited input power to the instrument meant the coolers may have to be set no higher than 140 watts each. If the cooling fluid temperature (nominally set between 16℃ and 24℃) were to be delivered on the warm extreme, the coolers would operate with a skin temperature of about 40℃, which would also reduce the cooler refrigeration capacity. A third concern was that the mechanical layout of the coolers within the instrument placed the compressors on the instrument floor necessitating long transfer lines to reach their respective cooler cold heads. Cumulatively, these effects could result in insufficient cooling power from the coolers to maintain the detector less than 65K. As shown in Figure 4, thermal modeling suggested that the performance degradation due to the extended transfer line length (effects of flow losses and increased void volume) could reduce the

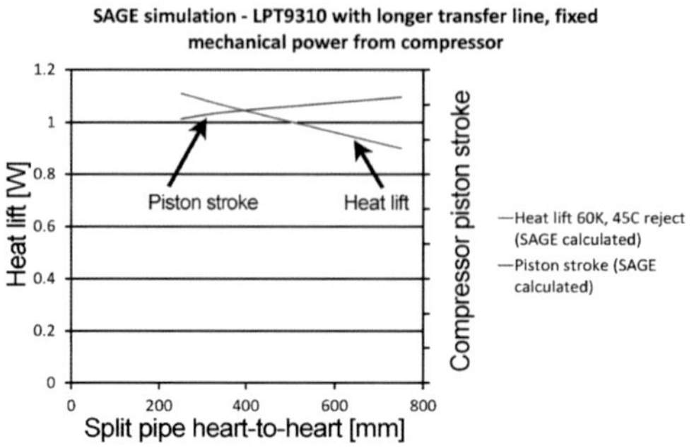

Figure 4. SAGE simulation of LPT9310 with longer transfer line.

refrigeration capacity of the coolers by an additional 200 mW. These performance limiting factors and the uncertainties convinced JPL that they needed Thales to implement a number of material upgrades to enhance LPT9310 cooler performances, as discussed in the following sections.

TRANSFER LINE PERFORMANCE

The requirements on the cooler for ECOSTRESS are a departure from what is typically required of the LPT9310. The required transfer line is significantly longer than the maximum specified value. Furthermore, the required cryogenic cold tip temperature of 60 K is lower than the operating point where the LPT9310 is typically used. (80 K). During initial discussions, the influence of the longer transfer line on cryogenic cold tip performance as well as the compressor stroke were modeled in the commercially-available SAGE® software. The result can be seen in Figure 4.

In order to verify that sufficient margin would be present with respect to compressor piston stroke, a test cooler was built with a long temporary transfer line. The standard LPT9310 specification limits the allowable center-to-center distance between the compressor and the pulse-tube to 500 mm, while distances exceeding 700 mm were expected. A test cooler was therefore built with a 750 mm center-to-center distance (straight transfer line). Sufficient margin for compressor piston stroke was found without the need for an increase in fill pressure.

ADDITIONAL SCREENING ON COMPRESSOR

The LPT9310 compressor can be expected to have an intrinsic high reliability, due to the flexure bearing and moving magnet design. A well-aligned flexure-bearing compressor can be expected to be near-free of wear, which eliminates compressor wear as a factor in cryocooler reliability. The standard parts inspection and manufacturing processes for such a compressor incorporate several steps to ensure that the desired wear-free piston motion is achieved. For the LPT9310 units delivered for ECOSTRESS, a number of additional tests were performed to verify this.

Two key ingredients are required for wear-free operation of a compressor. First, the flexure bearing needs to provide the as-designed high radial stiffness over its operating life. Second, the compressor piston needs to be aligned inside the cylinder such that the motion of the piston is in parallel to the cylinder, maintaining a constant radial gap over its entire stroke.

Even though the inspection criteria for flexure springs are well-understood and no faulty flexures have inadvertently passed the current Thales inspection criteria, flexures from the same production batch as those used in the compressors were subjected to 10^7 cycles of 110% of the physical stroke limit (end-stop stroke), or approximately 170% of the nominal steady-state stroke, as an additional test. This test is not normally performed for off-the-shelf coolers, but is routinely done for space compressors where additional screening is required, such as the Thales LPTC compressor. The flexure fatigue test cabinet can be seen in Figure 5.

In order to ensure that the alignment remained unchanged throughout testing, a ring-down test was performed after several steps in assembly and test. In this test, a DC current is briefly run through a motor coil in order to move a piston to the extreme position. The DC current is then removed and the motor back EMF is measured to monitor the damped oscillation of the piston. The ring-down test is used as a diagnostic test to monitor compressor performance after key steps in the cooler assembly and test sequence, such as:

- After additional run-in hours
- After proof pressure testing
- After acceptance-level random vibration testing

An example ring-down result plot can be seen in Figure 6.

ADDITIONAL COOLER-LEVEL TESTING

A number of additional assembly and test steps were performed on the coolers to ensure workmanship. An additional 168 hours of full-power operation were accumulated on each cooler. An additional bake-out at 100℃ was performed with the cooler internals vacuum pumped, to remove any residual contamination. The coolers were subjected to a proof pressure test to verify all welds and brazes. Finally, the EM coolers were subjected to acceptance-level random vibration, based on the "General Environmental Verification Standard (GEVS) For GSFC Flight Programs and Projects". The test configuration is seen in Figure 7. For the upgraded coolers described later, an updated spectrum was provided, representative of the expected launch loads on the cooler, as seen in Figure 8.

FURTHER ANALYSIS AND JUSTIFICATION

As has been pointed out before, the LPT9310 is a commercial product not developed to any applicable space standards. Analyzing the cooler with respect to specific standards will therefore result in a number of non-compliances. One specific area of attention is the properties of the LPT9310 as an item of pressurized hardware. As the cryocooler was not designed towards the proof pressure factor of safety typically required for flight (1.5X MEOP), a reduced pressure was used for proof testing. In order to provide evidence of the suitability of the cooler design, burst tests were performed on a separate pulse-tube as well as a compressor. Both compressor and pulse tube burst at the expected locations (see Figure 9) and met applicable burst pressure requirements, with burst pressures exceeding 100 bar in both cases, which provides a comfortable margin above the required burst pressure of 2X MEOP (Maximum Expected Operating Pressure) required.

EM TESTING OF STANDARD LPT9310 COOLERS

Upon delivery of the three ECOSTRESS EM coolers to JPL, cooler S/N 1280, with the longest transfer line (731 mm), was set up for thermal vacuum testing. Figure 10 shows the test setup used that was designed to simulate the operating conditions on the ISS. Prototype heat exchangers from MaxQ Technology, similar to the flight units, were bonded onto the compressor and expander with Nusil CV-2946 using 0.028 mm bond wire to guarantee the bond thickness. Braycote grease 601EF was lightly applied to the bonding surfaces prior to the Nusil in order to ensure that the bonds were not permanent.

The prototype heat exchangers were connected to a temperature and mass flow rate controlled fluid loop containing Fluorinert FC72 heat transfer fluid (the same as used on the ISS). The fluid

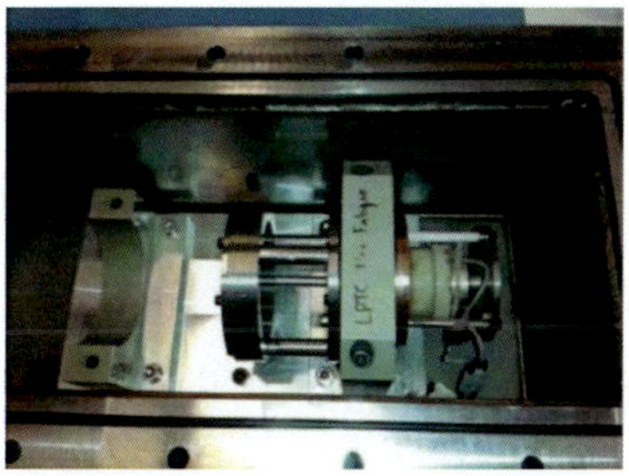

Figure 5. Flexure overstroke test.

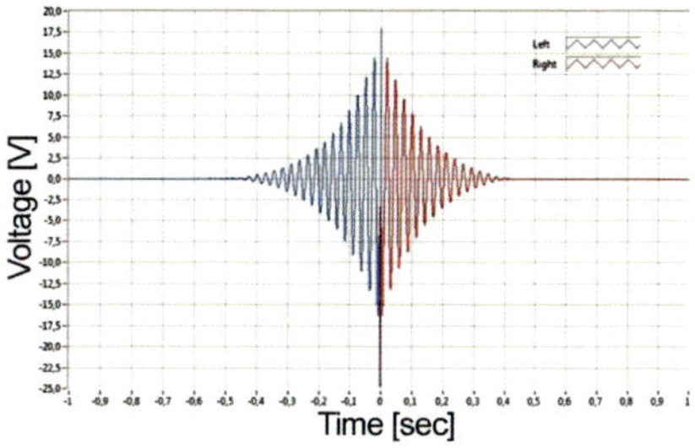

Figure 6. Ring-down test

Figure 7. Random vibration test.

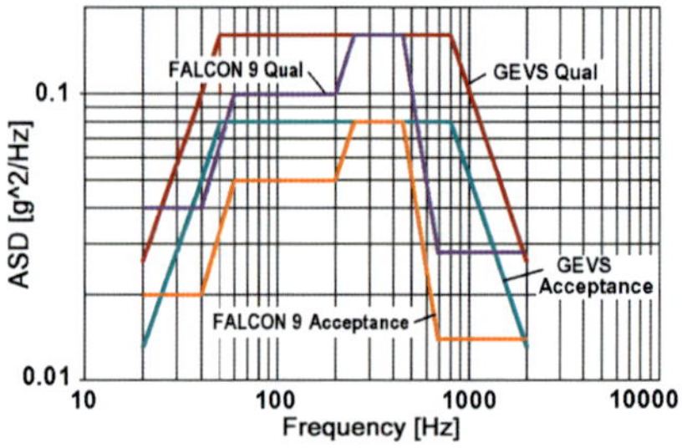

Figure 8. Random vibration spectra.

first passed through the expander heat exchangers and then the compressor heat exchangers. The fluid temperature was measured before the expander heat exchangers, between the expander and compressor, and after the compressor heat exchangers. The measurements were made by thermocouples fed directly into the fluid by means of T-connections. The fluid inlet temperature was defined as that measured before the expander heat exchangers.

The cryocooler was operated with Thales XPCDE 4865 drive electronics at 47 Hz. The input power to each compressor motor was measured using Yokogawa WT310 power meters. In addition, the heat rejection from the compressor and the expander was calculated from the product of the FC72 mass flow rate, the temperature dependent specific heat, and the temperature drop between the inlet and outlet of the heat exchangers. The sum of these gave an independent heat rejection measurement that closely agreed with the sum of the measured cooler input power and cold block heater power.

Cooler S/N 1280 was performance tested in thermal vacuum under worst case scenario ISS conditions, with the anticipated low FC72 fluid flow rate of 155 kg/hr, and at higher flow rate up to 182 kg/hr, and for several different possible FC72 fluid inlet temperature scenarios. Figure 11 shows the heat lift vs. the compressor input power for a fluid loop mass flow rate of 155 kg/hr, a 24°C fluid inlet temperature, and a 60 K cold tip. The heat lift increased nearly linearly with increasing compressor input power and reached a maximum of 1.68 W for 170 W of input power

Figure 12 shows the heat lift vs. cold tip temperature for two different compressor input powers with a 155 kg/hr fluid mass flow rate and a 24°C fluid inlet temperature. For a cold tip temperature of 60 K, the heat lift with 170 W of input power was nearly 0.4 W larger than with 140 W input power.

Figure 13 shows the sensitivity of the cooler heat lift as a function of the FC72 fluid inlet temperature for two different FC72 mass flow rates while the cooler operated with 140 W input power and with a 60 K cold tip temperature. For a given fluid inlet temperature, the heat lift at 182 kg/hr was slightly larger than at 155 kg/hr.

The long transfer line length (731 mm) of cooler S/N1280 had a much smaller effect on the performance than anticipated. The S/N 1280 was capable of 1.12 W of cooling at 60 K with a 40°C

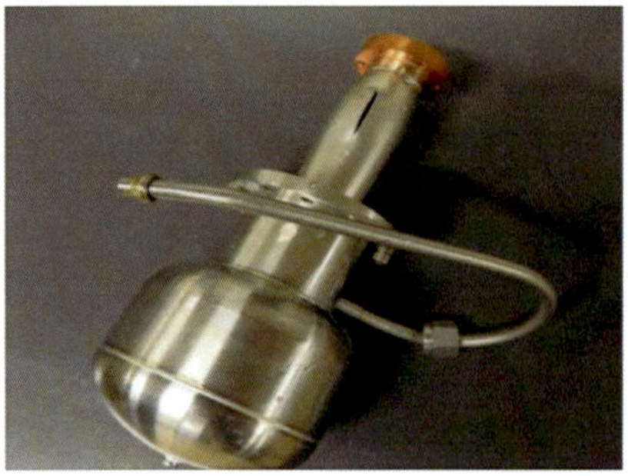

Figure 9. Burst test, standard pulse tube.

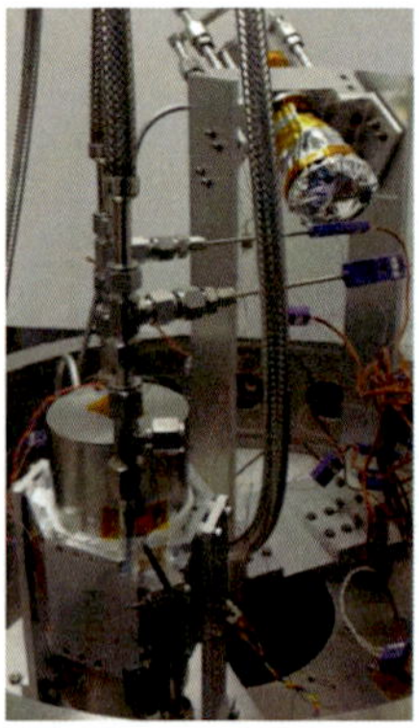

Figure 10. EM cooler TVAC test setup at JPL.

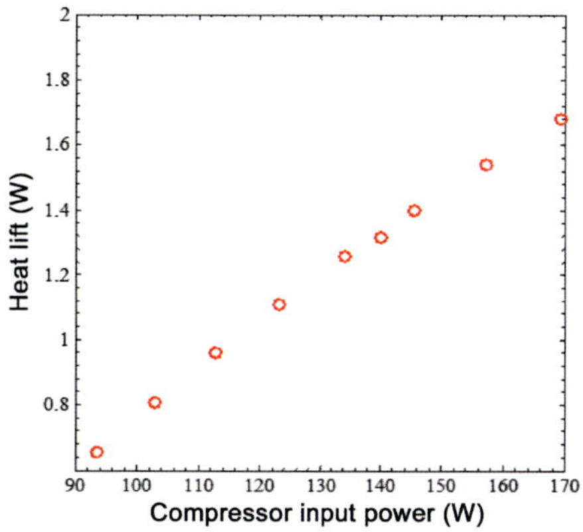

Figure 11. Heat lift vs. compressor input power for 155 kg/hr mass flow rate, 60 K cold tip, and 24°C fluid inlet temperature

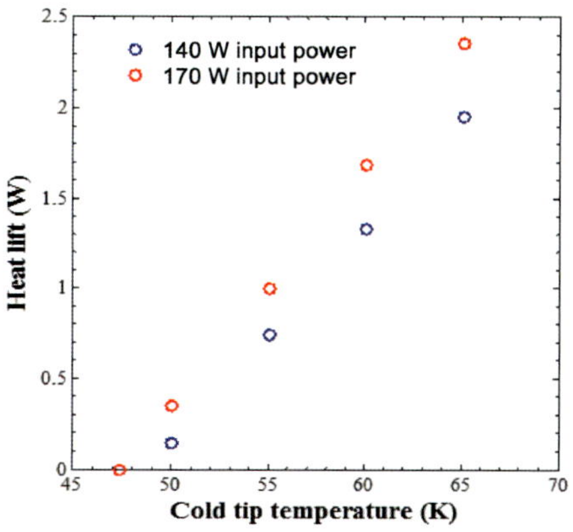

Figure 12. Heat lift vs. cold tip temperature for 155 kg/hr fluid flow rate and 24°C fluid inlet temperature at various compressor input powers.

compressor temperature measured on either side of the heat exchanger. In comparison, a standard LPT9310/17, with a transfer line length of 180 mm and operating under the same conditions, produced 1.19 W of cooling, suggesting that the S/N1280 cooler had only a performance reduction of 70 mW.

With the 140 watt input power limitation on the ECOSTRESS coolers the two coolers for the 60-K focal plane do provide sufficient cooling as designed, however there is an inadequate performance margin for the instrument. As such it was deemed imperative to find a performance enhancement design for ECOSTRESS.

LPT9310 HIGH PERFORMANCE

As the specified performance of the off-the-shelf LPT9310 shows insufficient margin versus the requirement, options were examined to build LPT9310 units with increased performance without invalidating the (off-the-shelf) heritage of the design. Two constraints were present:

- Schedule: only low-risk "first time right" changes, as the ECOSTRESS project schedule does not allow for an iterative design
- Interface: The upgraded cold heads needed to be usable as form-fit drop-in replacements for the existing "EM" coolers

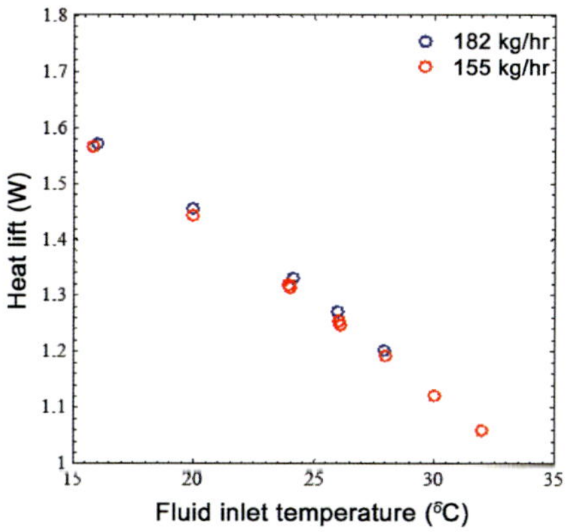

Figure 13. Heat lift vs. fluid inlet temperature for various mass flow rates with 140 W compressor input power and 60 K cold tip.

Table 1. Trade-off for improvement options.

Design option	Interface impact	Process and schedule risk	Selected
Optimized regenerator design	None	Low – Clear heritage present	Yes
Low conductance material between hot and cold	None	Medium – connection processes changed, but heritage present	Yes – outer tube (Titanium alloy)
High-conductance material on warm side	None / Low	High – significant impact on design, limited heritage, metal seals needed	No
Optimized cold heat exchanger	Medium	Medium – limited heritage, risk of design iteration	No
Optimized warm heat exchanger	Medium (potential)	High	No
Optimized inertance	High	High	No

Several material upgrades were tested in the past, but were discarded in the series-produced design because the cost increase outweighed the benefit for most (industrial, non-space) applications. However, because the ECOSTRESS focal plane array requires cooling at 60 K, a number of previously-discarded upgrade options were expected to provide a significant performance benefit. A trade-off was performed to select which upgrades to implement. Options for increasing the performances of a COTS pulse-tube cooler were described previously [6], these design options are repeated in Table 1 and the trade-off result is presented for each option.

It was decided to implement an optimized regenerator composition as well as a Titanium alloy outer tube. The former increases regenerator efficiency at 60 K, while the latter significantly reduces the parasitic loss.

The effect of performing these upgrades can be seen in a comparative measurement, Figure 14. Depending on the input power and reject temperature, the observed improvement was anywhere between 700 and 1000 mW at 60 K. Of this improvement, somewhere between 400 and 500 mW is explained by the reduced conduction loss through the outer tube (change to Titanium alloy), while the remainder is attributed to the increased regenerator efficiency.

A delta qualification program was performed on the LPT9310-HP. Apart from verifying the performance of the upgraded cold head, special attention was paid to all manufacturing (bonding) processes that were changed. Sample-level tests were performed, including tensile strength tests, and a pulse-tube was subjected to a fatigue cycle, including 8 proof pressure cycles, 50 pressure cycles to MEOP, and 80 cooldown-warmup cycles. The number of cycles was based on the fatigue cycle guidelines prescribed by ESA [7]. The pulse tube was subjected to qualification level random vibration testing (envelope spectrum encompassing the protoflight qualification levels for the Falcon 9) after which it was verified to be functional. Finally, a burst test was performed on the LPT9310 design, which resulted in a burst value of 140 bar.

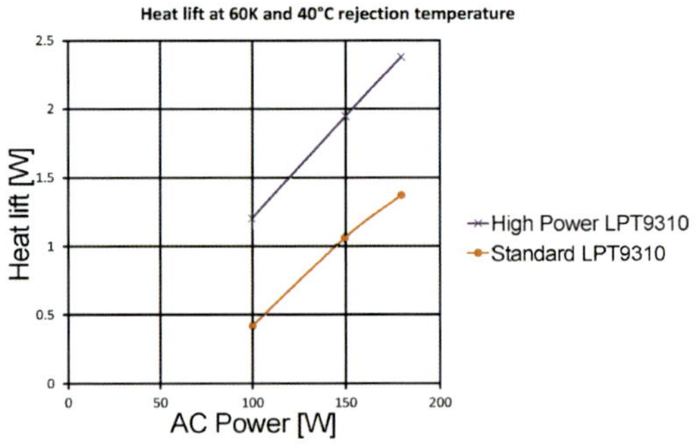

Figure 14. Performance LPT9310-HP vs standard.

CONCLUSION

Testing was performed on standard Thales LPT9310 coolers, which were delivered as engineering models to be used in JPL's ECOSTRESS instrument.

An upgraded high-performance version of the LPT9310 was subsequently built, with upgrades selected based on minimizing risk to schedule as well as avoiding changes to interface dimensions. A qualification test campaign was completed successfully, with the upgraded design meeting qualification-level robustness requirements after being subjected to fatigue cycling as well as providing the required efficiency increase.

Flight models are currently in production and will be delivered to JPL July 2016, with launch planned for 2018.

Future work is considered to investigate an upgraded LPT9310 with a true designed-for-space cryocooler compressor, such as the Thales LPTC compressor.

ACKNOWLEDGMENT

Some of the work described in this paper was carried out at the Jet Propulsion Laboratory, California Institute of Technology, under contract with the National Aeronautics and Space Administration. Government sponsorship is acknowledged.

REFERENCES

1. Trollier, T., Tanchon, Buquet, J., Ravex, A., Charles, I., Coynel, A., Duband, L., Ercolani, E., Guillemet, L., Mullié, J., Dam, J., Benschop, T., Linder, M., "Design of a Large Heat Lift 40 K to 80 K Pulse Tube Cryocooler for Space Applications," *Cryocoolers 14*, ICC Press, Boulder, CO (2007), pp.75-82.

2. Trollier, T., Ravex, A., Crespi, P., Mullié, J., Bruins, P., Benschop, T., Linder, M., "High Capacity Flexure Bearing Stirling Cryocooler On-Board the ISS," *Cryocoolers 12*, Springer US (2003), pp 165-172.

3. Willems, D., Arts, R., Douwen, J., "State-of-the art cryocooler solutions for HPGe detectors," Canberra website, *http://www.canberra.com/literature/detectors/tech_papers/CP5-plus_C48083.pdf* (2015).

4. Paine, C.G., "Thermal Performance Testing of Two Thales 9310 Pulse-tube Cryocoolers for PHyTIR," *Adv. in Cryogenic Engineering*, Vol. 59, Amer. Institute of Physics, Melville, NY (2014), p. 1800.

5. Johnson, D.L., McKinley, I.M., Rodriguez, J.I., Tseng, H., Carroll, B.A., "Characterization Testing of the Thales LPT9310 Pulse Tube Cooler," *Cryocoolers 18*, ICC Press, Boulder, CO (2015), pp. 125-133.

6. Arts, R., Willems, D., Mullié, J., Benschop, T., "Advantages of high-frequency Pulse-tube technology and its applications in infrared sensing," *SPIE Defense+ Security,* International Society for Optics and Photonics (2016), pp. 982108-982108.

7. ESA-ESTEC, "Structural design and verification of pressurized hardware", ECSS-E-ST-32-02C Rev. 1

Small Scale Cooler: Extending Space Developed Technology into Adjacent Markets

P. Iredale, C. F. Cheuk, N. Hardy, S. Barclay, M. Crook[1], G. Gilley[1] and S. Brown[1]

Honeywell Hymatic, Redditch, UK
[1]STFC, Rutherford Appleton Laboratory, Didcot, UK

ABSTRACT

Honeywell Hymatic in collaboration with Science & Technology Facilities Council (STFC), Rutherford Appleton Laboratory (RAL) have further developed the prototype design of the Small Scale Cooler (SSC) (Figure 1), presented at the Space Cryogenics workshop at ESTEC 2013 [1], to a production ready system capable of serving the original market i.e. space, but in addition serving applications such as EO/IR and radioisotope detection. The aim is to provide a SSC with low mass, cooling power of 0.75W at 77K and with high reliability built upon heritage design language.

This work builds upon the achievements of the initial prototype build by STFC with particular attention being paid to optimizing the overall cooler system efficiency, adapting the design for manufacture at production volume and accommodating differing system integration requirements, whilst maintaining a level of commonality suitable for multiple applications.

Productization activities include full system drawing release, definition of manufacturing and process methodology to assure product quality, development of tooling and assembly equipment, and procedures to build and test coolers.

Figure 1. Photograph of the Small Scale Cooler next to a UK one penny coin for scale.

Cryocoolers 19, edited by S.D. Miller and R.G. Ross, Jr.

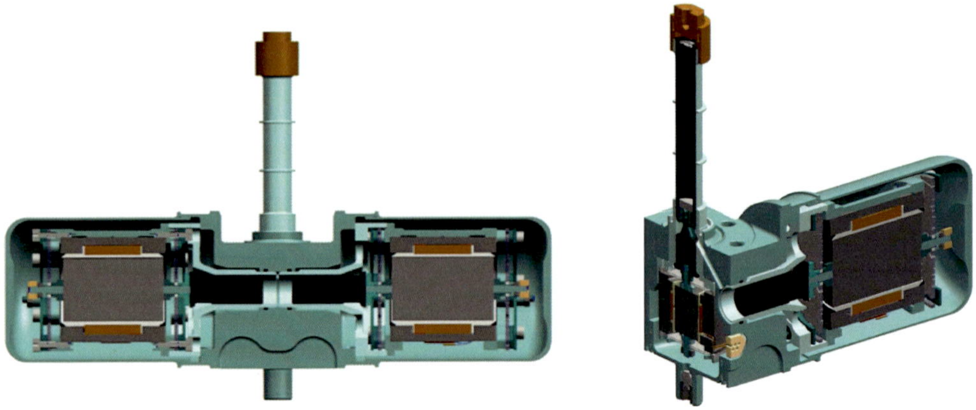

Figure 2. Cross sectional arrangements of the SSC showing the back-to-back compressor design with integral cold-head arrangement. Details of the motor arrangements have been omitted.

INTRODUCTION

Honeywell Hymatic has over 60 years experience in manufacturing JT coolers, Cryocoolers and Space Compressors (as used by Northrop Grumman e.g. the HEC Compressor). In 2013 Honeywell Hymatic signed a license with RAL to industrialize a Small Scale Cooler (SSC) for space and adjacent markets. Hymatic delivered a prototype SSC to an industrial customer for integration to a sensor and performance evaluation. Positive feedback and suggestions on improvements were received as a consequence. Continuous operation in the customers device, >12 months and counting, has resulted in no change in performance. This paper highlights the design characteristics of the SSC and the process of putting the cooler into production. The cross section of the SSC is shown in Figure 2.

DESIGN FEATURES OF THE SSC

The primary design features are as follows:

- Dual opposed compressor design for balancing.
- Displacer mechanism is driven by a small motor to allow for phase adjustment.
- Design based on RAL Space heritage and long life technology such as true clearance frictionless seals and flexure bearings.
- All Titanium welded body design for Helium hermetic gas retention and low mass.
- New moving magnet motor for both the compressor and displacer drive. Axially magnetized magnetic pair configuration obviating the need for inner iron cores.

The design parameters and prototype measured performance are as follows:

- Mass: 620g.
- Size: 152mm X 55mm X 102mm.
- Power Input: 22W.
- Lift: 500mW @ 77K (21°C ambient).
- Closed Loop Temperature Stability: ±5mK 10mins, ±30mK 1hr.

THE NEW MOTOR

The new motor design has advantages and disadvantages over conventional designs that must be considered.

Advantages:

- Lower part count and thus reduced bill of material costs. Lower part costs as a consequence of not having a moving coil former which introduces complexity. Additionally there is no radial magnetized magnet assembly that introduces component and process costs.
- Low magnetic interference with minimal external field.
- Stationary coil with no moving electrical connections.
- Enhanced design flexibility; to adjust frequency and easily scalable.
- Compact.

Disadvantages:

- Eddy current loss needs to be carefully controlled in order to minimize motor losses. This may be done by careful material selection.
- Radial magnetic force requires the careful selection and design of suspension springs with high radial stiffness.

VARIATION OF RADIAL FORCE AND MOTOR CONSTANT WITH RADIAL GAP

The moving magnets are located in an air gap defined by the magnet assembly on the inside, and the yoke on the outside. The magnetic forces are trying to make the moving magnet eccentric radially in the air gap, and these forces are opposed by the mechanical spring forces which are trying to center it. The radial magnetic force decreases with increasing radial gap according to an inverse square law which can be seen in Figure 3. It should be noted that the maximum radial force occurs at the mid-stroke.

From Figure 3 which shows the radial magnetic force at mid stroke when the magnet assembly is displaced eccentrically by a typical assembly clearance of 20 micron, it can be seen that a larger radial gap between the motor and yoke is beneficial as it reduces the magnetic side load. Figure 4 shows that there is a 13% decrease in motor constant when the radial gap is increased from 0.2mm to 1.2mm with a corresponding reduction in radial force by a factor of 10. It is important to consider this during the design of the motor in order to ensure that a true clearance can be maintained between the piston and cylinder.

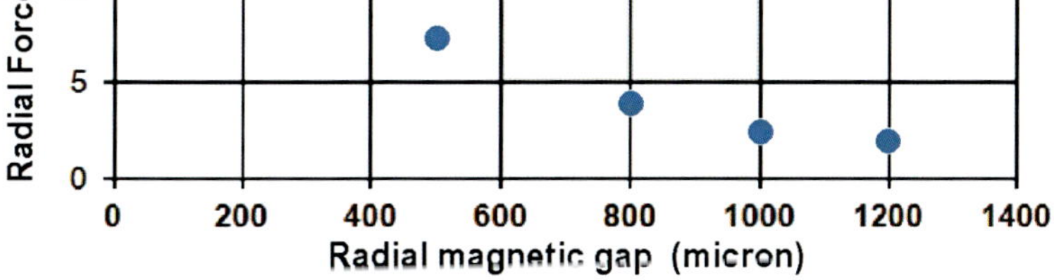

Figure 3. Plot of radial magnetic force vs radial magnetic gap. Data courtesy of RAL.

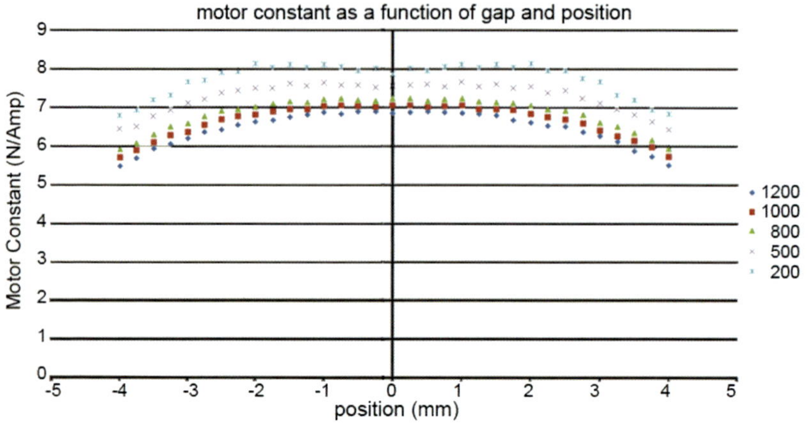

Figure 4. Plot of motor constant as a function of gap and position. Data courtesy of RAL.

Table 1 shows the radial stiffness due to the magnetic forces and the mechanical springs as a function of axial position. The radial stiffness is taken at a radial displacement of 0.22mm which is the radial bump stop position of the magnet assembly. Figure 5 shows the margin between magnetic and mechanical spring rates for the minimum number of springs, which is 2.

Table 1. Radial Spring Rates: Typical Motor Configuration

Axial Displacement (mm)	Radial Spring Rate at 0.22 mm radial offset (N/mm)			Factor of Safety
	Magnetic	Mechanical (2 springs)	Mechanical (6 springs)	
0	127.9	273.70		
3.3	93.1	138.75	416.25	4.47
3.96		111.76		
5		69.23		

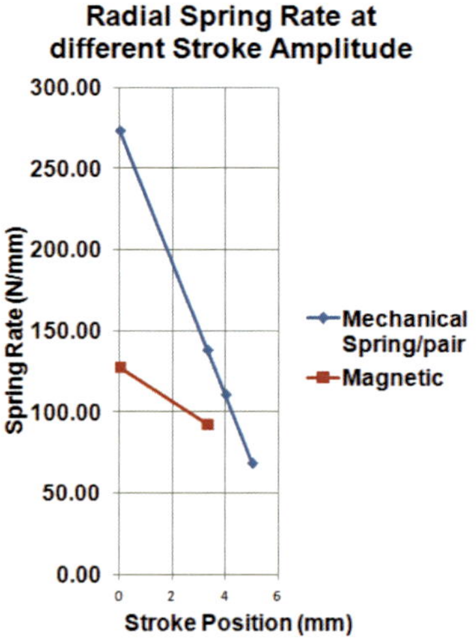

Figure 5. Plot displays how the mechanical radial spring rate has been designed such that it exceeds the magnetic radial spring rate. This is for 2 mechanical springs.

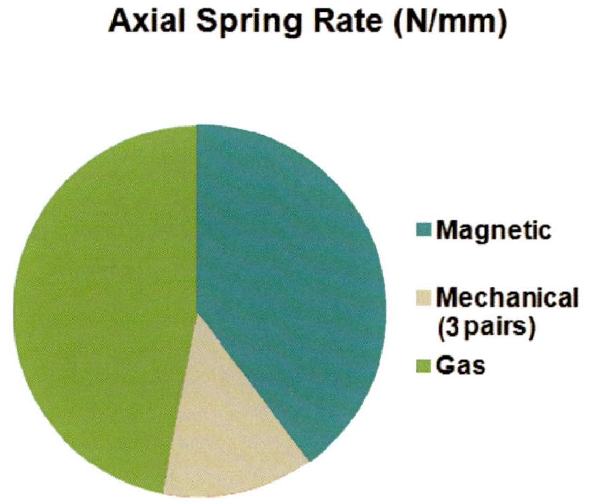

Figure 6. Proportional representation of motor axial spring rates.

AXIAL MAGNETIC SPRING RATE

The axial magnetic spring is a significant contributor to the total axial spring stiffness, the proportions of which can be demonstrated in Table 2 and Figure 6. With minor modifications to the motor configuration the desired spring rate can be varied whilst minimizing the change in motor force. This spring rate can be tailored to be positive, negative or zero as seen in Figure 7. With the contribution of the axial magnetic spring, this design lends itself to high operating frequencies due to the high additional spring rates.

Table 2. Axial Spring Rates: Typical Motor Configuration

Magnetic	N/mm	6.56
Mechanical (6 springs)	N/mm	2.16
Magnetic + mechanical	N/mm	8.72
Gas Spring	N/mm	7.74
Total	N/mm	16.45
Moving Mass	gram	49.23
Natural Frequency	Hz	92.00

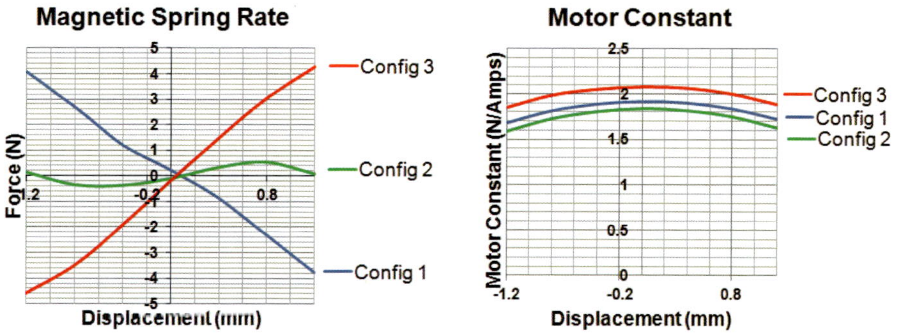

Figure 7. Shows magnetic spring rates for varying motor configurations and their corresponding motor constants. Data courtesy of RAL.

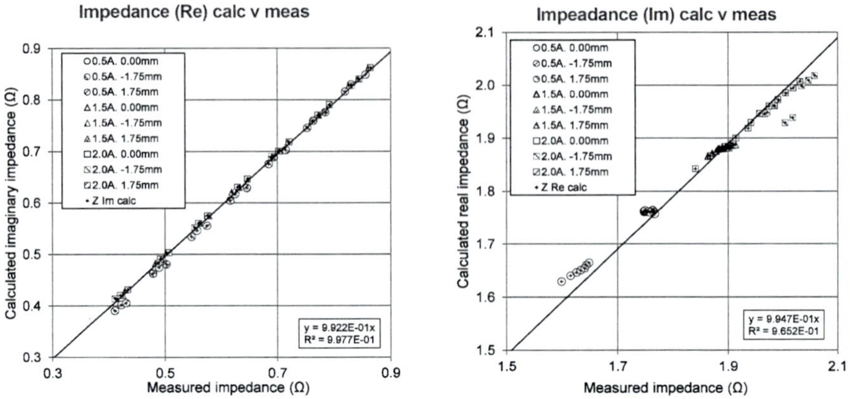

Figure 8. A comparison between measured total impedance and that derived from static analysis which displays a good agreement over a wide range of operating conditions. (f=60Hz-120Hz, I=0.5A-2.0A, x=-1.75mm-+1.75mm). Data courtesy of RAL.

SMALL SCALE COOLER MOTOR LOSS ANALYSIS

RAL developed a series of electrical models to understand the motor losses. The stator core was represented by a parallel inductance and resistance, and the total impedance included the winding resistance. Core losses arose from the 'core resistance' which was a function of geometry and operating conditions. Dynamic losses arose from the magnetic (back emf) damping. Static measurements (motor stalled) at various positions as a function of current and frequency gave the functional form of the core resistance, and also core inductance. Fits to simple functions could be made. Figure 8 shows the comparisons between measured impedance and the static analysis providing a good correlation over varying operating conditions. Dynamic measurements as a function of frequency and stroke included the back emf damping. Back emf damping was verified by free oscillation decay. Static, dynamic and Joule loss contributions could be separated, the results of which can be seen in Figure 9.

As a result the position waveform could be derived from the knowledge of the total impedance under particular operating conditions. This was compared to actual measurements using a laser displacement sensor as seen in Figure 10.

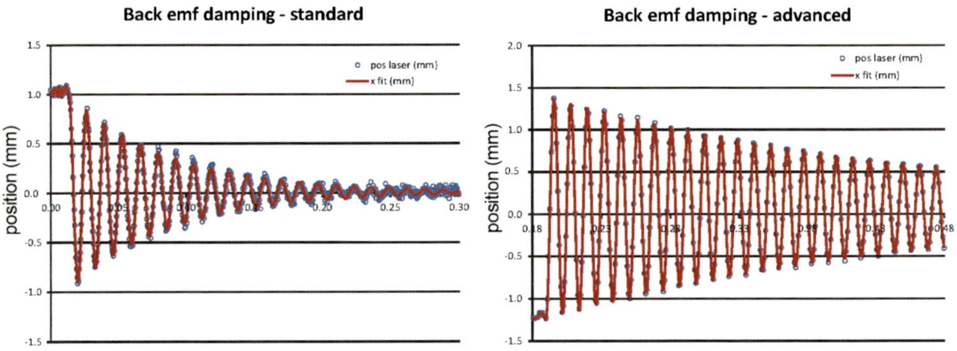

Figure 9. Dynamic measurements for Eddy loss reduction. Standard geometry – damping 1.6Ns/m. Advanced geometry – damping 0.36Ns/m. Data courtesy of RAL.

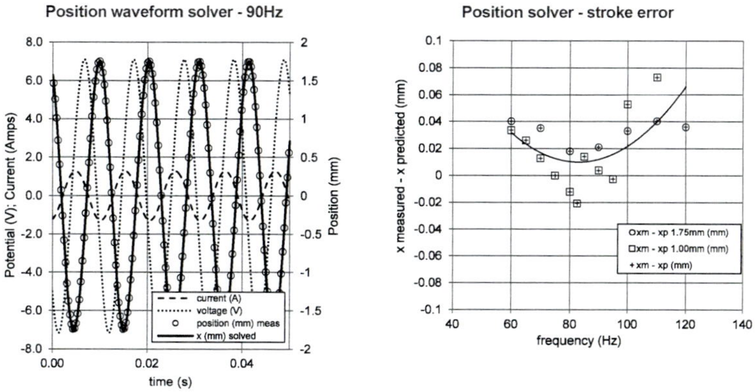

Figure 10. Comparison between measured position using a laser displacement sensor and derived position. Data courtesy of RAL.

PRODUCTIONIZATION PERFORMANCE ENHANCEMENT & PRODUCT RELIABILITY

As a consequence of the RAL development activities for the SSC and customer feedback from performance evaluations of the prototype, the following design modifications were implemented for the production model in order to match the cooler to the requirements of our industrial customer:

- Improved thermal management – modification of the centre plate design and heat sinking.
- Increasing the compressor stroke amplitude to 3.3mm for faster cool down and higher performance capacity.
- Improvements for the cold head phase control to further facilitate faster cool down. This focused on displacer modifications.
- Reduction of eddy current loss by interruption of the eddy current flow path.

The primary aim of these modifications is to drive the performance up to approximately 750mW at 77K (21°C ambient) from the current performance of 500mW.

Design for reliability has taken lessons learnt from Hymatic heritage products used for space applications (such as the HEC compressors used by Northrop Grumman). Spring design has been taken through FEA stress analysis at over stroke positions as shown in Figure 11, in addition to practical testing whereby the following has been undertaken:

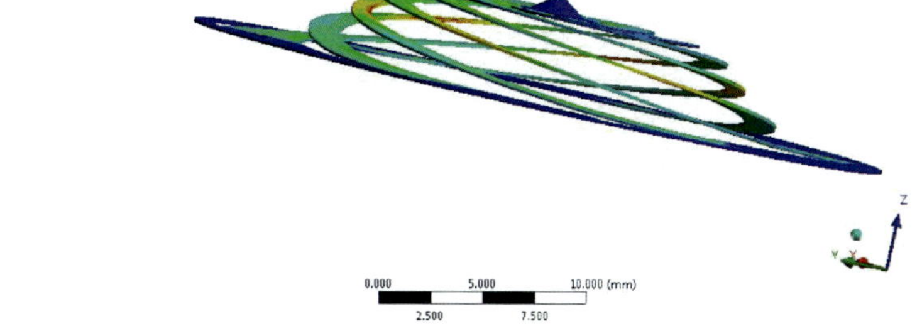

Figure 11. FEA performed on the compressor spring using ANSYS Workbench V16.2.

Compressor Spring Testing:

- 25% over stroke from mechanical stop position.
- 5mm amplitude during test.
- Fatigue testing completed to 800 million cycles.

Displacer Spring Testing:

- 25% over stroke as above.
- 1.8mm amplitude during test.
- Fatigue testing completed to 800 million cycles.

LINEARITY OF PISTON MOVEMENT

Linearity of movement of the piston throughout its stroke has been measured to be less than 8μm; however the following has been noted during assembly:

Sources of error and control:

- Piston movement is controlled by a pair of spring stacks at either end of the motor. The spring mounting faces must be parallel and concentric.
- Unbalanced radial magnetic forces cause piston radial misalignment. The magnetic and spiral spring stiffness must be matched to minimize the radial deflection. The magnet assembly must be centralized in the yoke at the zero force position.

Implementation:

- The manufacturing method for the spring mounting faces controls geometric tolerances and the sequence ensures that this is done at minimal cost.
- Piston linearity of movement is measured in 4 planes to ensure process consistency.
- The magnet is centralized in the yoke using a pair of XY force transducers monitoring forces less than 0.2N.

FLAWLESS LAUNCH, PRODUCT DEVELOPMENT, AND QUALIFICATION

Honeywell Hymatic is currently establishing a production line for the manufacture of approximately 30 SSCs per month. The production line is being established using Honeywell 'Flawless Launch' philosophies:

- Product designed for manufacturing and assembly (producibility). Whereby producibility is seen on an equal footing to technical requirements.
- Zero customer escapes, i.e. no failures of the product at the customer. Zero defect mindset quality culture.
- Early manufacturing and supplier engagement in the product development process.
- Lean manufacturing and de-skilling of operations, mistake proof (Poka-Yoke) tooling, design for Six Sigma (DFSS) methodology.
- Rolling Throughput Yield requirement of 99% for low rate initial production. Rolling Throughput Yield is a combination of First Pass Yield's within the assembly and test of a product.

The main objectives of these initiatives are that of customer satisfaction, competitive advantage, and maintaining ultra high reliability as seen on space products.

Qualification testing is now being planned on the initial production standard units in order to meet the requirements of our customers. This includes high and low temperature operation, temperature cycling at storage conditions (followed by leak and performance checks), performance testing in three orthogonal orientations, random vibration and shock, exported vibration, cool down characterization, cold finger side load deflection tests, and finally drop test survivability (2m onto concrete with the cold finger protected).

ACKNOWLEDGMENTS

Honeywell Hymatic would like to acknowledge the Science & Technology Facilities Council, Rutherford Appleton Laboratory for their continued support on the development of the Small Scale Cooler initially developed by RAL under European Space Agency funding. Honeywell Hymatic would also like to thank C F Cheuk for his focus on the development of the SSC and his many years of dedicated support; we wish him a long and happy retirement come July 8th 2016.

REFERENCES

1. Crook M. "Progress Towards a Small Scale Cooler for Space Applications," *5th European Space Cryogenics Workshop*, Noordwijk, Netherlands (2013).

Overview of Micro-Miniature Stirling Cryocoolers for High Temperature Applications

X. P. Chen[1], H. Sun[1], X. L. Nie[1], Z. H. Gan[2,3]

[1]Kunming Institute of Physics, Kunming, P.R. China, 650223
[2]Cryogenics Lab., Institute of Refrigeration and Cryogenics, Zhejiang University, Hangzhou, P. R. China, 310027
[3]Key Lab. of Ref. & Cryo. Technology of Zhejiang Province, Hangzhou, P.R. China, 310027

ABSTRACT

Micro-miniature Stirling cryocoolers for high temperature applications have bee ndeveloped recently to meet market demands. This paper compares micro-miniature Stirling coolers for Higher Operation Temperature applications with a miniature cooler a 2nd FPA base on four cooler subsystems. Thermal dynamics is inherent impetus to size, weight and power, so the regenerator is the key component of the cooler. This paper also presents a novel micro-miniature cooler for Higher Operation Temperature.

INTRODUCTION

Miniature Stirling cryocoolers are widely used in an infrared detector assembly. Since 1980, miniature Stirling refrigerators have been miniaturization with higher efficiency. Micro integral rotary Stirling cryocooler for 2nd generation Focal Plane Array (FPA) operating at 77K-85K usually weigh about 400 g, and split linear Stirling cryocooler weigh 1 kg. Based on miniaturization and optimization performance, micro Stirling cryocooler life has become a critical parameter. The Mean Time To Failure (MTTF) of a linear cryocooler employing a flexure bearing and moving magnet motor is more than 20,000 hours, e.g., THALES and AIM products. The MTTF of a linear cryocooler employing flexure bearing and moving coil motor is 15,000 hours, e.g., DRS products. The MTTF of a rotary cooler exceeds 10,000 hours, e.g., RICOR and THALES [1]. The data is based upon experimental results in laboratory test involving a few sample. The actual MTTF is affected by the environment that the cryocoolers are exposed to. RICOR has tracked their fielded coolers and performed statistical analysis of the K508. MTTF for this cooler reached about 33,000 hours for coolers with static observation, 12,000 hours for hand held devices, and 5,376 hours for armored vehicle [2]. A cold tip temperature of 77K results in a heavy load on the cooler and a reduced life. However, the life of the cryocooler is expected to increase by reducing the load in response to a higher operating temperature.

Higher operation temperature (HOT) FPA detector improves the operating temperature and is beneficial to reduce the load of cooling system, and will benefit the coolers MTTF. Much research

Cryocoolers 19, edited by S.D. Miller and R.G. Ross, Jr.

on high temperature infrared detector have be published in recent years. This trend is driven by the Size, Weigh and Power (SWaP)[3] concept. SWaP[3] is proposed based on the high performance and low cost. The concept means reducing the Size, Weight, and Power with a higher Performance, and lower Price. On the basis of a mature detector structure and state of the art chip preparation, high performance often requires smaller dark current reduced by lower temperature. So it is difficult to reduce the size, weight and power consumption of coolers. HOT detectors rely on new structures and methods of production improving operating temperature [3]. HOT is a promising method to achieve a reduced SWaP.

The 3rd generation MCT focal plane array should work at higher temperature and without losing performance as compared to the 77K performance. This stimulates High Operating Temperature (HOT) MCT detector development. Operating temperature has just an effect on thermal-induced free carriers resulting in noise, and no effect on other performance, such as cut-off wave length and quantum efficiency [4]. A HOT MCT chip can be realized by a new fabrication method and novel structures which suppress thermal-induced carriers. Since 1999, HOT MCT chips truly became commercially viable products, operating at 80K, 90K, 120K, 150K, 175K, 200K [5]. For FPA operating at 77K, the parasitic heat load is the main load. Parasitic loads decrease as the operating temperature rises, and the cooling load also decreases. HOT enables the novel miniature Stirling cooler to reduce Size, Weight, Power, and achieve a longer life of cooler for HOT application.

PRINCIPLES OF MINIATURE STIRLING COOLER FOR HOT

Efficiency of Stirling cooler is the result of four subsystems in total. That is, the overall efficiency is a product of thermodynamic cycle efficiency, mechanical efficiency of drive gear, motor efficiency and electronic work transforming efficiency of control and drive electronic device.

$$\eta_{cooler}=\eta_{cycle}\times\eta_{machine}\times\eta_{motor}\times\eta_{CDE} \quad (1)$$

It is expected to significantly reduce the power consumption of the detector assembly by improving the thermodynamic cycle efficiency with a higher operating temperature. The main reason for thermodynamic efficiency improvement is,

(1) The Coefficient of Performance of a HOT Stirling cooler increases as the operating temperature of the FPA (Te) increases, that is

$$COP_{Stirling}=T_e/(T_{com}-T_e) \quad (2)$$

Since the heat rejected temperature of the compressor (T_{com}) is atmosphere temperature, COP will increase as Te increases.

(2) The specific heat capacity variation of the matrix filled in the regenerator will decrease in a small temperature range. The ratio of the specific heat capacity between the matrix and the gas is greater than a 77K cooler. The larger ratio is beneficial to attenuate temperature hysteresis and improve the heat exchange efficiency in the regenerator. The actual cycle is closer to the ideal reverse Stirling cycle. With a shorter regenerator, there is greater influence by the hysteresis loop. The configuration should be considered for a HOT Stirling cooler.

Comparison Miniature Stirling Cooler for HOT and 2nd Generation FPA

With a cold finger temperature of 95K-200K, the thermodynamic cycle efficiency of a HOT miniature Stilrling cooler is higher than that of 80K coolers for second generation FPA. The reduction in overall efficiency translates into mechanical efficiency, motor efficiency and cooler driver electronics (CDE) efficiency.

Mechanical Efficiency

The unavoidable friction from bearings, piston and other parts in a rotary cooler wastes the partial drive power. The proportion of the friction consumption to total power is higher than those cooler for 80K application. However, linear compressor that rely on contact friction by flexure bearings have an advantage in mechanical efficiency to rotary cooler. The higher operating temperature, the greater the advantage of the linear Stirling cooler mechanical efficiency.

Motor Efficiency

The brushless DC motor for a HOT rotary Sitrling cooler reaches about 70% - 90% efficiency which is comparable to those motor for 80K cooler. The efficiency of a linear motors is individually determined by the design that obviously affects the uniform distribution of the magnetic field.

Frequency of a linear motor for a HOT application rises as a result of lightweight spring oscillator, and this frequency is higher than that for 2nd generation FPA at 80K. Linear coolers by Raytheon for 2nd generation operate at 50-60Hz [6-12], coolers by COBHAM at 54-60Hz [13]. COBHAM LC1076 cooler for HOT operates at 85Hz [14] and RICOR K527 for HOT at 75Hz by comparison.

CDE efficiency

The efficiency of CDE is mainly determined by the efficiency of power transformation module. The novel CDE offers potential for improving efficiency up to 90% in a recent report [15].

Comparison Rotary Cooler and Linear Cooler for HOT

Rotary coolers have several characteristics. They depending upon the rotation speed to regulate cooling capacity, the stroke of piston is constant, and the motor speed which is determined by the input voltage does not involve the thermodynamic cycle and the mechanics. To maintain a specific cryogenic temperature, the rotation speed is regulated and monitored and adjusted in a timely manner for the specified heat load. The principle of the rotary cooler regulation mode is useful for two operating modes, the rapid cool-down mode and the regulation mode for a nominal voltage. A booster promotes the motor speed in cool-down mode to offer the extra cooling capacity to shorten the cool-down time. Nominal input voltage for the optimized frequency increases the motor efficiency.

The operating frequency of a linear compressor is fixed by the natural frequency of spring oscillator. The cooling power of a linear cooler is tuned through changing the linear motor input voltage that dominates the stroke of compressor piston. The stroke of the piston is a key parameter for the thermodynamics cycle and mechanism of the linear cooler. The input frequency and voltage amplitude are inflexible.

The characteristics that the regulation loop bases on the operating frequency and the frequency can be decoupled from thermodynamics and mechanism provides flexibility to Integrated Detector Dewar Cooler Assembly (IDDCA) employed rotary Stirling cooler. Rapid cool-down is compatible with high thermodynamic efficiency through two work mode switch. On the other hand, the linear cooler has some merits that no fraction between the piston and the sleeve extend the lifetime and suppress acoustic noise.

Structure Features on Miniature Stirling Cooler for HOT

The coolers for HOT FPA are significantly smaller than existing products. The reason is that the parasitic load from the dewar decreases as the high operating temperature. The electronics power consumption for a HOT FPA is consistent with the reduction in parasitic load. Cooler miniaturization relies on the smaller swept volume and the smaller motor. The novel product has a small diameter and length for the linear compressor, e.g., 30 mm in diameter and 60 mm in length for the opposed piston compressor, or 40 mm in length for single piston as seen in Table 1. In addition, the regenerator and the cold finger can be reduced due to an increase in thermodynamic efficiency. The cold finger is usually small as shown in Table 1 and Table 2, e.g., 1/4 inch in diameter and 30-40 mm in length. A rotary cooler for HOT can cool down in 3-4 minutes as a result of a booster. It provides an alternative for SWaP thermal imaging system.

Main Aspect of Miniature Cooler for HOT

The cooling capacity (Q) is proportional to the swept volume of the compressor (V_{com}), the charging pressure (P_m), the cold end temperature (T_e) and the operating frequency (f)[19],

Table 1. Typical linear Stirling cooler and Stirling Pulse-Tube cooler for HOT application

	AIM SX020 [16]	COBHAM LC1076 [15]	THALES UP8197	RICOR K527 [17]	Lockheed Martin STAR [18]
Profile			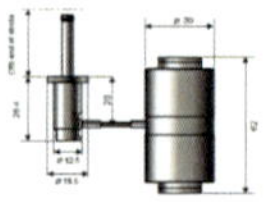	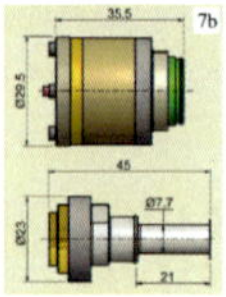	
Cooling capacity	0.5W@23°C /160K	0.4W@23°C /120K	1.2W@23°C /150K		0.69W@23°C /150K
Compressor	Φ26.5×57.6mm	Φ28.2×53.6mm	Φ30×62mm	Φ29.5×35.5mm	Φ32×90mm
Cold finger	Φ6×27.28 /46.5mm	Φ5×32.1mm	Φ6×30mm	Φ7.6×21 /45mm	Φ9.7×61mm Φ42×50mm reservoir
Weight	240g	260g	250g	140g	328g
Cooldown time		5 min (175J@120K)			8 min (150K)
Temperature range	-54°C~+85°C	-40°C~+71°C			-15°C~+55°C

Table 2. Typical rotary Stirling cooler for HOT application

	RICOR K562S [19]	RICOR K580	THALES RM1
Profile			
Cooling capacity	0.3W@71°C /110K	0.5W@71°C /150K	420mW@20°C /110K
Compressor	34.2×38.5×59.5mm	35×48.2×54.8mm	49.5×50.5×72mm
Cold finger	Φ7.6×45mm	Φ6.1×31mm	Φ6.1×38.5mm
Cooldown time	4 min (160J@110K)	3 min (210J@150K)	4.5 min (110J@110K)
Input power (regulation)	3W@150mW	2W@200mW	2.9W@100mW
Temperature range	-40°C~+71°C	-40°C~+71°C	-40°C~+71°C
Typical assembly	Hot-Pelican MW(640×512) (15μm) 、Kinglet、EPSILON MW(384×288)(15μm)		MW 384×288(15μm)

$$Q= (f/2)\, T_e\, P_m\, V_{com} \times 10^{-5}\ (W) \tag{3}$$

The product of the charging pressure and the swept volume has a positive correlation to the compressor mechanical load which has an effect on the machine lifetime. The high charging pressure is unsuitable for HOT Stirling cooler.

The pressure drop through the generator is negligible. It is thought that the volume of gas in the compression chamber is roughly equal to the volume of gas in the expansion chamber, that

$$V_{com} / V_e = T_{com} / T_e \tag{4}$$

The radius of the compressor piston (R) and the radius of the displacer (r) in the integral rotary cooler has the following relationship,

$$R/r= (T_{com} / T_e)^{1/2} \tag{5}$$

Isothermal model Stirling refrigeration cycle in cold storage caused by the relative dead volume is from 1.8 to 2.5.

The relative dead volume (s) is empirically assumed 1.8 - 2.5 in classic Schmidt's model for Stirling cyrocooler.

Figure 1. Miniature cooler for HOT and novel CDE

$$s=A_{reg}L_{reg}\Psi/V_{com} \tag{6}$$

It is worthy to note that the thermal gradient along the cold finger decreases in the HOT miniature Stirling cooler. It is achievable that the novel miniature coolers employ a shorter regenerator and cold finger. A short regenerator us beneficial to reduce the relative dead volume. When the regenerator is shorter, the temperature hysteresis loop becomes more serious. The improvement in COP may not eliminate the temperature hysteresis loop effect. It is important to balance the pressure drop and thermal exchange in the regenerator for a miniature cooler.

MINIATURE COOLER DESCRIPTION IN KIP

The novel miniature cooler is developed for the increasing demand ranging from 120K to 150K in KIP. The principle design of the novel cooler is low charging pressure for the reliability of the bearing and the piston coating. Another design principle is the optimization of the armature and air gap which balances the efficiency and the avoidance of outgassing.

Figure 1 shows the profile of the miniature cooler with a 1/4 inch diameter × 30 mm length. The mass of the cooler is 200 g. At the time of this writing, the prototype is currently being performance tested for optimization. Figure 1 shows the new digital electronics with booster developed for the novel cooler used for HOT application. This electronics has an adjustable booster and 30mK accuracy.

SUMMARY

- Rotary Stirling cooler for HOT is flexible for a the thermal imaging system, in light of the compromise between rapid cool-down and SWaP.
- Linear Stirling cooler for HOT is advantageous for the coefficient of performance, especially higher than 150K.
- Miniaturization of flexure bearing combined with the high frequency linear motor offers the potential for a linear cooler.
- The preliminary rotary Stirling cryocooler and novel electronics with booster are developed in KIP.

ACKNOWLEDGMENT

This work is financially supported by the National Natural Science Foundation of China (No. 51376157).

REFERENCES

1. Chen, X.P., Xia, M., Wang, L.Y. and Gan, Z.H., "The Reliability Development of Miniature Stirling Cryocoolers," *Cryocoolers 18,* ICC Press, Boulder (2014), pp. 601-607.

2. Pundak, N., Porat, Z., Barak, M., Zur, Y. and Pasternak, G., "Field reliability of Ricor microcoolers," *SPIE Defense, Security, and Sensing,* International Society for Optics and Photonics (2009), pp.72983I:1-4.

3. Ye, Z.H., Chen, Y.Y., Zhang, P., "Overview of Latest Technologies of HgCdTe Infrared Photoelectric Detectors," *INFRARED*, vol. 35, no. 2 (2014), pp. 1-8.

4. Wang, Y.F., Liu, P., "On the High Operating Temperature Mercury Cadmium Telluride Infrared Detector (I),"*INFRARED,* vol. 35, no. 9 (2014), pp. 1-5.

5. Wang, Y.F., Liu, P., "On the High Operating Temperature Mercury Cadmium Telluride Infrared Detector (II)," *INFRARED,* vol. 35, no. 10 (2014), pp. 7-13.

6. Raytheon Vision Systems, "DFOISR Ref: 04-S-0945(7049)," (2006), [2007-05], www.raytheon.com.

7 Raytheon Vision Systems, "DFOISR Ref: 04-S-0946(7050)," (2006), [2007-05], www.raytheon.com.

8 Raytheon Vision Systems. "DFOISR Ref: 04-S-0947(7070)," (2006), [2007-05], www.raytheon.com.

9 Raytheon Vision Systems. "DFOISR Ref: 04-S-0948(7062)," (2006), [2007-05], www.raytheon.com.

10 Raytheon Vision Systems. "DFOISR Ref: 04-S-0957(7052)," (2006), [2007-05], www.raytheon.com.

11 Raytheon Vision Systems. "DFOISR Ref: 04-S-0958(7060)," (2006), [2007-05], www.raytheon.com.

12 Raytheon Vision Systems. "DTS99-00539_Rev.7100(7055)," (2006), [2007-05], www.raytheon.com.

13 Carleton Life Support Systems Inc. "ADV10616 REV A(CC1067)," (2013), [2015-06], www.cobham.com./lifesupport.

14 Squires, M., "Cobham Microcooler for High Temperature Applications," *Cryocoolers 18*, ICC Press, Boulder (2014), pp.73-77.

15 Pilvelait, B., Zagarola, M., Finger, W., Bingham, R., Kaszeta, R. and Olson, J.R., "Micro-size Cryocooler Control Electronics," *Cryocoolers 18*, ICC Press, Boulder (2014), pp. 351-357.

16 Mai, M., Rosenhagen, C., Ruehlich, I., "Development of Single Piston Moving Magnet Cryocooler SX020," *Cryocoolers 18*, ICC Press, Boulder (2014), pp. 65-71.

17 Veprik, A., Riabzev, S., Avishay, N., Oster, D. and Tuitto, A., "Linear Cryogenic Coolers for Hot Infrared Detectors," *Cryocoolers 17*, ICC Press, Boulder (2012), pp. 61-70.

18 Nast, T.C., Roth, E., Olson, J.R., Champagne, P. and Frank, D., "Qualification of Lockheed Martin Micro Pulse Tube Cryocooler to TRL6," *Cryocoolers 18,* ICC Press, Boulder (2014), pp. 45-50.

19 Filis, A., Haim, Z.B., Pundak, N. and Broyde, R. "Micro Miniature Rotary Stirling Cryocooler for Compact, Lightweight and Low Power Thermal Imaging Systems," *SPIE Defense, Security, and Sensing*, vol. 7298 (2009), pp.729818:1-9.

Numerical Investigation of Environmental Temperature on Performance of Integral Crank Driven Stirling Cryocooler

Y. J. Hong, J. Ko, H. B. Kim, S. J. Park

Korea Institute of Machinery & Materials
Daejeon, 31403, Korea

ABSTRACT

Typical performance factors of Stirling cryocoolers are cool-down time, cooling capacity, power consumption, and efficiency. These factors depend on operating conditions of the Stirling cryocooler such as the charge pressure of helium gas, the rotational speed of the driving mechanism, and the thermal environment. The cooling capacity and power consumption are directly related to its thermal environment. The efficiency of the Stirling cryocooler increases with increasing the cooling temperature, while the efficiency decreases with increasing the environmental temperature. In this study, effects of the environmental temperature on performances of the integral crank driven Stirling cryocooler were investigated by numerical analyses using the Sage software. The simulation model takes into account thermodynamic losses due to regenerator inefficiency, pressure drop, shuttle heat transfer and solid conduction. Contributions of each loss to the cooling capacity with changes of the environmental temperature are shown as a result.

INTRODUCTION

Stirling cryocoolers have been widely used for cooling of infrared detectors (InSb, HgCdTe, etc.) and HTS devices to cryogenic temperature. The Stirling cryocooler with a rotary type compressor is applicable to cooling a device for compact mobile thermal imaging systems, because the cryocooler has a compact structure and is light weight.

Typical performance factors of the Stirling cryocooler are cool-down time, cooling capacity, power consumption and efficiency. These factors depend on operating conditions of the cryocooler, such as the charge pressure of the working fluid, rotational speed of driving mechanism, and thermal environment. The cooling capacity and power consumption are directly related to its thermal environment. In general, the efficiency of the Stirling cryocooler increases with increasing cooling temperature, while the efficiency decreases with increasing environmental temperature [1, 2].

Operation processes of the Stirling cryocooler are based on a cyclic compression and an expansion of working fluid helium. The working fluid is filled in the cryocooler at given conditions of temperature and pressure. Cyclic mean pressure of the working fluid in the cryocooler changes with the temperature of the cold end and the environment.

In general, the Stirling cryocooler has small cooling capacity and high power consumption at elevated environmental temperatures. Otherwise, at low environmental temperatures, it has

Cryocoolers 19, edited by S.D. Miller and R.G. Ross, Jr.

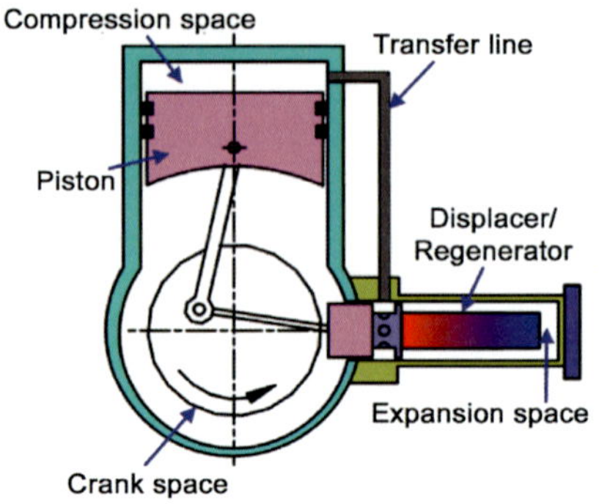

Figure 1. Schematic diagram of a Stirling cryocooler

the other tendency. In this study, effects of the environmental temperature on performances of the integral crank driven Stirling cryocooler were investigated by numerical analyses using the Sage software [3]. P-V work in the compression and expansion space, with contribution of losses in the expansion space are discussed.

SIMULATION MODEL OF A STIRLING CRYOCOOLER

A schematic diagram of the Stirling cryocooler is shown in Figure 1. The integral Stirling cryocooler consists of a motor, a crank-rod mechanism, a compression piston, a regenerator, a displacer, a compression space, an expansion space and a crank space. The Stirling cryocooler has a cooling capacity of 0.5 W at 80 K. The rotational speed of the motor is in the range of 20 to 70 RPS [4].

The Sage software is used for simulations of the performance of the Stirling cryocooler. Figure 2 shows the root level of the Sage model of the Stirling cryocooler. In order to know the cooling capacity, we need to calculate losses in the expansion space of the Stirling cryocooler. Thermal conduction losses occur along the wall of the expansion cylinder, the regenerator sleeve, and the regenerator. Enthalpy flow losses due to imperfect heat transfer arise in the regenerator. Enthalpy flow losses also arise in the clearance between the expansion cylinder and the regenerator sleeve. In the Sage model, the thermal conduction loss and the enthalpy flow loss are taken into account. In the simulation, the cold end temperature is 80 K and rejection temperatures of the cryocooler are equal to the environmental temperature. Efficiency of the motor is assumed to be constant. Radiation loss is not considered in the calculation of the cooling capacity.

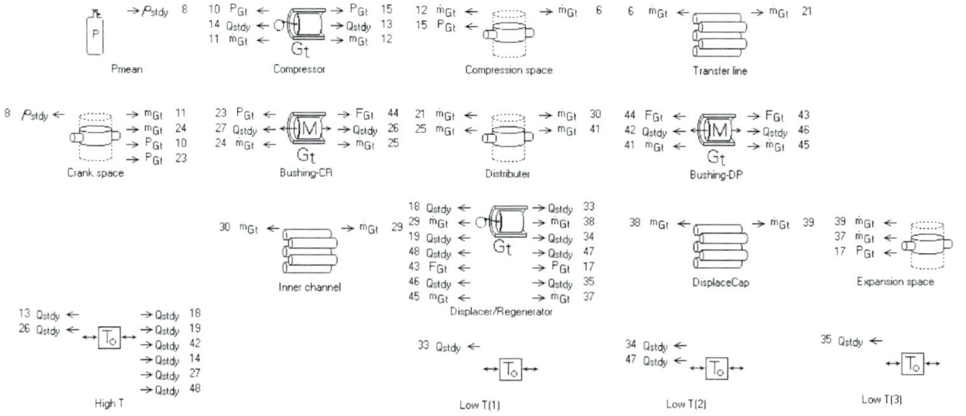

Figure 2. Sage model of the Stirling cryocooler

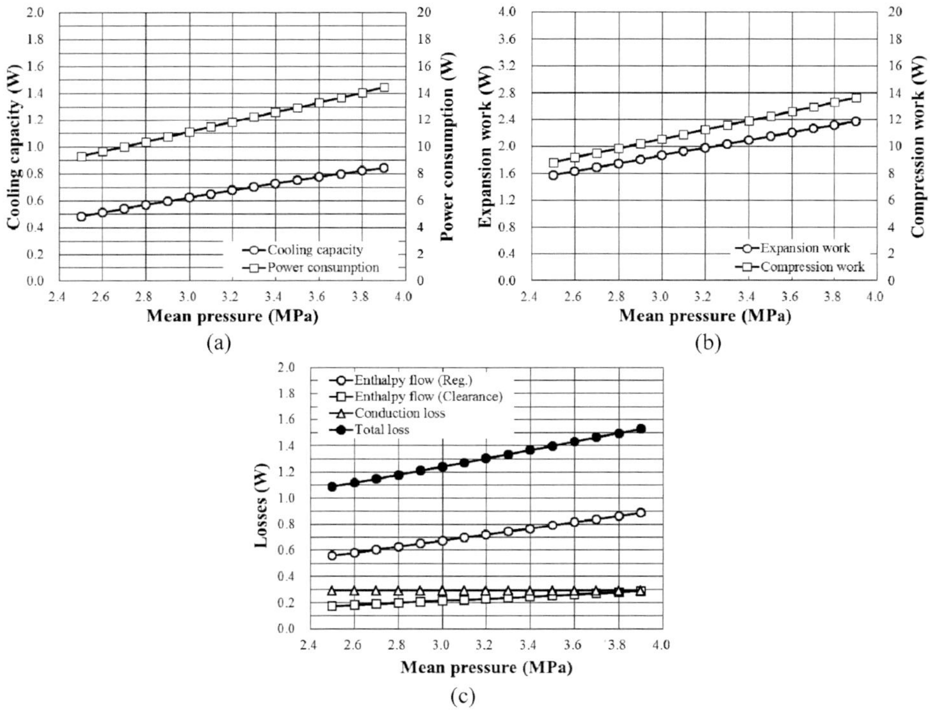

Figure 3. Performance of the Stirling cryocooler with different mean pressures; (a) cooling capacity and power consumption, (b) expansion and compression work, and (c) losses

SIMULATION RESULTS

Effects of Mean Pressure and Rotational Speed

Figure 3 shows simulation results for the performance of the Stirling cryocooler at a rotational speed of 60 RPS and an environmental temperature of 300 K. The cooling capacity and power consumption increases as the cyclic mean pressure of helium gas increases; which is mainly caused by increases of the expansion work and compression work. The total loss is about 60 to 70% of the expansion work; and a major loss in the expansion space is the enthalpy flow through the regenerator. Enthalpy losses from the regenerator and the clearance increase as the mean pressure increases. This must have been due to increases of the mass flow rate through the regenerator and the clearance. On the other hand, the conduction loss remains steady due to no changes in the cold end temperature or the rejection temperature.

As shown in Figure 3, at the mean pressure of 2.5 MPa, the cooling capacity and the power consumption of the cryocooler are 0.48 W and 9.3 W, respectively. At a mean pressure of 3.9 MPa, the cooling capacity and the power consumption of the cryocooler are 0.84 W and 14.5 W, respectively. The increase of 56% in the mean pressure induces an increase of 56% in the power consumption and an increase of 75% in the cooling capacity.

Figure 4 shows predicted performances of the cryocooler over a range of rotational speeds from 20 to 80 RPS at the mean pressure of 3.5 MPa and an environmental temperature of 300 K. The Sage model predicts no cooling capacity at 20 RPS from a power consumption of 4.5 W, because the total loss exceeds the expansion work. Above 30 RPS, the cooling capacity and the power consumption increase with the rotational speed. The fraction of the total loss in the expansion work decreases with the increases of the rotational speed. At the low rotational speed, the enthalpy flow loss through the regenerator and the conduction loss have similar contributions for the total loss. At the high rotational speed, the enthalpy flow loss through the regenerator has significant effects on the total loss.

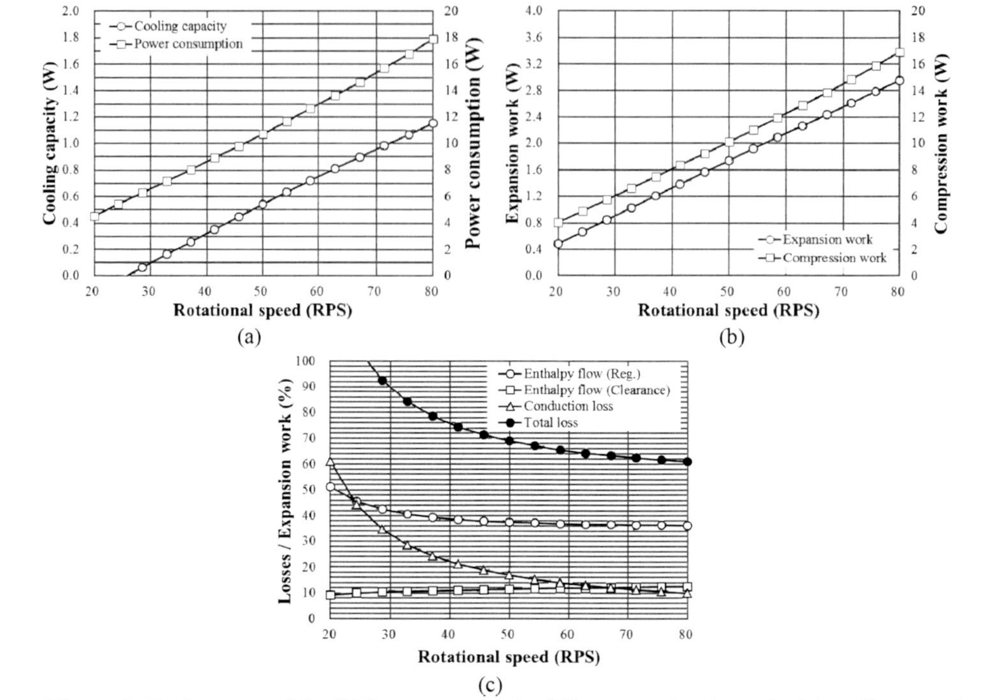

Figure 4. Performance of the Stirling cryocooler in different rotational speeds; (a) cooling capacity and power consumption, (b) expansion and compression work, and (c) losses

Effects of Environmental Temperature

In the Stirling cryocooler, the cryocooler is filled with helium gas at given conditions of temperature and pressure. So, the cyclic mean pressure of helium in the cryocooler changes with density changes due to the variation of temperatures of the cold end and the environment.

Figure 5 shows predicted performances of the cryocooler for cases over a range of environmental temperatures from 240 to 340 K at a mean pressure of 3.5 MPa and a rotational speed of 60 RPS. In the simulation, variations of the cyclic mean pressure of helium is not considered. The Sage model predicts that the Stirling cryocooler has a small cooling capacity and a high power consumption at the elevated environmental temperature. At the environmental temperature of 340 K, the cooling capacity and the power consumption are 0.56 W and 13.6 W, respectively. At an environmental temperature of 240 K, the cooling capacity and the power consumption are 1.04 W and 12.0 W, respectively. These phenomena are mainly caused by the decrease of the expansion work and the increase of the compression work.

As shown in Figure 5, enthalpy flow loss through the regenerator and conduction loss increase as the environmental temperature increases. The precentage of the total loss in the expansion work increase with the environmental temperature by up to 72.5% at 340 K.

Figure 6 shows P-V diagrams of the compression and the expansion space of the cryocooler with variations of the mean pressure due to the change of the environmental temperature. Simulations are performed for cases over a range of environmental temperatures from 240 to 340 K at a rotational speed of 60 RPS. The charging condition at a temperature of 290 K and a pressure of 3.5 MPa. The cyclic mean pressure is determined by using isothermal processes with ideal gas assumptions [5].

As shown in Figure 6, P-V diagrams show a shift to the high pressure region as environmental temperature increases. Mean pressures, at environmental temperatures of 240, 300, and 340 K, are 2.67, 3.28, and 3.68 MPa, respectively. The results show that spans of the pressure variation increase slightly as the environmental temperature increase.

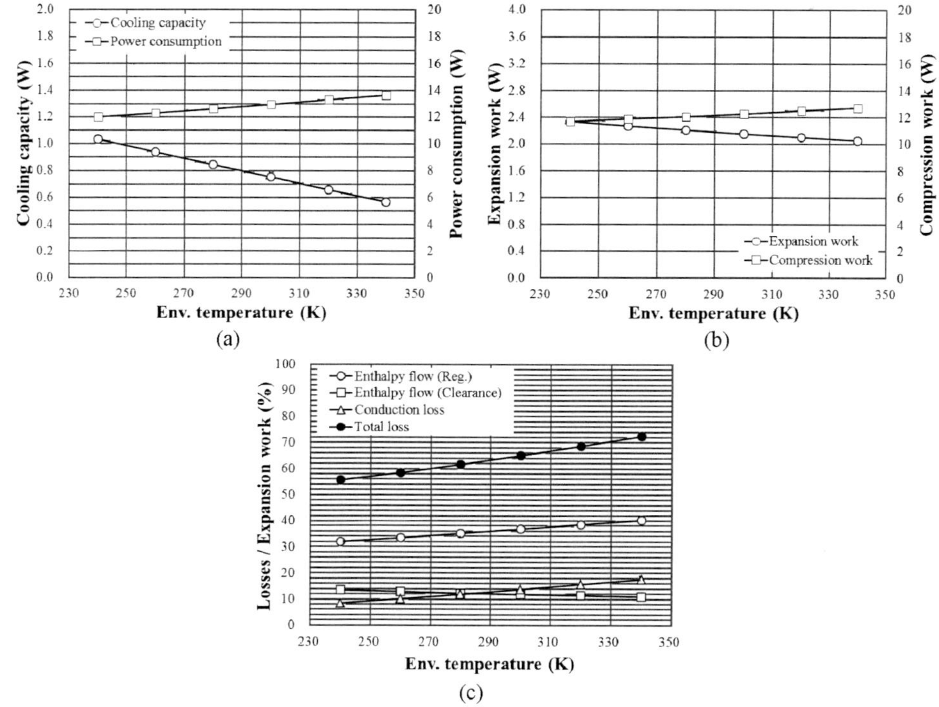

Figure 5. Performance of the Stirling cryocooler with different environmental temperatures; (a) cooling capacity and power consumption, (b) expansion and compression work, and (c) losses

Figure 7 shows predicted performances of the cryocooler at the charge pressure of 3.5 MPa and a rotational speed of 60 RPS. At an environmental temperature of 340 K, the cooling capacity and the power consumption are 0.62 W and 14.4 W, respectively. At an environmental temperature of 240 K, the cooling capacity and power consumption are 0.80 W and 9.1 W, respectively. These results show the effects of variations of the mean pressure on the performance of the Stirling cryocooler. In comparison with Figure 5, at elevated temperature, the elevated mean pressure results in increases of the cooling capacity and power consumption. On the other hand, at the low temperature, the lower mean pressure results in decreases of the cooling capacity and power consumption. Also shown in Figure 7, the compression and expansion work increases as the environment temperature increases.

The enthalpy flow loss through the regenerator and the conduction loss increase as the environmental temperature increases. The percentage of the total loss in the expansion work increases with

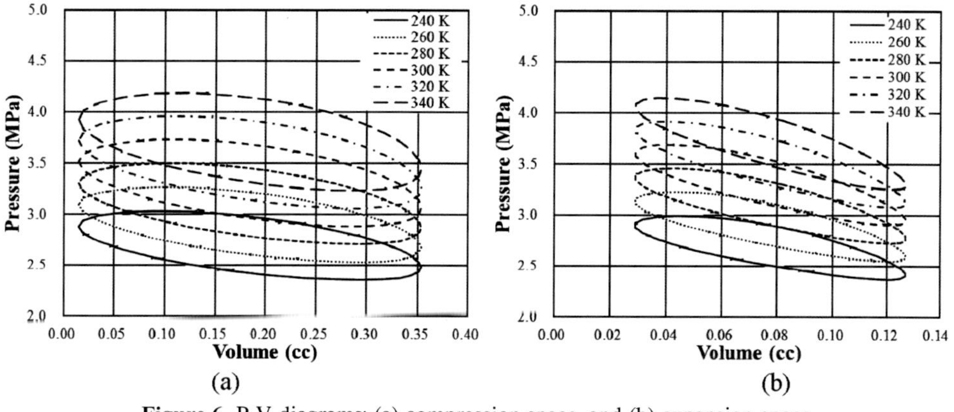

Figure 6. P-V diagrams; (a) compression space, and (b) expansion space

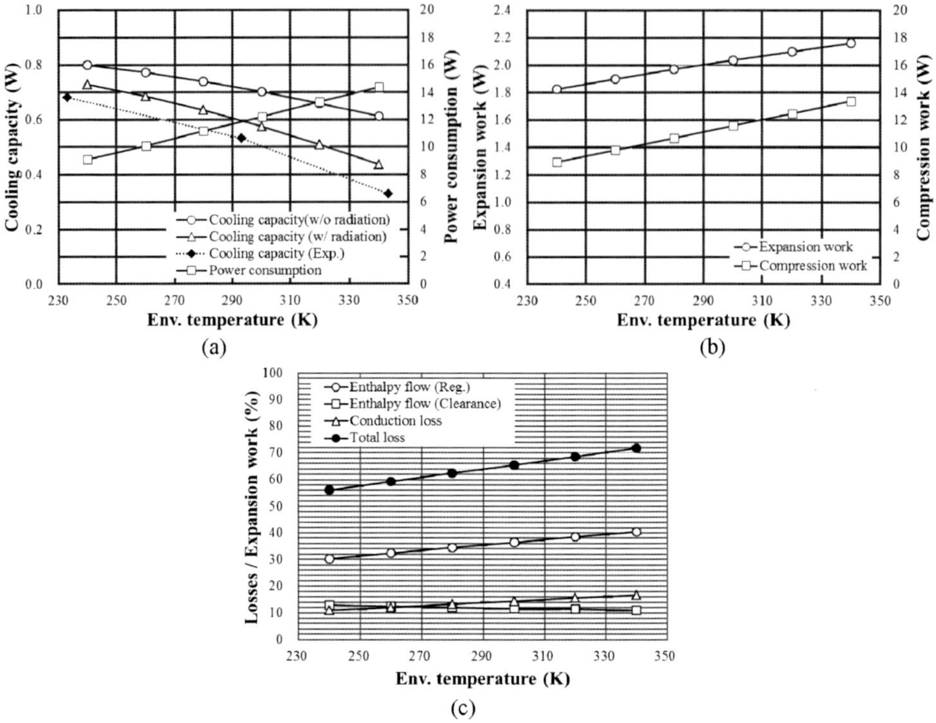

Figure 7. Performance of the Stirling cryocooler with different environmental temperatures; (a) cooling capacity and power consumption, (b) expansion and compression work, and (c) losses

the environmental temperature to 71.7% at 340 K. The total loss at the environmental temperatures of 240, 300, and 340 K are 56.2, 65.5, and 71.7% of the expansion work, respectively. A major loss in the expansion space is the enthalpy flow through the regenerator, which exceeds half of the total loss in the Stirling cryocooler.

In order to compare simulation results with the cooling capacity from experiments, the radiation loss should be considered in the predicted cooling capacity from the Sage model. As shown in Figure 7, the cooling capacity with the radiation loss is lower than the predicted cooling capacity without the radiation loss. Hence, the radiation loss was estimated from the boil-off measurement. The cooling capacity from the simulation and the experiment are slightly different, but exhibit the same tendency. The variation of the mean pressure, which arises from the change of the environmental temperature, should be considered to predict the performance of the Stirling cryocooler.

SUMMARY

In this study, the effects of the environmental temperature on performances of an integral crank driven Stirling cryocooler were investigated by numerical analyses using the Sage software. The variation of the mean pressure, which arises from the change of the environmental temperature, should be considered to predict the performance of the Stirling cryocooler. The enthalpy flow loss through the regenerator and the conduction loss increase as the environmental temperature increases. The percentage of the total loss in the expansion work increases as the environmental temperature increases, and a major loss in the expansion space is the enthalpy flow through the regenerator.

ACKNOWLEDGMENT

This research was supported by the Space Core Technology Development Program of the National Research Foundation (NRF) funded by the Ministry of Science, ICT & Future Planning (2015M1A3A3A03027288).

REFERENCES

1. Walker, G., *Miniature Refrigerators for Cryogenic Sensors and Cold Electronics,* Clarendon Press, Oxford (1989), pp. 89-120.
2. Donabedian, M., *Spacecraft Thermal Control Handboolk, Volume II: Cryogenics,* The Aerospace Press, El Segundo (2002), pp. 121-129.
3. Gedeon, D., *Sage User's Guide,* Gedeon Associates, Athens (2013)
4. Hong, Y. J., Kim. H. B., Park, S. J., "Effects of the environmental temperature on the performance of the Stirling cryocooler," *Journal of KIASC*, vol. 11, no. 3 (2009), pp. 65-68 (In Korean).
5. https://www.ohio.edu/mechanical/stirling/

Global Vibration Control of Split Stirling Linear Cryogenic Cooler

A. Veprik[1], A. Tuito[2]

[1] SemiConductor Devices, POB 2250, Haifa, 31021, Israel
[2] Israel Ministry of Defense, Kirya, Tel Aviv, 64734, Israel

ABSTRACT

High operating temperature, low size, weight and power (SWAP) infrared imagers often rely on miniature split Stirling linear cryocoolers composed of an electro-dynamically driven compressor and a pneumatically driven expander. For compactness, the two compoents are mounted side-by-side to a common rigid frame and connected by a configurable transfer line.

Vibration export produced by such a cryocooler may be thought of as a pair of tonal and coherent forces resulting from the almost sinusoidal imbalanced reciprocation of the moving assemblies inside the compressor and expander. The cooler-induced vibration of the infrared imager is, therefore, comprised of a coupled angular and translational components manifesting themselves in the form of angular line of sight jitter and translational defocusing.

The authors present a Multimodal Tuned Dynamic Absorber, featuring translational and tilting dynamic modes, the frequencies which are essentially tuned to the fixed driving frequency. Dynamic analysis shows that the dynamic reactions (force and moment) produced by such a dynamic absorber may simultaneously attenuate both translational and angular components of cryocooler-induced vibration, and improve the quality of the imagery.

INTRODUCTION

Split Stirling linear micro-miniature cryocoolers find use in advanced infrared (IR) imagers. Over the last few years major vendors, such as DRS, AIM, L-3, Cobham and Ricor developed and fielded new models for low cost (SWAP) cryocoolers in support of high operating temperature (150K and above) IR focal plane arrays.

It is worth noting that DRS [1,2], L3 [3] and Cobham [4] are following the traditional dual-piston (low vibration) compressor design concept, while Ricor [5-12] and AIM [13,14] favor a single-piston compressor offering lower SWAP and manufacturing expenses.

This decision has been substantiated by the choice of a relatively high driving frequency (100 Hz and above) leading to more compact design, smaller strokes and lower weight for the moving assemblies, resulting in lower the vibration export magnitudes. Along with these lines, the visual effect of high-frequency interference is less harmful. This favorable combination makes SWAP single-piston compressors adequate for the relatively heavy or fixedly mounted electro-optic payloads [15, 16].

In the case of inherently more vibration sensitive lightweight hand-held and gyro-stabilized payloads, the magnitudes of the cooler-induced vibration may be reduced by a payload design. In

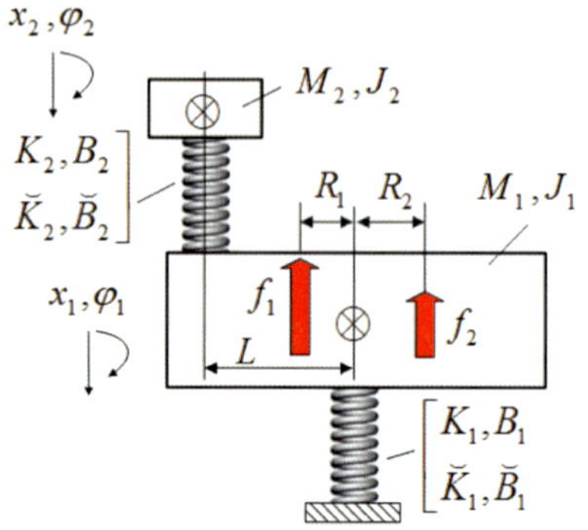

Figure 1. Dynamic model

gyro-stabilized applications, for example, the compressor and expander may be placed to minimize the moment about each gimbal axis.

When the design constraints prevent the prefered packaging or the payloads are sensitive to vibration, the cryocooler induced vibration may be efficiently attenuated to the acceptable levels using a Tuned Dynamic Absorber (TDA) mounted in-line with the compressor (major source of vibration export) to produce a counterbalancing force. This configuration will ultimately eliminate the translational dynamic response at the attachment point. Usually, it is advised to make such a TDA in the form of an undamped "mass-spring" single degree of freedom (SDOF) translational system, where the frequencies of tilting, in-plane translations, rotation about the axis are well separated from the frequency of the axial translational mode, which needs to be tuned exactly to the driving frequency (or vice versa), see [17,18]. Examples are known of cryogenically cooled long-range thermal weapon sights (TWS) or hand-held imagers, relying on the optional TDA [19,20].

Although the vibration export produced by the expander is relatively small, it cannot be disregarded in the light-weight hand held and gyrostabilized electro-optic applications. In these cases, the in-line mounting of the compressor and the expander units is favorable. Single-axis consolidation of the exported vibration induced by the compressor and the expander allows for effective use of a single, inline mounted translational TDA. Unfortunately, this option is not always practical due to the packaging constraints. For compactness, the expander and compressor are usually packaged side-by-side. In this case, the expander-induced exported vibration may produce a moment about the payload center of gravity resulting in an angular line-of-sight jitter and a translational defocusing.

The authors present a patent pending tunable Multimodal Tuned Dynamic Absorber (MTDA) with the translational frequencies and the two tilting mode frequencies are essentially tuned to the driving frequency. Dynamic analysis shows that the dynamic reactions (force and moment) produced by such a dynamic absorber are capable of simultaneously attenuatiing the translational and angular components of cryocooler induced vibration.

DYNAMIC MODEL, EQUATIONS OF MOTION AND ANALYSIS

Figure 1 shows the simplified four-degrees of freedom dynamic model, where the electro-optical payload having mass M_1 and moment of inertia J_1 is flexibly supported from the stationary base using a viscoelastic member featuring the linear and angular stiffness, and the damping $K_1, B_1; \breve{K}_1, \breve{B}_1$, respectively.

It is further assumed that both the compressor and the expander are rigidly mounted to the payload frame at distances of R_1 and R_2 with respect to the center of gravity. The moving assemblies' reciprocating inside the compressor and expander have masses $m_{1,2}$. Their reciprocation with (fixed) angular driving frequency ω, magnitudes $\delta_{1,2}$ and constant phase lag φ result in a tonal force export, which may be thought of as a pair of coherent forces $f_1 = m_1\delta_1\omega^2 \sin(\omega t)$ and $f_2 = m_2\delta_2\omega^2 \sin(\omega t + \varphi)$, respectively.

The MTDA is schematically represented as a lumped body of mass M_2 and moment of inertia J_2, which is supported from the payload frame at a distance L with respect to the center of gravity using viscoelastic member featuring linear and angular stiffness and damping $K_2, B_2; \breve{K}_2, \breve{B}_2$, respectively. The translation and tilt responses are x_1, φ_1 for the payload frame and x_2, φ_2 for the MTDA, respectively.

Differential equations of motion in time domain are as follows:

$$
\begin{aligned}
&M_1\ddot{x}_1 + K_1x_1 + K_2\left(x_1 + L\varphi_1 - x_2\right) + B_1\dot{x}_1 + B_2\left(\dot{x}_1 + L\dot{\varphi}_1 - \dot{x}_2\right) = f_1 + f_2 \\
&J_1\ddot{\varphi}_1 + \breve{K}_1\varphi_1 + K_2L\left(x_1 + L\varphi_1 - x_2\right) + \breve{K}_2\left(\varphi_1 - \varphi_2\right) + \breve{B}_1\dot{\varphi}_1 + B_2L\left(\dot{x}_1 + L\dot{\varphi}_1 - \dot{x}_2\right) + \breve{B}_2\left(\dot{\varphi}_1 - \dot{\varphi}_2\right) = f_1R_1 - f_2R_2 \\
&M_2\ddot{x}_2 - K_2\left(x_1 + L\varphi_1 - x_2\right) - B_2\left(\dot{x}_1 + L\dot{\varphi}_1 - \dot{x}_2\right) = 0 \\
&J_2\ddot{\varphi}_2 - \breve{K}_2\left(\varphi_1 - \varphi_2\right) - \breve{B}_2\left(\dot{\varphi}_1 - \dot{\varphi}_2\right) = 0
\end{aligned} \tag{1}
$$

Complex Fourier transform, $G(j\omega) = \int_{-\infty}^{\infty} g(t)e^{-j\omega t}dt$, where $j = \sqrt{-1}$ is complex unity, yields a transition from time into a frequency domain $x_{1,2}(t) \Leftrightarrow X_{1,2}(j\omega)$, $\varphi_{1,2}(t) \Leftrightarrow \Phi_{1,2}(j\omega)$, $f_{1,2}(t) \Leftrightarrow F_{1,2}(j\omega)$. Substitution into (1) yields the set of 4 linear algebraic equations:

$$
\begin{aligned}
&\left[-\omega^2 M_1 + \left(K_1 + j\omega B_1\right) + \left(K_2 + j\omega B_2\right)\right]X_1 + \left(K_2 + j\omega B_2\right)\left(L\Phi_1 - X_2\right) = F_1 + F_2 \\
&\left[-\omega^2 J_1 + \left(\breve{K}_1 + j\omega\breve{B}_1\right) + \left(\breve{K}_2 + j\omega\breve{B}_2\right) + \left(K_2 + j\omega B_2\right)L^2\right]\Phi_1 + \left(K_2 + j\omega B_2\right)L\left(X_1 - X_2\right) - \\
&-\left(\breve{K}_2 + j\omega\breve{B}_2\right)\Phi_2 = F_1R_1 - F_2R_2 \\
&\left[-\omega^2 M_2 + \left(K_2 + j\omega B_2\right)\right]X_2 - \left(K_2 + j\omega B_2\right)\left(X_1 + L\Phi_1\right) = 0 \\
&\left(-\omega^2 J_2 + \breve{K}_2 + j\omega\breve{B}_2\right)\Phi_2 - \left(\breve{K}_2 + j\omega\breve{B}_2\right)\Phi_1 = 0
\end{aligned} \tag{2}
$$

From the third and fourth equations of (2)

$$
\Phi_1 = \frac{\left(-\omega^2 J_2 + \breve{K}_2 + j\omega\breve{B}_2\right)}{\breve{K}_2 + j\omega\breve{B}_2}\Phi_2;\ X_1 = \frac{\left(-\omega^2 M_2 + K_2 + j\omega B_2\right)}{K_2 + j\omega B_2}X_2 - \frac{\left(-\omega^2 J_2 + \breve{K}_2 + j\omega\breve{B}_2\right)}{\breve{K}_2 + j\omega\breve{B}_2}L\Phi_2 \tag{3}
$$

Substituting (3) into (1) eliminates X_1 and Φ_1 and yields a set of linear equations in X_2 and Φ_2

$$
\begin{aligned}
&\mathbf{\Lambda}_{11}X_2 + \mathbf{\Lambda}_{12}L\Phi_2 = F_1 + F_2; \\
&\mathbf{\Lambda}_{21}X_2 + \mathbf{\Lambda}_{22}\Phi_2 = F_1R_1 + F_2R_2
\end{aligned} \tag{4}
$$

where

$$
\mathbf{\Lambda}_{11} = \frac{\left(-\omega^2 M_2 + K_2 + j\omega B_2\right)\left[-\omega^2 M_1 + K_1 + K_2 + j\omega B_1 + j\omega B_2\right] - \left(K_2 + j\omega B_2\right)^2}{K_2 + j\omega B_2};
$$

$$
\mathbf{\Lambda}_{12} = -\frac{\left(-\omega^2 J_2 + \breve{K}_2 + j\omega\breve{B}_2\right)\left[-\omega^2 M_1 + K_1 + j\omega B_1\right]}{\breve{K}_2 + j\omega\breve{B}_2}L;\ \mathbf{\Lambda}_{21} = -\omega^2 M_2 L;
$$

$$
\mathbf{\Lambda}_{22} = \frac{\left[-\omega^2 J_1 + \breve{K}_1 + \breve{K}_1 + j\omega\breve{B}_2 + j\omega\breve{B}_1\right]\left(-\omega^2 J_2 + \breve{K}_2 + j\omega\breve{B}_2\right) - \left(\breve{K}_2 + j\omega\breve{B}_2\right)^2}{\breve{K}_2 + j\omega\breve{B}_2}
$$

Solution to (4) may be obtained by, say, using Kramer's rule

$$
X_2 = \frac{\mathbf{\Delta}_{X_2}}{\mathbf{\Delta}};\Phi_2 = \frac{\mathbf{\Delta}_{\Phi_2}}{\mathbf{\Delta}} \tag{5}
$$

where determinants are

$$
\mathbf{\Delta} = \begin{vmatrix} \mathbf{\Lambda}_{11} & \mathbf{\Lambda}_{12} \\ \mathbf{\Lambda}_{21} & \mathbf{\Lambda}_{22} \end{vmatrix};\ \mathbf{\Delta}_{X_2} = \begin{vmatrix} F_1 + F_2 & \mathbf{\Lambda}_{12} \\ F_1R_1 + F_2R_2 & \mathbf{\Lambda}_{22} \end{vmatrix};\ \mathbf{\Delta}_{\Phi_2} = \begin{vmatrix} \mathbf{\Lambda}_{11} & F_1 + F \\ \mathbf{\Lambda}_{21} & F_1R_1 + F_2R_2 \end{vmatrix}
$$

Substituting (5) into (3) yields

$$\Phi_1 = \frac{\left(-\omega^2 J_2 + \breve{K}_2 + j\omega\breve{B}_2\right)}{\breve{K}_2 + j\omega\breve{B}_2}\frac{\Delta_{\Phi_2}}{\Delta};\; X_1 = \frac{\left(-\omega^2 M_2 + K_2 + j\omega B_2\right)}{K_2 + j\omega B_2}\frac{\Delta_{X_2}}{\Delta} - \frac{\left(-\omega^2 J_2 + \breve{K}_2 + j\omega\breve{B}_2\right)}{\breve{K}_2 + j\omega\breve{B}_2} L\frac{\Delta_{\Phi_2}}{\Delta} \quad (6)$$

Assuming the case of ideally tuned, MTDA $\left(K_2 = \omega^2 M_2;\; \breve{K}_2 = \omega^2 J_2;\; \breve{B}_2 \to 0;\; B_2 \to 0\right)$ from (5), we find finite values for translational and angular responses of MTDA at the driving frequency; these are:

$$X_2 \approx -\left(F_1 + F_2\right)/K_2;\; \Phi_2 \approx \left[-\left(F_1R_1 + F_2R_2\right) + L\left(F_1 + F_2\right)\right]/\breve{K}_2 \quad (7)$$

From (7), the force of dynamic reaction exerted by MTDA upon the payload frame is $\mathbf{Q} \approx K_2X_2 = -\left(F_1 + F_2\right)$, thus evidently counterbalancing aggregate force applied by the cryocooler. Along with these lines, the moment of dynamic reaction exerted by MTDA upon the payload frame is $\mathbf{T} \approx \Phi_2\breve{K}_2 = -\left(F_1R_1 - F_2R_2\right) + L\left(F_1 + F_2\right)$, thus evidently counterbalancing aggregate moment produced by the cryocooler along with the moment resulted from the MTDA translational motion.

Since X_2 and Φ_2 in (7) are finite, monotonous and almost constant functions in close vicinity to the driving frequency, we can use the below approximations

$$\Phi_1 \approx \frac{\left[-\left(F_1R_1 + F_2R_2\right) + L\left(F_1 + F_2\right)\right]}{\breve{K}_2^2}\left(-\omega^2 J_2 + \breve{K}_2 + j\omega\breve{B}_2\right); \quad (8)$$

$$X_1 \approx -\frac{\left(F_1 + F_2\right)}{K_2^2}\left(-\omega^2 M_2 + K_2 + j\omega B_2\right) - \frac{\left[-\left(F_1R_1 + F_2R_2\right) + L\left(F_1 + F_2\right)\right]L}{\breve{K}_2^2}\left(-\omega^2 J_2 + \breve{K}_2 + j\omega\breve{B}_2\right)$$

From (8), $\Phi_1 = 0$ and $X_1 = 0$ at the driving frequency, provided $K_2 = \omega^2 M_2$, $\breve{K}_2 = \omega^2 J_2$ and $\breve{B}_2 = B_2 = 0$. Also, from (8) the best geometrical location for the MTDA mounting is defined. Recalling that forces F_1 and F_2 are, generally speaking, complex numbers, we calculate the module of complex moment $\mathbf{T} = -\left(F_1R_1 - F_2R_2\right) + L\left(F_1 + F_2\right)$ and find the optimum value

$$L_{opt} = \frac{|F_1|^2 R_1 + |F_2|^2 R_2}{|F_2|^2 + 2\,\mathrm{Re}\left(F_2\right)|F_1| + |F_1|^2} \quad (9)$$

which minimizes its magnitude, where notation $\mathrm{Re}(*)$ means real part. It is worth noting that for the split Stirling linear cryocooler, the magnitude ratio F_2/F_1 is almost constant over the wide range of working conditions typical of the temperature control operational mode. From (9), the optimal position of MTDA will be, therefore, almost invariant over the range of working conditions.

For example, in the typical case $\varphi = 90\deg$ and $\mathrm{Re}\left(F_2\right) = 0$, thus

$$L_{opt} = \frac{|F_1|^2 R_1 + |F_2|^2 R_2}{|F_2|^2 + |F_1|^2};\; \mathbf{T}_{\min} = \frac{|F_2||F_1|}{|F_2|^2 + |F_1|^2}\sqrt{|F_1|^2\left(R_1 - R_2\right)^2 + |F_2|^2\left(R_1 + R_2\right)^2} \quad (10)$$

Figure 2 portrays the typical dependence of the moment $\mathbf{T}$ on the position of MTDA. The numerical values used in this example are: $|F_2| = 3N$; $|F_1| = 13N$; $\varphi = 90\deg$; $R_1 = 35mm$ and $R_2 = -10mm$. From (10), when $L_{opt} = 32.7mm$ and $\mathbf{T}_{\min} = 0.13Nm$, this correlates well with the graph in Figure 2. This indicates the theoretical possibility of simultaneously attenuating the translation and tilting responses of the payload frame using a properly located, undamped single MTDA with matched translation and tilt frequencies.

In the most general case when $B_2 \neq 0$; $\breve{B}_2 \neq 0$, the entire analysis will be much more complicated. The numerical example below provides a practical insight into the attainable performance.

DYNAMIC DESIGN OF MUTIMODAL TDA

In this numerical example, the following set of variables will be used: $M_1 = 1\ kg$; $M_2 = 0.03\ kg$; $J_1 = 0.0033\,kg \cdot m^2$; $J_2 = 0.000099\,kg \cdot m^2$. It is worth noting that such a MTDA adds only 3% to the total weight of the payload, the moment of inertia of such a MTDA is 3% of this payload. The payload is assumed to have a low frequency mount, thus mimicking the case of a hand-held infrared imager. Translational and tilting frequencies are assumed to be 10 Hz and damping ratios of 15%, respectively. The driving frequency is 105 Hz. The nominal frequencies of translational and tilting

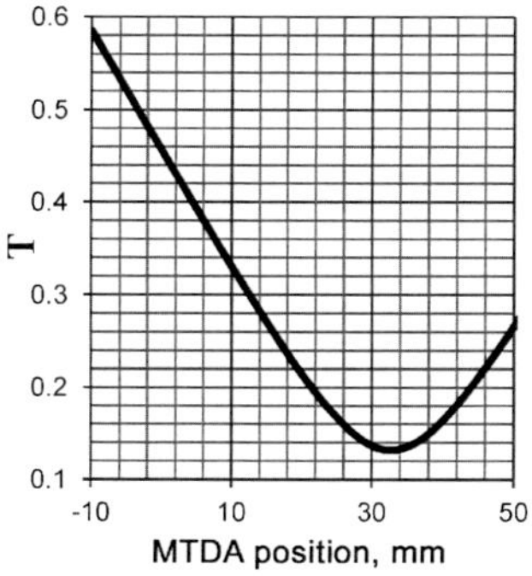

Figure 2. Optimal placement of MTDA

modes of MTDA will be equal to the driving frequency. The typical damping ratios for both modes will be conservatively assumed to be as low as 0.03%. The distances from the center of gravity are $R_1 = 0.035m$ and $R_2 = -0.01m$. Typical magnitudes for the force export are: $\|F_1\| = 13\ N$ and $\|F_2\| = 3N$, and the phase lag is assumed to be 90 deg.

In the reference case with no MTDA attached, the magnitudes of translation at the cold tip location and the tilt response of the payload are $27\mu m$ and $301\mu rad$, respectively. The magnitude of translation response is comparable to the typical focus depth and the magnitude of tilt response which is definitely higher than a typical acceptable value of $25\mu rad$. This substantiates the need for using a MTDA. Further, we consider the prior single degree of freedom (SDOF) TDA having only translational mode. The frequency of which is exactly tuned to the driving frequency. Translational and tilting response of the payload will be calculated at different TDA locations ranging from 10 mm to 50 mm relative to the payload center of gravity. Finally, we consider the two degrees of freedom (TDOF) MTDA featuring translational and tilting modes tuned exactly to the driving frequency. Translational and tilting responses of the payload will be calculated at different TDA locations ranging from 10 mm to 50 mm relative to the payload center of gravity.

Figures 3a and 3b compare the payload translation and tilt responses in the above three cases. In particular, the dashed lines represent the reference case with no TDA, double lines portray the case of thr SDOF TDA and the thick lines portray the case of TDOF TDA. From Figure 3, using SDOF TDA reduces both translation and tilt at the cold head location, the attenuation ratio is dependent on the TDA position with regard to the payload center of gravity. The optimal position is 33 mm, this is in accordance with Figure 2. It is clearly seen that SDOF TDA, even placed at the optimal position, is not capable of completely neutralizing the cooler-induced vibration. Such a TDA yields only a 4.6-fold and 9-fold attenuation of tilt and translation response at the location of cold head, respectively. Although the translational response is quite negligible in comparison with

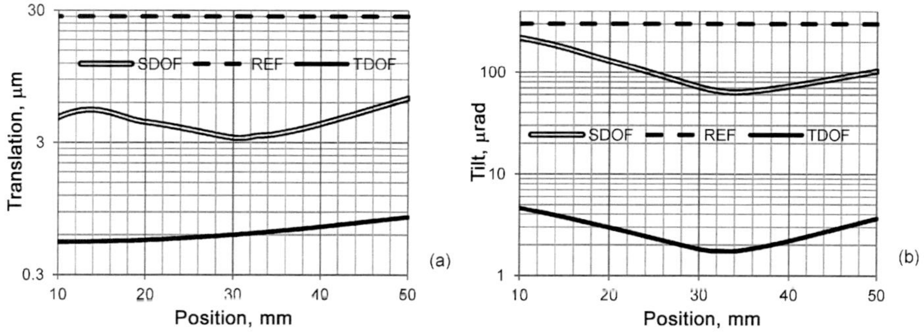

Figure 3. Translation and tilt responses of payload

the typical depth of focus, the achieved 65 μrad magnitude of the payload tilt does not meet the requirement of 25 μrad typical of a hand-held application.

The TDOF TDA is different from SDOF TDA. It yields an essential improvement: the attenuation ratios are much higher: 45-fold for translation and 176-fold for the tilt and the attained 1.7 μrad tilt magnitude does meet the most stringent requirements. It is important to notice, that these improvements have been achieved without adding weight to the MTDA – just by making favorable use of its rotational inertia. It is also important to notice that the performance may be further improved by using an MTDA having a higher moment of inertia. Along these lines, optimal placement of TDA approximately corresponds to the location of compressor which allows placing the TDA in-line with compressor axis.

In the previous reasoning, we assumed that the resonant frequencies of translational and tilting modes were tuned exactly to the driving frequency, namely to 105 Hz. It goes without saying that in practice such accurate tuning will be very complicated. Therefore, prior to advising the mechanical design of MTDA we need to study the influence of frequency detuning on attainable performance. To start with, we will assume that the translational mode is tuned to the driving frequency 105 Hz exactly and vary the frequency of the tilting mode over the range 104.5-105.5 Hz.

Figure 4a shows the dependences of the magnitude of tilt and translation on the value of the tilting frequency at fixed translation frequency. The MTDA has been mounted inline with compressor at $L = 35mm$. From Figure 4a, tuning the tilting frequency results in essential attenuation of the tilt response, but this has very little influence on the translation response. Similarly, at the fixed and tuned tilting frequency, the tuning translation frequency results in essential attenuation of the translation response, but has very minor influence on the angular response (see Figure 4b). From Figure 4, the anti-resonant notches are wide enough to perform accurate frequency tuning and matching.

MECHANICAL DESIGN OF MUTIMODAL TDA

Figure 5 shows a diagrammatic view of the side-by-side split Stirling cryogenic cooler comprised of compressor 1 and expander 2 interconnected by the transfer line 3, whereupon the MTDA includes a primary proof ring 4 which is supported from one side by the flexural bearing 5 having central anchor for mounting to compressor housing 1. The secondary proof ring 6 is coaxial and displaceable along the primary proof ring; the outer diameter of the primary ring is tightly and slidably matched to the inner diameter of the secondary proof ring. In this preferable embodiment, the flexural bearing is made in the form of a circular planar spring with 4 symmetrical spiral slots.

In this embodiment, the translational frequency depends on the weight of the aggregate proof mass and axial spring rate of the flexural bearing. Along with these lines, the tilting frequencies depend on the angular spring rate of the said flexural bearing and moment of inertia of the aggregate proof mass. By finite elements design of the flexural bearing and proof rings, the said frequencies of the translation and tilting modes may be equalized and made essentially equal the fixed working frequency typical of the cryocooler.

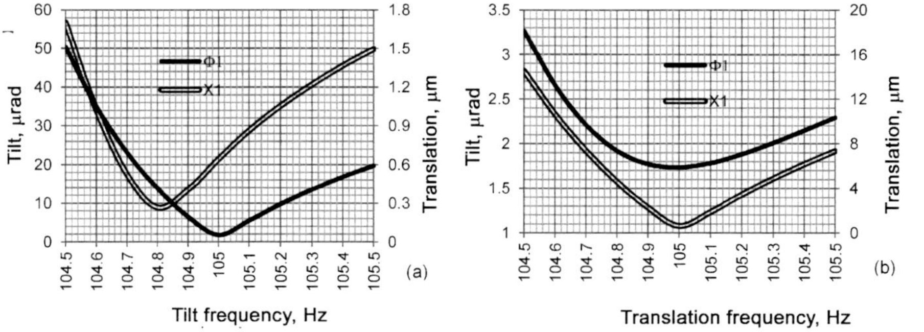

Figure 4. Translation and tilt response of payload at different frequency detuning.

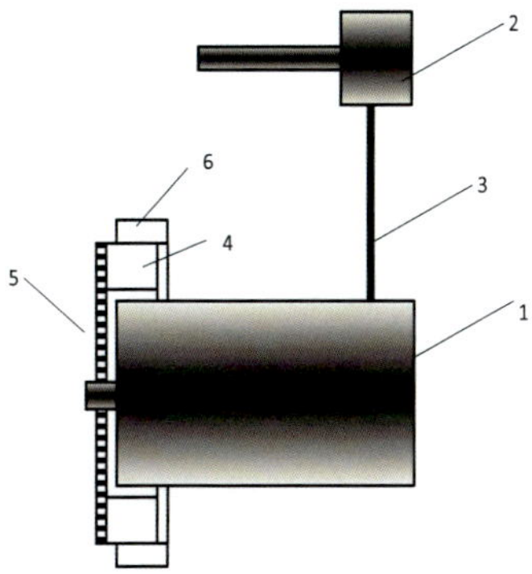

Figure 5. MTDA concept

Figure 6 shows the flexural bearing (a) and self-explanatory exploded view (b) of actual CAD design of such a TDA.

The modal analysis has been performed using the ANSYS 16.2 Finite Element software. Figure 7 shows the first 6 modal shapes: rotation about TDA axis (a), translation along the TDA axis (b), tilting modes (c,d) and in-plane modes (e,f). The frequencies of tilting and translation modes are 105 Hz. This has been achieved by virtually displacing the correction ring into particular position, it will be loosely set to "zero". The frequency of rotational mode is 65 Hz, the frequencies of the in-plane modes are 195 Hz.

Figure 8 shows the dependence of translation and tilt frequencies on the position of the secondary proof ring over the range from -0.5 mm to +0.5 mm relative to the above "zero" position. From Figure 8, the translation frequency (105 Hz) is not affected by the position of corrective proof ring while tilting frequencies vary simultaneously from 104.4 Hz to 105.5 Hz. It is also seen that at "zero" position all three frequencies are practically the same.

In practice, however, because of the manufacturing tolerances, on the spot fine frequency tuning is needed. This is achievable in two stages. First by precisely regulating the driving frequency and minimizing dynamic translation, the driving frequency is matched to the frequency of translation mode. Second by axially displacing the secondary proof ring relatively to the primary proof ring, this will alter the moment of inertia of the aggregate proof mass without altering its weight, and minimizing the angular dynamic response. The tilting frequencies may be accurately adjusted without affecting the translation frequency, to which the driving frequency has been accurately matched during the first stage.

Figure 9 shows the variation of angular and translational responses during the above explained tuning procedure performed with MTDA having initial translational and tilting frequencies 105 Hz and 105.4 Hz, respectively. Firstl by varying the driving frequency from 104.5 to 105.5 Hz, we find the minimum of the translation occurring at approximately 105 Hz. Figure 9a shows dependence

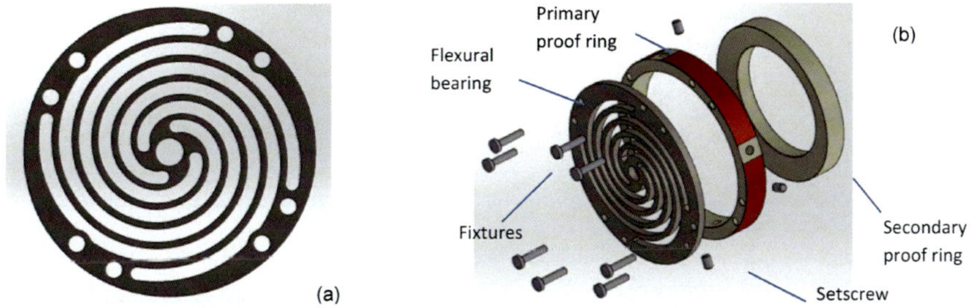

Figure 6. Design of MTDA

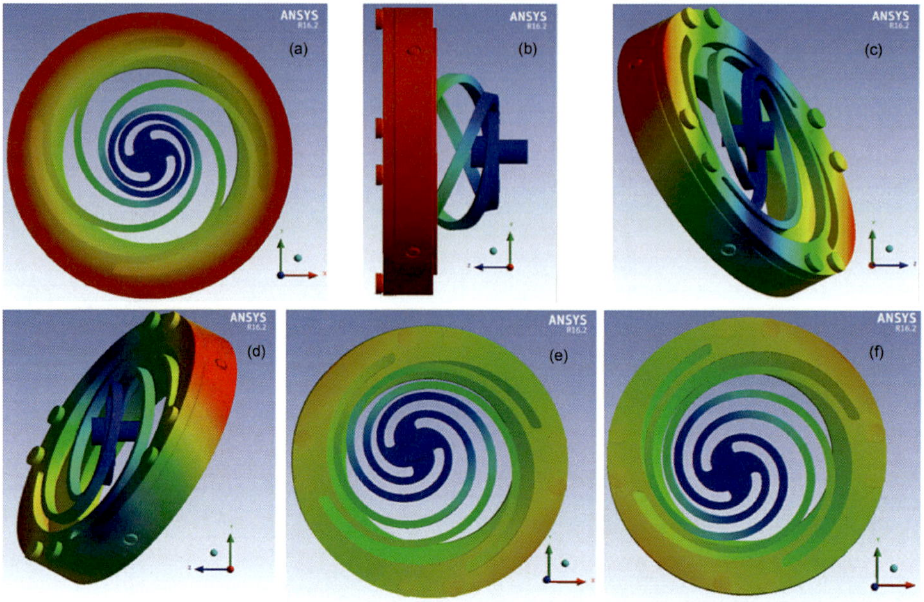

Figure 7. Dynamic modes

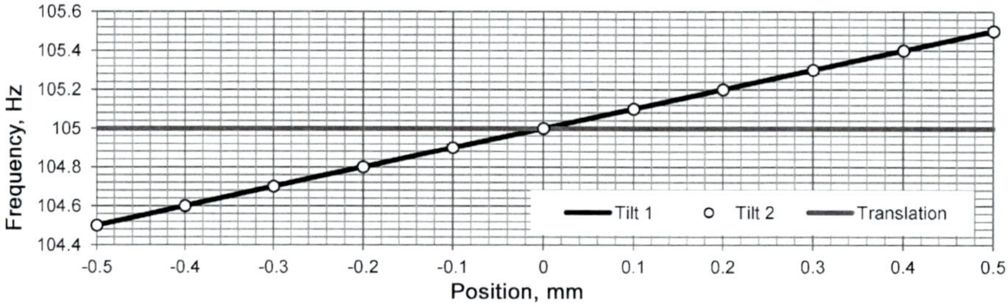

Figure 8. Resonant frequencies at different positions of corrective proof ring

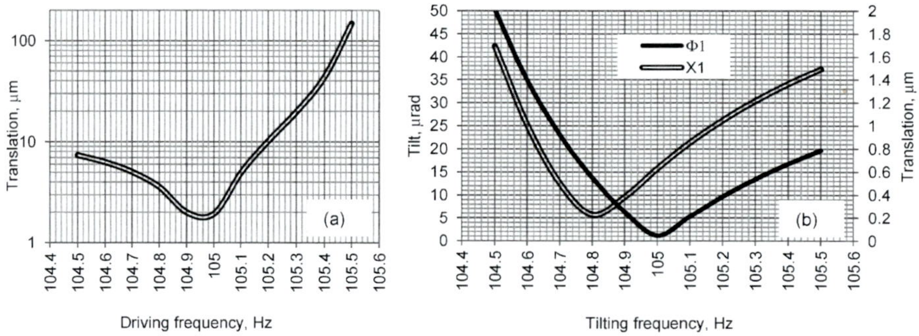

Figure 9. Scenario of tuning procedure

of payload displacement on the driving frequency with minimum of 1.93*mm* at 105Hz. Second at a fixed driving frequency of 105Hz, we axially displace the secondary proof ring and monitor the angular response using, say, an angular accelerometer. Figure 9b shows the appropriate dependence and optimum position of the secondary proof ring. From Figure 9b before tuning, the secondary

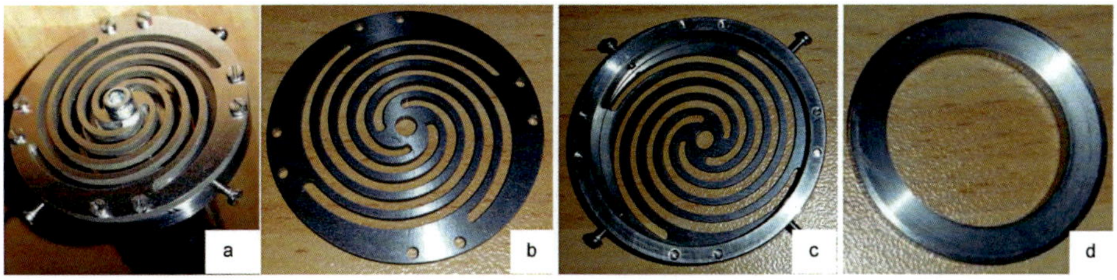

Figure 10. MTDA parts and assembly

proof ring is displaced from the optimum position by -0.5 mm; the linear response is 1.93 mm and angular response is 51 mrad. In the optimum position the angular response is 1.2 mrad and linear response is 0.6 mm. It is interesting to note that linear response is practically independent on the axial position of the correction proof ring.

FEASIBILITY STUDY: FINE MODAL TUNING

Figure 10 pictures the assembled (a) MTDA comprises of (b) flexural bearing, (c) primary proof ring with mounted flexural bearing and (d) secondary corrective proof ring.

Figure 11 shows the experimental setup for accelerance evaluation. The MTDA 1 is mounted upon the proof rod 2; the instrumented modal hammer 3 (Bruel&Kjaer Type 8206) is used for applying force (input) and accelerometer 4 (Endevko Type 752A12) is used for measuring resulting acceleration (output). The data acquisition and processing has been performed using dual-channel signal analyzer (Data Physics QUATTRO, not shown). As explained in [21], the location of typical antiresonance frequency in the module of accelerance of the combined system indicates the value of the resonant frequency of MTDA.

In Figure 12, the superimposed are spectra of moduli of translation and tilt accelerances. The typical anti-resonant notches corresponding to the translational and tilting frequencies of MTDA in both graphs are observed in the vicinity of the desired frequency 105 Hz. The secondary anti-resonant notch in the tilt accelerance corresponds to the in-plane mode at 193 Hz. By fine axially displacing the secondary proof ring precise matching the resonant frequencies is possible, as shown in Figure 13, where zoomed portions of the above accelerance are portrayed at different positions of the secondary proof ring, namely at -0.5 mm, 0 mm and +0.5 mm. It is worth noticing that the value of translational frequency is not affected by a position of the secondary ring.

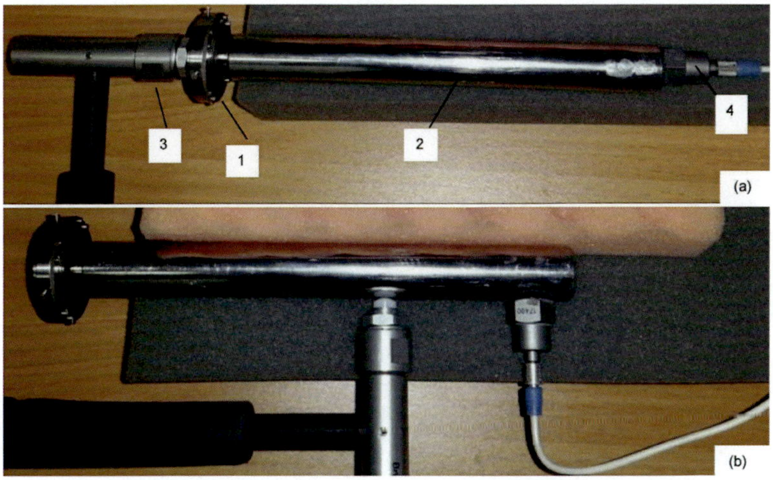

Figure 11. Evaluation of translation (a) and tilt (b) accelerances

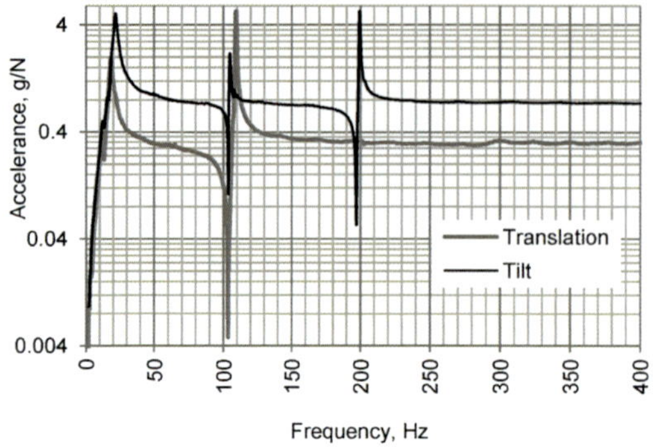

Figure 12. Tilt and translation accelerances

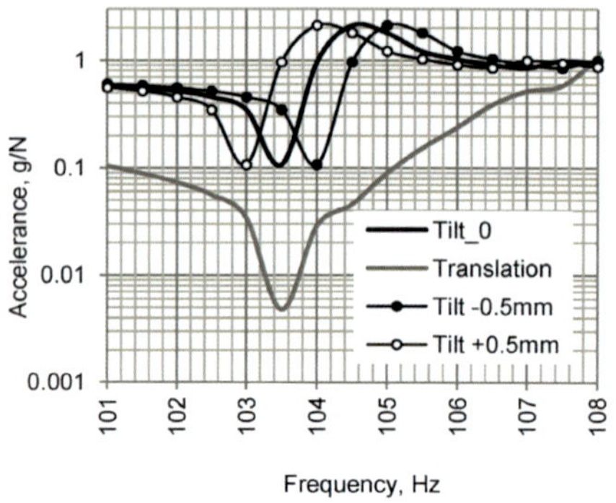

Figure 13. Fine frequency tuning

PRELIMINARY TESTING

Figure 14 shows the experimental setup. The miniature vibration exciter 1 is fixedly clamped on the top of vibration mounted platform 2 weighting approximately 2 kg. The MTTDA 3 is mounted collinearly but with essential (30mm) offset relatively to the exciter axis. The driving tonal signal is generated by the Data Physics QUATTRO signal analyzer 4 and amplified by the Ingenia Neptune digital servo drive (www.ingeniamc.com) 5. The Endevko Type 752A12 accelerometer 6 is mounted inline (b) and perpendicular to the shaker axis (c) and is used for measuring resulting dynamic response.

Figure 15 compares vibration spectra in axial (a) and perpendicular (b) directions with and with no MTDA . From Figure 15, in spite of the essential offset and small mass ratio, the impressive 70-fold vibration attenuation ratio has been achieved in both directions. In a tuned condition, the cooperative dual-mode operation of MTDA may be visually observed, as portrayed in Figure 16.

CONCLUSIONS AND FUTUTRE WORK

The analytical study has shown that MTDA has improved potential for essential attenuation of cryocooler induced line of sight jitter. The preliminary experimentation supported the approach to precise modal tuning. The authors are planning the full-scale experimental feasibility study.

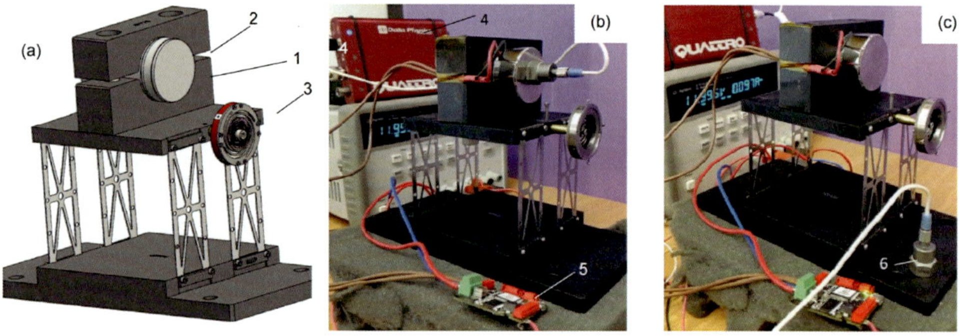

Figure 14. Experimental setup

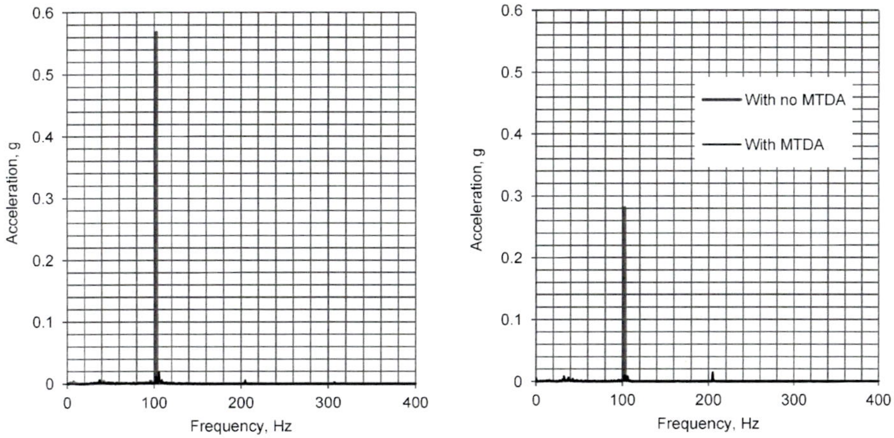

Figure 15. Experimental setup and attained performance

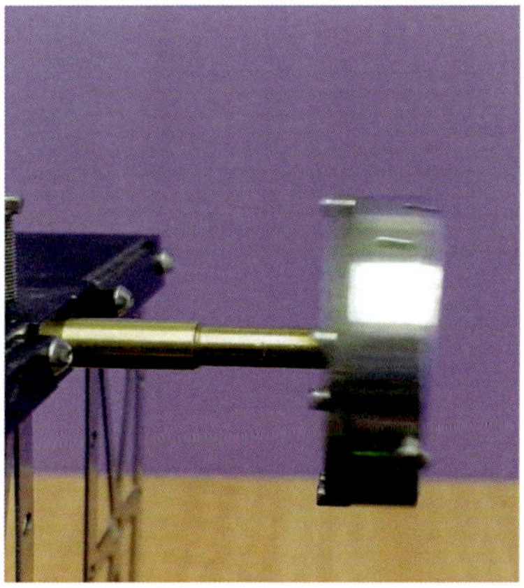

Figure 16. MTDA translation and tilt

REFERENCES

1. Rawlings, R. and Averitt, G., "A linear drive cryocooler for ultra-small infrared sensor systems," *Proc. SPIE* 9070, 90702R (2014).
2. http://www.drs.com/Products/RSTA/PDF/CD64012MWU.pdf
3. http://www.cinele.com/images/datasheets/ir_products/cryogenic_coolers3.pdf
4. http://www.cobham.com/about-cobham/mission-systems/about-us/life-support/davenport products/ cryocooling-systems.aspx
5. http://www.ricor.com/products/split-linear/k527/
6. Veprik, A., Babitsky, V., Pundak, N. and Riabzev, S., "Vibration Control of Linear Split Stirling Cryogenic Cooler for Airborne Infrared Application", *Shock and Vibration*, 7 (6), 363-379 (2000)
7. Veprik, A., Vilenchik, H., Pundak, N., "Microminiature linear split Stirling cryogenic cooler for portable infrared imagers," *Cryocoolers 14*, ICC Press, Boulder, CO (2007), pp. 105-115.
8. Veprik, A., Vilenchik, H., Riabzev, S., Pundak, N., "Novel microminiature linear split Stirling cryogenic cooler for portable infrared imagers," *Proc. SPIE*, 6542, 65422F (2007)
9. Veprik, A., Zechtzer, S. and Pundak N., "Compact linear split Stirling cryogenic cooler for high temperature infrared imagers," *Cryocoolers 16*, ICC Press, Boulder, CO (2011), pp. 121-132.

10. Veprik, A., Zechtzer, S. and Pundak N., "Split Stirling linear cryogenic cooler for a new generation of high temperature infrared imagers," *Proc. SPIE* 7660, 76602K (2010).

11. Veprik, A., Zechtzer, S., Pundak, N., Kirkconnell, C., Freeman, J. and Riabzev, S., "Adaptation of the low cost and low power tactical split Stirling cryogenic cooler for aerospace applications," *Proc. SPIE* 8012, 80122J (2011).

12. Veprik A., Riabzev, S., Avishay, N., Oster, D., and Tuitto, A., "Linear cryogenic coolers for HOT infrared detectors," *Proc. SPIE*. 8353, 83531V (2012)

13. Rühlich, I., Mai, M., Rosenhagen, C. et al., "Compact high-efficiency linear cryocooler in single-piston moving magnet design for HOT detectors," *Proc. SPIE* 8353, 83531T (2012)

14. Rühlich, I., Mai, M., Withopf, A. et al., "AIM cryocooler developments for HOT detectors," *Proc. SPIE* 9070, 90702P (2014)

15. Ashcroft, A.P., Richardson, L., Harji, J., and Giuliano, F., "Half-TV Format MWIR Sensor Module Incorporating Proximity Electronics," *Proc. SPIE* 8353, 83530J-1 (2012)

16. http://www.finmeccanicausa.com/capabilities/Eclipse.aspx.

17. Ross, B. and Olan, R.W., "Passive balance system for machines", USA Patent 5895033 A (1996)

18. Rosenhagen, C. and Rühlich, I., "Compensating oscillation device," European Patent WO 2014206542 A1 (2014)

19. http://www.aim-ir.com/fileadmin/files/Data_Sheets_Thermal_sights/2013_neu/2013_AIM_ daten-blatt_A4_HuntIR-K2_engl.pdf

20. http://elbitsystems.com/Elbitmain/files/LILY-L.pdf

21. Veprik, A., Tuitto, A., "Tuned dynamic absorber for split Stirling cryogenic cooler," *Proc. SPIE* 9821, Tri-Technology Device Refrigeration (TTDR), 98210F (2016)

Raytheon High-Frequency Pulse Tube Cryocooler Testing

T. Conrad, R. Yates, B. Schaefer, D. Kuo

Raytheon Space and Airborne Systems
El Segundo, CA 90025

ABSTRACT

Since the 1970s, Raytheon has developed, built, tested and integrated high performance cryocoolers. Our versatile designs for single and multi-stage cryocoolers provide reliable operation for temperatures from 10 to 200 kelvin with power levels ranging from 50 W to nearly 600 W. These cryocoolers incorporate clearance seals, flexure suspensions, hermetic housings and dynamic balancing to provide long service life and reliable operation in all relevant environments. Recently, Raytheon has developed an advanced regenerator technology capable of operating efficiently at high frequencies and outperforming traditional screen regenerators. The Raytheon Advanced Miniature (RAM-100) cryocooler, a flight packaged, high frequency, single stage pulse tube cooler with an integrated surge volume and inertance tube, has been designed for use with this regenerator. Design details and experimentally measured performance of two iterations of the RAM cryocooler are presented in this paper.

INTRODUCTION

Raytheon's Advanced Miniature (RAM) cryocooler was developed to provide the high reliability and thermodynamic efficiency typical of a space cryocooler with a lower cost and lead time than such systems have traditionally required [1]. This has been accomplished by emphasizing simplified and streamlined manufacturing processes in the RAM design. This approach is similar to that taken for the Raytheon Dual-Use Cryocooler (DUC) [2], but for the RAM design performance, robustness, and ease of integration were prioritized above pure ease of manufacturing. Additionally, the RAM cooler was developed to take advantage of the systematic benefits of a high operating frequency. These include a reduction of lower frequency exported disturbance, to which space systems are often sensitive, and a decrease in the relative size and mass of cryocooler components.

When dealing with cryocooler exported disturbance, the higher operating frequency of the RAM compared to other similar capacity cryocoolers results in several benefits for a systems integrator. First, the increased fundamental frequency results in fewer harmonics in the frequency range where structures and instruments are typically sensitive to disturbances. Additionally, soft-mount isolation systems are generally more effective at attenuating vibration as the source frequency increases. With a higher frequency source, these soft-mounts can be made stiffer for the same attenuation roll-off, resulting in more survivable soft-mounts and possibly eliminating the need for launch-locks.

To date, Raytheon has built and tested two flight-capable RAM Thermo-Mechanical Units (TMUs) that are able to operate between 90 and 100 Hz and deliver these operational benefits. They incorporate the Raytheon advanced regenerator [3], which has the potential to outperform

Cryocoolers 19, edited by S.D. Miller and R.G. Ross, Jr.

traditional regenerators in this frequency range, and have undergone extensive analysis during the design to ensure that they are robust and reliable.

MECHANICAL DESIGN AND SPECIFICATIONS

The RAM cryocooler is a split pulse tube design as shown in the photograph of Figure 1. The compressor is an Oxford-style design featuring dual opposed pistons for dynamic balancing, linear motors, a flexure suspension and non-contacting clearance seals. The expander module incorporates a concentric pulse tube and regenerator and an integrated surge volume and inertance tube. Additional aspects of both module designs are discussed in the paragraphs below.

The compressor architecture of the RAM cryocooler draws upon several recent design efforts, including the Low Temperature Raytheon Stirling/Pulse Tube hybrid (LT-RSP2) production compressor [4]. The RAM compressor incorporates clearance gap seals and flexure suspended moving mechanisms similar to other Raytheon space cryocoolers, but does away with piston position sensors in order to minimize package size and electronics complexity. Instead, piston position sensing and centering is performed using a technique based on the back-EMF of the motors.

The thermal, structural, and electrical interfaces of the compressor have been designed to simplify integration of the RAM cryocooler into an objective system. Heat rejection is accomplished through a single flange perpendicular to the mounting interface, which greatly simplifies attachment of heat pipes or thermal straps. The housing design allows simple three bolt mounting and incorporation of load washers while the RAM electronics are capable of utilizing either load washers or chassis-mounted accelerometers for vibration control feedback, allowing flexibility depending on the integration requirements. Industry standard connectors are installed for straightforward power and telemetry wiring. For robustness every sealed interface, with the exception of the transfer-line connecting the compressor and expander, is hermetically welded to eliminate the potential for leaks through handling or radiation exposure.

The RAM expander also incorporates a number of manufacturing and thermodynamic improvements, capitalizing on several years of Raytheon IRAD advancements. The cold-end tip incorporates two flat mounting interfaces, creating predictable, consistent thermal-strap and instrumentation conduction paths. Heat rejection is accomplished through either or both of the expander body sides, which incorporate bolt patterns for direct connection of heat pipes or thermal straps. The seamless design of the expander includes no external tubing or joints for inertance tube/surge volume attachments and the expander is designed to be mounted using a built-in vacuum interface, ensuring easy integration with instrument dewar structures.

Unit production and assembly of the RAM TMU is straight-forward and efficient, minimizing unnecessary critical path steps. Fabrication of the initial unit was completed very quickly with assembly of the compressor module requiring less than a week, including all curing, bake-out, and

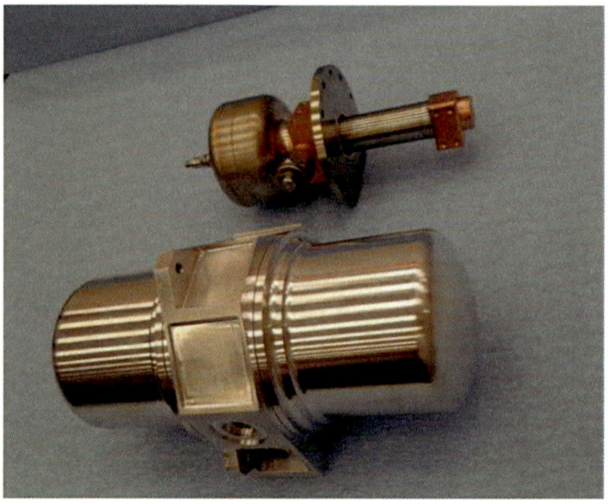

Figure 1. Photograph of RAM-100 Compressor and Expander.

Table 1. Comparison of RAM and DUC dimensions and masses

DUC TMU	Compressor	Expander	RAM TMU	Compressor	Expander
Mass	3.9 kg	1.4 kg	Mass:	2.8 kg	0.9 kg
Package Size			**Package Size**		
Length:	8.2 in	5.4 in	Length:	7.9 in	6.3 in
Width:	4.5 in	4 in	Width:	3.75 in	3.4 in
Height:	4.5 in	4 in	Height:	3.75 in	3.4 in

alignment processes. In quantity production, assembly is expected to be completed in a single day for this compressor design. Initial checkout and acceptance procedures were also completed quickly and without incident. The expander module, was designed with a simplified braze/weld schedule, removing several manufacturing steps found in previously designed coolers; this resulted in a significant decrease in assembly time and reduction of critical-path supply risks.

Size and Mass

The increased operating frequency of the RAM cryocooler relative to its predecessor, the Raytheon DUC, resulted in reductions size and mass for both the compressor and expander modules. These reductions, totaling approximately 1.6 kg of mass, were accomplished while integrating thermal, electrical, and structural interfaces which were mostly not incorporated into the DUC design. The lone dimension to grow from the DUC to the RAM was the expander length, which increased due to the integration of the surge volume and inertance tube into the expander body.

Structural Analysis

Extensive structural analysis was performed during the development of the RAM cryocooler in order to ensure robustness and reliability. This included industry standard pressure vessel, modal, thermo-elastic, and radiation analyses as well as comprehensive random vibration, quasi-static loading, shock, and combined loading analyses. To guarantee that the RAM-100 would meet most future shock and vibration requirements, a 300 g inertial load was used for the purposes of verifying shock and as a bounding case on random vibration analysis. Any shock event will be strictly dependent on the particular system in which the event happens, however several design features of the RAM-100 compressor and expander result in a relative lack of susceptibility to a shock environment. The TMU components were also examined against a random vibration profile of .2 g^2/Hz across all resonant frequencies. In all of these analysis cases, the compressor and expander of the RAM-100 cryocooler showed positive margins.

Electronics

The RAM cryocooler is designed for compatibility with the Raytheon Cryocooler Thermal Control Electronics (CTCE) as well as third party, tactical style cryocooler electronics.

MECHANICAL DESIGN AND SPECIFICATIONS

Initial thermodynamic characterization of the RAM cryocooler was completed in 2015, the results of which are shown in the Ross plot in Figure 2. This initial testing was performed at a 91 Hz operating frequency with a heat reject temperature of 295 K. Because the first planned application for the RAM cooler requires cooling at ~150 K, the first expander included a higher flow regenerator more suitable for warmer cold tip temperatures. As a result, the first RAM cooler performed well at cold tip temperatures of approximately 100 K and above but exhibited reduced performance at lower temperatures. The less restrictive regenerator likely also resulted in the observed reductions in the optimal operating frequency and maximum input power, both due to a reduction in the pressure ratio and resulting gas spring contribution to the compressor resonant frequency.

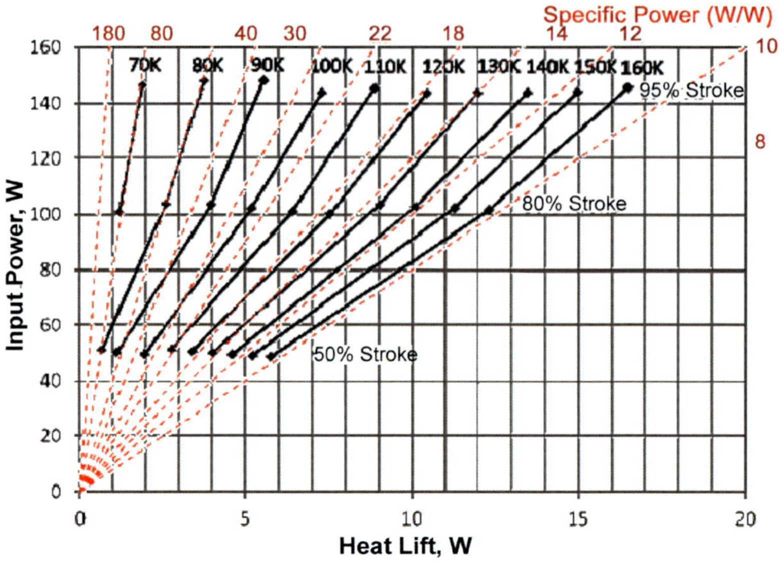

Figure 2. Ross plot for RAM cryocooler with regenerator optimized for cooling above 100 K.

A brassboard RAM expander has also been constructed which incorporates a regenerator optimized for lower temperature operation. Testing of this expander with a production RAM compressor was performed in early 2016; the results of this testing are shown as a Ross plot in Figure 3. This expander significantly outperformed the previous expander, which had been optimized for higher temperature operation, at temperatures below approximately 100 K and achieved a no-load temperature of 42 K. The brassboard testing was performed at an operating frequency of 100 Hz and a reject temperature of 300 K. A production version of this lower temperature expander, to be completed later this year, will incorporate the same design features, geometry, and robustness as the current unit; the only changes to be made are to the design parameters of the regenerator. Due

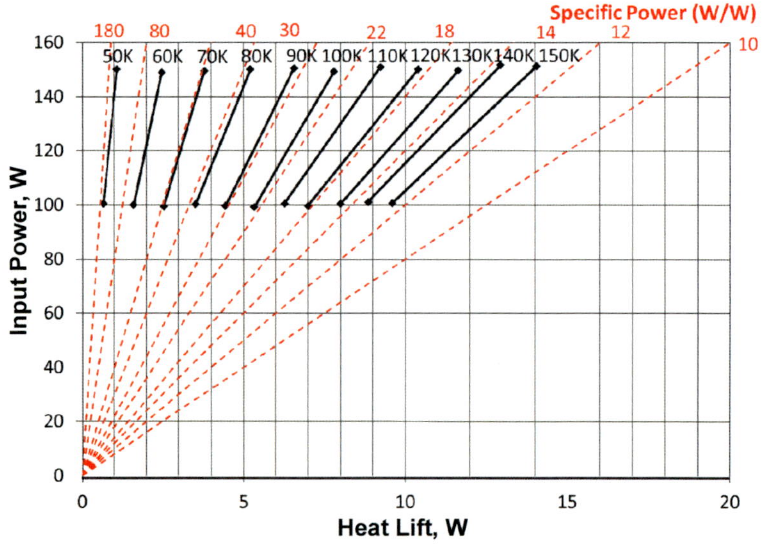

Figure 3. Ross plot for brassboard RAM cryocooler with regenerator optimized for cooling below 100 K.

to slight geometry differences between the brassboard and planned production unit, the production expander is predicted to operate optimally at 100 Hz.

CONCLUSIONS

The RAM cryocooler has been developed as a miniaturized, flight ready TMU capable of capitalizing on the system benefits of high frequency operation. It incorporates many legacy design features from other Raytheon space cryocoolers and has been designed with robustness, reliability, and ease of integration in mind. Initial testing of a production RAM cryocooler has been performed along with additional testing of a brassboard expander optimized for lower cold tip temperatures.

ACKNOWLEDGMENT

The authors would like to thank the Air Force Research Laboratory for their broad support of the DUC system development which preceded and led to the RAM cryocooler.

REFERENCES

1. Conrad, T., Schaefer, B., Yates, R., Bellis, L., Bruckman, D., Im, Y., Pillar, M., and Barr, M., "Raytheon Productized High Frequency Cryocooler," *Cryocoolers 18*, ICC Press, Boulder, CO (2015), pp. 135-139.

2. Hon, R., Kesler, C., Sigurdson, D., "Integrated Testing of a Complete Low Cost Space Cryocooler System," Cryocoolers 15, ICC Press, Boulder, CO (2009), pp. 125-132.

3. Schaefer, B.R., Bellis, L., Ellis, M.J., and Conrad, T.J., "Advanced Regenerator Testing in the Raytheon Dual-Use Cryocooler," *Adv. in Cryogenic Engineering*, Vol. 59, Amer. Institute of Physics, Melville, NY (2014), p. 570

4. Schaefer, B.R., Bellis, L., Conrad, T., Bruckman, D., Yates, R., Pillar, M., and Barr, M., "Raytheon Low Temperature RSP2 Production Program," *Cryocoolers 18*, ICC Press, Boulder, CO (2015), pp 19-25.

Qualification of a European Large Pulse Tube Cooler System for Space Applications

T. Wiertz, J. Urbano

Air Liquide Advanced Technologies
Sassenage, France

ABSTRACT

Since 2010, Air Liquide Advanced Technologies has been going through the process of preparing and conducting the qualification of the Large Pulse Tube Cooler (LPTC) for earth observation and meteorological infrared instruments. It is the first European space cooler based on pulse tube technology. The LPTC consists of a compressor, a pulse tube and the drive electronics. It ensures cooling of infrared detectors at temperatures between 50K and 80K with cooling power between 2W and 7W.

The tasks addressed during qualification include product industrialization; verification engineering; clean room facilities set up; test bench design, manufacturing and acceptance; test engineering; qualification test conducting and test data processing. Qualification scope not only covers the integrated system, but also the process and the components.

Space coolers are exposed to qualification risks because of their complexity. In fact, they are among the pieces of space hardware addressing the largest spectrum of technical disciplines: mechanics, heat transfer, fluid, thermodynamics, fracture mechanics, mechanisms, micro-vibrations, power electrics, analog and digital electronics, EMC, radiation, etc.

Thanks to the involvement of the highly skilled space coolers team, we are now achieving successful qualification of the LPTC for its first application program. The events and achievements of this qualification, together with lessons learned will be presented.

INTRODUCTION, A BRIEF HISTORY OF AIR LIQUIDE SPACE COOLERS

Air Liquide began space cryogenics activities in the early 1960's and has been a key design and manufacturing player of cryogenic propellant tanks for all generations of Ariane launch vehicles. Capitalizing on its cryogenic launch vehicle activities, Air Liquide began working on orbital cryogenics projects in the 1990's.

This effort led to three main achievements on Eurpoean missions and contributions to the ISS:

- MELFI freezer onboard the ISS. Air Liquide designed and qualified the Turbo Brayton machine cooling the MELFI to -80 °C. Today, three of these cryocoolers are onboard the ISS, providing reliable cooling for almost ten years.
- HERSCHEL Infrared Space Observatory. Air Liquide performed the detailed design, manufacturing and testing of the superfluid helium cryostat for Herschel, its cryogenic lines, thermal links and cold shields.

Cryocoolers 19, edited by S.D. Miller and R.G. Ross, Jr.

- PLANCK cosmic microwave background mapping satellite. Air Liquide engineered and qualified the 0.1K dilution cooler, its cryostat and launch lock device, and its gas system.

Capitalizing on these successes, Air Liquide began developing CRYOSYSTEM, a 90K cooler for the ISS. This cooler was based on a Stirling cooler provided by Thales Cryogenics BV. Development went on to a successful PDR and extensive tests on a Development Model (DM). Reduction of ISS funding and Columbia Shuttle tragedy led to the termination of the project.

At the same time, need for several infrared instruments onboard future European missions emerged and ESA and CNES R&D effort were engaged towards the development of new coolers based on pulse tube technology. Two products emerged from this work:

- MPTC (Miniature Pulse Tube Cooler), a cooler capable of 50 W of electrical input power, well adapted for 80K – 150K applications.
- LPTC (Large Pulse Tube Cooler), a cooler capable of 160W electrical power, well adapted to 50K – 80K applications.

Both coolers incorporate a compressor designed and manufactured by Thales Cryogenics BV and a pulse tube based on CEA/SBT heritage. Space applications finally came for the LPTC for which Air Liquide now holds 3 orders totaling 24 flight cooler systems.

LPTC COOLER DESIGN DESCRIPTION

LPTC consists of a complete cooler system, including a mechanical cooler, drive electronics, harness and sensors. The mechanical cooler is a split configuration with a double split pipe (compressor – cold finger split pipe and inertance pipe). Inertance and buffer have been placed around one of the compressor motors for better compactness. Split pipe geometry can be adapted to each customer configuration. Figure 1 shows cryocooler main parts, excluding the harness.

The pulse tube is a coaxial design for compactness and for ease of the thermal interface. The regenerator tube is protected by a passive cold shield and mechanically supported by a launch support. The hot thermal interface is a flat surface opposite the cold end. The pulse tube contains a redundant cryogenic temperature sensor embedded in the cold tip.

The compressor design is dual opposed pistons. The pistons are driven by linear moving magnet motors, with external static coils which are not in contact with the Helium. Such a design eliminates the main contamination source and the need of electrical feed-throughs, which are a major risk of

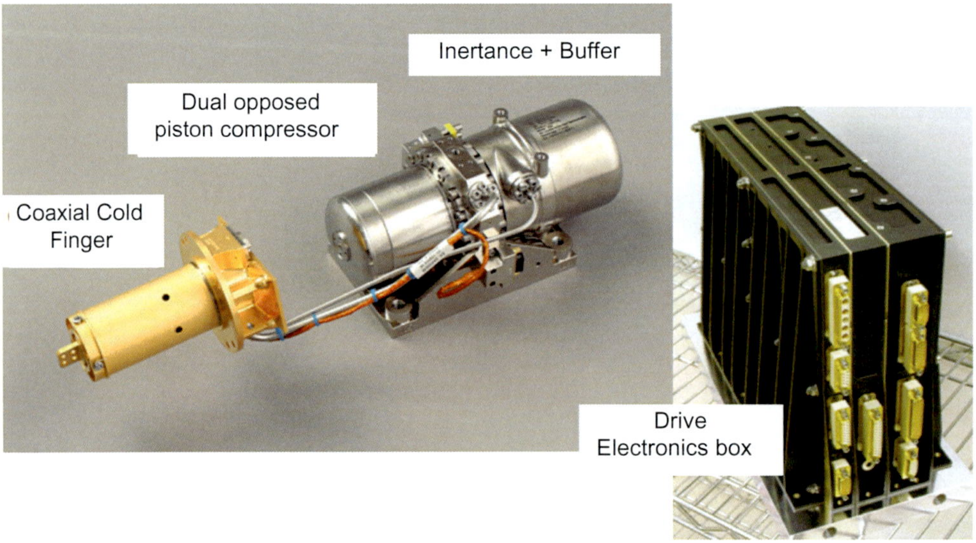

Figure 1. LPTC system overview

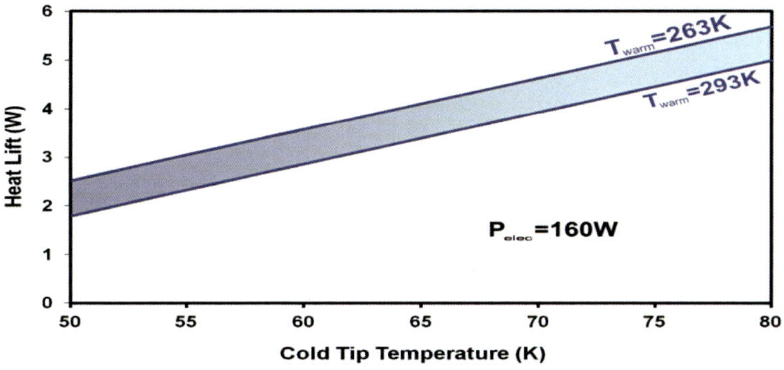

Figure 2. LPTC cryogenic performance

leakage. The use of static coils suppresses the risk of failure due to a breaking of flying lead connections and wires. Compressor design was optimized to limit radiated magnetic field.

Drive electronics design can be highly dependent on the satellite interfaces (power bus, communication, etc.). So far, two different electronic units have been developed.

Maximum power supply to the compressor is 160 Wrms. With this power setting, and depending on the reject temperature, the cryogenic power is between 2 W and 7 W, as depicted in Figure 2. Together with this performance, emphasis was put on the following performances of the system:

- Temperature stability control loop through active power adaptation. This feature allows optimal efficiency and satellite power consumption together with very stable cold tip conditions and extremely good reaction to external conditions changes. This feature is particularly interesting for low earth orbit payloads performing observation maneuvers.
- Lifetime. The system was designed to a lifetime covering at least 15 years of storage followed by 10 years of continuous operation. A redundant system offers 98% reliability over a 10 years operation.
- Active micro vibration reduction. Like temperature control, micro vibration reduction is achieved through an active control loop able to react to any change in the cooler operating conditions, thus offering optimal availability of a vibration free environment to the payload.

QUALIFICATION CHALLENGES

Qualification of the LPTC represents several challenges for the Air Liquide team. Three orders have been placed for different payloads in a short time frame. The schedule obliges Air Liquide to run development and qualification in parallel. One of the developments is aimed at the Meteosat Third Generation (MTG) run by ESA, which requires a full qualification. This represents a constraint but also an advantage for the other programs. It was decided to achieve qualification on two programs only and to supply the third with an off the shelf coolers, based on the existing qualified cooler. This Qualification By Similarity (QBS) policy will be applied to future orders.

The main challenge lies in the scope of the provided system. It is the first time that our team is responsible for the whole system, including drive electronics. This not only consists of the electronics control unit qualification, but it also addresses the entire operational aspect of the system, including temperature and vibration control loops, operational procedures, TM/TC and data management.

Space cryogenic cooler qualification is not simple in that it implies a lot of technical disciplines. Some of these were new to the team, mainly giga cycle fatigue and micro vibration. These add to all other technical challenges posed by thermo-mechanical structures. On top of these fields, all know how related to electronic components and electronic design was also new to the team.

PRELIMINARY DEVELOPMENTS AND ENGINEERING MODELS

ESA and CNES strongly supported early development of a European pulse tube cooler. Conscious of the most critical aspects of a new technology qualification, they promoted work emphasizing the lifetime demonstration and micro vibration reduction. This came in addition to more classical R&D work, dealing with cryogenic performance optimization and preliminary environment tests.

Two mechanical coolers were placed in endurance testing in 2007. A third was placed in endurance testing in 2014, and the first qualification model should join the program very shortly. The main challenge of endurance testing is to keep the coolers running. Hopefully, they are reliable and can run smoothly for years. In fact, smooth running of the test bench and control of test environment are the main challenges of endurance testing. In order to be able to notice a drift in cooler efficiency, one needs to be able to run it in very constant conditions during long runs. The refurbished endurance test room and revamped test benches allow us running the coolers continuously, with just one week maintenance per year. Endurance test room is shown on Figure 3. Improved environment control in the test room allows easier monitoring of the cooler performance. With more than eleven years of operation cumulated on three coolers, we have so far noticed no performance degradation.

PROCESS AND COMPONENTS QUALIFICATION

For the purpose of Meteosat Third Generation, one of the two programs on which LPTC qualification is achieved, all processes and components qualification have to comply with the European standards, ECSS for design rules and process and ESCC for components. These standards are particularly demanding but provide a very solid and versatile reference for the product.

Qualification activities on process and components often include engineering of qualification tests, design and manufacturing of qualification parts and jigs, qualification samples preparation and testing. Many tests, particularly those involving gas and cryogenics could be done on Air Liquide testing facilities: thermal shocks, burst tests, strength tests at low temperature, etc.

In several cases we had to deal with suppliers not familiar with space qualifications and make sure that we could implement process and part traceability. This was an additional difficulty and requires a reinforced follow up of these suppliers all through the program, including flight models production phase.

Main process and components qualification needed for the purpose of LPTC qualification were:

- Vacuum brazing and EB welding for the cold finger
- Gold plating on aluminum, titanium and copper parts
- Metallic seals
- Temperature sensors
- Hermetic connectors
- Force transducers

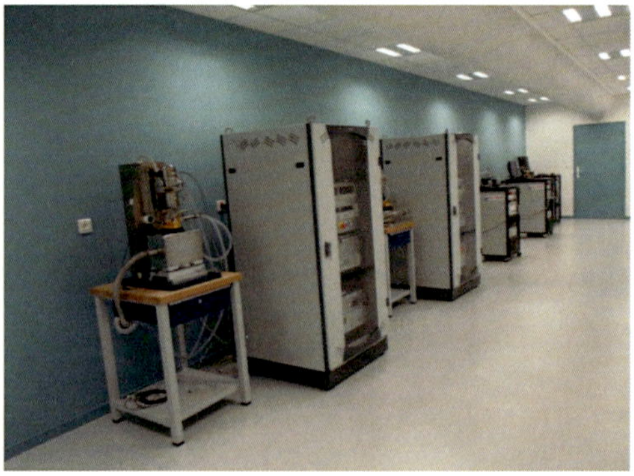

Figure 3. Endurance test room

Figure 4. Giga cycle tests on regenerator tubes

Giga cycle qualification

Giga cycle regime is a special field of fatigue analysis, where the classical S-N curves do not apply. It was shown that fatigue threshold can be significantly lower than 10^7 cycle S-N curve asymptote. The regenerator tube is subjected to pressure waves throughout the operating life of the cooler. For a 10 year required lifetime, this represents several tens of billions cycles. So giga cycle fatigue regime is applicable.

An accelerated test was performed on three regenerator tube assemblies, using piezo actuators to apply an axial load to the regenerator tube. Test set up is shown on Figure 4. To cope with the absence of circumferential load cycles, axial load was increased. Test acceleration obtained this way allowed applying 10 year life within 4 months testing time. The assemblies used for testing allowed validating the regenerator tube itself (thin titanium tube), brazing of the cold tip and EB welding of the tube flange for giga cycle strength. No failure occurred on the 3 tested samples.

SUBSYSTEMS QUALIFICATION

For the development of LPTC system, two main parts were subcontracted:

- Compressor assembly was contracted to Thales Cryogenics BV
- Electronics was contracted to SITAEL from Italy or CRISA from Spain, depending on the program application.

Customer requirement specifications were flown down to suppliers and sub assemblies were subjected to a completed PDR – CDR – Qualification Review development cycle. Suppliers were also responsible for qualification of their process and parts.

Main final qualification tests consist of mechanical vibration and shock, and Thermal Vacuum Test (TVT). Generally, these tests are also done at system level. They require complex test facilities not available in house. In order to optimize schedule and reduce risks, some smart arrangements have been made for performing these tests at sub system level e.g.

- Vibration tests at compressor level were split between sub system and system level, with only one test performed at compressor level to check main resonance mode. This shortcut was possible thanks to earlier tests performed on compressor engineering models, which brought great confidence in a positive outcome. Shock tests were also done only at system level.
- For electronics, vibration tests and TVT tests were done using the same facilities and the same test team as for system tests. Vibration tests on electronics are not repeated at system level. Electronics vibration test set up is shown on Figure 5.

QUALIFICATION ACTIVITIES AT SYSTEM LEVEL

Qualification tests campaign at system level represents the most problematic part of the qualification. For LPTC system, the retained programmatic approach for qualification was an EQM model.

EQM model is identical to flight model, with some possible relaxation on component screening. It is not refurbishable to flight standard.

AIT facilities

Last orbital hardware assembled and tested at Air Liquide was HERSCHEL – PLANCK material, which was delivered long ago. With the forthcoming AIT of a large quantity of flight coolers, the installations used for HERSCHEL had to be deeply refurbished and extended. Moreover test benches dedicated to cooler performance were designed and built.

AIT facilities now consist in a $65m^2$ ISO5 clean room, with molecular contamination control and humidity regulation. This clean room is populated with one TVT test bench and two micro vibration test benches. Also included in the clean room are bake ovens, bake out chamber, parasitic heat loss measurement bench and clean, dry storage cabinets. Two flight models have been delivered so far. To sustain the foreseen AIT ramp up, an extension of clean room to $110m^2$ is under way, with additional TVT and micro vibration test benches.

Qualification test campaign

Qualification test campaign at system level consisted in running test in the following order:

- Reference cryogenic and micro vibration test (mechanical cooler only)
- Vibration tests at mechanical cooler level
- Cross check to reference test
- Electronics incoming tests
- Reference cryogenic test at system level
- TVT tests
- Cross check to reference test
- EMC tests
- Control loops settings
- Final cryogenic and micro vibration performance test

This test campaign differs from the flight acceptance by little, but was conducted more thoroughly, e.g. by performing the tests over a wider operating frequency range, and by checking more functionalities in the system. It was also longer to perform because most tests procedures were run for the first time.

Environment test campaigns were rather uneventful, except incidents reported here below. Vibration tests were performed externally. To maintain cleanliness of the system, it was assembled

Figure 5. Vibration tests on the electronics box (with prototype cooler)

inside the clean room with the vibration jig. TVT test was performed with electronics, applying 5 independent temperature profiles on the compressor, cold finger, electronics, general radiation environment and cold finger radiation environment. TVT test set up is shown on Figure 6. Performance check before and after these environment tests showed no drift of the performances due to application of extreme conditions.

A big part of the work, not necessarily expected, was the functionalities verification and the settings of the regulation loops. Because LPTC is aimed at very diverse applications, functionalities offered to the end user and performance of the control loops were both maximized. The number of parameter accessible to the user for cooler operation exceeds 80. Qualification of the cooler implies to verify all functionalities and to explore the settings of control loops and fully understand the sensitivity of the system to user parameters. A few examples:

- Test scenarios had to be devised to test system protections against excessive voltage or current
- Temperature regulation is ensured by adapting compressor power to cope with varying external conditions (reject temperature variations, heat lift variations). Vibration reduction active control directly reacts to power variations triggered by temperature regulation. Such interaction between both regulation loop needs to be taken into account when optimizing performance
- Electrical parameters vary slightly from one cooler to another. Voltage protections then need to be precisely set in order to optimize performance (e.g. minimizing cool down time) while ensuring reliable protection of the cooler.

Incidents we had to face during qualification

The number of tests and system complexity are such that there is invariably some time, something that will go wrong during qualification. Although engineering tests have prepared the qualification and lowered the risks, one must expect incidents during the qualification and be ready to react. Two interruptions occurred during the qualification campaign:

- During vibration tests, we experienced pipe failure due to unexpectedly high dynamic response. A design modification was decided and tested at component level. It was then applied to the qualification model and the test campaign was resumed with a complementary vibration test.

Figure 6. TVT test bench with qualification model (vacuum vessel not visible)

- Shortly after TVT test, a capacitor in the electronics experienced arcing and was completely vaporized. No single point failure design made that this event went undetected for some time. It was finally detected during handling of the electronics box. The box had to be sent back for opening and repair.

In both cases, the engineering teams' responses to the incident were quick and efficient. Nevertheless, qualification schedule was extended by several months due to both incidents.

CONCLUSIONS

Qualification of the LPTC represented several challenges for Air Liquide. It was a technological premiere. It represented an extension of our scope of work compared to former projects with inclusion of electronic control unit,. A large quantity of orders for the product made qualification time critical and to be run in parallel with preparation of flight hardware AIT ramp up.

Despite several incidents, qualification is now complete on the first application program, opening the path to flight models delivery, which occurred very shortly afterwards.

A great tribute has to be paid to our engineering and qualification test team, and to the teams of our suppliers for this successful achievement.

Second qualification campaign on Meteosat Third Generation coolers is now on the rails and will allow achievement of a broad qualification scope for the LPTC.

REFERENCES

1. L. Sentis, J. Delmas, P. Camus, G. Guyot, Y. Blanc "Cryogenic tests of a 0.1K Dilution Cooler for Planck HFI," *Cryocoolers 13*, Springer US (2005), p. 533.
2. T. Trollier, C. Aubry, J. Butterworth, S. Martha, A. Seidel, L. De Parolis "Cryogenic Tests of a Development Model for the 90K Freezer of the International Space Station," *Cryocoolers 13*, Springer US (2005), p. 661.
3. T. Trollier, A. Ravex, I. Charles, L. Duband, J. Mullié, P. Bruins, T. Benschop, M. Linder "Performance Tests Results of a Miniature 50 to 80K Pulse Tube Cooler," *Cryocoolers 13*, Springer US (2005), p. 93.
4. T. Trollier, J. Tanchon, J. Buquet, A. Ravex "Status of Air Liquide Space Pulse Tube Cryocoolers" *Cryocoolers 15*, ICC Press, Boulder, CO (2009), p. 115.
5. J. Tanchon, T. Trollier, Y. Icart, A. Ravex, C. Chassaing, J. Butterworth, C. Daniel, R. Briet "150K Pulse Tube Cooler for Micro Satellites," *Cryocoolers 17*, ICC Press, Boulder, CO (2013), p. 53.
6. J. Butterworth, G. Aigouy, C. Chassaing, B. Debray, A. Huguet "Pulse Tube Coolers for Meteosat Third Generation," *Adv. in Cryogenic Engineering*, Vol. 59, Amer. Institute of Physics, Melville, NY (2014), p. 520.
7. C. Bathias, L. Drouillac, P. Le François "How and why the fatigue S-N curve does not approach a horizontal asymptote," *International Journal of Fatigue*, Vol. 23 (2001), p. 143.

LPT6510 Pulse Tube Cooler for 60-150 K Applications

R. Arts[1], J. Mullié[1], J. Tanchon[2], T. Trollier[2]

[1]Thales Cryogenics B.V., Eindhoven, The Netherlands
[2]Absolut System SAS, Seyssinet-Pariset, France

ABSTRACT

For a number of future instruments, such as the OCO-3 instrument, there is an anticipated need for a high-efficiency, small-scale pulse-tube cryocooler for the 60-150 K operating regime.

The LPT6510 cryocooler, based on the Thales Cryogenics MPTC compressor originally developed under ESA TRP and the Absolut System SSC80 pulse-tube cold finger, was designed to fill this niche in an elegant manner. In this paper, an update is given regarding the ongoing development of this cryocooler.

Key performance data as well as an overview of the Manufacturing, Assembly, Integration and Test (MAIT) processes as compared to the LPTC compressor heritage is shown. A novel new design for the buffer-inertance assembly will be presented. The presentation will conclude with a discussion of various aspects of the integration. This will include the heat sink and mounting design as well as the induced vibration characteristics of this cooler when operated with and without active vibration control.

INTRODUCTION

The need for a compact pulse-tube cryocooler for 60-150 K space applications has long been anticipated, which resulted in the Thales MPTC compressor being developed under ESA funding [1]. Whereas the MPTC cooler developed under this ESA contract has not been used in satellite programs to date, a commercial spin-off has been produced as the Thales LPT9510 [2]. This cooler has subsequently been series-produced and is even regarded as a viable option for low-cost flight missions, such as TacSat-3 [3]. This has resulted in attention from several interested parties in using the LPT9510 for low-cost flight programs [4, 5, 7].

However, when examining the performances of the LPT9510 it becomes apparent that the cooler was developed as a compromise between cost and efficiency. For space applications where power budgets are tight, a more efficient alternative is preferred.

Therefore Thales Cryogenics B.V. has partnered with Absolut System SAS to develop a designed-for-space, high efficiency pulse-tube cooler, based on the MPTC compressor developed under ESA funding (Figure 1), and the SSC80 pulse-tube developed originally for a high-efficiency terrestrial astronomy application (Figure 2).

DESIGN OUTLINE

Early in the development, a working point was defined for the cryocooler. As the development was not performed against a specific requirement, a generalized requirement was formulated. A

Figure 1. MPTC compressor

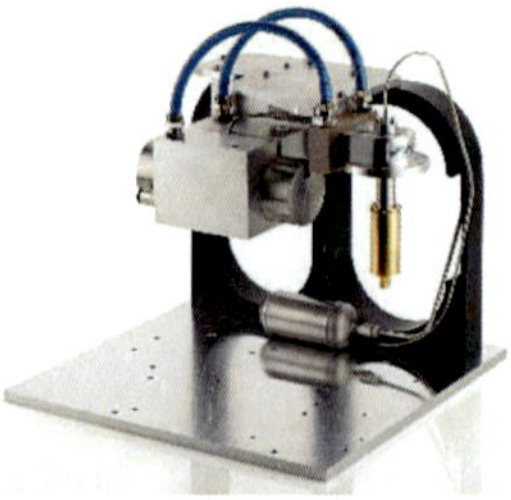

Figure 2. SSC80 pulse-tube for IPAG

working point of 100-110 K was initially targeted for the optimization of efficiency, with a maximum heat lift of 3 W and a system mass below 3 kg. Exported vibrations should be in the 100 mN range for most satellite applications, with the use of an active vibration reduction (AVR).

In order to determine the design parameters, several trade-off studies were performed. For the cryocooler compressor, this resulted largely in the same choices as those initially made for the MPTC compressor: an aluminum center part, with Titanium alloy motor shells.

Another aspect of the trade-off was the position of the inertance tube and buffer volume of the pulse-tube cold finger. This resulted in a design where the inertance and buffer volume were integrated into the compressor.

Finally, the choice was made to mount the cold finger directly on the compressor, rather than using a transfer line between them. This has the advantage of eliminating an interface, as mechanical mounting and heat sinking can all be done through the aluminum center part of the compressor. Furthermore, removing the transfer line has an advantage for cryogenic performance.

For applications where positioning freedom of the cold head with respect to the compressor is required, a specific version with a transfer line can be produced. This is not needed in many space applications, as a thermal strap is typically used to connect the cold finger tip to the object to be cooled, providing freedom in positioning and mechanical isolation.

An artist impression of the proposed design can be seen in Figure 3.

BREADBOARD DEMONSTRATOR

In order to validate the design concept, a breadboard demonstrator model was built based on existing building blocks with minimal modifications. A photograph of this demonstrator can be seen in Figure 4.

A test campaign was conducted to provide preliminary validation of all key performances. First, the heat lift and efficiency were tested. All performances obtained were measured with multi-layer insulation around the pulse-tube and heat sinking through the compressor baseplate. Results can be seen in Figure 5 and Figure 6.

A well-known effect in pulse-tube coolers is the sensitivity of performance to the warm side temperature of the cold finger. As the heat sinking concept was changed compared to a standard split pulse-tube cooler, the temperature difference between the compressor base plate and the pulse-

Figure 3. Artist impression of final system.

Figure 4. Demonstrator model.

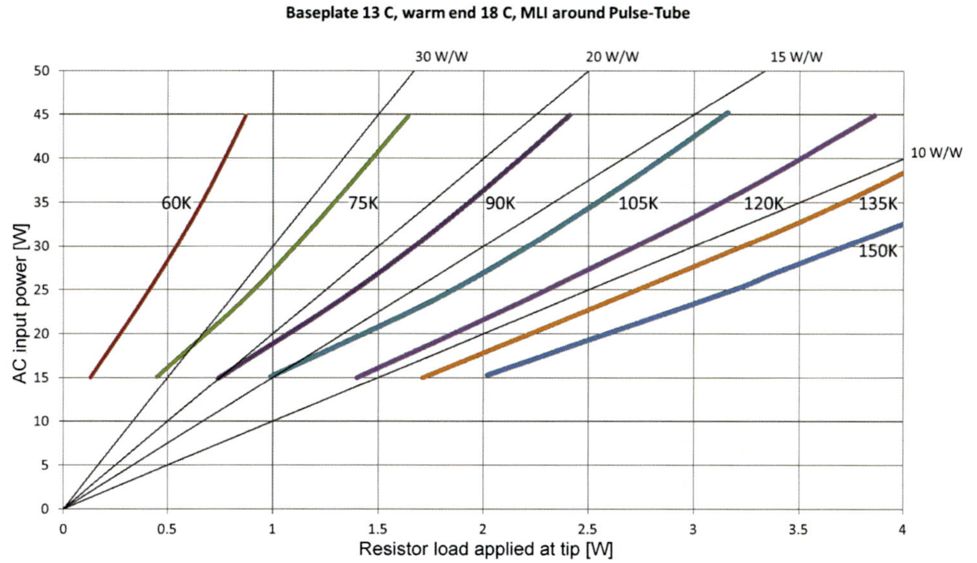

Figure 5. Heat lift at 13 C base plate temperature

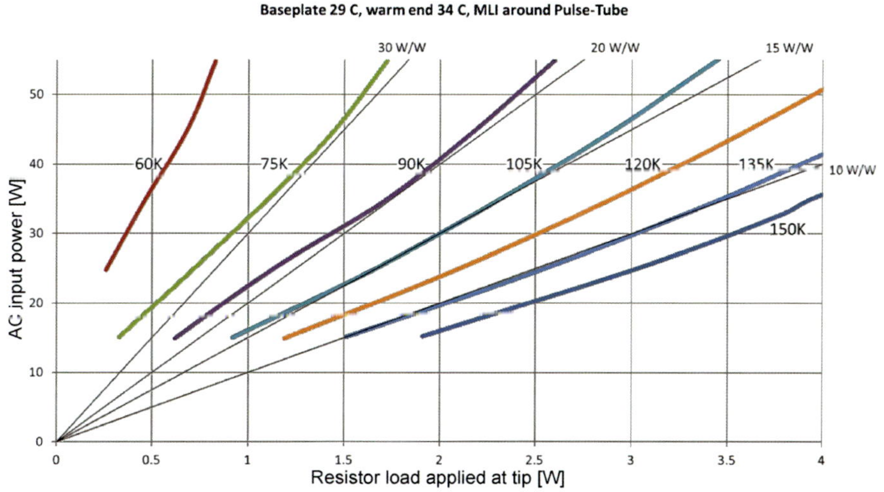

Figure 6. Heat lift at 29 C base plate temperature.

tube warm side was measured. The test was done in a climate chamber at normal atmosphere, with the entire cooler wrapped in multi-layer insulation to ensure heat rejection through the base plate. The result can be seen in Figure 7. It can be seen that, even with a breadboard setup based on the modified existing hardware, the thermal contact from the cold head to the compressor center part is sufficient to ensure a ΔT of 5 K at 45 W input to the cooler. In the final design, it is expected that this still can be improved.

A key requirement for any designed-for-space cryocooler is the exported vibration. As an initial characterization of the exported vibrations signature, the breadboard cooler was suspended from springs and fitted with a three-axis accelerometer. The vibration signature of the cooler was measured with a standard drive signal and with active vibration cancellation, in both cases using the Thales CDE7232 bench-top drive electronics. The measurement data can be seen in Figure 8 and Figure 9. It can be seen that using active vibration cancellation, the exported vibrations could

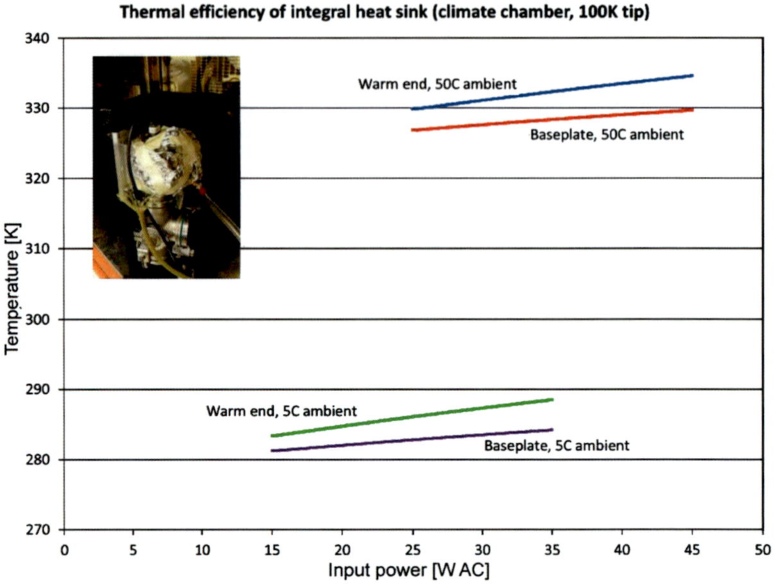

Figure 7. Efficiency of heat sinking through base plate

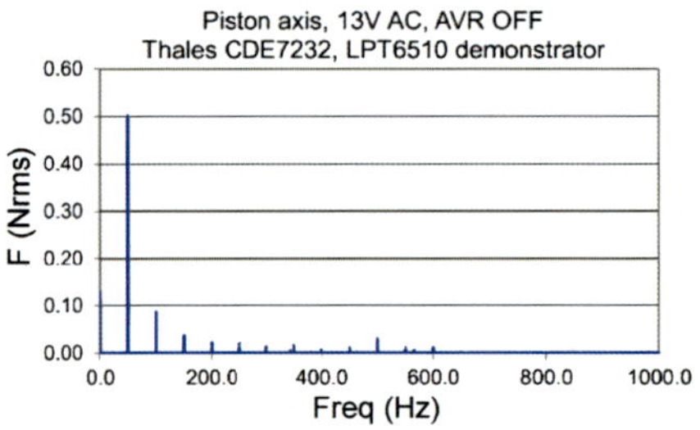

Figure 8. Exported vibrations (compressor axis), AVR OFF.

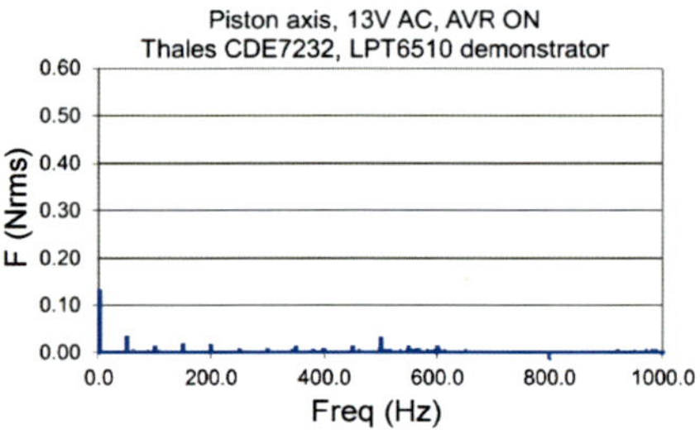

Figure 9. Exported vibrations (compressor axis), AVR ON.

be reduced well below 100 mN in the compressor piston axis. The vibrations exported in off-axis directions were in the 100 mN range.

As neither partner in the development of the LPT6510 currently has radiation-hardened space qualified drive electronics available, the compatibility of the LPT6510 breadboard with off-the-shelf radiation-hardened cooler drive electronics was assessed, by performing a functional test of the cooler when driven with the Iris HP-LCCE drive electronics [6]. The LPT6510 was successfully driven with both sets of electronics, the Thales bench-top CDE7232 and the Iris HP-LCCE brassboard model, with no special effort required to operate the cooler with the HP-LCCE.

As several minor changes are still planned for the final design, no environmental tests have yet been performed. However, the breadboard model is fully representative of the cryogenic performance of the final design.

CONCLUSION AND OUTLOOK

A number of minor design changes (parts and materials) are still required in order to bring the LPT6510 to true flight standard, but the base design as tested in the breadboard demonstrator shows good efficiency, low vibrations and a compact design convenient for system integrators.

Custom variants based on the base design as well as specific product qualification tests can be performed for specific requirements. Thales and Absolut invite feedback on the presented performances from all interested parties.

ACKNOWLEDGEMENTS

All work was performed using internal funding at Thales and Absolut. The authors thank Jet Propulsion Laboratories for giving their feedback on the proposed design and specifications of the LPT6510 cooler. We furthermore thank Iris Technology Corporation for loaning a prototype HP-LCCE cooler drive electronics to us for initial functional testing. Finally, we would like to acknowledge ESA for funding the initial development of the MPTC compressor that served as the basis for the compressor of the LPT6510.

REFERENCES

1. Trollier, T., Ravex, A., Charles, I., Coynel, A., Duband, L., Mullié, J., Bruins, P., Benschop, T., Linder, M., "Miniature 50 to 80 K Pulse Tube Cooler for Space Applications," *Cryocoolers 12*, Springer US, pp.165-172.

2. Mullié, J., Bruins, P., Charles, I., Coynel, A., Duband, L., "Development of the LPT9510 1W concentric pulse-tube,"*Adv. in Cryogenic Engineering*, Vol. 51, Amer. Institute of Physics, Melville, NY (2006), pp. 1332-1339.

3. Straight, S., et al, "Tactical Satellite-3 Mission Overview and Initial Lessons Learned," *24th Annual AIAA/USU Conference on Small Satellites* (2010).

4. Johnson, D., Rodriguez, J., Carroll, B., Bustamante, J., Kirkconnell, C., Luong, T., Murphy, J., Haley, M., "Integrated testing of the Thales LPT9510 pulse tube cooler and the Iris LCEE electronics," *Adv. in Cryogenic Engineering*, Vol. 59, Amer. Institute of Physics, Melville, NY (2014), pp. 1806.

5. Johnson, D., McKinley, I., Rodriguez, J., "Flight Qualification Testing of the Thales LPT9510 Pulse Tube Cooler," *Cryocoolers 18*, ICC Press, Boulder, CO (2015), pp 115.

6. Kirkconnell, C., Luong, T., Shaw, L., Moore, R., Ellis, M., "Flight Qualification Testing of Cryocooler Electronics," *Cryocoolers 18*, ICC Press, Boulder, CO (2015), pp 339.

7. Arts, R., Benschop, T., Bruins, P., Rijks, T., de Jonge, G., Trollier, T., "Development of Cost-effective Cryocoolers for Space," *Cryocoolers 18,* ICC Press, Boulder, CO (2015), pp. 595-600.

Status of AIM Space Cryocoolers

M. Mai, S. Zehner, A. Withopf, I. Rühlich

AIM Infrarot Module GmbH
Heilbronn, Germany

ABSTRACT

The growing market of commercial space applications and satellite programs has significantly increased the demand for infrared detectors. High performance Infrared (IR) detector systems are mostly equipped with Stirling or pulse tube coolers to provide sufficient cryogenic cooling for maximum electro-optical performance. The focus for these cryocoolers is on reliability and life time, thermal performance, and also on cost. To fulfill these demands AIM has focused on Integrated Detector Cooler Assemblies (IDCA) with long life cryocoolers similar to those being used for tactical applications.

Design of this cooler family is based on all-welded compressors featuring flexure bearing and moving magnet technology with customer-specific thermo-mechanical and electrical interfaces. In distinction to most of the tactical applications, a pulse tube is used instead of a Stirling coldfinger.

Some years ago, AIM delivered a MW-IR-IDCA with a SF400 pulse tube cooler for a Korean Space program. This unit was launched in March 2015 and is the first European pulse tube cooler in space.

An overview of ongoing AIM space cryocooler development activities is presented. This includes the latest cooling performance maps, environmental tests, and life test results of the SF-cooler family. The status of life time testing with > 65,000 real-time hours at elevated temperature and high power operation is also presented.

The reliability of an overall system can be increased by incorporating redundant cryocoolers. For a standard, compact IDCA a fully redundant cooling system cannot be achieved. Therefore a switching valve has been developed to allow operation of one pulse tube from a redundant pair of compressors. This activity was supported by the German DLR Space Agency in a study "Reliability and redundancy concepts for space cryocoolers used in IDCA solutions." The improvement of this valve is the topic of phase II. The valve was optimized in regard to its reliability and lifetime. An overview of these activities is also presented.

INTRODUCTION

AIM started the development of a long-life cooler with flexure bearing and moving magnet technology several years ago [2, 3]. The main focus during this development was to improve the lifetime of tactical coolers. Additionally, AIM developed a pulse tube cooler in cooperation with the University of Giessen [4, 5]. When AIM, also an IR system manufacturer, started with the development of infrared detector systems for low Earth orbit operation, it was decided to build up Integrated Detector Cooler Assemblies (IDCA). Thus, a space qualified cryocooler also was needed.

Cryocoolers 19, edited by S.D. Miller and R.G. Ross, Jr.

Table 1. Comparison of standard linear (moving coil) and flexure bearing (moving magnet) compressor design.

Category	Percentage change of SF design based on SL compressor
Number of parts	-56 %
Thread joints	-71 %
Glued joints	-60 %
Welded joints	+275 %
Materials with potential for outgassing into Helium vessel	**-97 %**

To fulfill typical space mission requirements in terms of reliability, mechanical robustness, and irradiation hardness, the AIM developed pulse tube space cryocoolers with flexure bearing moving magnet compressor technology based on a tactical long life cooler with the same basic design. AIMs Long Wave Infrared (LW-IR) detector system is based on the SF400 flexure bearing compressor. For a Short Wave Infrared (SW-IR) detector system AIM has chosen the smaller SF100 compressor.

To fulfill typical space mission requirements in terms of reliability, mechanical robustness, and irradiation hardness, a radiation hardened, redundant cooler electronics had to be developed.

The following sections give an overview of the design of our cooling system and the environmental qualification of both cooling systems and the cooler electronics.

TACTICAL LONG-LIFE COOLER DESIGN

Design of Flexure Bearing Moving Magnet Compressors

More than ten years ago, AIM started its activities to develop long life coolers. The MTTF of a standard linear cooler (SL-Design) was around 10,000 hours at that time. The main lifetime limiting factor for these coolers was the contamination of the working gas, primarily caused by the coil inside the vacuum vessel. Also the wearout mechanism of the piston coating and of the Stirling expander contributed to a limited life of the cryocooler.

For the design of a long life cooler the desire was to entirely eliminate these mechanisms. Thus, AIM decided to develop a new compressor of a moving magnet flexure bearing design. The moving magnet design allows the positioning of the coils outside the helium vessel. Beside the removal of this contamination source, there is no need for an electrical feedthrough, a source for statistical failures. The elimination of a flying wire further helps to decrease the statistical failure rate.

To reduce the amount of organic material inside the compressor to a minimum it was necessary to change the connection technology. Minimum use of glued and threaded joints helps to minimize the contamination potential. Table 1 shows a comparison of the standard linear (moving coil) and the flexure bearing moving magnet design.

In order to prevent abrasion of the piston coating, the long life cooler series uses flexure bearing technology with clearance seals. The idea is to centre the piston within its sleeve to eliminate contact and wearout. Of course, this is not an easy task considering a nominal piston gap of only a few microns. To realize this concept it was necessary to develop a computerized alignment system with sufficient accuracy.

Design of Pulse Tube Coldfinger

One key requirement for the development of a long-life tactical cooler was the desire for a form fit function replacement of a standard linear Stirling cooler. For the coldfinger, a compatibility to the

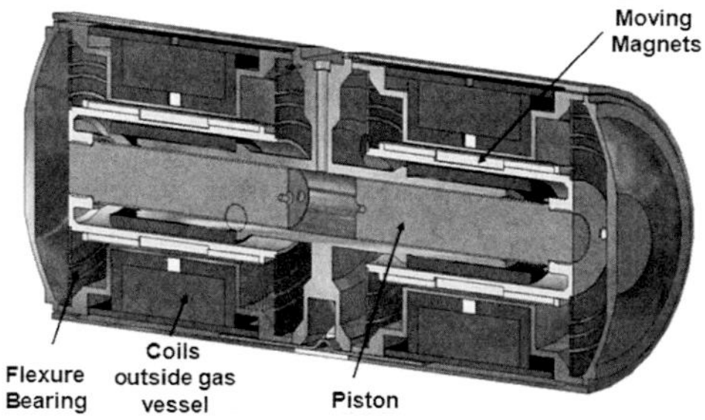

Figure 1. Principle design of AIM flexure bearing moving magnet compressors [2].

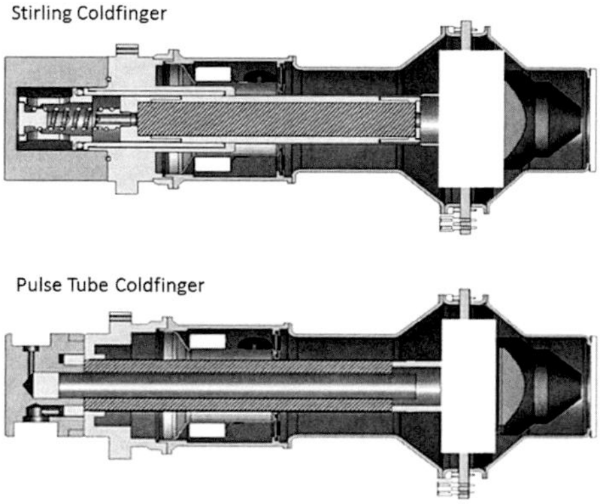

Figure 2. Comparison of AIM Stirling coldfinger and pulse tube coldfinger in AIM standard ½" dewar.

AIM standard ½" dewar was required. For the long-life cooler, a redesign of the Stirling coldfinger was reasonable. The use of high performance plastics reduced the wear and tear. The careful selection of construction materials and connection technologies led to a lifetime in the range of 20,000 to 30,000 hours.

For an ultra-long life cooler, the use of a pulse-tube coldfinger is necessary. In cooperation with the University of Gießen, AIM developed a compact coaxial pulse tube coldfinger that fully fits into the AIM standard ½" dewar. Without any organic materials or moving parts, the pulse-tube coldfinger is designed for lifetimes above 60,000 h. So far, no life limiting effect is known or observed for such pulse tube coldfingers.

SPACE SPECIFIC DESIGN

Once a cryocooler has the capability to meet the lifetime requirement of a space mission, the main difference between a tactical and a space cooler is its interfaces. For a tactical cooler, besides the mechanical clamping, an additional interface to mount a fan is a common approach. For low power applications sometimes just the mechanical interface is used for heat removal. However, space coolers typically need a suitable interface for the mounting of a thermal link or a heat pipe, and in addition, there are specific requirements for the mechanical interface. Depending on the thermal management concept, conductive cooling through this mechanical interface may not be desired.

For different space programs AIM has developed, in close collaboration with the customer, a reasonable thermal and mechanical interface. Figure 3 shows a comparison between two space pulse-tube coldfingers. They are based on the same thermodynamic design of the coldfinger. The difference is just the thermo-mechanical interfaces. The thermal interfaces are matched to the program-specific heat pipes. The dewar flange is adapted to the outer dewar used in the particular program.

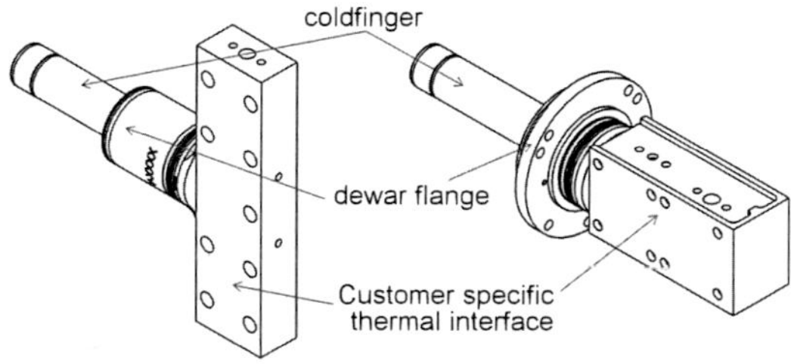

Figure 3. Thermal and mechanical interface of pulse tube coldfinger for different applications.

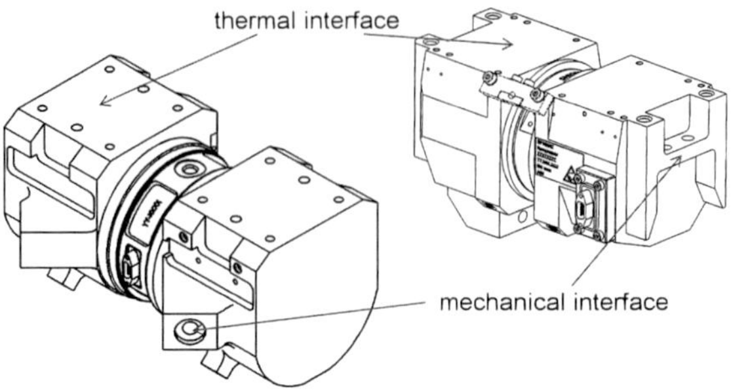

Figure 4. Thermal and mechanical interface of SF400 for a Korean space program (left) and SF100 for a German space program (right).

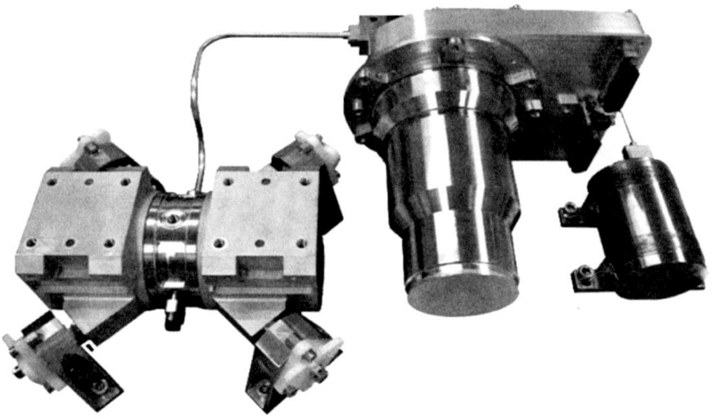

Figure 5. AIM MWIR detector with SF400 flexure bearing compressor + ½" pulse tube cold finger.

The interfaces of the compressor are matched to the requirements of its corresponding program. While clamping of the compressor is standard for tactical cryocoolers, space coolers need a more specific mechanical interface. Also, the thermal interface is more demanding due to the vacuum in space. To benefit from the maturity of tactical coolers, only the outer housing was changed to provide reasonable space interfaces. Figure 4 shows the thermal and mechanical interfaces of the SF400 and the SF100 in their space version.

KOREAN SPACE PROGRAM

Some years ago AIM delivered the flight model of an integrated detector cooler assembly for a Korean space program (see Figure 5). Due to the huge size of the detector and cold shield of this mid wave detector system, the SF400 flexure bearing compressor was chosen to provide enough performance margin. The cooler was optimized for an operating temperature of 80 K and includes the adapted interfaces outlined above. The total weight of the mechanical cryocooler is 4.70 kg. The cooler system was delivered with a fully redundant cooler control electronics (Figure 6) with an additional weight of 3.44 kg.

The cryocooler and the cooler control electronics have been qualified as part of the IDCA qualification program. Details of the qualification and also performance data of the cooling system can be found in [1]. A picture taken from the Korean satellite, which was launched in March 2015, is shown in Figure 7.

Figure 6. AIM cooler drive electronics (fully redundant)

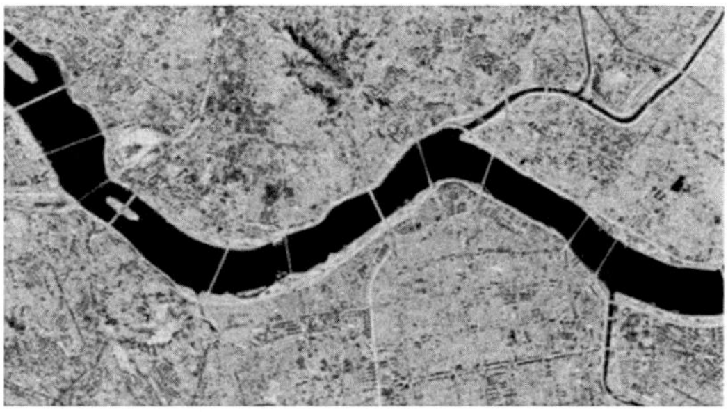

Figure 7. Picture of Soul taken from Korean satellite (Source: KARI)

GERMAN HYPERSPECTRAL IMAGING INSTRUMENT

For this instrument a short wave infrared detector will be used. Because of the wave length of the detector, the operating temperature of the cooler is significantly higher than the operating temperature of the mid wave system. The cooler for this system is optimized for operation at 150 K. Due to the lower input power needed for cooling in this temperature range, AIM selected the smaller SF100 compressor for this application. Figure 8 shows the SF100 cooler integrated into the IDCA mounted on a transportation plate.

With the customer-specific interfaces, the total weight of the mechanical cryocooler is 3.00 kg. The cooler control electronics are based on the electronics used for the Korean program. Due to customer requirements, an additional Tele-measurement and command capability was implemented. Including these additional capabilities, the weight of the electronics is still below 3.50 kg.

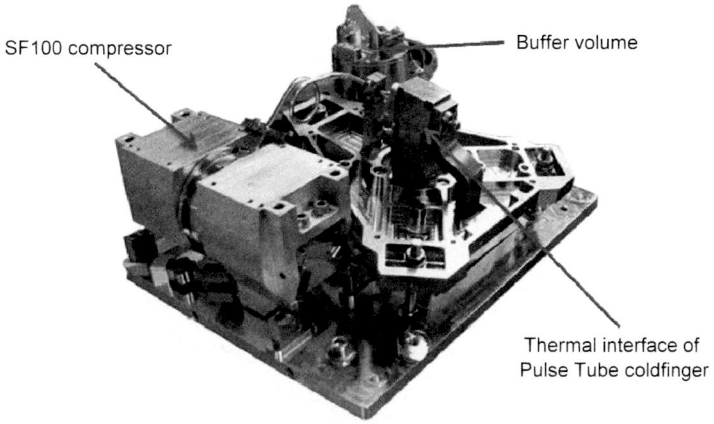

Figure 8. AIM SWIR detector with SF100 flexure bearing compressor with ½" pulse tube cold finger

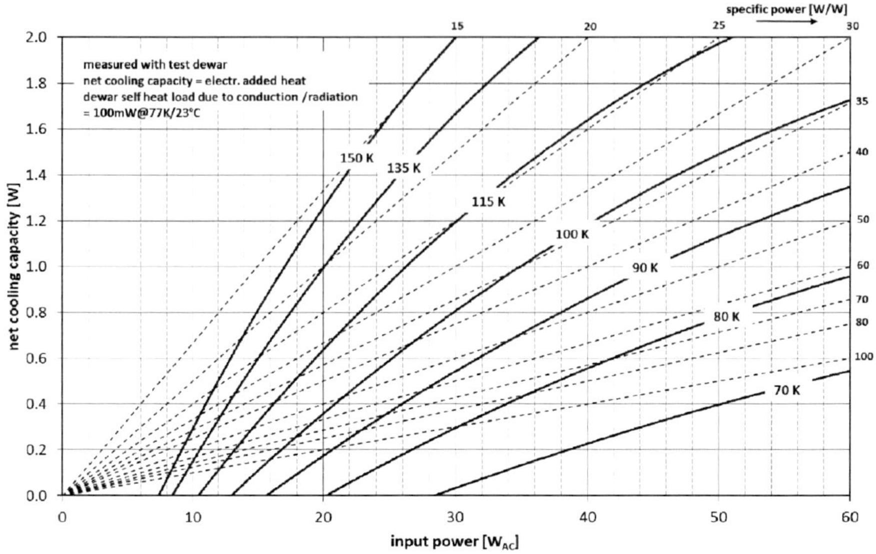

Figure 9. Performance data for the SF100 pulse tube cooler.

On a cooler level, the SF100 has passed launch vibration and thermal vacuum pretests. The final qualification of the cooler will be performed at the IDCA level. The QM of the IDCA is currently being built up. The launch for this system is scheduled for 2018. Figure 9 shows the thermal performance data.

COOLING SYSTEM LIFETIME VERIFICATION

Long lifetime is one of the main requirements of a space cooler. Thus, lifetime tests are mandatory. As part of system qualification AIM performs lifetime tests for different coolers. All coolers operate in open-loop-mode with a fixed input power of at least 80% of the rated power. Thus, the input power is 1.5 to three times higher than the steady state power predicted for nominal operating conditions. The heat removal is realized by natural convection only. Thermal interface temperatures are on the order of 50°C, and thus roughly 25 K above the nominal reject temperature at the system level.

Cooling performance and helium leakage tests are performed on a regular basis to check performance and life time stability. In addition, high temperature operating cycles are being performed regularly. Figure 10 shows, by way of example, the lifetime test of the SF400 cooler in bench-top operation.

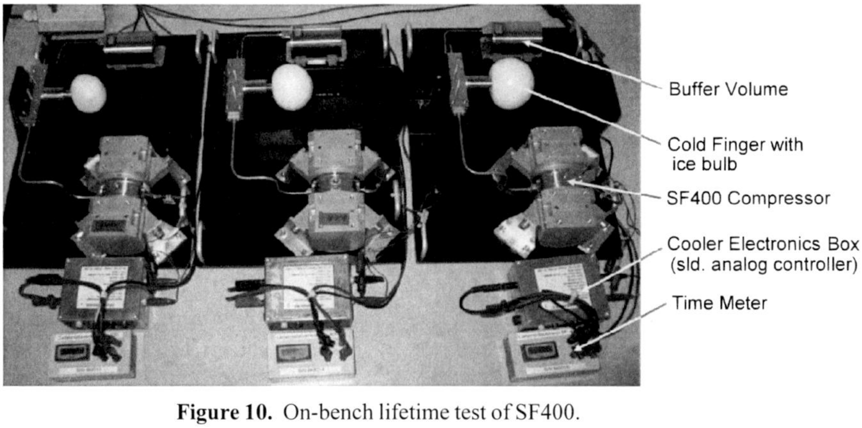

Figure 10. On-bench lifetime test of SF400.

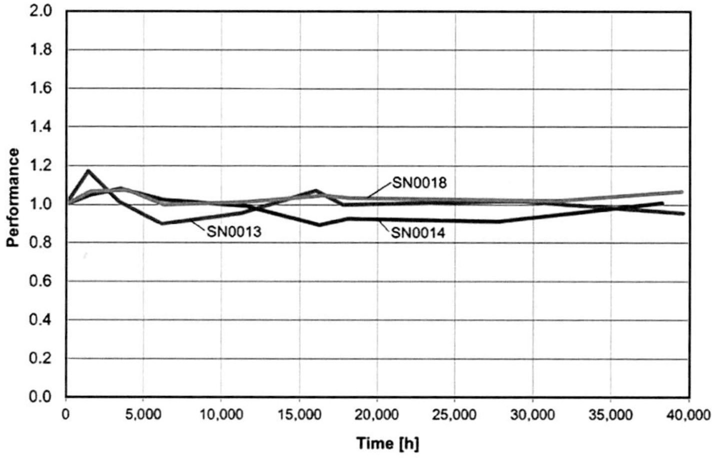

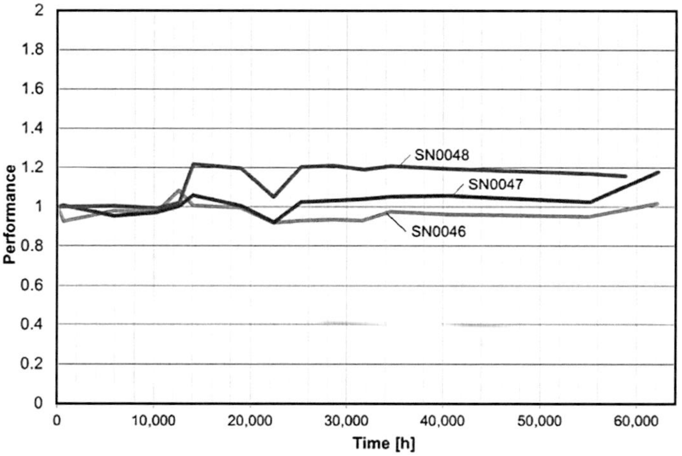

Figure 11. Relative performance over lifetime of SF400 (top) and SF100 (bottom), both with pulse tube coldfinger.

The relative performance of SF100 and SF400 over lifetime is shown in Figure 11. Relative performance is the ratio of cooling capacity divided by cooling capacity at the start of lifetime test for the nominal operating point. Deviations can be attributed to changing measurement equipment over time. First of all the integration into the measurement dewar and the calibration of the copper thermal mass results in variations. Table 2 shows the status of coolers in test with achieved life time operating hours to date.

Table 2. Status Lifetime test of AIM Long-Life cooler (May 2016).

Compressor type	Coldfinger type	No. of coolers	Operating hours	Status
SF070	¼" Stirling	1	~63 700	Ongoing, minor performance degradation observed yet
SF070	¼" Stirling	3	~35 200	
SF100	13mm Stirling	2	~34 500	Stopped due to displacer wear out after 10% Performance degradation
SF100	½" Pulse Tube	3	~64 800	Ongoing, no performance degradation observed yet
SF400	½" Pulse Tube	3	~41 800	Ongoing, no performance degradation observed yet

Figure 12. AIM pulse tube cooler with redundant compressors connected via switching valve.

FURTHER ACTIVITIES

Compressor Redundancy

As described above, reliability is the predominant requirement for space cryocoolers. Even for cryocoolers being designed to achieve lifetimes in excess of the mission life, single failures may lead to an early failure of the instrument. A common way to increase the reliability of space systems is redundancy, which is mainly used for cooler drive electronics. But for integrated detector cooler assemblies cooler redundancy is very demanding. Integrating a second pulse tube in the dewar will result in loss of some of the key advantages that are attractive for this IDCA approach, namely compactness, thermal efficiency, and usage of standard proven parts. In the scope of a DLR funded project, AIM investigated redundancy concepts for IDCA systems.

Advantages and disadvantages of various systems have been analyzed and assessed in detail. Due to the fact that the pulse tube has a very low statistical failure rate and no known ageing phenomena, use of only a redundant compressor was judged as a viable solution.

To realize compressor redundancy, an additional switching valve is required. There are a lot of requirements that need to be fulfilled by this valve. It has to have a very low leakage rate, internal as well as external. Internal leakage between the two compressors would increase the power consumption of the cooler. An external leakage rate would decrease the lifetime of the cooling system due to loss of the working gas. Also, organic materials would decrease the lifetime due to contamination of the working gas.

The development of a demonstrator of a suitable switching valve was part of the first study (Fig. 12). Analysis of weak points and optimization of the demonstrator was part of a second study, also funded by the DLR.

Performance Optimization

Besides reliability, the performance of space cryocoolers is one of the main requirements. Higher efficiency decreases the power consumption and thus also the heat to be removed by a radiator.

During development of the AIM pulse tube, a main requirement was to fit into the AIM standard dewar. If this requirement were given up, a considerable performance improvement would be possible. Simulations done so far predict an increase in efficiency of about 25%. With the lower PV-power needed with this 25% improvement, it would be possible to substitute a smaller compressor for the same application. The SF100 compressor would be able to provide sufficient cooling for midwave IR applications down to 80 K. The SF400 compressor even becomes a suitable solution for application down to 65 K. Besides decreasing the power consumption of the cryocooler system, the weight of the cooling system can also be reduced.

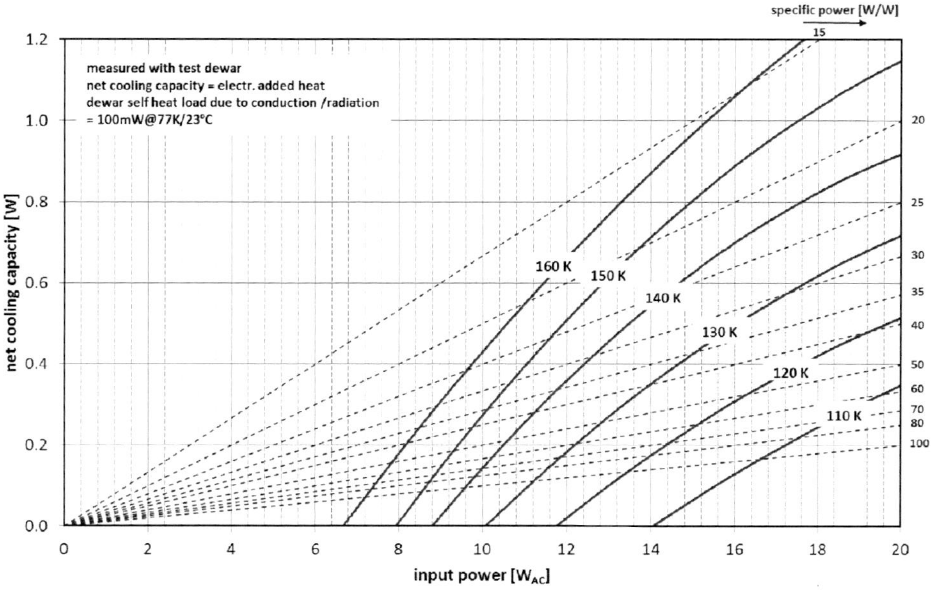

Figure 13. Performance data of SF070 with AIM pulse tube coldfinger.

Pulse Tube Cooler with SF070

Within the last decade there have been numerous activities to increase the operating temperature of infrared detectors. In the past, for example, a standard mid-wave detector operated at about 80 K. A state-of-the-art High Operating Temperature (HOT) mid-wave IR detector now allows operation at 140 K and above. With increased operating temperature the needed PV-power and thus compressor size can be decreased.

Several years ago, AIM developed the SF070 [2], based on the same technologies (moving magnet, flexure bearing) like the SF100 and the SF400. This compressor also can be used to provide the necessary PV-power for the AIM pulse tube coldfinger. Figure 13 shows a performance mapping for this cooler. For use as a space cooler, the thermal and mechanical interfaces need to be adapted.

SUMMARY AND CONCLUSIONS

In recent years, AIM has been involved in several space cryogenics applications. Space cryocoolers for different programs have been developed. A MW-IR-IDCA with AIM space cryocooler was successfully qualified, delivered, and launched in 2015. The qualification model of a SWIR-IDCA, also utilizing an AIM space cryocooler, is currently being assembled. Launch for this module is scheduled for 2018. Additionally, AIM has developed a switching valve for space cryocoolers. This valve allows a redundant compressor to be incorporated within an IDCA system.

Further activities are related to improvement of efficiency and reduction of weight. Simulations predict improvements in efficiency on the order of 25% by optimization of the pulse tube coldfinger. This also allows switching to a smaller compressor for the same cooling capacity due to the less PV-power needed.

Performance mapping of the SF070 compressor with the same AIM pulse tube coldfinger demonstrated that this is also a viable solution to provide sufficient PV-power for the AIM pulse tube coldfinger. Using the same technologies as the SF100 and SF400 coolers, this cooler can also meet typical requirements of space missions.

REFERENCES

1. S. Zehner, M. Mai, A. Withopf, I. Rühlich, „ AIM Space Cryocooling System Qualification," *Cryocoolers 17*, ICC Press, Boulder, CO (2012), pp. 85-92.
2. M. Mai, I. Ruehlich, C. Rosenhagen, Th. Wiedmann "Development of the Miniature Flexure Bearing Cryocooler SF070," *Cryocoolers 15*, Kluwer Academic/Plenum Publishers, New York (2009), pp. 133-138.
3. Rühlich, I., Mai, M., Wiedmann, Th., Rosenhagen, C., "Flexure Bearing Compressor in the One Watt Linear (OWL) Envelope," *Proc. SPIE*, Vol. 6542 (2007), pp. 654221-1 to -7.
4. K. Klundt, M. Dietrich, G. Thummes, C. Heiden, „Miniature Pulse Tube Cooler with linear compressor," *Infrared Technology and Applications XXVI, Proc. SPIE*, Vol 4130 (2000), pp. 394-405.
5. G. Thummes, L.W. Yang, "Development of Stirling-type Pulse Tube coolers driven by commercial linear compressors," *Infrared Technology and Applications XXVI, Proc. SPIE*, Vol 4820 (2003), pp. 1-14.

5 W at 77 K without Breaking the Bank

D. Wilcox, B. Rock, P. Spoor, J. Corey

Chart Industries - Qdrive
Troy, NY 12180

ABSTRACT

Chart Industries has developed a small cryocooler for mass production at a cost far less than prior state of the art, using a novel external-stator motor and magnetic-piston construction that eliminates electrical feedthroughs, springs, and flexures, combined with a coldfinger that eliminates all brazing and welding. This new acoustic Stirling cooler operates directly on 60 Hz AC power, with direct forced air rejection to ambient (no other support services or equipment is required). Developed for local liquefaction of medical gases, this breakthrough in affordable, small cryocooling is now available in initial samples for any application. This paper presents the novel construction, price estimates, performance maps, and ongoing development of this cooler.

INTRODUCTION

Chart Industries has undertaken the endeavor of developing a novel, low-cost cryocooler to address the need for local, on-demand, cheap liquefaction of medical gases. This cryocooler, combined with Chart's cryogenic storage dewars and gas concentrators can provide consumers with a compact, on-site gas liquefaction appliance that is the perfect combination of quality, performance and affordable. Given the intended marketplace, cost is the single biggest challenge associated with the acoustic Stirling (or 'pulse-tube') cryocooler. In order to meet the strict cost requirement making use of existing, 'off-the-shelf' cryocoolers is simply not an option. Therefore, a new external-stator based electro-acoustic motor[1] was devised to significantly reduce the production time, complexity, and cost of this motor compared to traditional flexure or spring-based reciprocating motors. This motor concept further reduces the cryocoolers estimated cost by eliminating the need for electrical feed-throughs, as its electrical connections, which provide power to the motors, are moved outside of the pressure vessel. In addition to the above mentioned cost saving features, this cryocooler makes use of no copper components, with the exception of the motor coils, or braze joints and uses only a limited number of motor tube and inertance coil weld joints to achieve a cost-effective, helium tight pressure vessel. The low cost cryocooler design coupled with the prototype performance test results that will be discussed, make this cryocooler an ideal option for researchers and OEMs looking to achieve 5 to 12 watts of cooling in the 77K to 100K range without 'breaking their bank.'

LOW COST CRYOCOOLER COMPOSITION

Basic Cryocooler Components

The cryocooler makes use of two, opposed, external-stator motors that are directly coupled to its coldfinger. The motors alternately compress and expand the working helium gas, 'driving' the coldfinger. Each motor consists of a thin-walled motor tube which serves the purpose of the pressure vessel wall, a reciprocating magnet core-piston, an iron stator assembly including wound bobbins and a back volume. Figure 1 shows a solid model of the motor.

The coaxial coldhead portion of the cryocooler consists of a stacked-plate warm heat exchanger (or aftercooler) which serves as the main heat rejection point, a coiled inertance tube, a compliance tank and a coldfinger. The coldfinger itself houses the regenerator, which is composed of low cost stainless-steel, random fiber filter material, a slotted/finned cold heat exchanger, and a phenolic (Garolite 10) thermal buffer tube with integrated, sintered screen or wire-knit puck flow straighteners at each end. Figure 2 shows the basic low cost cryocooler components. The solid model shown in Figure 2 represents something close to the preferred embodiment of a product (or OEM system) minus integrated air or water cooling.

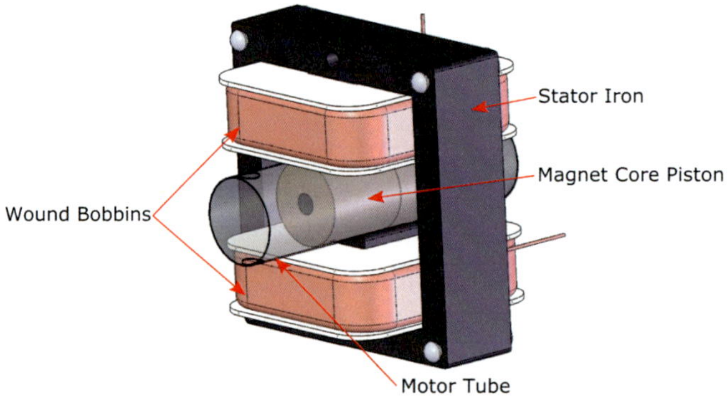

Figure 1. Novel magnetic reciprocating piston, external-stator motor used in the low cost cryocooler. The motor 'tube' is made transparent here to show the relationship between the motors external stator and magnet core-piston. Two of these motors comprise the acoustic drive or pressure wave generator portion of the cryocooler and drive its coldfinger (acoustic load).

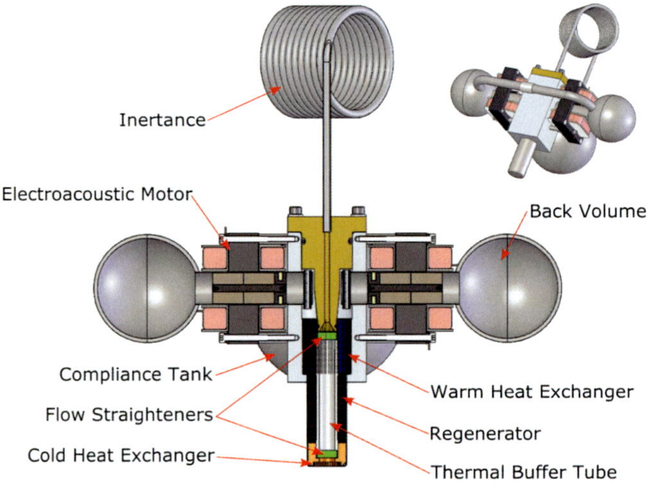

Figure 2. Section view with inset solid model (top right) of the low cost cryocooler. In the sectioned view the compliance tank is hidden from view. Neither water nor air heat reject options are shown attached to the warm heat exchanger. Space volume is approximately 38 cm x 38 cm x 26 cm.

Cryocooler Cost Estimate

Based on the manufacturing cost analysis that we have carried out, which includes direct material costs, labor costs and manufacturing overhead, and the estimated market volume, we believe when this cryocooler becomes commercially available, its sale price will be in the $3000 range, if not less. This price includes integrated air or water cooling based on the end users specifications.

EXPERIMENTAL RESULTS TO DATE

In addition to achieving its target production cost we believe that the cryocooler also has to be able to achieve a cooling capacity of ~5W at ~77K in order for it to be commercially viable for a small liquefaction appliance. Figure 3 shows a prototype cryocooler in an experimental performance load test set up. Figure 4 shows a close up of the prototype cryocooler on a vacuum can with certain visible components identified.

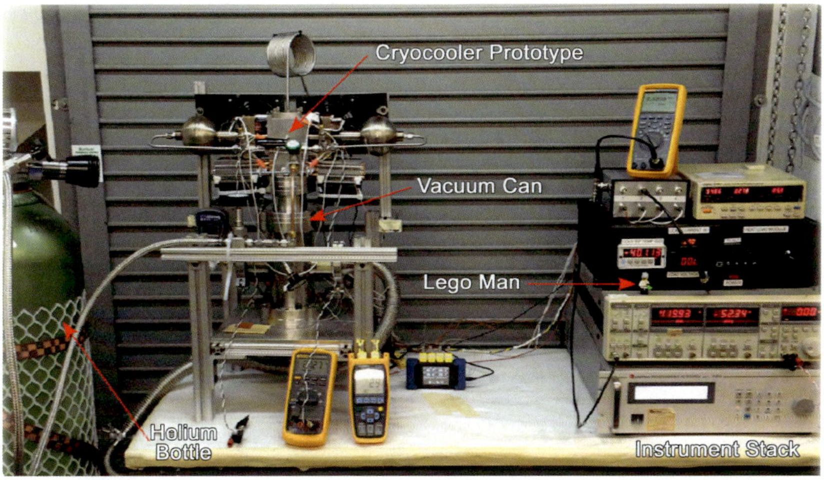

Figure 3. A prototype cryocooler in test. Note Lego Man on instrument stack for scale.

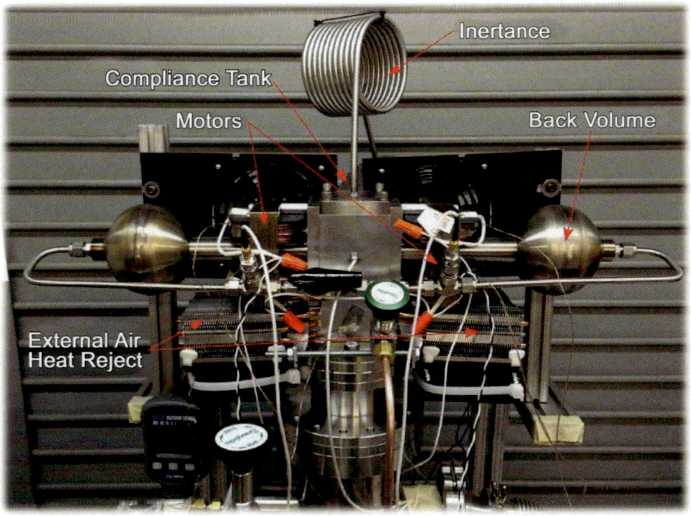

Figure 4. Close up of experimental cryocooler prototype. The coldfinger is not visible as it is inserted into a vacuum can for testing.

Analysis of available data

Before integrating the cryocooler into a full liquefaction appliance, it is important to evaluate its cooling power to assure that it will meet the performance target of 5W at 77K. The experimental test set up pictured in Figure 3 is used to make a series of heat load measurements at various operating conditions. Figure 5 shows two, two point load curves for the cryocooler with electrical inputs of 200We and 250We. The performance is measured at 77K and 100K as these are the approximate liquefaction points of nitrogen at Standard Temperature and Pressure (STP) and oxygen at 18 psig, respectively. We then measured the cryocoolers heat load at several other electrical input points, while maintaining a coldfinger tip temperature of 77K. Figure 6 shows the measured power curves. If we assume that the curves in Figure 6 are linear, we can conclude that the cryocooler should be able to achieve ~7W of cooling at 77K and over 14W of cooling at 100K with an electrical input of 350We, which is the allocated electrical input power limit for the project.

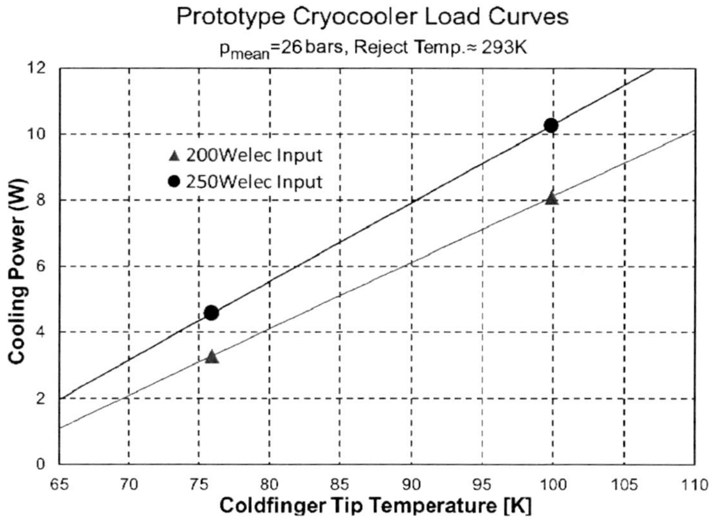

Figure 5. Load curve for prototype cryocooler.

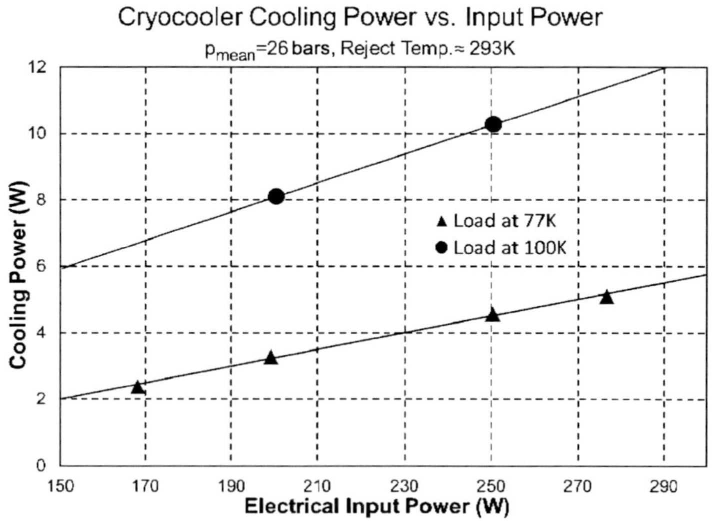

Figure 6. Cooling capacity of the prototype cryocooler versus electrical input power.

CONCLUSIONS

Chart, Inc. has successfully constructed a low cost cryocooler that is able to achieve 5W of cooling at 77K and won't 'break the bank' to purchase. While these test results are very encouraging there is still more experimental work that needs to be carried out to fully characterize the prototypes performance. Additional power curve testing (up to an input of 350We), testing at higher reject temperatures (up to 351K) and evaluating the heat leak of the experimental set up still needs to be completed. Additionally, the construction and performance testing of several more prototypes to confirm repeatability of the design, its manufacturing methods, and performance must be carried out. Furthermore, oxygen and nitrogen liquefaction testing to determine the necessary cooling capacity based on desired appliance specific production rates (liters/day, gallons/day, etc.) also needs to be completed. Development work on a liquefaction appliance is currently underway.

REFERENCES

1. Corey, J.A., Douglas, A.W., "Cryocooler with magnetic reciprocating piston," U.S. Patent Application 14/450,142.

Progress in Development of High Capacity Stirling Cryocooler Using a Linear Compressor

J. Ko, H. Yeom, Y.J. Hong, H. Kim, S. In, S.J. Park

Korea Institute of Machinery & Materials
Daejeon, Korea(S), 34103

ABSTRACT

A high capacity Stirling cryocooler is being developed for HTS electric power devices. The developed cryocooler is driven by a dual-opposed type linear compressor and has gamma-type configuration. The first prototype was fabricated and tested and the results of the cooling performance tests were presented at CEC 2015. In this study, a modified prototype is fabricated to improve its cooling performance. The thrust constant of a linear motor is increased and a regenerator material is changed from screen mesh to random fiber.

In the cooling performance tests, the cold head temperature is measured at varying heat loads with a fixed input current. The results show a no-load temperature of 43.3 K, a cooling capacity of 650 W at 76.8 K with 8.76 kW electric power input and 21.6 % of Carnot COP. Turn down performance is also measured by adjusting input voltage for a fixed cold head temperature.

INTRODUCTION

High Temperature Superconductor (HTS) electric power devices such as HTS cables and HTS Fault Current Limiter (HTS-FCL) are undergoing field testing after the R&D phase in Korea[1-3]. They are expected to be commercialized in the near future, and a reliable and efficient high capacity cryocooler is indispensable for their commercialization. A crank-driven type Stirling cryocooler is commonly used for some kW class cryogenic cooling systems. It is commercially available and technically matured but, still has problems of vibration and frequent maintenance due to a crank-cam driving mechanism and the use of lubricating oil. A Stirling cryocooler driven by a dual-opposed linear compressor can solve these problems because it has the advantage of oil-free operation and is inherently vibration-free.

Previously, we fabricated and tested the first prototype of the high capacity Stirling cryocooler driven by the linear compressor. The experimental results showed a no-load temperature of 47.8 K, a cooling capacity of 440 W at 78.1 K with 6.45 kW of electric power input and 19.4 % of Carnot COP[4]. In this study, we modify the first prototype to improve the cooling performance. The dynamic behavior, the symmetric characteristics due to use of two linear motors, the cooling capacity and the efficiency are investigated from the cooling performance tests when the input current to a cryocooler is fixed. The variation of the cooling performance is measured from a turn down operation with a fixed cold head temperature.

Cryocoolers 19, edited by S.D. Miller and R.G. Ross, Jr.

FABRICATION AND EXPERIMENTAL SETUP

The stator of the first prototype had a 260 coil turns. Its thrust characteristic is designed to have 6 kW input power at 60 Hz operation. We changed the operating frequency to 45 Hz after the performance test of the first prototype. Coil turns were increased to 292 turns to increase the thrust constant without a change to the dimension of the linear motor.

The regenerator has dimensions of 86, 130, and 80 mm for the inner diameter, outer diameter and length, respectively. A random fiber of SS 316L is used as the regenerator material instead of a stainless steel screen mesh in the modified cryocooler. The cold-end and the warm-end heat exchangers have a slit-type configuration and the heat is rejected by cooling water at warm-end heat exchanger. Figure 1 shows the modified cryocooler and the installation of heaters and temperature sensors at the cold head. To apply a heat load and to measure the spatial temperature distribution, 24 cartridge heaters of 50 W(Lakeshore, HTR-50) are uniformly inserted and 8 temperature sensors (Lakeshore, DT-670A-CU) are attached.

Figure 2 shows the electric connection for driving the cryocooler and measurements during tests. Two linear motors are electrically connected in parallel and an AC power supply (EXTECH, 6530) supplies electric power to the linear compressor. A resonant capacitor is connected between the AC power supply and each linear motor for power factor correction. Several capacitors are used to satisfy the desired capacitance and to withstand the voltage. Voltage, current, power and power factor are measured with the power analyzer (Yokogawa, WT1800). Electric heat is supplied by an AC power supply (Chroma, 61504) for the heat load tests. Accelerometers(PCB, 353B18) are attached to the piston and the displacer inside the cryocooler and the measured acceleration is con-

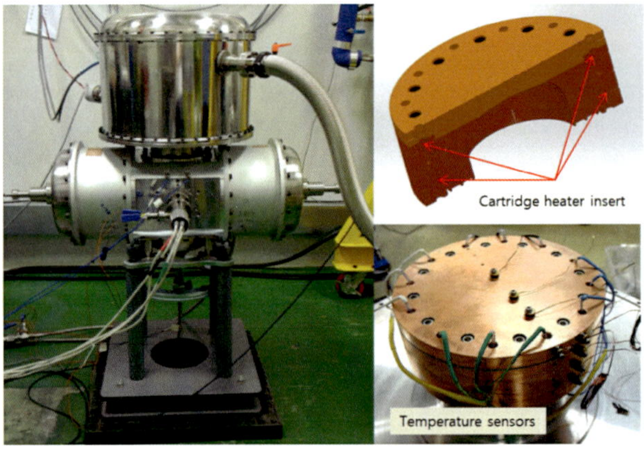

Figure 1. Photo of fabricated Stirling cryocooler and cold-head.

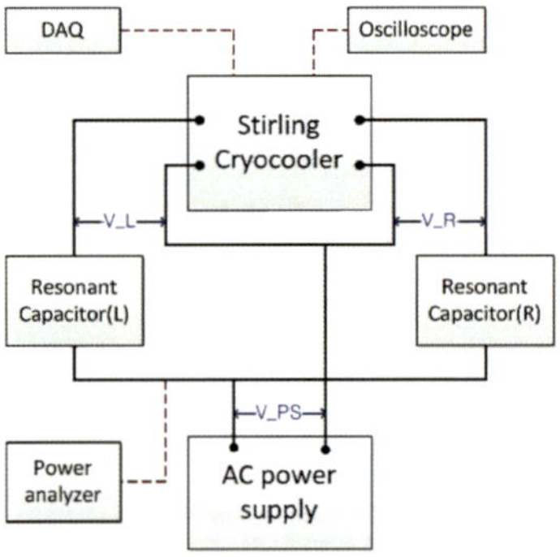

Figure 2. Schematic diagram of driving circuit and measurement.

verted to a displacement. Dynamic pressure sensors (PCB, 112A21) are installed to measure the dynamic pressure in the compression space. Waveforms for the acceleration, the dynamic pressure, the input current and the voltage are recorded to the oscilloscope (Yokogawa, DLM4038).

EXPERIMENTAL RESULTS

Cooling Performance Test

In the cooling performance test, the input voltage to the cryocooler is adjusted to a fixed input current of 36.7Arms. Figure 3 shows the measured cold head temperature of each point at steady state. The temperature deviation from the eight (8) spatially distributed sensors are within 1 K for all cases. Input voltage and power factor measured at the AC power supply (V_PS) and each linear motor (V_L, V_R) are shown in Figure 4. Driving voltage gradually increases with increasing cold head temperature. Voltage at the linear motor is higher than at AC power supply. The voltage difference between the two linear motors is less than 5 %. The power factor at the AC power supply is nearly 1.0, whereas value at linear motor is close to 0.35, which confirms that the resonant capacitor worked well.

Figure 5 and 6 show the displacement of piston and displacer and dynamic pressure in the compression space. Amplitude of piston, displacer and compression space pressure decrease as decreasing cold head temperature. The phase difference between input current and piston displacement(ϕ_{ixp}) decreases slightly as the cold head temperature increases. However, the phase difference between displacer and piston displacement(ϕ_{xdxp}) is almost constant. The phase difference between the pressure and the piston displacement(ϕ_{Pxp}) decreases as the cold head temperature increases. It means that the gas spring effect becomes stronger rather than gas damping effect with increasing cold head temperature.

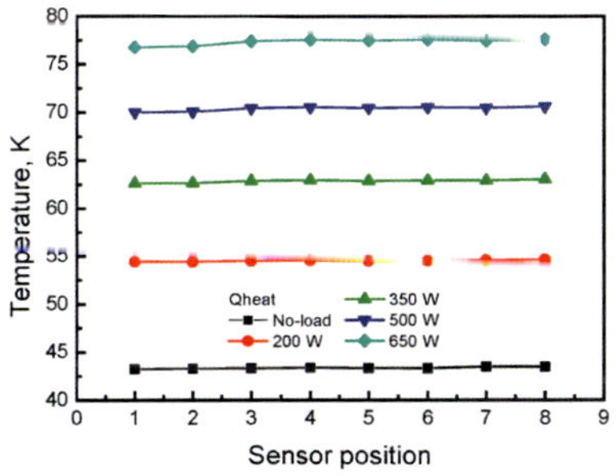

Figure 3. Spatial distribution of cold head temperature.

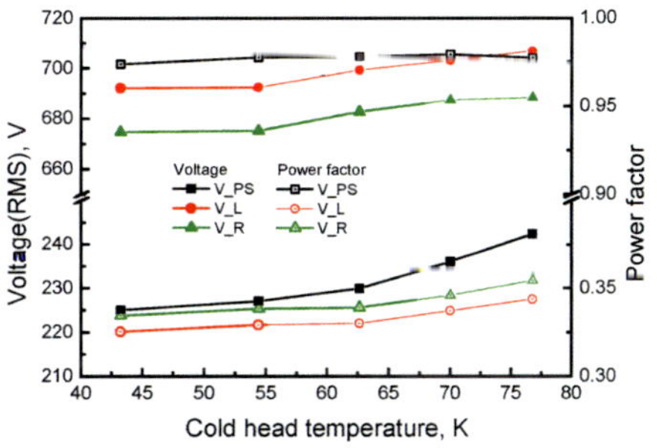

Figure 4. Voltage and power factor in driving circuit

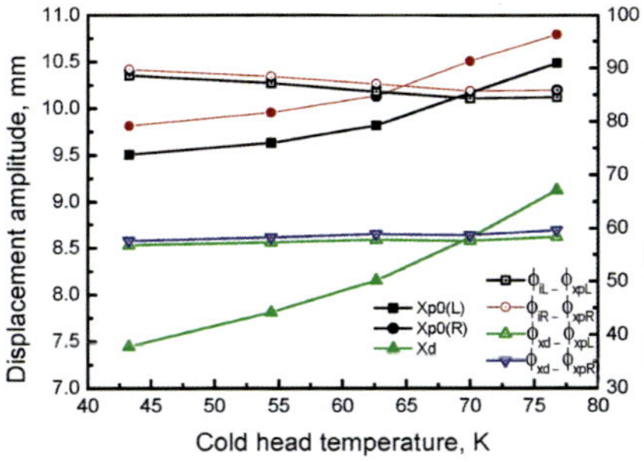

Figure 5. Displacement of piston and displacer.

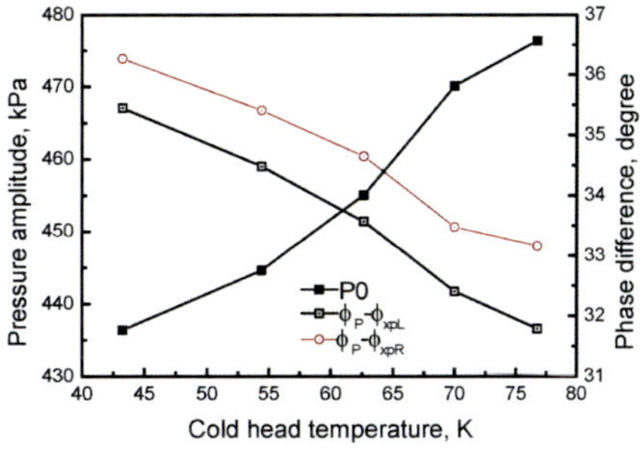

Figure 6. Pressure at compression space.

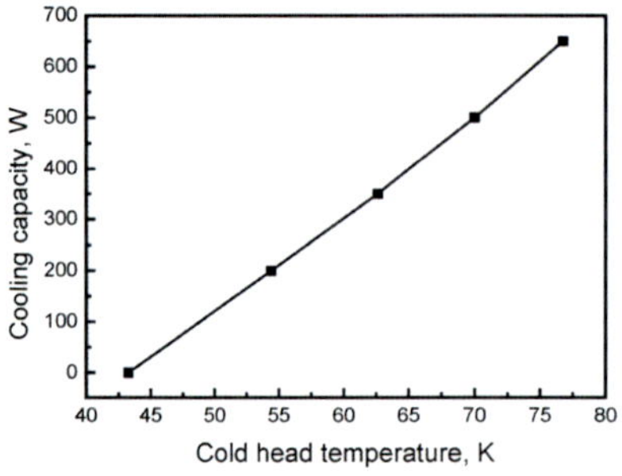

Figure 7. Cooling capacity.

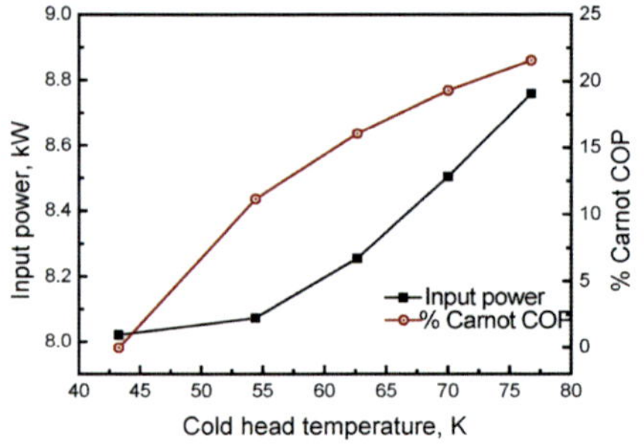

Figure 8. Input power and % Carnot COP.

Figure 7 and 8 show the measured the cooling capacity, the input power and the efficiency of the developed cryocooler. The no-load temperature is 43.4 K and the cooling capacity is 650 W at 76.8 K with 8.76 kW electric power input. The input power and % Carnot COP increases as cold head temperature increase.

Turn Down Operation

A turn down performance is informative to a cryocooler user. The developed cryocooler is tested with varying input voltages at a fixed cold head temperature of 77.9 K. To investigate the applicability to an LNG re-liquefaction system, the test is carried out at a cold head temperature of 110 K without any modification. Figure 9 and 10 show the results of the turn down operation. It measured a cooling capacity of 533 W with 3.33 kW of electric input power and 27.7 % of Carnot COP at 110 K operation. At that condition, the input voltage and current is 170 V and 19.7 Arms value, respectively. Figure 10 shows a decrease in efficiency with the turn down operation.

SUMMARY

In this study, the previously developed high capacity Stirling cryocooler driven by a linear compressor was modified to improve the cooling performance. Dynamic and electric symmetry are confirmed from the experimental results. The results of the cooling performance test with a fixed input current show a no-load temperature of 43.3 K, a cooling capacity of 650 W at 76.8 K with 8.76 kW electric power input and 21.6 % of Carnot COP. The turn down test was run while the cold head temperature is maintained at 77.9 K and 110 K to demonstrate the applicability to LNG re-liquefaction and HTS. The developed cryocooler is applicable to LN_2 cooling for HTS devices and also LNG re-liquefaction.

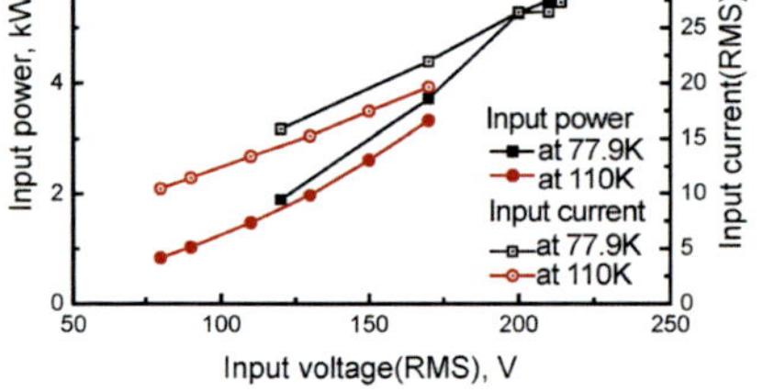

Figure 9. Turn down operation: input power and current.

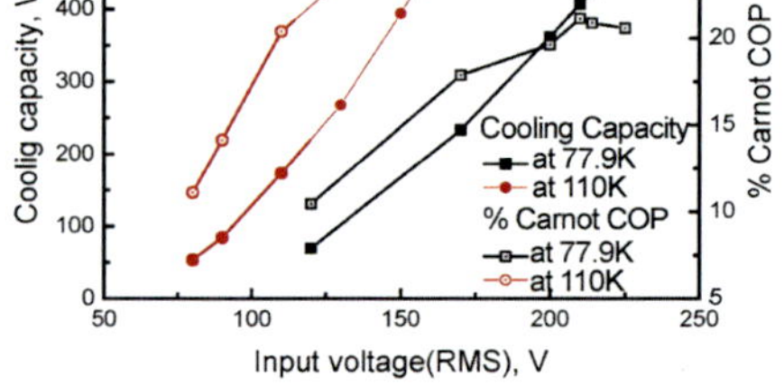

Figure 10. Turn down operation: cooling capacity and efficiency.

ACKNOWLEDGMENT

This work was supported by the Korea Institute of Machinery & Material(KIMM) and the Ministry of Science, ICT & Future Planning(MSIP).

REFERENCES

1. Ryu, C., Jang, H., Choi, C., Kim, Y., Kim, H., "Current status of demonstration and comercialization of HTS cable system in grid in Korea," *Applied Superconductivity and Electromagnetic Devices(ASEMD), 2013 IEEE International Conference*, pp. 539-542.
2. Kim, H., Lee, J.Y., Kim, H.R., Yang, S.E., Yu, S.D., Kim, W.S., Hyun, O.B., Ko, J. and Yeom, H., "An effect of HTS wire configuration on quench recovery time in a resistive SFCL," *IEEE Transactions on Applied Superconductivity* 23 (3) (2013), pp. 5604104-5604104.
3. Lee, S., Yoon, J., Yang, B., Moon, Y. and Lee, B., "Analysis model development and specification proposal of 154kV SFCL for the application to a live grid in South Korea," *Physica C: Superconductivity* 504 (2014), pp. 148-152.
4. Ko, J., Kim, H., Hong, Y.J., Yeom, H., In, S., Park, S.J., "Development of 1 kW Stirling cryocooler using a linear compressor," *IOP Conference series: Materials Science and Engineering*, vol. 101 (2015) 012092.

Investigation of Gravitational Effects in Pulse Tube Cryocoolers Using 3-D CFD Simulations

T. Fang[1], T. I. Mulcahey[2], P.S. Spoor[3], M.D. Perrella[1], T. J. Conrad[4], S. M. Ghiaasiaan[1]

[1]Georgia Tech Cryo Lab, G.W. Woodruff School of Mechanical Engineering, Georgia Institute of Technology, Atlanta, GA 30332 USA

[2]CSA Medical Inc., Lexington, Massachusetts 02421

[3]Chart, Biomedical Division, Chart Inc., Troy, New York 12188

[4]Raytheon Space and Airborne Systems
El Segundo, CA 90245 USA

ABSTRACT

Stirling-type pulse tube cryocoolers (PTC) are increasingly used in tactical applications as well as ground testing of space systems. Some PTCs exhibit sensitivity to gravitational orientation and often lose significant cooling performance unless situated with the cold end pointing downward. Previous investigations have indicated that some coolers exhibit sensitivity while others do not; however, a reliable method of predicting the level of sensitivity during the design process has not been developed. In this study, we attempt to derive a semi-analytical method that can be used to ensure that a PTC will remain functional in adverse static tilt conditions. The development of the method is based on experimentally-validated 3-D computational fluid dynamics (CFD) simulations that identify relationships between pulse tube geometry and operating conditions including frequency, mass flow rate, pressure ratio, mass-pressure phase, hot and cold end temperatures, and static tilt angle. The validation of the computational model is based on experimental data related to a number of pulse tube cryocoolers. Simulation results and experimental data are gathered to yield the relationship between inclined PTC performance and pulse tube convection numbers (N_{PTC}). The data hereby can offer guideline for PTC designs.

INTRODUCTION

Stirling-type pulse tube cryocoolers (PTC) are especially attractive for some applications because of their mechanical simplicity and long life reliability. Pulse tubes are critical components of which the primary function is to create phase lag of mass flow to pressure wave. Many PTCs experience significant losses of cooling power when the cold end is not pointing downward due to convective effects occurring in the pulse tube [1-3]. This phenomenon has been experimentally and numerically studied. Swift and Backhaus [4, 5] experimentally analyzed gravity orientation

Cryocoolers 19, edited by S.D. Miller and R.G. Ross, Jr.

sensitivities of PTCs and analytically formulated a theory to explain the phenomenon and offer guidelines to PTC designs. Berryhill and Spoor [6] found that their experimental results generally agreed with Swift and Backhaus's [4, 5] theory. But their work suggests that the safety factor in the theory may need to be modified in order to ensure reliable inclined condition performance.

Meanwhile, computational fluid dynamics (CFD) approaches have been used to model pulse tube flow and thermal physics as well. Taylor [7, 8] developed a two-dimensional, axisymmetric CFD model as a predictive tool of PTC designs. Some 3-D PTC CFD simulation approaches have also been implemented and experimentally validated [9-11]. Some parametric studies have been conducted based on 3-D CFD models [12].

In this study, which is a follow-up to an earlier investigation [9, 12] based on more data and analysis, a half-symmetric 3-D CFD model is used in component level pulse tube parametric analysis. Simulation and experimental results are gathered to exhibit the relationship between inclined PTC performance and pulse tube convection numbers (N_{PTC}).

PULSE TUBE THERMODYNAMICS AND CONVECTIVE INSTABILITY

The thermodynamics and convective instability of pulse tubes are briefly reviewed in this section. More detail can be found in [9, 12]. The relevant energy flows, and their directions, are noted in Figure 1 and consistent with a cycle-averaged condition. In Figure 1, boundary planes one (1) and two (2) represent the cold and warm ends, respectively.

Pulse Tube Thermodynamics

The pulse tube converts the Gibbs free energy to an enthalpy flow (travels from cold to warm). The Gibbs energy, also referred as the acoustic (PV) power and is expressed as,

$$\left\langle P\dot{V}\right\rangle = \frac{1}{2} P_d \dot{V}_d \cos\left(\theta_{m-p}\right) \tag{1}$$

where P_d is the dynamic pressure amplitude, $\dot{V}_d$ is the volume flow amplitude, and θ_{m-p} is the phase between mass flow and pressure. In an ideal pulse tube, the conversion is perfect and the enthalpy flow is equal to the Gibbs energy. In an ideal pulse tube with using the remaining terms in Figure 1, it can be shown that the total net energy flow in the pulse tube at a cyclic condition is,

$$\left\langle \dot{E}_{PT,net}\right\rangle = \left\langle \dot{H}\right\rangle - \left\langle \dot{Q}_g\right\rangle - \left\langle \dot{Q}_w\right\rangle \tag{2}$$

where $\left\langle \dot{H}\right\rangle$ is the pulse tube enthalpy flow, $\left\langle \dot{Q}_g\right\rangle$ is the conduction in the working fluid, and $\left\langle \dot{Q}_w\right\rangle$ is the conduction in the solid wall of the pulse tube. In practice, the acoustic power to enthalpy conversion is not perfect due to numerous complex loss mechanisms that include all forms of streaming, shuttle heat transport, turbulent mixing and jetting, and gravitationally induced instability. All these mechanisms reduce the available enthalpy flow in the pulse tube and subsequently the cooling power. Minimizing these losses has been the subject of numerous empirical and numerical studies over the years. The mechanisms responsible for all these losses are relatively well understood, with the exception of gravitationally-induced instability.

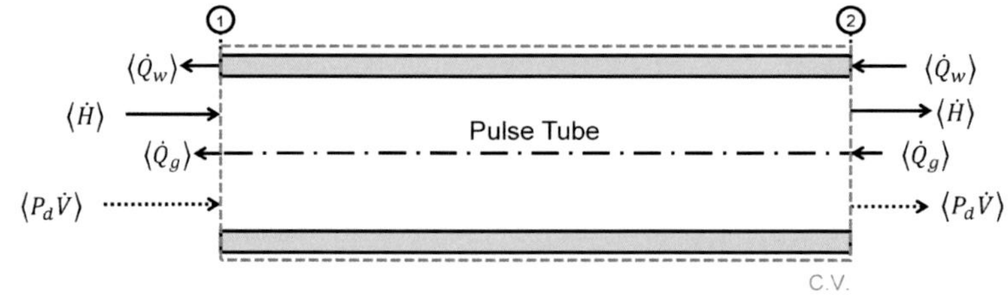

Figure 1. Illustration showing the relevant energy flows for a pulse tube. PV power is included in the energy flow diagram with dashed arrows for reference. The temperature gradient is cold to hot from left to right, i.e. from 1→2.

The gravitationally-induced stability loss in pulse tube coolers causes severe cooling power degradation when the pulse tube cryocooler is operated in a configuration where the temperature gradient (cold below warm) is not aligned with the gravitational field. Once the cooler deviates from the ideal gradient orientation with respect to gravity, the flow tends to become unstable, and an undesirable circulation cell develops when instability prevails.

In this work, we quantify the loss associated with the tilt effect as,

$$|L(\theta)| = \frac{\langle \dot{E}_{PT,net}(0)\rangle - \langle \dot{E}_{PT,net}(\theta)\rangle}{\langle \dot{E}_{PT,net}(0)\rangle} \tag{3}$$

where $\langle \dot{E}_{PT,net}(0)\rangle$ is the net pulse tube energy flow at zero degrees deviation from ideal and $\langle \dot{E}_{PT,net}(\theta)\rangle$ (i.e., when the cold tip is above the warm end of the cryocooler, and the axis of the cryocooler is vertical) is the net pulse tube energy flow at an angle θ removed from zero. A clear understanding of this loss and its parametric dependence on various design and operational parameters is required to enable the use of modern high-frequency pulse tube coolers in applications where operation under non-ideal cooler-gravitation alignment in expected.

Convective Instability

A classic example of convective instability arises in a quiescent enclosure initially containing quiescent fluid, in which the top and bottom surfaces are at differing temperatures and the vertical surfaces are adiabatic. For any specific no-trivial tilt angle at some critical temperature differential, the buoyant forces in the fluid overcome viscous forces and buoyncy-driven cells develop. The strength of these cells depends on the severity of the imbalance between buoyancy and viscous forces and is characterized using the Rayleigh number. The phenomenology of instability in a pulse tube is to some extent similar. In the pulse tube, unlike an otherwise quiescent fluid, there is a bulk velocity arising from the oscillating flow field. In this case the convection stability criteria can be related to a ratio of the inertial forces (and not viscous forces) in the oscillating fluid relative to the buoyant forces in the fluid.

The convective instability and conditions leading to its occurrence in pulse tube can of course be modeled using high-fidelity computational fluid dynamic (CFD) simulations, provided that such simulations are set up with correct boundary conditions and closure relations. Such CFD simulations are complex and computationally demanding given that they involve periodic multi-dimensional flow. Simpler, semi-analytical models suitable for iterative design and optimization calculations are thus needed.

Swift and Backhaus [4, 5] modeled the convective instability in a pulse tube based on an analogy with an inverted pendulum, noting that a statically unbalanced inverted pendulum can be dynamically stabilized using correct dynamic excitation at its base. This observation is consistent with experimental evidence related to high frequency pulse tube cryocoolers. Experiment has shown that increased operating frequency in pulse tube cryocoolers mitigates the convective instability caused by tilt from ideal vertical configuration. Using a semi-analytical model Swift and Backhaus [5] argued that the forthcoming non-dimensional pulse tube convection number could be used as a criterion to determine the vulnerability of a pulse tube cryocooler to the adverse effect of non-ideal tilt angle,

$$N_{PTC} = \frac{\omega^2 a^2}{g(\alpha_s D\sin\theta - L\cos\theta)}\sqrt{\left(\frac{\Delta T}{T_{avg}}\right)} \tag{4}$$

where ω is the angular frequency associated with the flow pulsation, a is the mass flow amplitude, g is gravitational acceleration, α_s is empirical fitting parameter, D and L are the pulse tube diameter and length, respectively, θ is the tilt angle from vertical, ΔT is the temperature difference between hot and cold, and T_{avg} is the average gas temperature in the pulse tube.

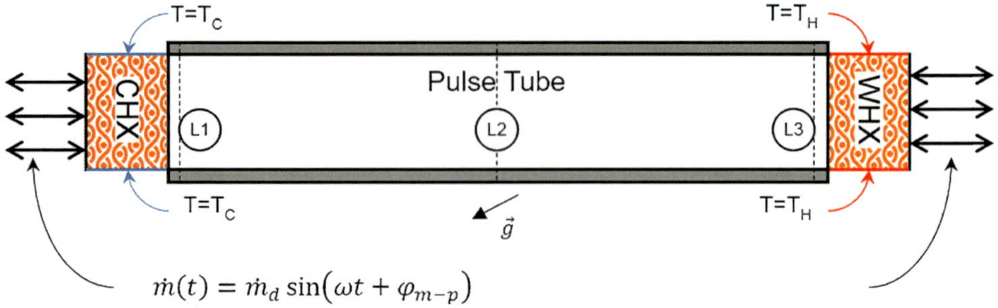

Figure 2. Schematic view of the computational domain along the plane of symmetry. Patterned sections represent porous domains. Solid filled wall section represents solid domain. Dashed lines represent computational monitor planes. Double headed arrows represent oscillatory boundary conditions.

Simulation Methods

A brief overview of the key characteristics of the ANSYS CFD [13] simulation methods used in this study will be given here. For further details of the simulation methods the reader is directed to [9], which describes two distinct methods for simulating the convective losses: a full-system approach which simulates an entire pulse tube cooler system, and an isolated model that includes only the pulse tube and adjacent heat exchangers. Figure 2 depicts an isolated system. The argument for the limitation of the simulation domain to the pulse tube and its adjacent heat exchangers is that these are the key components that are most sensitive to gravity. All simulations are performed in 3-D half-symmetry, with the symmetry plane coincident with the plane in which the gravity vector varies. It was shown earlier that an isolated model introduces only approximately 6% error to the predictions, while reducing the number of CPU's required for parallel processing by a factor of four and simultaneously reducing the simulation time by a factor of three. Validation of this model against a set of commercial cryocoolers is given in [10], where error was reported as a percentage of the net energy flow across the pulse tube. An average error of 3.7% and a maximum error of 8.2% were reported using this modeling methodology, with largest errors occurring when the cooler was driven below its design input power, resulting in very low Reynolds numbers and therefore was most unstable. Numerical simulation error magnitudes were in the range of approximately 100 mW, indicating that the simulation method is most aptly suited for pulse tube coolers with larger net cooling capacity.

An open system modeling method is used for the isolated domain consisting of the cold heat exchange (CHX), pulse tube, and warm heat exchanger (WHX) (See Figure 2.) A sinusoidal mass flow boundary condition with specified temperature, magnitude and phase angle relative to pressure oscillation is applied at the entrance to the CHX (the left side of the system shown in Fig. 2). A sinusoidal pressure with specified amplitude and frequency ω provides closure to the fluid domain at the warm heat exchangers. Isothermal boundary conditions are imposed at the cold and warm heat exchangers. Monitoring surfaces are defined across the pulse tube locations denoted L1, L2 and L3, which record the thermodynamic and hydrodynamic parameters required to compute and isolate energy flows across the domain. Conjugate heat transfer with the pulse tube wall is included to simulate thermally driven boundary phenomena that may impact off-axis performance. Gas properties are obtained from the NIST real gas models reported in the REFPROP database [14]. The heat exchangers are modeled as porous domains with hydrodynamic parameters based on those reported for packed screens in oscillatory flow [15].

The domain is spatially discretized using approximately 75,000 ordered mesh elements, the minimum quantity found to yield grid-independence with discretization error less than 0.5% of the net pulse tube energy flow. Inflation meshing was used within the fluid domain to capture boundary effects. The transient simulation was carried out to periodic steady state using time steps that yielded 400 samples per period of oscillation. Typical simulations required approximately 30,000 time steps for the solution to converge for convectively stable configurations. Simulations were carried out on high performance computing (HPC) cluster resources in order to compute the solution in

12 parallel processes, with a simulation duration of approximately 255 hours per case. In some instances 24 CPUs were used for each case, which reduced simulation time to 155 hours.

RESULTS

We examined 145 simulated pulse tube conditions with driving frequency ranging from 25 Hz to 60 Hz, cold end temperatures from 4 K to 80 K, rejection temperatures of 20 K to 300 K, mass flow rates from 1 g/s to 4 g/s, and mass flow phase angles of -30° to +30° relative to pressure. Two computational domains were used with length/diameter aspect ratio of eight (8), commonly used as a design minimum, and four (4), which is often required for larger capacity pulse tubes as cross-sectional area scales with mass flow while length does not. Experimentally measured off-axis performance for three commercial cryocoolers are also included in the analysis to add to the body of data from which to draw conclusions.

Upon completion, each simulation was post-processed to determine the net energy flow across the pulse tube at the cold end according to Equation 1, adjacent to the cold heat exchanger but within the gas domain outside the transition from porous to open flow channel. The energy flow data at varying inclination angle θ was compared to the energy flow in the vertical orientation via Equation 3, yielding the normalized energy flow loss $|L|$, for which a value of zero indicates no reduction in cooling performance at the specified angle, a value of one indicates complete loss of net cooling power, and a value greater than one indicates that the pulse transfers energy from the warm end to the cold end. Normalized energy flow loss is plotted in Figures 3, 4 and 5 as a function of the tilt angle, with the oscillatory Reynolds number Re^{δ} as a parameter. The Reynolds number based on Stokes' boundary layer thickness is used to characterize the flow and ranges from 43 (very laminar) to 350. The oscillatory Reynolds number at which transition to intermittent turbulence occurs is approximately 500 [16].

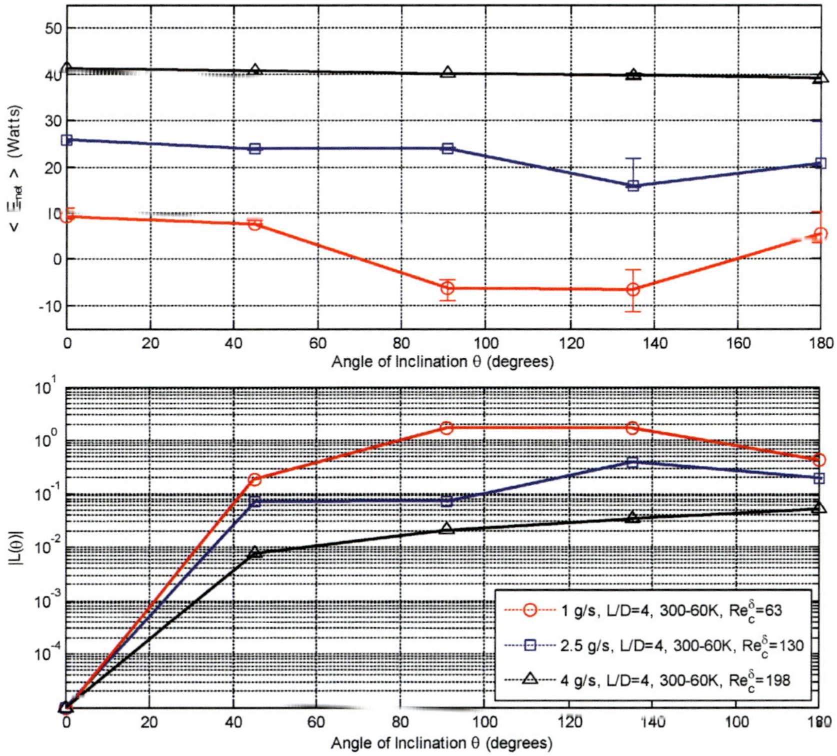

Figure 3. Net pulse tube energy flow and normalized loss coefficient for pulse tube with L/D=4 aspect ratio driven at 60 Hz at various mass flows in five discrete orientations with respect to gravity. Phase angle of -30° (mass lags pressure) and pressure ratio of 1.2 operating between 300 K and 60 K.

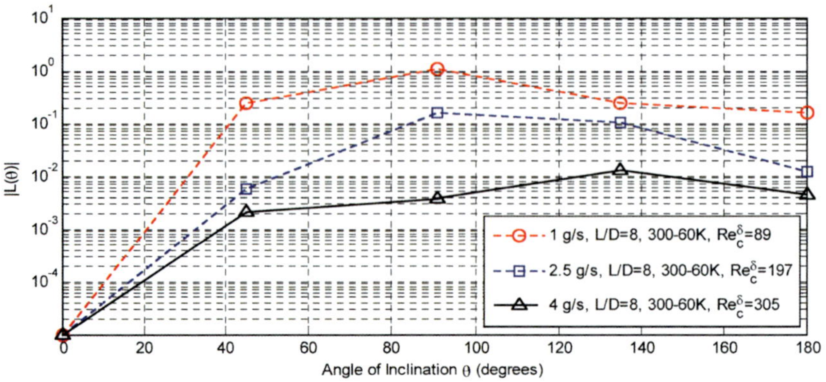

Figure 4. Net pulse tube energy flow and normalized loss coefficient for pulse tube with L/D=8 aspect ratio driven at 60 Hz at various mass flows in five discrete orientations with respect to gravity. Phase angle of -30° (mass lags pressure) and pressure ratio of 1.2 operating between 300 K and 60 K.

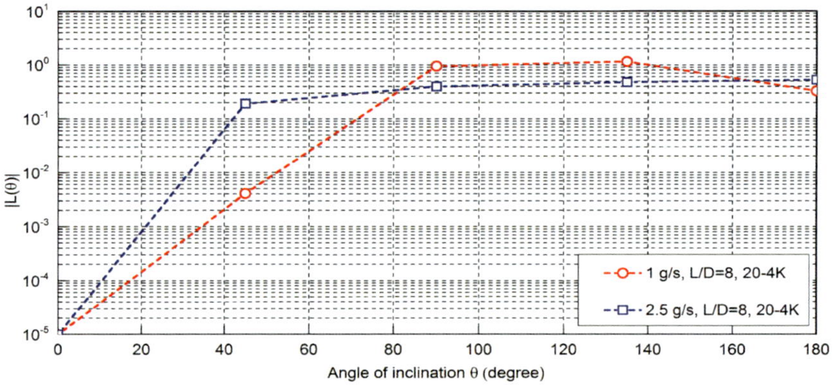

Figure 5. Net pulse tube energy flow and normalized loss coefficient for pulse tube with L/D=8 aspect ratio driven at 60 Hz at various mass flows in five discrete orientations with respect to gravity. Phase angle of -30° (mass lags pressure) and pressure ratio of 1.2 operating between 20 K and 4 K.

Figure 3 presents representative energy flow loss data as a function of the mass flow amplitude, and therefore the Reynolds number, for a pulse tube with length-to-diameter aspect ratio of four. The pressure used in all simulations was 2.0 MPa with a pressure amplitude of 400 kPa. The mass flow amplitude is directly related to the input power, and therefore the efficiency decreases significantly for higher mass flows at inclination angles above 45°. The error bars displayed in the net energy (4) under the typical pulse tube design condition that mass flow lags pressure by 30°. The average flow plots in Figures 3 and 4 are used for unstable cases to show the maximum and minimum energy flows observed over the final 0.25 s of simulated flow time. Larger error bars indicate a more unstable flow regime.

Figure 4 compares performance for a pulse tube with identical operating conditions to the results reported in Figure 3 and identical pulse tube gas volume but with an aspect ratio of eight (8). In this case, only a single operating point (1 g/s, 90° orientation) resulted in catastrophic loss of cooling, and an increase in the mass flow rate to 4 g/s reduced orientation sensitivity by two orders of magnitude.

Figure 5 shows normalized cooling of the pulse tube in the same working condition as shown in Fig. 4, except the pulse tube cryocooler operates between 20 K and 4 K. In this case, the performance of the pulse tube was more sensitive to gravity orientation than when the pulse tube operated between 300 K and 60 K temperatures. Two points (1 g/s, 90° and 135° orientation) shown catastrophic loss of cooling. Other points also presented more loss of cooling in comparison with Fig. 4.

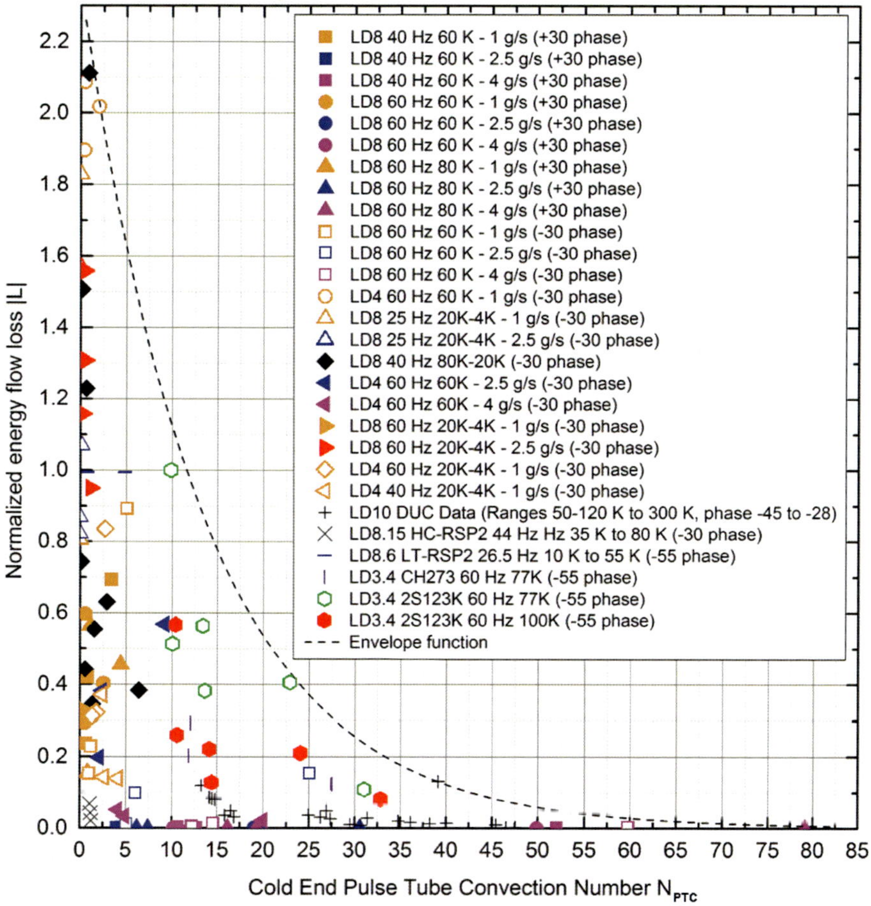

Figure 6. |L| vs N_{PTC} results. Legend entries indicate the aspect ratio (LD), driving frequency, cold end temperature, mass flow, and phase angle of mass flow relative to pressure (dash line indicates the envelop function).

In extreme low temperatures (in this case 20 K to 4 K), inclination effects are more severe because of the stronger convection resulted from larger variances in density.

In order to assimilate all data sets collected, the pulse tube convection number N_{PTC} defined by Equation 4 was computed for each case. Figure 6 shows the normalized energy flow loss |L| as a function of the computed pulse tube convection number N_{PTC} for all numerically simulated and experimentally measured data collected in this study. There is considerable scatter in the data depicted in Figure 6. This is understandable in view of the simplicity of the model and the complexity of the thermo-fluid phenomena in a pulse tube. Thus, the pulse tube convection number should only be used as an order of magnitude indicator of the orientation sensitivity. The calculation of the expected loss in energy flow of a pulse tube cryocooler may need a detailed CFD simulation.

The convective flow patterns can be qualitatively different for different conditions and N_{PTC} may not be able to fully characterize a coolers' performance. Figure 7 compares the flow patterns at different inclination angles. The sample case had the parameters of LD8, 60 Hz, 1 g/s mass flow, -30° phase lag of mass to pressure, and warm and cold heat exchanger temperatures of 20 K and 4 K, respectively. The temperature gradients in the middle part of the pulse tube indicated clear convective flow patterns for 90°, 135° and 180° conditions. However, the patterns were not the same at different angles, even in terms of their overall structure. Figure 8 depicts the convective flow patterns for 90° and 180° tilt angles. Contours of several time steps within a cooling cycle are shown in

Figure 7. Sample temperature contours (LD8, 60 Hz, 1 g/s, 20K-4K, -30 phase) at inclination angle of 0, 45, 90, 135 and 180 degrees.

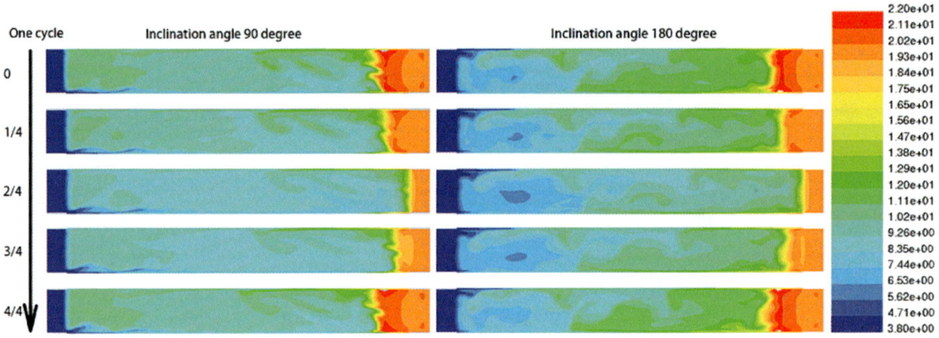

Figure 8. Temperature contours (LD8, 60 Hz, 1 g/s, 20K-4K, -30 phase) of 90 and 180 degrees inclination angle at 0/4, 1/4, 2/4, 3/4 and 4/4 point of a full cooling cycle.

this figure. Flow patterns for the same cases were shown in Fig. 7. At 90° inclination, there was a strong convective circulation that persisted during the entire cooling cycle and could not be broke down by oscillation. On the other hand, the convective flow inside the pulse tube at 180° is chaotic, indicating no clear flow circulation pattern, and instead suggests relatively localized mixing near the warm end of the pulse tube. Convective flow was not traveling from one end of the pulse tube to the other end, as could be observed in the case of 90° tilt, for example. As a result, the thermal stratification at 180° tilt angle was better than the thermal stratification at 90° tilt angle, which implied weaker undesirable convective heat transfer at 180° tilt angle. As a result, as noted in Figs. 3, 4 and 5, the 180° cases generally had a smaller cooling loss than the 90° case, even though N_{PTC} at 180° is smaller than at 90°.

DISCUSSION

In general, observation of the data presented in Figure 6 yields interesting trends. There is notable evidence that use of higher operating frequency, when possible, is advisable. Furthermore, larger aspect ratio pulse tubes generally perform better off-axis for temperatures above ~60 K. For temperatures below ~15 K, off-axis effects are noticeably more severe as expected due to the larger disparity in density between cold and warm ends when compared to more moderate temperature cases. Finally, larger mass flow amplitudes generally serve to reduce the off-axis sensitivity. This is attributed to larger inertial forces, which are stabilizing when compared to the buoyancy forces.

All of the observed cases reported in Figure 6 report orientation sensitivity in terms of normalized energy flow loss below a bounding function which is fitted to the largest observed loss in over the full range studied. These extreme points are fitted with an exponential decay function of the form shown in Equation 5. All points simulated and measured fall below the cutoff line described by Equation 5.

$$|L(\theta)|_{max} \leq 2.3446e^{-0.0738N_{PTC}} \quad (5)$$

The utility of the data reported herein is the ability to determine the expected order of magnitude loss as a function of a calculated pulse tube convection number for a given theoretical design. This factor can be incorporated in a scoping level study to avoid designs that indicate undesirable orientation sensitivity. Once a prospective design is developed, it should be examined using the above described 3-D CFD methodology at the worst anticipated angle of inclination to verify the level of sensitivity prior to fabrication.

CONCLUSIONS

Three-dimensional CFD simulation has been successfully applied to predict performance losses due to gravity-driven convective instability in pulse tube cryocoolers over a wide range of operating characteristics and two geometries. The results were characterized in terms of a nondimensional loss of net energy flow as well as the nondimensional pulse tube convection number. The results indicate that the upper limit of the magnitude of cooling power loss decays exponentially with increasing pulse tube convection number and sensitivity is minimized by applying the design which yields the highest laminar Reynolds number in the pulse tube. The reported relationship in this article should be used for scoping purposes only. Complete thermodynamic designs should rely on detailed 3-D CFD-assisted simulations using correct boundary conditions and closure relations.

REFERENCES

1. Thummes, G., M. Schreiber, R. Landgraf, and C. Heiden, "Convective heat losses in pulse tube coolers: effect of pulse tube inclination," *Cryocoolers 9*, Plenum Publishers, New York (1997), pp. 393-402.
2. Ross, Ronald G., and Dean L. Johnson. "Effect of gravity orientation on the thermal performance of Stirling-type pulse tube cryocoolers," *Cryogenics,* vol. 44, no. 6 (2004), pp. 403-408.
3. Shiraishi, M., Takamatsu, K., Murakami, M., & Nakano, A., "Dependence of convective secondary flow on inclination angle in an inclined pulse tube refrigerator revealed by visualization," *Cryogenics,* vol. 44, no. 2 (2004), pp. 101-107.
4. Swift, G.W. and Backhaus, S., "Why high-frequency pulse tubes can be tipped," *Cryocoolers 16*, ICC Press, Boulder, CO (2011), pp. 183-192.
5. Swift, G.W. and Backhaus, S., "The pulse tube and the pendulum," *Journal of the Acoustical Society of America*, vol. 126, no. 5 (2009), pp. 2273-2284.
6. Berryhill, A. and Spoor, P.S., "High-frequency pulse tubes can't always be tipped," *Advances in Cryogenic Engineering*, vol. 1434, (2012), pp. 1593-1599.
7. Taylor, R.P., "Development and Experimental Validation of a Pulse-Tube Design Tool Using Computational Fluid Dynamics," PhD thesis, The University of Wisconsin-Madison, 2009.
8. Taylor, R.P., Nellis, G. F., and Klein, S. A., "Optimal pulse tube design using computational fluid dynamics," *Advances in Cryogenic Engineering* (J. G. Weisand, I., ed.), American Institute of Physics, vol. 53B (2007), pp. 1445-1451.
9. Mulcahey, T.I., "Convective Instability of Oscillatory Flow in Pulse Tube Cryocoolers due to Asymmetric Gravitational Body Force," PhD thesis, Georgia Institute of Technology, (2014).
10. Mulcahey, T.I., Conrad, T.J., Ghiaasiaan, S.M., and Pathak, M.G., "Investigation of gravitational effects in pulse tube cryocoolers using 3-D CFD," *Advances in Cryogenic Engineering: Transactions of the Cryogenic Engineering Conference - CEC* (II, J. G. W., ed.), vol. 59, (2013).
11. Mulcahey, T. I., Conrad, T. J., and Ghiaasiaan, S. M., "CFD modeling of tilt induced cooling losses in inertance tube pulse tube cryocoolers," *Cryocoolers 17*, ICC Press, Boulder, CO (2012), pp. 143-150.

12. Mulcahey T.I., Taylor R.P., Ghiaasiaan S.M., Kirkconnell C.S., Pathak M.G., "Parametric Numerical Study of Off-Axis Pulse Tube Losses," *Cryocoolers 18*, ICC Press, Boulder, CO (2014), pp. 253-261.

13. ANSYS, ANSYS FLUENT User's Guide. No. Release 14.5, Canonsburg, PA 15317: Southpointe, (2012).

14. Lemmon, E.W. and Huber, M.L., REFPROP: NIST Materials Reference Properties Database: Version 9.1. NIST, (2013).

15. Cha, J.S., Ghiaasiaan, S.M., and Kirkconnell, C.S., "Measurement of anisotropic hydrodynamic parameters of pulse tube or Stirling cryocooler regenerators," *Advances in Cryogenic Engineering*, Vol. 51, Amer. Institute of Physics, Melville, NY (2006), pp. 1911-1918.

16. Akhavan, R., Kamm, R. D., and Shapiro, A. H., "An investigation of transition to turbulence in bounded oscillatory stokes flows part 1: experiments," *J. Fluid Mechanics*, vol. 225, (1991), pp. 395-422.

Electric-Mechanical-Acoustic Coupling Characteristics for Pulse Tube Cryocoolers

L. Y. Wang, Z. H. Gan, Y. X. Guo, B.Wang, S. H. Wang

Institute of Refrigeration and Cryogenics, Zhejiang University
Hangzhou 310027, P.R. China

Key Laboratory of Refrigeration and Cryogenic Technology of Zhejiang Province
Hangzhou 310027, P.R. China

ABSTRACT

A pulse tube cooler driven by a linear compressor can be equivalent to a network consisting of electric, mechanical and acoustic parts, among which the coupling is a critical issue for improving cooling performance. However, it is still a research area that remains poorly understood. In this work, we develop a coupled phasor diagram containing electric, mechanical and acoustic parts using a vector analysis method. The phasor analysis reveals the coupling characteristics inside a pulse tube cryocooler, especially the electric-to-acoustic conversion efficiency, the power factor, and so on. This is expected to help with the design and performance improvement of a pulse tube cryocooler.

INTRODUCTION

Since the Oxford-style linear compressor was introduced by Davey in 1981, it has become a key element that ensures the high reliability and high efficiency of many Stirling-type cryocoolers.[1,2] For analysis purposes, the compressor can be divided into three parts: the electric, the mechanical, and the acoustic impedance parts.[3,4] The operation of the linear compressor comprises complex transition processes among these three parts. Coordination and coupling among each part are important for the cryocooler to achieve its best performance. Previous studies have reported on the dynamic characteristics of linear compressors, and the acoustic impedance on the piston was modeled using an equivalent damping coefficient and a gas spring stiffness.[5,6] Wakeland used an equivalent circuit model to analyze the performance and optimized the acoustic resistance under the resonant condition.[3] Swift also derived the efficiency using acoustic impedance analyses, and optimized the piston area at resonance.[7] Radebaugh et al. discussed the acoustic impedance match based on a given commercial linear compressor.[8] Dai et al. analyzed the impedance matching principle between the linear compressor and the cold head from the perspective of energy balance and derived the criteria for both maximizing the efficiency and maximizing power output.[9] Although much has been learned, some topics remain poorly understood. How to choose the best working condition for a given linear compressor? Are there any possible coupling methods in case the linear compressor does not match the cooler? And, how about the coupling characteristics among the entire electric-mechanical-acoustic parts?

This paper tries to answer these questions. A specific impedance point, called the 'sweet spot' is proposed aiming to obtain the highest efficiency as well as the maximum power output at the same

Cryocoolers 19, edited by S.D. Miller and R.G. Ross, Jr.

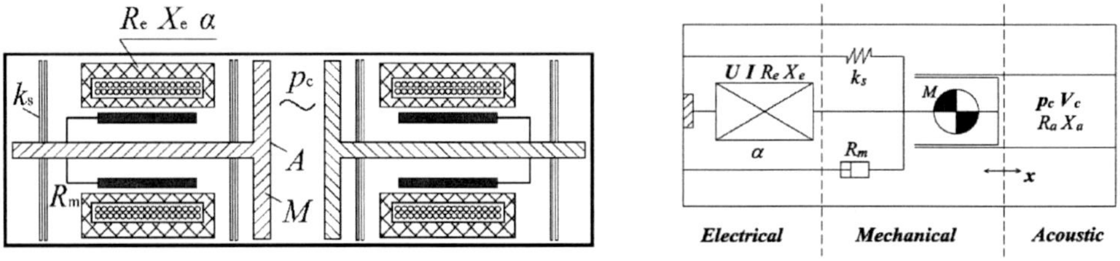

Figure 1. Schematic diagram and physical model of a linear compressor.

time. Some possible additional impedance matching methods are introduced based on an electrical impedance matching network. Finally, an electric-mechanical-acoustic coupling phasor diagram is introduced to analyze the coupling characteristics of the entire system.

'SWEET SPOT' FOR A LINEAR COMPRESSOR

The linear compressor shown in Fig. 1 is a kind of electrodynamic device that can be described by the following linear harmonic-approximation equations[7]:

$$\boldsymbol{U} = R_e \boldsymbol{I} + j\omega X_e \boldsymbol{I} + \alpha \boldsymbol{v} \tag{1}$$

$$M\boldsymbol{a} = \alpha \boldsymbol{I} - R_m \boldsymbol{v} - k_s \boldsymbol{x} - \boldsymbol{p}_c A \tag{2}$$

where $\boldsymbol{U}$ is the complex voltage across the electric terminals, with bold representing complex vectors, R_e is the electric resistance, $\boldsymbol{I}$ is the complex electric current, X_e is the electric inductance, α is the transduction coefficient which is the product of the magnetic field and the length of the wire in the field, $\boldsymbol{x}$ is the piston displacement, $\boldsymbol{v}$ is its velocity and $\boldsymbol{a}$ is its acceleration, V_c is the volume flow rate at the piston face, $\boldsymbol{p}_c$ is the pressure amplitude inside the compression volume, R_a is the acoustic resistance and X_a is the acoustic reactance, A is the piston area, R_m is the mechanical resistance, j equals to $\sqrt{-1}$, M is the moving mass, k_s is the mechanical spring stiffness. These two equations express Ohm's law and Newton's law, respectively.

We try to find a specific working condition for a given linear compressor to obtain its highest efficiency as well as maximum power output at the same time. To achieve the highest efficiency, both R_a and X_a should be fixed meeting the equations[3]:

$$A^2 X_a = \frac{k}{\omega} - m\omega \tag{3}$$

$$A^2 R_a = R_m \sqrt{\frac{\alpha^2}{R_e R_m} + 1} \tag{4}$$

where $\omega = 2\pi f$ is angular frequency. To achieve the maximum power output, the following equation should be satisfied which means that the current and the displacement reaches their limits at the same time[9]:

$$\frac{\alpha I_{\max}}{\omega x_{\max}} = A^2 R_a + R_m \tag{5}$$

Combining Equations (3)–(5), we obtain

$$\frac{\omega}{\alpha} R_m \left(\sqrt{\frac{\alpha^2}{R_e R_m} + 1} + 1 \right) = \frac{I_{\max}}{x_{\max}} \tag{6}$$

In Equation (6), there is only one parameter that can be adjusted for a given linear compressor, that is the frequency ω. Also, there is only one acoustic impedance point (R_a, X_a, f) to meet this requirement. This specific impedance point is called the 'sweet spot' for a linear compressor.

Taking the existing commercial linear compressor Qdrive 2S132W[10] in our lab for instance, its parameters are listed in Table 1.

Table 1. Parameters of CFIC 2S132W linear compressor (single motor).

Parameters	R_e (Ω)	k_s (N/m)	R_m (N·s/m)	α (N/A)	M (kg)	d_p (m)	I_{max} (A)	X_{max} (mm)
Value	0.671	44800	10.7	23.5	0.721	0.0394	12.02	6.0

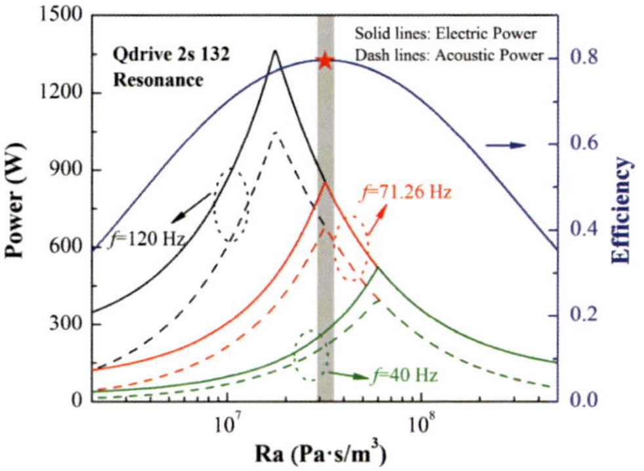

Figure 2. Efficiency and power for different frequencies under resonance of Qdrive 2S132W

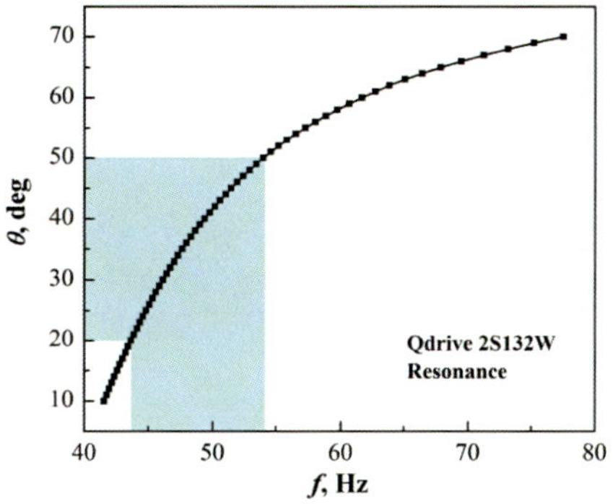

Figure 3. Phase angle vs. frequency under resonance of Qdrive 2S132W

Substituting these parameters from Table 1 into Equation (6), we get the specific frequency:

$$f = \frac{1}{2\pi}\frac{\alpha I_{max}}{x_{max} R_m\left(\sqrt{\frac{\alpha^2}{R_e R_m}+1}+1\right)} = 71.26 \text{ Hz}$$

Calculations for three different frequencies (40 Hz, 71.26 Hz and 120 Hz) under resonance are carried out as shown in Fig. 2. The efficiency curves are overlapped, which means the efficiency has nothing to do with frequency as long as resonance is achieved. As for the power output, for all three frequencies, when R_a is small (e.g. less that about 3.0e7 Pa·s/m^3 for 71.26 Hz), the displacement reaches its upper limit before the current does; that is, the maximum displacement limits the power input and the power output. When R_a is large (e.g. larger that about 3.0e7 Pa·s/m^3 for 71.26 Hz), the current reaches its upper limit before the displacement does. In this case, the maximum current limits the power input and the power output. As a result, there is a specific R_a to reach the maximum power output for three frequencies (R_a meeting Equation (5)), and the higher the frequency is, the higher the maximum power output will be among these frequencies. The 71.26 Hz frequency is the one where the highest efficiency accompanies the maximum power output; this is the frequency the linear compressor should work at so as to fulfill its capability.

However, we note that the rated working frequency of this compressor is 60 Hz[10], which was also used in our previous work.[11] We turn to the cryocooler side to see how this frequency affects the phase angle θ between mass flow and pressure wave at the compressor outlet. For the acoustic impedance we have:

$$X_a = -R_a \tan\theta \tag{7}$$

Combining the resonant Equation (3), we obtain the relation between f and θ:

$$f = \frac{1}{2\pi}\frac{1}{2M}\left[A^2 R_a \tan\theta + \sqrt{A^4 R_a^2 \tan^2\theta + 4Mk_s}\right] \tag{8}$$

To achieve the highest efficiency, Equation (4) should be satisfied. By substituting parameters from Table 1, we obtain the relation between f and θ, as shown in Fig. 3. It is found that θ increases as f increases. Generally, θ should be between 20~50 degrees (mass flow leading the pressure wave) for a cryocooler, which needs the f to be between about 43~54 Hz in this case shown from Fig. 3. But this frequency region is lower than the 'sweet' frequency 71.26 Hz, which means that

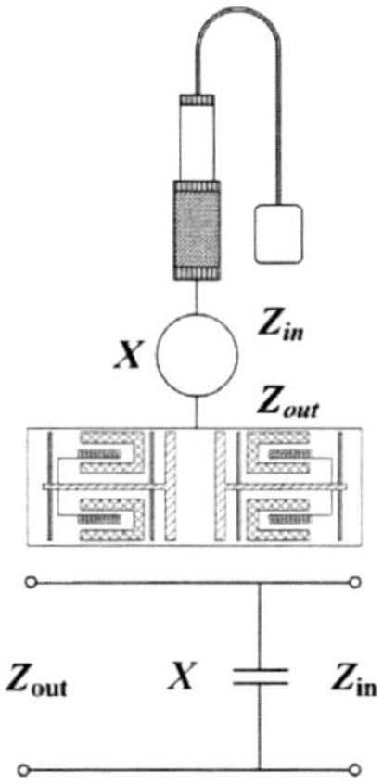

Figure 4. Extra volume matching and its equivalent circuit.

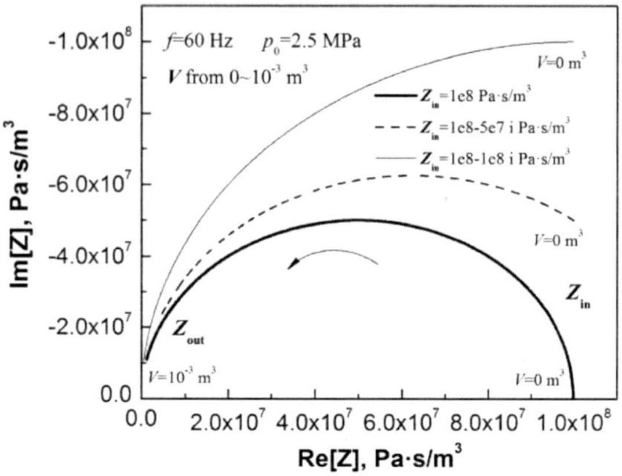

Figure 5. Effects of extra volume on Zout.

power output is limited as shown from Fig. 2. On the other hand, if driving the linear compressor at 71.26 Hz, θ will be much larger for the cryocooler according to Fig. 3. This is one of the cases that mismatch between the linear compressor and the cryocooler occurs; here extra matching methods are needed.

POSSIBLE IMPEDANCE MATCHING METHODS

Additional matching methods are needed to ensure high efficiency of both the linear compressor and the cryocooler, and no extra dissipation is expected to be introduced. The most simple way is to add a buffer volume between them (in series or in parallel). Figure 4 shows the structure and its equivalent circuit. Here Z_{in} is the acoustic impedance of the cooler, X is the reactance of the volume, and Z_{out} is the acoustic impedance at the compressor outlet. We have the equation:

$$\boldsymbol{Z}_{out} = \frac{1}{1/\boldsymbol{Z}_{in} + 1/\boldsymbol{X}} \tag{9}$$

We calculate the effects of volume (0~10^{-3} m^3) on Z_{out} for different Z_{in}, as shown in Fig. 5. It is seen that the extra volume can adjust the acoustic impedance in a quite wide range, but for each Z_{in}, Z_{out} will vary along a specific curve as the volume increases. This matching method can not realize the impedance match between an original Z_{in} and an arbitrary Z_{out}.

In the electronics field, an impedance matching network is widely used.[12] Here we introduce the simple L type double component impedance matching network. The four possible matching structures and their equivalent circuits are shown in Fig. 6, in which the buffer volume is analogous to electric capacitance and the inertance tube is analogous to electric inductance.

These matching networks have a much wider impedance adjusting range (other than a specific curve). Taking the first network in Fig. 6 as an example, it is described by the following equation:

$$\boldsymbol{Z}_{\text{out}} = \frac{1}{1/\boldsymbol{Z}_{\text{in}} + \mathrm{i}\omega C} + \mathrm{i}\omega L \tag{10}$$

where C is the acoustic capacitance of the volume, and L is the acoustic inertance of the inertance tube. Writing Z_{in} and Z_{out} in the forms:

$$\begin{aligned} \boldsymbol{Z}_{\text{in}} &= R_{\text{in}} + \mathrm{i}X_{\text{in}} \\ \boldsymbol{Z}_{\text{out}} &= R_{\text{out}} + \mathrm{i}\,X_{\text{out}} \end{aligned} \tag{11}$$

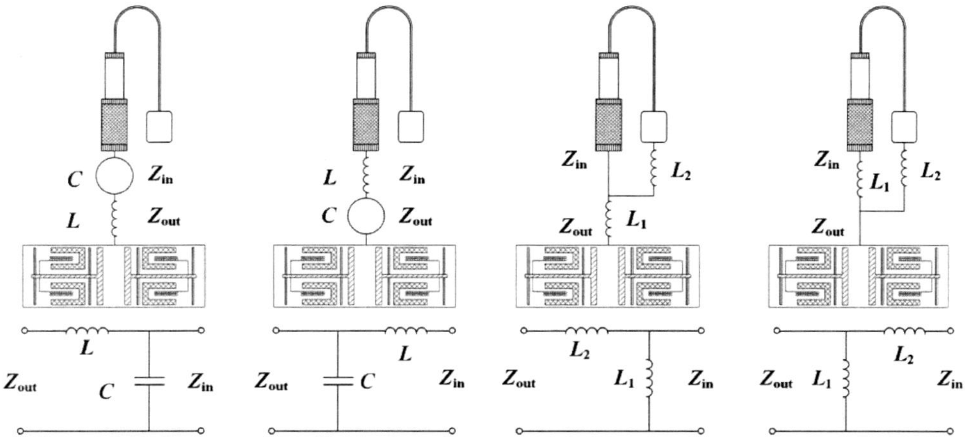

Figure 6. Four L type impedance matching networks and their equivalent circuits

By substituting Equation (11) into Equation (10), we can derive the required C and L:

$$C=\frac{1}{\omega}\frac{X_{\text{in}}\pm\sqrt{\frac{R_{\text{in}}}{R_{\text{out}}}\left(R_{\text{in}}^{2}+X_{\text{in}}^{2}\right)-R_{\text{in}}^{2}}}{R_{\text{in}}^{2}+X_{\text{in}}^{2}}$$
$$L=\frac{1}{\omega}\left[\frac{1}{\omega C}-\frac{R_{\text{out}}\left(1-\omega CX_{\text{in}}\right)}{\omega CX_{\text{in}}}+X_{\text{out}}\right] \tag{12}$$

That is, for any given Z_{in} and Z_{out}, Z_{out} could be achieved at the compressor outlet by using the L type impedance matching network with C and L calculated from Equation (12). There are still some forbidden zones that can not be reached, these are not discussed in this paper.

ELECTRIC-MECHANICAL-ACOUSTIC COUPLING

In a Stirling-type cryocoolers driven by a linear compressor, most parameters inside are sinusoidal. Thus, a phasor analysis provides a convenient way to describe the phase relationship between mass flow and pressure wave inside the cooler[13], the forces acting on the piston[14], as well as the voltage balance within the electric elements.[15]

For the voltage balance, Equation (1), the first term on the right side represents the voltage drop in the electric resistance that is in phase with current. The second term on the right side represents the voltage change due to electric inductance, which is 90° out of phase with current. The third term on the right side represents the induced voltage due to the piston movement, which is in phase with piston velocity. Figure 7(a) shows the phasors of each term in the complex frequency domain, in which $\boldsymbol{I}$ is chosen to be the x-axis. The sum of the above three vectors is the total voltage $\boldsymbol{U}$. It should be noted that the phase angle between $\boldsymbol{U}$ and $\boldsymbol{I}$, ψ, has a cosine value known as the power factor.

In the force balance, Equation (2), the inertial force $M\boldsymbol{a}$ is in phase with piston acceleration. For each term on the right side, the motor force $\alpha\boldsymbol{I}$ is in phase with current, the damping force $R_m\boldsymbol{v}$ is in phase with piston velocity, the mechanical spring force $k_s\boldsymbol{x}$ is in phase with piston displacement, and the gas force $\boldsymbol{p}_cA$ is in phase with pressure wave. Figure 7(b) shows the phasors of each term in the complex frequency domain, in which $\boldsymbol{x}$ is chosen to be the x-axis. Here the phase angle between $\boldsymbol{v}$ and $\boldsymbol{I}$, β, is a critical angle for a linear compressor. When $\beta = 0$, the motor force is minimized for the same power output, which means the highest efficiency is obtained; this condition is known as resonance. What's more, the phase angle between $\boldsymbol{v}$ and $\boldsymbol{p}_c$, φ, depends on the particular cold head attached to the compressor. For the cold head side, according to the equation for conservation of

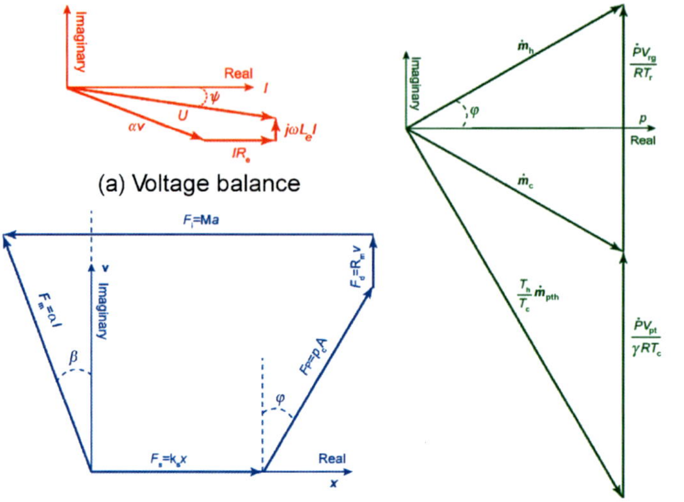

Figure 7. Phasor diagrams of electric, mechanical and acoustic parts.

mass, we have[13]:

$$\dot{m}_h = \dot{m}_c + \frac{\dot{P}V_{reg}}{RT_a} \qquad \dot{m}_c = \frac{T_{ht}}{T_c}\dot{m}_{ht} + \frac{\dot{P}V_{pt}}{\gamma RT_c} \tag{13}$$

where $\dot{m}$ is mass flow rate, $\dot{P}$ is the rate of change of pressure wave, V is the gas volume of the element, R is the gas constant per unit mass, T_a is the average temperature in the component, γ is the ratio of the specific heats, subscripts h, c, ht, reg, pt means hot end of the regenerator, cold end, hot end of the pulse tube, regenerator, pulse tube, respectively. According to Equation (13), we obtain the phasor diagram of the mass flow and pressure wave in the cold head, shown as Fig. 7(c). Here, to achieve the best performance of a regenerator, generally the optimum phase relationship is that $\boldsymbol{m}$ and $\boldsymbol{p}$ are in phase at approximately the midpoint of the regenerator.[16] Typically the flow at the cold end will lag the pressure by about 30°, whereas at the warm end the flow leads the pressure by about 30°. To meet such a phase relationship, the flow will always lag the pressure by about 60° at the warm end of the pulse tube, as shown in Fig. 7(c). Here the phase angle between $\boldsymbol{m}$ and $\boldsymbol{p}$ at the hot end of the regenerator, φ, is same as that in Fig. 7(b).

All the analyses above are in the same complex frequency domain. It would be helpful to combine the three phasor diagrams together, to see how the different parts couple with each other. Fig. 8 shows the joint phasor diagram of the entire electric–mechanical–acoustic parts in a cryocooler which is a combination of the three diagrams from Fig. 7. In this figure, the electric and mechanical parts are coupled by $\boldsymbol{v}$ and $\boldsymbol{I}$, while the mechanical and acoustic parts are coupled by $\boldsymbol{v}$ ($\dot{m}$) and $\boldsymbol{p}_c$. The relative magnitudes of different parts are only schematic.

Figure 8 is helpful in the design, analysis and optimization of a cryocooler, especially in the understanding of coupling among the three different parts. For example, on the one hand, in order to ensure good cooling performance of the cold head, the phase angle φ should be around 30°. This in turn requires a reasonable design of the linear compressor so as to maintain such a phase relationship between $\boldsymbol{v}$ and $\boldsymbol{p}_c$. On the other hand, the linear compressor should work around the resonance to get its highest efficiency; this needs the phase angle β to be near 0°. The requests above can be satisfied by adjusting the parameters in the compressor, e.g. M, k_s, A.

In some applications, people are concerned about the power factor (that is the phase angle ψ). It is noted from Figure 8 that $\psi = 0$ and $\beta = 0$ can not occur at the same time as long as $L_e \neq 0$, that is, power factor equals 1 doesn't mean resonance and vice versa.[4] Sometimes attention should also be paid to power factor to avoid too high a voltage. In this case, the compressor efficiency should be sacrificed a little ($\beta \neq 0$, and $\boldsymbol{v}$ leading $\boldsymbol{I}$).

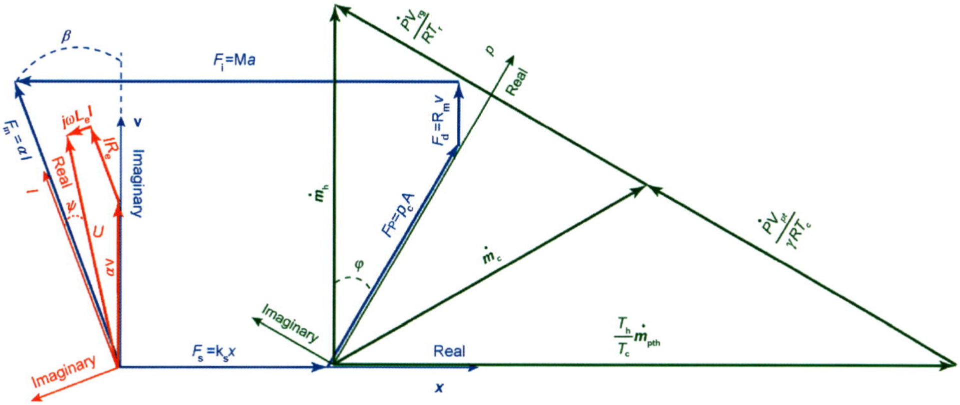

Figure 8. Joint electric-mechanical-acoustic coupling phasor diagram

All the discussions here are aimed at the phase angles in Fig. 8. Similar coupling analyses can also be done specific to magnitudes of each vector. Figure 8 reveals the intrinsic coupling mechanism between different parts, and provides references for reaching the best performance of a Stirling-type cryocooler.

CONCLUSIONS

Study on the coupling characteristics of a cryocooler is carried out. In order to fulfill the capability of a linear compressor, it should be operated around such a 'sweep spot' to obtain the highest efficiency and maximum power output at the same time. Several possible impedance matching methods in analogy with electrical matching networks are introduced if the compressor doesn't match the cold head. A joint electric-mechanical-acoustic coupling phasor diagram is achieved. It reveals the intrinsic coupling mechanism between different parts and provides references for reaching the best performance of a Stirling-type cryocooler.

ACKNOWLEDGMENTS

This work is financially supported by the National Natural Science Foundation of China (No. 51376157) and the Specialized Research Fund for the Doctoral Program of Higher Education of China (No. 20130101110098). The authors would like to express their gratitude to Dr. Ray Radebaugh for his helpful discussion at Zhejiang University.

REFERENCES

1. Davey, G., "The Oxford University Miniature Cryogenic Refrigerator," *International Conference on Advanced infrared detectors and systems*, (1981), pp. 39.

2. Davey, G., "Review of the Oxford Cryocooler," *Adv. in Cryogenic Engineering,* Vol. 35B, Plenum Publishing Corp., New York (1990), pp. 1423-1430.

3. Wakeland, R.S., "Use of Electrodynamic Drivers in Thermoacoustic Refrigerators," *Journal of the Acoustical Society of America*, vol. 107, no, 2 (2000), pp. 827-832.

4. Gan, Z.H., Wang L.Y., Zhao S.Y., Song Y.J., Wang W.W., and Wu Y.N., "Acoustic Impedance Characteristics of Linear Compressors," *Journal of Zhejiang University-Science A*, vol. 14, no, 7 (2013), pp. 494-503.

5. Koh, D.Y., Hong Y.J., Park S.J., Kim H.B., and Lee K.S., "A Study on the Linear Compressor Characteristics of the Stirling Cryocooler," *Cryogenics*, vol. 42, no, 6-7 (2002), pp. 427-432.

6. Chen, N., Tang Y.J., Wu Y.N., Chen X., and Xu L., "Study on Static and Dynamic Characteristics of Moving Magnet Linearcompressors," *Cryogenics*, vol. 47, no, (2007), pp. 457-467.

7. Swift, G., *Thermoacoustics: A Unifying Perspective for Some Engines and Refrigerators*. 2002, Los Alamos National Laboratory: Acoustical Society of America Publications.

8. Radebaugh, R., Garaway I., and Veprik A.M., "Development of Miniature, High Frequency Pulse Tube Cryocoolers," Proc. SPIE 7660, Infrared Technology and Applications XXXVI, 76602J (4 May 2010); doi: 10.1117/12.852766.

9. Dai, W., Luo E.C., Wang X.T., and Wu Z.H., "Impedance Match for Stirling Type Cryocoolers," *Cryogenics*, vol. 51, no, 4 (2011), pp. 168-172.

10. http://www.Chartindustries.Com/Industry/Industry-Products/Gas-Systems/Qdrive/Qdrive-Products/Pressure-Wave-Generators/Pressure-Wave-Generators-2s132w.

11. Wang, L.Y., Wu M., Sun X., and Gan Z.H., "A Cascade Pulse Tube Cooler Capable of Energy Recovery," *Applied Energy*, vol. 164, no, (2016), pp. 572-578.

12. Matthaei, G., Jones E.M.T., and Young L., *Microwave Filters, Impedance-Matching Networks, and Coupling Structures*. Artech Microwave Library. 1980: Artech House.

13. Radebaugh, R., "Thermodynamics of Regenerative Refrigerators," *Generation of Low Temperature and It's Applications*, (2003), pp. 1-20.

14. Marquardt, E. and Radebaugh R., "Design Equations and Scaling Laws for Linear Compressors with Flexure Springs," *7th International Cryocooler Conference Proceedings*, Air Force Phillips Laboratory Report PL-CP--93-1001, Kirtland Air Force Base, NM, April 1993, pp. 783-804.

15. Radebaugh, R., Lewis M., and Bradley P., "Verification of the Back-Emf Method for Piston Velocity Measurements," *Cryocoolers 17*, ICC Press, Boulder, CO (2012), pp. 357-365.

16. Radebaugh, R., Lewis M., Luo E., Pfotenhauer J.M., Nellis G.F., and Schunk L.A., "Inertance Tube Optimization for Pulse Tube Refrigerators," *Advances in Cryogenic Engineering, Vols 51A and B*, Amer. Institute of Physics, Melville, NY (2006), pp. 59-67.

Passive Mechanical Device for Phase Shifting in a Pulse Tube Cryocooler

D. Radchenko[1], G. Grossman[2]

[1]Ricor Cryogenic and Vacuum Systems, Ein Harod Ihud, Israel
[2]Technion – Israel Institute of Technology, Haifa 32000, Israel

ABSTRACT

A passive mechanical device for phase shifting was developed for a miniature Pulse Tube In-Line cryocooler developed earlier at the Rechler Cryogenic Laboratory, Technion-Israel Institute of Technology. A passive piston supported on flexure bearings was proposed for this mechanical device. In order to produce the desired phase shift in the regenerator, the piston mass, flexure bearings stiffness and damping coefficient of the mechanical system were obtained using a phasor analysis and a mechanical analogy, and by simulations using a commercial numerical solver for pulse tube cryocoolers (Sage®). In these calculations, other parameters of the existing cryocooler remained without changes. Simulation results were compared with those from the theoretical model. A prototype was manufactured and experiments were performed. The lowest temperature observed in the experiments was 158K. The discrepancy between theoretical results and experimental results was due to the limitations in the precision manufacturing and assembling of the passive mechanical phase shifter.

One of the difficulties of the instrumentation in the current experiment was measurement of the passive piston high frequency movement (at about 100 Hz) that had to be contact-less. It was proposed to use a laser beam and reflective surface mounted to the back side of the passive piston. The proposed measurement system was designed, manufactured and used in the experiments.

INTRODUCTION

A present worldwide trend focuses on the miniaturization of regenerative cryocoolers. It is evident that the pulse tube cryocooler is one of the best candidates for miniaturization. Downscaling the mechanical dimensions without a change in the operating parameters becomes problematic, as internal heat losses become dominant in the device. The method proposed to counteract these heat losses is to increase the fill pressure and the operating frequency of the miniature cryocoolers. A critical parameter for efficient operation is the proper phase shift between thr pressure and the mass flow rate at the different parts of the cryocooler.

One of the smallest pulse tube cryocoolers constructed to-date was designed at the Rechler Cryogenic Laboratory, Technion-Israel Institute of Technology by Sobol et al. [1]. An inertance tube and a reservoir were used for phase shifting in the cryocooler. This miniature pulse tube cryocooler achieved a no-load temperature of 99K, at a heat rejection temperature of 305K, and provided 400 mW cooling at

Cryocoolers 19, edited by S.D. Miller and R.G. Ross, Jr.

110K at an operating frequency of 103 Hz. The main constraint of this cryocooler was phase shift limitation using the inertance tube. In miniature devices, the impedance created by the inertance tube has a large resistive component, which limits its ability to produce the required phase shift.

A possible solution for increasing the phase shift in the pulse tube cryocoolers was proposed by Matsubara et al. [2]. The concept was to use a passive mechanical device instead of an inertance tube and a reservoir. Theoretically, any desired phase shift could be produced by such a passive mechanical device and it could be miniaturized without deterioration of its performance, in contrast to the inertance tube.

In 2012 Lewis, Bradley and Radebaugh from NIST [3] developed and tested a miniature pulse tube with actively controlled piston installed at the hot end of the cryocooler instead of the inertance tube. The phase shift was controlled by software that locked the phase between the compressor and the piston at the hot end. It was shown that the efficiency of the pulse tube cryocooler with actively controlled piston was better than that of the same cryocooler with the inertance tube. The improvement of the efficiency was attained because the use of a controlled warm expander made possible flexible control in optimization of phase angles.

In the current research, the inertance tube and reservoir of the existing cryocooler [1] were replaced by a passive mechanical device. Design parameters of the existing cryocooler remain unchanged. The overall length of the assembly of aftercooler, regenerator, cold heat exchanger, buffer tube and hot heat exchanger is 31 mm. The regenerator is filled with SS#635 mesh screen and its length is 12 mm. The proposed design of the mechanical phase shifter is a passive piston supported on flexure bearings. The passive piston diameter is 5 mm. The main design parameters selected to produce the desired phase shift using the passive piston are: piston mass (50 gm), overall flexure bearings stiffness (16000 N/m) and damping coefficient whose source is oscillating gas (helium) in a precise clearance gap (20 μm) between piston and cylinder. These design parameters were obtained using phasor theory and compared to the simulation of the pulse tube cryocooler with the proposed mechanism for phase shifting by commercial software Sage®. Final optimization of the design parameters was performed by Sage®.

For the comparison of the experimental and the theoretical model, it was necessary to measure the passive piston displacement at high frequencies. For this purpose, a contact-less displacement measurement system was designed, manufactured and implemented.

THEORETICAL ANALYSIS

Phasor Analysis for the Pulse Tube Cryocooler with Passive Piston

Phasor analysis is a common theoretical tool for investigation of different parameters in regenerative cryocoolers. Using such analysis eliminates time dependence from the equations which are transformed from the time-dependent space to the vector space and only phases and amplitudes of the flow oscillation become important. Radebaugh [4] introduced phasor analysis in a pulse tube refrigerator where the phase shifter was an orifice. The difference between [4] and the analysis introduced in this work is that the orifice and reservoir were replaced here by the passive piston (see Figure 1). The pressure phasor is permanently located along the real axis as a reference. Another difference is the following: in the regenerator the phase shift has to be close to zero for efficient performance of the pulse tube cryocooler. Thus, the phase angles 'upward' (counter-clockwise) from the regenerator are positive and the phase angles 'downward' (clockwise) from the regenerator are negative. Since the only difference between the pulse tube cryocooler in [4] and the one in this work is the passive piston, only the expressions of the mass flow in the passive piston would be introduced here. A more detailed analysis is given in [5].

The passive piston volume can be expressed as follows:

$$\vec{V}_p = V_{p0} + \frac{V_{p\_amp}}{2}\left[1+\sin\left(\omega t + \phi_p\right)\right] \qquad (1)$$

where: V_p is the volume of the passive piston; V_{p0} is the 'dead' volume; V_{p_amp} is the swept, or stroke volume, and $V_{p\_amp}/2$ is the amplitude of the swept volume.

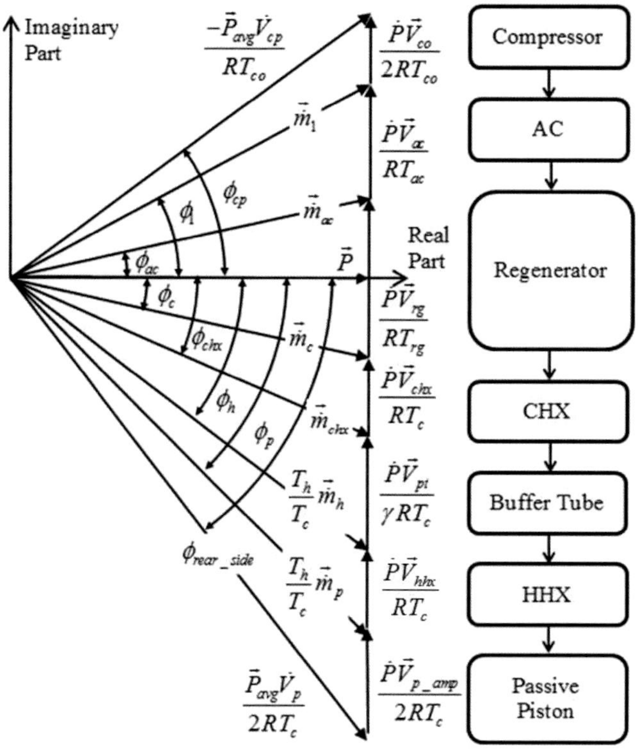

Figure 1. Schematic diagram for a Pulse Tube cryocooler equipped with a Passive Piston - in Vector space.

The passive piston mass conservation can be expressed as follows:

$$\vec{\dot{m}}_p = \frac{\dot{\vec{P}} V_{p\_amp}}{2RT_h} + \frac{P_a \dot{\vec{V}}}{RT_h} \tag{2}$$

where $\dot{m}_p$ is the mass flow, P_a is the mean pressure, T_h is the 'hot' temperature, $\dot{V}$ is the volumetric flow, and $\dot{P}$ is the pressure fluctuation in time.

Or,

$$\frac{T_h}{T_c}\vec{\dot{m}}_p = \frac{\dot{\vec{P}} V_{p\_amp}}{2RT_c} + \frac{P_a \dot{\vec{V}}}{RT_c} \tag{3}$$

where T_c is the 'cold' temperature.

The phasor analysis indicates the following: the optimum performance in the regenerator can be attained if the mass flow rate vector to the regenerator from the aftercooler would be equal to the mass flow rate vector from the cold heat exchanger. In this case these vectors would be symmetric about the real axis of the diagram. This causes the average mass flow to be in phase with the pressure in the regenerator. In the conventional Stirling cryocooler the same phasor arrangement would produce efficient performance of the device.

Therefore, from Figure 1, in order to obtain the optimal phase shift in the regenerator the following should hold simultaneously:

$$\begin{aligned} \phi_{ac} &= -\phi_c \\ |\dot{m}_{ac}| &= |\dot{m}_c| \end{aligned} \tag{4}$$

MECHANICAL ANALOGY

The solution obtained from the phasor analysis performed in the previous section does not include mechanical parameters that are necessary for the design of the passive piston. The proposed method for obtaining these design parameters is a mass-spring-damper analogy of the pulse tube cryocooler. The results from applying this method can be used together with results from the phasor analysis. The displacement of the gas acted upon by the compressor is represented by x_{cp} and that of the gas acted upon by the passive piston is represented by x_{pp} (which is assumed to be the same as the passive piston displacement). The gas in the pulse tube is represented as a spring with finite stiffness k_{pt} and negligible mass. The passive piston is represented as a mass, M_{pp}; spring stiffness, k_b, which can be a mechanical spring or a gas spring; and damping coefficient, c_{pp}. The compressor piston working area (the area interacting with the gas) is represented by A_{cp} and the passive piston working area is represented by A_{pp}. The pulse tube volume is represented as V_{pt} and the passive piston volume as V_b. A schematic mass-spring-damper model of the pulse tube cryocooler with a passive piston is shown at Figure 2.

The oscillation of gas in the pulse tube cryocooler can be described using this method by a second order ordinary differential equation (ODE):

$$M_{pp}\ddot{x}_{pp} + c_{pp}\dot{x}_{pp} + \left(k_b + k_{pt}\right)x_{pp} = k_{pt}\frac{A_{pp}}{A_{cp}}x_{cp} \tag{5}$$

where the ratio A_{pp}/A_{cp} represents an amplified ratio of spring stiffness in the pulse tube due to the different working areas of the compressor and the passive piston. The gas stiffness can be found from Hook's law and the expression for the polytropic compression.

Then, the equation (5) can be expressed as follows:

$$M_{pp}\ddot{x}_{pp} + c_{pp}\dot{x}_{pp} + x_{pp}k_{tot} = \frac{x_1 k_{pt} A_{pp}}{A_{cp}}\cos(\omega t) \tag{6}$$

where x_1 is the amplitude of the compressor piston displacement.

Initial conditions of the passive piston displacement and velocity are:

$$x_{pp}(0) = 0; \dot{x}_{pp}(0) = 0 \tag{7}$$

The solution for equation (6) is:

$$x_{pp}(t) = \tilde{A}\cos(\omega t) + \tilde{B}\sin(\omega t) \tag{8}$$

where

$$\tilde{A} = \frac{x_1 k_{pt} A_{pp}\left(k_{tot} - M_{pp}\omega^2\right)}{A_{cp}\left[\left(c_{pp}\omega\right)^2 + \left(k_{tot} - M_{pp}\omega^2\right)^2\right]}; \quad \tilde{B} = \frac{x_1 k_{pt} A_{pp} c_{pp}\omega}{A_{cp}\left[\left(c_{pp}\omega\right)^2 + \left(k_{tot} - M_{pp}\omega^2\right)^2\right]}$$

The mass flow rate of the gas due to the piston oscillation can be expressed as follows:

$$\vec{\dot{m}}_g = \rho A_{pp}\dot{x}_{pp} \tag{9}$$

where: $\dot{m}_g$ is the mass flow rate of the gas, and ρ is the density of the gas.

Therefore,

$$\vec{\dot{m}}_g(t) = \vec{\dot{m}}_p(t) = \rho A_{pp}\omega\left[-\tilde{A}\sin(\omega t) + \tilde{B}\cos(\omega t)\right] = \rho A_{pp}\omega\sqrt{\tilde{A}^2 + \tilde{B}^2}\sin\left(\omega t + \tilde{\phi}_{in}\right) \tag{10}$$

Or,

$$\vec{\dot{m}}_p(t) = \rho A_{pp}\omega\sqrt{\tilde{A}^2 + \tilde{B}^2}\cos\left(\omega t + \tilde{\phi}_{in} - \frac{\pi}{2}\right) \tag{11}$$

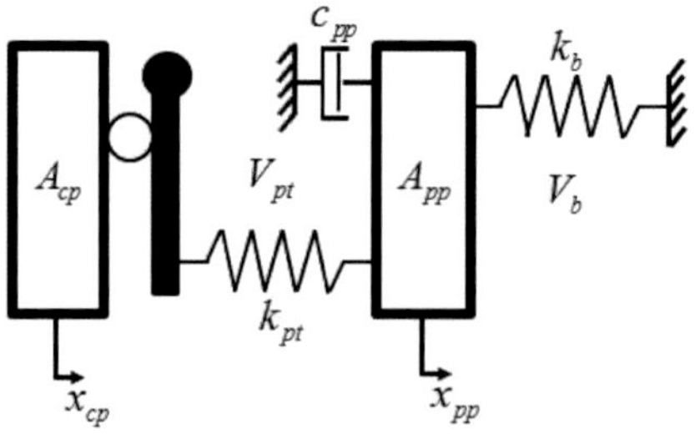

Figure 2. Schematic mass-spring-damper model of pulse tube cryocooler with a passive piston.

Where,

$$\tilde{\phi}_{in} = \arctan\left(-\frac{\tilde{B}}{\tilde{A}}\right)$$

The pressure of the piston on the gas, $\vec{P}_p$, can be found from the following:

$$\vec{P}_p = \sqrt{\tilde{C}^2 + \tilde{D}^2}\cos\left(\omega t + \phi_{pp} - \frac{\pi}{2}\right) \tag{12}$$

Where,

$$\phi_{pp} = \arctan\left(-\frac{\tilde{C}}{\tilde{D}}\right);\ \tilde{C} = \frac{k_{pt}x_1}{A_{cp}} - \tilde{A}\frac{k_{pt}}{A_{pp}};\ \tilde{D} = \tilde{B}\frac{k_{pt}}{A_{pp}}$$

In Figure 1 the reference for the real axis is pressure. In other words, the pressure vector has to be coincident with the x direction, namely the phase of the pressure has to be zero. Thus, it is necessary to shift the phase in equation (12) to be consistent with the discussion in the previous Section.

Therefore,

$$\vec{P}_p = -\omega\sqrt{\tilde{C}^2 + \tilde{D}^2}\sin(\omega t) \tag{13}$$

$$\vec{\dot{m}}_p(t) = \rho A_{pp}\omega\sqrt{\tilde{A}^2 + \tilde{B}^2}\cos\left(\omega t + \phi_{\dot{m}_p}\right) \tag{14}$$

where, $\phi_{\dot{m}_p} = \tilde{\phi}_{in} - \phi_{pp}$ is the phase shift of the mass flow rate. So,

$$\phi_{\dot{m}_p} = \arctan\left(-\frac{\tilde{B}}{\tilde{A}}\right) - \arctan\left(-\frac{\tilde{C}}{\tilde{D}}\right) \tag{15}$$

This result can now be substituted into the results from the previous section. In conclusion, the following should hold simultaneously for the optimal performance of the pulse tube cryocooler with passive piston:

$$\arctan\left(\frac{\tilde{G}}{\tilde{H}}\right) = -\arctan\left(\frac{\tilde{M}}{\tilde{N}}\right) \Rightarrow \frac{\tilde{G}}{\tilde{H}} = -\frac{\tilde{M}}{\tilde{N}} \tag{16}$$

and

$$\sqrt{\tilde{M}^2 + \tilde{N}^2} - \sqrt{\tilde{G}^2 + \tilde{H}^2} \Rightarrow \tilde{M}^2 + \tilde{N}^2 = \tilde{G}^2 + \tilde{H}^2 \tag{17}$$

Where,

$$\tilde{G} = \frac{P_{avg}}{RT_{co}}\frac{V_{co}}{2}\omega\sin\left(\phi_{pp}\right);\ \tilde{H} = \frac{P_{avg}}{RT_{co}}\frac{V_{co}}{2}\omega\cos\left(\phi_{pp}\right) + \left[\frac{V_{co}}{2} + V_{ac}\right]\frac{\omega\sqrt{\tilde{C}^2 + \tilde{D}^2}}{RT_{co}}$$

$$M = \frac{T_h}{T_c}\rho A_{pp}\omega\sqrt{\tilde{A}^2 + \tilde{B}^2}\cos\left(\phi_{\dot{m}_p}\right)$$

$$\tilde{N} = \frac{\omega\sqrt{\tilde{C}^2 + \tilde{D}^2}}{RT_c}\left[V_{hhx} + \frac{V_{pt}}{\gamma} + V_{chx}\right] - \frac{T_h}{T_c}\rho A_{pp}\omega\sqrt{\tilde{A}^2 + \tilde{B}^2}\sin\left(\phi_{\dot{m}_p}\right)$$

Equations (16) , (17) were solved for the passive piston mass, spring stiffness and damping coefficient that are included in the definition of $\tilde{G}, \tilde{H}, \tilde{M}, \tilde{N}$ by Matlab® software. In the following section these results would be compared with the numerical solution from Sage® software that is dedicated to solving and optimizing different configurations of pulse tube cryocooler.

PULSE TUBE CRYOCOOLER DESIGN IN SAGE®

Overview

For the Stirling-based cycle cryocoolers the most commercial software for analysis is Sage®, developed by Gedeon Associates [6]. Sage® has a visual interface in which different standard components such as pistons, and heat exchangers are assembled into a complete machine. The optimization possibility of the design is a great advantage of Sage®, which makes it a very useful tool for cryocooler design. The software employs theoretical models and empirical correlations in calculations of different parameters. The assembly has a modular form where each component of the system is organized logically inyo a hierarchical tree structure. In Figure 3 the root level editing window of Sage® with the final model of the Passive Piston Pulse Tube cryocooler is shown. It is based on the model of the earlier design developed by Sobol et al. [1].

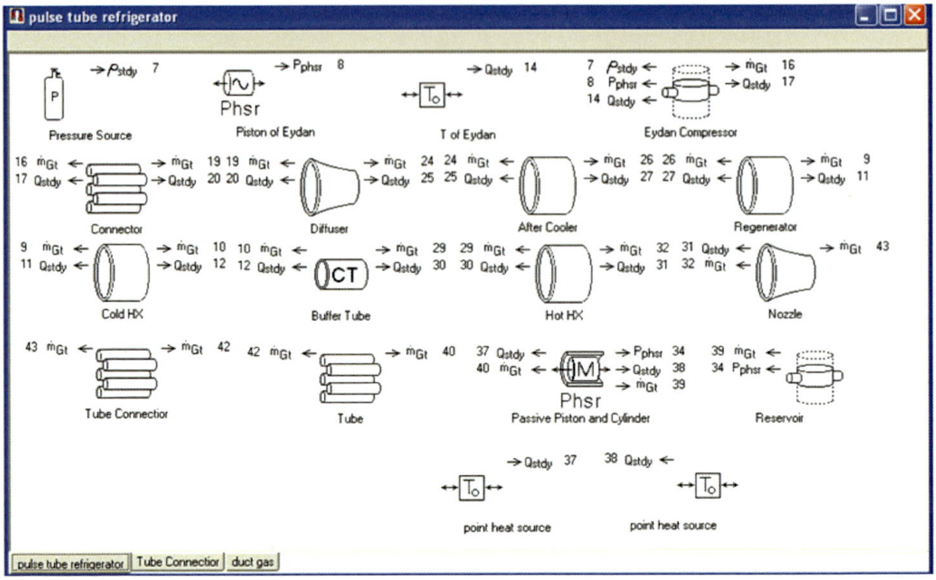

Figure 3. Root-level editing window of Sage® with the final model of the Passive Piston Pulse Tube cryocooler.

Optimization

Available parameters for optimization are limited because the present design of the Passive Piston Pulse Tube cryocooler is based on the already designed and manufactured Inertance Tube Pulse Tube cryocooler [1]. Most parameters such as dimensions, frequency, compressor piston displacement, mesh properties, frequency and temperatures of the previous design had to remain unchanged. The parameters available for optimization are passive piston physical parameters which determine piston dimensions such as piston mass, spring stiffness and damping coefficient. The objective function in the optimization was the maximum net cooling power at the cold end of the pulse tube model. The diameter of the piston in the phase shifter was selected to be 5 mm. The entire pulse tube cryocooler was analyzed to attain the objective function. Several combinations of optimized parameters were found. Due to the manufacturing restrictions, the values of parameters found in the optimization process were rounded off to acceptable values.

Simulation Results

The results from the Sage® optimization were compared to the results from the theoretical analysis. Clearance gap width was calculated only by Sage®. The damping coefficient and clearance gap in both solutions were constant and independent of piston mass at the examined range of piston mass values.

However, the spring stiffness behavior was different. In Figure 4 a linear fit was calculated for both solutions. The goodness of fit was about 1.0; therefore, it can be concluded that the spring stiffness had a linear behavior at the tested range of piston mass values. A linear fit function was also calculated and its graph was shown for the analytical solution from theoretical analysis, as well as for the Sage® solution. It is necessary to discuss the difference between the graph of analytical solution and of Sage® solution. Both solutions have a linear behavior. Also, it is quite simple to realize that analytical and numerical solutions are parallel. However, the shifting from the x axis of these two functions is quite different. The reason for such behavior is that the analytical model neglects the compressibility and the helium gas stiffness; therefore, the calculated spring stiffness in the case of the analytical solution also contains in it the real gas stiffness. In contrast, in the Sage® software the real gas compressibility and stiffness is taken into account, while the model is solved or optimized. Thus, the value of spring stiffness from the analytical solution is greater than the one

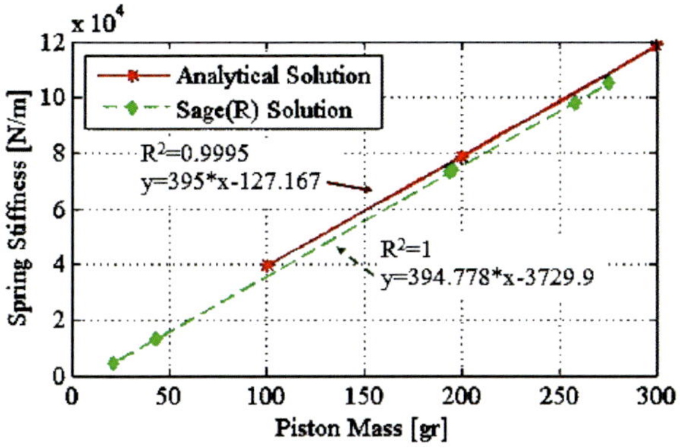

Figure 4. Spring stiffness vs. piston mass.

Table 1. Final parameters for Passive Piston model.

Piston Mass	Spring Stiffness	Damping Coefficient	Clearance Gap Width
50 gm	16000 N/m	1 x 10^{-3} N*s/m	20 μm

from the Sage® solution at the same mass value for each mass. Piston mass of 50 gm was chosen as design point for the passive piston design. In summary, final parameters for piston design are summarized Table 1.

EXPERIMENTAL SETUP

The parameters to be measured are the temperature, pressure and the passive piston displacement. Temperatures were measured at the aftercooler, cold heat exchanger and hot heat exchanger (see Figure 5). At the cold heat exchanger the temperature had to be measured with high precision; therefore, a diode gauge was employed, with accuracy of ±0.1K. Aftercooler and hot heat exchanger temperature accuracy are less critical; therefore, K-Type thermocouples were used. Pressure oscillations were measured at the interface between the compressor and aftercooler, buffer tube and hot heat exchanger, and at the inlet to the passive piston compression volume (see Figure 5).

Piston Displacement Measurement System

The main difficulty in the experiment was the measurement of the passive piston displacement without the physical contact of any mechanical part with the oscillating piston. The values of the displacement have to be measured at the same times as the pressure values at different cryocooler points. Using values of the passive piston displacement, the velocity of the passive piston can be calculated using differentiation by a finite differences method. The velocity of the passive piston can provide an estimate for the gas volumetric flow rate at the passive piston. This parameter is necessary for evaluation of the phase shift between the volumetric flow of the gas and pressure of the gas in the passive piston. The importance of this parameter was explained previously.

For piston displacement measurement an optical system was used. The operating principle of the system is as follows: a laser beam passes through the glass at the rear side of the cylinder cover and reflects from the reflective surface on the rear side of the piston; then the reflected beam passes back through the glass and impinges on the position sensitive detector (PSD). The laser beam main specifications are: power: 0.9mW, wavelength: 655 nm. The active area size of the PSD main specification is 1x12 mm. The PSD is based on a p n junction that produces current and voltage which depend on the location of the laser beam impingement. A schematic concept of the piston displacement measurement is shown in Figure 6 and its mounting to the passive piston assembly is shown in Figure 5. After the post-processing and amplification, the output voltage signal can be collected by a data acquisition program.

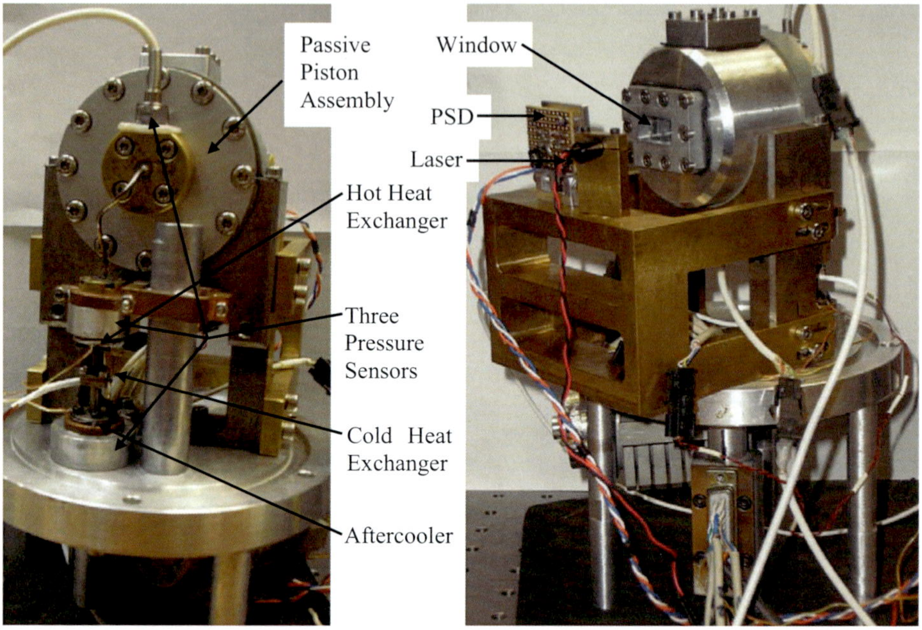

Figure 5. Cryocooler exclusive of vacuum chamber with passive piston displacement measurement system.

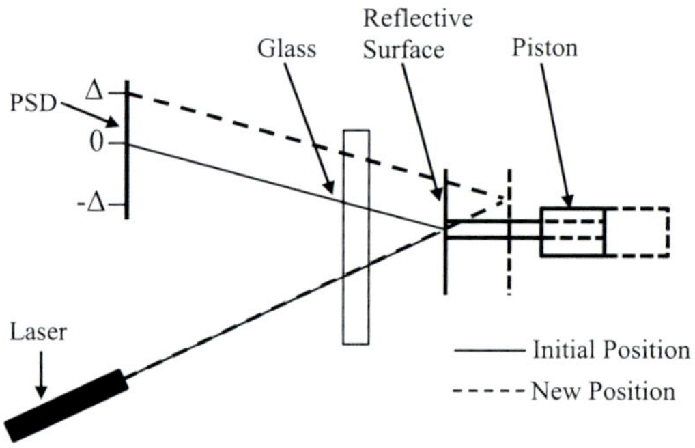

Figure 6. Schematic of piston displacement measurement.

In order to eliminate the convection heat transfer at the cold end it is convenient to perform the experiments with the cryocooler in a vacuum chamber. However, due to the design restrictions, the passive piston displacement could only be measured without the vacuum chamber.

RESULTS

The experiments were performed at the environmental temperature of 296K. Fill pressure in the refrigerator was set to 40 bar. Cryocooler performance was checked at several frequencies of the compressor near the design point of 100 Hz. The target pressure ratio was 1.3 (as set in the design). The temperature recorded was the lowest temperature obtained at the cold heat exchanger. In order to produce vacuum in the vacuum chamber a Varian® vacuum system was used.

Passive Piston Displacement Measurement Results

As mentioned before, due to design restrictions, the passive piston displacement measurement was performed without the vacuum chamber. Figure 7 shows the variation of pressure, piston dis-

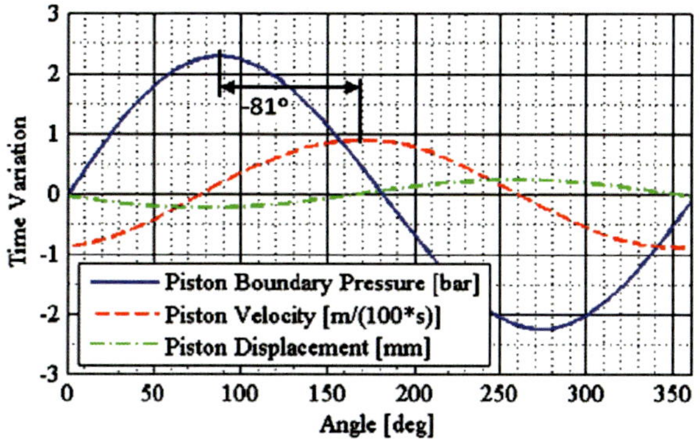

Figure 7. Pressure at the inlet to the passive piston, passive piston displacement and velocity at frequency of 120 [Hz].

placement, and piston velocity at the inlet to the passive piston at frequency of 120 Hz. The phase shift at the boundary of the passive piston between piston velocity and pressure oscillation, which is the same as the phase between mass flow of the gas and pressure oscillation, was about -81°.

Cryocooler Performance Measurement

As mentioned earlier, in order to decrease heat transfer by convection at the cold end, the experiment was performed in a vacuum chamber. The temperature at the cold heat exchanger was measured at several frequencies around the design point. The results are shown in Figure 8. The lowest temperature of 157.8K was reached at a frequency of 120 Hz. The pressure ratio at the aftercooler inlet was measured at the same frequencies (see Figure 8). The pressure ratio at high frequencies decreased from the design point of 1.3 to 1.2.

CONCLUSIONS

The present research has proved the viability of the concept of using a passive mechanical mechanism as a phase shifter of the flow in a pulse tube cryocooler. There was fine agreement between analytical and numerical solution by the Sage® software. However, the temperatures at the cold end reached in the experiments were higher than those reached with the cryocooler described in [1]. Also, the pressure ratio was lower than the design value. This was due to manufacturing and assembly inaccuracies that added parasitic friction between piston and cylinder. However, the phase shift between piston velocity and pressure oscillation that was measured at the inlet to the passive piston demonstrated the feasibility of the concept.

Passive piston displacement contactless measurement system was proposed and used in the

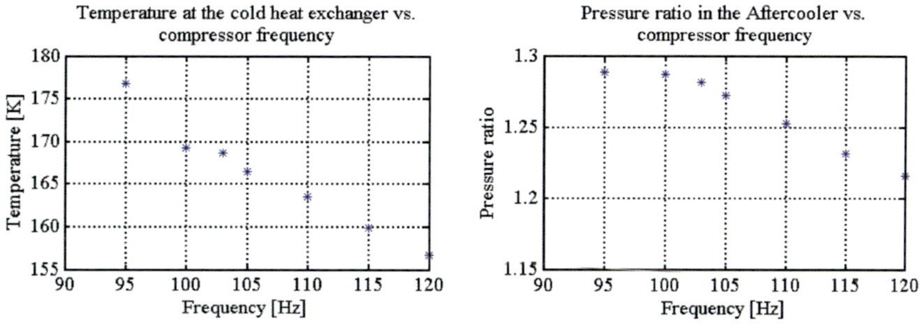

Figure 8. Results of the experiment in vacuum chamber.

experiments. The advantage of the proposed system was in the contactless concept and ability to measure small displacement values at high frequencies, due to the high accuracy of the PSD. The results of measurements helped to understand gas behavior in the pulse tube cryocooler operating at high frequencies.

ACKNOWLEDGEMENTS

The authors like to thank the Israel Ministry of Defense/MAFAT and the Rechler Family Foundation for their generous financial support, Ricor – Cryogenic and Vacuum Systems Ltd. for their support of the equipment for the experiments, Yossi Cohen and Zeev Hershkovitch for their efforts in manufacturing the system, Shlomo Weiss for his support with electronics, Sergey Sobol for his support in technical questions and Israel Rosinsky for his help in the experiments.

REFERENCES

1. Sobol, S., Katz, Y. and Grossman, G., "A Study of a Miniature In-Line Pulse Tube Cryocooler," *Cryocoolers 16*, ICC Press, Boulder, CO (2011), pp. 87-95

2. Matsubara, Y. and Miyake, A., "Alternative Methods of the Orifice Pulse Tube Refrigerator," *Proceedings of the 5th International Cryocooler Conference*, (1989), pp. 127-135

3. Lewis, M., Bradley, P. and Radebaugh, R., "Experiments with Linear Compressors for Phase Shifting in Pulse Tube Cryocoolers," *Adv. in Cryogenic Engineering*, Vol. 57, Amer. Institute of Physics, Melville, NY (2012), pp. 1600-1607

4. Radebaugh, R., "A Review of Pulse Tube Refrigeration," *Adv. in Cryogenic Engineering*, Vol. 35B, Springer US (1990), pp. 1191-1205

5. Radchenko, D., "Investigation of a Passive Mechanical Mechanism for Phase Shifting of the Flow in a Pulse Tube Cryocooler," M.Sc. Thesis, Faculty of Mechanical Engineering, Technion, 2015

6. Gedeon, D., "Sage® - Stirling Cycle Model Class Reference Guide," Gedeon Associates, Athens, OH, USA, April 2006

Simulation and Experimental Analysis of Pulse Tube Refrigerator with an Active Phase Controller

S. Yang, Y. N. Wu, Z. P. Zhou, Z. H. Jiang, A. K. Zhang

Shanghai Institute of Technical Physics
Chinese Academy of Sciences
Shanghai, 200083, P. R. China

ABSTRACT:

The pulse tube refrigerator (PTR) with an active phase controller (APC) is a novel type of pulse tube refrigerator, where the inertance tube and the reservoir are replaced by an APC. In this study, a numerical analysis of a PTR with an APC was performed using a two-dimensional axisymmetric Computational Fluid Dynamics (CFD). The influences of the phase and swept volume of APC on phase, mass flow rate, pressure ratio and the performance of PTR are discussed. Some experiments are carried out to validate the numerical results.

INTRODUCTION

As illustrated in Figure 1, in the PTR, the displacer piston (as found in the Stirling and Gifford-McMahon (GM) regenerative systems) is eliminated, the cold moving parts have been replaced by stationary components (orifice or inertance tube). In pulse tube refrigerators with small cooling capacity, the inertance tube cannot shift the phase between the mass flow and pressure to the ideal angle because of the low acoustic power at the inlet of the inertance tube[1-2]. In addition, the acoustic power is absorbed and dissipated into waste heat in the inertance tube, which causes a decrease in the efficiency of the PTR[3].

A key component to create a more efficient PTR is the phase shifter[4], which controls the phase shift between the mass flow and the pressure. In this study, a linear compressor, which is directly connected at the warm end of the pulse tube using a connecting tube, is used as a phase shifter (Figure 2)[5-6].

Figures 2 and 3 show an electrical analogy model of the pulse tube refrigerator with an inertance tube or APC. In this model, the pressure wave generator (PWG) and the active phase controller (APC) are analogous to electric motors, the regenerator is analogous to an electric resistance and the inertance and the reservoir are analogous to an electric capacity and electrical ground[7-8].

Comparing these two electrical analogy models of the pulse tube refrigerator with an inertance tube or APC, the advantage of the active phase controller pulse tube (APCPTR) is the convenience of controlling the phase between the pressure and the mass flow in the PTR. In addition, when the hot end of the regenerator and the back of the APC are connected to recover power from the gas in the hot end of the pulse tube, the system efficiency can be increased.

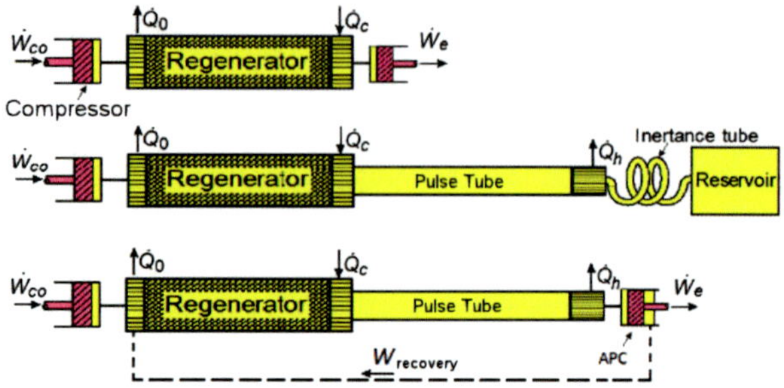

Figure 1. Schematic of three different configurations of cryocooler based on Stirling cycle

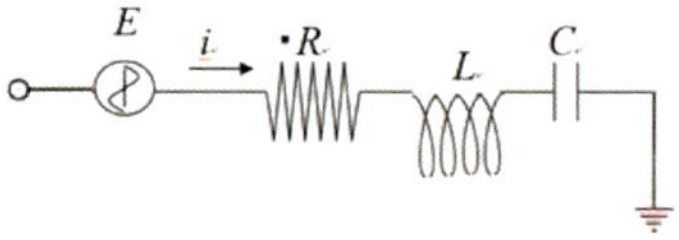

Figure 2. Electrical analogy of pulse tube refrigerator with inertance tube

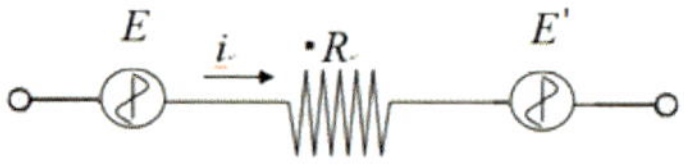

Figure 3. Electrical analogy of pulse tube refrigerator with APC

SIMULATION MODEL

In this study, FLUENT®, CFD modelling software, is employed to simulate the flow of fluid and the pressure distribution in the pulse tube refrigerator[9]. All of the details of modeling and meshing of the 2-D axis-symmetric pulse tube refrigerator model is done using the modeling software GAMBIT®. In FLUENT®, the conservation equations of mass, momentum and velocity are solved using a finite volume method. There are several turbulence models available in the code. The flow was calculated by the SIMPLE® method, and a second order upwind differential scheme was applied for the approximation of the convective terms. The implemented mesh involved about 350,000 cells (Figure 2).

A complete pulse tube refrigerator using a linear compressor as a phase controller consists of the compressor (1), the transfer line (2), the aftercooler (3), the regenerator (4), the cold-end heat exchanger (5), the pulse tube (6), the hot-and heat exchanger (7), the connecting tube (8), and the active phase controller (9) (see Figure 4).

The governing equations of Fluent® used to solve the mass, momentum, and energy for the working medium in this computational model are:

$$\frac{\partial \rho}{\partial t} + \nabla \cdot (\rho V) = 0 \tag{1}$$

$$\rho \frac{D\vec{V}}{Dt} = -\rho g \vec{n} - \nabla \vec{P} + \nabla \cdot \left\{ \mu \left[(\nabla \vec{v} + \nabla \vec{v}^t) - \frac{2}{3} \nabla \cdot \vec{v} I \right] \right\} \tag{2}$$

$$\rho c \left[\frac{\partial T}{\partial t} + (\vec{v} \cdot \nabla) T \right] = \nabla \cdot (k \nabla T) \tag{3}$$

The compressor and the APC are modeled using dynamic mesh. A User Defined Function (UDF) is written to track the motion of the pistons with respect to time.

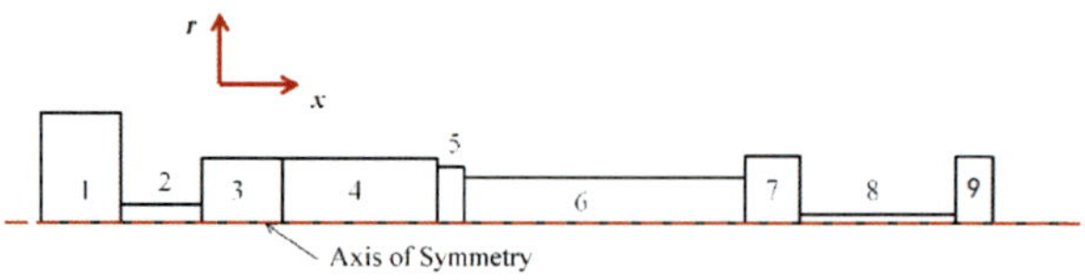

Figure 4. Schematic of the simulation model

The regenerator and the heat exchangers are modeled as porous regions; their viscous loss and the inertial loss are considered. In Fluent®, porous media are modeled by the addition of a momentum source term to the standard fluid flow equations. The source term is composed of two parts: a viscous loss term and an inertial loss term.[10-11]

$$s_i = -\left\{\sum_{j=1}^{3} D_{ij} u v_j + \sum_{j=1}^{3} C_{ij} \frac{1}{2} \rho |v| v_j\right\} \tag{4}$$

where s_i is the source term for the i the (x, y or z) momentum equation, $|v|$ is the magnitude of the velocity and D and C are prescribed matrices. This momentum sink contributes to the pressure gradient in the porous cell, creating a pressure drop that is proportional to the fluid velocity (or velocity squared) in the cell.

$$s_i = -\left\{\frac{\mu}{\alpha} v_i + C_2 \frac{1}{2} \rho |v| v_i\right\} \tag{5}$$

where α is the permeability and C_2 is the inertial resistance factor, simply specify D and C as diagonal matrices with $\frac{1}{\alpha}$ and C_2, respectively, on the diagonals (and zero for the other elements).

In the regenerator, the thermal conductivity and specific heat capacity of material are only functions of temperature.

The working fluid, helium, is modeled as an ideal gas; its viscosity and thermal conductivity are only functions of temperature.

The warm-end heat exchanger and the aftercooler are set to a constant temperature of 300K.

The working frequency of this PTR is 50 Hz. In this simulation, the time step of this transient model is 0.0002 second.

RESULTS AND DISCUSSION

Velocity in two kinds of pulse tube refrigerator in one cycle

In this study, a numerical simulation of 2 different PTRs were performed to identify the differences between a PTR with APC and a PTR with inertance tube. In the simulation, the swept volume of the piston of the PWG is 4.53 x 10^{-6} m^3, the swept volume of the APC is changing from 1.0625 x 10^{-6} m^3 to 3.1875 x 10^{-6} m^3, the phase between PWG (Pressure Wave Generator) and APC is changing from 70°to 140°(PWG-APC), and the temperature of cold heat exchanger is fixed at 100K.

Figures 5 and 6 show the flow of helium in the pulse tube refrigerator with an inertance tube and the pulse refrigerator with APC in one cycle. The CFD analysis shows that the APC which controls the phase between the mass flow and the pressure by changing the phase and the swept volume of APC can replace the inertance tube.[2]

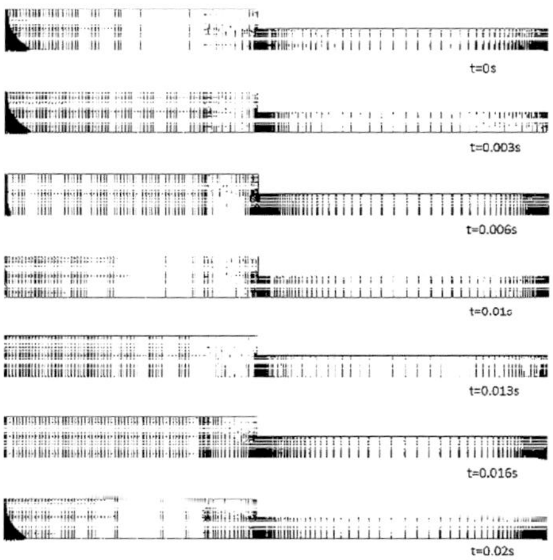

Figure 5. Snapshot of velocity vectors in pulse tube refrigerator with APC in one cycle

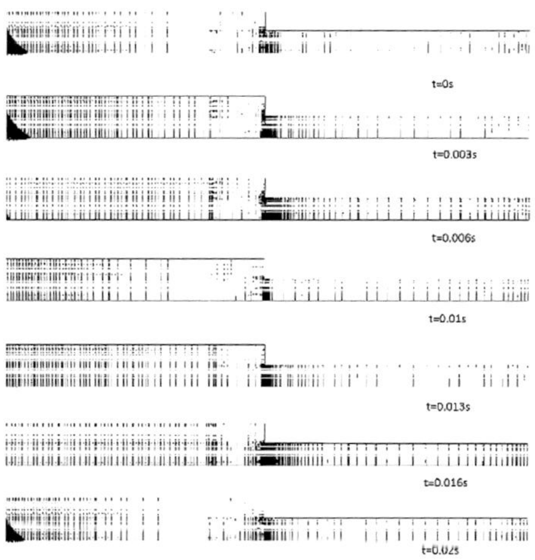

Figure 6. Snapshot of velocity vectors in pulse tube refrigerator with inertance tube in one cycle

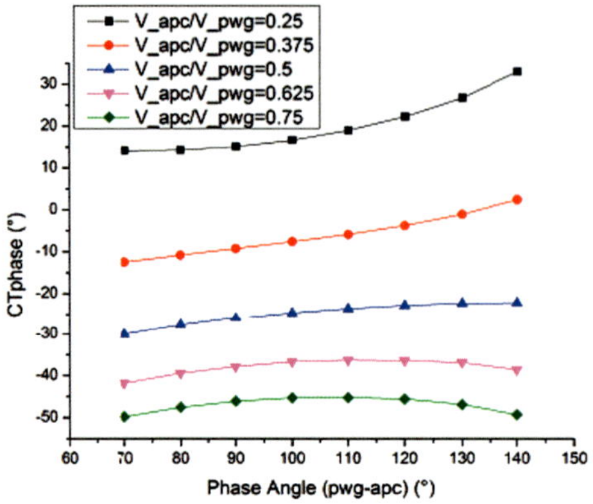

Figure 7. CT phase versus phase between PWG and APC

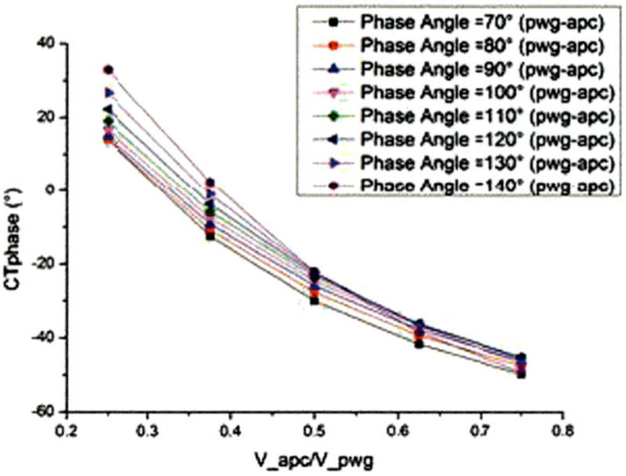

Figure 8. CT phase versus Vapc/Vpwg

Phase between the Pressure and the Axial Velocity (CTphase) at the Cold End of PTR

Figure 7 shows the phase between the pressure and the axial velocity at the cold end of PTR and the phase between PWG and APC are strongly correlated. Figure 8 shows the phase between the pressure and the axial velocity at the cold end of the pulse tube and swept volume of PWG and APC are strongly correlated. With the increase of the swept volume in APC, the phase between the pressure and the axial velocity at the cold end of the pulse tube is reduced.

Both swept volume and phase of the PWG and the APC have a great influence on the phase between the pressure and the axial velocity, the swept volume of APC plays a more important role in it.

Mass Flow Rate (Mc) at the Cold End of PTR

Figure 9 shows mass flow rate at the cold end of the pulse tube and the phase angle between the pistons in the compressor and the APC are strongly correlated. With the increase of the phase between the compressor and APC, the mass flow rate at the cold end of the pulse tube is reduced.

Figure 10 shows the phase between the pressure and the axial velocity at the cold end of the pulse tube and swept volume of two pistons are strongly correlated. With the increase of the swept volume in APC, the phase between the pressure and the axial velocity at the cold end of the pulse tube is increased.

Both swept volume and phase of the compressor and APC play a great influence on the phase between the pressure and the axial velocity; there exists an optimum phase angle between the two pistons and the optimum swept volume of the two pistons.

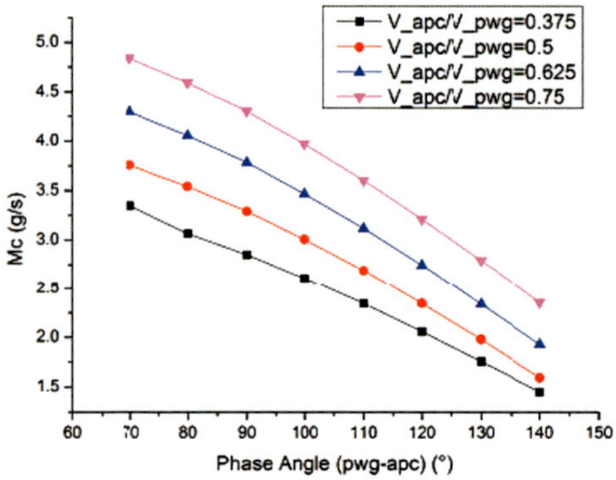

Figure 9. Mc phase between PWG and APC

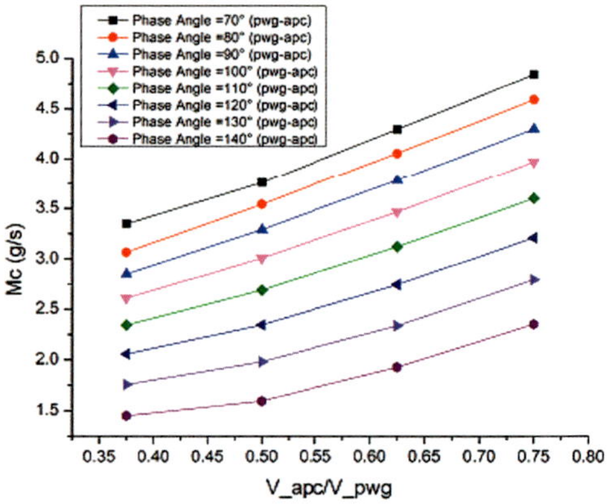

Figure 10. Mc versus Vapc/Vpwg

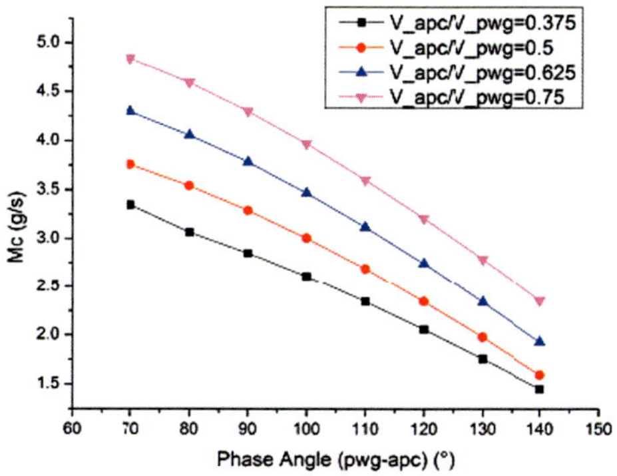

Figure 11. CTPr versus phase between PWG and APC

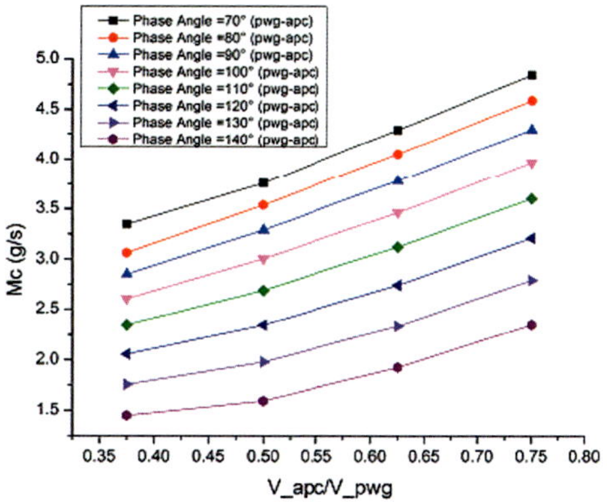

Figure 12. CTPr versus Vapc/Vpwg

Pressure Ratio at the Cold End of the Pulse Tube (CTPr)

Figure 11 shows the pressure ratio at the cold end of the pulse tube and the phase angle between the pistons in the compressor and the APC are strongly correlated. With the increase of the phase between the compressor and APC, the pressure ratio at the cold end of the pulse tube is increased.

Figure 12 shows the pressure ratio at the cold end of the pulse tube varying the swept volume of the APC. With an increase in the swept volume of the APC, the pressure ratio at the cold end of the pulse tube is increased. The increase in the pressure ratio at the cold end of the pulse tube produces an increase of the efficiency. With an increase in the swept volume in APC, the power supply increases. There exists an optimum phase angle between the two pistons and the optimum swept volume of the two pistons.

The simulation shows that the active phase controller can be applied to pulse tube refrigerators that have low acoustic power at the warm end, by changing the phase and the swept volume of compressor and APC.

Cooling Capacity (Qc) of PTR with APC

Figure 13 shows that the cooling capacity of the PTR with APC and the phase between the PWG and APC are strongly correlated; there exists an optimum phase for highest Qc.

Figure 14 shows the cooling capacity of the PTR with APC varying the swept volume of APC. With the increase of the swept volume of APC, the cooling capacity of the PTR is increased. With the increase of the swept volume in APC, the power supply increased as well; there exists an optimum phase and an optimum swept volume of the PWG and APC.

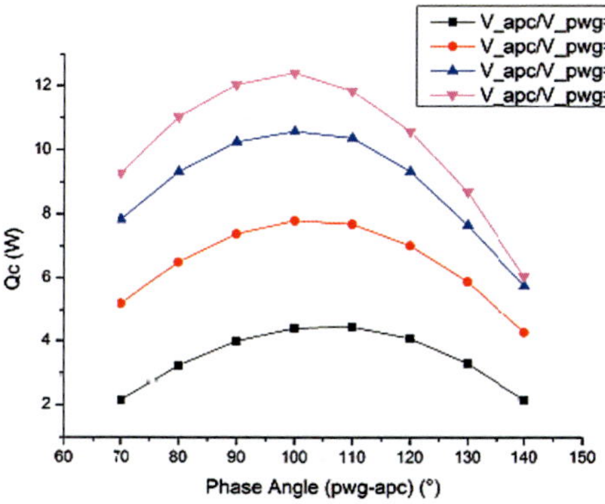

Figure 13. Qc versus phase between PWG and APC

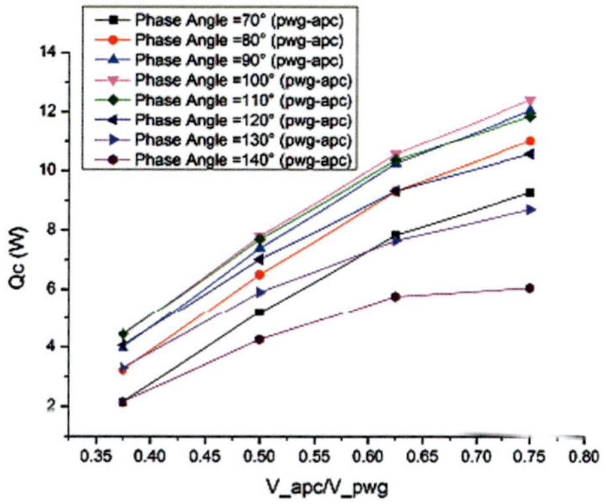

Figure 14. Qc versus Vapc/Vpwg

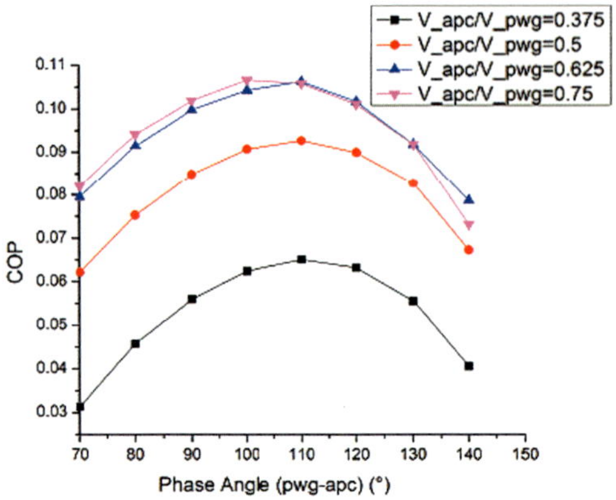

Figure 15. COP versus phase between PWG and APC

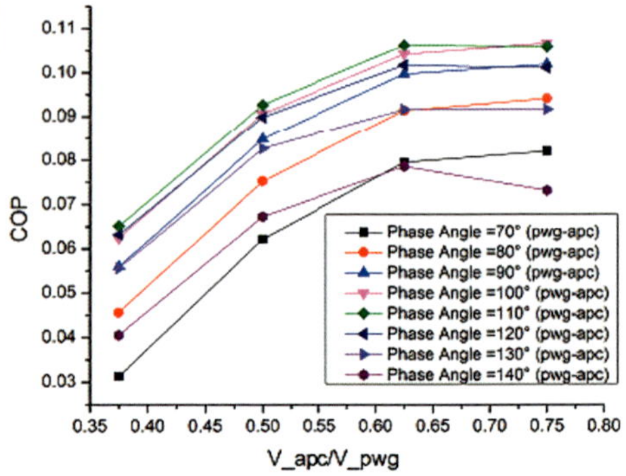

Figure 16. COP versus Vapc/Vpwg

COP of PTR with APC

Figures 15 and 16 show that as the phase angle between the PWG and the APC increases, the COP increases until the phase angle reaches a certain value. There exists an optimum phase and an optimum swept volume for the PWG and APC.

The quantity of cooling capacity depends on the phase between the pressure and the axial velocity, the mass flow rate and the pressure ratio at the cold end of the pulse tube. The COP of PTR depends on the cooling capacity and the PV power supplied by the PWG.

$$Q_c \propto \langle \dot{W} \rangle_t = \langle P\dot{V} \rangle_t = \frac{1}{2}|P||\dot{V}|\cos\theta \tag{6}$$

The simulation shows that, the phase and swept volumes of PWG and APC play an important influence on the phase angle of the phase between the pressure and the axial velocity at the cold heat exchanger of the pulse tube.

Experimental Setup

A photograph of the experimental setup is shown in Figure 17. The left side compressor is used as the pressure wave generator (PWG) and the right side compressor is used as the active phase controller (APC). The APC is directly connected to the warm end of the pulse tube using a stainless steel tube.

As shown in the schematic in Figure 18, the PWG and the APC are operated by amplified electric signals from a function generator. The power supply of the PWG and the APC consists of two power amplifiers. The phase angle between the PWG and APC is dominated by the signal generator. The displacement transducers are used to measure the displacement of the piston in PWG and APC.

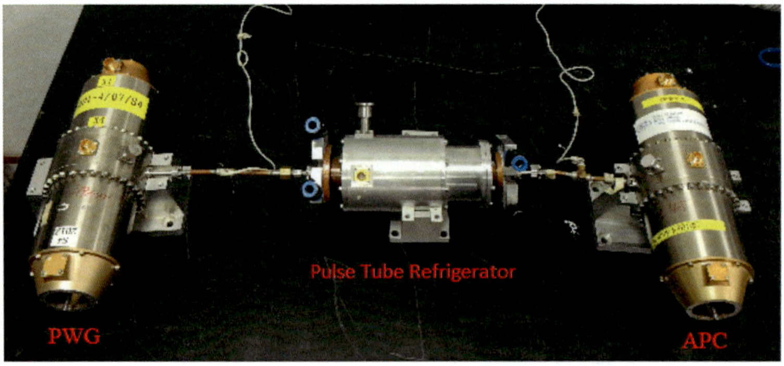

Figure 17. Photograph of experimental setup

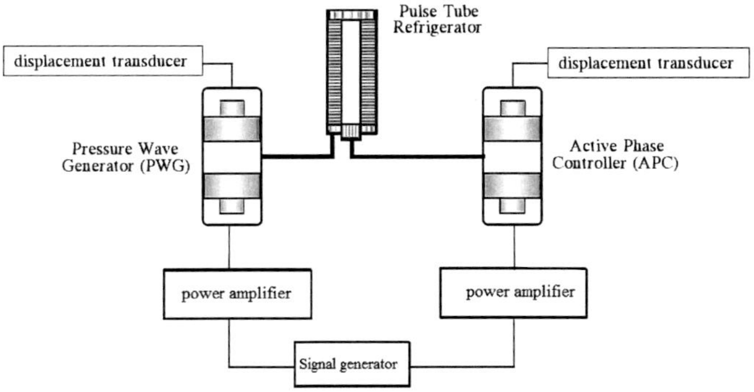

Figure 18. Schematic of experimental setup

All the experiments have been carried out at an operating frequency of 50 Hz and an initial charging pressure of helium gas of 3.2 MPa. The warm-end heat exchanger and the aftercooler are set to a constant temperature of 300 K. These experimental results are shown as follows.

EXPERIMENTAL RESULTS

Carnot Efficiency of PTR with APC

In this experiment, all the experiments have been carried out at an operating frequency of 50 Hz and an initial charging pressure of helium gas of 3.2 MPa. The warm-end heat exchanger and the aftercooler are set to a constant temperature of 300 K. The initial input power of the compressor of the PWG was 120 W, and the initial input power of the compressor of the APC was 15.5 W. The cold end temperature was fixed at 100 K.

Figure 19 shows the phase PWG and APC by varying the COP of the pulse tube refrigerator. The highest COP is 1.05×10^{-2}, when the phase angle between the PWG and APC is 110° (PWG-COP). The predicted performance of the PTR is in reasonable agreement with the experimental data.

No-Load Temperature of PTR with APC

In this experiment, all the experiments have been carried out at an operating frequency of 50 Hz and the initial charging pressure of helium gas is 3.2 MPa. The warm-end heat exchanger and the aftercooler are set to a constant temperature of 300 K. The amplitude of the compressor is 8 mm, and the amplitude of the APC is changed from 2 mm to 7 mm.

Figure 20 shows the temperature at the cold heat exchanger of the pulse tube refrigerator without a thermal load; the lowest temperature is 75.5 K, when amplitude of the APC was 7 mm. The predicted performance of the pulse tube refrigerator is in reasonable agreement with the experimental data.

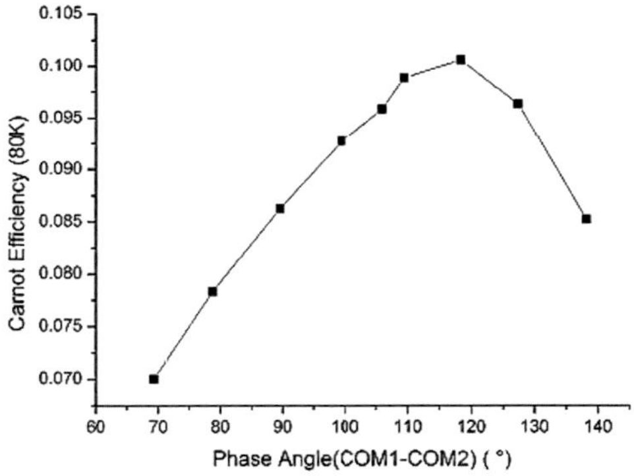

Figure 19. Carnot efficiency versus input power of PWG and APC

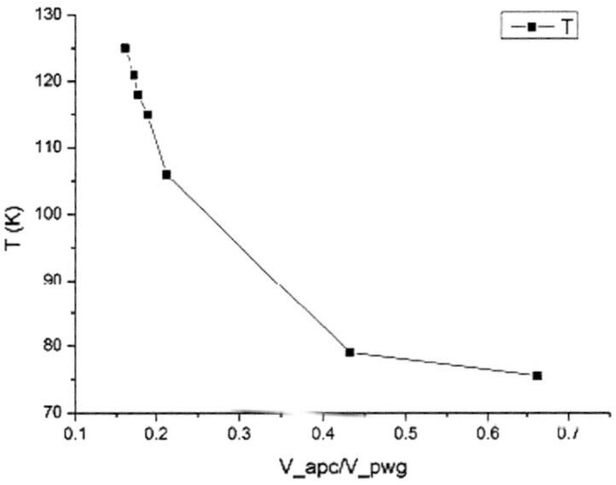

Figure 20. No-load temperature of PTR with APC versus V_{apc} / V_{pwg}

CONCLUSION

In this study, a numerical model of Stirling type pulse tube refrigerator with an active phase controller was performed using a two-dimensional axis-symmetric CFD. The influences of the phase and swept volume of the APC on the phase, mass flow, pressure ratio and the performance of PTR are discussed. The predicted performance of the pulse tube refrigerator is in reasonable agreement with the experimental data.

1. The Stirling type pulse tube refrigerator with the active phase controller has been experimentally investigated.
2. The investigations have demonstrated the capability of CFD models to predict the performance and discover the losses in the pulse tube refrigerator.
3. There exist an optimum phase between two pistons and the optimum swept volume of two pistons for the PTR to improve system efficiency.

ACKNOWLEDGMENTS

This work was supported by the Natural Science Foundation of Shanghai (NO. 16ZR1441500).

REFERENCES

1. Hu, J.Y., Ren, J., Luo, E.C. and Dai, W.,. "Study on the Inertance Tube and Double-inlet Phase Shifting Modes in Pulse Tube Refrigerators," *Energy Conversion and Management*, 52(2) (2011), pp. 1077-1085.

2. Antao, D.S., Farouk, B., "Numerical and Experimental Characterization of the Inertance Effect on Pulse Tube Refrigerator Performance," *International Journal of Heat and Mass Transfer*, 76 (2014), pp. 33-44.

3. Swift, G.W., Gardner, D.L., Backhaus,S., "Acoustic recovery of lost power in pulse tube refrigerators," *Journal of the Acoustical Society of America*, 105(2) (1999), pp. 711-724.

4. Li, Z., Gan, Z. and Qiu, L., "Cold Inertance Tube for 4 K Stirling Type Pulse Tube Cryocoolers," *Physics Procedia*, 67 (2015), pp. 451-455.

5. Brito, M., Peskett, G., "Numerical Model of Free Warm Expander Pulse Tube Cooler," *Cryogenics*, 41(10) (2001), pp. 751-755.

6. Brito, M., Peskett, G., "Experimental Analysis of Free Warm Expander Pulse Tube," *Cryogenics*, 41(10) (2001), pp. 757-762.

7. Riley, P.H., "Towards a Transient Simulation of Thermo-acoustic Engines Using an Electrical Analogy," *Procedia Engineering*, 56 (2013), pp. 821-828.

8. Bailly, Y., Nika, P., "Comparison of Two Models of a Double Inlet Miniature Pulse Tube Refrigerator: Part B Electrical Analogy," *Cryogenics*, 42(10) (2002), pp. 605-615.

9. Dang, H., Zhao, Y., "CFD Modeling and Experimental Verification of a Single-stage Coaxial Stirling-type Pulse Tube Cryocooler Without Either Double-inlet Or Multi-bypass Operating at 30–35 K Using Mixed Stainless Steel Mesh Regenerator Matrices," *Cryogenics*, 78 (2016), pp. 40-50.

10. Rout, S.K., Gupta, A.K., Choudhury, B.K., Sahoo, R.K. and Sarangi, S.K., "Influence of Porosity on the Performance of a Pulse Tube Refrigerator: a CFD Study," *Procedia Engineering*, 51 (2013), pp. 609-616.

11. Cha, J.S., Ghiaasiaan, S.M. and Kirkconnell, C.S., "Oscillatory Flow in Microporous Media Applied in Pulse – Tube and Stirling – Cycle Cryocooler Regenerators," *Experimental Thermal and Fluid Science*, 32(6) (2008), pp. 1264-1278.

Diaphragm-Type Mechanism for Passive Phase Shifting in Miniature PT Cryocooler

S. Sobol, G. Grossman

Technion – Israel Institute of Technology
Haifa, Israel 32000

ABSTRACT

A properly operating pulse tube (PT) cryocooler requires flow-to-pressure phase shifting at the hot end to provide close to zero phase within the regenerator and the cold heat exchanger. Typical optimum flow-to-pressure phase at the hot end is around -60 degrees; thus, the phase shifting mechanism should provide a distinct inertial behavior with a certain capability to dissipate acoustic power.

The most common method of creating the proper phase shift involves the use of an inertance tube (IT), a long and thin tube with a reservoir attached to its far end. However, as a PT cryocooler is scaled down, the IT efficiency degrades due to the rapid increase of the resistive fluid impedance. Additionally, miniaturization of the cryocooler is generally accompanied by an increase in the operating frequency, which, in turn, magnifies the ineffectiveness of the IT further. A combination of the IT with a bypass flow tubing is also not sufficient for adequate shifting of the flow-to-pressure phase in ultra-miniature PT cryocoolers.

In fact, the only available method of creating the proper phase in a miniature high-frequency PT cryocooler entails the use of a warm mechanical expander. We propose to construct an expander consisting of a passively oscillating mass suspended on soft silicone diaphragms, which are frictionless, to provide a perfect dynamic seal with ease of miniaturization, and to perform the necessary function of a viscous damper. The research includes an analytical model, numerical simulations and an actual prototype of the proposed phase shifting mechanism tested on our MTSa miniature PT cryocooler.

INTRODUCTION

In recent years, many efforts have been made worldwide to miniaturize a pulse tube (PT) cryocooler, while retaining an adequate cooling capacity at lowest possible temperature. The corresponding miniaturization should be accompanied by increasing the filling pressure, the pressure ratio and the operating frequency. R. Radebaugh in 2006 showed that a PT can theoretically be operate effectively at frequencies up to 1000 Hz with a screen type regenerator of 7 mm length only [1]. However, no one has been able to approach these values practically. One of the obstacles to the PT miniaturization is the inability to produce a proper flow-to-pressure phase shifting mechanism of the appropriate miniature size.

A properly operating PT cryocooler requires very little flow-to-pressure phase within the regenerator and the cold heat exchanger. The corresponding phase at the hot end is obtained at about -60°, which is equivalent to a phase of -150° of the gas travel relative to the pressure wave. The

Cryocoolers 19, edited by S.D. Miller and R.G. Ross, Jr.

phase much under -90° indicates a distinct inertial behavior of the planned phase shifting mechanism, which should also have a certain capability to dissipate acoustic power.

The most common passive method of the flow-to-pressure phase shifting involves the use of a inertance tube (IT), a long and thin tube with a reservoir attached to its far end. The IT is a very simple and reliable mechanism, which is also very effective in medium and large scale PT cryocoolers. However, as the PT is scaled down, the effectiveness of the IT degrades due to a rapid increase of the resistive fluid impedance, which is inversely proportional to the tube diameter to the 5th power, whereas the inertance component changes just as the 2nd power [2]. Additionally, the flow resistance increases further with the frequency, the filling pressure and the pressure ratio, while the inertance effect is independent on these parameters. As a result, the maximum flow-to-pressure phase that the IT is able to create drops to zero as the PT cryocooler is miniaturized while retaining a nominal cooling capacity. The combination of the IT with a bypass flow tubing may improve the phase by about 10 degrees, but it is definitely not sufficient for the proper operation of an ultra-miniature PT cryocooler.

In fact, to date the only available method of creating the proper phase in a miniature high-frequency PT cryocooler entails the use of a mechanical warm expander (WE). Active WE systems, employing the inversely operating pressure oscillator as a phase shifting mechanism were investigated in the past [3], [4]; however, this solution is expensive and hard to miniaturize. Passive WE systems were theoretically considered [3], [5]. We propose to implement a passive expander consisting of a pressure driven oscillating mass suspended on soft silicone diaphragms, which are frictionless, provide a perfect dynamic seal, are easy to miniaturize, and contribute to the required damping. Unlike the IT, the mechanical oscillator has no inherent size or frequency dependent restrictions. Therefore, a properly designed passive expander should be able to supply the required flow-to-pressure phase at any operational frequency regardless of the PT dimensions and pressure amplitude.

THEORY

Figure 1 demonstrates schematically the proposed passive warm expander (WE). The oscillating mass of the system consists of two pistons with cross sectional areas A_1 and A_2, linked together by a rigid heavy rod of mass m. The pistons partition the interior space of the system housing into three volumes: front volume (V_f), intermediate volume (V_i) and buffer volume (V_b). The corresponding absolute pressures within the volumes are given by P_f, P_i and P_b. The average pressure P_0 is uniform along the system and is equal to the filling pressure of the PT. The front volume is connected to the hot end of the PT through the transfer line, and must provide the proper flow phase and amplitude to the cryocooler.

The linked pistons are supported axially by a mechanical spring k_m, which allows axial displacement x relative to the equilibrium point. Since the spring is intended to represent the stiffness of elastomeric diaphragms, the hysteretic damping coefficient η is also taken into consideration. Piston A_2 includes apertures filled by porous material, which allow for the gas to flow from the buffer volume to the intermediate volume and vice versa with a certain pressure drop. A proper design of the gas passaging might supply a linear flow to pressure drop relation, which should result in creating a linear viscous damping mechanism within the system.

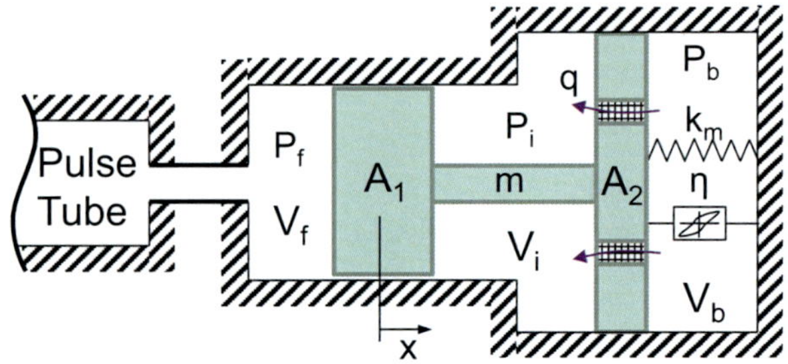

Figure 1. Schematic of the passive warm expander.

Linear Model of the Passive Warm Expander

The governing equations of the system are given by (1) and (2):

$$m\ddot{x} + k_m (1 + i\eta) x = (P_f - P_i) A_1 + (P_i - P_b) A_2 \tag{1}$$

$$q = -C_V (P_i - P_b) \tag{2}$$

Where, $i = \sqrt{-1}$, q is the volumetric flow rate through the apertures in piston A_2, C_V is the linear flow coefficient. Under the assumption of small pressure ratios, the mass conservation equation within the adiabatic buffer volume can be expressed by (3) [6]:

$$-\frac{P_0}{RT_0}\mathbf{q} = \frac{V_{b0}\dot{\mathbf{P}}_\mathbf{b}}{\gamma RT_0} + \frac{P_0\dot{\mathbf{V}}_\mathbf{b}}{RT_0} \tag{3}$$

Where, γ is the adiabatic constant, R is the gas constant, and T_0 is the time average temperature.

Eq. (3) is given in vector notation, where the first term on the right hand side of (3) represents the mass accumulation within the volume, and the second term represents the volume change rate due to displacement of the pistons. All three vectors are assumed to be purely sinusoidal with a uniform angular velocity ω. In this case Eq. (3) can be simplified as follows:

$$-P_0\mathbf{q} = \omega\left(\frac{1}{\gamma}V_{b0}\mathbf{P}_\mathbf{b} + P_0\mathbf{V}_\mathbf{b}\right) \tag{4}$$

Where $\mathbf{P}_b$ and $\mathbf{V}_b$ are pressure and volume sinusoidal functions, substituted in (4) instead of their respective time derivatives in (3). Now, we would like to consider a very special case, where the pressure amplitude is about two orders of magnitude lower than the filling pressure, and, on the other hand, the volume amplitude is about half the mean buffer volume V_{b0}:

$$\begin{cases} \dfrac{|\mathbf{P}_\mathbf{b}|}{P_0} \sim 0.01 \\ \dfrac{|\mathbf{V}_\mathbf{b}|}{V_{b0}} \sim 0.5 \end{cases} \Rightarrow \quad \left|\frac{1}{\gamma}V_{b0}\mathbf{P}_\mathbf{b}\right| \ll |P_0\mathbf{V}_\mathbf{b}| \tag{5}$$

In this special case the gas accumulation term in (3) is neglected relative to the volume change term, and thus can be omitted. As a result, the following algebraic relation for q is obtained:

$$q \approx -\dot{V}_b = A_2\dot{x} \tag{6}$$

Substitution (6) of (2) into provides a relation between the pressure drop on the A_2 piston and the piston velocity:

$$P_i - P_b = -\frac{A_2\dot{x}}{C_V} \tag{7}$$

Substitution of (7) into (1) provides the modified motion equation of the pistons:

$$m\ddot{x} + c\dot{x} + k_m(1 + i\eta)x = (P_f - P_i)A_1 \tag{8}$$

Where:

$$c = \frac{A_2^2}{C_V} \tag{9}$$

The resulting relation (8) is actually a classic linear motion equation driven by oscillating pressure force, and having additional hysteretic damping. For the pressure given by (10), the piston displacement and the front volume magnitude can be expressed in general by (11) :

$$P_f - P_i = P_1 \cos\omega t \tag{10}$$

$$\begin{matrix} x = x_1\cos(\omega t + \theta) \\ V_f = V_0 + V_1\cos(\omega t + \theta) \end{matrix}, \quad V_1 = A_1 x_1 \tag{11}$$

where x_1 and V_1 are piston and volume amplitudes respectively, and θ is actually the gas travel-to-pressure phase. A general solution for the volume amplitude and the corresponding phase is the following:

$$\begin{cases} V_1 = \dfrac{P_1 A_1^2}{\sqrt{\left(\omega c + \eta k_m\right)^2 + \left(\omega^2 m - k_m\right)^2}} \\ \tan\theta = \dfrac{\omega c + \eta k_m}{\omega^2 m - k_m} \end{cases} \tag{12}$$

Matching the Passive Warm Expander to the PT Cryocooler

We consider a specific pulse tube cryocooler, designed to operate at a frequency $f_0=\omega_0/2\pi$ with a pressure amplitude P_1. A separate analysis of the PT shows that a phase shifting mechanism connected to the hot end should supply the swept volume amplitude of V_1 and the travel-to-pressure phase of θ_0 for achieving the proper PT operation. The phase shifting mechanism is implemented by a passive WE described schematically in Figure 1, and the WE oscillations are governed by Eq. (12). The pistons are suspended on elastic diaphragms, which possess total axial stiffness k_m and the corresponding hysteretic damping coefficient ηk_m. Cross sectional areas of the pistons can be chosen arbitrarily.

The first objective refers to estimation of a mass and a viscous damping coefficient as a function of A_1, k_m and η, which would create the proper dynamics of the WE matching the PT requirements. For this purpose, Eq. (12) should be solved for m and c, while it is known in advance that θ_0 should be below -90°. The appropriate solution is given by Eq. (13):

$$\begin{cases} m = \dfrac{k_g + k_m}{\omega_0^2} \\ c = \dfrac{k_g\left|\tan\theta_0\right| - \eta k_m}{\omega_0} \end{cases}, \quad -180° < \theta_0 < -90° \tag{13}$$

Where:

$$k_g = \frac{P_1 A_1^2 \left|\cos\theta_0\right|}{V_1} \tag{14}$$

One should notice that the stiffness k_g defined in (14) is actually the PT gas stiffness measured on the A_1 piston boundary.

In order to verify the results obtained in (13), we first modified the SAGE™ model of our MTSa miniature PT cryocooler [7] for operating with a passive piston instead of the original IT system. Figure 2 partially demonstrates a root level of the SAGE™ model, where the bottom row components represent the structure of the passive expander. The schematic of the SAGE model expander differs from the one shown in Figure 1 by the absence of the secondary piston A_2. Instead, there is a single buffer volume and a linear damper connected in parallel to the oscillating mass and the mechanical spring as shown in Figure 2 in a separate window.

MTSa [7] was designed to operate at 100 Hz with a filling pressure of 40 bar and a pressure ratio of 1.3. Due to the pressure drops along the cryocooler, the average pressure amplitude within the hot heat exchanger is obtained at 4.45 bar. According to previous simulations we know that the optimum swept volume amplitude and the travel-to-pressure phase at the hot end should be 23.6 mm^3 and -145°, respectively.

We chose a passive piston of 5 mm diameter, which should result in a 1.2 mm piston amplitude, k_m = 3 N/mm and η = 0. In this case the gas stiffness is obtained at 5.95 N/mm according to (14). According to (14), the oscillating mass and the damping coefficient should be 22.5 g and 6.8 N/m, respectively. These analytically obtained parameters (A_1, m, c and k_m) were substituted into the SAGE model, which was run at various frequencies while retaining the nominal pressure amplitude at the hot end. The resulting piston amplitude and phase are shown in Figure 3 together with the analytical frequency response. It can be clearly seen that both the analytical and the SAGE models satisfy perfectly the required 1.2 mm amplitude and -145° phase at 100 Hz.

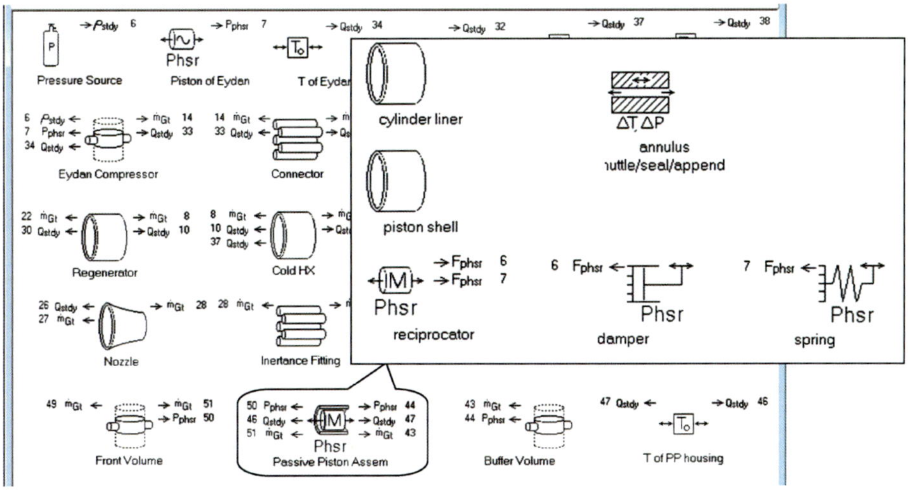

Figure 2. SAGE model of the MTSa PT with a passive warm expander.

PRELIMINARY PROTOTYPE

Viscous Damping Mechanism

Implementation of the viscous damper is based on Eq. (9), and should not violate the conditions given in (5). A linear flow-to-pressure-drop relation can be achieved using mesh stacks with sufficiently small gas passages, which ensure a laminar flow regime. A suitable candidate was found in a random fiber stainless steel mesh of 0.4 mm thickness with a porosity of 0.75. The mesh was cut into discs of 2 mm diameter for further arrangement within the A_2 apertures. Figure 4 shows the enlarged disk illustration together with a measured flow capacity of a single disk, where the examined fluid was pure nitrogen at room conditions. According to the figure, the flow curve is perfectly linear for pressure drops up to 0.3 bar at least, where the flow reaches 2.5e-5 m^3/s. The linear flow coefficient of a single disk is estimated at 8.4334e-10 $(m^3/s)/Pa$.

Because of some design constrains, the effective diameter of the secondary piston A_2 was set to 14 mm. According to Eq. (6) the maximum flow rate at 100 Hz with the piston amplitude of 1.2 mm is estimated at 1.16e-4 m^3/s. This value exceeds by about four times the linear flow range of a single disk; hence, in order to ensure small pressure drops, at least four single disks should be ar-

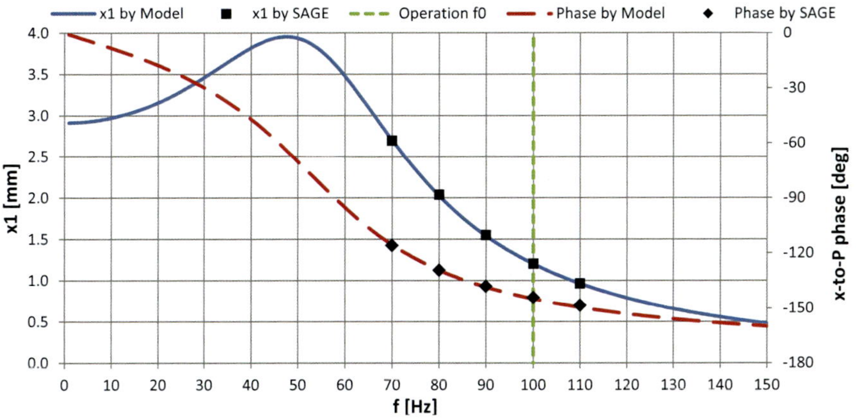

Figure 3. SAGE™ vs. analytical model frequency response.

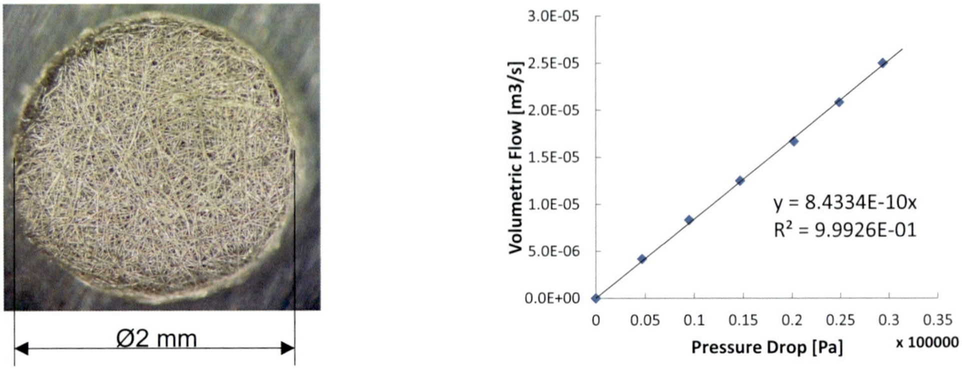

Figure 4. Illustration and flow capacity of a single random mesh disk.

ranged across the A_2 piston in parallel. According to Eq. (9), arrangement of four disks in parallel should result in 7.0 N/m for the viscous damping coefficient. In the previous section of the paper we considered a case where the required viscous damping in the absence of the hysteresis was estimated at 6.8 N/m, which is quite close to the value obtained here.

CAD Model

Figure 5 demonstrates exploded and section views of the preliminary WE assembly. The A_1 and A_2 pistons are molded together with the corresponding cylinders by 25 shore-A adhesive silicone. The pistons and the cylinders were machined with patterns of small apertures for enlarging the contact areas, thus improving the silicone to metal adhesion. The mutual axis was tapped by M2 thread for simplicity in fastening of the mass element, the A_2 piston and the mirror located on the right hand end of the axis. The mirror is utilized for the real time measurement of the axial displacement using an external laser-detector system. The buffer volume of the housing is covered by a glass window, which allows the laser beam to pass freely in both directions regardless of the

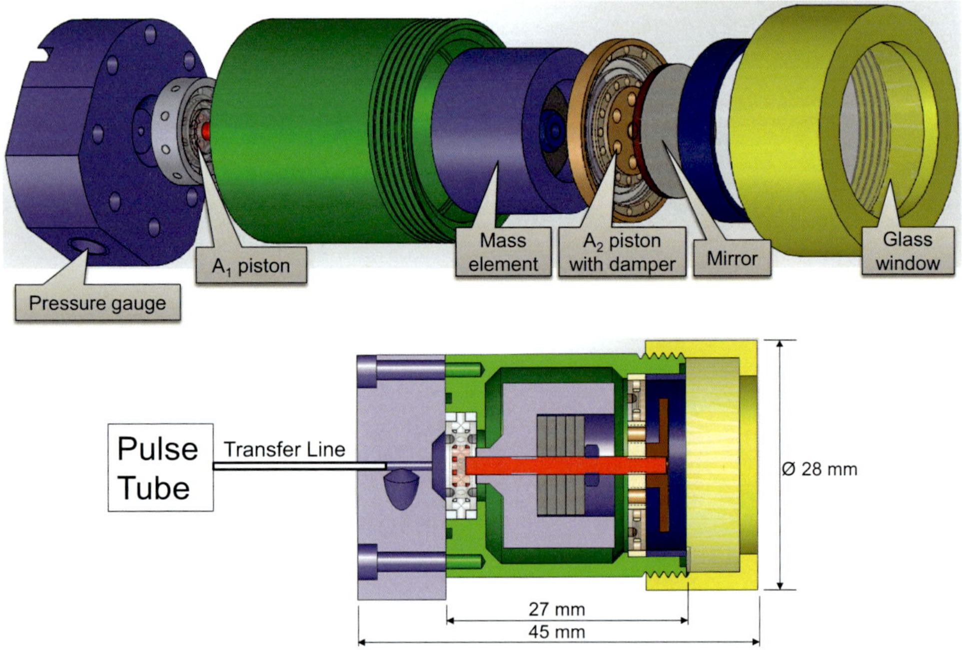

Figure 5. CAD model of the preliminary warm expander.

axis instantaneous location. The front side of the housing is covered by a thick flange for allowing pressure gauge installation.

The total length of the WE system is 45 mm. The length can be decreased by about 10 mm with the pressure and displacement measurement systems eliminated. In this case the overall size of the WE housing would be dictated mostly by the mass element dimensions. According to Eq. (13) the mass and its corresponding dimensions are intended to decrease as the operating frequency increases and/or as the silicone diaphragms create lower axial stiffness. Currently, the oscillating mass is adjustable between 24 and 32 gram according to the number of the lead disks added to the main mass element.

Silicone Diaphragms

Advantages of the diaphragm type suspensions over the commonly used gas-gap sealing methods are the following: they provide frictionless motion, are a perfect dynamic seal, are easy to miniaturize, and do not require high precision and assist in damping due to the natural hysteresis of polymers. Additionally, well treated silicones possess very good temperature stability and produce very little outgassing. On the other hand, the diaphragms may produce the enhanced axial stiffness; the effective diameter may vary with the pressure drop; fatigue problem is more topical; and the non-linear effects are quite significant. The influence of the latter is magnified while operating in the vicinity of the resonance frequency; however, the non-linear effects can be reduced drastically by applying sufficient damping on the oscillator.

Figure 6 demonstrates an example of a numerical simulation of the A_1 piston diaphragm, where the piston is displaced downwards by 1 mm, and pressure of 4 bar is applied on the bottom surfaces. Mechanical properties of the silicone were specified by hyper elastic Neo-Hookean material model with an initial shear modulus of 0.22 MPa and the incompressibility parameter of 17.7 GPa^{-1}. The maximum equivalent strain was obtained at 77.7%, while the maximum allowable strain of the specific silicone should exceed 300%. The equivalent piston diameter is obtained at 5.4 mm. No-pressure simulation of the diaphragm provided estimation of the axial stiffness, which was obtained increasing from 4.4 to 4.7 N/mm as a function of the piston displacement.

Experiment Setup

A preliminary prototype of the proposed passive WE has been constructed and connected to our miniature MTSa PT cryocooler through 0.7 mm inner diameter, 100 mm length transfer line. Figure 7 illustrates the experimental setup, with the WE installed vertically, and the front volume rotated upwards. The WE housing shares a rigid basis with a travel measurement system, consisting of a laser module and a Position Sensitive Detector (PSD) unit. Both are adjustable for a proper angle and location relative to the mirror within the WE housing.

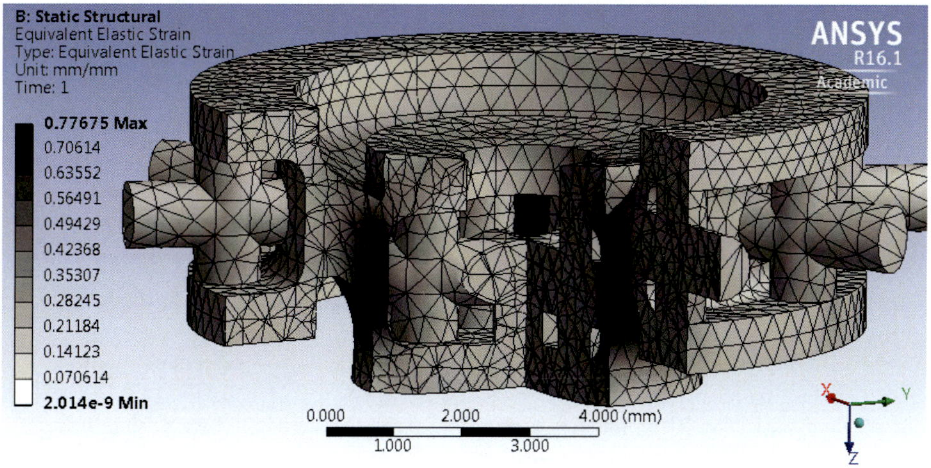

Figure 6. Simulation on A_1 piston diaphragm with $\Delta x = 1$ mm, $\Delta P = 4$ bar.

Figure 7. Experiment setup.

Four pressure gauges are installed along the PT-WE system: in the aftercooler, in the hot heat exchanger, in the front and the intermediate volumes of the WE. The pressure wave is supplied by a pressure oscillator located under the illustrated setup, outside the vacuum chamber.

RESULTS AND DISCUSSION

The experiments were conducted with the mass element set to the maximum possible 32 g due to the enhanced axial stiffness of the silicone diaphragms obtained after the molding. The total static axial spring constant due to both diaphragms elasticity was estimated at about 12 N/mm, while the initial prediction based on numerical simulations was about 8 N/mm. The viscous damping mechanism included six random fiber disks, as described previously, in parallel arrangement. Pressure ratio at the front volume of the WE system was retained for all the examined frequencies at nearly 1.1, which is equivalent to 1.85 bar amplitude over the filling pressure of 40 bar.

Figure 8 (top row) summarizes piston amplitudes and phases relative to the pressure difference on A_1 at various operating frequencies. The smooth curves represent the best fit theoretical model response according to Eq. (8). It can be clearly seen that at least one additional motion mode was excited at 130 Hz in the experiment. The mode has distorted the predicted amplitude curve; however, it had a minor effect on the phase. Apparently, the phase response should provide more reliable information on the axial travel mode than the amplitude.

The phase response curve crosses -90° at 102 Hz, which should be the natural frequency of the axial motion. Therefore, taking into account the oscillating mass of 32 g, the effective dynamic spring constant k_m is estimated at 13 N/mm, which is about 8% larger than the measured static stiffness. Since the hysteretic damping constant of the utilized silicone was estimated previously at about 0.1, the only underestimated parameter in Eq. (8) remained the viscous damping coefficient. The best fit between the theoretical model and the measured data was obtained for c = 6.5 Ns/m, which is quite close to the one obtained in the previous chapter of the paper despite a different fluid and different pressure.

A bottom row of plots in Figure 8 shows the time scaled measurements of the pistons travel and the appropriate pressures at the frequency of 140 Hz, which provided a phase lag of 145° required for operation of our MTSa PT cryocooler. Certainly, this frequency is too distant from the nominal 100 Hz, and the amplitude at 140 Hz is still affected by the undesired mode; thus it is unreasonable to conduct the cooling capacity experiments on MTSa at these conditions. One should notice that

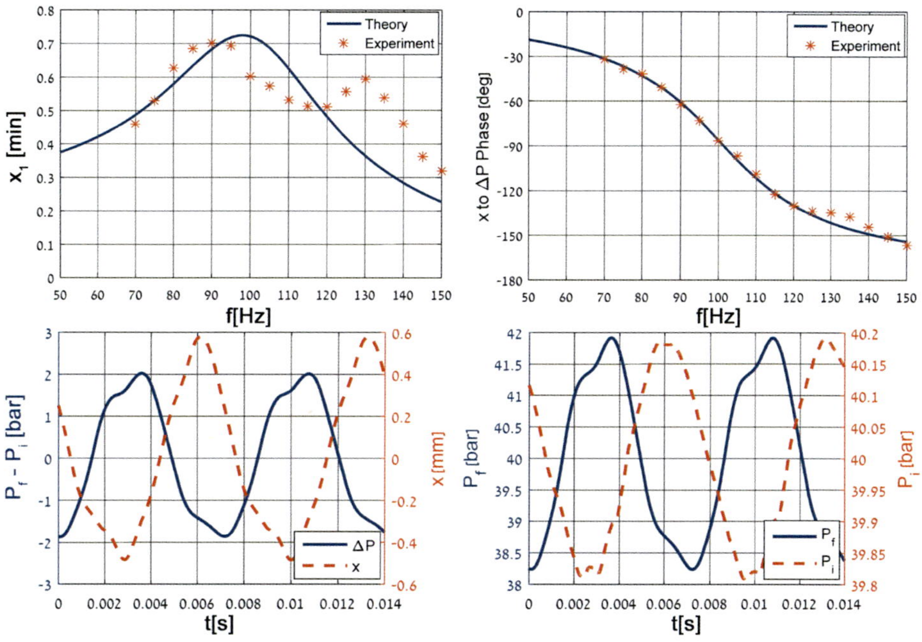

Figure 8. Top – Frequency response of the prototype and the corresponding theoretical model; Bottom – Measured pressures and displacement at 140 Hz.

the intermediate pressure P_i is obtained with about ten times lower amplitude than that of the front pressure P_f, and, therefore, has a minor effect on x-to-P_f phase and the pressure force applied on A_l piston.

SUMMARY AND CONCLUSIONS

A simplified analytical model was developed for designing a passive warm expander (WE), with the intention to provide a proper flow-to-pressure phase shifting for PT cryocoolers regardless of the PT size, operating frequency and pressure ratio. The model estimates the mass and damping coefficient values for the WE oscillator as functions of the arbitrarily chosen spring constant, the passive piston diameter and the PT requirements. The viscous damping is determined by the secondary piston diameter and the linear flow coefficient obtained from the gas flow capacity through the secondary piston apertures.

The analytical model demonstrated a perfect agreement with the SAGE™ model, where the damping mechanism was not implemented entirely but represented by artificial viscous damping component with a constant coefficient. The models have verified that our MTSa PT cryocooler operating at 100 Hz with pressure ratio 1.3 requires for optimum operation the WE oscillator of 23.5 gr with damping coefficient of 6.8 Ns/m, when the passive piston diameter is set to 5 mm and the spring constant is 3 N/mm.

A preliminary prototype of the passive WE system was designed, fabricated and tested at various frequencies. The overall dimensions of the WE system are Ø28x45 mm, where about 10 mm of the length are extraneous due to pressure and displacement measurement systems. The prototype includes two rigidly linked pistons of 5.4 and 14.0 mm diameters suspended on silicone diaphragms, a mass element of 32 gr, and viscous damping mechanism, which produces a linear damping coefficient of 6.5 Ns/m.

The WE prototype proved the ability to create any flow-to-pressure phase appropriate for a PT cryocooler. The implemented viscous damping mechanism demonstrated a good agreement with the theoretical model. Furthermore, the actual damping coefficient is amenable to easy tuning by changing the amount and/or the arrangement of the mesh disks within the apertures across the A_2 piston.

Unfortunately, several technical drawbacks did not enable perfect matching between the WE system and our MTSa PT cryocooler. First, material properties of the silicone utilized for the diaphragms were estimated inaccurately; thus, the total measured axial spring constant was about 60% larger than the calculated one. Additional stiffness increase was obtained in dynamic operation due to the pressure drop applied on the secondary piston and the total gas compressibility in the intermediate and the buffer volumes. Finally, the axial spring constant turned out almost twice larger than the required one for the 32 gram mass, and, as a result, the natural frequency of the oscillator was shifted by a factor of 1.4. Consequently, the desired -145° of travel-to-pressure phase was obtained at 140 Hz instead of at the nominal 100 Hz.

The second drawback refers to excitation of additional motion modes of the oscillator in the vicinity of the operating frequencies. At least one parasitic mode was detected at 130 Hz; however, it affected the measured amplitudes in a wide range around. Obviously, the WE system requires a full modal analysis and a subsequent dynamic redesign, which would shift the undesired modes far away from the operational range of frequencies.

ACKNOWLEDGMENT

The generous financial support of the Rechler Family, MAFAT and the Technion is gratefully acknowledged.

REFERENCES

1. Radebaugh, R. and O'Gallagher, A., "Regenerator operation at very high frequencies for microcryocoolers," *Adv. in Cryogenic Engineering*, Vol. 51, Amer. Institute of Physics, Melville, NY (2006), pp. 1919-1928.

2. Radebaugh, R., Lewis, M., Luo, E., Pfotenhauer, J.M., Nellis, G.F. and Schunk, L.A., "Inertance Tube Optimization for Pulse Tube Refrigerators," *Adv. in Cryogenic Engineering*, Vol. 51, Amer. Institute of Physics, Melville, NY (2006), pp. 59-67.

3. Brito, M.C. and Peskett, G.D., "Experimental Analysis Of Free Warm Expander Pulse Tube," *Cryogenics,* vol. 41 (2001), pp. 757–762

4. Lewis, M., Bradley, P. and Radebaugh, R., "Experiments With Linear Compressors For Phase Shifting In Pulse Tube Crycoolers," *Adv. in Cryogenic Engineering*, Vol. 57, Amer. Institute of Physics, Melville, NY (2012).

5. Zhu, S. and Nogawa, M., "Pulse Tube Stirling Machine with Warm Gas-Driven Displacer," *Cryogenics,* vol. 50 (2010), pp. 320–330.

6. Radebaugh, R., "A Review of Pulse Tube Refrigeration," *Adv. in Cryogenic Engineering*, vol. 35, Springer US (1990), pp. 1191-1205.

7. Sobol, S., Katz, Y. and Grossman, G., "A Study Of A Miniature In-Line Pulse Tube Cryocooler," *Cryocoolers 16*, ICC Press, Boulder, CO (2011), pp. 87-95

Recent Progress with an Adjustable Inertance Tube

W.J. Zhou, J.M. Pfotenhauer, G.F. Nellis

University of Wisconsin - Madison
Madison, WI 53706

ABSTRACT

The cylindrical threaded adjustable inertance tube creates an acoustic element using the engagement of two threads. It is therefore possible to change the characteristics of the element easily and during operation by adjusting the rotational position of the outer screw thread. The threaded adjustable inertance tube has the capability to shift the phase angle between pressure and flow between 0 and -30 degrees, as demonstrated by previously reported experimental test results and supported by predictions made using the distributed component model. However, the performance of the current iteration of the device is limited by the leakage paths that are inherent between the engaged threads of the two screws. Since the working fluid flows through the leak to the average pressure in the channel formed by the root of the two screws, it is reasonable to connect these two flow channels in parallel to eliminate the leakage effect. Therefore, this paper reports the performance of the parallel threaded adjustable inertance tube using the appropriately modified distributed component model by eliminating the leaks terms and introducing another inertance tube in parallel with the original one. The preliminary modeling results suggest that the phase angle shift ability of the modified device will increase from -65 degrees to +40 degrees by using two inertance tubes in parallel.

INTRODUCTION

In previous reports [1-4], the development of an adjustable inertance tube for Stirling-type pulse tube refrigerators has been chronicled including analyses and measurements with linear, threaded conical, and threaded cylindrical adjustable inertance tubes. The present report continues the account with a update on the work-in-progress, and presents an encouraging calculation suggesting that an adjustable inertance tube can shift the phase angle between the pressure and flow oscillations in a pulse tube from -60 degrees to +40 degrees, altogether a shift of nearly 100 degrees.

Various models exist for designing inertance tubes that predict the resultant phase angle between the pressure and flow oscillations at the connection between the pulse tube and inertance tube as a function of the diameter and length of the inertance tube [5-7]. However, fabrication and assembly of all the components in a pulse tube refrigerator can result in a different phase angle during operation than desired. In most cases the phase angle during operation is unknown, but if the cooling performance is less than expected, it is reasonable to expect that the performance can be improved by adjusting the phase angle. Furthermore, an adjustable phase shifting mechanism can be useful for various operating modes, such as cooldown vs. steady state operation.

Cryocoolers 19, edited by S.D. Miller and R.G. Ross, Jr.

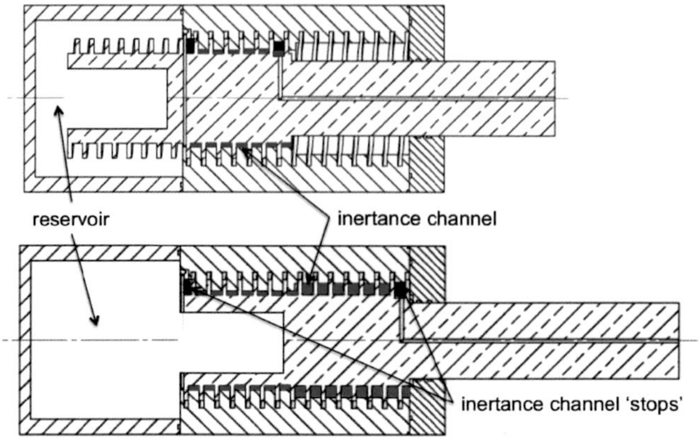

Figure 1. Threaded cylindrical adjustable inertance tube. The solid shaded regions highlight the helical channel between the root of the inner thread and tip of the outer thread that define the intertance tube or 'channel'. The top figure displays the most compact setting where the length and hydraulic diameter of the inertance channel are 1.74 m and 7.4 mm respectively, while the bottom displays the most extended setting, adding 1.86 m of a channel with a hydraulic diameter of 14.5 mm to the overall length.

CYLINDRICAL ADJUSTABLE INERTANCE TUBE

Potential and Challenge

The adjustable inertance tube that presents the most promising ability to shift the phase over a large range with minimal impact on the associated acoustic power is the cylindrical threaded version. A recent report [4] demonstrated that the geometry shown in Figure 1 has the potential to shift the phase angle from +60 degrees to – 40 degrees. Here the helical channel that exists as the gap between the root of the inner screw and the thread-tip of the outer screw forms the inertance tube. The gas flow passage extends from the pulse tube through the centerline passage visible at the right side in Figure 1, behind the solid stop at the right end of the inner thread, along the helical inertance channel, and finally through a hole behind the stop at the left end of the outer thread, into the reservoir. The phase between the pressure and flow oscillation at the pulse-tube connection to the inertance tube shifts as a result of increasing the length of the inertance channel and at the same time increasing the contribution of the channel with a larger cross sectional area. However, a problematic feature of the geometry results from the gas leakage path between the threads; the leak significantly decreases the phase shifting potential of the inertance channel.

As reported [4], the modified version of the distributed component model for the cylindrical threaded inertance tube design that includes the effects of leaks between the threads also demonstrates an equivalent degradation in phase shifting potential. In the same report, an effort to eliminate the leakage path between the threads by applying vacuum grease to the threads of the inner screw provides significant, but incomplete, recovery of the phase shifting potential.

Using the thickness of the leakage path between the threads as an adjustable parameter, the modified distributed component model matches the measured length- and diameter-dependent phase shift at an operating frequency of 60 Hz with a gap size of 15.5 microns. Agreement between the model and measurements with the same thickness of the leak-gap degrades as the frequency is reduced. At the lowest measured frequency of 27 Hz, the deviation between the model and experiment ranges from -27 degrees to + 36 degrees dependent on the given length.

The model and measurement of the associated acoustic power, while in the same quantitative range, are also not in good agreement. In view of the significant degradation in the phase-shifting performance that is associated with the thread-to-thread leak, various options have been explored to mitigate the leak. The most promising option, depicted in Figure 2 uses the helical channel formed by

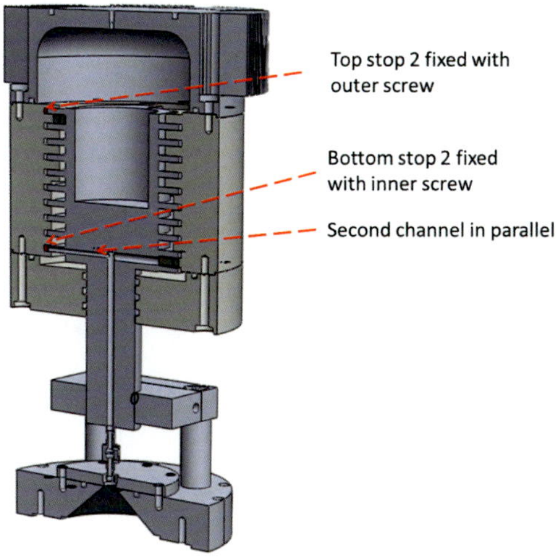

Figure 2. Cylindrical adjustable inertance tube. Two parallel helical channels are defined by 1) the gap between the root of the inner thread and tip of the outer thread, and 2) the gap between the root of the outer thread and tip of the inner thread.

the gap between the root of the outer screw and the thread-tip of the inner screw as an inertance tube operating in parallel with the original inertance channel. Spring-loaded stops, one fixed with the outer screw near the connection to the reservoir, and the other fixed with the inner screw near the connection to the pulse tube, define the ends for both of the parallel inertance channels. As the outer screw is turned from its most extended location, to its most compact location with respect to the inner screw, the effective diameter of the inner inertance channel changes from 11 mm to 7 mm while the diameter of the outer inertance channel remains constant at 2 mm. Over the same range, the length of both channels changes from 3.6 m to 1.74 m.

A passage connecting the inner and outer inertance channels at the pulse-tube end, and passages connecting both channels to the reservoir at the other end guarantee that the pressure is identical for both channels at those two locations. Since the geometry-defined inertance and compliance features of the two parallel channels are not the same, the pressures at any location along the helical length of the channels are not expected to be identical. However, as shown in Figure 3, the calculated pressure profiles for the two parallel channels are very similar.

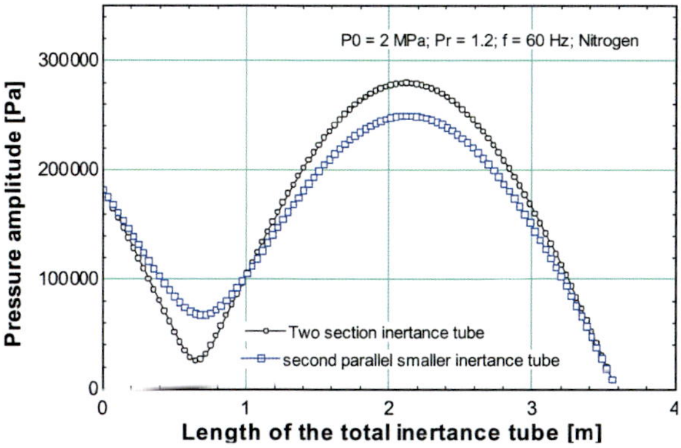

Figure 3. The Pressure profile along the length of the parallel helical channels. The 0 m location is the pulse tube end while the 3.6 m location is at the reservoir end.

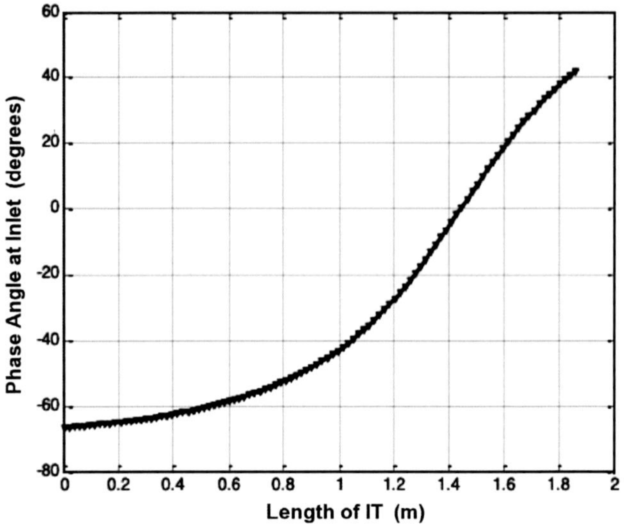

Figure 4. Anticipated phase shift produced by the parallel inertance channels as a function of the added length when the outer screw is turned from the compact location to the extended location.

In view of the small pressure difference between the two channels at any location, one may expect that the gas leak between the two channels is quite small. The resulting phase shift produced by the adjustable cylindrical inertance has been calculated using the modified distributed component model, including the leak between the threads, and is displayed in Figure 4. As shown there, the phase between the pressure and flow oscillations is expected to range from -65° to +40°, a change of more than 100° as the cylindrical threaded screw is turned from its most compact to its most extended configuration. During the fall and winter of 2016-2017 the required modifications will be made to the existing cylindrical screw so that it functions as two parallel inertance channels, and its operating characteristics will be measured.

SUMMARY

A progress update regarding the development of an adjustable inertance tube reveals that the cylindrical threaded design provides a large potential phase shifting ability. The leak between the channel formed by the gap between the root of the inner thread and tip of the outer thread and the channel formed by the gap between the root of the outer thread and tip of the inner thread can be mitigated by connecting the two helical inertance channels at both the pulse tube and reservoir ends, thereby creating a set of parallel interance tubes. The anticipated phase shift of the parallel inertance channel arrangement will be larger than 100 degrees.

REFERENCES

1. Pfotenhauer, J.M., Steiner, T.J. and Qiu, L.M. "Continuously Variable Inertance Tubes for Pulse Tube Refrigerators," in *Cryocoolers 16*, S. Miller, ed., ICC Press (2010), pp. 275-280.

2. Zhou, W.J., Pfotenhauer, J.M., "Real Time Phase Shifting with an Adjustable Inertance Tube," *Cryocoolers 17*, S. Miller, ed., ICC Press (2012), pp. 229-238.

3. Zhou, W.J., Pfotenhauer, J.M., and Nellis, G.F., "Developing a Cylindrical Adjustable Inertance Tube for Pulse Tube Cryocoolers," *Cryocoolers 18*, S. Miller, ed., ICC Press (2014), pp. 237-243.

4. Zhou, W.J., Pfotenhauer, J.M., Nellis, G.F., and Liu, S.Y., "Performance of an Adjustable, Threaded Inertance Tube," *Advances in Cryogenic Engineering*, vol. 61, IOP (2015).

5. Luo, E., R. Radebaugh, and M. Lewis, "Inertance Tube Models and Their Experimental Verification," *AIP Conference Proceedings*, vol. 710 (1) (2004), pp. 1485-1492.

6. Schunk, L.O., G.F. Nellis, and J.M. Pfotenhauer, Experimental Investigation and Modeling of Inertance Tubes. *Journal of Fluids Engineering*, vol. 127(5) (2005), pp. 1029-1037.

7. Swift, Greg, "Thermoacoustics: A Unifying Perspective for Some Engines and Refrigerators," *The Journal of the Acoustical Society of America,* vol. 113 (5) (May 2003), pp. 2379-2381.

Experimental Investigation of Piston-Free Expander Accompanied by Cold Linear Compressor

J. Park, S. Jeong

Cryogenic Engineering Laboratory, Mechanical Engineering Dept.,
School of Mechanical and Aerospace Engineering,
Korea Advanced Institute of Science and Technology, Daejeon, Korea

ABSTRACT

This research focuses on the study of a Stirling-type pulse tube refrigerator (PTR) designed to operate with a cold compressor, where 'cold' means much colder than ambient room temperature including cryogenic operating temperatures. Thus, the PV power is directly generated at a cryogenic temperature and transmitted to the cold expansion volume without undergoing a separate precooling process. Implementing a cold reservoir as a heat rejecter and regulating the entire operating temperature range of the PTR under sub-ambient temperature will enable the PTR to operate efficiently in cold environments like space.

As a proof of concept, this experimental investigation was conducted with a cold linear compressor operating at 80 K; whereby a no-load temperature of 18.7 K was achieved. The PTR including the cold linear compressor is expected to be further optimized at the system level for operation between 80 K and 20 K at the resonant state. This paper describes both preliminary test results and further discussions for an 'optimized' PTR system.

WHY DO WE FOCUS ON COLD COMPRESSION?

The main reason for cold compression is to improve compression efficiency and to get a 'lower temperature' at the cold expansion space. In order to get a 'lower temperature', many researchers have worked towards cascade pulse tube refrigerators (PTRs) introducing a precooling stage [1-3] or multistage cooling [4, 5]. Let us imagine that the pressure-volume work (PV work) is generated at a 'cold (sub-ambient temperature)' state. As the thermodynamic state of the working gas becomes close to its saturation curve, the specific volume decreases and incompressibility increases. The work consumption rate is represented by the following equation [6]:

$$\dot{W}_{PdV} = \int P dv \tag{1}$$

The reduced specific volume of the working gas enables the work consumption in pressurizing it to be decreased. The generated PV work is then transmitted to a 'colder' state with a precooling process. The work reduction due to the compression temperature ratio to the temperature where the working gas expands, is much smaller relative to that of a typical ambient-temperature Stirling-type PTR. This fact implies that the available mechanical work (PV work) transmitted to the expansion volume

Cryocoolers 19, edited by S.D. Miller and R.G. Ross, Jr.

is effectively convertible to increased cooling. Therefore, the cold-Stirling-type PTR can readily reach a lower target temperature with larger refrigeration potential than a typical ambient-temperature single-stage Stirling-type PTR. For a cold reservoir one can utilize the technically-mature liquefaction cycles widely-available in the gas liquefaction industry, i.e. liquefied natural gas (LNG), liquid nitrogen (LN_2) and liquid oxygen (LOX), etc. This concept can even be extended to liquid hydrogen (LH_2) and liquid helium (LHe) temperatures.

In order to implement a cold-Stirling-type PTR (or other regenerative-type refrigerator) for the large-scale gas industry, a fundamental investigation of the important characteristics must be carried out first [4, 7-9]. Finally, in the process of converting electrical to mechanical compression of the working gas, losses due to Joule heating are generated. When the linear compressor operates at a cryogenic temperature, it is possible to utilize high temperature superconductors (HTS) instead of typical metallic conductor coils for the linear motor. A cold compressor implemented with a HTS linear motor can intuitively eliminate the ohmic loss so that a higher motor efficiency can be expected. Nevertheless, since the heat generated due to the alternating current (AC) loss at cryogenic temperature generally creates significant burden, it has to be minimized for efficient motor operation. Several advantages of the 'cold compressor with piston-free cold expander' include:

(i) High density of the working gas; reducing compression work to the working gas
(ii) Cascade-like refrigeration system; increasing thermal efficiency
(iii) Superconducting enabling environment; increasing in electrical efficiency

In this paper, in order to demonstrate the concept of 'cold compression', a new cryogenically compatible linear compressor was developed. The cold linear compressor stably produces PV work around LN_2 temperature, and the efficiency of the electric input to the mechanical work conversion is estimated to exceed approximately 80% over a broad frequency range. As an expansion device, our pulsating tube was mated with a multilayered regenerator to widen the temperature span between the compression space where the PV work is generated and the cold-end where the working gas expands. At an operating frequency of 50 Hz, a lowest no-load temperature of 18.7 K was recorded. This paper also describes further investigations in two areas: first, to improve the resonance matching between the mechanical compressor and thermal efficiency needs of the cold end; this included a dynamic analysis of the whole PTR system with the commercial software (SAGE). The second investigation involved optimization of the multilayered regenerator with a sophisticated regenerator model (REGEN).

COLD LINEAR COMPRESSOR

As noted in Fig. 1, a 'single acting' and 'moving magnet' type linear compressor was developed for cryogenic temperature usage. The piston and the cylinder body were made of the same material (aluminum) and their surfaces were coated with Teflon. The piston has a 'gas bearing' to mitigate mechanical friction between the piston and the cylinder wall. We also installed a multi-flexure piston support (ASTM304) to maintain a neutral position of the piston. Radially magnetized permanent magnet segments were bonded with Stycast 2850FT epoxy on a stainless-steel guide that was mechanically connected with the piston. Inner and outer iron yokes were laminated for minimizing eddy current generation, and the inter gap between them was precisely adjusted to avoid conflict with the permanent magnet unit at cryogenic temperatures. We monitored the mean temperature in the compression space by installing a thermometer (DT-670, Lakeshore). In order to measure the effective mechanical power generated by the linear compressor, a dynamic pressure transducer (601A, Kistler) and an accelerometer (8730AE500M8, Kistler) were also mounted in the compression space and on the piston, respectively.

As noted in Fig. 1, all the compressor components including the electric coil were sealed in a hermetic stainless steel housing that was cooled by being submerged in liquid nitrogen (LN_2).

CRYOSTAT

The regenerator shell is 22.3 mm in diameter, 0.3 mm in thickness and 40 mm high. We selected a combination of stainless steel mesh screens (#400) and antimonial lead spheres ($Pb_{96}Sb_4$, 75 μm)

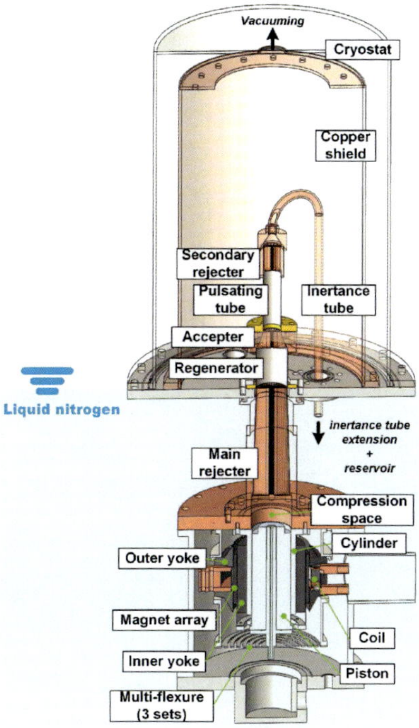

Figure 1. Rendering of the developed Stirling-type PTR system with its cold linear compressor.

after considering the thermal penetration depth of the working gas and of the regenerative material itself at the target operating temperatures. The lead spheres, with a filled length of 10 mm, were positioned nearest to the cold-end. Strictly speaking, the portion of each regenerative material in this first regenerator was a preliminary estimate, but had not yet been optimized. To provide for operational diagnostics, we attached E-type thermocouples on the outer surface of the regenerator shell (their opposite junctions were submerged in LN_2 for improved measurement accuracy). The thermocouples provide temperatures along the regenerator's axial direction that are useful to determine the effective thermal conduction through the matrix.

The pulsating tube canister (the pulse tube) is 12.7 mm in diameter, 0.3 mm in thickness, and 75 mm long. At the cold side of the pulsating tube and the regenerator, we installed an oxygen-free high conductivity (OFHC) copper cold-end heat exchanger (Accepter). Due to the difference in the tube diameters of the pulsating tube and the regenerator, it provided a smooth connection between them. To provide heat exchange surfaces, we carved several channels using electric discharge machining (EDM) within the copper. Other heat exchangers (main and secondary rejecters) also adopted this 'slit-type heat exchanger.' This 'slit-type heat exchanger' has several characteristics; it can have axisymmetric geometry with respect to the flowing direction, smooth channel connection, and no thermal contact resistance within the heat exchanger body. Nevertheless, it can create a flow maldistribution problem, so we installed flow straighteners at three locations of the PTR (at both ends of the pulsating tube and at the phase controller entrance) [10]. These consist of few layers of copper mesh screens (#200, totally 3 mm in thickness). The heat exchangers were designed and fabricated not to exceed a 1 K temperature difference between the working gas and the heat exchanger body at the corresponding target heat lift (100 W for the main heatsink rejector and 10 W for both the secondary heatsink rejector and the coldend heat exchanger) [11]. On the outer surfaces of both the coldend heat exchanger and the secondary heatsink, resistance thermometers (Cernox RTD, Lakeshore) were placed. An inertance tube (3.6 mm in inner diameter and 1.5 m in length) and a reservoir (volume of 500 cc) were connected in series to the secondary heatsink. A dynamic pressure trans-

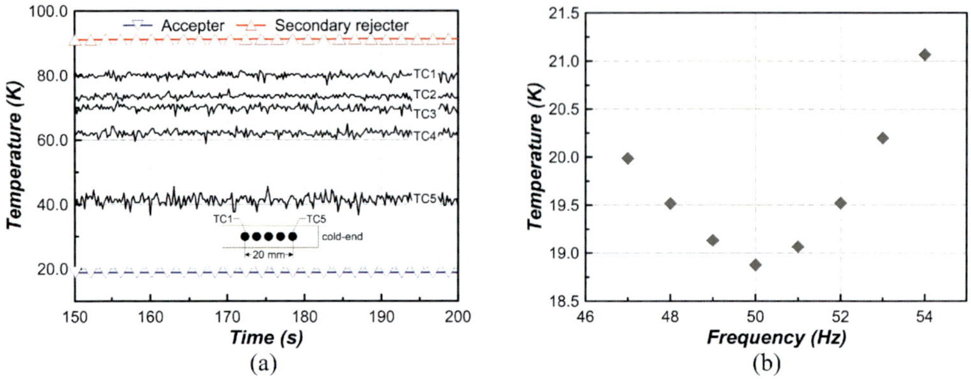

Figure 2. (a) Steady-state temperature distribution at a frequency of 50 Hz (b) frequency dependency of the coldend temperature (without thermal load).

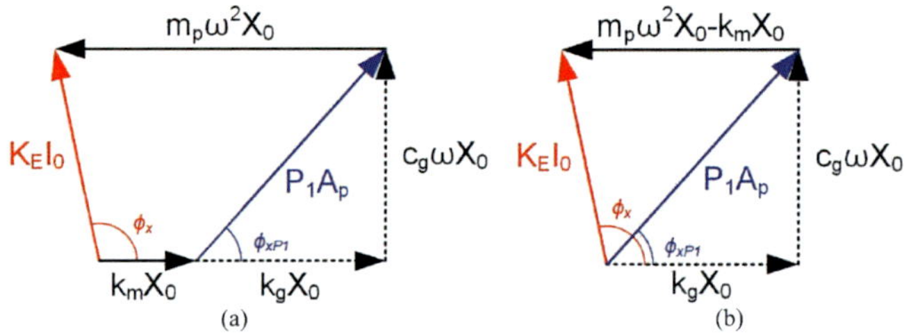

Figure 3. (a) Force phasor diagram of the linear compressor and (b) the modified one.

ducer (601A, Kistler) was also mounted to observe the pressure at the location where the working gas expands. A copper shield that was conductively cooled by LN_2 blocks thermal radiation to the cold parts. The secondary heatsink, moreover, was thermally connected to the copper shield by means of copper braid wires. All the cold parts are thermally insulated in vacuum with a copper radiation shield cooled by LN_2. The WHX, furthermore, is thermally anchored on the surface of the copper radiation shield by means of copper braid wires. We implemented a dynamic absorber to cancel the mechanical vibration that the single-acting piston generates. This consists of a flexure and a balancing mass that were manually tuned for the operating frequency.

EXPERIMENTAL RESULTS AND DISCUSSIONS

Figure 2(a) shows the temperature distribution at steady-state with the operating frequency of 50 Hz. In this case, a no-load temperature of 18.7 K was achieved, and it took 10 minutes to cool down from 80 K. Figure 2(b) shows the no-load temperature variation with the operating frequency.

In Figure 3(a), we introduce a force phasor diagram to determine the phase angle between the pressure and the piston displacement. Since the inertia and the spring force terms are in phase with each other, the phasor diagram can be simplified as depicted in Fig. 3(b). As we know the magnitudes of the physical variables as tabulated in Table 1, we can find the phase angle of the pressure relative to that of the piston displacement by applying the law of cosines. Consequently, we can evaluate the PV work in the compression space by using the following equation:

$$\dot{W}_{PdV1} = f\int P_1 dV = \pi f\left|P_1\right|\left|X_0\right|A_p\sin\left(\phi_{xP1}\right) \qquad (2)$$

where, f is the operating frequency (Hz), P_1 is the dynamic pressure magnitude at the compression space (Pa), X_0 is the piston displacement (from the neutral position to its maximum peak, m), A_p is the cross-sectional area of the piston (m^2) and ϕ_{xP1} is the phase angle of the pressure relative to that of the piston displacement (°). Figure 4(a) shows the relative phase angle and the resultant PV power as estimated through the phasor analysis. Finally, we can claim that the cold linear compressor

Table 1. The measured physical variables and the pressure phase angle relative to that of the piston displacement and the corresponding PV work at the compression space

f (Hz)	X_0 (mm)	P_1 (kPa)	Analytically predicted from measured data with force phasor diagram	
			ϕ_{xP1} (°)	W_{PdV1} (W)
47	4.3	210.8	37.3	46.0
48	4.5	217.4	36.3	50.3
49	4.9	229.0	35.1	57.8
50	5.3	242.5	32.4	61.3
51	5.6	249.1	31.1	65.8
52	5.8	253.7	29.5	68.3
53	6.0	254.5	28.1	68.6
54	6.1	255.5	26.8	68.2

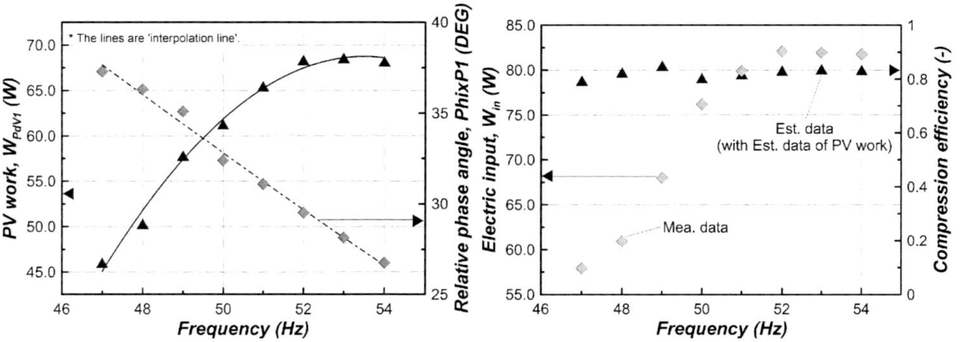

(a) (b)

Figure 4. (a) The expected pressure phase angle relative to that of the piston displacement and the resultant PV work at the compression space and (b) the corresponding compression efficiency to the electric input.

stably generates PV power with the compression efficiency of over 80%, as represented in Fig. 4(b). The compression efficiency can be defined as the equation:

$$\eta_{comp.} = \frac{\dot{W}_{PdV1}}{\dot{W}_{in}} \tag{3}$$

$$\dot{W}_{in} = f\int (V \cdot I)dt = \frac{1}{2}V_0 I_0 \cos\phi_v \tag{4}$$

where, W_{in} is the electric input power (W) defined by Equation (4), V_0 is the peak voltage across the electric coil (V), I_0 is the peak current through the electric coil (A) and ϕ_v is the phase angle of the voltage relative to that of the current flow (°).

FURTHER IMPROVEMENT IN TWO WAYS

As discussed in the last section, the lowest temperature at the coldend was recorded to be 18.7 K at an operating frequency of 50 Hz. Can we say that the 'above results' are the best? The answer is definitely 'no'. Therefore, we have tried to improve the performance of the PTR in two ways as follows:

(i) Resonance matching between the thermal and mechanical systems
(ii) Regenerator optimization

For the first step, we need to analyze the physical variables (piston displacement and dynamic pressure amplitude) and whether the PTR properly operates near its resonance or not. In order to simulate the physical variables, a commercial software (SAGE) was used, as it considers the thermo-hydraulic governing equations for the working gas as well as the thermal properties of the solid

Table 2. Geometrical dimensions and operating parameters of the PTR system.

		Sim. data (i) & Exp. data	Sim. data (ii)
Linear compressor	Piston mass, m_p	235 g	260 g
	Mechanical stiffness, k_m	2,500 N m^{-1}	0 N m^{-1}
	Motor constant, K_E	4.8 N A^{-1}	
	Electric resistance, R_c	0.05 Ω	
	Electric inductance, L_c	1.9 mH	
Pulse tube components	Compression space	16,430 mm^3	
	Regenerator	22.2 mm (O.D.), 0.3 mm (T), 40 mm (H) *30 mm-portion; #400 stainless steel mesh (porosity: 0.65)* *10 mm-portion; 75 μm-diameter lead sphere (porosity: 0.35)*	
	Pulse tube	12.7 mm (O.D.), 0.3 mm (T), 75 mm (H)	
	Heat exchangers	Main rejecter: 9,600 mm^3; Secondary rejecter: 1,570 mm^3; Accepter: 1,310 mm^3	
	Phase controller	Inertance tube: 3.6 mm (I.D), 1.6 m (L); Reservoir: 500,000 mm^3	
Operating parameters	Operating temperature	Main rejecter: 80 K; Secondary rejecter: 90 K; Accepter: 20 K	
	Charging pressure	2 MPa	
	Input current	9.7 $A_{rms.}$	

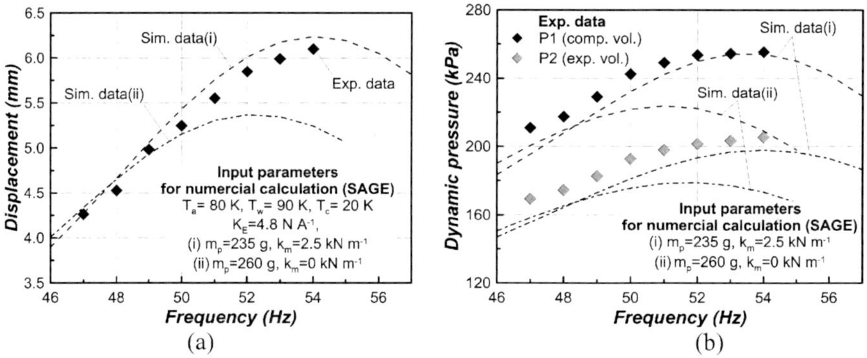

Figure 5. Comparison of (a) the piston displacement and (b) the dynamic pressure amplitude between the present results and the expected values with the linear compressor modifications

regenerator elements. The Sage model also includes the equations of motion (EOM) that describe the dynamic characteristics of the cold compressor's piston. As noted in the previous section, it seems that the frequency of 50 Hz (thermally best one) is not well matched to the compressor resonance when it is operated at cryogenic temperatures (its resonance is much higher than 50 Hz). At this point, we decided to reduce the natural frequency of the cold linear compressor by increasing the mass of the moving piston. Furthermore, it was determined that eliminating a flexure was advantageous for improving the compressor's mechanical reliability at cryogenic temperatures.

Table 2 shows the geometrical dimensions of the PTR system and the operating parameters. The values in the column 'Sim. data (i) & Exp. data' are designated for the apparatus currently constructed and tested in this research paper. The values for the further analysis with the modified compressor are listed in the last column 'Sim. data (ii).' Figure 5 shows a comparison of the piston displacement and the dynamic pressure amplitudes in the PTR system between the present results (both experimental and numerical) and the expected values with the proposed compressor modifications. The results obviously signify that the peak point is shifted from the 'high frequency region' to near 50 Hz.

Next, various methods can be pursued to characterize the regenerator thermo-hydraulic performance with its geometry and operating conditions. In this analysis, REGEN was adopted to analyze the thermal performance of the regenerator at the coldend with respect to the regenerative material

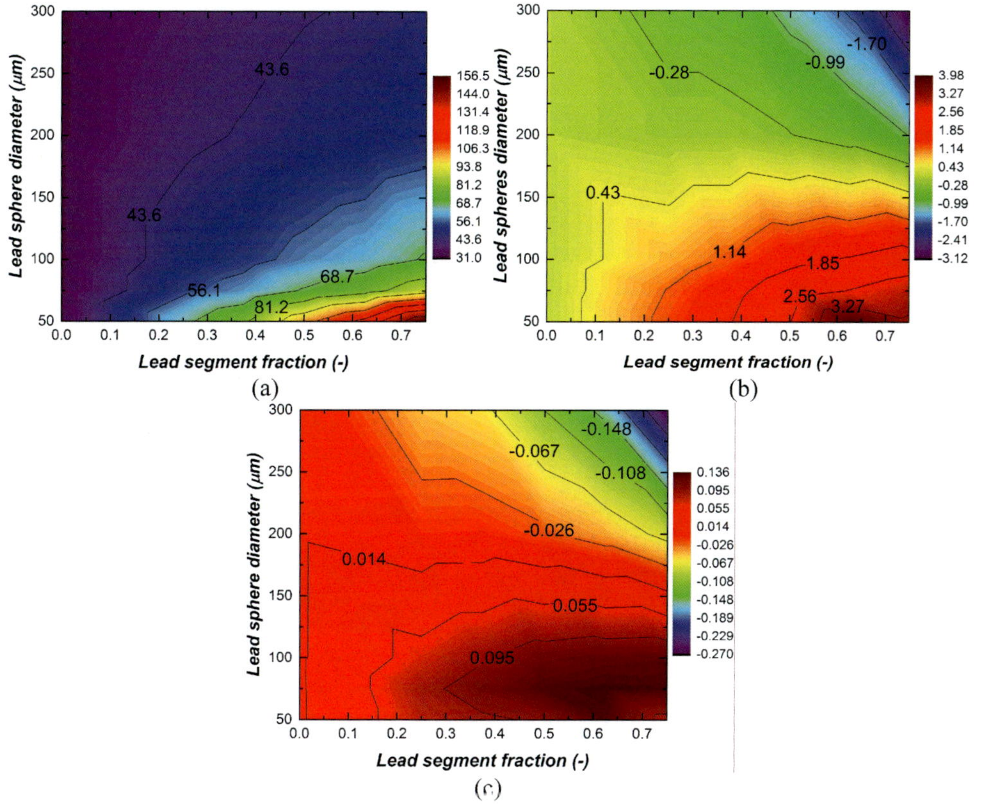

Figure 6. Performance maps of the multilayer regenerator with the fixed length of the regenerator (40 mm), (a) the required PV work at the warm-end of the regenerator, (b) the cooling power and (c) the COP of the PTR system

packing type. As mentioned in the previous section, the regenerator consists of two regenerative materials (stainless steel mesh screens and tiny lead spheres). Thus, the packing fraction and the particle size of each material should be optimized; this is especially true for the lead spheres. Also, the total length of the regenerator should be optimized. The optimization procedure was as follows:

(i) Determine the particle size of the lead spheres with the fixed regenerator length

(ii) Determine the total length of the regenerator with the selected size of the lead spheres

Since, as shown in the temperature distribution along the regenerator, Fig. 2(b), the thermal duty on the lead segment is exceedingly high relative to that of the total regenerator. Thus, the total amount of lead spheres should be increased beyond the current status. Table 3 lists the fixed and the controlling parameters for the optimization process. The fixed parameters, especially the mass flow rate, the pressure ratio, and the phase angle between them, are based on the computation results with the compressor modifications 'Sim. data (ii).' The operating frequency, therefore, is set to be 50 Hz. Figure 6 shows maps with respect to the particle size of the lead spheres and their portion (length fraction to the length of the regenerator of 40 mm). Figure 6(a) shows how much PV work is required at the warm-end of the regenerator when it operates with the input parameters tabulated in Table 3 and the specified geometric configurations. Figure 6(b) also presents the cooling power at the cold end of the regenerator. The cooling power includes auxiliary losses ($Q_{aux.\ losses}$) that the PTR system generates, i.e. shuttle heat transfer in the pulsating tube and conductive losses. Since we assume that the operating conditions are identical for all the calculations, the aforementioned losses can be roughly calculated [12]. The shuttle and the conduction thermal losses are estimated to be 0.96 W and 0.23 W, respectively. The COP is defined

Table 3. Operational parameters of regenerator (input parameters of REGEN).

	Warm-end temperature, T_a (= temp. of the main rejecter)	80 K
	Cold-end temperature, T_c (= temp. of the accepter)	20 K
	Operating frequency, f	50 Hz
Fixed parameter	Mean pressure, P_m	2 MPa
	Pressure ratio at the cold-end, $P_r=(P_m+P_2)/(P_m-P_2)$	1.2
	Inner diameter of regenerator, D_r	21.6 mm
	Mass flow rate at the cold-end, m_c	6 g s^{-1}
	Mass phase angle relative to that of the pressure at the cold end, □$_{mP2}$	-15°
	Particle size of the lead spheres	50~300 μm
Controlling parameter	Length of the regenerator	40~70 mm
	Length fraction of the lead matrix to the total regenerator length	0~0.75

as the ratio of the effective cooling power (Q_c-$Q_{aux.\ losses}$) divided by the PV work at the warm-end of the regenerator and multiplied by the temperature ratio between them (T_r=T_a/T_c), as follows:.

$$COP = \frac{\dot{Q}_c - \dot{Q}_{aux.losses}}{\dot{W}_{PdV1}} T_r \tag{5}$$

From the map in Figure 6(c), which represents the coefficient of performance (COP), it appears that the 75 μm-diameter lead spheres tested in this paper was a good choice. As mentioned above, nevertheless, the map indicates the maximum COP when the length fraction of the lead matrix becomes approximately 60% (24 mm in equivalent length of the lead segment) not at 25% for the case of 40 mm-regenerator length. Although we did not measure the cooling power experimentally, the computation results predict that the current PTR system with some compressor modifications can achieve a cooling power level of 1.1 W with a COP of 0.087. Next, we have tried to optimize the length of the regenerator with the selected size of the lead sphere of 75 μm, as shown in Fig. 7.

Figure 7(c) clearly indicates the optimized regenerator dimension of 65 mm in length with the length fraction of the lead matrix of 40% (26 mm in equivalent length of the lead segment). The computation results also confirm that the equivalent length of the lead segment remains almost constant even if the total length of the regenerator is changed. As shown in Fig. 7(b), modifications, including further optimization on the regenerator length, can greatly enhance the cooling performance, i.e. a cooling power of 4.05 W with the COP of 0.17. From Fig. 7(a), we can also expect the input power of 119.5 W with the compression efficiency of 80%.

CONCLUSIONS

The development and test of a Stirling-type PTR system with a cold linear compressor has been discussed in this paper. We have tried to analyze the performance of the preliminary cold PTR system, and consequently an 'optimized system' has been suggested for maximizing its potential from the tested PTR system. The operational issues and main features regarding the Stirling-type PTR system are listed as follows.

(i) The cold linear compressor stably generates the mechanical work (PV work) at liquid nitrogen (LN_2) temperature.

(ii) The PV work at the compression space is indirectly deduced by a simple force phasor analysis. The resultant electrical to mechanical work conversion efficiency is estimated to exceed 80% for the entire frequency range.

(iii) The lowest temperature at the cold-end of the PTR system is recorded to be 18.7 K.

(iv) Through the dynamic analysis on the whole PTR system, the results obviously show that the peaks of the physical variables are shifted from the 'high frequency region' to near the target operating frequency of 50 Hz.

(v) For the lead matrix, the particle size of 75 μm-diameter is proper but its total amount should be increased to achieve the full potential of the PTR system.

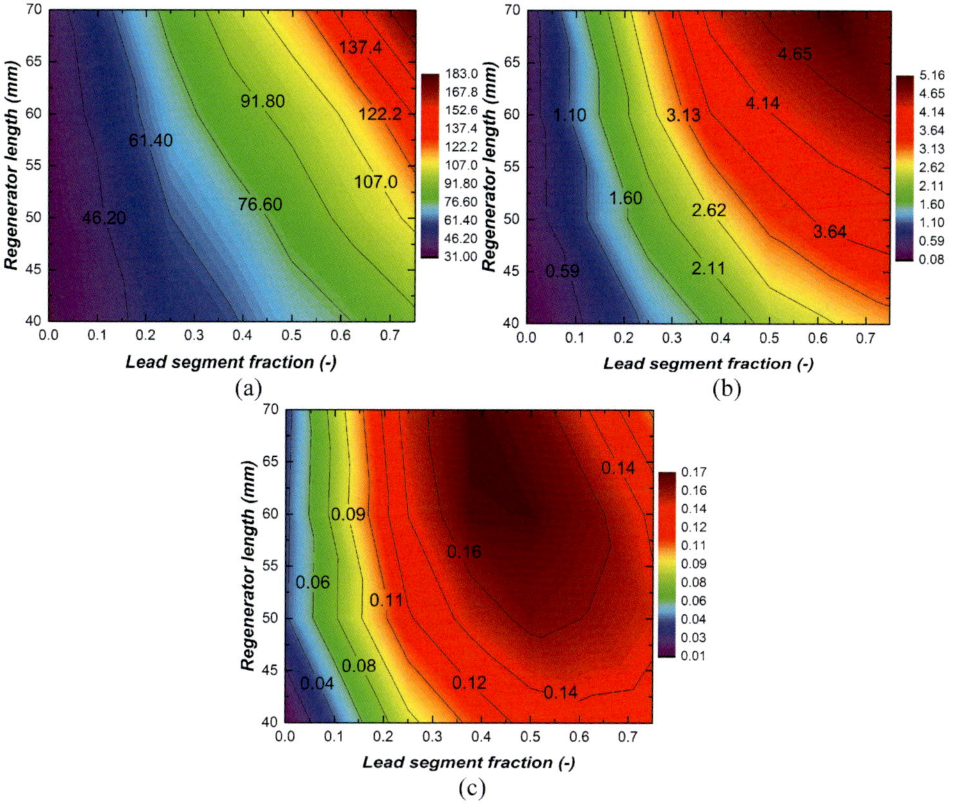

Figure 7. Regenerator's total length dependency of the thermal performance, (a) the required PV work at the warm-end of the regenerator, (b) the cooling power and (c) the COP of the PTR system.

With the modifications for both the cold linear compressor and the regenerator, an enhanced cooling performance, e.g.. cooling power of 4.1 W with the COP of 0.17, is expected. The overall experimental and analytical results signify that this 'piston-free colder expander' is viable and useful in many potential applications such as space and gas liquefaction industries.

ACKNOWLEDGMENT

This research was supported by the Converging Research Center Program through the Ministry of Science, ICT and Future Planning, Korea (2014M3C1A8048836). We also wish to thank National Institute of Standards and Technology (NIST) Cryogenics Technologies Group for providing REGEN3.3 program.

REFERENCES

1. J. M. Poncet, I. Charles, A. Gauthier, and T. Trollier, "Low Temperature High Frequency Pulse Tube Cooler Using Precooling," *Cryocoolers 13*, R. Ross, Jr., Ed., Springer US (2005), pp. 149-155.

2. G. Thummes, S. Bender, and C. Heiden, "Approaching the ^{4}He lambda line with a liquid nitrogen precooled two-stage pulse tube refrigerator," *Cryogenics,* vol. 36 (1996), pp. 709-711.

3. Z. Li, Y. Jiang, Z. Gan, L. Qiu, and J. Chen, "Performance of a precooled 4 K Stirling type high frequency pulse tube cryocooler with Gd_2O_2S," *Journal of Zhejiang University SCIENCE A,* vol. 15 (2014), pp. 508-516.

4. M. Dietrich and G. Thummes, "Two-stage high frequency pulse tube cooler for refrigeration at 25 K," *Cryogenics,* vol. 50 (2010), pp. 281-286.

5. Z. H. Gan, B. Y. Fan, Y. Z. Wu, L. M. Qiu, X. J. Zhang, and G. B. Chen, "A two-stage Stirling-type pulse tube cryocooler with a cold inertance tube," *Cryogenics,* vol. 50 (2010), pp. 426-431.

6. B. M. A. Cengel Y. A., *Thermodynamics*, 6 ed.: McGraw Hill, 2004.

7. E. Ercolani, J. M. Poncet, I. Charles, L. Duband, J. Tanchon, T. Trollier*, et al.*, "Design and prototyping of a large capacity high frequency pulse tube," *Cryogenics,* vol. 48 (2008), pp. 439-447.

8. M. Dietrich, L. W. Yang, and G. Thummes, "High-power Stirling-type pulse tube cryocooler: Observation and reduction of regenerator temperature-inhomogeneities," *Cryogenics,* vol. 47 (2007), pp. 306-314.

9. D. M. Sun, M. Dietrich, and G. Thummes, "High-power Stirling-type pulse tube cooler working below 30 K," *Cryogenics,* vol. 49 (2009), pp. 457-462.

10. T. Ki and S. Jeong, "Stirling-type pulse tube refrigerator with slit-type heat exchangers for HTS superconducting motor," *Cryogenics,* vol. 51 (2011), pp. 341-346.

11. T. Ki and S. Jeong, "Optimal design of the pulse tube refrigerator with slit-type heat exchangers," *Cryogenics,* vol. 50 (2010), pp. 608-614.

12. J. Jung and S. Jeong, "Surface Heat Pumping Loss in a Pulse Tube Refrigerator," *Cryocoolers 12,* R.G. Ross, Jr. Ed., Springer US (2003), pp. 371-378.

Cascading Three Pulse Tube Coolers with Work Recovery

L. Y. Wang, Z. H. Gan, Q. Y. Zhao, Z. Y. Jin, Y.R. Song

Institute of Refrigeration and Cryogenics,
Zhejiang University,
Hangzhou 310027, P.R.China

Key Laboratory of Refrigeration and Cryogenic Tech. of Zhejiang Province,
Hangzhou 310027, P.R.China

ABSTRACT

A cascade pulse tube cooler (PTC) consists of sub-PTCs that are connected by transmission tubes in between; it can recover the PV work at the warm end of the pulse tube to drive the subsequent stages. The more cascading stages it has, the closer the efficiency will approach Carnot efficiency. In this paper, a cascade PTC consisting of three sub-PTCs is presented, and experimental results show that while working at 233 K, the cooling powers at the three stages are 164.9 W, 70.7 W and 18.0 W, respectively. This results in a total cooling power of 253.6 W, 39.9% over a PTC without cascading.

INTRODUCTION

A pulse tube cryocooler (PTC) cannot work with Carnot efficiency due basically to the expansion work that has to be dissipated thermally at the warm end of the pulse tube. This dissipation is especially prominent at high cooling capacity or at high temperatures, as it reduces the COP and limits the application of PTCs above 120 K. This intrinsic limitation limits its capability of reaching Carnot efficiency even in the ideal case. That is, the ideal efficiency of the PTC is only T_c/T_h, which is lower than the Carnot efficiency $T_c/(T_h-T_c)$.[1] Therefore, how to recover this amount of dissipated work becomes a critical issue in a highly efficient PTC.

Previous studies have been carried out to recover this amount of PV work: one way is by introducing moving parts,[2-5] and another is by conducting loop configurations.[6-9] Also, G. Swift proposed a quarter-wavelength pulse tube cooler in which a second pulse tube cooler was added after the quarter-wavelength pulse tube to recover the expansion work, this idea provides a new way to recover the PV work without introducing either moving parts or streaming.[10,11]

We started our program studying cascade PTC with work recovery in 2012. In 2013, our single stage PTC reached 181.3 W cooling power at 233 K.[12] In 2014, our two-stage cascade PTC obtained 241.6 W cooling power (175.0 W from the first stage and 66.6 W from the second stage).[13,14]

In the work reported here, we build a three-stage cascade PTC.

Cryocoolers 19, edited by S.D. Miller and R.G. Ross, Jr.

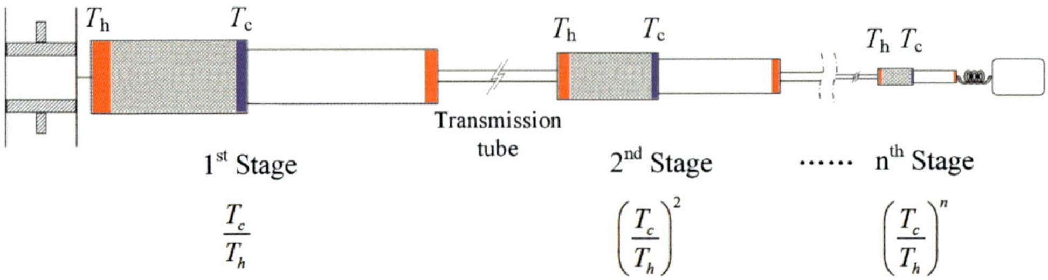

Figure 1. Schematic diagram of a multi-stage cascade PTC

THEORETICAL ANALYSIS

Figure 1 shows a schematic diagram of a multi-stage cascade PTC. It consists of n-stages of sub-PTCs with transmission tubes in between (here n is a positive integer number not less than 2). Each sub-PTC works at the same hot end and cold end temperatures T_h and T_c, respectively. The linear compressor provides the original driving power, the acoustic power at the outlet of the former stage is transferred to the latter one by the transmission tube. Assuming that there is no loss in each stage of sub-PTC, then the cooling efficiency of each stage is T_c/T_h. We can obtain the total cooling efficiency of the whole system:[13,14]

$$COP_n = \frac{Q_{c1}+Q_{c2}+\cdots+Q_{cn}}{E_1} = \frac{T_c}{T_h} + \left(\frac{T_c}{T_h}\right)^2 + \left(\frac{T_c}{T_h}\right)^3 + \cdots + \left(\frac{T_c}{T_h}\right)^n = \frac{T_c}{T_h - T_c}\left[1-\left(\frac{T_c}{T_h}\right)^n\right] \tag{1}$$

when n goes to infinity, the COP will become:

$$COP_{n\to\infty} = \frac{T_c}{T_h - T_c} \tag{2}$$

That is, the efficiency of an infinite stage cascade PTC will be the Carnot efficiency. This theoretical analysis with ideal hypothesis is not only beautiful but also meaningful, because there may be many different ways similar to this to get to Carnot efficiency for a refrigerator. Equation (2) in this paper shows us one of such refrigeration methods which approaches Carnot efficiency step by step, and it is theoretically realizable learning that each term in Equation (2) corresponds to each stage of sub-PTC in Fig. 1.

A three stage cascade PTC was designed based on REGEN[15] and Sage[16]. In this work, we used an existing linear compressor of model CFIC 2s132 (the rated power is 500 W at 60 Hz, 2.5 MPa charging pressure) to obtain a relatively high cooling power, and a cooling temperature of 233 K (-40°C) was chosen for the first step.

The key point in designing a cascade PTC is to ensure good phase relations inside each stage. Figure 2 shows the phasor diagram of the entire cascade cooler, in which three vectors labeled p_1, p_2 and p_3 represent the pressure waves in the 1st stage, 2nd stage, and 3rd stage, respectively. The vectors with open-headed arrows represent the mass flow in the 1st stage, the vectors with blunt-headed arrows represent that in the 2nd stage, while the vectors with sharp pointed arrows represent that in the 3rd stage. It is shown that the reversal of phase relation between pressure wave and mass flow by the transmission tubes plays an important role, so that appropriate phase relations inside all three stages are satisfied at the same time. That is, at the warm end of the regenerator, the mass flow leads the pressure wave; while at the cold end of the regenerator, the pressure wave leads the mass flow. The pressure wave of the 1st stage is turned by 167.6 degrees while entering into the 2nd stage; at the same time, the mass flow is turned by 112.0 degrees. Similarly, the pressure wave of the 2nd stage is turned by 162.0 degrees while entering into the 3rd stage; at the same time the mass flow is turned by 97.7 degrees.

Table 1 lists the simulation results of the cascade PTC based on a Sage model. It includes three operation modes: single-stage operation, two-stage cascade operation, and three-stage cascade

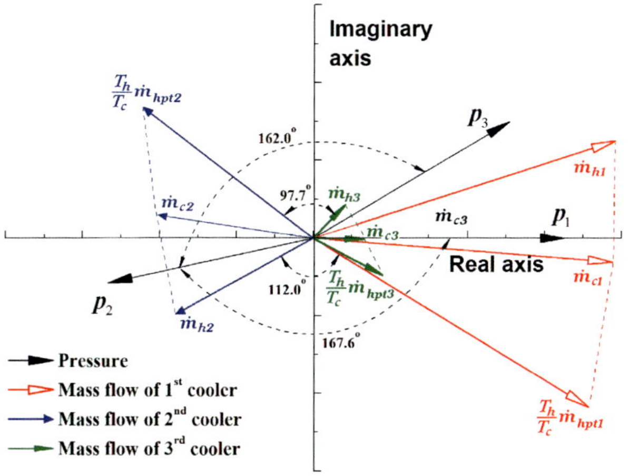

Figure 2. Phasor diagram of the three-stage cascade PTC

Table 1. Calculation results of three-stage cascade PTC based on Sage code

	Single stage[12]	Two-stage cascade[13,14]			Three-stage cascade			
		1st	2nd	In total	1st	2nd	3rd	In total
Electric power input / W	500							
Acoustic power input / W	389.9	379.1	125.7	379.1	348.9	143.3	46.0	348.9
Cooling power at 233 K / W	196.3	184.5	65.1	249.6	172.3	75.4	24.1	271.8
COP	**0.3926**	**0.4992**			**0.5436**			
Percent of Carnot / %	11.29	14.35			15.63			

operation, all with the same electric power input of 500 W. The three-stage cascade PTC is expected to achieve a total cooling power of 271.8 W at 233 K; the corresponding cooling efficiency will be improved 38.5% compared with the single stage PTC and 8.9% compared with the two-stage cascade PTC.

EXPERIMENTAL SYSTEM

To demonstrate the feasibility of the basic idea, experiments were carried out. A three-stage cascade PTC setup was designed, as shown in Figs. 3 and 4. The system consists of a linear compressor, a 1st stage PTC, a transmission tube I, a 2nd stage PTC, a transmission tube II, a 3rd stage PTC. Each PTC consists of an aftercooler, a regenerator, a cold heat exchanger, a pulse tube, and a secondary hot heat exchanger. The 1st stage PTC and the 2nd stage PTC are connected by a 7 m long transmission tube with inner diameter of 14.2 mm; while the 2nd stage PTC and the 3rd stage PTC are connected by a 6.3 m long transmission tube with inner diameter of 10.0 mm.

The CFIC 2s132 model linear compressor is driven by an AC power supply to regulate the frequency and power output. The four hot end heat exchangers located at the warm ends of the regenerators and pulse tubes are cooled by water at room temperature. Non-vacuum expanded pearlite is used for thermal insulation.

Table 2 lists the main parameters of the cascade PTC. The fill matrix in the three regenerators is a 200 mesh stainless steel with a porosity of 0.6704. The measurement system includes measurements of temperature, pressure and cooling power. Two rhodium-iron resistance thermometers are mounted on the cold end of the 1st stage PTC, two platinum resistance thermometers are mounted on the cold end of the 2nd stage PTC, while three rhodium-iron resistance thermometers are mounted on the cold end of the 3rd stage PTC, all calibrated with accuracy of ±0.1 K. Four pressure sensors are employed to measure the pressures at the compressor back space, the inlet of the 1st stage, the inlet of the 2nd stage, and the inlet of the 3rd stage, as shown as P_1, P_2, P_3 and P_4 in Fig. 3. For P_1, P_3 and P_4, KISTLER 211B3 piezoelectric pressure transducers are used to measure the amplitude and phase of the oscillating pressure; for P2, an Entran EPX piezoresistive pressure transducer is used to

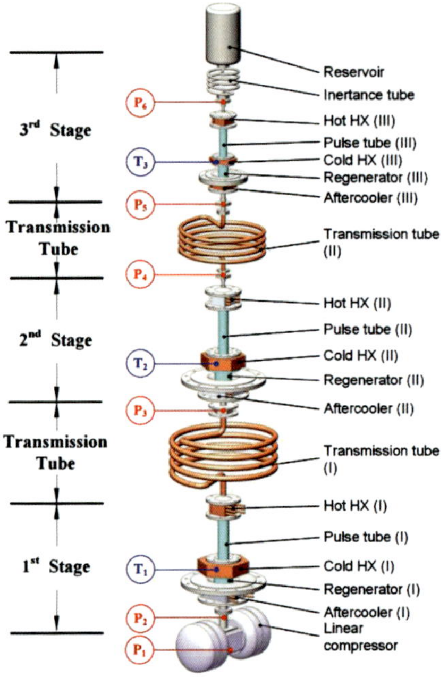

Figure 3. Schematic drawing of three-stage cascade PTC

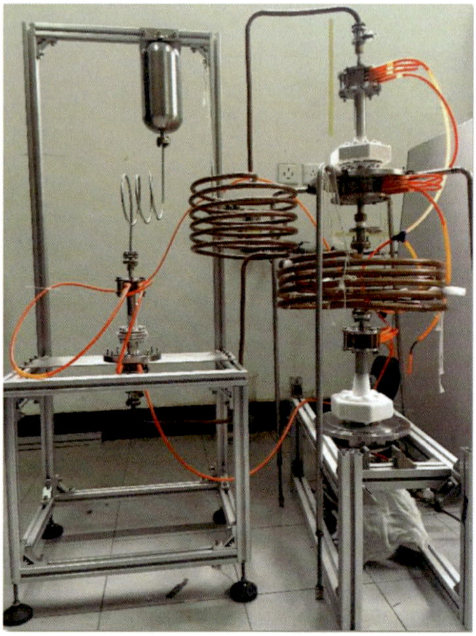

Figure 4. Experimental setup of three-stage cascade PTC

Table 2. Main parameters of the three-stage cascade PTC

	Part Name	Dimension
1st Stage	Regenerator	53.7 mm i.d., 37.7 mm long
	Pulse tube	30.5 mm i.d., 134.3 mm long
	Inertance tube (in single-stage operation)	8 mm i.d., 2.67 m long
	Reservoir (in single-stage operation)	450 cm^3
Transmission tube I		14.2 mm i.d., 7 m long
2nd Stage	Regenerator	47.6 mm i.d., 48 mm long
	Pulse tube	27 mm i.d., 150 mm long
	Inertance tube (in two-stage cascade operation)	6 mm i.d., 1.4 m long
	Reservoir (in two-stage cascade operation)	1000 cm^3
Transmission tube II		10.0 mm i.d., 6.3 m long
3rd Stage	Regenerator	27.0 mm i.d., 53 mm long
	Pulse tube	19.6 mm i.d., 100 mm long
	Inertance tube	4 mm i.d., 1.3 m long
	Reservoir	1000 cm^3

measure the dynamic as well as the mean pressure. Real-time data is collected using LabVIEW in a PC. To measure the cooling power, four 50 Ω resistors capable of providing 200 W heating power are installed on the cold end of the 1st stage, three 50 Ω resistors capable of providing 150 W heating power are installed on the cold end of the 2nd stage, and another three 500 Ω resistors capable of providing 29.4 W heating power are installed on the cold end of the 3rd stage.

RESULTS AND DISCUSSION

During the preliminary test, after about 7.5 hours, the temperatures of the three PTCs come to steadystate; their final temperatures were 144.8 K, 146.0 K and 188.5 K, respectively. If compared with single-stage operation (final temperature of 99.7 K) or two-stage cascade operation (120.8 K and 127.2 K, respectively), these are higher than before. This is mainly due to the sacrifice of the 1st and the 2nd cooler, e.g. their phasor diagrams are compressed to some degree. We compare the

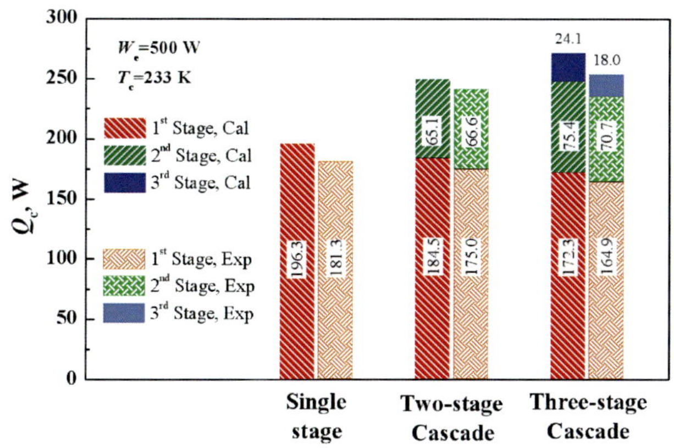

Figure 5. Comparison of cooling power

cooling capacities among three operations modes. Those are: singlestage operation, two-stage cascade operation, and three-stage cascade operation. Here the electric power input to the linear compressor is fixed at 500 W, and the cold end temperature is fixed at 233 K. As shown in Fig. 5, the single-stage PTC can supply a cooling power of 181.3 W, while for the two-stage cascade PTC, the cooling power of the 1st and 2nd stages are 175.0 W and 66.6 W, thus 241.6 W cooling capacity in total. In three-stage operation, the cooling powers of the three stages are 164.9 W, 70.7 W and 18.0 W, making a total cooling power of 253.6 W. The cooling efficiency is improved by 39.9% compared with single stage PTC and 5.0% compared with the two-stage cascade PTC. This positively demonstrates the feasibility of the cascade concept. In addition, the cooling capacity agrees quite well with the calculation results as shown in Figure 5; this verifies our design. In this work, the cooling temperature is higher than the cryogenic temperature ($\leq$120 K) reflecting the existing conditions in our lab. But for most cryogenic applications such as HTS and LNG, lower temperatures and even higher cooling capacities are required; in such cases, the cascade PTC can be more attractive.

It should be noticed that although extra cooling power can be obtained, the 1st stage PTC is always playing a dominant role. It is found that the 1st stage cooler deteriorates as the cascading stage number increases. Its no-load temperature goes up, and its cooling capacity goes down as shown from Fig. 5 — from both calculation and experiment. This means the 1st stage cooler sacrifices itself to some extent. Figure 6 shows the calculated phasor diagrams of the 1st stage PTC under three operating modes. We can see that the phasor diagram of the 1st stage is compressed more and more in both two and three-stage cascade operation. For the phase span between the mass flow at the warm end of the regenerator and the warm end of the pulse tube, this value is about 79° in single-stage operation, which is reduced to about 61° in two-stage cascade operation, and even further reduced to about 49° in three-stage cascade operation. As a result, the cooling performance of the 1st stage declines. From this point of view, the number of cascading stages maybe not the more the better in practice. It is beneficial before the loss of the former stages become even larger than the benefit of the additional cascading stages.

The pressure waves at different positions shown in Fig. 3 are measured at 233 K under cascade operation, as shown in Fig. 7. The pressure ratios in the 1st, 2nd and 3rd stages are 1.140, 1.164 and 1.157, respectively. It is shown that the phase angle between P1 and P2 is 138.0°, for the mass flow in the compression space and P1 should be 90° out of phase. This indicates 48.0° between the mass flow and the pressure wave at the entrance to the 1st stage. What is more important, it is also shown in Fig. 6 that the measured phase difference between three sub-PTCs, those are 170.8° between 1st and 2nd stage, while 165.3° between 2nd and 3rd stage, which coincide well with the calculated 167.6° and 162.0° (see Fig. 3). This verifies the phase reversion function of the transmission tubes, which is a key design in the whole system.

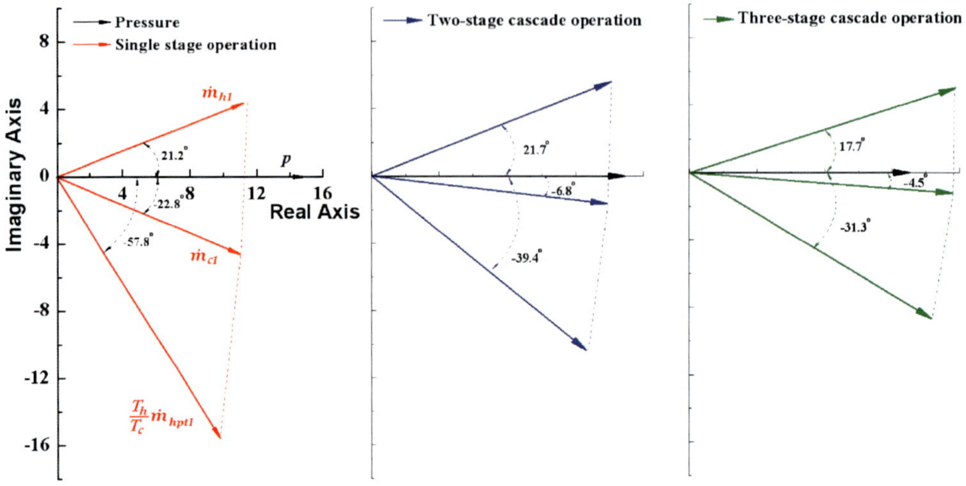

Figure 6. Calculated phasor diagrams of the 1st stage PTC under three different operating modes

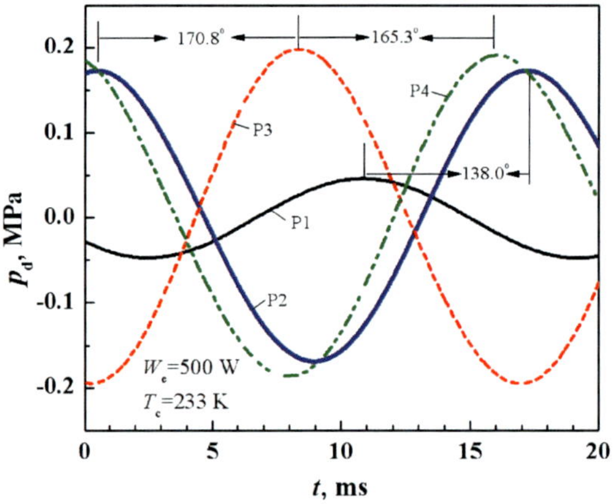

Figure 7. Measured pressure waves

CONCLUSIONS

A three-stage cascade PTC has been designed and tested. These preliminary tests show that the cooling powers of the three stages are 164.9 W, 70.7 W and 18.0 W, respectively, making a total cooling power of 253.6 W at 233K. The cooling efficiency is improved by 39.9% compared with a single stage PTC and 5.0% compared with a two-stage cascade PTC. This demonstrates the concept of cascade PTCs with work recovery, and lays a good foundation for much lower temperatures with potential use in HTS and LNG applications.

ACKNOWLEDGEMENTS

This work is financially supported by the National Natural Science Foundation of China (No. 51376157) and the Specialized Research Fund for the Doctoral Program of Higher Education of China (No. 20130101110098).

REFERENCES

1. Kittel, P., "Ideal Orifice Pulse Tube Refrigerator Performance," *Cryogenics*, vol. 32, no. 9 (1992), pp. 843-844.

2. Matsubara, Y., Miyake, A., "Alternative Methods of The Orifice Pulse Tube Refrigerator," *Cryocoolers 5, Proceedings of the International Cryocooler Conference*, Monterey, CA, August 18-19, 1988, Conference Chaired by P. Lindquist, AFWAL/FDSG, Wright-Patterson AFB, OH, pp. 127-135.

3. Ishizaki, Y., Ishizaki, E., "Experimental Performance of Modified Pulse Tube Refrigerator Below 80 K Down to 23 K," *7th International Cryocooler Conference Proceedings*, Air Force Phillips Laboratory Report PL-CP--93-1001, Kirtland Air Force Base, NM, April 1993, pp.140-146.

4. Ishizaki, Y., Ishizaki, E., "Prototype of Pulse Tube Refrigerator for Practical Use," *Advances in Cryogenic Engineering*, vol. 39B, Plenum Publishing Corp., New York (1994), pp. 1433-1439.

5. Ki, T., Jeong, S., "Design and Analysis of Compact Work-Recovery Phase Shifter for Pulse Tube Refrigerator," *Cryogenics*, vol. 52, no. 2-3 (2012), pp. 105-110.

6. Swift, G.W., Gardner, D.L., Backhaus, S., "Acoustic Recovery of Lost Power in Pulse Tube Refrigerators," *J Acoust Soc Am*, vol. 105, no. 2 (1999), pp. 711–24.

7. Zhu, S., Nogawa, M., Inoue, T., "Numerical Simulation of A Step-Piston Type Series Two- Stage Pulse Tube Refrigerator," *Cryogenics*, vol. 47, no. 9–10 (2007), pp. 483-489.

8. Zhu, S., "Step Piston Pulse Tube Refrigerator," *Cryogenics,* vol. 64, no. 0 (2014), pp. 63-69.

9. Hu, J.Y., Luo, E.C., Zhang, L.M., Wang, X.T., Dai, W., "A Double-Acting Thermoacoustic Cryocooler for High Temperature Superconducting Electric Power Grids," *Applied Energy*, vol. 112, no. 0 (2013), pp. 1166-1170.

10. Swift, G.W., Gardner, D.L., Backhaus, S.N., "Quarter-Wave Pulse Tube," *Cryogenics*, vol. 51, no. 10 (2011), pp. 575-583.

11. Swift, G.W., Gardner, D.L., Backhaus, S., "Staging Two Coolers Through A Quarter-Wave Tube," *Crocoolers 17*, International Cryocooler Conference Inc., Boulder (2012), pp. 179-185.

12. Zhu, J.K., Song, Y.J., Wang, L.Y., Huang, X.Q., Gan, Z.H., "A Cascade Pulse Tube Cooler With Work Recovery," *Advances in Cryogenics Engineering*, vol. 59, Amer. Institute of Physics, Melville, NY (2014), pp. 1417-1423.

13. Wang, L.Y., Wu, M., Sun, X., Gan, Z.H., "A Cascade Pulse Tube Cooler Capable of Energy Recovery," *Applied Energy*, vol. 164 (2016), pp. 572-578

14. Wang, L.Y., Wu, M., Zhu, J.K., Jin, Z.Y., Sun, X., Gan Z.H., "Study on A Cascade Pulse Tube Cooler With Energy Recovery: New Method for Approaching Carnot," *IOP Conf. Series: Materials Science and Engineering*, vol. 101 (2015) 012182:1-8.

15. Gary, J., Daney, D.E., Radebaugh, R., "A Computational Model for a Regenerator," *Proc. of the Third Cryocooler Conference*, NBS Special Publication 698, National Bureau of Standards, Boulder, CO, (1985), pp. 199-211.

16. Gedeon, D., *Sage User's Guide*. Gedeon Associates, Athens (2013).

Gas Spring Effect in a Displacer Pulse Tube Refrigerator

S. Zhu[1,2]

[1]Shanghai Key Lab of Vehicle Aerodynamics and Vehicle Thermal Management Systems, Tongji University, Shanghai, 201804, China
[2]Institute of Refrigeration and Cryogenics, School of Mechanical Engineering, Tongji University, Shanghai, 201804, China

ABSTRACT

Displacer type pulse tube refrigerator is a work recover type pulse tube refrigerator. It has high potential to achieve high efficiency. Gas spring in the displacer type pulse tube refrigerator is investigated with numerical simulation in this paper. It is found to be effective for increasing working frequency. With the increase of piston diameter and dead volume, the operation frequency can be increased without increasing the mechanical spring stiffness under the condition of the linear motor operating with current displacement ratio. The displacer rod diameter is not only a parameter to optimize the displacer stroke for the COP, but also is an effective parameter for decreasing the mechanical spring stiffness. Operation frequency increases with an increase in the piston diameter and the dead volume, without increasing mechanical spring stiffness when current displacement ratio is a constant.

INTRODUCTION

Pulse tube refrigerator can be with[1-6] or without[7-10] a warm displacer. The current trend is to develop a Stirling type pulse tube refrigerator with high frequency and high efficiency. High frequency means a high power density in the linear motor which can effectively decrease the weight and the linear motor size. The long development history of inertance tube and double inlet pulse tube refrigerator shows that high efficiency needs a work recover type phase shifter. Warm displacer type pulse tube refrigerator, whose theoretical efficiency is the same as the Stirling refrigerator, is the simplest type of work recover refrigerator. One of the development tendency is to increase the operation frequency[11,12]. Higher frequency means higher stiffness of the spring and lighter of the moving part of the linear motor. Weight of the moving part is limited by material and structure, which is difficult to loss. In brief, increasing stiffness of mechanical spring means to increase the number of flexible springs, which is a high cost method. Increasing gas spring stiffness without increase mechanical spring stiffness of the linear compressor is a realistic method. The gas spring effect of the displacer rod also should be considered for decreasing mechanical spring stiffness of the displacer. In this paper, a numerical simulation is used to study the gas spring effect of the linear motor and displacer.

Cryocoolers 19, edited by S.D. Miller and R.G. Ross, Jr.

STRUCTURE

Figure 1 is the schematic of the warm displacer type pulse tube refrigerator. It includes a compressor, a cold head and a warm displacer. The compressor includes a linear motor, a piston and a dead volume. The cold head includes an after cooler, a regenerator, a cold heat exchanger and a pulse tube. The displacer has a rod which is connected to a spring in the displacer buffer. The displacer forms the displacer front space which is connected to the warm end of the pulse tube and the displacer back space which is connected to the compression space.

The linear compressor supplies power, generates pressure oscillation and oscillation gas flow in the refrigerator. The oscillation gas flow generates a pressure difference across both end of the displacer due to the pressure drop across the regenerator and heat exchangers. There is also a pressure difference across both ends of the rod due to the pressure difference between the displacer buffer and displacer back space, which causes the displacer to oscillate. Regardless of recovering the expansion power from the pulse tube, the displacer acts as a phase shifter to control the gas flow in the regenerator to let the regenerator reach a higher efficiency. The rod has two functions, one is to supply additional driving force for the displacer, another is to be a gas spring due to the pressure difference at both ends of the rod.

Basic data of the refrigerator is found in Table 1. During the numerical simulation, the piston weight and displacer weight is not changed. The resonant point of the linear motor is adjusted by the piston diameter and dead volume. The displacer natural frequency is adjusted by the displacer spring stiffness or rod diameter.

NUMERICAL METHOD

The numerical method found in S. W. Zhu, et. al.[9] is used for this simulation. The linear motor force current relation is assumed to be linear which is available for the moving coil and the moving magnet structure. This method is originally developed for the simulation of the double inlet pulse tube refrigerator[4], and improved for the simulation of the inertance tube pulse tube refrigerator[5], and some other types of pulse tube refrigerators. Past developments such as double inlet, inertance tube and active buffer pulse tube refrigerator, show that it is effective for the mechanism study of a new type of pulse tube refrigerator. The invention of the double inlet pulse tube refrigerator is partly due to the result of this numerical method.

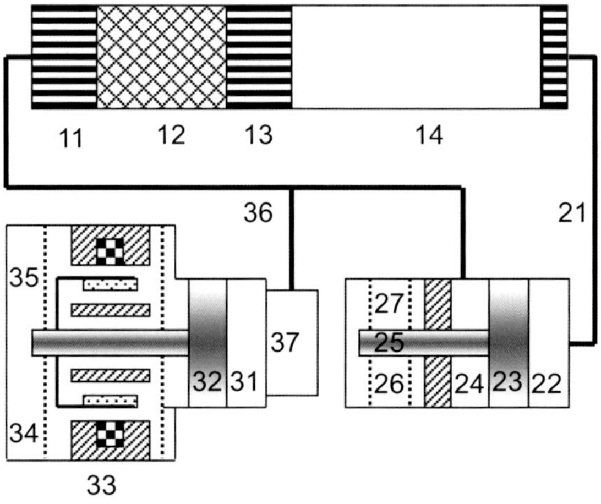

Figure 1. Displacer pulse tube refrigerator with rod: 11. warm heat exchanger 12. regenerator 13. cold heat exchanger 14. pulse tube 21. displacer connecting tube 22. displacer front space 23. displacer 24. displacer back space 25. displacer rod 26. displacer spring 27. displacer buffer 31. compression space 32. compressor piston 33. linear motor 34. motor spring 35. motor house 36. compressor connecting tube 37. dead volume

Table 1. Basic data of the refrigerator

Regenerator	Φ80mm×50mm, wire diameter 0.025mm, porosity 0.7
Pulse tube	Φ30mm×150mm
Motor	Spring stiffness 160N/mm, piston weight 2.9kg, motor force factor 100N/A, voltage 500V
Displacer	Displacer diameter 70mm, rod diameter 35mm, weight 0.4kg, spring stiffness 150N/mm
Operation condition	Room temperature 300K, refrigeration temperature 77K, charge pressure 2MPa, operation frequency 100Hz

PISTON DIAMETER EFFECT

In order to let a pulse tube refrigerator whose natural frequency is lower than operation frequency to match the resonant point without increasing the mechanical spring, the piston diameter should be enlarged to increase the gas spring stiffness. With the increase in the piston diameter, the compressor will supply extra gas to the cold head. In order to balance the extra gas, a dead volume is needed. Figure 2a shows the piston diameter effect to the dead volume(Vd) of the compressor. Figure 2b shows the piston diameter effect to the current displacement ratio of the linear motor(I/Xp), and the current (I). Figure 2c shows the piston diameter effect to the displacement of the piston(Xp), the displacer(Xd), and the phase angle difference (ϕ) between the piston and displacer. Figure 2d shows the piston diameter effect to the input power(W), and the cooling power(Q). Figure 2e shows the piston diameter effect to the COP and motor efficiency(Em). With the piston diameter expanding, the dead volume and piston displacement increase; and the current, current displacement ratio, phase angle difference between the piston and displacer decrease for the same performance. There is a peak for the displacer displacement, the motor input power, the cooling power, and the COP when the piston diameter increasing. Motor efficiency increases with piston diameter increasing as well.

For a given linear motor, there is an optimal current and displacement. There is a ratio of the optimum current divided by the optimum piston displacement which is named optimum current displacement ratio. Finding the best match of the linear motor becomes a search to find the relation between the optimum displacement ratio and the piston diameter. Figure 2 shows that the optimum displacement ratio decreases with an increasing piston diameter. Hence, there is a piston diameter with which the optimum current displacement ratio can be reached.

ROD DIAMETER EFFECT

The linear motor can reach high frequency with increasing piston diameter. The displacer also can operate at a high frequency with a larger rod diameter, which is demonstrated in Figure 3. As shown in Figure 3, piston diameter is 152mm, when rod diameter is increased from 5mm to 45mm, though there is a peak of COP, the difference is rather small in a wide range, the spring stiffness of the displacer(Kd) almost decreases about 100 N/mm when the rod diameter increases from 5mm to 45mm. Due to the increasing rod diameter, the dead volume of the compressor has to be increased a little in order to adjust the resonant point of the linear motor (Figure 3a). The current displacement ratio and current increase slightly (Figure 3b). The piston displacement basically has no variation, the displacer displacement has few increasing (Figure 3c). The phase angle difference between the piston and displacer, input power and cooling power increase (Figure 3d). The motor efficiency almost remains unchanged (Figure 3e).

Rod diameter has a strong effect of decreasing mechanical spring stiffness, and has a slight influence on the others, thereby, we may design the displacer with a slightly larger rod diameter for decreasing the mechanical spring.

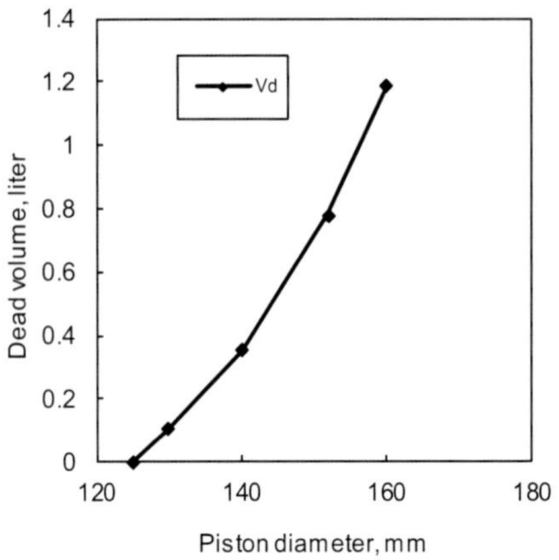

Figure 2a. Dead volume vs. piston diameter

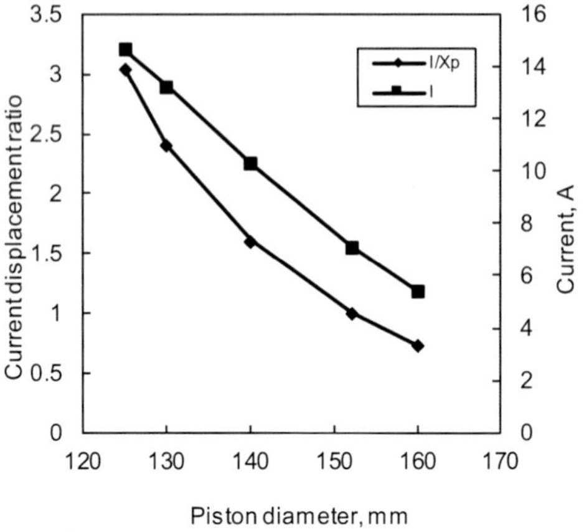

Figure 2b. Current diplacement ratio and current vs. piston diameter

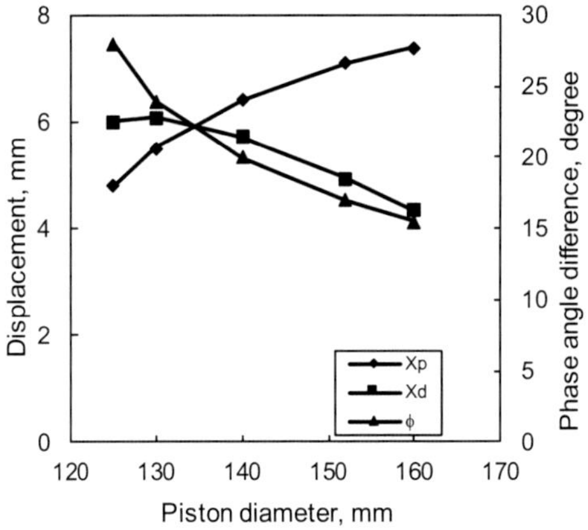

Figure 2c. Phase angle diffference and displacement of piston and displacer vs. piston diameter

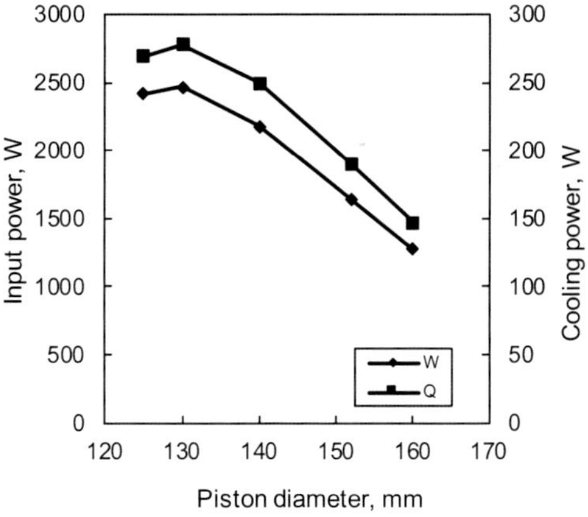

Figure 2d. Input power and cooling power vs. piston diameter

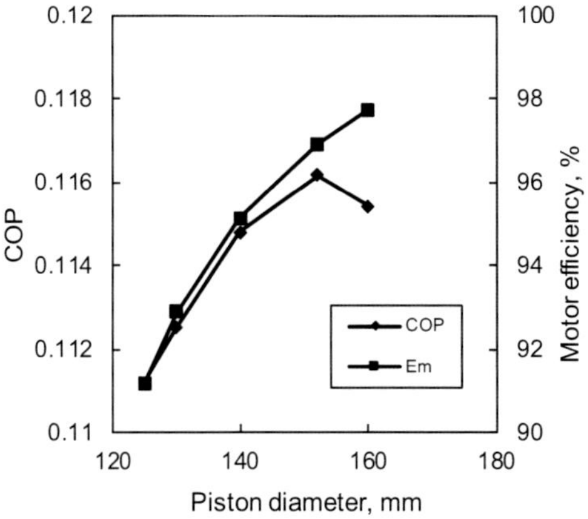

Figure 2e. COP and motor efficiency vs. pistion diameter

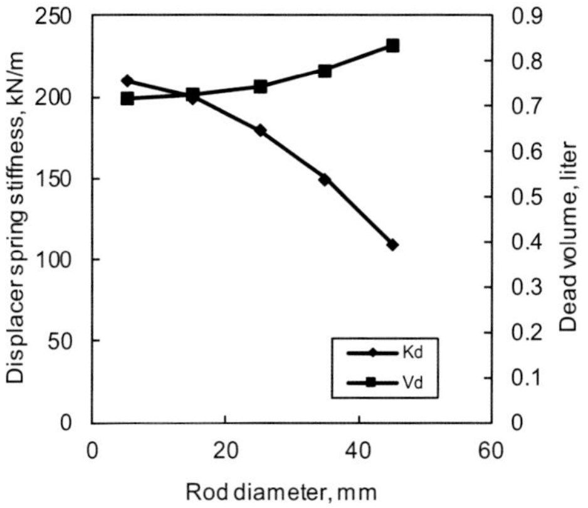

Figure 3a. Dead volume and displacr stiffness vs. rod diameter

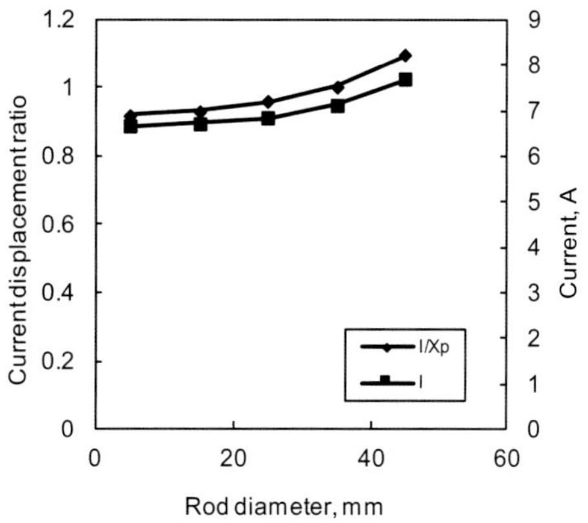

Figure 3b. Current displacement ratio and current vs. rod diameter

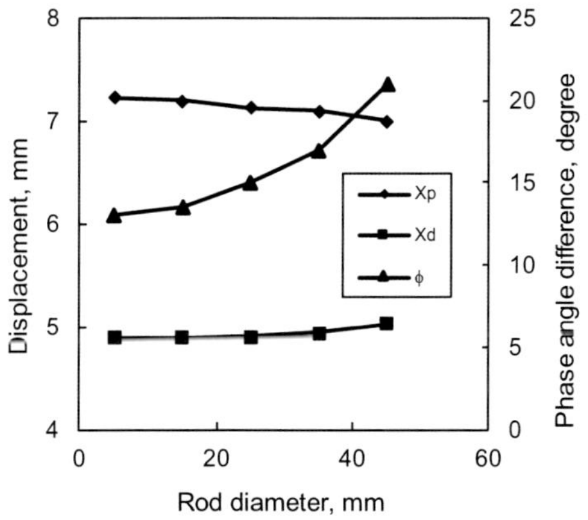

Figure 3c. Phase angle diffference and displacement of piston and displacer vs. rod diameter

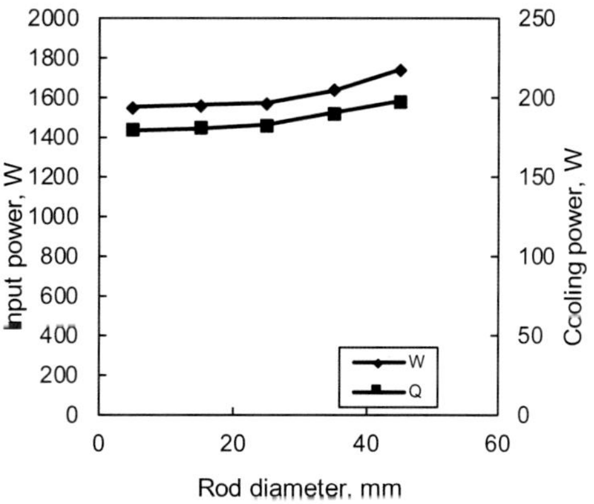

Figure 3d. Input power and cooling power vs. rod diameter

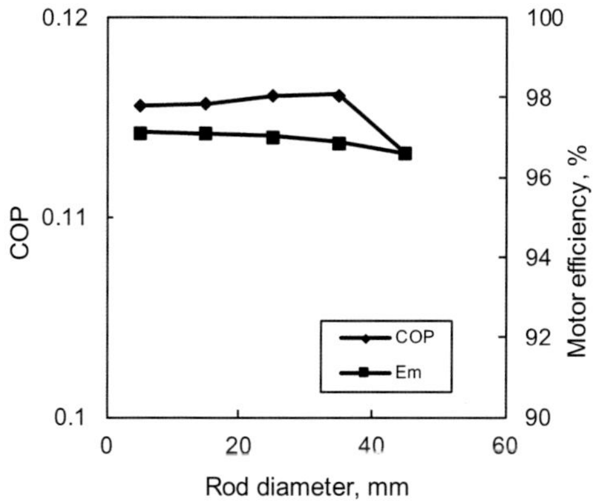

Figure 3e. COP and motor efficiency vs. rod diameter

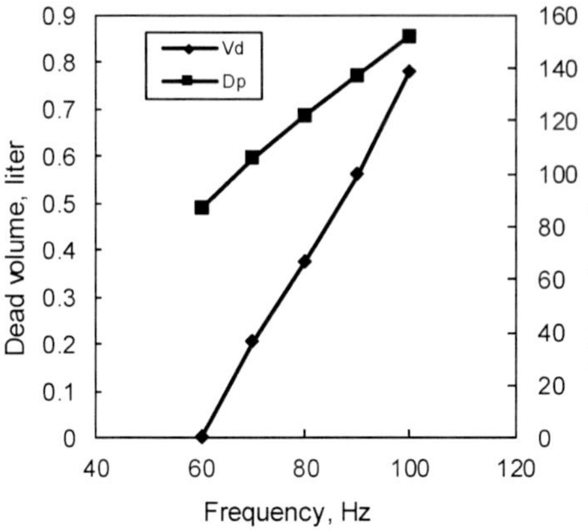

Figure 4a. Dead volume and piston diameter vs. frequency

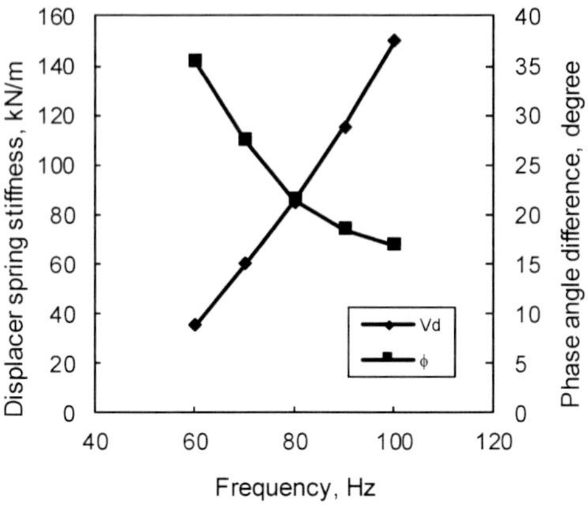

Figure 4b. Displacer spring stiffness and phase angle difference vs. frequency

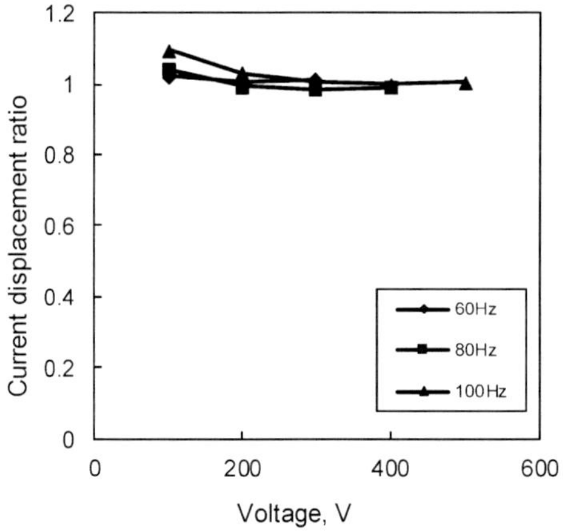

Figure 5a. Current displacement ratio vs. frequency and voltage

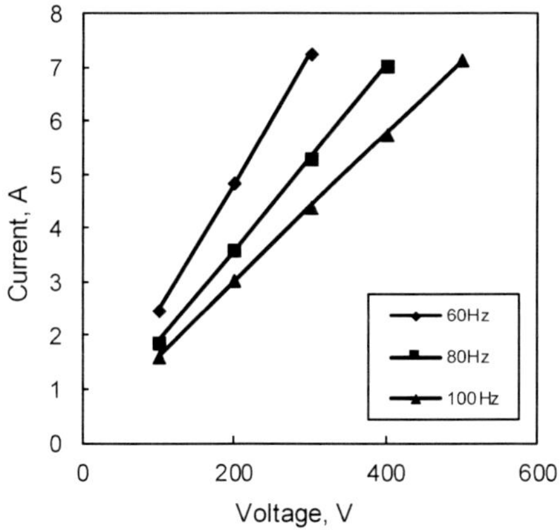

Figure 5b. Current vs. frequency and voltage

FREQUENCY EFFECT

Assuming the rated current displacement ratio is 1A/mm, displacer spring stiffness can be changed to fit the frequency change, the piston diameter and compressor dead volume can be adjusted to let linear motor operate at resonant point. The frequency effect is shown in Figure 4. Figure 4 shows that the piston diameter, compressor dead volume (4a), displacer spring stiffness (4b) increases with an increase in frequency. The phase angle difference between the piston and displacer decrease. Figure 5 shows the current displacement ratio (5a), current (5b), input power (5c), cooling power (5d) and COP change (5e) with the voltage. There is a peak COP for each frequency. At higher frequency, the COP is lower. Current is almost increases linearly with the an increase of voltage. However, the current displacement ratio almost does not change when the voltage increases.

Figure 5e shows that the cold head should be redesigned for a higher operation frequency. Otherwise, the efficiency decreases with an increase of operation frequency. Figure 5a shows that the current displacement ratio is a weak function of voltage for a rather wide range. So, in Figure 2, we can use constant voltage to find the relation between the current displacement ratio and the piston diameter.

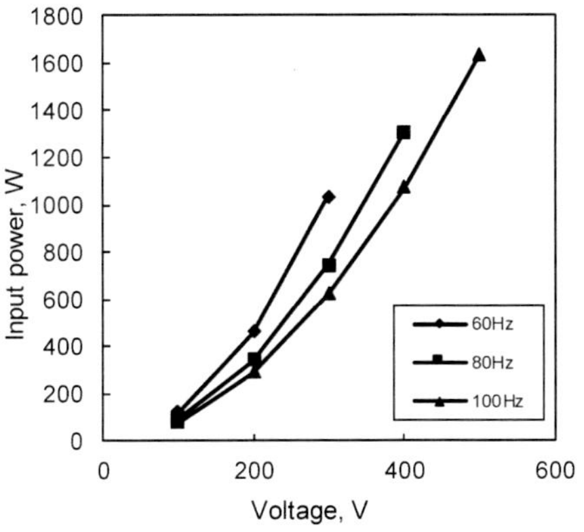

Figure 5c. Input power vs. frequency and voltage

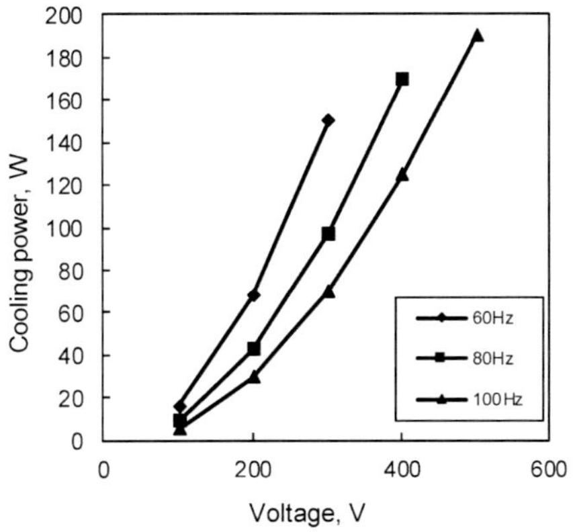

Figure 5d. Cooling power vs. frequency and voltage

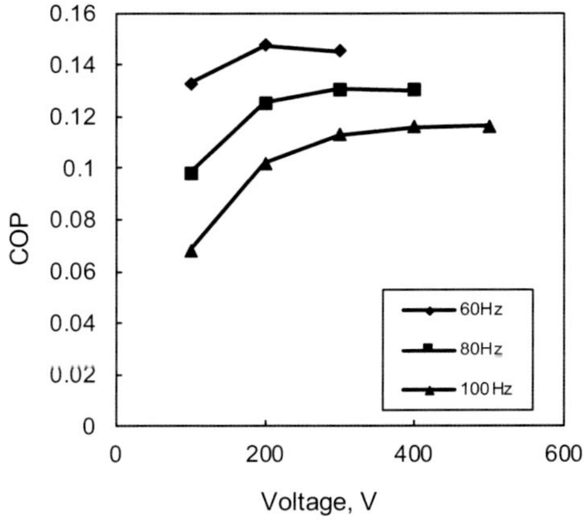

Figure 5e. COP vs. frequency and voltage

CONCLUSION

For a displacer type pulse tube refrigerator, operating at some frequency, the dead volume should be increased to keep the linear motor at its resonant point. When the piston diameter is increased, the current displacement ratio decreases with the piston diameter increasing, which means that there is a piston diameter that the linear motor can be operated at resonant point with the optimum current displacement ratio. An increase in the rod diameter can decrease the mechanical spring stiffness at same operation frequency. The rod diameter influences the motor matching slightly with the cold head. Operation frequency increases with an increase in the piston diameter and the dead volume without increasing mechanical spring stiffness at the optimum current displacement ratio.

ACKNOWLEDGMENT

This work is supported by the National Natural Science Foundation of China (No. 51476117).

REFERENCES

1. Gifford WE, Longsworth RC, "Pulse-tube refrigeration," *Journal of Engineering for Industry* 86, no. 3 (1964), pp. 264-268.
2. Mikulin EI, Tarasov AA, Shkrebyonock MP, "Low temperature expansion pulse tubes, " *Adv. in Cryogenic Engineering*, Vol. 29. Springer US (1984), pp. 629-637.

3. Radebaugh, R., Zimmerman, J., Smith, D.R. and Louie, B., A comparison of three types of pulse tube refrigerators: new method for reaching 60 K, *Adv. in Cryogenic Engineering*, Vol. 31, Springer US (1986), pp. 779-789.

4. Zhu SW, Wu PY, Chen ZQ, "Double inlet pulse tube refrigerators: an important improvement," *Cryogenics*, 30, no. 6 (1990), pp. 514-520.

5. Zhu, S.W., Zhou, S.L., Yoshimura, N. and Matsubara, Y., "Phase shift effect of the long neck tube for pulse tube refrigerator," *Cryocoolers 9*, Springer US (1997), pp. 269-278.

6. Hu, J.Y., Dai, W., Luo, E.C., Wang, X.T., Huang, Y., "Development of high efficiency Stirling-type pulse tube cryocoolers," *Cryogenics*, 50, no. 9 (2010), pp. 603–607.

7. Radebaugh, R., "Development of the pulse tube refrigerator as an efficient and reliable cryocooler," *Proceedings of Institute of Refrigeration, vol. 96,* (2000), pp. 11-31.

8. Matsubara, Y., "Future trend of pulse tube cryocooler research," *Proceedings of the twentieth international cryogenic engineering conference,* Beijing, China (2004), pp.189–96.

9. Zhu, S.W., Nogawa, M., "Pulse tube stirling machine with warm gas-driven displacer," *Cryogenics*, 50, no. 5 (2010), pp. 320-330.

10. Wang, X.T., Zhang, Y.B., Li, H.B., Dai, W., Chen, S., Lei, G., Luo, E.C., "A high efficiency hybrid stirling-pulse tube cyocooler," *AIP Advances* 5, No. 3 (2015), p. 037127.

11. Petach, M., Waterman, M., Pruitt, G., Tward, E.. "High frequency coaxial pulse tube microcooler," Cryocoolers 15, ICC Press, Boulder, CO (2009), pp. 97–103

12. Wu, Y.Z., Gan, Z.H., Qiu, L.M., Chen, J., Li, Z.P. and Weisend, J.G., "Study on a single-stage 120 hz pulse tube cryocooler," *Adv. in Cryogenic Engineering*, Vol. 55, Amer. Institute of Physics, Melville, NY (2010), p. 175.

Smart Energy Compact Chiller for Helium Compressors

J. Gardiner, T. Sayles, K. Knight, J. Diederichs, J. Sloan, S. Spagna

Quantum Design Inc.
San Diego, CA 92121

ABSTRACT

Recent advances in smart energy compressors for helium cryocoolers have shown to significantly reduce power consumption during operational lifetime, especially when the system is cold or in "standby" mode. However, these compressors typically require at least a 3 ton industrial chiller or larger for adequate chilled water cooling.

This paper introduces an updated version of the Quantum Design (QD) smart energy chiller dedicated to operation with a variable speed, water cooled helium compressor (QD model HLC 4500). Running at fixed speeds (60 Hz capsule, 60 Hz head) with a Gifford-McMahon (G-M) cold head or pulse tube, cooling power is found to be similar to the cooling power of a standard compressor and industrial chiller with power savings of up to 2 kW at 60 Hz. Additionally, the variable speed compressor can be adjusted to reduce power consumption when cooling power needs are less which results in potential power savings of up to 5 kW.

The chiller features dedicated cooling of the helium and oil circuits that leads to increased efficiency and a smaller footprint than other commercially available chillers. A weatherized housing allows for flexible installation options including operation outdoors in a variety of climates.

INTRODUCTION

In 2012, Quantum Design introduced a "Modular Architecture" option for the HLC 4500[1] water cooled helium compressor that allowed a simple modification of the standard compressor to provide a compact solution for customers without access to a chilled water source[2]. While the new design retains the core feature of separate brazed-plate heat exchangers (BPHX), to optimize cooling, new improvements have been implemented such as a simplified cooling scheme, a two stage digital controller to control fan and refrigerator power and optimization for outdoor installation.

Previous studies have shown that reducing the speed of the compressor scroll capsule can significantly reduce overall power consumption on Gifford-McMahon (G-M) cold heads[3,4] and pulse tubes[5,6]. By varying the head and capsule speed of the compressor, cooling power can be optimized while taking advantage of the reduced power draw at lower speeds. Further power savings come from utilizing the chiller refrigerator to only cool the helium circuit rather than unnecessarily overcooling of the oil circuit as standard in a traditional industrial chiller solution.

Cryocoolers 19, edited by S.D. Miller and R.G. Ross, Jr.

RADIATOR DESIGN UPDATE

Figure 1 shows the updated configuration of the compressor and chiller interface. Previously, the helium circuit featured dedicated copper lines running between the condenser unit and helium BHPX, which had the advantage of directly cooling the helium gas with R134a but complicated the installation as the distance between indoor and outdoor units can extend up to 30 meters. this will add significant cost in materials and requires a certified refrigeration technician for installation. By diverting some of the water/glycol mixture cooled by the radiator to the refrigerator (controlled by a manually controlled globe valve) for further cooling, the interface between the indoor and outdoor unit is now simplified to three rubber hoses. The smaller diameter (1/4" ID) helium supply hose carries refrigerator chilled water to the helium BHPX while the larger diameter (3/4" ID) oil supply hose carries radiator cooled water to the oil BHPX. A single ¾" ID rubber hose returns heater water back to the chiller for cooling. A 1/3 HP centrifugal pump keeps the water continuously pumping with discharge pressure of approximately 20 psig and flow of 4 gallons a minute when 30 meter lines are used.

Another objective of the redesign was to ensure that the chiller can operate a distance up to 30 meters from the indoor unit at an outside ambient temperature up to 45°C while keeping the helium temperature optimally between 15 to 28°C with a maximum of 35°C as specified by G-M and Pulse Tube manufacturers[7,8] Operating at higher helium temperatures can potentially reduce the effective cooling power of the system. Also, previous studies have also shown that oil carryover from the compressor coalescing filter to the adsorber is substantially higher at helium temperatures above 30°C requiring more frequent service[4]. To accomplish this design objective, the following hardware updates were implemented: a U-shaped radiator to increase heat exchanged area, a larger fan with increased air flow, a larger refrigerator unit to increase cooling capacity and a larger pump to accommodate the pressure drops of 30 meter hoses (Figure 2). Additionally, a protective environmental coating is added to the finished refrigerator assembly and all electronics are contained within a NEMA 4X enclosure with liquid tight cord grips for each cable (no external connectors). The overall footprint of the chiller is 22 x 22 inches while a standard 3 ton industrial chiller takes up over double the floor space[9].

THERMAL PERFORMANCE EXPERIMENTAL RESULTS

The updated HLC 4500 chiller is designed to operate under a wide range of ambient temperatures from -30°C to 45°C which makes it suitable for outdoor installation in most climates. Figure 3 shows the effect on the compressor temperatures while running the chiller at the upper range of allowable ambient temperatures. From the data, we see that while running the chiller at the most thermally stressful condition (45°C ambient temperature and full compressor scroll speed of 60 Hz) the helium temperature stays below the 35°C recommended for cold heads and below the 90°C maximum helium discharge (capsule) recommended by the helium scroll compressor manufacturer.

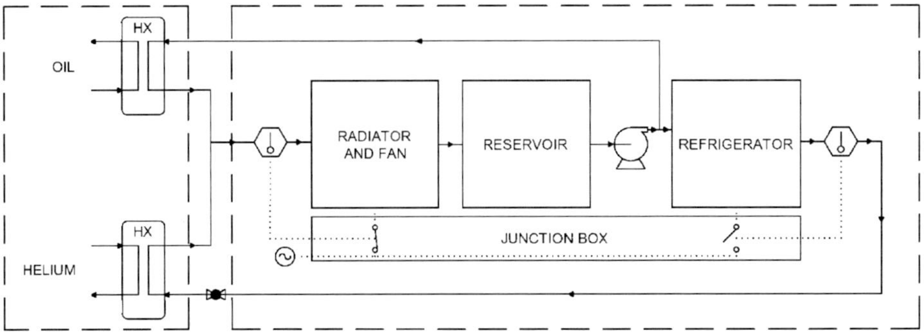

Figure 1. Updated configuration of water cooled compressor and chiller

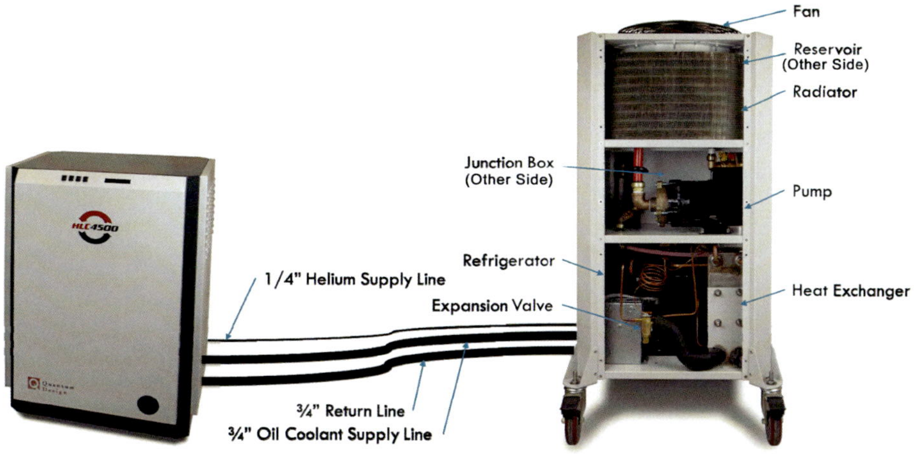

Figure 2. Updated configuration of water cooled compressor and chiller

Should the chiller ambient temperature raise above 45°C, the chiller refrigerator unit features a built in thermostat that will shutoff power to the refrigerator to prevent damage to the refrigeration compressor. Additionally, the HLC 4500 compressor features adaptive speed control that will reduce the compressor speed to keep the compressor running if the capsule temperature reaches 90°C. These adaptive features can also keep the compressor running at reduced cooling power if there is reduced chilled water flow (such as a pinched rubber hose) or air flow (such as significant debris build up on air filters).

When the full cooling power of the compressor is not needed and an energy savings is desired, the compressor can be run at reduced scroll capsule and head speeds. Running the compressor at reduced speeds also reduces capsule, oil and helium temperatures, which along with the shorter duty cycle of the wear components of the compressor and head, can reduce wear and lengthen the lifetime of major system components. Figure 4 shows the reduced capsule temperature range of 50°C to 60°C and reduced helium temperature of 17°C to 25°C. The automated temperature controller of the updated chiller design also eliminates the need to throttle the flow of chilled water when compressor speeds are changed or ambient temperatures shift, allowing for a more hassle free experience for the end user.

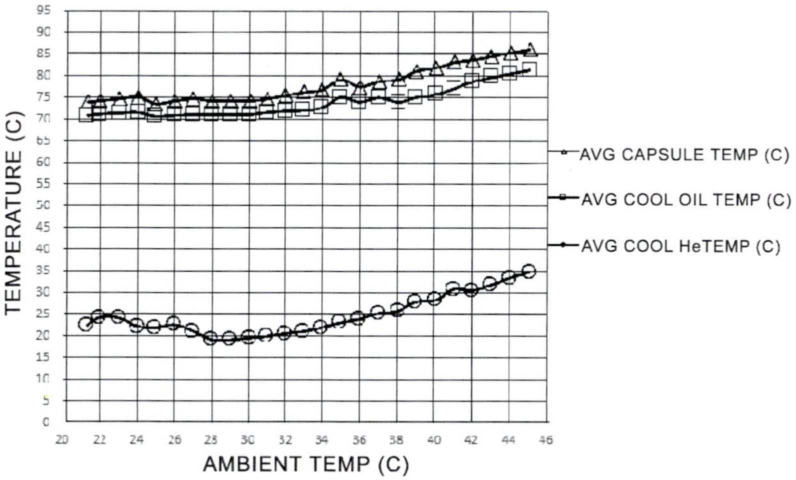

Figure 3. Thermal testing results at 60 Hz capsule speed

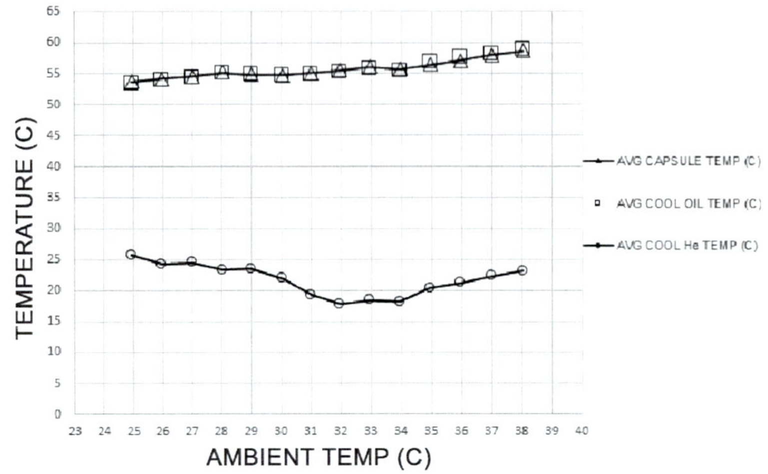

Figure 4. Thermal testing results at 30 Hz capsule speed

POWER DRAW EXPERIMENTAL RESULTS

The HLC 4500 and chiller combo was run with several RDK-415D[9] G-M 1.5W cold heads and flow impedance fixtures, with a static helium charge pressure of 1.5 MPa, at various speeds while cooling down and while cold. Head speed was kept at 60 Hz. The actual power draw of the compressor and chiller was measured with a 3 phase Fluke 437 power meter. Figure 5 shows that the actual power draw of the HLC 4500 compressor varied greatly depending on compressor

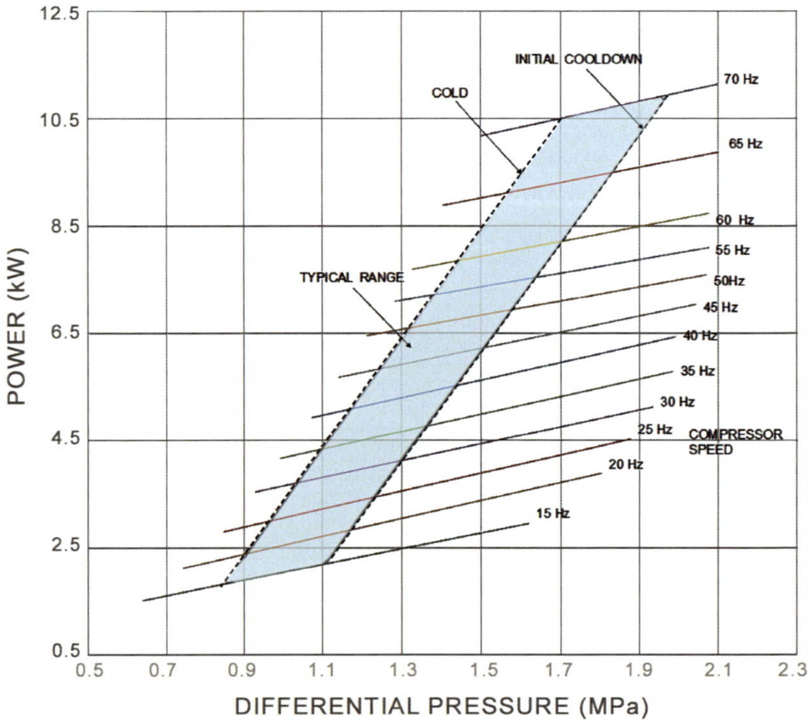

Figure 5. HLC 4500 compressor actual power draw at various speeds

scroll speed and differential pressure (Supply Pressure - Return Pressure) of the compressor. During cool down at a given speed, the differential pressure typically was about 0.3 MPa larger than the differential pressure once the system was cold at 4.2K leading to a power reduction of about 0.5 kW once cold.

The chiller power draw varies depending on which chiller components are energized as shown in Figure 6. At lower ambient temperatures, the fan and refrigerator power will cycle to maintain the helium temperature in the desired 20 – 28°C range giving the chiller an actual power draw of 0.4 to 1.3 kW depending on cooling needs. The HLC 4500 and chiller paired with an average 1.5 W cold head has a maximum power draw of 11.8 kW during cool down at 70 Hz in higher ambient temperatures (above 30°C) and a minimum power draw of 2.1 kW while cold at 15 Hz and running in a lower ambient temperature (below 30°C). By comparison, a Sumitomo F70 compressor has a maximum measured power draw of 9.0 kW when charged to 1.5 MPa, running at 60 Hz and cooling down a 1.5 W cold head. However, the F-70 requires an external chilled water source, typically a minimum of a 3 ton industrial chiller with a maximum power draw of 3.5 kW[10] bringing the total max power draw up to 12.5 kW. Once cold the minimum power draw of the F70 and 3 ton industrial chiller combo is reduced to approximately 11.0 kW. From these results we see that the potential power savings of the HLC 4500 and chiller can be significant, especially if the application does not require the full cooling power capacity of the cold head. At 60 Hz and in cold state, the combined power draw of the HLC 4500 and chiller is approximately 9.0 kW.

COOLING POWER EXPERIMENTAL RESULTS

The HLC 4500 and chiller combo was run with a SRP-082B2S[11] 0.9W 4.2K class pulse tube at various compressor and remote valve speeds. Figure 7 shows the HLC 4500 achieved 0.88 W 2nd stage cooling power at a capsule speed of 71 Hz and remote valve speed of 62 Hz. The F70 compressor was able to achieve 0.88 W 2nd stage cooling at 60/60 Hz due to the larger Copeland scroll capsule[12] used by the F70 compared to the Hitachi scroll capsule[13] utilized by the HLC 4500.

The HLC 4500 allows the capsule and head speeds to be independently set so that cooling power can be optimized and we found that generally for a given capsule speed ideally the head speed should be several Hz lower to maximize cooling power on the pulse tube as shown in Figure 8. At a 40 Hz capsule speed, cooling power is reduced to 0.4 W but overall power draw is reduced to 6.3 kW – an approximate 5 kW power savings compared with the F70 and industrial chiller combination. Also from the data we see that cooling power falls off substantially at capsule speeds less than 40 Hz limiting the practical use of the pulse tube below 40 Hz and limiting further significant power savings.

Next the HLC 4500 and chiller combo was run with a RDK-415D cold head at various capsule and head speeds. The results of these tests are seen in Figure 8. Figure 9 shows a 2nd stage cooling

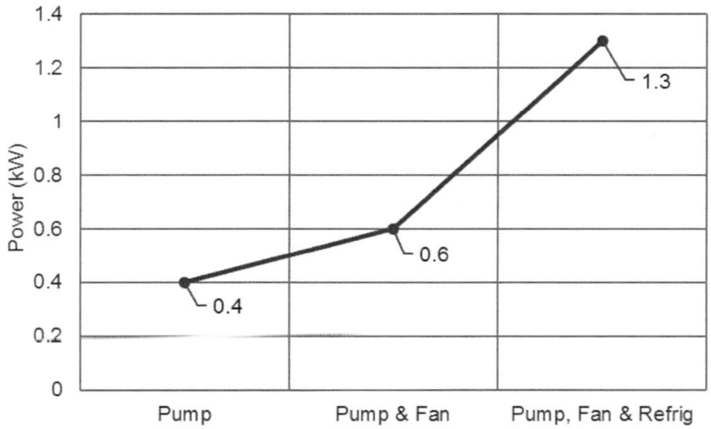

Figure 6. HLC 4500 chiller actual power draw by operational state

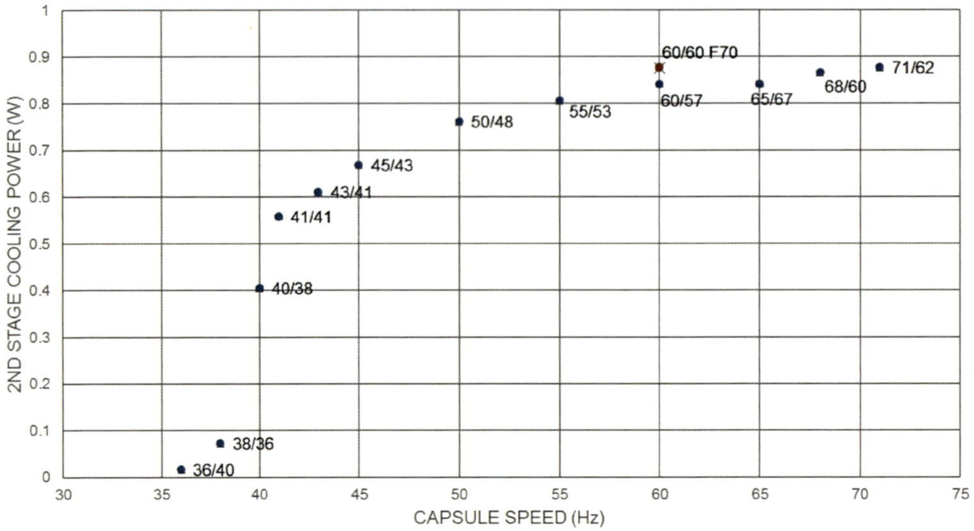

Figure 7. Cooling power with 4.2 K pulse tube at tuned (capsule/remote valve) speeds

power of 1.95 W at 66 Hz. Unlike the pulse tube, the G-M cooling power follows capsule speed more linearly – allowing for practical application at speeds down to 15 Hz. At 15 Hz capsule speed the available 2nd stage cooling power is still a significant 0.3 W – an approximate 8 kW savings over an F70 compressor and industrial chiller combo. Running the compressor at a capsule speed less than 15 Hz is not recommend (nor allowed) on the HLC 4500 as there is not sufficient differential

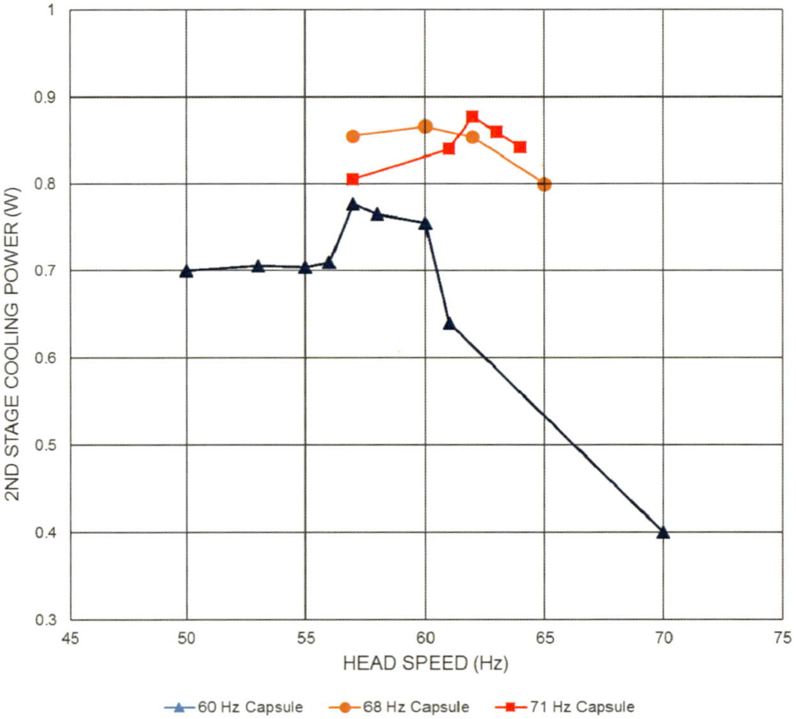

Figure 8. Head speed "tuning" to optimize cooling power at a given capsule speed

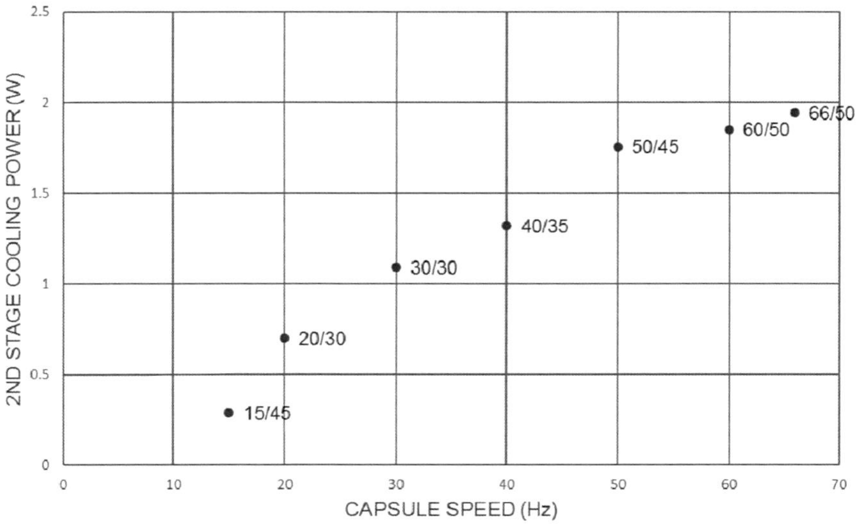

Figure 9. Cooling power with 4.2 K G-M at tuned (capsule/head) speeds

pressures at lower speeds to adequately return oil to the capsule from the coalescing filter. Like the pulse tube, keeping the G-M head speed several Hz lower than a given capsule speed yields maximum 2nd stage cooling power. Depending on the compressor configuration, the upper limit for capsule speed ranges from 66 to 71 Hz and is a result of the maximum capsule inverter rating of 18 amps. Unless ultimate cooling power is the top application priority, the additional power draw and wear on moving parts should also be considered when determining the maximum allowed capsule speed.

CONCLUSIONS

The updated modular architecture of the HLC 4500 helium compressor and chiller provides a novel smart energy helium compressor solution by enabling the end user to "dynamically tune" capsule and head speeds, allowing for optimization of cooling power needs and total power draw. Reducing compressor speeds also helps lengthen the service life of major system components such as scrolls, bearings and moving seals. Further testing planned includes: expanded thermal testing in cold environments, cold head tuning refinement, testing at capsule speeds above 70 Hz (larger inverter) and optimization of 1st stage cooling power. Compressor speeds and parameters can easily be remotely adjusted via CAN bus offering further automation and refinement of compressor diagnostics and control. Further software development is planned to streamline the graphical user interface and offer real time display of power draw and cooling power.

REFERENCES

1. Model HLC 4500: a 7.5W class liquid helium compressor manufactured by Quantum Design Inc.
2. Diederichs, J., Simmonds, M.B., Sayles, T. and Spagna, S., "A Modular Architecture of Helium Compressors Larger than 2.5 kW," *Cryocoolers 17*, ICC Press, Boulder, CO (2013), pp. 367-370.
3. Sayles, T., Martien, D., Diederichs, J., Sloan, J., Bird, D. and Spagna, S., "Enhanced Helium Compressor Operation for Sensitive Measuring Instrumentation," *Cryocoolers 16*, ICC Press, Boulder, CO (2011), pp. 371-373.
4. Chialvo, C., Sayles, T., Diederichs, J., Spagna, S., Gabal, M., Sesé, J. and Rillo, C., "Smart Energy Compressors for Helium Liquefiers," *Cryocoolers 18*, ICC Press, Boulder, CO (2015), pp. 551-554.
5. Wang C., "Dynamic Operation of a 4 K Pulse Tube Cryocooler with Inverter Compressors," *Cryocoolers 16*, ICC Press, Boulder, CO (2011), pp. 65-70

6. Wang, C., Beyer, A., Cosco, J., Lichtenwalter, B. and Brown, E., "Energy Efficient Operations of 4 K Pulse Tube Cryocoolers," *Cryocoolers 18*, ICC Press, Boulder, CO (2014), pp. 203-210

7. Model F70: a 7.5 kW class liquid cooled compressor manufactured by SHI Cryogenics Inc.

8. Model PT410: a 1.0 W 4K class helium cryogenic refrigerator manufactured by Cryomech Inc.

9. Model RDK-415D: a 1.5W 4K class helium cryogenic refrigerator manufactured by SHI Cryogenics Inc.

10. Model ACI-3: a 3 ton air-cooled chiller manufactured by Pacific Rim International.

11. Model RP-082B2S: a 0.9W 4K class helium cryogenic refrigerator manufactured by Cryomech Inc.

12. Model ZCH72: a horizontal helium cryogenic compressor manufactured by Emerson Climate Technologies.

13. Model S603DH: a vertical helium cryogenic compressor manufactured by Hitachi Appliances, Inc.

Dry Compact 1 kW Compressor for GM and Pulse Tube Cryocoolers

J. Höhne

Pressure Wave Systems GmbH
Munich, Germany

ABSTRACT

Pressure Wave Systems GmbH has developed a dry and compact 1 kW compressor system for the operation with Gifford-McMahon and pulse tube cold heads.

The compression unit consists of two metal bellow based "cylinders" of 500 cm^3 driven by electro-hydraulics. As compared to flexure-bearing compressors, the system runs at a much lower frequency between 0.5 and 2 Hz. The compressor is air-cooled, low-vibration, orientation-independent and uses single-phase electrical power. Technology, performance and application of the compressor system will be discussed.

INTRODUCTION

Recent advances in superconducting sensor and electronics technology e.g. (single) photon detectors create a demand for commercial low power, compact and mobile 4K and sub-4K cooling systems. Today, to our knowledge the smallest commercial 4K cold head on the market is the Sumitomo RDK-101 with around 100 mW cooling power at 4.2K and a no-load temperature of around 2.5K [1]. It is available with a 1.2 kW single-phase air-cooled compressor (CNA-11). This unit is well suited for many of the applications mentioned above. Nevertheless, many of these applications are currently run with no thermal load at the cold head base temperature of around 2.5K or actively heated at much higher temperatures up to 10K. As a consequence, a smaller cold head / compressor combination with reduced cooling power resulting in reduced electrical input power in the range of 500 W or smaller would be sufficient. For this and other applications we have developed a new dry compressor technology that can be scaled to serve the input power requirements described above.

TECHNOLOGY

Our compressor technology is based on a metal bellow driven by electro-hydraulics. Figures 1a and 1b illustrate the working principle. When the electric motor drives the hydraulic gear pump to move the hydraulic oil towards the metal bellow, the bellow and the gas inside are compressed (Figure 1a). Moving oil in the other direction (Figure 1b) allows the bellow and the gas to relax.

This set-up already has many advantages compared to the conventional scroll technology: the working gas only gets in contact with metal surfaces. Oil separator technology as used in scroll systems is not needed. Furthermore, our metal bellows are guaranteed for 80 million cycles. This corresponds to 5 years of continuous running at a frequency of 0.5 Hz. The gear pump itself is a

Cryocoolers 19, edited by S.D. Miller and R.G. Ross, Jr.

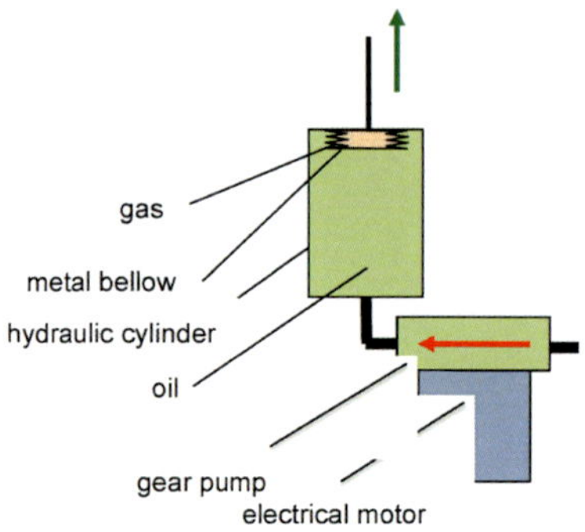

Figure 1a. Compression of the metal bellow and resulting compression of the gas volume

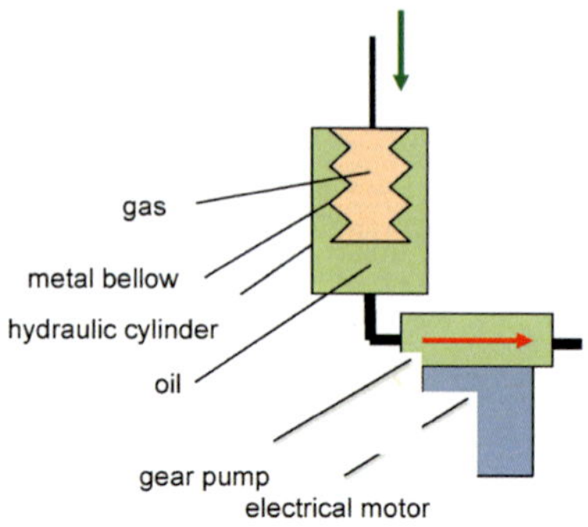

Figure 1b. Relaxation of the metal bellow and resulting relaxation of the gas volume

standard part and has shown the same reliability as needed for the bellow. We expect to exceed the maintenance intervals of conventional scroll units with this technology.

COMPRESSOR DESIGN

To make efficient use of this technology it is favorable to arrange the two bellows connected to one gear pump and drive the oil back and forth between the two bellows as shown in Figure 2. This has the additional advantage that the oil is always kept in a closed loop. In this way contamination of the hydraulic oil can be ruled out.

Two check valves on each cylinder control the gas flow. Our proprietary check valves have been tested to over 100 million cycles which translates to 6 years of continuous running time at 0.5 Hz. Buffer volumes in the range of liters are needed to keep the pressure fluctuations at reasonable values. Commercially available flex-lines (3 m) have been used to connect the compressor to the cold head.

Figure 3a shows a picture of the actual compression unit. In Figure 3b the complete 19" rack-mountable compressor package is shown.

PERFORMANCE / TEST RESULTS

The compressor was run with the Sumitomo RDK-101 cold head as depicted in Figure 2. The cold head motor was driven by the compressor electronics. Tests were performed with 230V / 50 Hz European grid voltage and frequency with electrical input powers of 1000 W, 750 W and 500 W. Figure 4a shows a typical cool down curve with 750 W electrical input power. The first and second

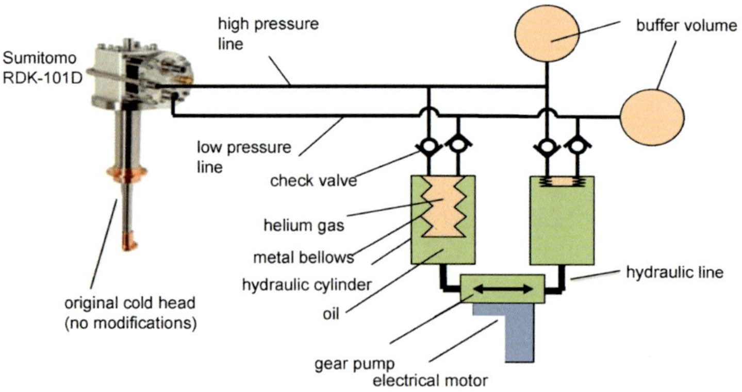

Figure 2. Compressor set-up

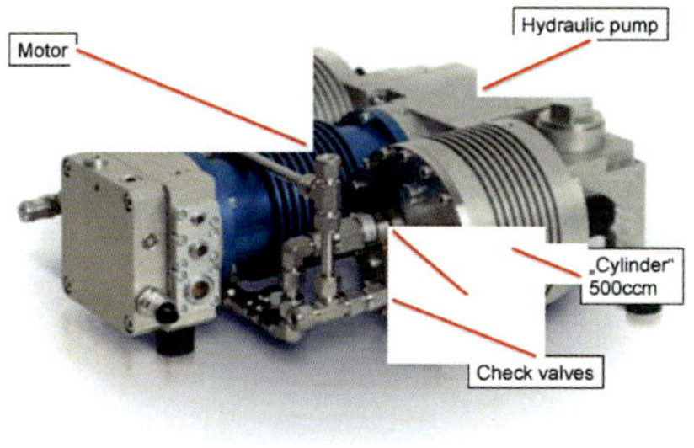

Figure 3a. Compression unit

Figure 3b. 19" compressor package

stage of the cold head reached no-load temperatures of 43K and 2.5K, respectively. Figure 4b shows the no-load temperatures for the three cases.

For 1000 W and 750 W the cold head still reaches its nominal no-load temperature although with increased 1st stage temperature in the case of 750 W. For 500 W of electrical input power a no-load temperature of 6K is reached. This would still be sufficient to run a sensor at 10K with some remaining cooling power.

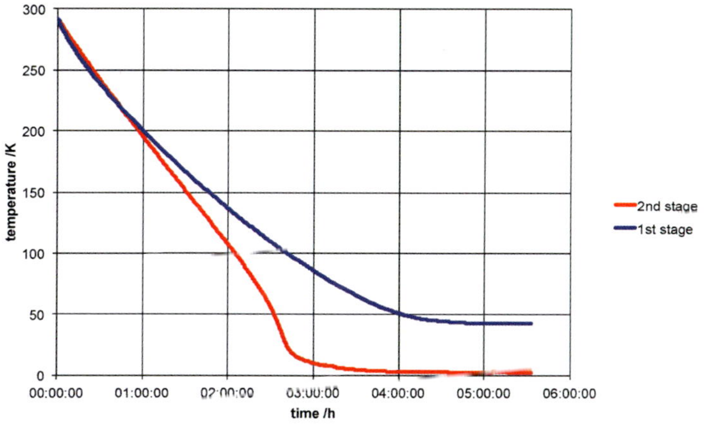

Figure 4a. Typical cool down with 750W of electrical input power

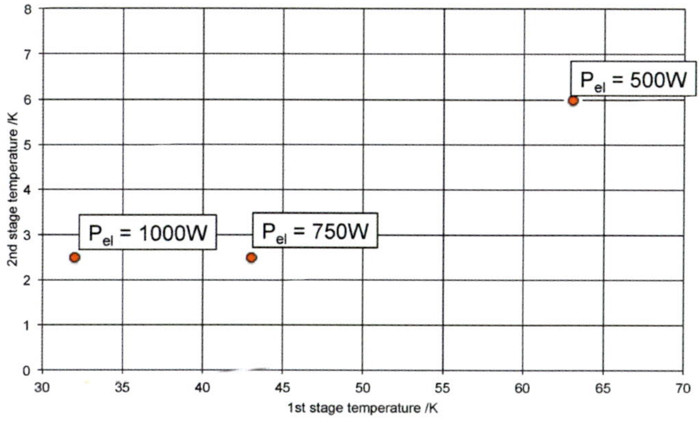

Figure 4b. No-load temperatures for different input powers

CONCLUSIONS AND OUTLOOK

A dry and compact 1 kW compressor has successfully been built and tested with a commercially available GM cold head. Desired base temperatures of around 2.5K could be reached with reduced electrical input power down to 750 W.

The performance results give rise to the assumption that low frequency 4K GM and pulse tube cryocoolers can be run with smaller input powers (e.g. 500 W or less) given the availability of smaller 4K cold heads. This would enable a much broader use of this technology and strengthen the acceptance of cryocoolers in daily use applications.

ACKNOWLEDGMENT

The author would like to thank M. Bühler from Low Temperature Solutions UG, Ismaning, for the many discussions. The challenging ideas and requirements of Vikas Anant from Photon Spot, Inc. have strongly influenced this work. The speed and accuracy in making mechanical parts by G. Töpelt is greatly appreciated. This work was happily supported by WMH, FH and JH.

REFERENCES

1. Sumitomo Heavy Industries, Ltd., Cryocooler Product Catalogue, 05/12

Testing of a Closed-Cycle Cryocooler Compressor for Deployment Aboard SOFIA Aircraft

S. Rosner[1], B. Helvensteijn, A. Kashani, L. Hofland, J. Wang, F. Pichay, R. Rowan, A. Gee, J. Brown, C. Risacher[2]

MEI Company, Moffett Field, CA 94035, USA
[1]SETI Institute, Mountain View, CA 94043, USA
[2]Max Planck Institut für Radioastronomie (MPIfR), Bonn, Germany

ABSTRACT

This paper outlines the regimen of developmental and verification tests performed on a Sumitomo Heavy Industries (SHI) model CSA-71A cryocooler compressor intended for installation in the Stratospheric Observatory For Infrared Astronomy (SOFIA) main cabin to support the upGREAT science instrument.

The first series of tests were performed on a large moving platform at the NASA Ames Research Center's Vertical Motion Simulator (VMS), with low frequency (1Hz) bouncing of the compressor to various g-force levels of 0.1g to 0.5g, at 0.1g intervals, simulating turbulence events characterized during typical missions. The next series involved tilting the running compressor over a range of angles to be expected in flight on board SOFIA. After this sequence the compressor was mounted on a shake table where it was subjected to 3-axes random vibrations from 10 to 2000Hz over a range of g forces.

For the last sequence of tests the compressor was placed inside an environmental chamber to simulate the range of conditions required by NASA for compliance with acceptance criteria established for both mission assurance and airworthiness for flight on board SOFIA.

INTRODUCTION

SOFIA Background

SOFIA is a joint project of NASA and the German Space Agency (DLR). It is an airborne infrared observatory featuring an open port 2.5m gyrostabilized and vibration isolated reflecting telescope integrated into a highly modified Boeing 747-SP aircraft. Flying at altitudes of up to 13.7km (45,000ft), SOFIA is able to make observations above about 99% of the atmospheric water vapor and most clouds, opening a window to the universe in the mid to far-infrared portion of the spectrum that is not available to ground-based observatories, even those on the highest mountaintops.

The SOFIA Program Office is located at the NASA Ames Research Center (ARC) at Moffett Field, CA, while flight operations are nominally based out of the NASA Armstrong Flight Research Center (AFRC) B703, in the Mojave Desert city of Palmdale, CA. An airborne observatory is that it can indeed be strategically deployed to remote locations to enable observations that cannot be achieved from a fixed location, and SOFIA is routinely deployed to Christchurch, New Zealand during the Southern

Cryocoolers 19, edited by S.D. Miller and R.G. Ross, Jr.

Figure 1. The SOFIA Boeing 747-SP, shown here on the tarmac at NASA Ames Research Center (ARC) at Moffett Field, CA, parked just forward of its predecessor, the Kuiper Airborne Observatory (KAO) Lockheed C-141A.

Hemisphere winter to observe targets in the southern skies and to take advantage of the exceptionally cold and dry conditions. SOFIA was deployed to New Zealand at the time of the ICC19 (2016).

SOFIA is the successor to several generations of earlier airborne infrared observatories operated by NASA, which over the years included a Convair CV-990 named "Galileo" (infrared solar and planetary observations from 1967 through 1969), a Learjet with a 12" open port reflecting telescope (1968 through the 1970's), and more recently a Lockheed C-141A Starlifter with a 36" open port reflecting telescope which flew missions from 1974 to 1995 as the "Kuiper Airborne Observatory" (KAO) [1]. See Figure 1 for a photograph showing SOFIA and KAO observatories.

Science instruments that observe in infrared wavelengths must maintain very low and stable temperatures for their focal plane detector arrays, often at or even below 4.2K. While closed-cycle cryocooler (CCC) technologies are ubiquitous in various medical and industrial applications and are commonly used in ground-based observatories. All SOFIA science instruments, like those that flew aboard earlier airborne observatories, have previously relied on expendable liquid cryogens (LN2 and LHe) to cool their cryostats to the required cryogenic temperatures. The SOFIA observatory set the stage for technological advancement towards using a CCC system.

The Science Case for Use of Closed-Cycle Cryocoolers Aboard SOFIA

A number of considerations supported the development of a CCC system qualified for use as an observatory mission system aboard the SOFIA aircraft. These include airworthiness and flight safety issues surrounding the storage and use of volumes of LHe, the increasing cost of LHe, lead-times and logistical complexities associated with procuring and delivering enough LHe to support extended observing campaigns (including deployment to remote bases of operations), and the value of providing infrastructure and support facilities comparable to those to which world-class observatories PI teams and instrumentalists have become accustomed at other.

In addition, the use of CCC systems has enabled the development, commissioning, and use of state-of-the-art detector technologies, such as those being developed and used with the GREAT instrument, designed and built by the Max Planck Institut für Radioastronomie (MPIfR) in Bonn, Germany. GREAT features a modular design that allows two distinct channels, housed in separate and interchangeable cryostats, to be integrated with the telescope on a common structural, optical and electronics "bus" such that they both simultaneously observe the same sky position. Between 2011 and early 2015, four channels were flown and commissioned for use with GREAT on SOFIA, all single-pixel heterodyne THz detectors cooled with liquid cryogens.

The next generation of GREAT channels, appropriately dubbed upGREAT, features a significantly higher number of detector pixels for enhanced observing efficiency, with a corresponding increase in power dissipation from associated active electronics. Volumetric constraints in the modular cryostats preclude use of LN2 and LHe cryogen volumes adequate for hold times at 4.2K as needed to support a full night of observing, further driving the requirement for use of a CCC system.

The upGREAT Low Frequency Array (LFA) channel that was flown and commissioned in May 2015 is a 14 pixel, dual polarization heterodyne receiver array that covers the 1.9-2.5THz range [2]. A 7 pixel High Frequency Array (HFA). which will observe at 4.745THz, is presently in development by MPIfR and is scheduled for commissioning in Fall 2016.

Encouraged by the excellent performance, successful operations, and commissioning of the upGREAT LFA channel and Cryocooler System aboard SOFIA, MPIfR is presently also developing another next-generation 4GREAT cryostat, which will use CCC pulse tube technology, and when combined with the upGREAT HFA channel, will replace all four of the 1st generation liquid cryogen cooled channels. All of these upGREAT and 4GREAT channel developments were made possible by the adoption of CCC technologies, and the development and integration of the enabling CCC infrastructure aboard the airborne SOFIA observatory.

The MPIfR upGREAT team specified a customized two-stage pulse tube cold head from transMIT GmbH (Giessen, Germany), based on its heat sinking capacity, thermal performance with relative insensitivity to tilt angles (important due to normal operational variations of the SOFIA telescope elevation angle), high reliability, and low vibration levels imparted to the instrument cryostat.

NOVEL AIRBORNE APPLICATION OF CCC SYSTEM FOR ASTRONOMY

Development of Closed-Cycle Cryocooler Infrastructure for SOFIA

In close coordination with the SOFIA Science Instrument and Mission Systems development groups, candidate CCC He compressors were evaluated to drive the transMIT pulse tube cold head, with the primary consideration being adequate integrated system thermal performance within a total power budget (for 2 compressors) of 20kVA (3-phase, 400Hz) aircraft power. Several trade studies were conducted, evaluating use of air-cooled versus water-cooled compressors, and use of 60Hz compressors with external frequency converters versus R&D needed to produce a 400Hz compressor. An air-cooled, 60Hz CSA-71A compressor from Sumitomo Heavy Industries (SHI) was selected, despite the fact that this compressor is qualified by SHI for use only with their own Gifford-McMahon (GM) technology cold heads.

The MPIfR upGREAT team already had experience using this compressor in their laboratory as well as in some high elevation ground-based observatories, but significant additional developmental testing was needed to show that it would perform acceptably in our dynamic airborne environment. For example, the SHI documentation included several cautionary statements advising against the use of the compressor downstream of inverters (e.g., frequency converters), and compressed He line runs approaching 40m, considerably longer than the standard 10 to 20m lengths, would be necessary, due to constraints surrounding the location of the compressor in the aircraft cabin, and the need for the flexible lines to traverse a cable wrap with an adequate "service loop" to allow the full range of motion of the telescope.

Typical industrial applications for CCC systems are relatively "static" with the compressors either on casters for portable use on level floors, or mounted to a level concrete pad for a more permanent installation. The technical documentation for the selected SHI compressor specifies a maximum tilt of ±15° for storage / shipping, and only ±5° during operation. In contrast, the SOFIA 747-SP aircraft routinely sees pitch angle variations from about 1° nose down while on the tarmac, to nearly 20° nose up during take-off. The accelerations associated with take-off and landing, as well as turns during ground maneuvers, also impart resultant effective tilt angles that must be considered for this dynamic application. Roll angles, e.g., during banked turns, though routinely higher than the pitch angle variations, are not generally an issue due to the dynamic properties of "coordinated turns," in which centripetal acceleration effectively nulls the roll angle. Furthermore, an aircraft mounted compressor needs to be able to withstand shock loads, such as those during landing, as well as vibrational loads and airborne turbulence that are well outside the nominal operational environments envelope for these compressors.

Coordination with SHI technical staff indicated that their primary concerns associated with a scroll compressor in such operating conditions were, overheating of the sealed compressor/motor capsule resulting in possible permanent damage, temporary reduction in compressor performance, and reduction in effectiveness of the oil mist separator, leading to loading of the oil mist adsorber canister with potential break-through and migration into the downstream lines. Oil fouling of the cold head, though deemed a very low probability, would almost certainly curtail a flight campaign and require extensive and time-consuming clean-up, and therefore carried a high enough consequence to warrant study. Armed with these concerns, the development team based at NASA ARC devised a development analysis and testing regimen to inform the design, as well as features to mitigate environmental perturbations and associated risks to mission success.

The first step involved characterization of the dynamic environment. Flight data (1Hz rate) representing both nominal and exceptional flight conditions was retrieved from the SOFIA archives, and analyzed to determine quasi-steady load factors on all 3 axes, as well as the resultant effective tilt vector angle. High frequency triaxial vibration measurements were acquired and analyzed to determine both the composite g_{rms} acceleration levels, as well as the power spectral densities (PSDs) from 1 to 2000Hz, characterizing the frequency content and amplitude of shock and turbulence loads. Careful review of these measurements and analyses led to a compressor mount design featuring cable coil vibration and shock isolators, as well as a damped pitch-axis gimbal to keep the compressor level regardless of the attitude of the aircraft and longitudinal accelerations during take-off and landing events.

Developmental Test Hardware and Testing Regimen

An existing test cryostat with its set of internal radiation shields was modified in order to accommodate the geometry of the mounting flange and heat sinks of a SHI RDK-408D2 GM cold head which was procured with the compressor. The headplate of the cryostat was outfitted with wiring feedthroughs and pressure relief valves. Four LakeShore diode thermometers (1 calibrated, 3 uncalibrated) were installed on the 2nd stage heat sink, and wire-wound heating elements were installed on the 1st and 2nd stages of the cold head.

The data acquisition system (DAS) was configured with a data rate to support statistical analyses (Mean, Standard Deviation, Skew) and anti-aliasing, to ensure that temperature oscillations caused by low frequency pressure cycling of the cold head could be effectively filtered out as noise while quantifying the non-oscillatory temperature trends.

This test setup was designed to allow the thermal performance of the integrated CCC system to be characterized and monitored with varying thermal loads, while subjecting the SHI CSA-71A compressor to the regimen of environmental tests organized into test sequences and test cases. In addition to monitoring of the integrated system thermal performance, the internal oil mist adsorber within the SHI CSA-71A compressor was removed and weighed between the test sequences to determine if and to what extent these dynamic stimuli can be correlated with increased rates of oil capture.

The details of these test operations and sequences are documented in several formal test procedures and test reports and are not repeated here but rather summarized in the following paragraphs.

Following initial setup and preliminary baseline performance test runs in the NASA Ames Cryolab, the compressor and test cryostat with the integrated cold head was transported to the Vertical Motion Simulator (VMS), with its large moving platform capable of simulating the low frequency (~1Hz) "bouncing" motion characteristic of typical flight turbulence or "chop." During this test, the He compressor was subjected to simulated turbulence ranging from 0.1g to 0.5 g in 0.1g intervals, for durations of 20 minutes at each level. At the higher g levels, the motionwas applied in bursts, each limited to 10 seconds, to avoid exciting resonance at the natural frequencies of the VMS facility structure and building.

The next series of test sequences back in the Cryolab was designed to study the effects of 2 axes tilting on compressor and integrated CCC system performance. In this test, the thermal performance of the system was measured while the compressor was tilted in both lateral and fore-aft directions (corresponding to aircraft pitch and roll axes) up to the maximum specified tilt level of ±5°.

Random vibration testing of the compressor – now mounted to its shock isolated pallet structure was conducted in the NASA ARC Engineering Evaluation Laboratory (EEL), in accordance with

Figure 2. The instrumented test cryostat with the SHI RDK-408D2 GM cold head is shown in the foreground of the left panel, with the SHI CSA-71A He compressor shown in the background on the moving platform of the NASA ARC Vertical Motion Simulator (VMS) to simulate low-frequency turbulence events. In the right panel, the compressor is shown mounted via its shock isolation mounts to the vibration table in the NASA ARC Engineering Evaluation Lab (EEL) configured for Z axis random vibration environmental acceptance testing. (Not shown here are the test setups for the tilt evaluation testing or the thermal vacuum environmental acceptance testing.)

NASA AFRC airworthiness and mission assurance environmental acceptance standards for equipment mounted in the pressurized cabin of a transport category turbojet aircraft such as the SOFIA 747-SP. During this test, the compressor was subjected to random vibration for 20 minutes in each primary axis, with a composite level of $0.78g_{rms}$ and spectral content per a prescribed PSD profile ranging from 10 to 2000Hz in all axes, and the full -6dB attenuation factor allowed for equipment with mass of $\geq$ 160lbm ($\geq$ 72.7kg).

Figure 2 shows the test setups for the low-frequency turbulence developmental testing on the VMS platform, and Z-axis random vibration environmental acceptance testing on the EEL vibration table.

Finally, thermal-vacuum chamber testing was conducted as a formal verification to assure compliance with NASA AFRC environmental acceptance airworthiness and mission assurance standards. During this portion of the testing, the compressor was operated within a test chamber capable of the specified range from 0°F to 160°F, and pressures corresponding to altitudes from sea level to 50,000ft. Based largely on concerns voiced by compressor manufacturer SHI related to the operating and storage temperature specifications of some electrolytic capacitors within the compressor's internal power supply, the Cryocooler System project successfully pursued a deviation from the AFRC standard, such that the compressor was to be subjected to only 130°F, while considerably lower than the nominal upper limit of 160°F is still quite high relative to even the highest temperatures that could reasonably be expected aboard SOFIA, even considering extended daytime ground operations out on the flight line during the hot Mojave Desert summer. Because the SOFIA power bus powering the Cryocooler System is protected by a decompression relay that immediately disconnects power in the event of a cabin decompression event, the compressor was shut off for test sequences in which the pressure altitude was ≥ 20,000ft. Given the importance of cooling air mass flow for rejecting the waste heat from air-cooled compressors, particularly when operating at elevated ambient temperatures compounded by the reduced pressures characteristic of a pressurized aircraft cabin at cruising altitudes, some additional attention was focused on the effects of partial obstructions to the cooling air inlet and outlet vents.

Though beyound the scope of this paper, it is also worth mentioning that in addition to the environmental acceptance testing outlined above (colloquially referred to as "Shake & Bake" testing), low-level (0.125g) 3-axes 5 to 150Hz sine sweep modal surveys of the compressor on the vibration table were also conducted, with the compressor/motor capsule and other heavy subassemblies instrumented with accelerometers to identify any resonances, with particular attention to any natural frequencies and resulting magnifications below 20Hz, as excitation of these lower frequencies are not effectively attenuated by the selected shock and vibration isolation mounts. Several structural modifications were made to stiffen such subassemblies and raise their natural frequencies to above 20Hz. Additional proof and burst pressure testing of pressurized systems was also conducted to satisfy FAA qualification and acceptance criteria for pneumatic systems.

Summary of Results, Discussion and Conclusions

As with the details of the various test procedures and sequences, a quantitative analysis of the measurements and statistical parameters is outside the scope of this paper but are summarized in the paragraphs below.

One very important finding was that despite the concerns voiced by SHI technical staff, the operations and thermal performance of the integrated CCC system appeared to be largely insensitive to most of the dynamic and environmental conditions to which the compressor was subjected during this testing regimen. The low frequency simulated turbulence, compressor tilting, and 3-axiesrandom vibration testing all resulted in barely detectible performance effects, considering the temperature readout precision and noise.

The one test in which a statistically significant perturbation in thermal performance was noted (manifested as a readily measurable increase in the mean temperature of the cold head 2nd stage heat sink) was the environmental acceptance testing of the compressor operating in the thermal vacuum chamber. However, correlation of the observed temperature increases with the variations in positioning of the compressor within the constrained interior dimensions of the test chamber supports the conclusion that even modest obstructions in the compressor's cooling airflow had a more significant effect than either the ambient temperature or pressure on the thermal performance of the integrated system.

Additional mitigating features implemented in the design include augmented instrumentation of the compressor to enhance visibility into the nominal operating temperatures and pressures and identify any trends that might indicate an incipient failure, coupled with a programmable logic controller (PLC)-based control system to monitor these conditions as well as housekeeping signals from the compressor, with programmable warning and alarm thresholds for these parameters, and logic to shut down the compressor if deemed necessary or prudent. An external oil mist adsorber canister and molecular sieve filter were installed in the high pressure (supply) line, as well as a sight glass between the internal and external adsorbers, to provide a visual indication of any oil mist that somehow made it past the internal adsorber to migrate outside of the compressor.

Though some of these tests and mitigations may well have erred on the side of caution, they provided the development team, the SOFIA Program, and PI team stakeholders the confidence to proceed, as well as quantitative dynamic limits which were used to establish compressor cut-off thresholds that were programmed into the PLC-based controller.

The result is a robust and reliable system that, despite operations well outside the nominal, meets specified environmental conditions, has now performed flawlessly on several flight series, including the successful commissioning of the upGREAT LFA channel and several observing campaigns. The Phase 1 Cryocooler System is presently aboard SOFIA in Christchurch, New Zealand for its first Southern Hemisphere deployment, where it continues to operate flawlessly in support of observations with the upGREAT LFA channel. Figure 3 depicts the Phase 1 Cryocooler System and the GREAT instrument (with the upGREAT LFA channel) installed and operating aboard SOFIA.

Figure 3. The Phase 1 Cryocooler System SHI CSA-71A He compressor shown installed here in the foreground on the SOFIA main cabin in its shock isolated pitch-axis gimbal mount, with the GREAT Science Instrument in the background, mounted to the SOFIA Telescope instrument mounting flange. The Programmable Logic Controller (PLC) based Cryocooler System control system with the touchscreen Graphical User Interface (GUI) is mounted to the guard rail surrounding the gimbal mounted compressor.

At the time of this publication, the SOFIA Program is investigating options for development of a Phase 2 CCC system, with two He compressors which will support simultaneous and parallel operation of two cold heads, and allow concurrent operation of both the upGREAT LFA and HFA channels, as well as the upGREAT HFA and 4GREAT channels. Unlike the Phase 1 Cryocooler System, which is installed aboard SOFIA temporarily when needed by a Science Instrument, the Phase 2 Cryocooler System will represent a permanently installed observatory mission system.

It is notable that both of the candidate 3rd generation SOFIA Science Instruments selected by NASA early in 2016 for competitive Phase A Instrument Concept Studies and follow-on development to support integration, test and commissioning flights planned for summer of 2018 will also use CCC technology in lieu of expendable liquid cryogens for cooling to cryogenic temperatures.

ACKNOWLEDGMENT

The authors would like to acknowledge and thank the SOFIA Program Office and staff at both NASA ARC and NASA AFRC for their financial and technical support of the development of this pathfinder airborne CCC system. We would also like to recognize the German Space Agency (DLR) and the Deutsches SOFIA Institut (DSI) for their partnership in this endeavor, hardware procurement, and fabrication support.

REFERENCES

1. W. W. Dolci, "Milestones in Airborne Astronomy: From the 1920's to the Present," *AIAA,* paper no. 975609, (1997).
2. C. Risacher et al., "First Supra-THz Heterodyne Array Receivers for Astronomy with the SOFIA Observatory," *IEEE Trans. on THz Sci. and Technol.*, vol. 6, issue 2 (2016), pp 199-211.

A Low Maintenance and Vibration 4 K GM Cryo-Refrigerator with Independent Variable Speed and Valve Timing

D. Snow, J. Lawton, R. Taff, J. Diederichs, J. Gardiner, M. B. Simmonds, J. Hamilton, S. Spagna

Quantum Design Inc.
San Diego, CA 92121

ABSTRACT

Over the course of developing a small scale cryogen-free physical property measurement system (VersaLab™) we have learned a great deal about the strengths and weaknesses of commercially available Gifford-McMahnon (G-M) cryocoolers with regard to our specific application. This has motivated us to develop a novel low vibration G-M cryocooler with a nominal cooling power of 0.25 W at 4.2 K. The system features an independently controlled valve assembly providing dynamic control of the intake and exhaust cycles of the refrigeration gas exchange between the cryocooler and the compressor. Some unique features of the valve assembly design are its significantly lower power consumption and its low frictional forces; which translates into improved durability and component longevity over existent designs. This, combined with the ability to dynamically control the frequency and the motion profile of the displacer and the precise control of the gas flow functions (independently of the displacer mechanics) allows for optimization of the cryocooler performance with regard to cooling power, efficiency, and vibration levels. Additionally, all moving components and seals have been designed for very low wear. Collaborations with other manufacturers and researchers have also provided us with insights into additional optimizations that may be desirable for other cryogenic instruments and platforms.

INTRODUCTION

The VersaLab™ Physical Property Measurement System (PPMS©) provides for a compact, cryogen-free temperature/magnetic field platform (50K -400K and +/- 3 tesla) that can host a variety of integrated measurements for the characterization of the physical properties of materials [1]. These include DC and AC magnetometry, heat capacity, thermal and electronic transport, as well as numerous possibilities for custom experimentation. The VersaLab™ is a completely cryogen-free system that uses less than one "K" type helium gas cylinder per year for sample temperature exchange and has a small 1 square meter footprint for the entire system (cryostat and air-cooled compressor). Since its inception, it became apparent to us that the "orientation-free" operation of G-M refrigerators could be used advantageously to minimize the size of the VersaLab™ cryostat, making it virtually portable. In addition, the availability of commercial G-M cryocoolers; with a smaller physical size and reduced cooling power (nominally 0.1 W at 4.2 K), perfectly balance the size and weight requirements for a portable system while still maintaining an adequate cooling budget to operate the cryostat's cryogenic sub-systems (i.e. superconducting magnet, sample chamber and cryopump). Figure 1 highlights the cooling strategy adopted in the VersaLab™ cryostat: the second stage of the cryocooler (Figure 1 (a)) provides cooling to operate a small 3T Niobium-Titanium (NbTi) conduction cooled superconducting solenoid, while the first stage (Figure 1 (b)) services a variable temperature sample environment from 50 K to 400 K. The reduced power consumption of the helium compressor; when used in conjunction with these compact cryocoolers make it an obvious choice for designing a cryostat that would meet most common physics and materials science undergraduate laboratory space and power requirements while maintaining the maximum flexibility in performance.

Subsequently it became apparent that commercially available G-M cryocoolers systems of this type were essentially a scaled-down and economical version of their larger and more powerful industrial grade coolers that typically deliver 1.0 to 1.5 W at 4.2 K. In practice, this heritage has proven to be problematic in achieving critically important requirements (of sensitive measurement instrumentation) such as high temperature stability, low vibration, low-power consumption, and long life. In general, G-M cryocoolers have earned their nickname "thumper" due to the incoming rush of gas that slams down through the displacers and consequently pushes on the drive rod. This combination of forces exercises a backlash in the mechanical linkages connecting the motor, scotch yoke, and displacers to create a characteristic thump or clack. The mechanical impulse is rich in harmonics, creating a comb of noise peaks roughly 1 Hz apart in the vibration spectrum. In the past, the valves that control the helium gas inlet and exhaust functions are actuated by cams mounted on a rotating crank that leave little opportunity for adjustment and timing. These adjustments may be employed in strategies to reduce the forces on the displacers and/or optimize the cooling provided in each cycle. Since the forces in the scaled-down version of the G-M cryocoolers are substantially lower than their larger counterparts we felt that a greater reduction of vibrations could be possible if the gas flow, linkages and bearings could be made to operate more smoothly and in concert with one another.

In this paper, we describe a novel "instrument" grade G-M cryocooler driven by a variable speed "energy smart" compressor [2, 3] that is specifically designed for the VersaLab™ and similar laboratory apparatus where energy efficiency, long life and low vibrations are of interest. Notable features of the cryocooler are the independent computer control of the displacer movement and a specially designed "spool" valve adapted for dynamic control of the intake and exhaust flow of the helium within the cryocooler. The valve is adapted for resilient and durable operation over an extended period of time. We show that this enhanced control can be used to beneficially optimize the cryocooler performance with regard to cooling power, efficiency, and vibration levels.

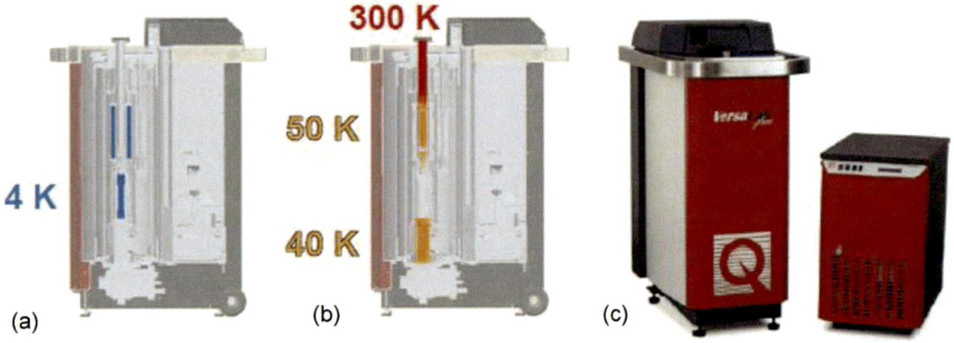

Figure 1. (a) A 3 T superconducting magnet is cooled by the 4 K stage using solid conduction cooling through high conductivity copper, (b) a thermal switch connects the sample chamber to the 40 K stage and cools the bottom of the sample chamber down to 50 K, (c) the complete system is shown with the small variable speed compressor for the G-M cryo-refrigerator.

The variable speed compressor can; by varying the scroll capsule speed, significantly reduce the overall power consumption for the system and deliver cooling power on "demand" according to the dynamic requirements (i.e., heat loads) within the system. For instance, when first cooling down the VersaLab™, the compressor and cryocooler will run at high speed to deliver maximum cooling power to the first and second stage. Then; once the system is cold and measuring a sample at a fixed temperature and magnetic field, the compressor and cryocooler will run at "low" speed to deliver minimal cooling power to the stages as required by the parasitic heat loads in the system. This saves wear on both head and compressor while reducing power consumption.

CRYOCOOLER SYSTEM DESCRIPTION

Figure 2 shows the physical configuration of the GA-1 cryocooler and details of the internal construction of the 2.5 kW class HAC900S variable speed compressor. The GA-1 is a two stage cryocooler that uses the G-M refrigeration cycle. The first and second stage cylinder dimensions are approximately 46 mm and 20 mm in diameter and 129 mm and 115 mm in length respectively. The displacers are driven by a synchronous 5-phase motor that is controlled by a computer controlled Pulse Width Modulated (PWM) circuit that feeds a half-bridge driver for each phase of the motor. The "spool" valve includes a linear armature driven by an induction motor, whose position and speed can be tracked and changed using microprocessor-based electronics. The valve assembly provides key benefits over previous designs. For instance, the armature is "floating" within a shuttle valve cylinder, which reduces the frictional contribution to power consumption, working lifetime and vibration. In general, the valve consumes a mere 0.1 W as opposed to the more conventional rotary valves that require about 20 W to operate. The 1st stage flange provides a cooling power of 4 W at 40 K and the 2nd stage flange provides a 0.25 W cooling capacity at 4.2 K. The displacers are constructed of phenolic and stainless steel, respectively, and filled with optimized amounts of phosphor bronze screens, lead and rare-earth materials.

Figure 2 (b) shows the details of the HAC900S. This compressor is air-cooled and runs on a 200-240 VAC 50/60 Hz power. Power consumption ranges from 800 W to 2 kW (~0.65 P.F.) depending on the capsule speed and the supply/return helium pressures. A commercially available inverter drives the compressor scroll capsule and a specially designed 5-phase inverter drives the cryocooler motor. Since the capsule motor is started slowly by the inverter, there is essentially

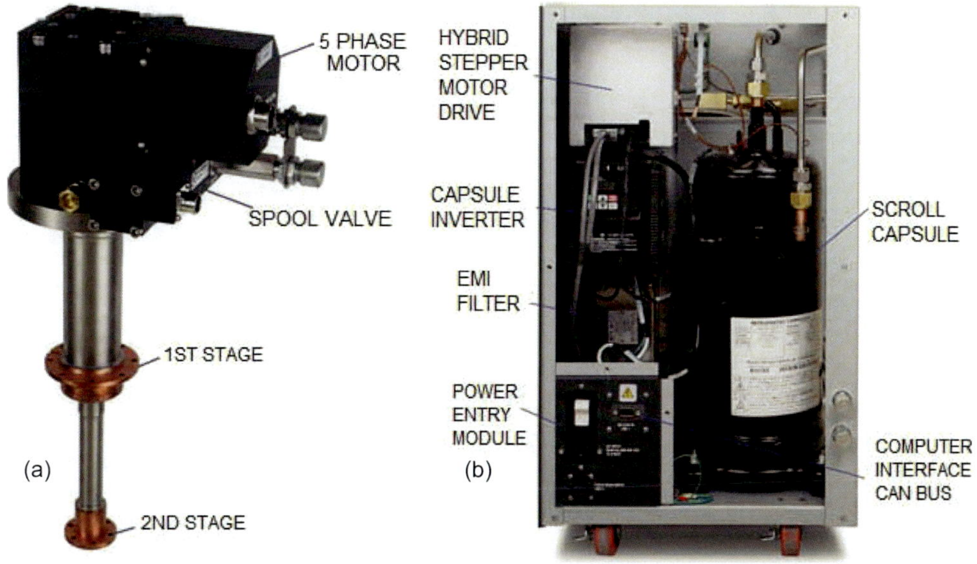

Figure 2. (a) The GA-1 cryo-refrigerator and, (b) details of the HAC900S variable speed compressor.

no surge-current or mechanical shudder at power-on. Ambient air is drawn into the front section where is cools the oil-coalescing filter and charcoal adsorber before being forced over the radiator coils, capsule, and inverters. Air exhaust, hose fittings, and electrical connectors are all at the rear panel. This configuration provides optimal cooling for the components that remove oil and vapor from the helium. Other features of the HAC900S compressor are all-stainless construction, oversized charcoal adsorber, electronic pressure, temperature, and oil-level sensors, and sophisticated firmware to control the state-engine and detect/report subtle fault conditions. The compressor has a full-featured Control Area Network (CAN/CANOpen) bus interface so that it integrates fully with the other modules in our instrument. We have also provided the hardware for an RS-232 interface to support OEM applications and a Software Development Kit (SDK), which enables third party programs to control compressor speed and query the HAC900S for diagnostic data.

Convenient speed control is the great side-benefit of this electronic drive system. We have found that the GA-1 cryocooler motor can be dynamically operated with input frequencies between 30 Hz and 70 Hz to achieve different system goals. At the low-frequency end, we can implement a "stand-by" mode that will extend the product lifetime and reduce power consumption. The system can be operational again after only a few minutes rather than many hours. At the high-frequency end, we can roughly double the cooling capacity of the head for transient high-load conditions such as rapid cooling of the sample-chamber in our VersaLab™ instrument. To properly accommodate such changes to the head speed, the compressor itself must be able to adjust its gas output over a range of at least 2:1.

Independent Refrigeration Cycle Control

Figure 3 depicts the principle of operation of the independent refrigeration cycle control. The displacers are driven with a reciprocating drive rod by use of a scotch yoke. The exact position of the displacer stroke and valve are independently operated by electric motors and tracked by two separate encoders. In practice, the displacer position provides for a "clock" signal of the refrigerator cycle providing a reference signal to be used by programmed electronics for the valve control. Figure 4 shows a typical oscilloscope waveform of the displacer and valve positions in a typical refrigeration/heat cycle. The electronics dynamically control the inlet/exhaust timing, dwell and

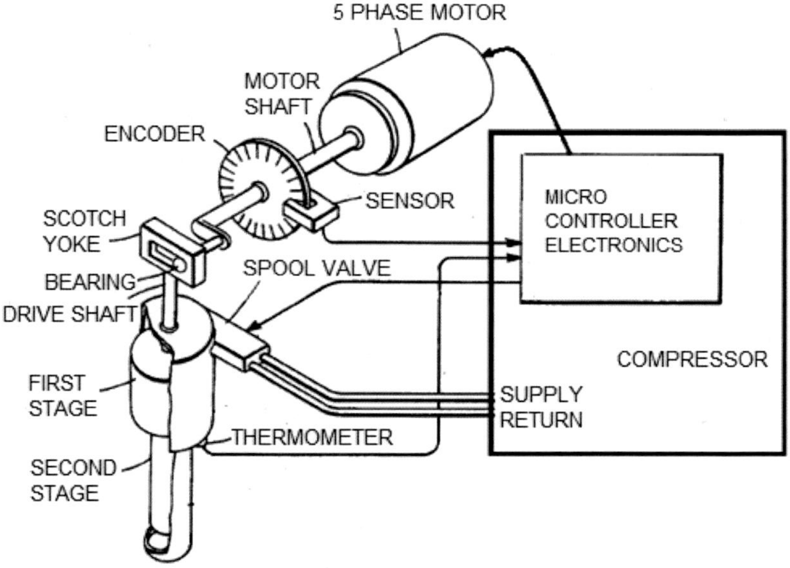

Figure 3. Control systems for the independent refrigeration cycle.

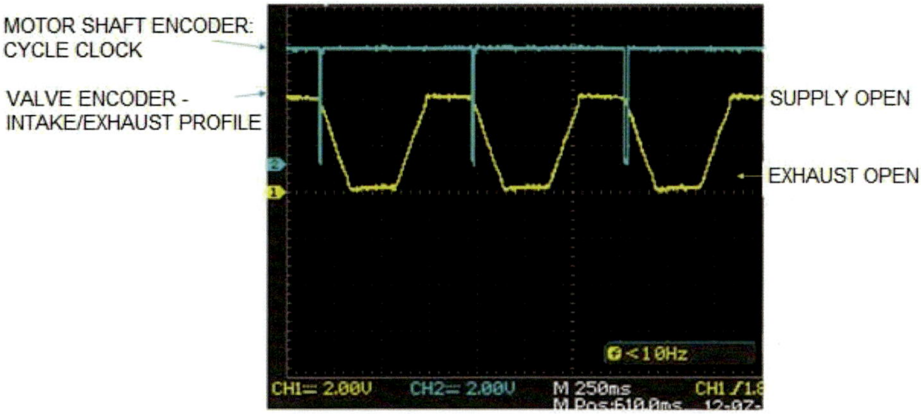

Figure 4. Independent control of the refrigeration/heat cycle.

flow areas relative to predetermined positions of the displacer. Unlike previous mechanisms, the valve can supply high pressure helium gas from the compressor to the cylinder and exhaust the helium gas back to the low pressure side of the compressor "on the fly" and in a very precise way, depending on the cooling requirements of the VersaLab™. This capability enables the system to be optimized with respect to cooling power and with added precision. Pressure changes within the cryocooler can be managed such that common problems including noise and losses due to fluid-friction are minimized. Such a valve, not being dependent on displacer mechanics, may also prove to be useful with modern pulse-tube cryocoolers.

EXPERIMENTAL RESULTS

Agile Temperature Control

The dynamic control of valve timing using a microprocessor allows for the GA-1 cryocooler to provide cooling and warming of the first and seconds stages in a "on-demand" basis. This form of temperature control is attained without the use of external heaters. In practice, this is accomplished by reversing the phase of the intake and exhaust functions during the refrigeration cycle. This feature is particularly desirable if a fast cycle of the cryocooler system is required for the regeneration of the cryo-pump at 77 K and subsequent cool down. Previously this might require the entire cryocooler system to the shut down for a several hour warm up. In our system, the cycle can be achieved in 30-40 minutes. Figure 5 shows data warming and cooling curves attained by varying the Phase Delay (PD) at a high speed. With no heat load, it takes 70 minutes for the first stage to reach 40 K and about 60 minutes for the second stage to achieve temperatures slightly below 2.5 K. From the base temperature, heating of the first and second stages can be simply achieved by varying the phase delay of the intake and exhaust functions during the refrigeration cycle. The data shows that different warming rates can be achieved by solely changing the phase delay at a fixed speed.

Load Maps

The performance of the GA-1 cryocooler driven by the HAC900S compressor was characterized by a number of conventional load maps with heat inputs up to 4 W applied to the first stage and loads up 0.2 W applied to the second stage. Figure 6 reveals the rich load map landscape obtained using different compressor and cryocooler input power frequencies. The data is taken at different cold head and capsule speeds. Of interest in this data is the exquisite control of the first and second stage cooling capacities as a function of the input frequency power for the capsule and cold head. During the "High" cooling mode the first stage cooling of the refrigerator is optimized to its lowest temperature. This mode of operation is particularly important in a liquefaction system where the gas to be liquefied needs to precooled before it can be liquefied at the second stage. The

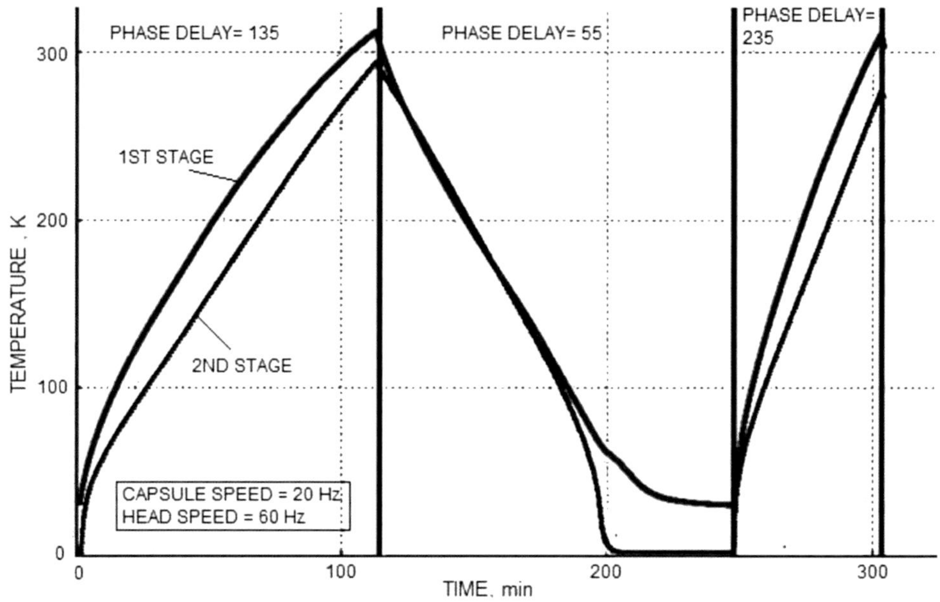

Figure 5. Warming and cooling curves.

"Low" cooling mode on the other hand maintains low temperatures at the second stage but reduces the cooling power at the first stage. This mode can be used for reducing power consumption, when the system is in "Stand-by" or idle.

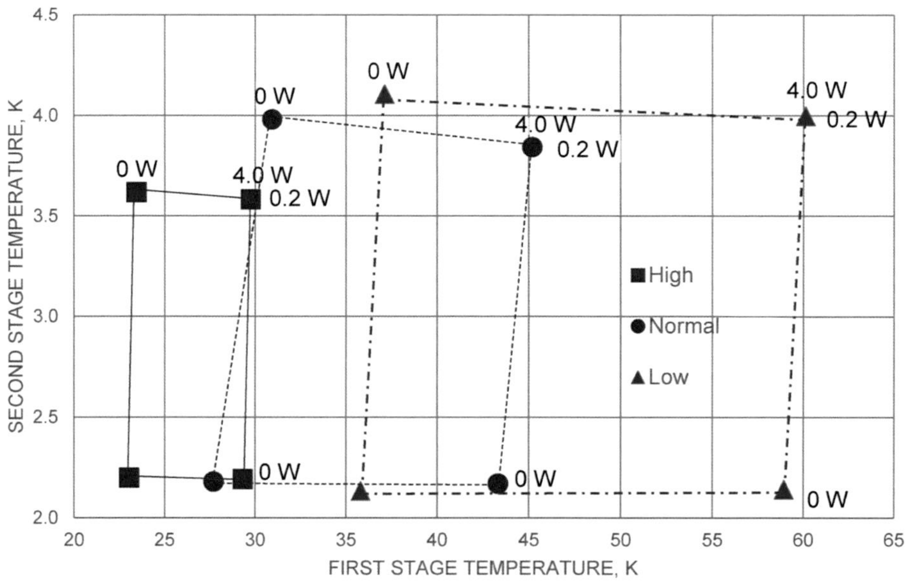

Figure 6. Cooling load maps of the two stage cold head with the HAC900S compressor.

Power Draw

Variation of the power draw versus input frequency is shown in Figure 7. The measurements were performed using a 3-phase Fluke 437 digital meter at three capsule speeds 30 Hz, 18 Hz and 13 Hz. The data shows that the power draw is dependent on the compressor capsule speed and to a lesser degree on the differential pressure (Supply – Return Pressure) of the compressor. At 30 Hz the average power consumption is approximately 2 kW, while at 13 Hz is closer to 1 KW. For a given speed during cool down was about 0.3 MPa larger than the differential pressure once the system was cold at its base temperature leading to a power reduction of typically 0.5 kW. In the VersaLab™ these lower speeds are attained when the magnetic field from the superconducting magnet is zero, and the temperature is stable. This idle state provides the user with substantial power savings over the course of even one year.

Vibration Spectrum

The vibrations of the G-M cryocooler were measured and compared to a commercial off-the-shelf unit of the same size (see Figure 8). To ensure accurate measurements of the vibration spectra, the cryocooler was isolated from external vibrational sources and care was taken to eliminate any dampening effects that the fixture itself might cause during the measurement. We accomplished this by forming a "crack" resistant concrete block weighing approximately 250 lbs standing on four high crush vibration dampening feet. Within the block is a cylindrical hole that accommodates the outer vacuum jacket; employed in load map experiments, without contacting the inner walls of the concrete. To further isolate the cryocooler from vibrations in the floor we bolted it to an aluminum plate through the four screws that normally hold the vacuum jacket. This plate rests on top of the concrete block separated with four Iso-damp grommets. The geophone used is the Geospace GS-11D 4.5Hz [4] and it is securely attached to the cryocooler at the top cover plate. This fixture secures the geophone by use of a Delrin® cap that is bolted to an aluminum base with four screws that are finger tightened to inhibit its ability to move within the fixture but without compressing the housing of the geophone. It is then bolted to the cryocooler using two of the screws holes that are used to hold the top plate. The geophone is wired to a 100x preamp through a shielded twisted pair cable

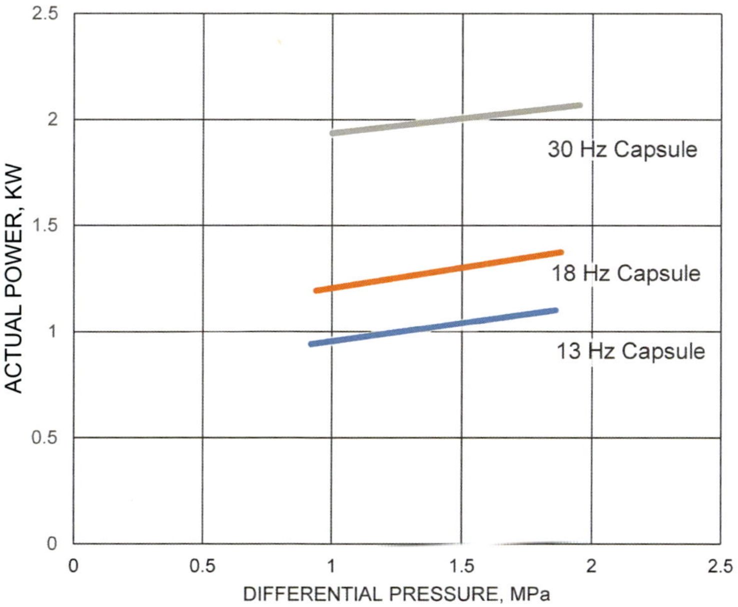

Figure 7. Power draw measurements of the GA-1 and HAC900S compressor.

and from the preamp to the spectrum analyzer with a standard BNC cable. The spectrum analyzer measurements are taken with a HP3561A in 200 Hz sections from 0-1kHz at 10 RMS; resulting in a bandwidth of 0.5Hz. The calculation to convert from dbV (from the spectrum analyzer) to meters of displacement is a transfer function that mimics a critically dampened spring and was verified at multiple frequencies as well as frequencies below its natural frequency to confirm our low frequency measurements (at 2 and 3 Hz). With the cryocoolers being driven at a 60 Hz motor input frequency, the vibration spectra indicate that the GA-1 has a lower vibration spectrum by at least one order of magnitude up to 1 KHz (gray curve). Experiments currently in progress, further indicate that the vibrations will continue to decrease at lower speeds.

Wear and Refurbishment

In general, the official recommended interval between head refurbishments is 10,000 hours for most commercially available G-M cryocoolers. This is only 14 months and refurbishing costs are typically several thousand dollars. This is one of the reasons, we operate the cooler in our VersaLab™ at reduced drive levels. It has been our experience that these heads typically run substantially longer than this period, even at standard drive levels. Unless down-time must be avoided at all costs, it is generally acceptable to run a head until it either begins making disturbing noises or fails to provide adequate cooling power.

We have looked at the typical wear patterns of several G-M cryocoolers and found several common themes. The greatest wear usually happens in the head, not in the displacer seals.

The plastic bushing on the motor eccentric (that drives the scotch yoke) shows the highest wear of all parts and this directly creates a loud clacking noise in the head (see previous comments about linkage slop). The next major wear items are the slide-bushings that constrain the scotch yoke to move vertically. As these wear, the drive rod moves transversely and tears up the gas seal that is riding on it.

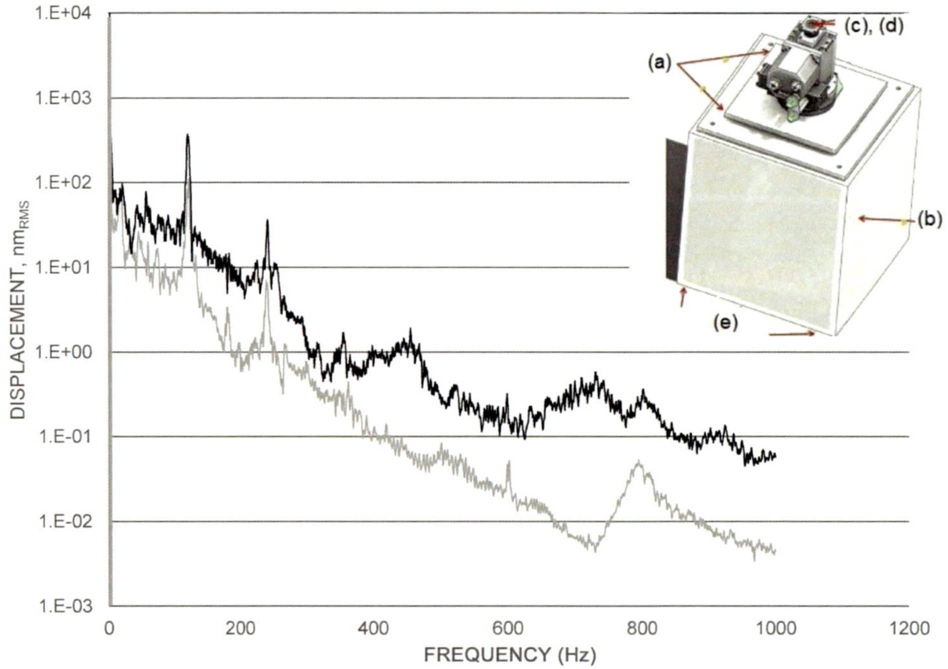

Figure 8. Vertical vibration spectra for the GA-1 cryo-refrigerator (Grey) and an "off the shelf" commercial cryo-refrigerator (Black). (Inset) Vibration test rig comprising of: (a) Iso-damp grommets, (b) Concrete block, (c) Geophone GS-11D, (d) HP3561A Spectrum Analyzer and, (e) Vibration isolation feet.

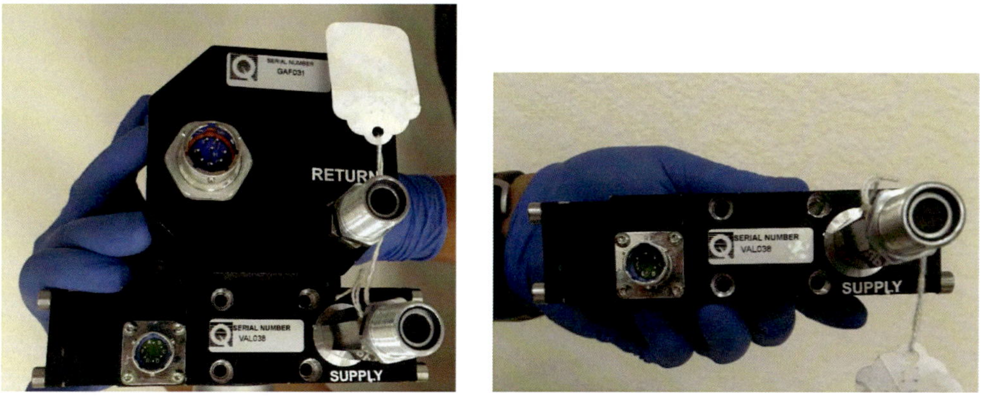

Figure 9. Replacement of the valve mechanism in GA-1 is accomplished by simply venting the system and unbolting four bolts.

In the GA-1 all the gas seals have been designed to provide a minimal amount of friction during operation. In addition, one of the benefits of an independent valve assembly, which is not mechanically linked to the displacer motion, is the ease of servicing the unit. Figure 9 shows the removal of the GA-1 spool valve by venting the cryocooler and unbolting four bolts that mate the valve to the main housing.

CONCLUSION

In this paper, we described a novel cryocooler and "smart energy" compressor system that dynamically controls the intake and exhaust functions of the refrigeration cycle. This enhanced control allows for optimizing the operation of the cryocooler system with respect to vibrations, cooling power, efficiency and service life.

ACKNOWLEDGMENT

We would like to thank Rich Reineman at GWR and Alex Woidtke at S2 Corportation for insightful discussions about the GA-1 cryocooler.

REFERENCES

1. Model Physical Property Measurement System (PPMS) VersaLab manufactured by Quantum Design Inc., San Diego, CA. For information and curricula, visit: education.qdusa.com.
2. Sayles T., Martien D., Diederichs J., Sloan J., Bird D., Spagna S., "Enhanced Helium Compressor Operation for sensitive Measuring Instrumentation," *Cryocoolers 16*, ICC Press, Boulder, CO (2011), pp. 371-373.
3. Chialvo C., Sayles T., Diederichs J. Spagna S., Gabal M., Sesé J., Rillo C., "Smart Energy Compressors for Helium Liquefiers," *Cryocoolers 18*, ICC Press, Boulder, CO (2015), pp. 551-554.
4. Model GS-11D manufactured by Geospace Technologies, Houston, TX.

Development Status of a High Cooling Capacity Single Stage GM Cryocooler

Q. Bao[1], M. Xu[1], and K. Yamada[2]

[1]Technology Research Center, Sumitomo Heavy Industries, Ltd.
Nishitokyo-city, Tokyo, Japan 188-8585
[2]Precision Equipment Group, Sumitomo Heavy Industries, Ltd.
Nishitokyo-city, Tokyo, Japan 188-8585

ABSTRACT

Continuous improvement of high temperature superconducting (HTS) materials has lead to a thriving development of various applications including superconducting motor, superconducting power transmission cable and superconducting power generator. Those applications, in common, require a high-capacity and highly reliable cooling solution to keep the superconducting material at the proper temperature.

To meet such a requirement, a high capacity GM cryocooler which can provide about 650 W cooling power at 80 K[1] was developed. Since the physical size of this prototype unit is considerably larger compared with current available GM cryocoolers, parameters including regenerator size and valve timing should be redesigned carefully in order to get an optimized performance.

An unsteady-flow, 1-D simulation method was used to predict the behavior of the prototype cryocooler, and the result was used to form the guideline for the basic design. After the first prototype unit became available, the design was improved by feeding back the experiment results to the prototype units. In this paper, both the simulation method and experimental results will be reported.

INTRODUCTION

Since its invention in 1950s[2], Gifford-McMahon (GM) cryocooler is known as one of the most reliable commercial cryocooler for multiple applications including cryopumps, cooling of the radiation shield for an MRI, and cooling of infrared sensors for the military[3]. Most of the development of the GM cryocooler in last decade focused on increasing cooling performance near 4 K, performance optimization in the 4-30K temperature region and increasing the Carnot efficiency. Development of high heat capacity materials played an important role in this stage. In the 1990s, the development of rare earth materials made it possible for a GM cryocoolers to reach a temperatures near 4K. Recently, by utilizing a new rare earth material, Gd_2O_2S (GOS)[4], a prototype compact GM cryocooler which can reach 2.1K under no-load conditions was developed[5-6].

On the other hand, rapid growth of high temperature superconducting (HTS) material applications brings new challenges to the current cryocoolers. In contrast to the 4K region applications, HTS applications often require much more cooling power with relatively high cooling temperature of 80K is acceptable. Applications including HTS superconducting motors, power transmission lines and power generators are usually considered to consume 10^2 W to 10^4 W cooling power (per

Cryocoolers 19, edited by S.D. Miller and R.G. Ross, Jr.

case). Most of these applications are currently using direct liquid nitrogen cooling or Turbo-Brayton cryocoolers. Though having the potential to significantly reduce the cost and space requirement, current commercial GM cryocoolers lack the suitable cooling power which is crucial in HTS applications. In the current market, even the largest GM cryocooler (scotch-yoke driven) can only provide about 100 W at 30K (with a 7 kW compressor unit)[7].

To fill the gap for high cooling capacity, a single stage GM cryocooler prototype unit is developed. After the design optimization of valve size, valve timing and regenerator material arrangement, a typical cooling capacity of about 650 W at 80K was achieved with an input power of about 13 kW. In this paper, the latest experiment results and optimization method will be described.

GENERAL DESIGN ASPECTS

A typical single stage, scotch-yoke driven GM type cryocooler design was chosen to be the start point of our development. As shown in Figure 1, the pressure wave is generated by a typical rotary valve. Pressure gauges which were used to monitor the compression and expansion room pressure are only utilized in the experiment. A 200 W brushless electric motor with controllable rotation speed is used to drive the motion of scotch-yoke. Figure 2 shows the more realistic external view of the current prototype unit. The physical size of this unit is about 680 mm in height and 640 mm in width.

The rotary valve is designed to be changeable so the operation timing and opening area can be easily optimized. Also the displacer is designed to be reopened for the arrangement of metal mesh inside to be changed.

In order to roughly determine the physical size of the expansion room, displacer and to achieve loss data for further optimization, a 1-D transient simulation was performed to show the principle performance prediction.

SIMULATION METHOD AND RESULTS

Simulation Method Description

A one-dimensional, transient, finite volume method simulation analysis was carried out to estimate the main loss in the prototype cryocooler. The compressibility of helium is taken into count by the continuity equation:

$$\frac{\partial \rho}{\partial t} + \nabla \cdot (\rho \vec{U}) = 0 \tag{1}$$

where ρ represents the density of helium, and $\vec{U}$ is the velocity vector.

The momentum equation is simplified to only consider the flow resistance caused by metal mesh in the displacer.

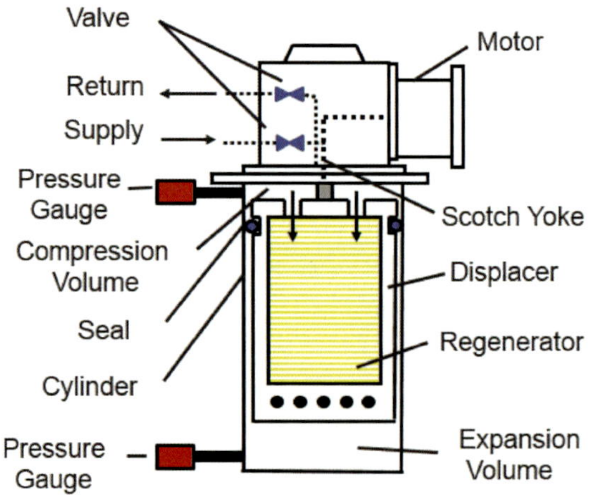

Figure 1. A simplified diagram of the prototype cryocooler.

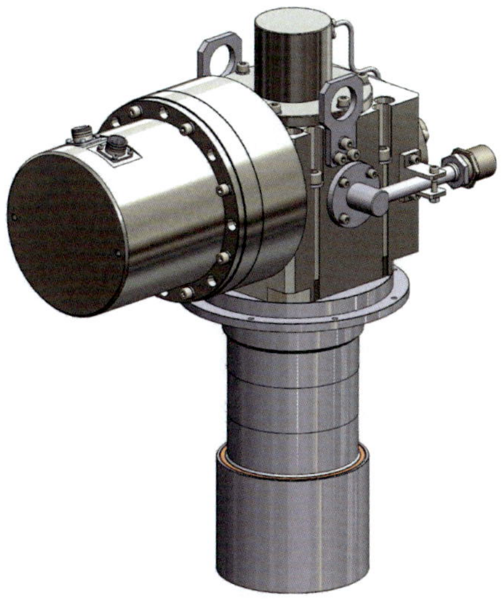

Figure 2. A 3-D drawing of the external view of current prototype unit.

$$\nabla p = -\frac{1}{2}\frac{f_r}{d_h}\rho\left|\vec{U}\right|^2\frac{\vec{U}}{|\vec{U}|} \qquad (2)$$

In Eq. 2, p represents the pressure of helium under consideration. f_r represents the Darcy friction coefficient posed by metal mesh and d_h is the hydraulic diameter of the calculated area. The pressure drop occurring in other components such as valve and connection pipes is modeled by a simpler calculation.

The energy equation of helium is:

$$\frac{\partial(\rho h)}{\partial t} + \nabla\cdot(\rho\vec{U}h) + h_f(T - T_s) = 0 \qquad (3)$$

where h is the enthalpy of helium and h_f stands for heat transfer coefficient between helium and metal mesh. T and T_s are the temperature of helium and mesh, respectively. The cooling capacity is calculated by a similar equation, except that the heat transfer is between the helium and the cold end heat exchanger.

Since the heat conduction loss through the wall of the cylinder, the shuttle loss and the radiation heat loss are not included in this calculation, additional analysis is needed to interpret the simulation results.

Analysis of Simulation Results

The simulation results for the cooling capacity and losses are summarized in Table 1. The cooling capacity is derived from the simulation results of the cooling power (as described in previous section), with additional consideration towards shuttle loss, thermal conduction loss through cylinder wall and radiation loss.

Regenerator loss is obviously the largest part among all losses. Therefore, reducing the regenerator loss should be one of the first priorities during the prototype development.

Pumping loss and shuttle loss, according to our experience, are sensitive to clearance size between the displacer and the expander cylinder. Actual loss in the experiment may be affected by dimension deviation, mechanical vibration and even surface abrasion. Since only a simplified model was used in simulation, the results of those terms may be different from the experiment.

Thermal conduction and heat radiation are also calculated separately from the simulation. Different from the pumping and the shuttle loss, these terms are considered to be rather consistent with the actual conditions. However, thermal conduction and radiation from ambient environment are usually difficult to reduce as long as the physical size is not significantly changed.

Table 1. Simulation Results of Cooling Capacity and Losses.

Item	Results (W)
PV power	853.4
Regenerator loss	91.5
Pumping loss (including pulse effect, minus means cooling effect)	-34.6
Shuttle loss	33.0
Thermal conduction loss	14.0
Radiation loss	2.4
Net cooling capacity	707.6

The calculated net cooling capacity of 707.6 W is considered to be a good starting point since it's considerably closer to the 650 W cooling capacity development target, and has a reasonable margin to cover the possible calculation error in future development.

EXPERIMENT SETUP AND RESULTS

Based on the principle design parameters derived from the simulation, a prototype unit was manufactured for further optimization. The test setup, optimization of valve timing and regenerator, and a typical cooling load-map and cool-down process will be described in the following sections.

Experiment Setup

Heat load is generated by 4 cartridge heaters attached to the cold head and the temperature on cold head is measured by a PtCO (Chino) sensor, as shown in Figure 3.

In order to provide sufficient helium flow to the expander, two helium compressor units (F70) were connected in parallel to the prototype unit. Typical total power consumption of compressor is about 7 kW x 2 = 14 kW. It should also be noted that F70 compressor itself is water-cooled and the power consumption of water chiller is not included here.

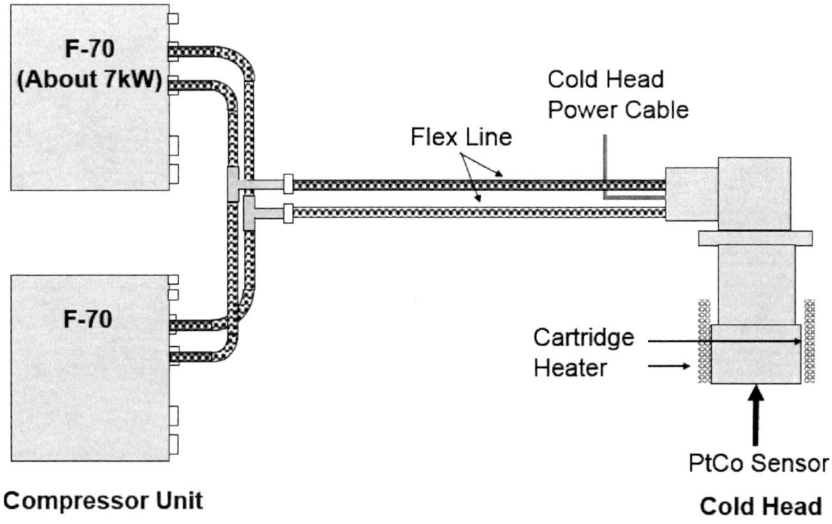

Figure 3. A schematic drawing of the experiment setup.

Valve Timing Optimization

Optimization value for the valve timing usually depends on the target temperature region, helium mass flow rate and characteristics of the compressor unit. Although there is already sufficient knowledge of compressor unit, the 80K temperature region is unfamiliar and we have no experience with the relatively large helium mass flow rate (compared with formal 4K and 20K GM cryocooler products). A comparison experiment was conducted to show the effect of expansion timing on the cooling capacity.

As shown in Table 2, the temperature of the cold head under 650 W heat load is only slightly improved with longer expansion timing, which means the expansion timing is already close to a local optimized point.

Regenerator Optimization

The regenerator optimization focused on the arrangement and amount of metal meshes (wire screens) filled in the displacer. Three kinds of copper and stainless meshes with different mesh size are used in this experiment. Considerable cooling capacity improvement was achieved by altering the amount of each kind of mesh, which is consistent with the simulation prediction described in previous section. Table 3 shows part of the results.

Cold head temperature under 650 W in the "Base line" case is different from the previous valve timing optimization experimental results because there were other minor modifications of the inner structure. Starting from the "Base line" case, increasing the amount of coarse meshes by 10% improved the cooling capacity by about 25-35 W. The best performance was almost exactly 650 W at 80K when the amount of coarse meshes is further increased.

Cooling Load-map and Cooling Down Process

A typical cooling load-map is shown in Figure 4. With all the optimization performed, a typical cooling temperature was 79.8K under 650 W heat load. The power consumption of the compressor at this point was about 13 kW (6.5kW x 2) The gap between this result and previous simulation analysis (707.6 W) is mainly caused by the calculation error in pumping loss, shuttle loss and lack of consideration to the characteristic of compressor unit.

A typical cooling down process is shown in Figure 5. The temperature on the cold head stage usually reaches its stable state within 1 hour under a no-load conditions. Minor modification including valve timing and regenerator material may affect the required cool-down time, but in principle the required time should be similar.

Table 2. Optimization Experiment Results for Valve Timing.

Item	Temperature under 650 W (K)
Case 1 (original timing based on simulation)	87.6
Case 2 (expansion timing + 1.1 %)	87.5
Case 3 (expansion timing + 3.3 %)	87.3

Table 3. Optimization Experiment Results for Regenerator Material.

Item	Temperature under 650 W (K)
Case 1 (Base line)	84.3
Case 2 (Portion of coarse meshes + 10 %)	82.3
Case 3 (Portion of coarse meshes + 20 %)	79.8

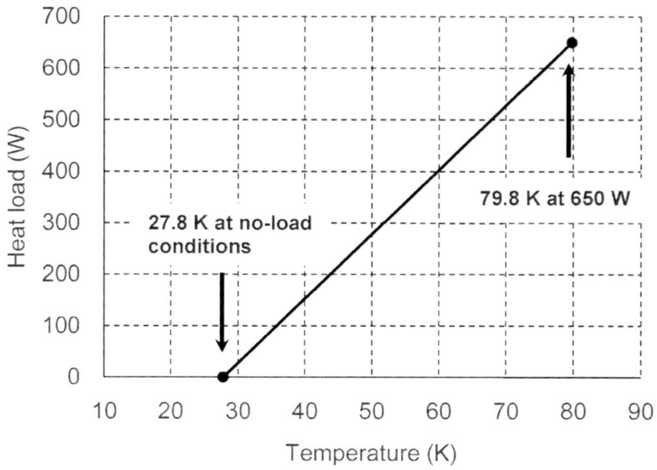

Figure 4. A typical cooling load-map.

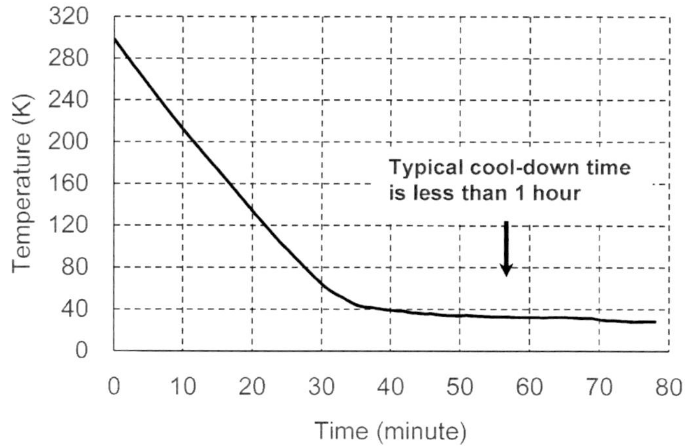

Figure 5. A typical cool-down process from room temperature.

CONCLUSION AND FUTURE WORK

A high capacity, single stage GM cryocooler prototype unit is developed to provide an alternative option in the 80K temperature region. The initial design was based on traditional scotch-yoke driven GM cryocooler structure with estimated parameters obtained from simulation method. After a prototype unit became available, further improvements were made mainly towards the valve timing and regenerator material based on experimental results.

The current prototype unit is about 680 mm x 640 mm in physical size. With 13~14 kW compressor power, a typical cooling capacity of 650 W at 80K was achieved. Typical cooling-down time from room temperature is under 1 hour.

As future work, more reliability and stability test will be taken to discover potential problem during long term operation.

REFERENCES

1. Gifford-McMahon Cryorefrigerators: http://www.cryomech.com/products/cryorefrigerators/gifford/

2. Gifford, W.E., and McMahon, H.O., "A new low-temperature gas expansion cycle," *Adv. in Cryogenic Eng.*, vol. 5, Springer US (1960), pp. 354-372.

3. Radenbaugh, R., "Refrigeration for superconductors," *Proceedings of the IEEE*, vol. 92, no.10 (2004), pp. 1719-1734.

4. Numazawa, T., Yanagitani, T., Nozawa, H., Ikeya, Y., Li, R. and Satoh, T., "A New Ceramic Magnetic Regenerator Material for 4 K Cryocoolers," *Cryocoolers 12*, Springer US (2003), pp.473-481

5. Xu, M., Bao, Q., Tsuchiya, A. and Li, R., "Development of compact 2K GM cryocoolers," *Physics Procedia*, vol. 67 (2015), pp. 491-496.

6. Bao, Q., Xu, M.Y., Tsuchiya, A. and Li, R., "Recent development status of compact 2 K GM cryocoolers," *IOP Conf. Series: Materials Science and Engineering*, vol. 101, no. 1 (2015), p 01236.

7. Yamada, K., "Development of a large cooling capacity single stage GM cryocooler," *Cryogenics*, vol. 63, (2014), pp. 110-113.

A 4 K Pulse Tube Cryocooler with Large Cooling Capacity

C. Wang

Cryomech, Inc.
Syracuse, NY, USA 13211

ABSTRACT

We are developing a large cooling capacity pulse tube cryocooler to be used for applications such as cryogen-free dry dilution refrigerators and large superconducting magnets, etc. The pulse tube cryocooler with standard design, model PT420S, is driven by a compressor, Cryomech model CP1114, with an input power of ~11.5 kW. After optimizing the PT420S design, it provides, simultaneously, a cooling capacity of 2.01 W at 4.2 K on the 2nd stage and 52.6 W at 45 K on the 1st stage. By modifying the first stage regenerator and flow controls, a large cooling capacity of 83 W at 45 K and 1.85 W at 4.2K could be provided. A low vibration design with a remote motor, Model PT420RM, can provide 1.76 W at 4.2K and 53W at 45 K.

INTRODUCTION

4 K pulse tube cryocoolers have opened many challenging applications which require low vibration, high reliability and long mean time between maintenance. For example, they cool superconducting magnets and SQUIDs; pre-cool dilution refrigerators, He3 adsorption coolers and adiabatic de-magnetization refrigerators for sub-Kelvin cooling[1,2]. Some 4 K pulse tube cryocoolers have been used for laboratory scale helium liquefiers and reliquefiers to re-liquefy the boil off from open liquid helium cryostats[3]. Many Cryomech 4 K pulse tube cryocoolers have operated continuously for more than 50,000 hours in the field without cold head maintenance.

Currently, the largest 4 K pulse tube cryocooler can provide 1.5 W at 4.2 K. Some large superconducting magnets have to use multiple 4 K pulse cryocoolers[4]. Furthermore, some applications of dry dilution refrigerator demand a large 4 K pulse tube cryocooler to increase cooling capacity at a temperature of ~10 mK. Some laboratories also require helium liquefiers with higher liquefaction rate.

In response to these market demands, Cryomech is developing a larger 4 K pulse tube cryocooler to provide a cooling capacity of ~2 W at 4.2 K. This paper presents some results of the development.

Cryocoolers 19, edited by S.D. Miller and R.G. Ross, Jr.

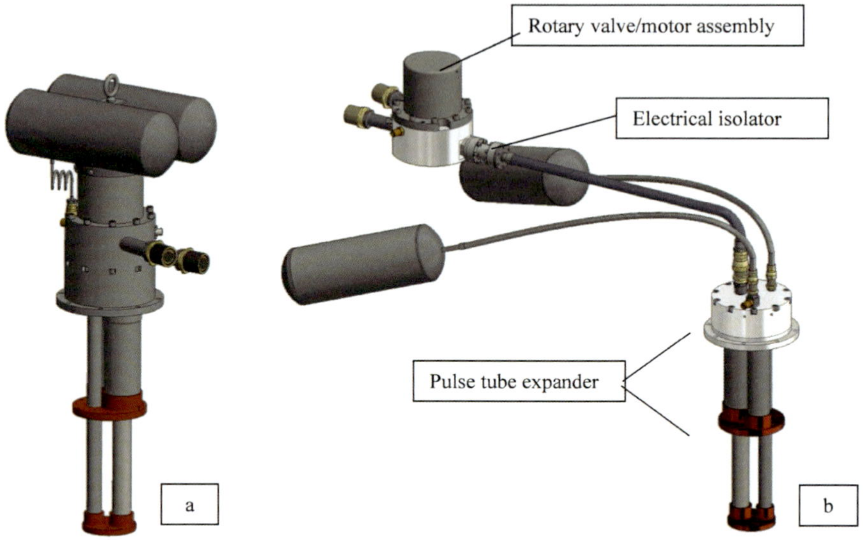

Figure 1. Large 4 K pulse tube cryocoolers. (a) standard design, model PT420S; (b) remote motor design, model PT420RM.

SYSTEM DESIGN

We are developing the 4 K pulse tube cryocooler with two types of rotary valve integration. 1. Standard design with the rotary valve/motor assembly integrated on the warm end of the pulse tube expander, See figure 1 (a), Model PT420S; 2. Remote motor design with the rotary valve/motor assembly separated from the pulse tube expander, see figure 1 (b), Model PT420RM. The standard design of the 4 K pulse tube cryocooler had been described in reference [1].

In the remote motor design, the rotary valve/motor assembly is connected to the pulse tube expander through a 61 cm long stainless steel flexible line. An electrical isolator is installed on the flexible line to isolate electrical noise generated by the compressor and cold head motor. Two reservoir volumes are connected to the pulse tube expander through one meter long stainless steel flexible lines. The remote motor model can be used for extremely vibration-sensitive applications.

Both the PT420S and PT420RM are driven by the same compressor, Cryomech model CP1114, with an input power of ~11.5 kW.

PERFORMANCE AND DISCUSSION

Performance Optimization

The development of the larger 4 K pulse tube cryocooler was started with the standard cold head model PT420S by modifying an existing 4 K pulse tube cryocooler, model PT415, which provides 1.5 W at 4.2 K. The new system is driven by a larger compressor model, CP1114.

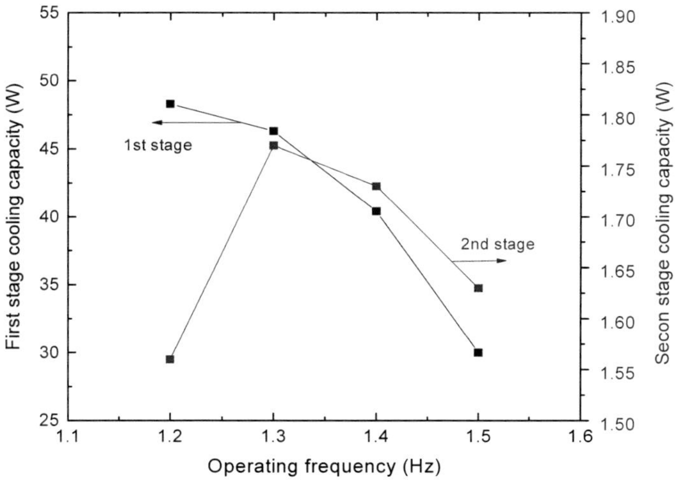

Figure 2. Performance optimization at different operating frequencies

The performance of the 1st stage at 45 K and 2nd stage at 4.2 K has been tested at several operating frequencies. The results are given in Figure 2. The 2nd stage provides the best performance of 1.77 W at 4.2 K with a frequency of 1.3 Hz. It also can also provide 1.73 W at 4.2 K with a 1.4 Hz operating frequency. All existing 4 K pulse tube cryocoolers at Cryomech operate at 1.4 Hz. Considering production compatibility, an operating frequency of 1.4 Hz was chosen for the PT420. The following improvements are all obtained at the frequency of 1.4 Hz.

Table 1 lists the performance improvements seen with a few modifications. Using new heat exchangers for the 1st and 2nd stage, the performance of the PT420S was improved to have 1.88 W on the 2nd stage and 41.4 W on the 1st stage. The highest 2nd stage capacity of 2.01 W at 4.2 K was achieved by increasing the sizes of both 1st and 2nd stage pulse tube volumes. Modifying the 1st stage regenerator gives the best 1st stage performance of 83W at 45 K. We obtained two results for the best 1st stage performance or the best 2nd stage performance with the No.3 design or No. 4 design.

Figure 3 shows the cooling capacity map for the best 2nd stage performance implementing the No. 3 design. It achieves bottom temperatures of 31.5 K on the 1st stage and 2.83 K on the 2nd stage. It can simultaneously provide 52.6 W at 45 K and 2.01 W at 4.2 K with an input power of 11.42 kW.

Table 1. Performance improvements of various designs

Cooling performance	No.1 design	No. 2 design	No. 3 design	No. 4 design
1st stage at 45 K	40.6 W	41.4 W	52.6 W	83.0 W
2nd stage at 4.2 K	1.73 W	1.88 W	2.01 W	1.85 W

No. 1, original PT415; No. 2, New 1st and 2nd stage heat exchanger; No. 3, increase both 1st and 2nd stage pulse tube volume; No. 4, new 1st stage regenerator.

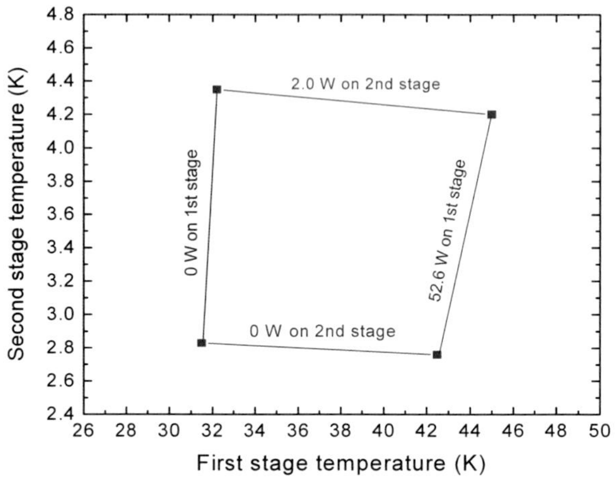

Figure 3. Cooling capacity map for the best 2nd stage cooling capacity

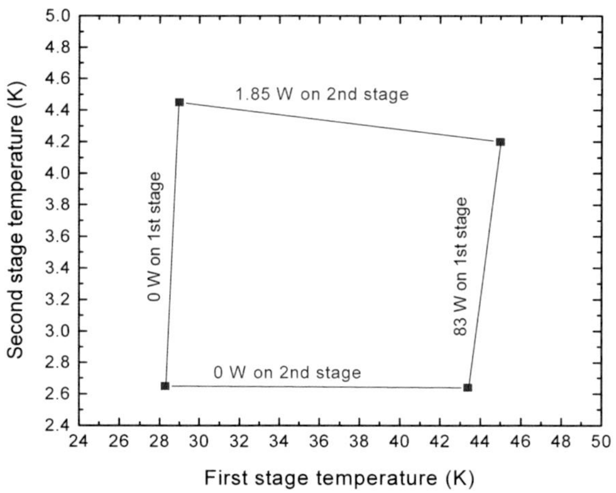

Figure 4. Cooling capacity map with a new regenerator

Figure 4 shows the cooling capacity map for the best 1st stage performance with No. 4 design. It has a lower bottom temperature of 28.3 K for the 1st stage and 2.65 K for the 2nd stage. The No. 4 design achieved lower temperatures on both stages than the No.3 design. The cryocooler can simultaneously provide 83 W at 45 K and 1.85 W with an input power of 11.47 kW.

Figure 5 shows a cool-down time of the PT420S with the No.4 design. The cool-down speed is measured with a 60 W heat load on the 1st stage and a 1.8 W load on the 2nd stage. It takes 133 min for the 1st stage to reach a bottom temperature of 38.7 K and 105 min for the 2nd stage to reach a bottom temperature of 4.15 K.

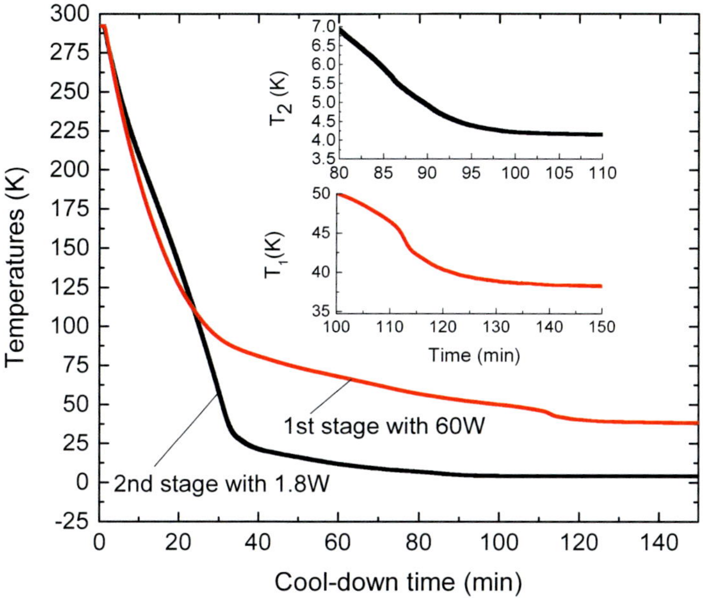

Figure 5. Cool-down curve of the PT420S

Performance of the remote motor model, PT420RM

Based on the improved performance of the PT420S, two remote motor pulse tube cryocoolers, Model PT420RM, were built for an OEM customer. They have comparable performance on both 1st and 2nd stages. Table 2 lists their performance. The cooling load map of #1 system is given in figure 6. It has bottom temperatures of 29.0 K on the 1st stage and 2.73 K on the 2nd stage. It provides 53 W at 45 K and 1.76 W at 4.2 K simultaneously with a power input of 11.34 kW. Comparing the results of the PT420RM with that of the PT420S, the 1st stage cooling capacity is significantly reduced from 83 W to 53 W at 45 K. This loss will be investigated later on.

Table 2. Performance of two PT420-RM systems

Cooling performance	#1 system	#2 system
1st stage at 45 K	53.0 W	52.6 W
2nd stage at 4.2 K	1.76 W	1.74 W

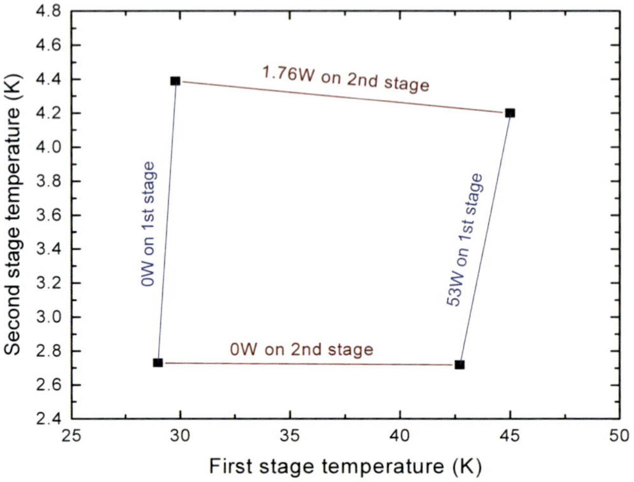

Figure 6. Cooling load map of the PT420RM

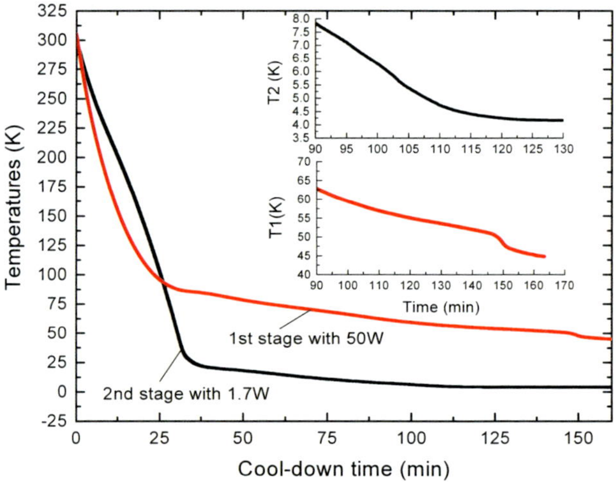

Figure 7. Cool-down curve of the PT420RM

Figure 7 shows the cool-down curve of the PT420RM with 50W on the 1st stage and 1.7 W on the 2nd stage. The PT420RM, with lower heat loads on the both stages, has a slower cooling down rate than the PT420S. It takes 125 min for the 2nd stage to reach a temperature of 4.14 K and 160 min for the 1st stage to reach 44.7 K.

Development of the PT420 pulse tube cryocooler is still ongoing to achieve better performance and a faster cool-down rate.

CONCLUSION

A high cooling capacity 4 K pulse tube cryocooler, with options of the standard and remote motor cold heads, is being developed at Cryomech, Inc. So far, this cryocooler can provide 2.01 W at 4.2 K and 52.6 W at 45 K using a standard cold head. Using a low vibration remote motor cold head, it can provide 1.76 W at 4.2 K and 53 W at 45 K. Developments are ongoing to provide a final product with increased performance.

ACKNOWLEDGMENT

Author would like to thank J. Cosco for his help in testing and B. Zerkle for building and assembling the system.

REFERENCES

1. Wang, C., "Characteristic of 4 K Pulse Tube Cryocooler in Application," *Proceeding of ICEC20*, Elsevier Publisher (2005), pp. 265-268.
2. Uhlig, K., "Dry dilution refrigerator with pulse-tube precooling," *Cryogenics,* Vol. 44 (2004), p. 53.
3. Wang, C., "A helium re-liquefier for recovery and liquefying helium vapor from cryostat," *Advances in Cryogenic Engineering,* Vol.55A, AIP publisher, Melville, NY (2010), pp. 687-694.
4. Green, M.A. and Wang, S.T., "Tests of four PT415 coolers installed in the drop-in mode," *Proceedings of ICEC-22*, Published by KIASC, Seoul Korea (2008), pp. 105-110.

A Double Pipe Regenerator for a 4 K Gifford-McMahon Cryocooler

S. Masuyama[1] and T. Numazawa[2]

[1]National Institute of Technology, Oshima College, Oshima, Japan
[2]National Institute for Materials Science, Tsukuba, Japan

ABSTRACT

A new regenerator structure named a *double pipe regenerator* was experimentally investigated. This structure was adapted to the second stage regenerator of a 4K Gifford-McMahon cryocooler to improve the cooling efficiency at 4.2 K. The double pipe regenerator consists of a regenerator housing and a stainless steel pipe. This pipe with a thin wall is inserted in the co-axial direction into the regenerator housing. The second stage regenerator was divided into two parts by two kinds of regenerator materials of lead (Pb) and holmium copper 2 ($HoCu_2$) spheres. These filling volume rate were 50%, respectively. Only the Pb part has the double pipe regenerator in consideration of the temperature distribution and helium properties.

The experimental results show that the cooling power was affected by the pipe size and the first stage temperature. The maximum cooling power at 4.2 K achieves 1.33 W with an input electric power of 7.3 kW. This is an improvement in the cooling power by 7%, compared with a conventional two-layer structure. Detailed experimental results will be shown in this paper.

INTRODUCTION

Regenerative 4K cryocoolers, Gifford-McMahon (G-M) and G-M type pulse tube cryocoolers, are widely used for advanced application, such as helium gas liquefaction and superconducting systems. An operation method of these cryocoolers is very easy, because users press the start button only. The cryocoolers are driven by the fixed operating frequency, and get to 4K level automatically. It seems apparently a complete machine. However, the cooling efficiency is insufficient. Much electrical input power is required to maintain the cooling temperatures of 4K level. The coefficient of performance of %Carnot is approximately one percent.

To improve the cooling efficiency, many modifications have been carried out since G-M cryocoolers reached the 4K level in the 1990s. In particular, the regenerator efficiency has been a key focus because this is directly linked to the cooling performance. One effective method is that a material which has a large specific heat at temperatures around 4K is packed in the regenerator. Holmium copper ($HoCu_2$) and gadolinium oxysulfide (Gd_2O_2S) will be considered as representative materials. In recent years, holmium-erbium based nitride has been developed[1,2] as a new candidate for 4K regenerator materials. The specific heat of this nitride is greater than that of $HoCu_2$ at temperatures around 4K.

Another approach with a modified regenerator configuration has been presented by Masuyama et al.[3-5] This approach has a regenerator configuration of a cylindrical form with a constant diameter.

Cryocoolers 19, edited by S.D. Miller and R.G. Ross, Jr.

In general, it is modified by using a bakelite rod. The bakelite rod is inserted in the regenerator as a dummy volume. This method was applied to the second stage regenerator of a 4K G-M cryocooler. The experimental result proved that this method improved the second stage cooling power at 4.2K slightly as well as the first stage cooling power at 40K.

This paper describes a new regenerator structure named a *double pipe regenerator* for the second stage regenerator of a 4K G-M cryocooler. This structure derives from the above method. The double pipe regenerator has a stainless steel pipe inserted into the regenerator. The regenerator structure and experimental results will be presented in this paper.

G-M CRYOCOOLER AND REGENERTAR CONFIGURATIONS

Two-stage G-M Cryocooler

To confirm the effect of the double pipe regenerator, a conventional two-stage G-M cryocooler, RDK-408D2 (SHI), with a water cooled compressor, C300G (SUZUKISHOKAN), were prepared. The operating conditions are as follows: the operating frequency of 1.2 Hz, the input electric power of 7.3 kW, and the initial charging pressure of helium gas of 1.6 MPa. All the experiments have been carried out under these conditions. A schematic diagram of the two-stage G-M cryocooler is shown in Figure 1. Two calibrated silicon diode thermometers and two electric heaters are attached to the first and second stages, respectively. The second stage and cylinder are covered by a radiation shield which is cooled with the first stage. The G-M cryocooler is operated in a vacuum condition of less than 10^{-4} Pa.

Second Stage Regenerator Configurations

Schematic diagrams of the second stage regenerator (the part in which regenerator materials are installed) are shown in Figs. 2 and 3. A general two-layer regenerator (hereinafter called "two-layer") with $HoCu_2$ and lead (Pb) spheres is shown in Figure 2. The cold side is packed with 50% by volume of $HoCu_2$ spheres with a diameter of 0.15-0.35 mm ($HoCu_2$ part), and the warm side is packed by Pb spheres with a diameter of 0.212-0.3 mm (Pb part). The filling weight of $HoCu_2$ and Pb spheres is 310 g and 330 g, respectively. To separate these regenerator materials, a separator, which is made of packed fine wire stainless steel screens, is set in the middle. The total length of the $HoCu_2$ and Pb parts are 14 cm, and the inner diameter of the regenerator is 3 cm, so that the volume of the regenerator housing is 99 cm^3. Table 1 gives these dimensions for the second stage regenerator housing.

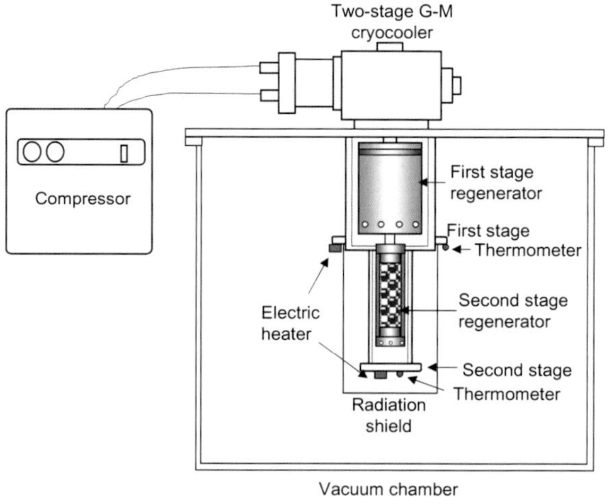

Figure 1. Schematic diagram of the two-stage G-M cryocooler and compressor

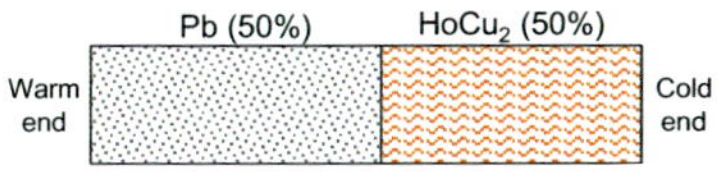

Figure 2. The second stage regenerator of two-layer with Pb and HoCu2 spheres

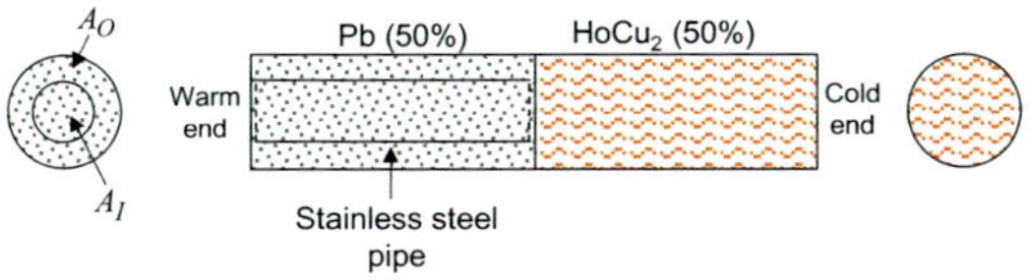

Figure 3. Double pipe regenerator with an inserted stainless steel pipe

Table 1. Dimensions of the second stage regenerator housing of the Pb part and the HoCu2 part.

Regenerator housing	Inner diameter [cm]	Length [cm]	Cross-sectionalarea[cm²]	Volume [cm³]
Pb part	3	7	7.1	49.5
$HoCu_2$ part	3	7	7.1	49.5

The double pipe regenerator with a stainless steel pipe is shown in Figure 3. The pipe is inserted in the co-axial direction of the Pb part. An photograph of the warm end is shown in Figure 4. In the experiments, three types of stainless steel pipes were prepared (length and thickness are constant). Table 2 presents the specifications of the pipes, the cross-sectional ratio, A_I / A_O, where A_I is the inside area, A_O is the outside area given in Figure 3, and the calculated thermal conduction loss of the pipe. The thermal conductivity of stainless steel from 60 K to 4.2 K was used. The thermal conduction loss of the double pipe regenerator, shown in Figure 3, does exist. However, using a thinner pipe is effective to decrease this thermal loss.

EXPERIMANTAL RESULTS OF COOLING POWER

The cool-down tests of the G-M cryocooler were started from the room temperature. The cool-down characteristics were almost the same for all the regenerator configurations. The cool-down time to the lowest temperature was approximately two hours.

To measure the cooling power, the temperature of the second stage was held at 4.2 K, and the temperature of the first stage was varied by adjusting the electric heater.

Figure 5 shows the results of the cooling power at 4.2 K as a function of the first stage temperature. In the case of the two-layer, the cooling power of 1.24 W remains constant between 40 K and 57 K. With the first stage temperatures of 24 K or 66 K, the cooling power is decreased to 1.1 W.

In the case of double pipe, all the pipes show the same cooling power at 24 K. After that, the cooling power increases with increasing first stage temperature until a peak is reached. Pipe S and L have similar characteristics. The maximum cooling power achieved ia 1.33 W at 50 K. It is an

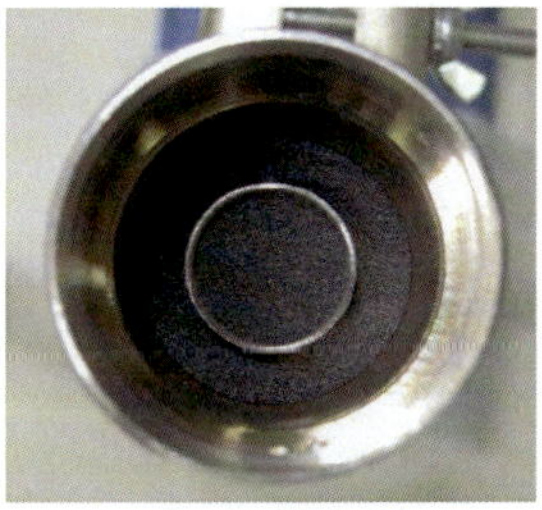

Figure 4. Photograph of the double pipe regenerator viewed from the warm end

Table 2. Specifications of three types of pipes, cross-sectional ratio, and calculated thermal conduction loss of stainless steel pipe.

Pipe type	Outer diameter [mm]	Length [mm]	Thickness [mm]	A_I / A_O	Thermal conduction loss* [mW]
S	10	67	0.5	0.10	44
M	15	67	0.5	0.50	81
L	22	67	0.5	1.1	100

* Thermal conductivity of stainless steel from 60 K to 4.2 K was used.

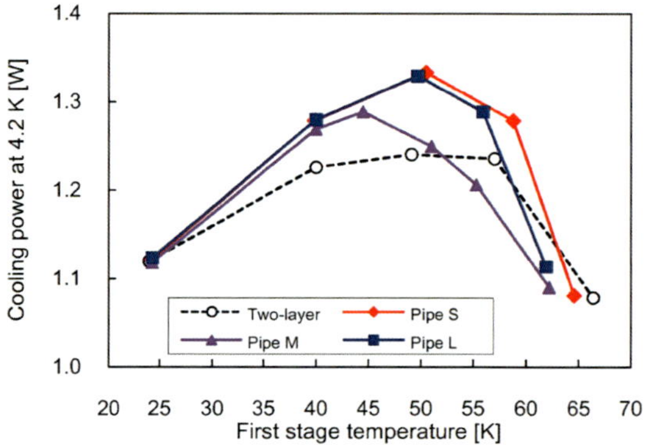

Figure 5. Comparison of the cooling power at 4.2 K for two-layer and double pipe regenerators

improvement by 7%, compared with the two-layer. In contrast, pipe M has a peak power of 1.29 W at 45 K. From the results, the double pipe regenerator is able to improve the second stage cooling power at 4.2 K. The cross-sectional ratio of 0.10 or 1.1 is optimum. Its effect depends on the first stage temperature and the specifications of the pipes. The reason for these intriguing results will be discussed in the next section.

Figure 6 presents the cooling power at 4.2 K as a function of the first stage power. With the first stage power of 59 W, pipe S and L achieve the maximum cooling power of 1.33 W.

DISCUSSION

In order to clarify the effect of the double pipe regenerator, the phenomena in the second stage regenerator including the helium properties are discussed in this section. Figure 7 shows

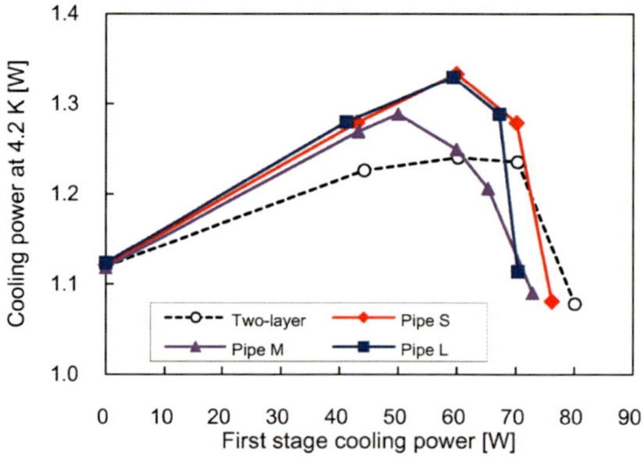

Figure 6. Cooling power at 4.2 K as a function of the first stage cooling power

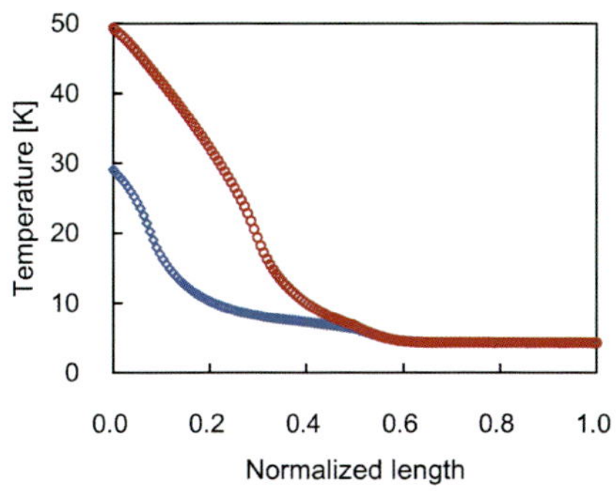

Figure 7. Simulated temperature distribution of the two-layer with Pb and HoCu2 spheres for two warm end temperatures of 30 K and 50 K

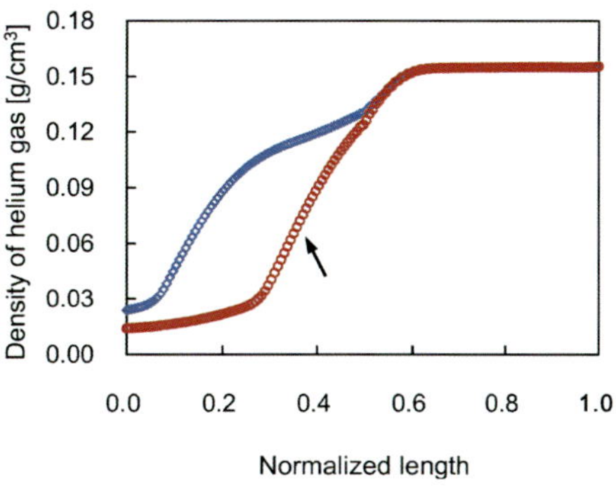

Figure 8. Calculated density of pressurized helium gas for two warm end temperatures

the simulated temperature distribution of the two-layer with Pb (50%) and $HoCu_2$ (50%) spheres. Two temperatures of 30 K and 50 K are considered for the warm end (normalized length equal to 0). The cold end (normalized length equal to 1) temperature is fixed at 4.2 K. Two curves have a large temperature difference at the warm end. In contrast, the cold end has the same characteristics.

From these temperature distributions, the density of helium gas was calculated at a mean pressure of 1.5 MPa. Figure 8 presents the calculation result for the two warm end temperatures. As shown, at a normalized length from 0 to 0.5, the density of the warm end of 50 K is smaller than that of 30 K. After that, two curves are in agreement at a high density. This density distribution means that increasing the warm end temperature from 30 K to 50 K decreases the amount of helium gas in the Pb part. Then, the decreased helium gas flows to the expansion volume through the $HoCu_2$ part. This effect leads to an improvement in the cooling power. Too high a warm end temperature, however, causes an increase in the enthalpy loss in the regenerator. As a result, the cooling power is decreased.

According to the experimental results shown in Figure 5, the cooling power increases from 24 K to 50 K for the two-layer and double pipe (pipe S and L) configurations are 11% and 19%, respectively. This difference is thought to be due to a disturbed flow of helium, which is not one-dimensional flow in the regenerator. From the density distribution, the disturbance in the Pb part is larger than that in the $HoCu_2$ part. In the case of the two-layer, the helium flow from the Pb part to the $HoCu_2$ part is suppressed by the disturbance. In contrast, in the case of the double pipe, the inserted pipe is considered to act as a rectifier which reduces the disturbance.

Figures 5 and 6 show the size effect of pipes. From the results, we propose a distribution of helium flow patterns which are divided into three in the Pb part as shown in Figure 9. The outer area has a disturbed flow because of the reflection from the wall. The middle area has an effective flow for regeneration. The inner area has a slow flow. In this inner area, the regenerative effect is considered to be small over the cycle.

The helium flow in the two-layer consists of the flow mixed by these three patterns that leads to a disturbed flow. In the case of the double pipe regenerator, the cooling power of pipe S and L is larger than that of pipe M at a first stage temperature of 50 K. It is thought that pipe S and L are inserted in the boundary layer of each area shown in Figure 9. Pipe S separates the inner and middle areas, and pipe L separates the outer and middle areas. Consequently, the helium flow is rectified that leads to an improvement in the cooling power. On the other hand, pipe M is inserted in the middle area. The divided middle areas, caused by inserting the pipe, mix the outer and inner areas and the effective flow area is decreased. This flow pattern makes a small improvement in the cooling power.

Compared with the Pb part, the whole area of the $HoCu_2$ part is considered to be an effective flow due to the high density of helium gas. To prove this proposal, a development of a measurement method regarding the helium flow patterns in the regenerator will be required.

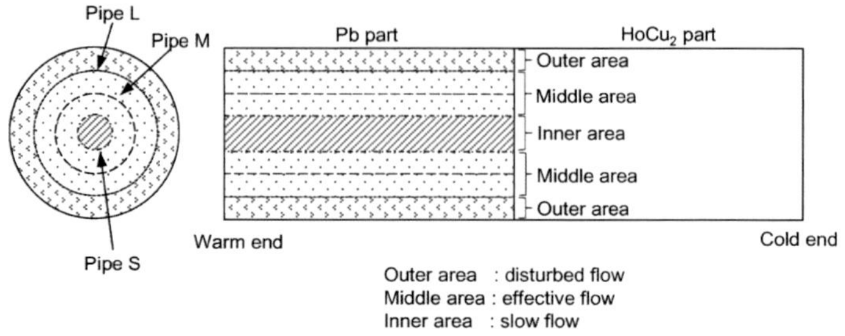

Figure 9. Proposed distribution of helium flow patterns in the Pb part

SUMMARY

A double pipe regenerator was developed to improve the cooling efficiency of a 4K G-M cryocooler. Three types of stainless steel pipes were inserted in the second stage regenerator to confirm its performance. The experimental results showed that the maximum cooling power of 1.33 W was achieved. It is an improvement in the cooling power by 7%, compared with the two-layer. The characteristics of the double pipe regenerator show that the cooling power depends on the first stage temperature, and the optimum cross-sectional ratio is 0.10 or 1.1. From these results, we proposed the helium flow pattern in the Pb part. The Pb part without the pipe has a disturbed flow. The inserted pipe is considered to act as a rectifier to reduce this disturbance.

The double pipe regenerator was proven to be an effective method to improve the cooling power with the two-layer (Pb-$HoCu_2$), and we will test it with a three-layer (Pb-$HoCu_2$-GOS) as a next step.

ACKNOWLEDGMENT

This study was supported by MEXT KAKENHI Grant-in-Aid for Scientific Research (C) 15K06693 and the NIFS Collaborative Research Program (NIFS15KECA034).

REFERENCES

1. Nakano T., Masuyama S., Hirayama Y., Izawa T., Nakagawa T., Fujimoto Y., Yamamoto T.A., Irie T., Nakamura, E. and Yamamoto, T. A., "ErN HoN Spherical Regenerator Material for 4K Cryocoolers," *Applied Physics Letters*, Vol. 101, No. 25 (2012), pp. 19081-19084.
2. Shoda K., Miyauchi T., Kobayashi H., Nakano T., Nakagawa T., Yamamoto T.A., Masuyama S., Fujimoto Y., Takata H., Kuriiwa T., and Yamamoto K., "Regenerator Material of Rare Earth Nitride ($Ho_xEr_{1-x}N$) for a 4K-GM Cryocooler," *Journal of Cryogenics and Superconductivity Society of Japan*, Vol. 50 (2015), pp. 132-136, (in Japanese).
3. Masuyama S., and Hiratsuka Y., "Effects of an Inserted Bakelite Rod in the 2nd Stage Regenerator of a Gifford-McMahon Cryocooler," *Journal of Cryogenics and Superconductivity Society of Japan*, Vol. 48 (2013), pp. 358-363, (in Japanese).
4. Masuyama S., Nagao T., and Numazawa T., "Magnetic Regenerator Material Economizing Method for 4K Gifford-McMahon Cryocoolers using Bakelite Rod," *Adv. in Cryogenic Engineering*, Vol. 59, Amer. Institute of Physics, Melville, NY (2014), pp. 1162-1168.
5. Masuyama S., Hiratsuka Y., and Numazawa T., "A Novel Approach to Optimize the 2nd Stage Regenerator Configuration of a 4K Gifford-McMahon Cryocooler," *Cryocoolers 18*, ICC Press, Boulder, CO (2015), pp. 271-276.

Size and Composition Optimization of $Er_xHo_{1-x}N$ Regenerators for 4K-GM Cryocoolers

T. Nakagawa[1], T. Miyauchi[1], T. Shiraishi[1], K. Shoda[1], T. A. Yamamoto[1], Y. Fujimoto[1], S. Masuyama[2]

[1]Osaka University, Suita, Osaka 5650871, Japan
[2]National Institute of Technology, Oshima College, Oshima, Yamaguchi 7422193, Japan

ABSTRACT

$Er_xHo_{1-x}N$ spheres were synthesized by nitriding Er_xHo_{1-x} metallic spheres with different sizes in the hot isostatic pressing method in a nitrogen sphere (1550°C and 195 MPa of 99.9999 % nitrogen). Cooling tests were performed using a commercially available two-stage GM cryocooler (Model SRDK-101D, Sumitomo Heavy Industries Ltd.) by loading $Er_xHo_{1-x}N$ spheres in the second stage regenerator column instead of $HoCu_2$. The optimal size of $Er_xHo_{1-x}N$ spheres was found to be approximately 150-200 μm. The most suitable composition was $Er_{0.75}Ho_{0.25}N$. When $Er_{0.75}Ho_{0.25}N$ with a size of 150-180 μm was used, the cooling power of the GM cryocooler at 4.2 K reached 0.266 W.

INTRODUCTION

Gifford-McMahon (GM) cryocoolers are cryogenic refrigerators widely used for helium and hydrogen liquefaction, and for systems using superconducting magnets such as magnetic resonance imaging diagnosis devices and magnetic levitation trains. Their performance and efficiency depend on the regenerator materials used. The most important requirement for the regenerator materials are high specific heat around the working temperatures. To improve the efficiency, the researchers and engineers are seeking advanced regenerator materials[1-4].

ErN and HoN are ferromagnetic with transition temperatures of 4 and 12 K, respectively[5,6]. They show large magnetic specific heat around their transition temperatures[5,7]. Nakano et al.[8] reported that they perform well as regenerators for 4K-GM cryocooler and the cooling power (CP) of the GM cryocooler at 4.2 K became high when the rare earth nitride spheres are a smaller size[8]. However, using nitride spheres smaller than 180 μm was not studied, so the optimal size of the nitride regenerator spheres has still been unknown.

Binary nitrides, $Er_xHo_{1-x}N$, are also promising regenerator materials because the peak temperature of the specific heat can be adjusted in the rage of 4 to 12 K by controlling the composition x[9]. Shoda et al.[10] measured the CP of the GM cryocooler using $Er_xHo_{1-x}N$ (x = 0.5, 0.75 and 1) regenerators. When $Er_{0.75}Ho_{0.25}N$ with a size of 180-212 μm was used, the CP of the GM cryocooler at 4.2 K (nominal CP at 4.2 K is 0.1 W) reached 0.206 W. However, the size and filling ratio of $Er_xHo_{1-x}N$ spheres to the second regenerator column were not unified in previous reports. The optimal composition of $Er_xHo_{1-x}N$ has not been clear.

Cryocoolers 19, edited by S.D. Miller and R.G. Ross, Jr.

The aim of this work is to find the most suitable composition and sphere diameter of $Er_xHo_{1-x}N$ as a regenerator material for a 4K GM cryocooler.

EXPERIMENTAL

$Er_xHo_{1-x}N$ (x = 0.75 and 0.5) spheres were synthesized by nitriding Er_xHo_{1-x} metallic spheres with different sizes in the hot isostatic pressing method in a nitrogen sphere (1550°C and 195 MPa of 99.9999 % nitrogen). Formation of $Er_xHo_{1-x}N$ was confirmed by the X-ray diffraction. The morphology of $Er_xHo_{1-x}N$ was observed by a scanning electron microscope (SEM). Cooling tests were performed using a commercially available two-stage GM cryocooler (Sumitomo Heavy Industries Ltd. Model SRDK-101D) with nominal cooling power of 0.1 W at 4.2 K. The operating conditions were the standard ones provided by supplier. The second regenerator column consists of packed Pb spheres, 200 μm diameter, together with the cryogenic magnetic regenerator. The initial regenerator material of $HoCu_2$ (180-300 μm in diameter) was replaced $Er_xHo_{1-x}N$ spheres with different sizes. The weights of the $Er_xHo_{1-x}N$ regenerator materials loaded in the column were measured. By varying the input power to a heater attached to the cold head and monitoring the equilibrium temperature, *T*, cooling power (CP) versus *T* curves were obtained. The lowest achieving temperature was determined when CP = 0. Size range, average size, filling amount and filling space ratio (7.32 cm^3 corresponds to 100 %) of $Er_xHo_{1-x}N$ spheres loaded to the second regenerator column are summarized in Table 1.

RESULTS AND DISCUSSION

The SEM images of $Er_{0.5}Ho_{0.5}N$ with different sizes are shown in Fig. 1. The figure indicates that the spherical $Er_{0.5}Ho_{0.5}N$, which was suitable for regenerators of cryocoolers, was obtained. Figure 2

Table 1. Composition, size range, average size, filling amount and filling space ratio of regenerator materials used in cooling tests of 4K-GM cryocooler.

Composition	Size range (μm)	Ave. size (μm)	Filling amount (g)	Filling space ratio (%)
$Er_{0.5}Ho_{0.5}N$	75-150	113	40	52
	180-212	196	41	54
	250-300	275	36	47
$Er_{0.75}Ho_{0.25}N$	150-180	175	40	52
	180-212	196	40	52
ErN [8]	180-212	196	41	53
	212-250	231	42	54
	250-300	275	41	54
$HoCu_2$	150-300	225	40	62

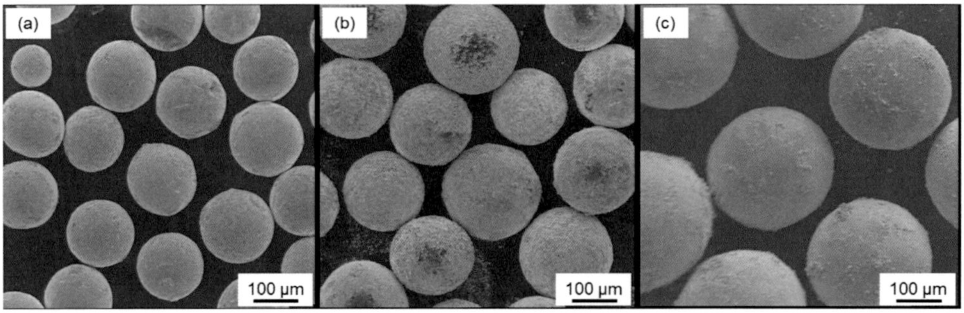

Figure 1. SEM images of Er0.5Ho0.5N regenerators with size of (a)75-150 mm, (b) 180-212 mm, (c) 250-300 mm.

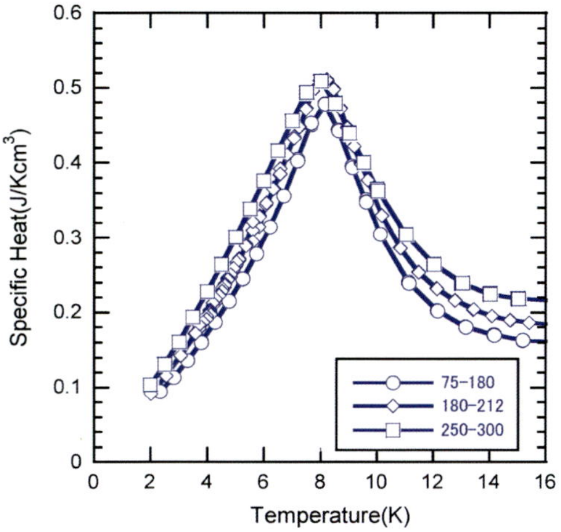

Figure 2. Temperature dependence of specific heats of Er0.5Ho0.5N with different sizes.

gives the temperature dependences of specific heat of $Er_{0.5}Ho_{0.5}N$ with different sizes. Regardless of the size, almost the same temperature dependence of the specific heat (peak temperature and amplitude) was obtained. No size dependency of the specific heat of the other compositions of $Er_xHo_{1-x}N$ was also observed.

Figure 3 shows the temperature dependence of CP of the GM cryocooler using $Er_{0.5}Ho_{0.5}N$ with different sizes as regenerators. This figure means that $Er_{0.5}Ho_{0.5}N$, regardless of the sizes, functions as regenerator materials of GM cryocooler which reaches 4.2 K. The CPs of the cryocooler at a temperature were changed when the different size of $Er_{0.5}Ho_{0.5}N$ regenerators were used. The 180-212 μm is the optimal size for $Er_{0.5}Ho_{0.5}N$ to obtain the highest CP at 4.2 K.

Figure 4 indicates the temperature dependence of CP for the cryocooler using $Er_{0.75}Ho_{0.25}N$. Higher CPs were observed compared to the $Er_{0.5}Ho_{0.5}N$ regenerators. The CPs of the cryocooler at different temperature using $HoCu_2$, which is employed in regenerators of commercially available

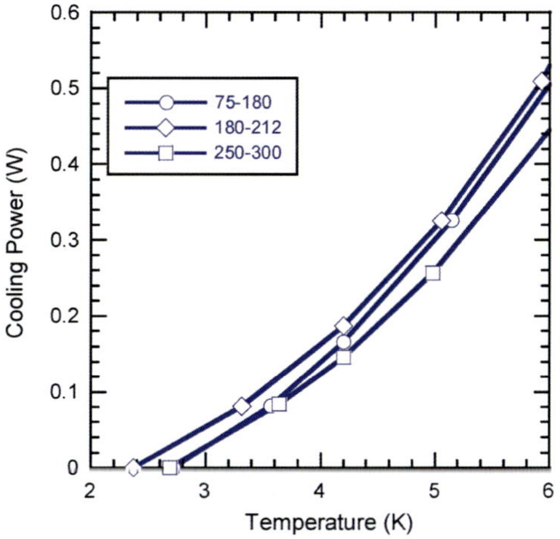

Figure 3. Cooling power of GM cryocooler using different size of Er0.5Ho0.5N regenerators.

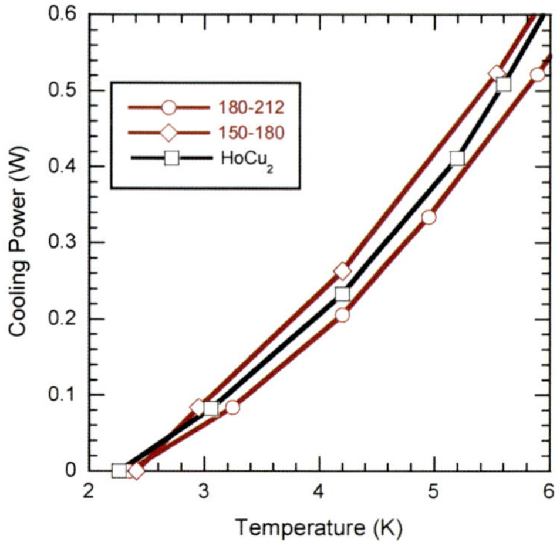

Figure 4. Cooling power of GM cryocooler using Er0.75Ho0.25N regenerators. The results of HoCu2 regenerator is also displayed for comparison.

4K GM cryocoolers, are also given in this figure. At a glance, the CP of $Er_{0.75}Ho_{0.25}N$ with 150-180 μm in diameter exceeds that of $HoCu_2$ (CP at 4.2 K is 0.233 W) can be seen.

Figure 5 shows the plots of the average sphere size vs. CP at 4.2 K for different compositions of $Er_xHo_{1-x}N$ (x = 1, 0.75 and 0.5). The data for ErN are previously reported[8]. The optimal size of $Er_xHo_{1-x}N$ spheres was found to be approximately 150-200 μm. When compared to the sphere diameter of 180-212 mm, which has a complete set of data by all composition, the composition $x = 0.75$ shows the highest CP. Therefore, the most suitable composition of $Er_xHo_{1-x}N$ is $x = 0.75$ as regenerator for 4K GM cryocooler. As shown in Table 1, filling space ratios of $Er_xHo_{1-x}N$ regenerators are less than that of $HoCu_2$. It is thought that the lower filling space ratios of $Er_xHo_{1-x}N$ are due to rough surface of them as seen in Figure 1. The filling space ratios of $Er_xHo_{1-x}N$ spheres will be improved by surface treatments. From the results of calculation by REGEN 3.3 code [11], CP is proportional to

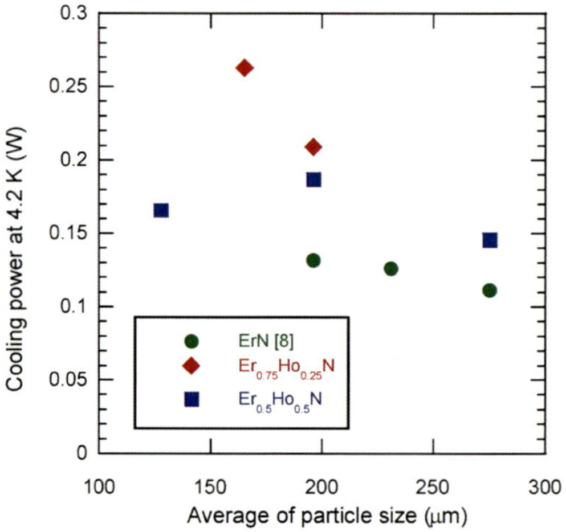

Figure 5. Average sphere size vs. CP at 4.2 K plots for the different composition of ErxHo1-xN (x = 1, 0.75 and 0.5)

the filling space ratio in the range from 50 to 70% [10]. If $Er_{0.75}Ho_{0.25}N$ with the size of 150-180 mm is filled by the same filling space ratio as $HoCu_2$, the CP of 4K GM cryocoolers will be increased to 0.317 W at 4.2 K.

CONCLUSIONS

To obtain the most suitable composition and the optimal sphere diameter of $Er_xHo_{1-x}N$ as regenerator materials for 4K GM cryocoolers, various size of $Er_xHo_{1-x}N$ (x = 1, 0.75 and 0.5) were prepared. $Er_xHo_{1-x}N$ sphere regenerators with different size were loaded into second regenerator column of 4K GM cryocooler with nominal cooling power of 0.1 W at 4.2 K. The temperature dependences of the cooling power were measured. It was found that 150-200 μm is the optimal size for $Er_xHo_{1-x}N$ spheres as regenerators of 4K-GM cryocoolers. Composition of the $Er_xHo_{1-x}N$ by which the highest cooling was obtained was x = 0.75. Using $Er_{0.75}Ho_{0.25}N$ with the size of 150-180 μm as regenerator, the cooling power of the GM cryocooler at 4.2 K reached 0.266 W. This value is higher than that in a case using $HoCu_2$.

REFERENCES

1. Sahashi, M., Tokai, Y., Kuriyama, T., Nakagome, H., Li, R., Ogawa, M. and Hashimoto, T., "New magnetic material R3T system with extremely large heat capacities used as heat regenerators," *Adv. in Cryogenic Engineering* 35 (1990), Springer US. pp. 1175-1182.
2. Bischof, J., Diviš, M., Svoboda, P. and Smetana, Z., The Specific Heat of HoCu2 in Magnetic Fields. *Physica Status Solidi (a)*, 114(2) (1989), pp. K229-K231.
3. Qiu, L. M., Numazawa, T., Thummes, G., " Performance improvement of a pulse tube cooler below 4 K by use of GdAlO 3 regenerator material," *Cryogenics* 41 (2001), pp. 693-696.
4. Numazawa, T., Satoh, T., " Cooling performance of a small GM cryocooler with a new ceramic magnetic regenerator material,"*Cryocooler* 12, Springer US (2003), pp. 397-402.
5. Nakagawa, T., T. Arakawa, et al., "Magnetocaloric effects of ferromagnetic erbium mononitride," *Journal of Alloys and Compounds*, Vol. 408-412, No. (2006), pp. 191-195..
6. Yamamoto, T. A., Nakagawa, T., Sako, K., Arakawa, T., Nitani, H., "Magnetocaloric effect of rare earth mono-nitrides, TbN and HoN," *Journal of Alloys and Compounds, Vol.* 376 (2004), pp. 17-22.
7. Nishio, S., Nakagawa, T., Arakawa, T., Tomioka, N., Yamamoto, T. A., Kusunose, T., Niihara, K., Numazawa, T., Kamiya, K., " Specific heat and thermal conductivity of HoN and ErN at cryogenic temperatures," *Journal of Applied Physics* 99(8) (2006).
8. Nakano, T., Masuyama, S., Hirayama, Y., Izawa, T., Nakagawa, T., Fujimoto, Y., Irie, T., Nakamura, E., Yamamoto, T. A., " ErN and HoN spherical regenerator materials for 4 K cryocoolers,"*Applied Physics Lett*ers 101 (25) (2012), 251908.
9. Nakano, T., Hirayama, Y., Izawa, T., Nakagawa, T., Yamamoto, T. A., Fujimoto, Y., Masuyama, S., Irie, T., Nakamura, E., " 4K-GM cryocooler performance and thermal conductivity of Ho x Er 1-x N," *Mater. Res. Soc. Symp. Proc.* 1492 Cambridge University Press (2013), pp. 53-58.
10. Shoda, K., Miyauchi, T., Kobayashi, H., Nakano, T., Nakagawa, T., Yamamoto, T. A., Masuyama, S., Fujimoto, Y., Takata, H., Kuriiwa, T., Yamakoto, K., *J. Cryo Super. Soc. Jpn.* 50 (2015), pp. 132-136.
11. REGEN 3.3 package; http://math.nist.gov/~AOGallagher/auxfiles/regenstart.html

Real Gas Effects on the COP of Regenerators Working at LowTemperatures

Q. Cao[1,2]

[1]Institute of Refrigeration and Cryogenics, School of Mechanical Engineering, Tongji University, Shanghai 201804, PR China
[2]Shanghai Key Laboratory of Vehicle Aerodynamics and Vehicle Thermal Management Systems, Tongji University, Shanghai 201804, PR China

ABSTRACT

Real gas properties impose profound effects on regenerators working at temperatures close to and below the critical point of the working fluid. The enthalpy flow associated with the pressure dependence, abbreviated as the pressure-induced enthalpy flow, generally affects the energy balance of regenerators. In this paper, the influence mechanism of the pressure-induced enthalpy flow on the COP has been analyzed. The main factor in keeping the energy balance of the regenerator is found to be the heat-associated enthalpy flow, which decreases the COP. Consequently, the temperature profile can be determined. The expression of the theoretical COP of the regenerator has been further derived. These discussions will contribute to the understanding of the working mechanism of the regenerator.

INTRODUCTION

Real gas properties impose profound effects on regenerators. The enthalpy flow associated with the pressure dependence, abbreviated as the pressure-induced enthalpy flow, is the significant real gas property, and generally induces profound heat losses at a temperature close to and below the critical point[1]. Regenerator temperature profiles that are far from linear have been observed numerically and experimentally since the 1990s[1-5]. This phenomenon has attracted great interest and significant progress has been achieved. The energy analysis and the expression of the COP has been discussed in previous study[6-8]. The important role of the heat-related enthalpy flow has identified based on the sources from open literature[9]. In this paper, the terms related to the loss mechanism will be clarified and the expression of the COP will also be derived.

SIGNIFICANCE OF THE PRESSURE-INDUCED ENTHALPY FLOW

The differential form of the specific pressure-induced enthalpy of real gases is expressed as

$$dh_p = v(1 - T\beta)dp \tag{1}$$

where the parameter β is the volume expansivity. In order to derive the magnitude of the pressure-induced enthalpy flow in the regenerator, some assumptions are made. A small pressure ratio is assumed, thus both the time-variation of volume expansivity and the time-variation of temperature

Cryocoolers 19, edited by S.D. Miller and R.G. Ross, Jr.

are regarded as constant during the integration of one cycle. The specific pressure-induced enthalpy is integrated over the pressure and the mean pressure is used as the reference pressure. The time-averaged pressure-induced enthalpy flow $\langle \dot{H}_p \rangle$ is calculated with the equation

$$\langle \dot{H}_p \rangle = (1 - T\beta)\frac{1}{\tau}\oint \dot{V}(p - p_0)dt \tag{2}$$

Compared with the expression of $\dot{P}\dot{V}$ power[10], $\langle \dot{H}_p \rangle$ is expressed as

$$\langle \dot{H}_p \rangle = (1 - T\beta)\langle pV \rangle \tag{3}$$

It is clear that the pressure-induced enthalpy flow is proportional to the PV power with a coefficient of (1-Tβ). In cases of relatively large pressure ratios, the volume expansivity and the temperature vary significantly with time in one cycle. The average product of the volume expansivity and the temperature should be used in the calculation.

The pressure-induced enthalpy flow usually accounts for a large percent of the PV power at temperatures close to and below the critical point. The parameter of ^4He[11] and ^3He[12] has been calculated, when the pressure is 1 MPa, (1-Tβ) reaches the lowest point of -96% at 8.7 K for ^4He[11], and reaches a value as high as 78% at 4.2 K, as shown in Figure 1. The real gas effect of ^3He is smaller than that of ^4He, but the parameter still reaches 69% at 4.2 K.

In order to estimate the variation of the pressure-induced enthalpy flow along the regenerator, the regenerator is assumed to be frictionless. Thus the PV power varies with the product of the compressibility factor and the temperature in the regenerator, as shown in the Eq. (4)

$$\frac{\langle pV \rangle_c}{\langle pV \rangle_h} = \frac{(ZT)_c}{(ZT)_h} \tag{4}$$

If the status at the hot end of the regenerator is used as the reference, the pressure-induced enthalpy flow at a certain temperature T inside the regenerator can be determined, and the difference of the pressure-induced enthalpy flow at two points of different temperature along the regenerator is calculated with the equation expressed as:

$$\Delta\langle \dot{H}_p \rangle = \Delta[(1 - T\beta)ZT](\langle pV \rangle_h/(ZT)_h) \tag{5}$$

The parameter [(1-Tβ)ZT] varies significantly with temperature, as shown in Figure 2. Three typical cases of PTCs[2, 13-15] are compared[9]. The maximum variation of the pressure-induced enthalpy flow within the working temperature range is estimated to account for 2.6, 1.6, and 1.2 times the PV power at the cold end for these three PTCs. Such a large pressure-induced enthalpy flow will influence the energy balance significantly.

ENERGY BALANCE IN THE REGENERATOR

By applying the first law of thermodynamics[10] to the regenerator section between heat exchangers at steady state, i.e. no heat exchange with external systems, the relationship of the heat conduction,

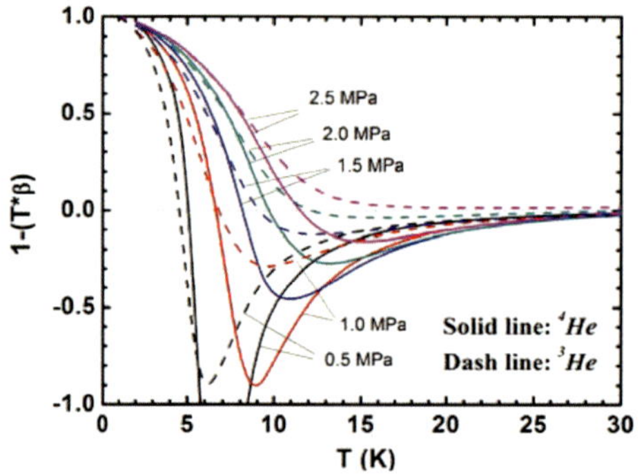

Figure 1. Dependence of parameter (1-Tβ) on temperature for ^4He and ^3He

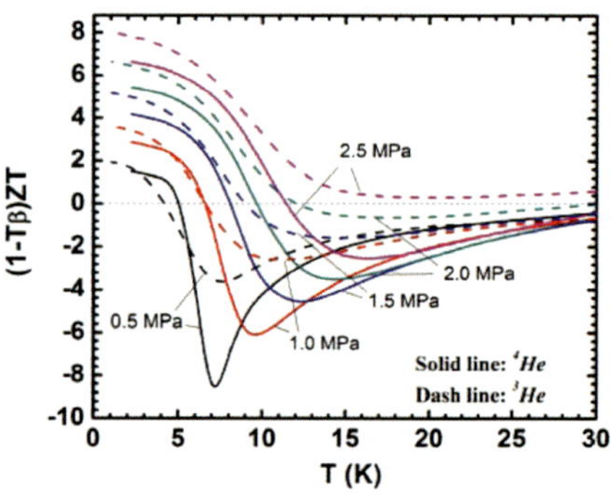

Figure 2. Dependence of parameter (1-Tβ)ZT on temperature for ^4He and ^3He

including that of gas, matrix and wall, the heat-associated enthalpy flow and the pressure-induced enthalpy flow is expressed as:

$$d\left(\langle \dot{H}_p \rangle + \langle \dot{H}_T \rangle + \dot{Q}_{cond}\right)/dx = 0 \tag{6}$$

If it is integrated over the length, the formula is rewritten as:

$$\langle \dot{H}_p \rangle + \langle \dot{H}_T \rangle + \dot{Q}_{cond} = \dot{Q}_{tot} \tag{7}$$

The total heat flow $\dot{Q}_{tot}$ must be kept constant over the length. As $\langle \dot{H}_p \rangle$ varies, $\dot{Q}_{cond}$ and/or $\langle \dot{H}_T \rangle$ must change accordingly in order to keep the energy balance.

The pressure-induced enthalpy flow usually dominates the total heat flow of the last-stage regenerator working between below 50 K and around 4.2 K. $\dot{Q}_{tot}$ should be larger than the maximum $\langle \dot{H}_p \rangle$ since neither $\dot{Q}_{cond}$ nor $\langle \dot{H}_T \rangle$ can be negative. The maximum $\langle \dot{H}_p \rangle$ usually occurs at the cold end in this temperature range, as shown in Figure 1. According to Eq. (7), the maximum of the sum of $\dot{Q}_{cond}$ and $\langle \dot{H}_T \rangle$ must be larger than the maximum variation of $\langle \dot{H}_p \rangle$ within the working temperature range, expressed as:

$$\left(\langle \dot{H}_T \rangle + \dot{Q}_{cond}\right)_{max} \geq \langle \dot{H}_p \rangle_{max} - \langle \dot{H}_p \rangle_{min} \tag{8}$$

Which factor of $\langle \dot{H}_T \rangle$ and $\dot{Q}_{cond}$ dominates the responsibility of the energy balance? According to the calculation[9] of three cases of PTCs , the maximum of the heat conduction is one to two-orders smaller than the maximum variation of the pressure-induced enthalpy flow in these cases. This conclusion can be extended to general cases. The only term that can keep the energy balance in practical circumstances is the heat-associated enthalpy flow. The heat-associated enthalpy flow can be in of the same magnitude of the pressure-induced enthalpy flow, and it is able to conserve the energy balance. Evidence has been found[9] from the numerical calculation for PTC in reference[2] and another PTC in reference[16]. In most cases, the heat conduction can be omitted since it constitutes a very small fraction. This point is used in the derivation in the following content.

FUNCTION OF THE REGENERATOR TEMPERATURE PROFILE

The temperature profile of regenerators deviates from linear to meet the requirement of energy balance. The relationship between the heat-associated enthalpy flow and the temperature gradient is proportional, which has been derived in a previous study[9]. The factor between the two terms is noted as k_T. This factor is derived to be:

$$k_T = -\frac{0.5946}{\pi^{1.3}} \frac{\left((1/\tau)(1-\emptyset)\rho_m c_{p,m} + h_{ht}\emptyset/d_h\right) c_{p,g} {d_h}^{1.3} Pr^{2/3} \cos\delta \langle PV \rangle^{0.65}}{\left((1/\tau)(1-\emptyset)\rho_m c_{p,m}\right) a_2 \mu^{0.3} \emptyset^{0.3} d_{reg}^{0.6} \left(Zv\cos(\varphi-\theta)\right)^{0.65}} \tag{9}$$

Considering the energy balance, and substituting the expression for $\langle \dot{H}_p \rangle$ and $\langle \dot{H}_T \rangle$, the governing equation for the temperature profile is expressed as:

$$\left(\langle pV \rangle_h / (ZT)_h\right) \frac{d[(1-T\beta)ZT]}{dx} - k_T \frac{d^2T}{dx^2} = 0 \tag{10}$$

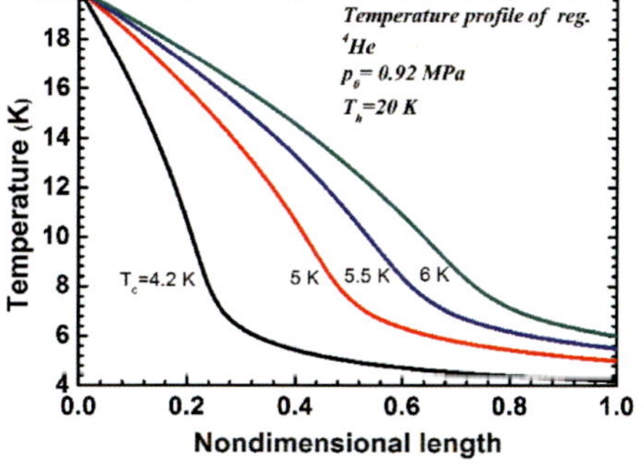

Figure 3. Calculated temperature distribution along the regenerator

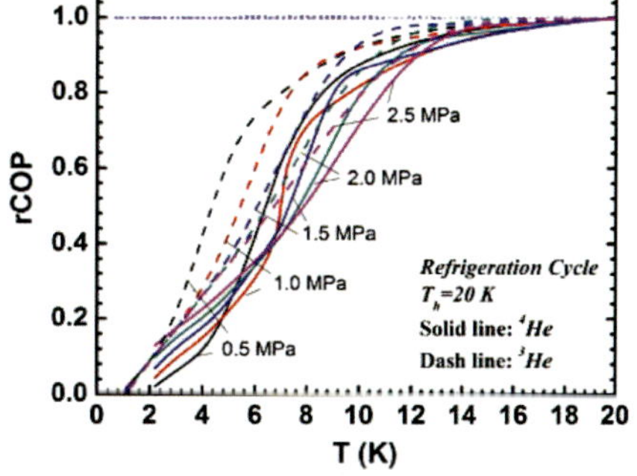

Figure 4. Dependence of the relative COP of the refrigerator cycle on the cold end temperature

An example of the temperature distribution is shown in Figure 3 based on the calculation of a PTC[14] with a hot end temperature of 20 K. The nonlinearity dominates as the cold end temperature approaches 4.2 K. The temperature profile is in accordance with previous experiment observation[4].

DERIVATION OF THE THEORETICAL COP

Since the total enthalpy flow is kept constant in the regenerator section between the heat exchangers, the heat-associated enthalpy flow at any location can be derived relative to the point when the pressure-induced enthalpy flow reaches a maximum of the regenerator section, as the equation shows:

$$\langle \dot{H}_p \rangle + \langle \dot{H}_T \rangle = \left(\langle \dot{H}_p \rangle + \langle \dot{H}_T \rangle \right)_{max\langle \dot{H}_p \rangle} \tag{11}$$

The most theoretical scenario of the regenerator, i.e. with least losses and not against the thermodynamic laws, can be ideal heat transfer when needed, the heat-associated enthalpy flow at that point equals to zero. Thus, the local heat-associated enthalpy flow can be rewritten as:

$$\langle \dot{\mathrm{H}}_{\mathrm{T}} \rangle = \max\langle \dot{\mathrm{H}}_{\mathrm{p}} \rangle - \langle \dot{\mathrm{H}}_{\mathrm{p}} \rangle \tag{12}$$

Substitute the equation to the expression for the refrigeration power available at the cold end[9]. The refrigeration power is expressed as

$$\dot{\mathrm{Q}}_{\mathrm{c}} = \langle \mathrm{pV} \rangle_{\mathrm{c}} - \max\langle \dot{\mathrm{H}}_{\mathrm{p}} \rangle \tag{13}$$

The COP of the regenerator is expressed as:

$$\mathrm{COP} = \frac{\langle \mathrm{pV} \rangle_{\mathrm{c}} - \max\langle \dot{\mathrm{H}}_{\mathrm{p}} \rangle}{\langle \mathrm{pV} \rangle_{\mathrm{h}} - \langle \mathrm{pV} \rangle_{\mathrm{c}}} \tag{14}$$

Substituting Eq. (3), (4) for the pressure-induced enthalpy flow and for the PV power, the COP can be rewritten as

$$COP = \frac{(ZT)_c - max[(1-\mathrm{T}\beta)ZT]}{(ZT)_h - (ZT)_c} \tag{15}$$

The relative COP, the COP divided by the Carnot COP, for ^{4}He and ^{3}He at the pressure range of 0.5-2.5 MPa is plotted in Figure 4. The hot end temperature is set to be 20 K. The COP is always smaller than the Carnot COP, because there is a heat loss named heat-associated enthalpy flow along the regenerator except the point with the maximum pressure-induced enthalpy flow.

Note that the theoretical COP for ^{4}He between a temperature below about 5 K and 20 K increases with the pressure. This is right in a vast pressure range of 0.1-10 MPa. The theoretical COP at 2.5 MPa is 50% higher than that at 1 MPa. Certainly there is a barrier of matrix heat capacity in practical cases. Still it means that the measured COP could be increased if the matrix can be improved.

This calculation is in accordance with the analysis in reference[6]. Although the expansion work of the pulse tube cryocooler is regarded to be dissipated in the calculation, the COP for the regenerator should be the same.

CONCLUSION

In this paper, the significance of the pressure-induced enthalpy flow and the main factor in keeping the energy balance of the regenerator have been discussed. Consequently, the temperature profile can be derived based on the energy equation. We have further derived the expression of the theoretical COP of the regenerator, taking into account of the real gas effects. These discussion may contribute to the understanding of the working mechanism of the regenerator.

ACKNOWLEDGMENT

This work is supported by the National Natural Science Foundation of China (No. 51506152) and "Yong Scholar Cultivating Program" of Tongji University (contract no.: 2014KJ034). The valuable discussion with Dr. P. Kittel, an emeritus member of NASA Ames Research Center and Prof. A.T.A.M. de Waele of Eindhoven University of Technology, on the topic of real gases is acknowledged. We are grateful to the relative discussion with Prof. Dr. S.W. Zhu from Tongji University,

Prof. Dr. Z.H. Gan and Prof. Dr. L.M. Qiu from Zhejiang University, Dr. R. Radebaugh from NIST. We also appreciate Dr. Y.H. Huang of Shanghai Jiao Tong University for providing the property data of ^{3}He.

REFERENCES

1. Radebaugh, R., Huang, Y., O'Gallagher, A., and Gary, J., "Optimization calculations for a 30 Hz, 4 K regenerator with helium-3 working fluid," *Adv. in Cryogenic Engineering*, Vol. 55, Amer. Institute of Physics, Melville, NY (2010), pp. 1581-1592.

2. Wang, C., "Numerical analysis of 4 K pulse tube coolers: Part I. Numerical simulation," *Cryogenics,* vol. 37, no.4 (1997), pp. 207-213.

3. Radebaugh, R., O'Gallagher, A., and Gary, J., "Regenerator behavior at 4 K: Effect of volume and porosity," *Adv. in Cryogenic Engineering*, Vol. 47B, Amer. Institute of Physics, Melville, NY (2002), pp. 961-968.

4. Lang, A., Hafner, H. U., and Heiden, C., "Systematic investigations of regenerators for 4.2K-refrigerators," *Adv. in Cryogenic Engineering*, Vol. 43B, Springer US (1998), pp. 1573-1580.

5. Seshake, H., Eda, T., Matsumoto, K., and Hashimoto, T., "Analysis of rare earth compound regenerators operating at 4 K," *Adv. in Cryogenic Engineering*, Vol. 37B, Springer US (1992), pp. 995-1001.

6. Radebaugh, R., Huang, Y., O'Gallagher, A., and Gary, J., "Calculated regenerator performance at 4 K with helium-4 and helium-3," *Adv. in Cryogenic Engineering*, Vol. 53, Amer. Institute of Physics, Melville, NY (2008), pp. 225-234.

7. Kittel, P., "Enthalpy, entropy, and exergy flows real gas effects in ideal pulse tube cryocoolers," *Adv. in Cryogenic Engineering*, Vol. 51, Amer. Institute of Physics, Melville, NY (2006), pp. 345-352.

8. Will, M. E. and de Waele, A. T. A. M., "Ideal pulse-tube refrigerators with real gases," *Journal of Applied Physics,* vol. 98, no.4 (2005): 044911.

9. Cao, Q., Qiu, L. M., and Gan, Z. H., "Real gas effects on the temperature profile of regenerators," *Cryogenics,* vol. 61, (2014), pp. 31-37

10. Radebaugh, R., "Thermodynamics of Regenerative Refrigerators," *Generation of Low Temperature and Its Applications*, Kamakura, Japan (2003), pp. 1-20.

11. Lemmon, E. W., McLinden, M., and Huber, M., "NIST Standard Reference Database 23, NIST Reference Fluid Thermodynamic and Transport Properties REFPROP," U.S. Dept. of Commerce (2007).

12. Huang, Y. H., Chen, G. B., and Arp, V. D., "Debye equation of state for fluid helium-3," *J. Chem. Phys.,* vol. 125, no.5 (2006): p. 054505.

13. Gao, J. L., Hiresaki, Y., and Matsubara, Y., "A Hybrid Two-Stage Refrigerator Operated at Temperatures Below 4K," *Adv. in Cryogenic Engineering*, Vol. 41B, Springer US (1996), pp. 1495-1502.

14. Cao, Q., Qiu, L. M., Zhi, X. Q., Han, L., Gan, Z. H., Zhang, X. B., Zhang, X. J., and Sun, D. M., "Impedance magnitude optimization of the regenerator in Stirling pulse tube cryocoolers working at liquid-helium temperatures," *Cryogenics,* vol. 58, (2013), pp. 38-44.

15. Xu, M. Y., de Waele, A., and Ju, Y. L., "A pulse tube refrigerator below 2 K," *Cryogenics,* vol. 39, no.10 (1999), pp. 865-869.

16. Wang, C., "Numerical analysis of 4 K pulse tube coolers: Part II. Performances and internal processes," *Cryogenics,* vol. 37, no.4 (1997), pp. 215-220.

Periodic Flow Hydrodynamic Resistance Parameters for Regenerator Filler Materials at Cryogenic Temperatures

M.D. Perrella and S.M. Ghiaasiaan
Georgia Institute of Technology
Atlanta, GA 30332

ABSTRACT

The regenerator is a critical component of all Stirling and Pulse Tube cryocoolers. It generally consists of a microporous metallic or rare-earth filler material within a container shell. The accurate modeling of the hydrodynamic and thermal behavior of different regenerator materials is crucial to the successful design and analysis of regenerative cryogenic systems. Due to the difficulty associated with experimental measurement at cryogenic temperatures, previous investigations that have used experimental measurements at steady and periodic flow conditions in conjunction with pore-level CFD analysis to determine the pertinent hydrodynamic parameters, namely the Darcy permeability and Forchheimer coefficients, have relied on experiments performed at ambient conditions. These results are assumed to be accurate for cryogenic temperatures since, for fully-developed flow, these parameters should depend only on the geometry of the porous medium (for Darcy permeability) and flow Reynolds number (for Forchheimer coefficient). There is a pressing need for the verification of the validity of the foregoing assumption. In this study, regenerators filled with spherical Er50Pr50 powder were assembled and tested under periodic helium flow at cryogenic temperatures. The mass flow and pressure drop data were correlated using a porous media CFD model to determine the Darcy Permeability and Forchheimer coefficients. These results are compared to the results of previous investigations at ambient temperature conditions, and the relevance of room-temperature models and correlations to cryogenic temperatures are critically assessed.

INTRODUCTION

Realistic prediction of the regenerator performance depends heavily upon the accurate modeling of the fluid-solid hydrodynamic and thermal interactions within the porous filler material of the regenerator. While it is theoretically possible to model the entire PTC system including the porous material on a microscopic level by directly solving the conservation equations, such an approach is computationally expensive and requires detailed knowledge of the microscopic geometry of the medium. The common approach is to convert the microscopic governing equations to the macroscopic level using volume averaging or a similar approach [1–5]. For the conservation of momentum under steady state conditions, for example, this yields the extended Darcy- Forchheimer model equation (1)

$$\langle \vec{u}_f \rangle = -\frac{\overline{\overline{K}}}{\mu_f} \cdot \left(\nabla \langle P_f \rangle^f - \rho_f \vec{g} \right) - \overline{\overline{F}} \cdot \langle \vec{u}_f \rangle \qquad (1)$$

Cryocoolers 19, edited by S.D. Miller and R.G. Ross, Jr.

where $\overline{\overline{K}}$ represents the Darcy permeability tensor and $\overline{\overline{F}}$ represents the Forchheimer inertial coefficient tensor, which are frequently referred to as hydrodynamic resistance parameters. These closure parameters are needed for CFD-assisted optimization of the regenerator design.

There are currently two main methods for determining the permeability and inertial coefficients for a particular medium which are pore-level CFD simulation and experimentation. In the first approach microscopic or pore-level CFD models of the regenerator are developed to determine the pressure drop characteristics by comparing the results to the macroscopic volume-averaged equations to determine the closure parameters. This technique has been employed successfully by multiple researchers including [6]. CFD simulations have also been used to model the heat transfer and thermal dispersion in porous media [7, 8]. This approach is limited because the real geometric configuration of a porous medium is often quite complicated and variable. Instead, experimental regenerator testing has been utilized extensively by researchers.

Using experiments and a one-dimensional porous media flow model Harvey [9] determined the hydrodynamic resistance parameters for metal foam, #400SS mesh, #325SS mesh, and 60μm perforated disks. Cha [10, 11] determined the steady and oscillatory flow closure parameters in the radial and axial directions for a variety of porous media using a CFDassisted methodology to iteratively determine the closure parameters. Cha also showed that oscillatory and steady flow closure parameters can be different. This method was later pursued by Clearman [12] and Landrum [13]. Other aspects of pulse tube cryocoolers (PTCs) were investigated by Conrad [14], Pathak [15], and Mulcahey [16].

Rare-earth alloy regenerator fillers suitable for applications in multistage PTCs where the regenerator of the final stage may operate at temperatures in the 10 – 20K range or lower include ErPr (erbium-praseodymium) alloy, which has a high heat capacity at and below 20K. Pathak [17, 18] performed experiments with a packed-bed of 69μm average diameter Er0.5Pr0.5 (alloys with 50% Er and 50% Pr) spheres in a prototypical regenerator for steady and oscillatory flows of helium at room temperature for a range of frequencies, charge pressures, and gas velocities. The Darcy permeability, which in general depends only on the geometry, remained relatively constant. The Forchheimer coefficient varied with the mass flow rate and was correlated in terms of the pore-based Reynolds number.

A key shortcoming of these and many other published studies are that their tests were conducted at room temperature. In theory, hydrodynamic resistance parameters should not be affected by the operating temperature if the geometric configuration of the porous medium remains the same, experimental verification of the forgoing assumption is needed. The objective of this work is thus to determine the Darcy permeability and Forchheimer inertial coefficients for Er0.5Pr0.5 powder, at cryogenic temperatures for a range of pressure and velocity amplitudes and compare the results to those obtained at ambient temperature conditions.

EXPERIMENTAL SETUP

The test section is shown schematically in Figure 1. The test section consists of a stainless regenerator sandwiched between a heat exchanger and a 7.24cc stainless steel surge volume with a Q-drive model 2S132W pressure wave generator (PWG) upstream connected via a 0.91m stainless steel transfer tube. The cold heat exchanger (CHX), is thermally attached to with the 1st stage of a Sumitomo model RDK-408D2 GM cryocooler with a copper bus bar. The first stage cold head of the Sumitomo GM cryocooler can deliver 50W of cooling at 43K, while the second stage provides 1W cooling power at 4.2K. The interior of the CHX is packed with #100 mesh copper screens (64.7% porosity). The regenerator is packed with Er0.5Pr0.5 micro spheres with 55μm mean diameter and 37% porosity. The ErPr powder is retained by #400 SS mesh screens at the inlet and outlet which are reinforced by two #60 SS screens. PCB piezotronics brand pressure sensors (models 102A05 and 102A10) are installed in specially designed ports on the upstream and downstream sides of the CHX and SV. The experimental setup and the GM cryocooler are mounted inside of a vacuum sealed dewar with modular feedthroughs (Figure 2). This dewar is capable of maintaining an insulating vacuum of 10^{-6} torr.

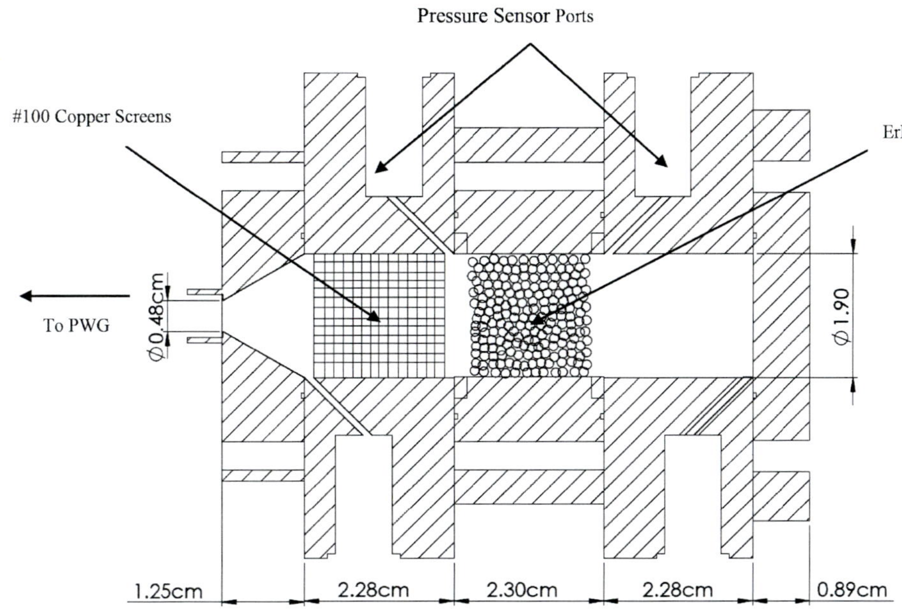

Figure 1. Experimental setup schematic

Instantaneous pressure measurements were made upstream and downstream of the regenerator test section in 2 second durations at a sample rate of 25.6kHz for a total of 51200 samples. The oscillatory pressure measurements were analyzed using Matlab's Fast Fourier Transform (FFT) capability for the fundamental frequency and the next four harmonics of the fundamental. Temperature was measured using a Cryo-con S900 silicon diode thermally mounted with cryogenic rated epoxy to the CHX. The mass flow between the regenerator test section and the surge volume (SV) is calculated by considering the instantaneous pressure and temperature within the SV and assuming that the surge volume is rigid and adiabatic. A simple analysis then leads to

$$\dot{m} = \frac{V}{gRT}\frac{dP}{dt} \tag{2}$$

The time derivative of the SV pressure can be determined numerically from the collected data to calculate the instantaneous mass flow rate into the SV. A FFT is then performed in order to develop an equation for mass flow rate as a function of time.

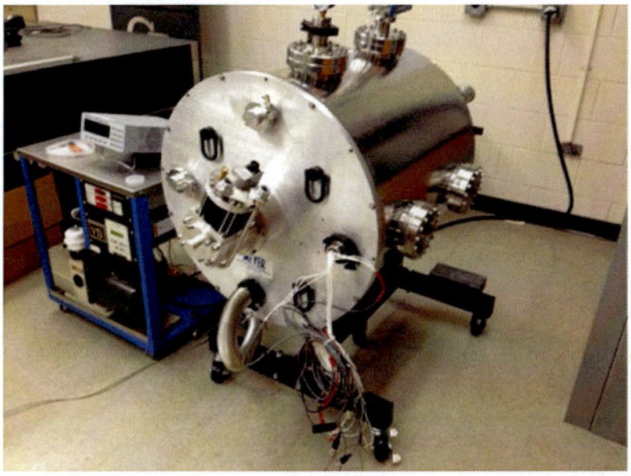

Figure 2. Cryogenic vacuum chamber with modular feedthroughs

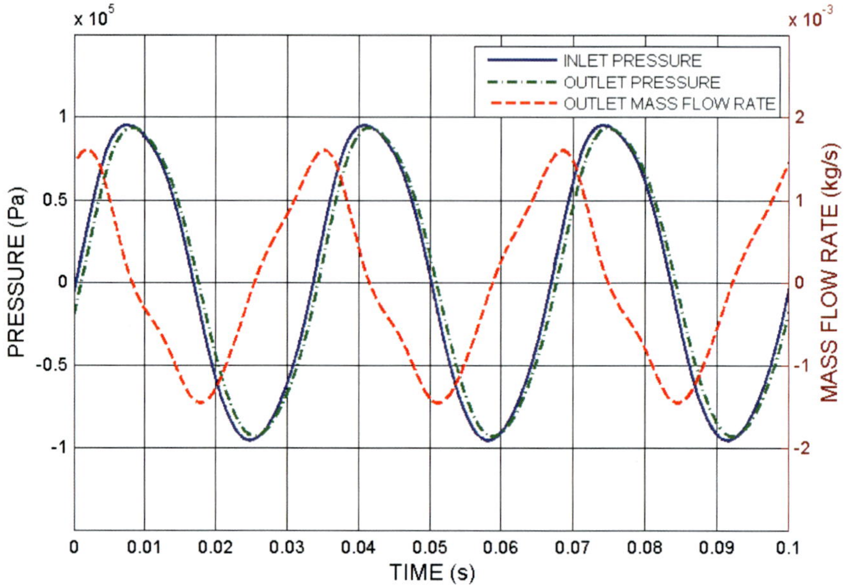

Figure 3. Inlet and outlet pressure oscillations with outlet mass flow rate at 75K and 30VAC PWG input

Some of the results are characterized using the pore based Reynolds number, which uses the square root of the permeability as the length scale.

$$\mathrm{Re}_K = \frac{|\dot{m}|K^{0.5}}{\mathrm{m}A} \tag{3}$$

Figure 3 shows typical upstream and downstream pressure oscillations along with the calculated mass flow rates for the same PWG input power at 75K. Figure 4 and Figure 5 show the amplitude of the instantaneous pressure drop across the regenerator versus the pore based Reynolds number for all samples. The center line represents a linear fit to the experimental data, and the bounding lines represent the upper and lower limits of the data based on the random uncertainty.

CFD ANALYSIS

The methodology has been discussed before [10, 18] and will be discussed briefly here. The CFD code ANSYS Fluent [19] is used to model the regenerator test section as a porous medium

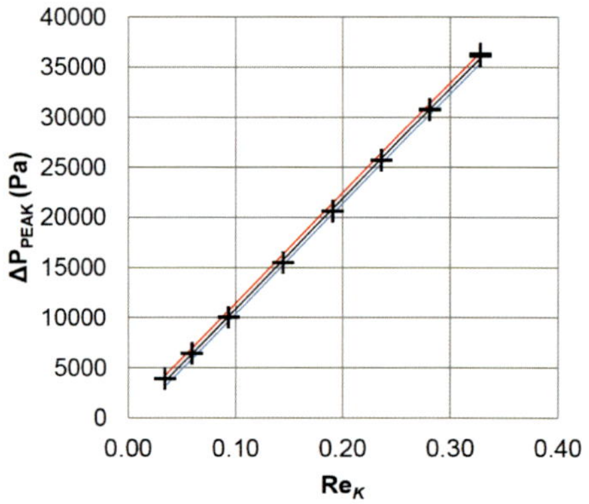

Figure 4. Regenerator pressure drop amplitude vs. Re_K at 300K

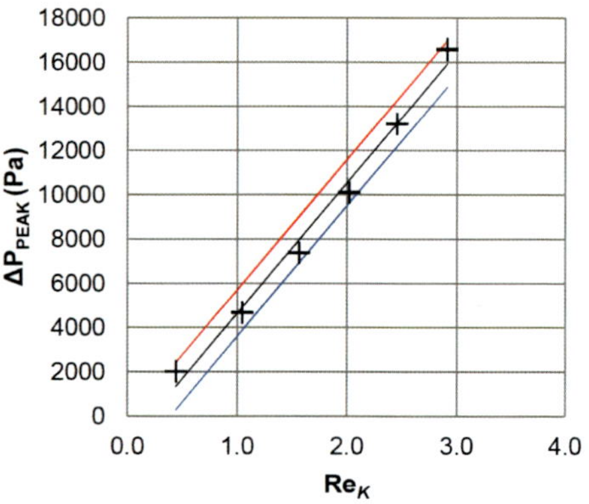

Figure 5. Regenerator pressure drop amplitude vs. Re_K at 75K

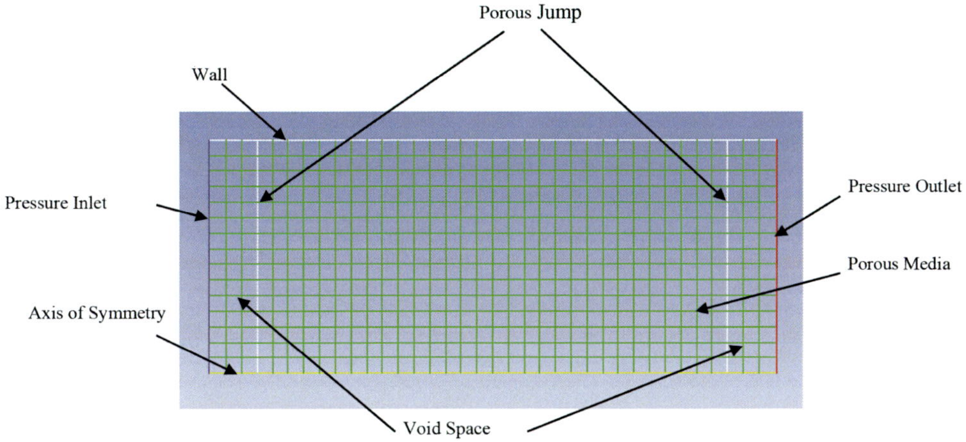

Figure 6. 2D axisymmetric geometry in Fluent

with helium as the working fluid. The governing equations for mass, momentum, and energy conservation in Fluent are as follows:

$$\frac{\partial \varepsilon\rho}{\partial t}+\nabla\cdot(\varepsilon\rho\,\vec{u})=0 \tag{4}$$

$$\frac{\partial(\varepsilon\rho\vec{u})}{\partial t}+\nabla\cdot(\varepsilon\rho\vec{u}\vec{u})+\varepsilon\nabla P-\nabla\cdot(\varepsilon\bar{\bar{\tau}})=-\left(\frac{\mu}{\bar{\bar{\beta}}}\cdot\vec{u}+\frac{\bar{\bar{C}}\rho}{2}\cdot\left|\vec{u}\right|\vec{u}\right) \tag{5}$$

$$\nabla\cdot[(\varepsilon k_f+(1-\varepsilon)k_s)\nabla T+(\bar{\bar{\tau}}\cdot\varepsilon\vec{u})]=\frac{\partial}{\partial t}(\varepsilon\rho_f E_f+(1-\varepsilon)\rho_s E_s)+\nabla\cdot(\varepsilon\vec{u}(\rho_f E_f+P)) \tag{6}$$

where $\bar{\bar{\beta}}$ (m^2) represents the viscous resistance tensor and $\bar{\bar{C}}$ (m^{-1}) the inertial resistance tensor of porous media. These equations are of course three-dimensional. The experiments involve axisymmetric flow, however, and 2-D flow simulations are performed using polar cylindrical coordinates. One can compare the momentum equation, Equation (5), with the 1-D momentum equation, Equation (1). If the porous medium is isotropic or the axial direction is a principle direction of the porous medium, then:

$$K=\varepsilon^2\beta \tag{7}$$

$$F=\frac{CK^{0.5}}{2\varepsilon^3} \tag{8}$$

For the packed bed of ErPr powder, isotropy is assumed. For flow through screens previous studies have shown that velocities in the radial direction are typically very small in comparison with the axial velocities and that CFD simulation results are insensitive to errors and inaccuracies in radial permeability and Forchheimer coefficients [10–13]. As a result, when axial resistance parameters are studied, for CFD simulations isotropic porous media are assumed.

The 2-D, axisymmetric geometry used in CFD simulations is shown in Figure 6. A mesh size of 0.635mm was used for all modeling based on mesh convergence and grid-independence tests. A user defined function (UDF) was created to apply the oscillatory pressure boundary conditions to the inlet and outlet. The #400 SS retaining mesh of the regenerator is modeled as a porous jump using the values of $\beta=2.5295\times10^{-11}\mathrm{m}^2$ and $C=120000\mathrm{m}^{-1}$ from Cha [10].

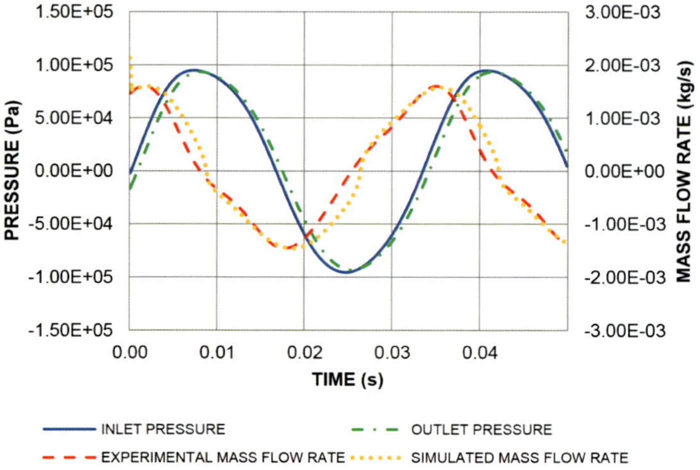

Figure 7. Simulated mass flow rate at 300K and 30VAC PWG input (Re_K=0.26)

Due to the inherent limitations of our measurement equipment, we were unable to measure the pressure drop across the regenerator within reasonable uncertainty limits at sufficiently low mass flow rates that would justify neglecting the inertial resistance. We therefore have used the permeability determined by Pathak [18], $K = 1.93\times10^{-11}\,\mathrm{m}^2$, which corresponds to a viscous resistance of $b = 1.41\times10^{-10}\,\mathrm{m}^2$. Subsequent parametric calculations confirmed that this assumption was justified. Once the proper boundary and operating conditions are supplied to CFD simulations, the inertial resistance is varied systematically until the simulated mass flow rate amplitude at the outlet matches the experimental value. Figure 7 and Figure 8 show some typical results. The simulations show excellent agreement with respect to both the magnitude and phase of the oscillatory mass flow rate.

RESULTS AND DISCUSSIONS

The results for all simulated data points are summarized in Figure 9 where the Forchheimer inertial resistance coefficient, *F*, is displayed as a function of the pore-based Reynolds number defined in Equation (3). The data in this study is also compared to the data of Pathak [18], which were obtained with Er0.5Pr0.5 powder at room temperature. It can be noted that the data obtained in the present study agree with Pathak's measurements very well in terms of trends. In terms of

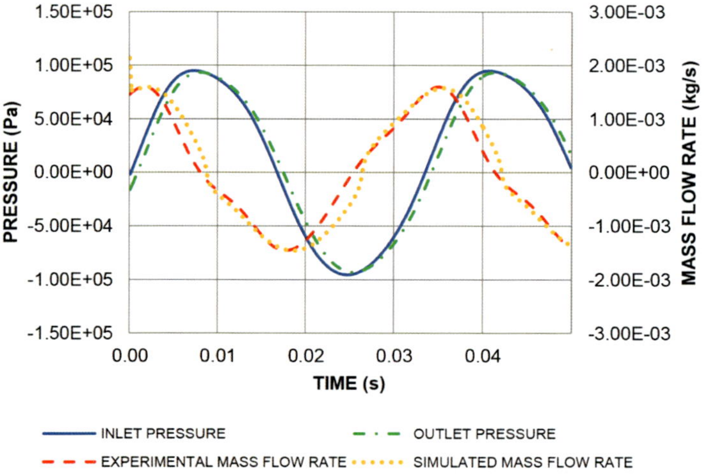

Figure 8. Simulated mass flow rate at 75K and 30VAC PWG input (ReK=2.71)

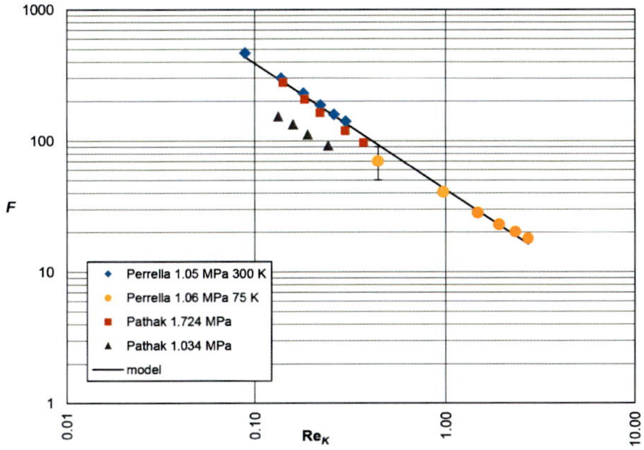

Figure 9. Variation of parameter F vs. pore-based Reynolds number, ReK

the magnitude of parameter F, the agreement between the two data sets is within a factor of two. The main reason for the disagreement could be the fact that in Pathak's experiments the Er0.5Pr0.5 particles were larger (63 to 75μm in diameter, compared with ≈50μm particle mean diameter) than in this study. More importantly, the results of this investigation show that the Darcy permeability and Forchheimer coefficient have similar values at 300K and 75K temperatures, suggesting that the coefficients may indeed be independent of temperature, as previously conjectured. The results also agree with the logarithmic relationship for Forchheimer coefficient derived by Pathak. The parameter F in this study was found to obey a power law relationship with pore scale Reynolds number, yielding the following correlation:

$$F = 42.48\,\mathrm{Re}_K^{-0.96} \tag{9}$$

A potential source of discrepancy between room temperature parameters and those measured at cryogenic temperatures is the possibility of subtle configurational changes in a regenerator, which is often assembled at room temperature, when its temperature is reduced to cryogenic levels. The apparent insensitivity of the solid-fluid momentum transport parameters in this investigation suggest that the effect of these configurational variations are negligible at least for packed powders.

While these findings are instructive, more work should be done to test the effects of varying operating frequencies, mean pressures, and even lower operating temperatures. More experiments addressing lower temperatures at as various frequencies and with different regenerator fillers are currently underway at Georgia Tech Cryo Lab.

CONCLUSION

In this study, a prototypical regenerator filled with spherical Er50Pr50 powder was assembled and tested under periodic helium flow at room (300K) and a cryogenic (75K) temperatures. The measured mass flow and pressure drop data were correlated with a porous media CFD model to determine the Darcy permeability and Forchheimer coefficients. The Darcy permeability, which depends only on the geometry of the porous medium, was taken from previous investigations using similar ErPr powder with a value of $K = 1.93\times10^{-11}\,\mathrm{m}^2$. The Forchheimer coefficients was found to decrease logarithmically with increasing pore-based Reynolds number according to Equation (9). The results of this analysis agree well with the results of previous investigations conducted at ambient conditions with slight discrepancies most likely due to the difference in ErPr particle size between the two studies. The values for Forchheimer coefficient in this analysis for temperatures of 300K and 75K at roughly the same mean pressure follow the same logarithmic trend and are well predicted by the same empirical correlation. Together, these findings suggest that the calculation of the porous media hydrodynamic resistance parameters is indeed independent of temperature as previously believed. Work is underway to strengthen this conclusion including experiments at lower temperatures with additional filler materials, mean pressures, and frequencies.

REFERENCES

1. Ochoa-Tapia, J. A., and Whitaker, S., "Momentum transfer at the boundary between a porous medium and a homogeneous fluid—I. Theoretical development," *Int. J. Heat Mass Transf.*, 38(14), (1995), pp. 2635–2646.
2. Ochoa-Tapia, J. A., and Whitaker, S., "Heat transfer at the boundary between a porous medium and a homogeneous fluid," *Int. J. Heat Mass Transf.*, 40(11), (1997), pp. 2691–2707.
3. Ochoa-Tapia, J. A., and Whitaker, S., "Momentum transfer at the boundary between a porous medium and a homogeneous fluid—II. Comparison with experiment," *Int. J. Heat Mass Transf.*, 38(14), (1995), pp. 2647–2655.
4. Whitaker, S., The method of Volume Averaging. Theory and Applications of Transport in Porous Media, *Kluwer Academic Publishers*., (1999)
5. Whitaker, S., "The Forchheimer equation: A theoretical development," *Transp. Porous Media*, 25(1), (1996), pp. 27–61.
6. Kim, S.-M., and Ghiaasiaan, S. M., "Numerical Modeling of Laminar Pulsating Flow in Porous Media," *J. Fluids Eng.*, 131(4), (2009), p. 041203.
7. Pathak, M. G., and Ghiaasiaan, S. M., "Convective heat transfer and thermal dispersion during laminar pulsating flow in porous media," *Int. J. Therm. Sci.*, 50(4), (2011), pp. 440–448.
8. Pathak, M. G., Mulcahey, T. I., and Ghiaasiaan, S. M., "Conjugate heat transfer during oscillatory laminar flow in porous media," *Int. J. Heat Mass Transf.*, 66, (2013), pp. 23–30.
9. Harvey, J. P., "Oscillatory compressible flow and heat transfer in porous media–application to cryocooler regenerators." PhD diss., Georgia Institute of Technology, 2003.
10. Cha, J. S., Ghiaasiaan, S. M., and Kirkconnell, C. S., "Oscillatory flow in microporous media applied in pulse - tube and Stirling - cycle cryocooler regenerators," *Exp. Therm. Fluid Sci.*, 32(6), (2008), pp. 1264–1278.
11. Cha, J., "Hydrodynamic parameters of micro porous media for steady and oscillatory flow: application to cryocooler regenerators," (2007).
12. Clearman, W. M., Cha, J. S., Ghiaasiaan, S. M., and Kirkconnell, C. S., "Anisotropic steady-flow hydrodynamic parameters of microporous media applied to pulse tube and Stirling cryocooler regenerators," *Cryogenics (Guildf).*, 48(3-4), (2008), pp. 112–121.
13. Landrum, E. C., Conrad, T. J., Ghiaasiaan, S. M., and Kirkconnell, C. S., "Hydrodynamic parameters of mesh fillers relevant to miniature regenerative cryocoolers," *Cryogenics (Guildf).*, 50(6-7), (2010), pp. 373–380.
14. Conrad, T. J., Ghiaasiaan, S. M., and Kirkconnell, C. S., "Simulation of Boundary Layer Effects in the Pulse Tube of a Miniature Cryocooler," *Cryocoolers 16*, ICC Press, Boulder, CO (2011), pp. 267–274.
15. Pathak, M. G., Ghiaasiaan, S. M., Radebaugh, R., Kashani, A., Feller, J., and Field, M., "The Design and Development of a High-Capacity Cryocooler Regenerator for Space Exploration," *AIAA SPACE 2012 Conference & Exposition*, Pasadena, CA (2012), pp. 1–8.
16. Mulcahey, T. I., Conrad, T. J., Ghiaasiaan, S. M., and Pathak, M. G., "Investigation of gravitational effects in pulse tube cryocoolers using 3-D CFD," *Adv. in Cryogenic Engineering*, Vol. 59, Amer. Institute of Physics, Melville, NY (2014), pp. 1002–1009.
17. Pathak, M. G., "Periodic flow physics in porous media of regenerative cryocoolers," (2013).
18. Pathak, M. G., Patel, V. C., Ghiaasiaan, S. M., Mulcahey, T. I., Helvensteijn, B. P., Kashani, a., and Feller, J. R., "Hydrodynamic parameters for ErPr cryocooler regenerator fillers under steady and periodic flow conditions," *Cryogenics (Guildf).*, 58, (2013), pp. 68–77.
19. "ANSYS Fluent," 2013

Heat Capacity Characterization of a 4 K Regenerator with non-Rare Earth Material

W. Chen, M. Jaeger, K. Rule

Creare LLC
Hanover, NH 03755

ABSTRACT

This paper presents separate effects test results of an advanced regenerator for a 4 K regenerative cryocooler. The regenerator is made of a non-rare earth material. It utilizes a unique approach to significantly enhance the regenerator's thermal capacity, which is the key to the efficient operation of a regenerative cryocooler. The regenerator employs a unique microchannel configuration that enables it to achieve high thermal performance at a low pressure drop. These performance features together will enable a regenerative cryocooler to achieve high thermal efficiency at low temperatures. To determine its effective heat capacity and heat transfer performance under prototypical conditions, the thermal response of the regenerator to a step temperature change in the inlet flow is characterized. This paper first describes the testing approach and the test setup. It then presents the measured heat capacity and the predicted values. Finally, the volumetric specific heat of the regenerator is compared with those of typical rare earth materials used in a 4 K regenerator.

INTRODUCTION

Superconductor digital electronics utilizing Josephson Junctions have very fast switching time and are able to operate at very high speeds in excess of 30 GHz for wideband, high frequency RF signal processing. A critical need for these superconductor electronics is reliable, efficient, and portable cryocoolers that can provide cooling at temperatures below 4 K to maintain the superconductivity of niobium-based microelectronics. Pulse tube cryocoolers are a promising technology for this application because there are no moving parts at their cold ends, minimizing mechanical vibration transmitted to cooled sensors and electronics and enhancing reliability. However, it is currently not possible for this type of cryocooler to reach temperatures below 4 K while maintaining acceptable thermal efficiency. The poor performance is mainly caused by the performance degradation in the regenerator at very low temperatures.

The losses in the regenerator account for more than 50% of the losses in a typical 4 K pulse tube cryocooler (Radebaugh et al. 2009). At low temperatures, one of the performance-limiting factors for a regenerator is its very limited thermal capacity. The specific heat of most conventional regenerator matrix materials falls off sharply with decreasing temperatures below about 40 K, proportional to the cube of the temperature ($Cp \propto T^3$). The decrease in specific heat substantially reduces the thermal mass of the regenerator at the cold end, limiting its capacity to cool the gas flowing to the cold expansion space. The low thermal capacity in the regenerator

Cryocoolers 19, edited by S.D. Miller and R.G. Ross, Jr.

causes a large temperature swing during periodic heat transfer and significantly reduces the efficiency of the regenerator.

The problem of limited thermal capacity in low-temperature regenerators cannot be solved by simply increasing the regenerator size. Increasing the size of a regenerator can indeed increase its heat transfer area and heat capacity. However, these benefits can be offset by the penalties associated with a larger regenerator void volume. A larger void volume increases compression heating and expansion cooling associated with the pressure oscillation in the regenerator. It also increases the gas flow rate at the warm end of the regenerator because of a larger dead volume.

Current research efforts have focused on utilizing rare-earth compounds that have relatively large magnetic heat capacities at low temperatures in the vicinity of their magnetic ordering transition temperatures. These compounds, however, are either very brittle (e.g., rare-earth/transition metal based intermetallic compounds) or very chemically reactive, limiting their practical use (Pecharsky et al. 1997). In addition, it is very difficult to use rare-earth compounds to produce small uniform spheres for packed beds or mesh screen with thin wires for a plate stack (Mérida and Barclay 1998). Furthermore, the peak heat capacity of each magnetic material only spans a very narrow temperature range. A hybrid matrix with layers of materials having different transition temperatures is required to achieve adequate heat capacity for sufficient performance.

REGENERATOR WITH ADSORBED HELIUM TO ENHANCE HEAT CAPACITY

The high heat capacity of the process helium in a pulse-tube cryocooler is the main reason why it is very challenging for a regenerator to achieve a high thermal efficiency. But the high heat capacity can also be the answer to the problem: adsorbed (trapped) helium can be used to store heat from the process gas (helium) passing through the regenerator. Porous adsorbents with high porosities, such as activated carbon, can adsorb significant amounts of helium at low temperatures. Moreover, the specific heat of helium molecules trapped in proper adsorbents remains very high. Consequently, the effective volumetric specific heat of helium in an adsorbent can be higher than those of rare-earth compounds (such as ErPr, GdRh, and ErNi) currently used in low-temperature regenerators. The specific heat of adsorbed helium and its macroscopic density in an adsorbent are discussed below.

Specific Heat of Adsorbed Helium Is Very High

The specific heat of helium depends strongly on the pore geometry and can be very high. The heat capacity of ^{4}He confined in adsorbents, such as molecular sieves, increases linearly with temperature (T) in the range between 3.25 K and 9 K (Wada et al. 1984). The adsorbed helium specific heat at temperature T can then be approximated as $(2.2 \times T)$ J/mol-K for $3.5 < T < 9$ K (Wada et al. 1984). It was also found that the specific heat of adsorbed helium can decrease as the channel dimensionality and pore size in the adsorbent material decrease (Konishi et al. 1993). The specific heat of helium molecules in an adsorbent can also decrease as the helium concentration in the adsorbent increases. Our study suggests that the minimum pore size of an optimum adsorbent should be larger than 20 Å (i.e., mesoporous adsorbents) to prevent a potential decrease in helium specific heat due to reduction in dimensionality.

Helium Densities in Adsorbents Can Be Very High

Typically, the macroscopic density of helium in an adsorbent is controlled by the porosity, pore size distribution of the adsorbent, and ambient pressure. For this particular application, the ambient helium pressure is very high, about 1 MPa in a typical pulse-tube cryocooler. This pressure is significantly higher than the capillary pressure associated with the porous structure. Consequently, the adsorbed helium densities in mesoporous adsorbents (20 Å $<$ pore size $<$ 500 Å) should be close to that in microporous adsorbents (pore size $<$ 20 Å). Using a mesoporous adsorbent prevents a potential decrease in helium specific heat. For these reasons, the helium adsorption density is largely controlled by the porosity of the adsorbent. Activated carbon can achieve a porosity higher than 60%, and a concentration of adsorbed helium higher than 0.1 g/cm^3 at 10 K (Duband et al. 1988).

MICROCHANNEL REGENERATOR WITH HELIUM ADSORBENT

Even though activated carbon can achieve very high helium adsorption capacities, its thermal conductivity is very low at temperatures below 20 K. This makes it difficult to transfer heat in or out of the helium/activated carbon mixture at high frequencies. Furthermore, thermally coupling activated carbon powders to a heat transfer surface is quite challenging. Due to these challenges, previous efforts to incorporate helium adsorbents into low-temperature regenerators have met with very limited success (Jones et al. 1987; Pron'ko et al. 1976). To overcome these problems, we developed a unique regenerator design with helium adsorbent in intimate thermal contact with convective transfer surfaces via a heat spreader. The helium adsorbent is a monolithic adsorbent directly integrated into the heat spreader surface by introducing micro pores in the heat spreader substrates. Consequently, there is no thermal resistance or thermal stresses at the interface between the adsorbent and the heat spreader. This approach also eliminates potential particulate contamination in the regenerator because the adsorbent is not a collection of loose particles.

The regenerator core consists of a stack of thin, micromachined silicon disks separated by thin polymer thermal insulators (Fig. 1). Each silicon disk has a thick layer of monolithic helium adsorbent at the center and an annular array of slots/fins farther out. The single crystal silicon disks serve as a heat spreader to effectively transfer heat from the process gas to the adsorbed helium in the helium adsorbent. When gas passes through the regenerator, thermal energy from the process gas first transfers to the fins near the outer rim of the disk, then spreads radially inward to the center base plate holding the adsorbent, and finally moves axially into the adsorbent, where a large amount of adsorbed helium provides a high heat capacity. Because the thermal diffusivity of silicon at low temperature is very high, thermal waves can easily pass from the fins to the center base plate.

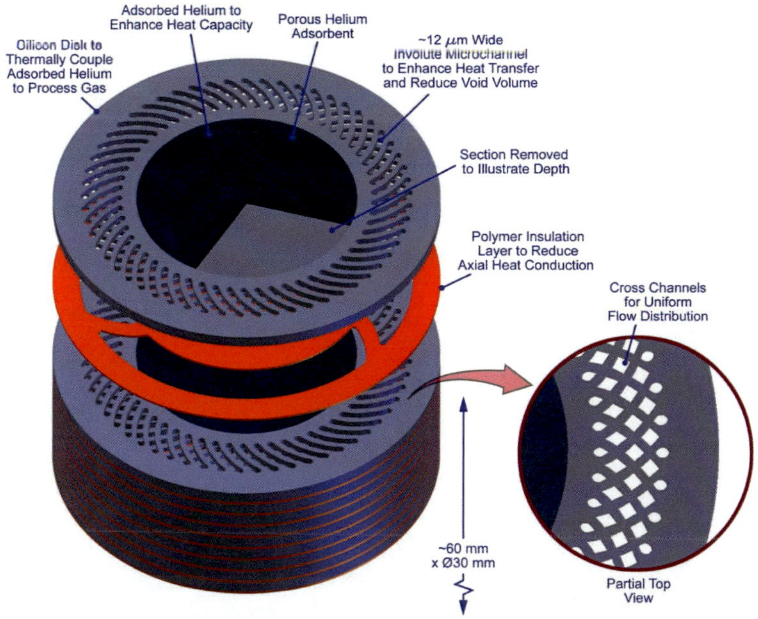

Figure 1. Configuration of a microchannel regenerator with adsorbed helium. Dimensions are not to scale for clarity.

Microchannels are machined in the silicon disks to form small flow paths for process helium, as shown in Fig. 2. Hundreds of involute slots are machined through the thickness of the wafer by Deep Reactive Ion Etching (DRIE). The involute spiral direction on one side of a disk is opposite to that on the other side. This allows the process gas to redistribute and helps to ensure that the flow distribution in the core is not sensitive to small variations in the channel geometry. The polymer spacer on each silicon disk reduces the conductive heat transfer along the regenerator axis. The composite disks are stacked together to form the regenerator core.

Brunauer-Emmett-Teller (BET) adsorption analysis shows that the pore surface area in the adsorbent is roughly 998 m^2/g, which is extremely high. The pore size distribution is shown in Fig. 3, suggesting the average pore diameter is roughly 3.5 nm, which is within the targeted range of 3 to 5 nm for this application.

The entire stack is housed inside a thin-walled titanium alloy shell. The stack is sandwiched between the hot end and cold flow distributors. One of the distributors includes a preload spring that maintains the composite disks in compression. The preload spring prevents the plates in the stack from separating from each other by gravity or vibrations and compensates for the mismatch in thermal expansion between the shell and the core.

(a) (b) (c)

Figure 2. Regenerator disks: (a) fin area of the silicon plate; (b) close-up view of the fin areal; (c) fracture surfaces of slot/fin features.

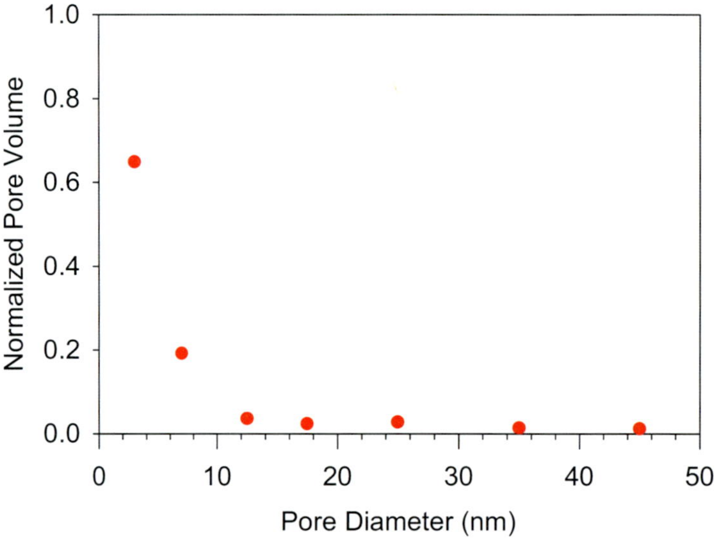

Figure 3. Pore size distribution for Creare's monolithic adsorbent.

OPERATION OF A HELIUM ADSORBENT REGENERATOR IN A CRYOCOOLER

The operation of a helium adsorbent regenerator in a pulse-tube cryocooler is very similar to a conventional regenerator. The main difference is during the cooldown process. In a cryocooler with a helium adsorbent regenerator, an appreciable amount of helium in the cryocooler will be adsorbed into the adsorbent during the cooldown process. This will cause the mean system pressure to decrease if no supplemental helium is supplied. This issue can be easily mitigated by slightly increasing the size of the helium reservoir or the cryocooler charge pressure to maintain its operating pressure after its cooldown. The helium adsorption during the cooldown process will also release additional adsorption heat into the regenerator, and thus increase the cooldown time. The estimated additional cooldown time is on the order of one minute. During steady state operation, however, there will be no noticeable difference between the operation of a helium adsorbent regenerator and a conventional regenerator.

The cycle period of pressure waves in a portable low-temperature pulse-tube cryocooler is very short, only about 50 ms (i.e., about 20 Hz). This cycle period is several orders of magnitude shorter than the adsorption and desorption time constant in an adsorbent, which is typically on the order of tens of seconds. Furthermore, the flow resistance between the adsorbent and process gas is quite high. Therefore, there will be no dynamic adsorption or desorption of helium in the adsorbents due to pressure oscillation or small local temperature swing in the regenerator within a pressure cycle. Here, it is the sensible energy of the adsorbed helium molecules being utilized, not the latent energy associated with adsorption and desorption.

REGENERATOR HEAT CAPACITY MEASUREMENT METHOD

We used a transient test technique to measure the regenerator heat capacity. We exposed the regenerator matrix to a step change in the inlet gas temperature and measured the temperature at the inlet and outlet as the thermal wave passed through the regenerator. The heat capacity is determined by the temporal profile of gas flow temperature at the inlet and outlet of the regenerator. The expected gas temperature at the regenerator outlet is illustrated in Fig. 4 when the gas inlet temperature has a "step" increase. The breakthrough time and the temporal profile of the outlet temperature are a function of the regenerator heat capacity, the flow heat capacity rate, and the thermal conductance between the gas flow and the matrix. Assuming a constant specific heat for the helium flow, the product of the cross hatched area between the two temperature profile curves (Fig. 4) and flow heat capacity rate $\dot{m}c_p$ is the energy absorbed by the regenerator. With the helium gas flow rate held constant, the regenerator heat capacity is proportional to the break through time, $\Delta t_{breakthrough}$.

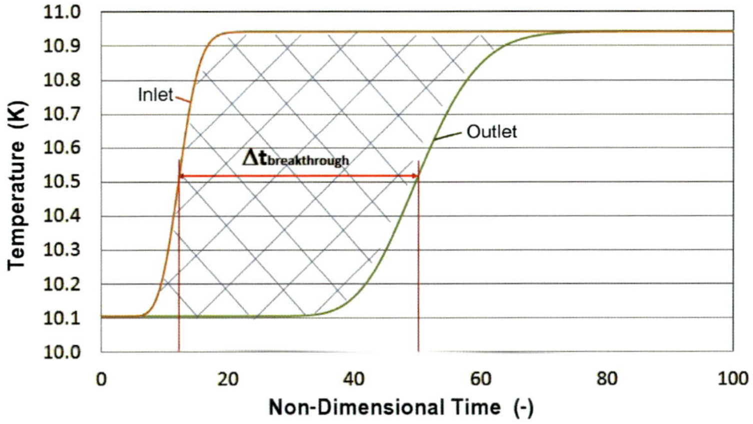

Figure 4. Illustration of gas temperature profiles at regenerator inlet and outlet as a function of time as thermal wave passes through regenerator.

The gas flow rate in this series of tests was selected so that the thermal wave breakthrough time is on the order of several seconds. This breakthrough time was long enough for our data acquisition system to resolve, but short enough to minimize the effects of parasitic heat leaks.

This test method was verified previously with a regenerator consisting of a stack of sintered stainless steel sheets.

TEST FACILITY

Fig. 5 shows a schematic of the test facility. Helium from a research-grade gas cylinder first passed through a mass flow meter, a gas cleaning subsystem (not shown) to remove contaminants, and then through a coil submerged in LN_2. It was cooled to about 100 K at the outlet of the coil. Next, it passed through a cold plate attached to a GM cold head to cool down to about 30 K and then through a Sumitomo cryocooler to further reduce its temperature to about 10 K. Finally, it passed through a heater and then entered the regenerator. At the beginning of each test, the heater was disabled and the entire regenerator was cooled down to a temperature very close to the temperature setpoint of the Sumitomo cold head. Once the regenerator temperature reached a steady state, the thin-film heater was turned on to rapidly raise the temperature of the helium flow entering the regenerator by 2 to 3 K. After the thermal wave passed through the regenerator, the heater was turned off, introducing another thermal wave through the regenerator, which cooled down the regenerator by 2 to 3 K back to the Stage 2 cold head temperature.

To reduce parasitic heating into the regenerator, we placed a cooling shroud around the regenerator. The cooling shroud was cooled by the exhaust flow from the regenerator. A relatively high-speed data sampling rate of 100 Hz was used to measure and record the transient gas flow temperatures at the inlet and outlet of the regenerator. We conducted a series of tests with regenerator inlet temperatures ranging from 10 to 22 K. The minimum test temperature was limited by the cooling capacity of the Stage 2 cold head. The heat load on the cold head was high because the test facility did not have a recuperator to reduce the heat load on the cold head. Tests were conducted at two different pressures to assess effects of the ambient helium pressure on the effective heat capacity of the adsorbed helium. The maximum pressure was limited to about 1.6 bar to limit the amount of helium in the regenerator dead volume.

Fig. 6 shows the regenerator sample used for characterizing the heat capacity of adsorbed helium/helium adsorbent. The regenerator consists of 80 plates with Kapton insulators in between the regenerator disks to reduce axial conduction. To minimize the effects of parasitic thermal mass, we located Swagelok fittings for temperature sensors, which have a relatively large thermal mass, at the two ends of the test apparatus where the fittings were not directly in the gas flow path. We also fabricated two thin sheathed Cernox sensors and used them to directly measure the gas flow temperatures at the inlet and outlet of the test stack to reduce the measurement response time. The Cernox sensors were located less than 0.25 in. away from the ends of the regenerator core.

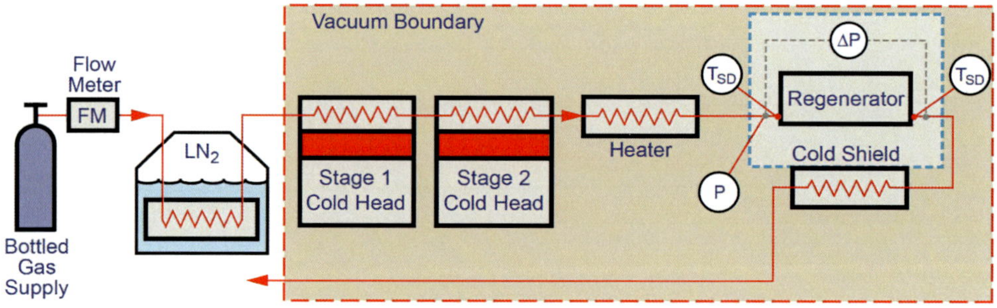

Figure 5. Test setup to characterize subscale regenerator heat capacity and thermal performance.

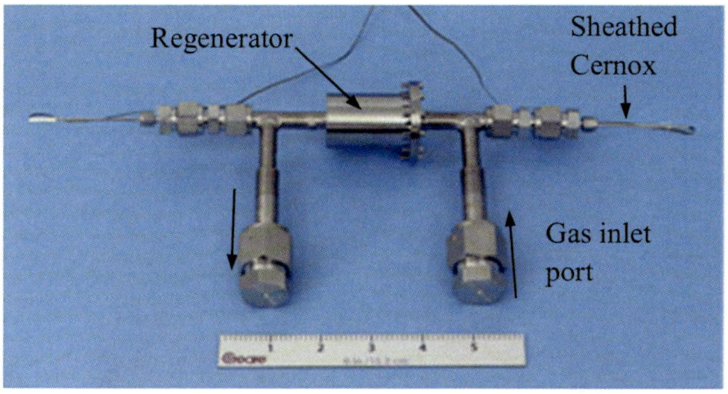

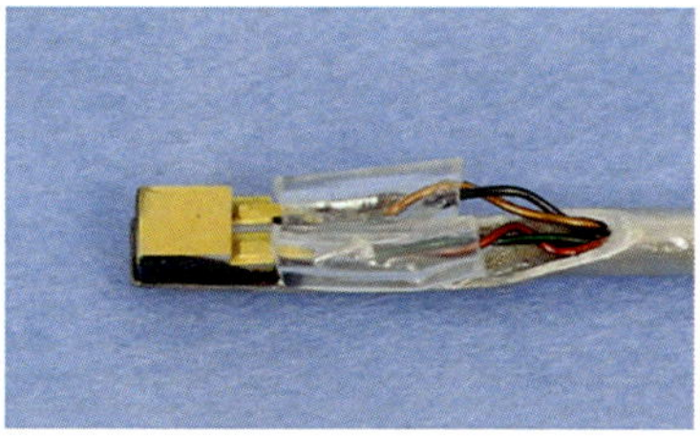

Sheathed SD Cernox with sensor exposed to gas flow

Figure 6. Regenerator sample with sheathed temperature sensors.

The regenerator test sample was designed to minimize the parasitic thermal mass for the core, which included the entire housing and manifold plates, as well as the dead volume inside the regenerator. At the nominal test temperature of 10 K, the resulting parasitic thermal mass was less than 10% of the core thermal mass, even if the entire parasitic mass participated (i.e., the thermal wave penetrated into all components and their temperatures all rose to the final temperature). The resulting dead volume was small enough that the resulting delay in thermal breakthrough was less than 2% of the actual breakthrough time.

DATA PROCESSING

Fig. 7 shows representative measured temperature profiles of helium flow entering and leaving the regenerator core during thermal wave breakthrough tests. Because of the participating parasitic thermal mass in the heater and the transfer line between the heater and the core, the temperature of the inlet gas took a few seconds to reach its steady state value. As the gas temperature started to rise rapidly at t = 281.7 s, the temperature of the helium exiting the core remained unchanged because the thermal wave had not reached the outlet of the stack. After t = 286.5 s, the outlet temperature began to rise, but at a rate slower than that of the inlet flow.

Because the thermal conductance (UA) of each plate was very high relative to the helium flow rate of 35 SLPM, the thermal effectiveness of each regenerator plate was very close to 1 (i.e., each plate had an NTU value close to 1). Therefore, the slow temperature rise after thermal breakthrough at the outlet flow was not caused by the limited heat transfer between the helium flow and the plates in the core. Two possible causes for the slow outlet flow temperature rise are the following:

1) The parasitic thermal mass of the inlet and outlet headers and other parts that are thermally connected to the headers. Because of the limited convective heat transfer between the helium flow and headers, very little parasitic thermal mass was participating at the thermal wave front. Therefore, the parasitic thermal mass did not delay the initial thermal breakthrough at the outlet, but only broadened the wave front.

2) Helium desorption associated with the temperature increase in the adsorbent. For these separate effects tests, the thermal wave breakthrough time was about 5 to10 s, which was long enough that a small fraction of adsorbed gas could be desorbed. The associated desorption heat could artificially increase the effective heat capability of the test core. The associated desorption heat would also broaden the thermal wave front (i.e., the active heat transfer zone) in the regenerator. However, it should not have had appreciable effect on the initial breakthrough time since desorption does not occur right at the wave front.

To exclude the effects of parasitic thermal mass and the desorption heat from the regenerator core heat capacity from the measurement, the initial thermal breakthrough time was used to estimate the core heat capacity, which was taken as the product of the helium flow heat capacity

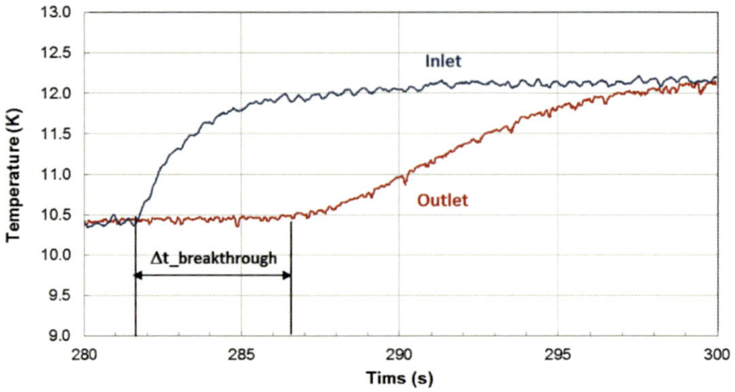

(a) Thermal Reponse with Increasing Inlet Temperature

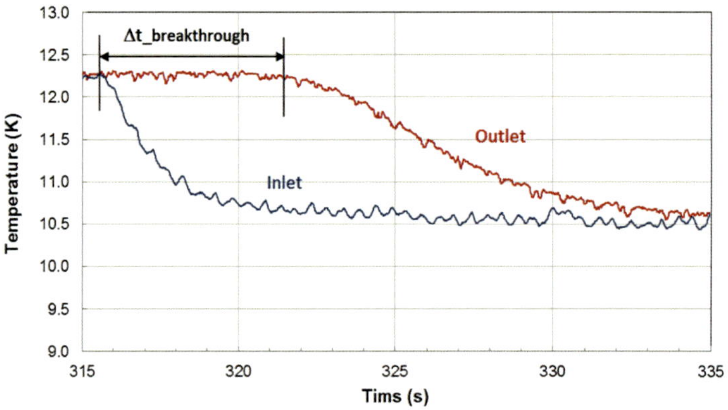

(b) Thermal Reponse with Decreasing Inlet Temperature

Figure 7. Measured regenerator inlet and outlet temperature profiles.

rate and initial breakthrough time. The initial breakthrough time was taken as the time difference for the temperatures of the outlet and inlet flows to change by 2.5% of the total temperature rise, as shown in Fig. 7.

TEST RESULTS AND DISCUSSIONS

Fig. 8 shows the measured volumetric specific heat of adsorbent/adsorbed helium composite based on the data processing approach discussed above. The measured volumetric specific heat values are comparable with those of typical rare earth materials used in low temperature regenerators. The volumetric specific heat remains very high in the entire test temperature range from 10 to 22 K. This is in contrast to the specific heat of a rare-earth material, which typically peaks over a narrow temperature range of a few Kelvins. The wide temperature span of high specific heat allows the use of the adsorbent/adsorbed helium to serve as the thermal storage media over the entire temperature range below 20 K.

The measured specific heat increases slightly when the ambient helium pressure is increased from 1.0 bar to 1.6 bar. It is worth noting that the helium pressures in these tests are significantly lower than typical mean pressures in low temperature cryocoolers, which are typically about 10 bar. Test data for helium adsorption in activated carbon at temperature above 10 K indicates that specific helium adsorption capacity still increases appreciably with the ambient pressure until the pressure is significantly higher than 20 bar. This suggests the volumetric specific heat of adsorbed helium in actual applications can be even higher than the values shown here.

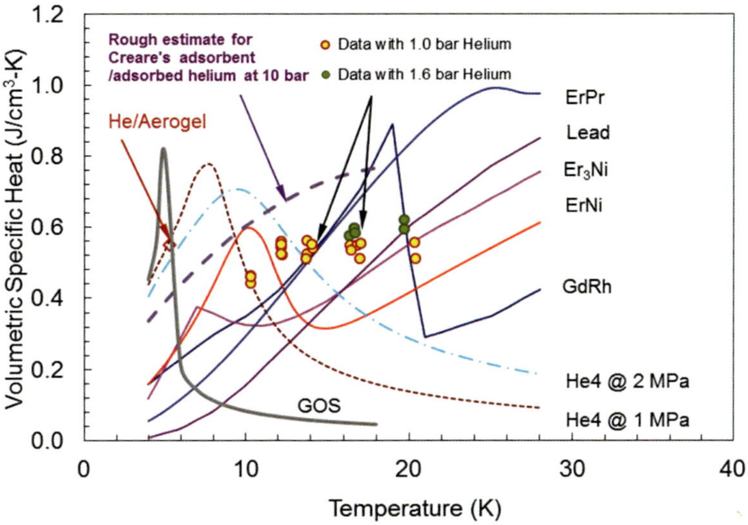

Figure 8. Comparison of measured volumetric specific heat of adsorbent/adsorbed helium with typical rare earth materials for low-temperature regenerators.

The measured specific heat is consistent with rough estimate values (Fig. 8). The volumetric specific heat of adsorbed helium is not well established in published literature. Its values are adsorbent dependent, and experimental data are scarcely available. To roughly estimate these values, the empirical correlation for adsorbed helium specific heat in Y zeolite, ($2.2 \times T$) J/mol-K for $3.5 < T < 9$ K, was used (Wada et al. 1984). The adsorbed helium density was taken as 0.202 (g/cm^3) (Steele 1956, Setoyama et al. 1996). An effective adsorbent porosity of 68% and adsorption isotherms similar to Chemviron SC2 (Hauer and Day 2002) were also assumed. The resulting rough values at 10 bar were established to guide the early design of the regenerator plates and their values were not expected to be very precise.

The measured values are also comparable to the reported volumetric specific heat of adsorbed helium in aerogel, which is 0.58 J/cm^3-K at 5.25K (Wong and Chan 1990). The pore size in their aerogel sample ranges from 20 Å to 2000 Å, and the reported helium density in the aerogel is about 0.08g/cm^3.

The specific heat measurement uncertainty resulting from instrumentation accuracy and data acquisition system resolution is relatively small, less than approximately 1.5%. The scattering of the test data is mainly caused by the uncertainty in determining the initial breakthrough point based on the recorded temperature profiles. The uncertainty is mainly due to the high temperature measurement noise caused by using low excitation current to avoid sensor self-heating, as well as small regenerator temperature drifts due to parasitic heating.

SUMMARY AND FUTURE WORK

A practical design and fabrication approach to use the sensible heat of adsorbed helium to enhance the heat capacity of a low-temperature regenerator has been developed. The resulting adsorbent is an integral part of the heat spreader and is in intimate thermal contact with convective transfer surfaces. This fabrication approach also avoids potential particulate contamination issues in the regenerator associated with use of loose adsorbent particles. Experimental results over the temperature range 10 to 20 K show that the volumetric specific heat of the adsorbent/adsorbed helium is comparable to typical rare earth materials used in low temperature regenerators. The adsorbent regenerator technology has strong potential to significantly enhance the performance of low-temperature regenerative cryocoolers.

Future research in this area should include assembling a Stirling or Pulse Tube cryocooler with the microchannel helium adsorbent regenerator, characterizing its thermal and fluid performance, and assessing its robustness for potential tactical and space applications. In addition to low temperature regenerators, the monolithic adsorbent technology also has applications in cryo-sorption pumps, sorption coolers, and sorption compressors for JT coolers.

ACKNOWLEDGMENT

The support and guidance from the Office of Navy Research are gratefully acknowledged.

REFERENCES

1. Duband, L., Chaussy, J. and Ravex, A., "High Pressure Adsorption Isotherms of Helium on Activated Charcoal," *Advances in Cryogenic Engineering*, vol. 33 (1988), pp. 1023–1030.

2. Hauer, V. and Day, Chr., "Cryosorbent Characterization of Activated Charcoal in the COOLSORP Facility," Final Report on Subtask 8 of Task VP1: Cryopump Development and Testing (ITER Task No. 448), Forschungszentrum Karlsruhe GmbH, Karlsruhe (2002).

3. Jones, J.A., Petrick, S.W., and Britcliffe, M.J., Cryogenic Regenerator Including Sarancarbon Heat Conduction Matrix, US Patent 4,835,973, 1987.

4. Konishi, K., Deguchi, H. and Takeda, K., "Heat Capacities of Helium in One- and Three-Dimensional Channels at Low Temperatures," *J Physics: Condensed Matter*, vol. 5, no. 11 (1993), pp. 1619–1632.

5. Mérida, W.R. and Barclay J.A., "Monolithic Regenerator Technology for Low Temperature (4 K) Gifford-McMahon Cryocoolers," *Advances in Cryogenic Engineering*, vol. 43, edited by P. Kittel, Plenum Press, New York, NY (1998), pp. 1597–1604.

6. Pecharsky, V.K., Gschneidner, K.A., Jr., McCallum, R.W. and Dennis, K.W., "Influence of Alloying on the Behavior and Properties of Er3Ni," *Cryocoolers 9*, edited by R. G. Ross, Jr., Plenum Press, New York, NY (1997), pp. 663–668.

7. Pron'ko, V.G., Amamchyan, R.G., Guilman, I.I. and Raygorodsky, A. I., International Cryogenic Engineering Conferences, Vol. 6, IPC Science and Technology Press (1976), pp. 86-88.

8. Radebaugh, R., Huang, Y., O'Gallagher, A. and Garya J., "Calculated Performance of Low-Porosity Regenerators at 4 K with 4He and 3He," *Cryocoolers 15*, ICC Press, Boulder, CO (2009), pp. 325–334.

9. Setoyama, N., Kaneko, K. and Rodriguez-Reinoso, F., "Ultramicropore Characterization of Microporous Carbons by Low Temperature Helium Adsorption," *J. Phys. Chem.*, vol. 100 (1996), pp. 10331–10336.

10. Steele, W.A., *J. Chem. Phys.*, vol. 25 (1956), p. 819.

11. Wada, N., Ito, T. and Watanabe, T., "A Semiquantum Liquid and a New Ordered Phase of 4He Adsorbed in Y Zeolite," *J. Physical Society of Japan*, vol. 53, no. 3 (1984), pp. 913–915.

12. Wong, A.P.Y. and Chan, M.H.W., "Liquid-Vapor Critical Point of 4He in Aerogel," *Phys. Rev. Lett.*, vol. 65 (1990), pp. 2567–2570.

James Webb Space Telescope Mid Infra-Red Instrument Cryocooler Electronics

D. Harvey[1], T. Flowers[1], N. Liu[1], K. Moore[1], D. Tran[1], P. Valenzuela[1], B. Franklin[2], D. Michaels[2]

[1]Northrop Grumman Aerospace Systems, Redondo Beach, CA, USA
[2]Jet Propulsion Laboratory, California Institute of Technology, Pasadena, CA, USA

ABSTRACT

The latest generation of long life, space pulse-tube cryocoolers require electronics capable of controlling self-induced vibration down to a fraction of a newton and coldhead temperature with high accuracy down to a few Kelvin. Other functions include engineering diagnostics, heater and valve control, telemetry and safety protection of the cryocooler subsystem against extreme environments and operational anomalies. The electronics are designed to survive the thermal, vibration, shock and radiation environment of launch and orbit, while providing a design life in excess of 10 years on-orbit. A number of our current generation high reliability radiation-hardened electronics units are deployed on-orbit on space flight payloads. This paper describes the features and performance of our latest flight electronics designed for the Pulse Tube and Joule-Thomson cryocoolers that providing 6K cooling for the James Webb Space Telescope (JWST) Mid Infra-Red Instrument (MIRI). The electronics are capable of highly accurate temperature measurement and control over the temperature range from 4 K to 15 K. Self-induced vibration is controlled to low levels on all harmonics up to the 16th. A unique active power filter controls peak-to-peak reflected ripple current on the primary power bus to a very low level. This paper provides a design overview and ground-test data for the MIRI JouleThomson and Pulse Tube flight Cryocooler Control Electronics (CCE) units.

INTRODUCTION

Northrop Grumman has been producing space cryocoolers and cryocooler control electronics for over 20 years, beginning with the AIRS Pulse Tube cryocooler system. The requirements for the electronics includes low mass, high efficiency, radiation hardness, high reliability, high temperature control accuracy and stability, the ability to control and minimize the vibration exported from the compressor, provide cryocooler safety protection and extensive on-orbit diagnostics. All of these objectives were met with the original space cryocooler electronics; however advances in electronic technology, along with experience gained manufacturing, testing, integrating and operating several space cryocooler systems (some on-orbit for fifteen years) provided an evolutionary path to a smaller, lower mass electronics package with improved performance and reliability; that design was presented in 2003.[1] Since that time, the design has been scaled up 2x in power, the precision PRT thermometry has been

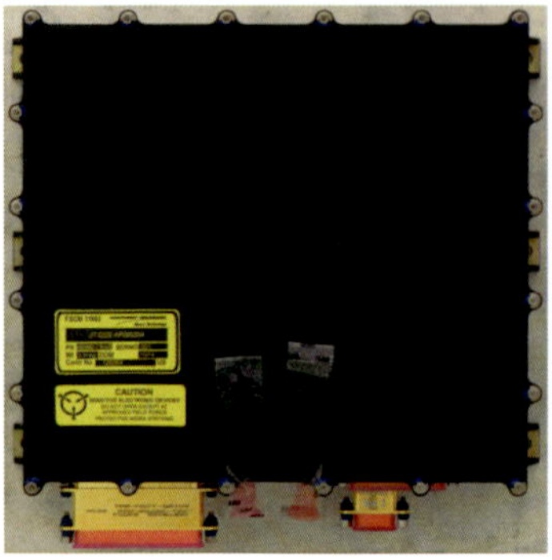

Figure 1. Flight JT Cryocooler Control Electronics.

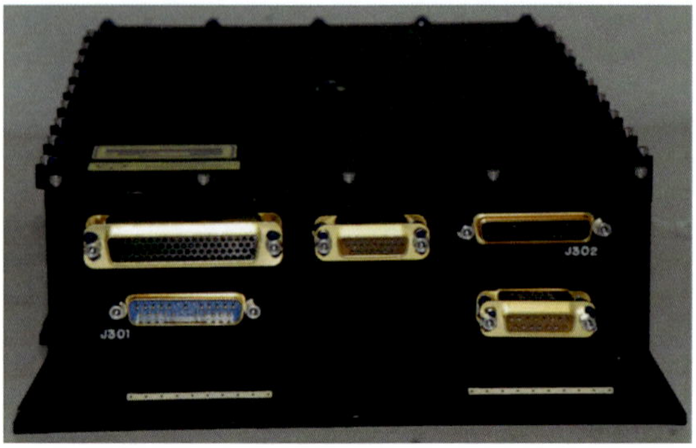

Figure 2. Flight PT Cryocooler Control Electronics.

replaced with precision Cernox[2] thermometry, heater and valve drive and control have been added, and a 1553B interface has been added.

In 2006, the Jet Propulsion Laboratory, as part of the JWST MIRI Cryocooler program, funded the development of two Cryocooler Control Electronics (CCE) units based on the second-generation Advanced Cryocooler Electronics (ACE). One CCE was designed to drive the Joule-Thomson (JT) recirculating compressor; the other was designed to drive the Pulse Tube (PT) compressor. Both CCEs were required to provide precision thermometry down to 4K and communicate with the ISIM (Integrated Science Instrument Module) Controller via a 1553 interface. The JT CCE retains the same power capability as the previous ACE CCE: 180 W. The PT CCE was required to provide up to 360 W output power to drive the PT compressor. Additionally, the PT CCE was required to control and drive five heaters and three latching valves. All other features and capabilities of the ACE (low mass per watt, high reliability, radiation hardness, extensive in-flight diagnostics and advanced vibration control) were retained. Photos of the completed JT and PT CCE units are shown in Figure 1 and Figure 2.

DESIGN DESCRIPTION

The JT and PT CCEs have similar architecture and use the same control board and flight software. The PT CCE's power section is scaled up 2x compared to the JT CCE. Both designs were based on the successful ACE (Advanced Cryocooler Electronics). Top-level design requirements are presented in Table 1. When the ACE was originally designed, we had modular scaling in mind since the HCC (High Capacity Cryocooler) was currently under development with low-temperature detector applications such as MIRI on the horizon. The MIRI PT cryocooler requires up to 325 W drive to meet MIRI cooling requirements, so it was a natural step to double the ACE power capability to 360W. Block diagrams of the JT and PT CCEs are shown in Figure 3 and Figure 4, respectively.

Each CCE is physically divided into two subassemblies: the power subassembly in the bottom contains the high-power DC-DC converters, switching power amplifiers and EMI filters; the control subassembly in the top contains the analog and digital data acquisition, signal processing, control circuitry, housekeeping converter(s), capacitor bank, and in the case of the PT CCE, the heater and valve drivers.

Similar to the ACE, the MIRI CCE's are a microcontroller-based system with all of the control loops implemented in software. 128 KB SRAM provides temporary code and data storage, while 8 KB PROM and 64 KB EEPROM provide non-volatile code and default parameter storage. All code is executed from radiation hardened SRAM. Boot and upload code are stored in PROM, with the balance of the software functions and user constants stored in EEPROM. The contents of the EEPROM can be uploaded on-orbit, providing a capability for

Table 1. Top-Level Design Requirements

Parameter	JT	PT	Comments
Bus Input Voltage	22 to 43VDC	22 to 43VDC	
Output Power	180W	360W	2 channels 10Ω/channel
Reflected Bus Ripple	<200mAp-p	<450mAp-p	at full power
Temperature Measurement Range	4K to 350K	4K to 350K	In 2 ranges
Temperature Control Range	5K to 250K	5K to 250K	
Temperature Measurement Accuracy	< ±10mK	< ±10mK	5K to 15K
	< ±1K	< ±1K	15K to 325K
Temperature Measurement Stability	< ±5mK	< ±5mK	Over 90 min interval
Vibration Control	<200mN	<200mN	On-axis, Adaptive control of each harmonic up to 16th
Cooler Drive Frequency Range	28Hz to 95Hz	28Hz to 95Hz	<0.1Hz increments
Command and Telemetry Interface	1553B	1553B	
Clock Sync Input	Yes	Yes	
Cooler Drive Sync Input	Yes	Yes	
Cooler Drive Sync Output	Yes	Yes	
Heater Drive Power	No	Up to 9W/ch	7 channels
Valve Drive	No	≥1Amp 100ms to 1s	6 channels, current-drive
User-Configurable On-Orbit	Yes	Yes	
Turn-Key Operation	Yes	Yes	
Firmware Uploadable On-Orbit	Yes	Yes	
Non-Volatile Storage	Yes	Yes	
Operating Temperature Range	-35C to +60C	-35C to +60C	Qual/Protoflight range, at full-power
Non-Operating TempRange	-35C to +70C	-35C to +70C	Qual/Protoflight range
Launch Vibration	14grms	14grms	
Total Dose Radiation Hardness	50 krad	50 krad	Can be spot-shielded for higher total dose
Size	7.85 × 9.20 × 3.00 inches	18.0 × 9.20 × 3.60 inches	Including mounting feet and filter-pin connectors
Mass	4.0 kg	9.5 kg	Including filter-pin connectors

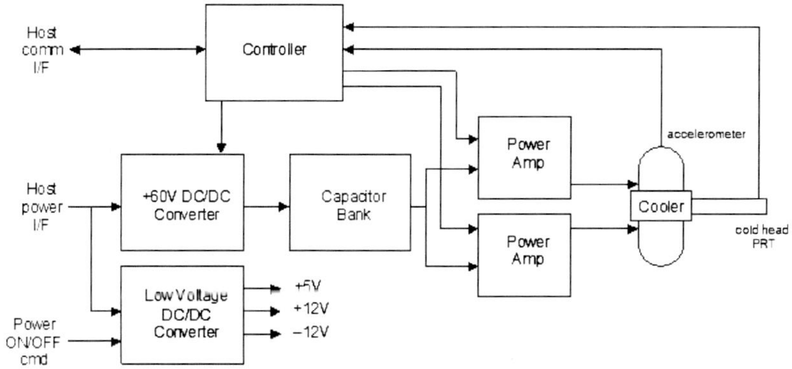

Figure 3. JT Cryocooler Control Electronics Block Diagram

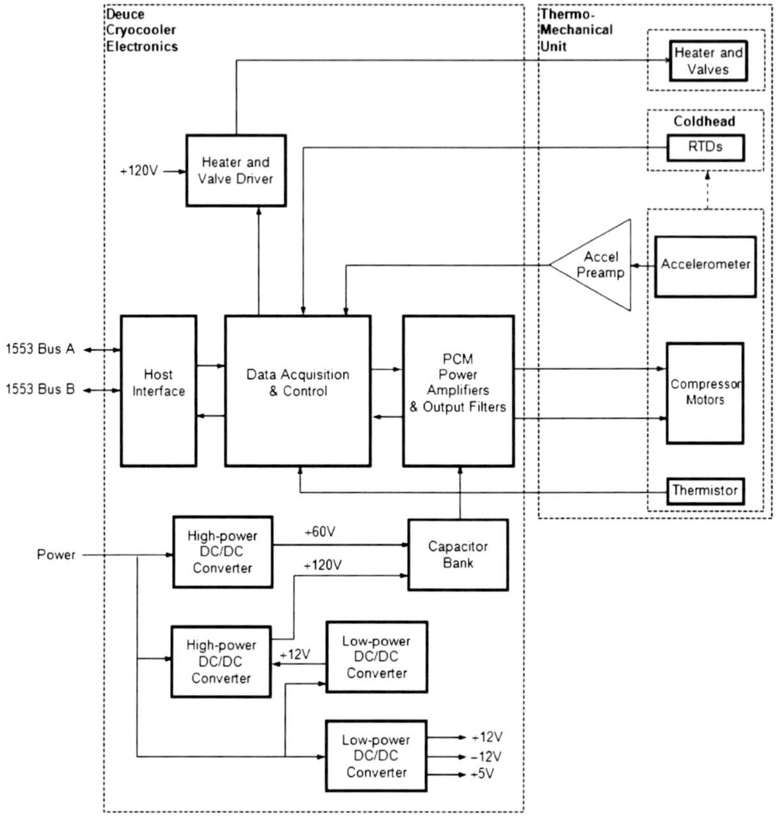

Figure 4. PT Cryocooler Control Electronics Block Diagram

operating parameter changes and software upgrades. The EEPROM also contains command macros, providing a user-configurable "turnkey" operation. A radiation-tolerant, Single Event Upset resistant FPGA provides various digital functions including address/data demultiplexing, address decoding, bus arbitration, DMA (Direct Memory Access), interrupt control, UART (Universal Asynchronous Receiver/Transmitter) serial communications, 1553B serial communications, heater control, valve control, analog data acquisition control, waveform look-up, digital-to-analog conversion, clock generation, synchronization and a watchdog timer. Synchronization between hardware and software activities is provided by a single interrupt that occurs once per cooler cycle (approx 30 Hz for PT, 90 Hz for JT). The data acquisition and waveform generation functions are synchronized to the cooler cycle interrupt. The software maintains a four second timer which is used to pace the software main loop. This four second timer is synchronized to the 1553 periodic broadcast message. Command decoding, telemetry transmission, mode changes, vibration control, temperature control and cooler safety functions are executed sequentially in the four second main-loop.

The precision temperature measurement function monitors up to seven Cernox RTDs (Resistance Thermometer Devices) mounted on the various cold heads in the MIRI cryocooler subsystem for the purpose of reporting and controlling cold head temperatures. The RTDs are measured using a 4-wire (Kelvin) connection, pulsed excitation and synchronized to the compressor drive waveform. Background subtraction removes DC errors, compressor drive pickup and thermocouple effects from the entire signal chain: RTD to A/D output. On-orbit end-to-end calibration enables absolute accuracy to within ±10mK. The on-orbit calibration is accomplished by sampling a precision resistor connected to the eighth Cernox channel. Separate Cernox RTD calibration and curve-fit coefficients for each channel are stored in EEPROM.

Vibration control is realized by sensing on-axis vibration with two redundant pairs of piezoelectric accelerometers mounted on the compressor assembly, two mounted on the JT compressor and two mounted on the PT compressor. Low-noise charge amplifiers are located on the compressor assembly to process the accelerometer signal and drive the cables to the JT and PT CCE's. The vibration signal is sampled synchronously and fed to a proprietary vibration control algorithm which continually adjusts the harmonic content of the compressor drive waveforms. The vibration control function is autonomous and adaptive, adjusting to variations in the compressors and structure dynamics.

Communication is provided over a (dual-redundant) 1553B interface. RS-422 synchronization clock inputs are provided for a 1 MHz reference clock and a cooler synchronization clock. An RS-422 synchronization clock output at the cooler drive frequency is also provided. The synchronization clocks are optional, and the CCEs functions normally in the absence of synchronization clocks. Fault management is enhanced by a RS-422 bi-level input (SAFE/HOLD) that provides the user a method of commanding the CCEs to Standby mode without using the 1553 communication channels. A RS-422 bi-level output (FAULT) provides the user a method of detecting cooler system safety exceptions without using the 1553 communication channels. A single-ended, bi-level POWER ON/OFF command enables the CCE's power converters. A passive caging relay monitor allows the user to observe the caging relay status without powering the CCEs.

A comprehensive diagnostic database is available for cryocooler system health monitoring and on-orbit debugging. Telemetry data includes cooler system mode, state and safety status, coldhead, compressor, heater and electronics temperatures, secondary power supply voltages, compressor voltage and current waveforms, vibration measurement parameters and vibration control parameters. In addition all areas of RAM can be read and written by the user for diagnostic and troubleshooting purposes.

A VCXO (voltage controlled crystal oscillator) generates a 16 MHz clock from which all internal clocks are derived. The 16 MHz clock is divided by 16 and phase-locked to the (optional) 1 MHz external reference clock.

The JT power subassembly contains a 200 W DC-DC converter, two switching power amplifiers, and EMI filters. The JT's DC-DC converter provides +60VDC to the capacitor bank and switching power amplifiers. The PT power subassembly contains two 200 W DC-DC converters, two switching power amplifiers, and EMI filters. The PT's DC-DC converters have their +60 VDC outputs wired in series to provide +120 VDC to the capacitor bank, switching power amplifiers and valve drive. The heater drives are powered off of the +60 VDC rail. The 60 V DC-DC converters are custom-designed modules that implement an active ripple control function. Each 60 V DC-DC converter is conservatively rated at 200 W output over the full range of bus voltages and operating temperatures. A smaller housekeeping converter provides +5 VDC and ±12 VDC to the control and power amplifier boards. An additional floating +12 VDC powers the secondary circuits in the upper 60 V DC-DC converter in the PT CCE. The 60 V DC-DC converters and the main housekeeping converter are synchronized to the CCE's 16 MHz master clock. The 60 V DC-DC converters are disabled when the CCEs are in Standby mode to conserve power. A single-ended, primary-referenced, bi-level command is provided to enable the main housekeeping converter, which in turn enables the 60 V DC-DC converters and auxiliary low-voltage converter. When all converters are disabled, the power consumption is less than 1W per CCE.

The power amplifiers are realized as H-bridge PCM switching amplifiers. Each JT channel is capable of delivering up to ±40 V peak, ±6 A peak or 90 W average. Each PT channel is capable of delivering up to ±72 V peak, ±9 A peak or 180 W average. Passive L-C filters smooth the power amplifier outputs and control EMI. Each amplifier channel contains a high-speed short-circuit current detector to protect the power amplifiers in the event of a shorted cable or load. The power amplifier design takes advantage of its inherent linearity by operating open-loop. The open-loop configuration ensures unconditional stability when driving the complex, non-linear load presented by the cryocooler compressors. Latching relays located on

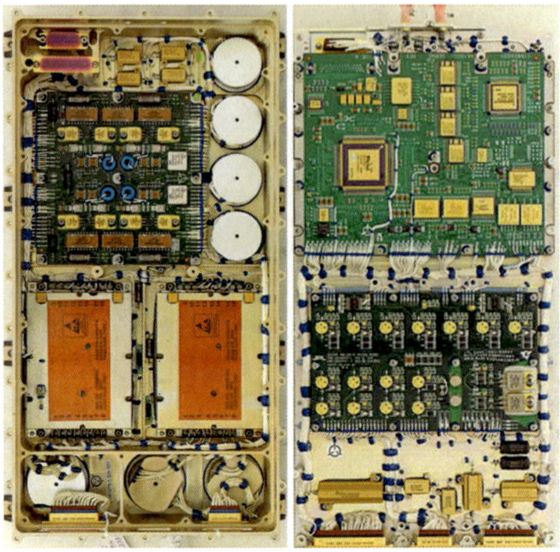

Figure 5. PT CCE subassemblies.

the power amplifier board cage (restrain) the cryocooler compressor pistons during launch by shorting the motor windings. These caging relays are dual-redundant and are controlled by user commands. In addition to 1553 telemetry indicating the caging relay status, a passive, bi-level output is provided on the communications connector to indicate caging relay status. This permits the user to monitor the caging relay status while the CCEs are powered off. The JT and PT power subassemblies are capable of delivering at least 180 W or 360 W, respectively, on a continuous basis to the cryocoolers.

The MIRI Cryocooler Subsystem contains a pair of latching cryogenic valves and six heaters. A heater and valve drive board has been designed and built at JPL and integrated into the PT CCE. Valves are all powered from the CCE's internal +120 VDC bus; heaters are powered from the +60 V bus. Because the valve winding DC resistance varies significantly over the range of operating temperatures, constant-current valve drive was selected. Series-redundant switches are incorporated into the valve and heater drivers to enhance reliability. The valves contain redundant coils which are driven by either the primary or secondary PT CCE. The CCE software can vary the valve drive pulse width under user command to accommodate faults such as shorted valve windings. Six heater drivers are provided for bang-bang control of valve heaters, decontamination heaters, and a He mass-flow sensor heater on the JT loop. Ballast resistors sized for each heater control the peak heater power. Average heater power is controlled by software bang-bang controllers. A solid-state circuit-breaker function is provided for each heater driver to protect against shorted heaters or cables.

Constructed similar to the ACE, the JT and PT CCE's are housed in an aluminum chassis machined from a single piece. Figure 5 shows the PT CCE internal configuration with the power subassembly on the left, and control subassembly on the right. Figure 6 shows the JT CCE internal configuration. 3-D solid modeling tools were used to design and layout the CCEs. Using these tools,

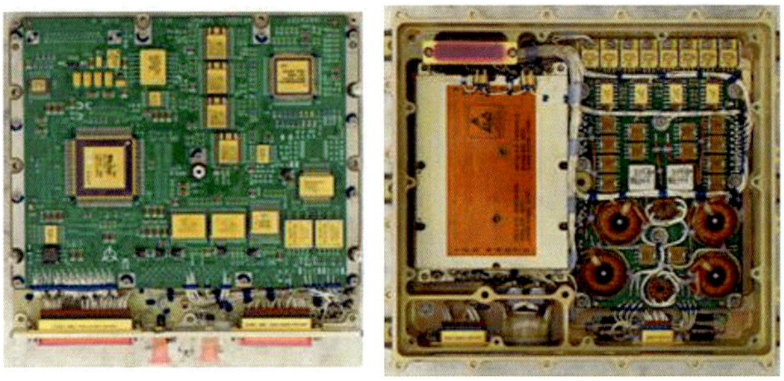

Figure 6. JT CCE subassemblies.

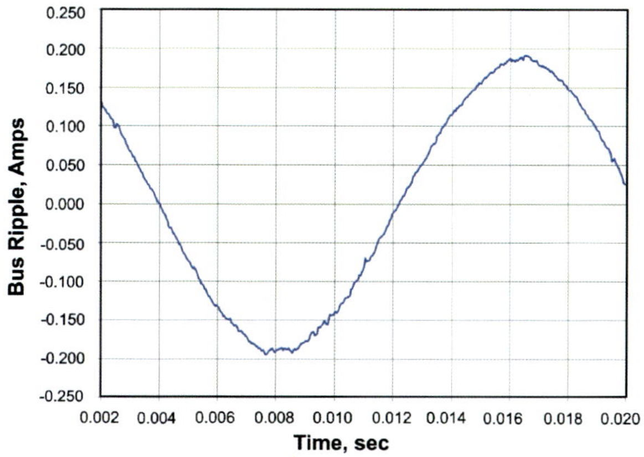

Figure 7. Flight PT bus current ripple

we were able to efficiently nest the various components, identify clearance and manufacturing issues early in the design process, accurately estimate mass, and perform thermal and structural analyses. The result was a CCE design that was easy to build and fit together perfectly on the first pass. All high power devices are mounted to the floor of the chassis, providing an excellent thermal path to the heat rejection surface. An internal shield plate forms an RF barrier inside the unit, separating the power and control subassemblies. The control subassembly is built up on the shield plate and then dropped into the unit. D-connectors located inside the unit connect the two subassemblies. An input filter cavity contains feedthrough filters, a common-mode choke, and the input power input connector. An output filter cavity contains feedthrough filters, two common-mode chokes, and the compressor drive connector. The shield plate closes off these RF-tight cavities.

D-connectors and two triaxial connectors provide interfaces to the user and the cryocooler: DC input power, communications, valve and heater drive, cryocooler compressor drive, and cryocooler instrumentation.

PERFORMANCE DATA

The flight CCEs were acceptance tested with resistive loads prior to integration with the MIRI cryocooler compressor assembly. Figure 7 shows the active ripple suppression at work: the PT CCE input ripple current is <400 mA p-p while producing full power (360 W) at 30.5 Hz with a 28 V bus. Figure 8 plots ripple current as a function of output power. Without ripple suppression, the input

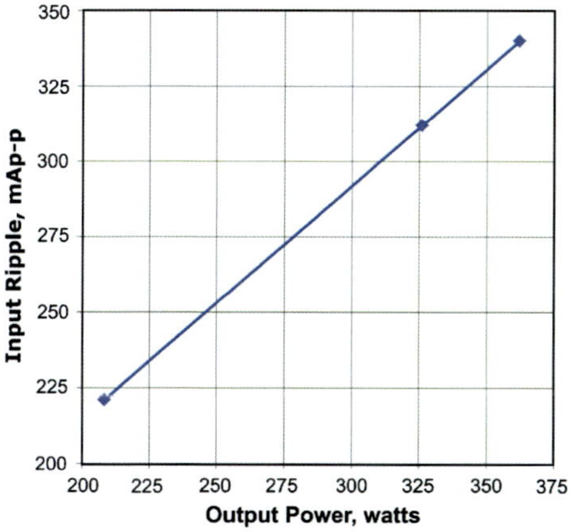

Figure 8. Flight PT bus current ripple vs output power.

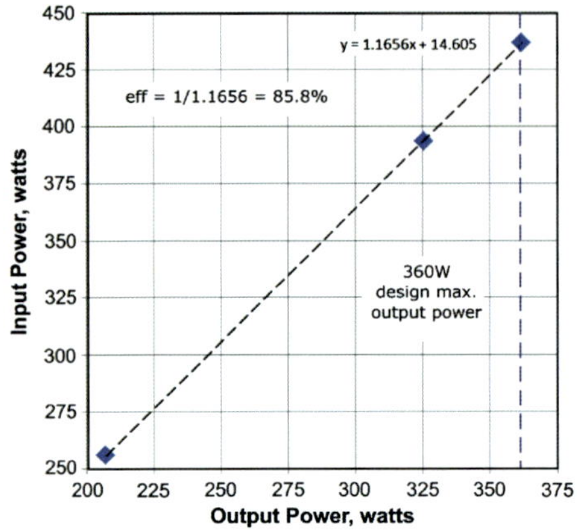

Figure 9. Flight PT CCE input power vs. output power.

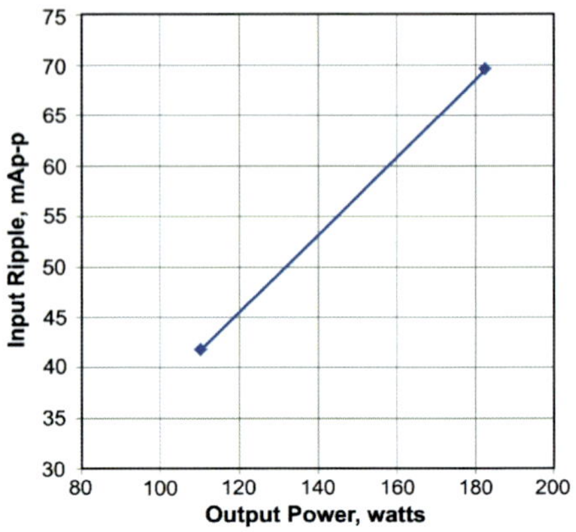

Figure 10. Flight JT CCE ripple current vs. output power.

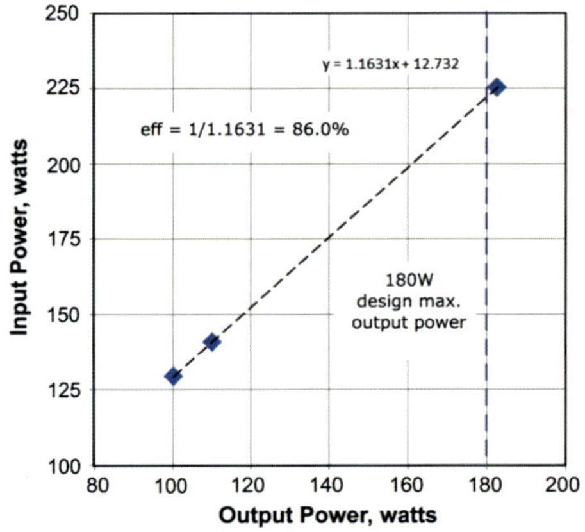

Figure 11. Flight JT CCE input power vs. output power.

current ripple would exceed 42 amps p-p. The PT CCE's active ripple suppression provides over 40 dB ripple attenuation at full power.

Figure 9 plots the PT CCE input power as a function of output power over a range of output power from 205 W to 365 W. The efficiency is flat at 85.8%. As shown in Figure 10, the JT CCE ripple current is <70 mA p-p while producing full power (180W) at 90Hz with a 28V bus. This is equivalent to 50 dB of attenuation.

Figure 11 plots the JT CCE input power as a function of output power over a range of output power from 100 W to 183 W. The efficiency is flat at 86.0%.

CONCLUSION

The MIRI Cryocooler Control Electronics meet all of their performance requirements for cooling the JWST MIRI detector to 6 K.

ACKNOWLEDGMENTS

This work was performed for the Jet Propulsion Laboratory, California Institute of Technology, sponsored by the United States Government under a Prime Contract between the California Institute of Technology and NASA. The heater and valve drive board design was carried out by Brian Franklin and Darren Michaels at the Jet Propulsion Laboratory.

The authors also acknowledge the significant contributions of the following Northrop Grumman Aerospace Systems personnel: Michael Petach, Huu Dang, Yen Loi, Hetmann Hsieh and Ray Fuentes.

REFERENCES

1. Harvey, D., Danial, A., Davis, T., Godden, J., Jackson, M., McCuskey, J., and Valenzuela, P., "Advanced Cryocooler Electronics for Space," *Cryogenics,* vol. 44, issue: 6-8, June-August 2004, pp. 589-593.

2. Cernox is a registered trademark of Lake Shore Cryotronics, Inc.

Challenges in Electronics Design and Qualification for Earth Observation Pulse Tube Space Cryocoolers

A. Gardelein[1], R. Gary[1], T. Wiertz[1], L. Fontani[2]

[1]Air Liquide Advanced Technologies, Sassenage, France
[2] SITAEL, Pisa, Italy

ABSTRACT

For more than 10 years, Air Liquide Advanced Technologies has developed a broad range of cryocooler systems to address diverse applications by translating into different performance ranges of cooling power, thermal stability, and mechanical vibration. In collaboration with SITAEL, the first cryocooler drive electronics unit was developed.

For Earth Observation applications, the critical performance metrics of a cryocooler system are the cooling efficiency, the temperature stability, and the ability to minimize the levels of exported vibration. Such performance requirements induce many challenges for electronics driving and controlling the cryocooler. These challenges cover domains such as power electronics, small signal acquisitions and digital logic design.

This paper will show the importance of associating the electronics design at the highest level of system design. One major contributor to system wall-plug efficiency is the electrical power factor of the cryocooler, which drives directly the maximum output of power supply and hence power components performance. Another major contributor on the digital side is the choice of the communications bus, and the control loop drive. The choice of technology to use to implement the control logic directly the impacts the overall system efficiency.

Eventually, the selection of electrical parts is mainly driven by two factors. Some of the critical components are either under ITAR or EAR99 restriction or are not qualified according to appropriate standards. The former concerns mainly FPGA, SRAM, 16-bits ADC, a ndPWM controllers. For the latter several components generally need a qualification program to be used on satellites, such as thermal sensors, hermetic connectors and force sensors.

INTRODUCTION

Air Liquide began its space activities in the 1960s with cryogenic propellant tanks for launch vehicles. In the late 1990s, orbital space cryogenics project were started and led to three big successes; MELFI turbo Brayton cooler for ISS, HERSCHEL cryostat and PLANCK dilution cooler. All those projects focused only on the thermo-mechanical, and gas management parts. It was only with the start of CRYOSYSTEM project that Air Liquide extended its involvement to driving electronics for space cryocoolers.

Unfortunately, CRYOSYSYTEM project was terminated after a successful PDR and did not lead to a full system development. It was only with the development of space coolers for earth

Cryocoolers 19, edited by S.D. Miller and R.G. Ross, Jr.

observation that Air Liquide pursued the development of a full cryocooler system. Today, LPTC coolers have successfully passed qualification campaign for the first French government application. Several orders have been placed making this product a great success in Europe. The development of the driving electronics of these coolers was one of the new challenges posed by this ambitious program.

CRYOCOOLER ELECTRONICS FUNCTIONS AND ARCHITECTURE

LPTC is a cooler aimed at a wide range of applications. Current orders already cover earth observation and weather observation, low orbit and geostationary orbit, imagers and interferometers. The diversity of the applications and users makes it difficult to settle on a limited functionality for the electronics. Instead, we made sure to incorporate the maximum number of functions and to provide for full user control for most of the system parameters. This was also deemed necessary as a risk reduction for the system development.

The main system functions to be implemented in the electronics are the following:

- Provide cold power to the cold tip.
- Ensure temperature regulation at the cold tip.
- Ensure exported micro vibration reduction.
- Report operating data.
- Sustain launch mechanical loads.
- Perform health check and report health check data.

Figure 1 shows LPTC system including mechanical cooler, driving electronics and vibration sensors. Figure 2 shows a block diagram of the cooler system with adjacent instrument environment.

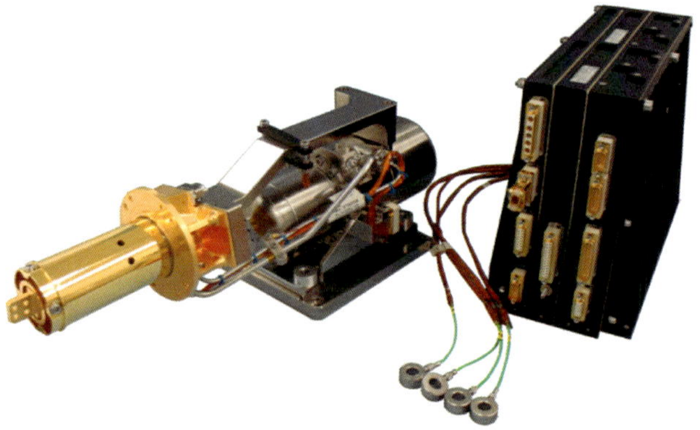

Figure 1. LPTC system overview

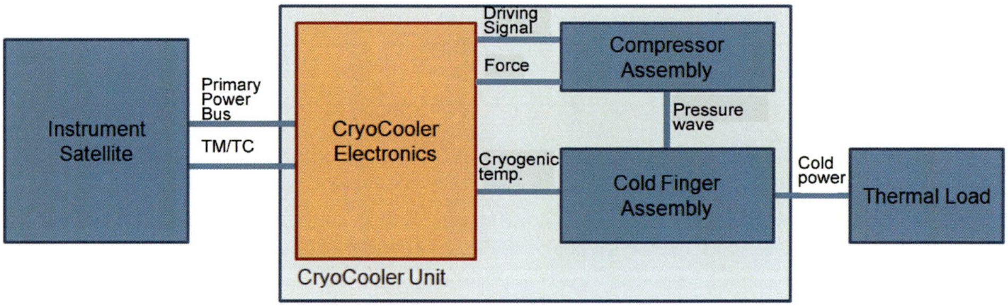

Figure 2. System block diagram

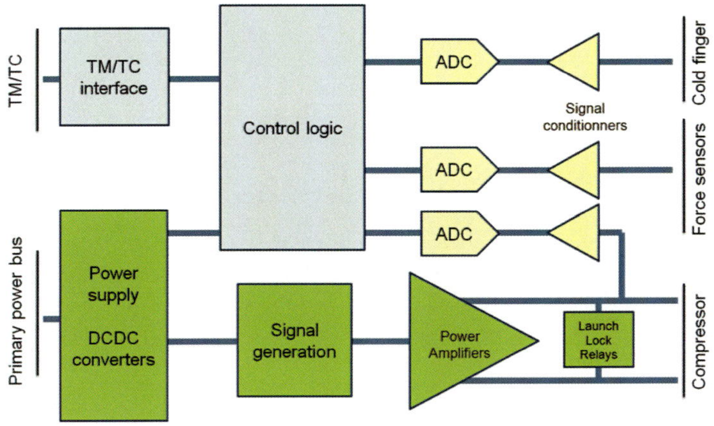

Figure 3. Driving electronics architecture

Fulfilling all of these functions and providing an interface with the instrument power bus and communication bus lead to an architecture incorporating several main blocks:

- Power conversion.
- Signal generation and amplification.
- TM/TC interface.
- Control logic.
- Sensor acquisition.

Figure 3 illustrates this architecture.

CHALLENGES FROM REQUIREMENTS

LPTC was designed and developed to provide a high level of reliability and performance. This strains the design of most parts of the system, and the electronics is no exception.

Performance requirements

The two main challenges of performance requirements are the temperature stability and the micro vibration reduction. These two functions have to be ensured for changing environmental conditions. Moreover, vibration reduction has to be ensured during cool down phase, to avoid perturbation on other satellite payload instruments.

Ensuring precise temperature and force regulation conditions the resolution of sensor acquisition chains. In our case, we seek a temperature stability better than 20mK. This requires temperature acquisition to have a resolution much better than a few mK over the whole user temperature required range. For exported forces, the target is to be able to reduce them below 50mN at operating frequency and harmonics below 500Hz. This again requires a high resolution of the sensor acquisition chain, below 10mN. On top of resolutions, accuracy and precision of the acquisition chains need also to be very good. For instance, a precision better than 1 Ω at 3 kΩ is necessary for temperature measurement.

Launch locking

A launch lock is necessary to protect compressor pistons from hitting endstops during launch. Customers require this function to be ensured with driving electronics OFF or ON. Passive (OFF) is achieved by shorting compressor coils on a resistance or RC circuit. Active locking involves an amplification circuit generating a counter motor force.

In both cases, energy is dissipated in the driving electronics but the satellite cooling system is not operating. Extra care must then be taken for the thermal control in this mode.

In the case of passive locking, the function needs to be triggered by relays operated once before launch. For reliability purpose, a series parallel combination of relays is implemented, which adds a considerable mass to the electronics. The challenge is also on relay qualification for the launch vibration environment.

Electrical interface

Depending e tsatellite, the electrical bus can be different. In the case of LPTC, we had up to 50% voltage difference between minimum and maximum supplied voltage to electronics. This, combined with requirements on low conducted emissions (CE) in common and differential modes, represents another challenge on the design, related to customer requirement. Besides the CE at high frequency, the challenge was how to reduce the CE at low frequency (mainly around 100 Hz) without impacting the mass. An active low pass filter, inside the CDE, was conceived, realized and successfully tested.

CHALLENGES FOR DESIGN

Designing a driving electronics for a cooler is always a challenge. In the case of electronics for the LPTC, two parts were a particular focus.

Power part

The main challenge on the power part comes from the fact that the cooler has a non optimal and varying power factor. The penalty of the low power factor on the efficiency is double. Parts with higher voltage rating are required, and these parts dissipate more. Moreover, reflected power needs to be absorbed. This requires filtering circuit to be added, but this cannot be done with a perfect efficiency. Table 1 illustrates the effect of voltage rating on MOSFET choice, showing the influence on dissipation.

Analog part

Temperature stability requirements are demanding to the design of the acquisition chain. Temperature sensors used in the LPTC are Cernox. An additional requirement is put on the maximum drift of the acquisition chain, which requires auto calibration to be implemented. Typical electrical requirements on the acquisition chain is 1Ω resistance resolution on a 3000Ω value. This translates into typical current probes of 100 μA and measurements of 100μV differences. To achieve this, 4 wire measurement is mandatory, but also high precision reference resistors for chain auto calibration. Design of instrumentation amplifier requires extra care.

For exported forces acquisition, we use piezo electric load washers. Typical sensitivity of the used sensor is 4 pC/N. The required resolution of force translates into measurement of currents as low as 1 pA. Achievement of this can be done only using low noise charge amplifiers. Downstream the charge amplifier, voltage measurement needs to be as low as 1mV.

Table 1. Effect of voltage rating on MOSFET dissipation

IRF P/N	Max voltage	RDSon	Max current
IRHMS57163SE	130	0.0145	45
IRHMS57160	100	0.0130	45
IRHMS57064	60	0.0060	45
IRHMS57Z60	30	0.0045	45

Force signal treatment is also a big task for the digital section. The LPTC is equipped with 4 load washers used to reconstruct 3 axis forces acting at the centre of the compressor. Induced vibration is encountered at driving frequencies and at its harmonics. This requires force signal to be transformed from time to frequency using pseudo FFT techniques. Such treatment occupies a large part of digital resources of the electronics.

EEE COMPONENTS

Several challenges come from the EEE components, technical and programmatic.

Technical challenges are related to components not in a qualified parts lists. Components such as temperature sensors, force transducers, hermetic connector and plug, coaxial connector for force transducers, are not part of European qualified parts lists. For those components, a qualification program had to be agreed and conducted. Although we did not fear any component not being suitable for application, agreement on the qualification program, formal conducting and discussion on potential test failures represents a long and not straight path. Qualification standards are often inspired from US MIL standards, requiring all environment test needed to use components in the field. Tests such as moisture, corrosion, cable pull strength are sometimes hard to fulfill. The question comes then to keep on trying to pass them, or to negotiate skipping them. In case of non skipping and failure, the subject comes again, when discussing deviation.

An additional difficulty comes from the fact that sometimes, the suppliers of the components are not familiar with space standards, which adds a product assurance additional burden to the work to be done for the qualification.

One interesting case was the piezo load washer. We went successfully through a complete qualification program for this component. It was only when reviewing the results with the customer that the question of the EEE category of the component was raised. After discussions, we concluded that the part being purely passive and not subjected to any external voltage or current, it could actually be non-EEE classified.

Some components such as relays required delta qualification because the applied environments were out of the qualification domains of the component. In our case random vibrations and shocks had to be played again at high levels to check the suitability of relays. This was actually a big risk of ending up with no technical solution. Test required to monitor micro openings of the relay during vibration. Figure 4 shows a test mounting for relay vibration qualification.

ENVIRONMENT AND QUALIFICATION CHALLENGES

Radiation

Radiation can be a problem for system availability because of SEU, SET, LET. The temperature and force regulation chains are complex and both involve a large number of parts. Studies were done to evaluate SEE scenarios and potential outages at system level.

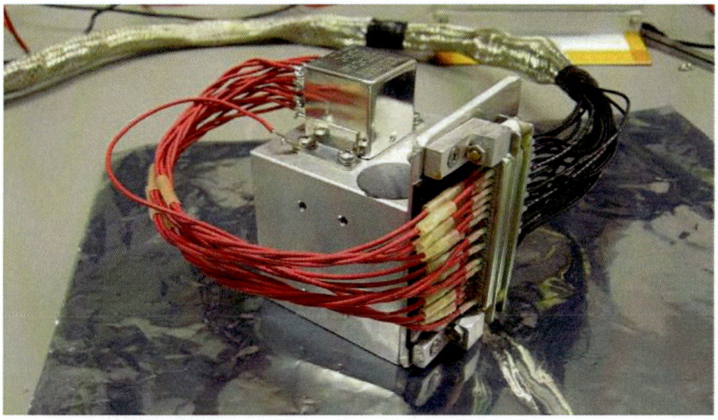

Figure 4. Relay mounting on vibration jig

Final demonstration was done through a test simulating an outage of temperature acquisition chain. For this test, we equipped the system with a special temperature harness and disconnected the sensor manually for al limited time. This was finally the simplest way to determine the effect of acquisition chain outage on the system and demonstrate how the system could handle it successfully.

Launch lock

Qualification of the launch lock at electronics level requires closing the loop of the launch locking function on a representative device by simulating the compressor. In order to assess actual dissipations in the system in a more representative way, it was asked also to perform qualification with a real mechanical cooler, applying vibration to the complete system. This of course adds to the difficulty and programmatic aspects because it requires:

- A large shaker to accommodate a much heavier load to be vibrated
- Adapted jig for the complete system
- Monitoring of compressor voltage and currents during vibration

Figure 5 illustrates the test set up used for launch lock qualification.

Test means

Qualification tests, environment or functional require a considerable amount of test means to be designed and manufactured. Most of them are not standard devices and require a lot of engineering to ensure adequate performance in the verification task:

- Passive cooler simulators (equivalent resistor circuits)
- Active cooler simulators (reactive circuits, representing potential values of cooler electrical properties, including variable power factor)
- Force sensor simulators (considering EMC)
- Temperature sensors simulators
- EGSE for system functional testing

PROGRAMMATIC CHALLENGES

Although first electronics developments occurred early in the LPTC technological maturity development program, the launch of several flight models qualification program, concurrent to the developments of electronics represents a very big challenge.

The complete system responsibility was a premiere for Air Liquide and we had to deal with early choices in the orientation of electronics fundamental design. One big challenge on the schedule is the lead time of components. Class 1 components were required for LPTC programs, due to reli-

Figure 5. Launch lock vibration qualification test

ability requirements. The downside is that lead time is one year for the most critical components, meaning that design choices gave to be made very early compared to qualification tests. This leaves no margin for design modification.

Other features such as VHDL code freezing need to be done very early, and any modification of the code takes several months to be validated. Again, having to handle a code modification late in the program is extremely difficult.

CONCLUSIONS AND PERSPECTIVE

Although electronics development had been started as early as mechanical cooler technology maturation, concurrent development and qualification of electronics and mechanical cooler is a challenging task. Technically, because cryocooler electronics package the power, analog and digital functionalities in the same box, the design and qualification engineering activities have to be run in parallel.

Air Liquide was responsible for the complete development of a cryocooler system for the first time. In collaboration with SITAEL, the Cooler Drive Electronics was developed. The first and complete demonstration model was designed and manufactured six years ago. Very high level requirements, combined with a high standard ESA qualification frame put a lot of pressure on the development and qualification activities, both technically and programmatically. Some conservative choices on system functionalities and their implementations had to be made. The same for component qualifications.

Future electronics developments will benefit quite a lot from this first experience. Electronics design fundamental choices cannot be easily disconnected from cooler requirements and satellite interfaces. The emphasis of future developments will be put on efficiency, cost and development and MAIT schedule.

Characterization Testing of Iris Cryocooler Electronics

Kerry Frohling

Iris Technology Corporation
Irvine, CA 92614

ABSTRACT

Two versions (100W and 200W) of the Low Cost Control Electronics (LCCE) have been developed by Iris Technology in collaboration with the Jet Propulsion Laboratory (JPL) to provide radiation hardened, high performance, modular, affordable cryocooler electronics. To date, the LCCEs have been used to drive and control a wide range of linear coolers, including the AIM SF100 (pulse tube and Stirling versions), Thales LPT9510, Thales LPT9310, Lockheed Martin Microcryocooler, Northrop Grumman Micro Pulse Tube Cooler, and the Ricor K527. Both of these new LCCEs include input ripple filters to mitigate the current ripple impressed on the input DC power and vibration mitigation to reduce the vibration exported by the cryocooler. This paper describes the performance testing of these LCCE devices, including test results and performance demonstrations of the LCCEs.

INTRODUCTION

This paper will concentrate on the results of acceptance testing of the LCCE-2. The LCCE-2 is a cryocooler-agnostic set of drive electronics, architected to support both traditional long-life space cryocoolers and tactical cryocoolers, providing the Payload Integrator with a wide range of radiation hard cryocooler system options. The LCCE-2 is modular, scalable, affordable, and provides the critical cryocooler operational functionality that fully envelopes the needs of a typical space mission (high efficiency motor drive, closed loop temperature control, input ripple filter, and vibration control). This design has been matured culminating in a flight-design Engineering Model immediately applicable for a wide range of missions. The LCCE products are also suitable for the control of other systems such as actuators, motors and heaters that require a controlled power drive.

The LCCE-2 specifically includes two enhancements that had not been part of previous LCCE efforts. These enhancements are an input ripple filter (IRF) to reduce the current ripple imparted onto the input power bus and a vibration control algorithm that is used to reduce the vibration produced by the compressor that could adversely affect overall system performance.

The LCCE-2 acceptance testing was performed with both a resistive load and a Thales LPT9510 cryocooler. Cryocooler vibration was measured with an Endevco 7703A-1000 accelerometer.

Cryocoolers 19, edited by S.D. Miller and R.G. Ross, Jr.

Table 1. Summary of Critical Requirements from LCCE-2 Performance Specification.

Parameter	Characteristic
Operating Input Voltage	22 to 35 VDC
Number of Motor Drives	2 independent motor drive outputs (each rated at > 50W)
Total Output Power	Continuous operation at 100 WAC and 28 VDC input voltage.
Efficiency	> 85%, with output power at 100 WAC and input voltage at 28 VDC.
Tare Power	< 2.2 W @28 VDC
Output Voltage Sinusoidal Symmetry	< 3% THD including harmonics up to 5 kHz, above 20 watts output power.
Operating Frequency	Control Output drive frequency shall be adjustable from 40 to 150 Hz with 0.1 Hz resolution.
Vibration Measurement	Measure cryocooler vibration at the fundamental (Fc), second harmonic (2 x Fc) and third harmonic (3 x Fc) of the drive frequency.
Input Current Ripple	< 200 mA peak-to-peak @ 100W output and 28 VDC input
Inrush Current	After application of +28V power the initial inrush (charging) current due to distributed capacitance, EMI filters, etc., shall be completed in 30 μs with its peak no greater than 25 Amps.
	The rate of change of inrush current after the initial application of +28V power shall not exceed 50 mA/μs.
	Steady state operation shall be attained within 100 ms from turn-on or the start operating mode change, except for motors.
Vibration Control	Measure cryocooler vibration at the fundamental (Fc), second harmonic (2 x Fc) and third harmonic (3 x Fc) of the drive frequency.
	Provide closed loop vibration control based on measured cryocooler vibration.
Temperature Sensor Set-point	Adjustable between 60K and 300K when using a Cernox CX-1080 sensor.
Temperature Sensor Sensitivity	< 0.1K (TBR) at 70K when using a Cernox CX-1080.

OVERVIEW OF LCCE-2 REQUIREMENTS

The detailed requirements for the LCCE-2 are given in Reference 1, and a summary of the critical performance criteria is provided in Table 1.

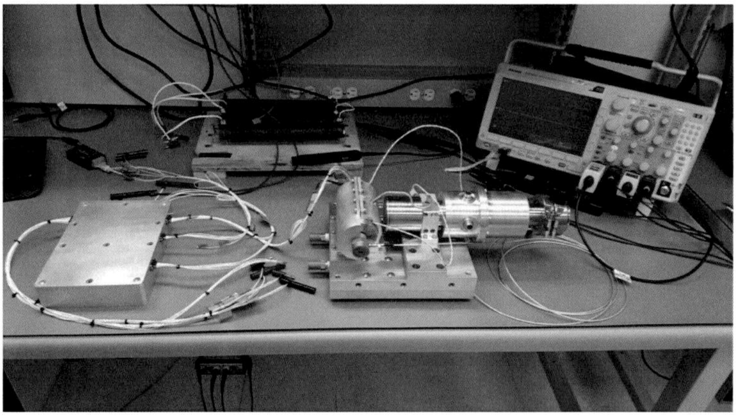

Figure 1. LCCE-2 Performance Qualification Test Setup.

LCCE-2 TESTING

LCCE-2 testing followed a standard build up. First circuit boards were tested stand-alone to validate performance prior to assembly into an enclosure. The assembled system was then subjected to more engineering tests to validate performance until it was determined to be ready for acceptance and qualification testing. The results presented in this paper have been excerpted from acceptance and qualification test results. Figure 1 shows the test setup for performance qualification testing.

LCCE-2 TEST RESULTS

The following paragraphs present the results of selected portions of the acceptance and qualification test results. Only tests with data that provide general interest or specific differentiation are presented herein.

Tare Power Test

The Tare Power test was performed at three voltages (22V, 28V, and 35V) under three conditions (Disabled, No Telemetry, and Telemetry ON). The results are shown in Table 2. When the LCCE-2 is "Disabled" the PWM controller is disable so none of the auxiliary voltages are produced, basically only the Input Ripple Filter is active. When the LCCE-2 is in the "No Telemetry" mode, all the voltages are present, but there is no communication and no activity in the CCE. When the LCCE-2 is in the "Telemetry ON" mode the CCE is inactive, but a telemetry message is being requested and sent every 500 ms. Note that the LCCE-2 is under the 2.2W requirement at 28V.

Input Voltage Range Test

The LCCE-2 was operated at three input voltage levels, 22V, 28V, and 35V. At each voltage power level, the LCCE-2 was programmed to output 110W into a resistive load. The output waveform was checked for sinusoidal characteristics and channel balance. This test was successful at all three input voltage levels.

Output Voltage Range Test

The peak output voltage is adjustable in 15.928mV steps from 0 to in excess of 22V. For the Output Voltage Range Test the LCCE-2 was operated at the worst case input voltage level of 22V. The LCCE-2 produced output waveforms at 5V (6.25W), 10V (25W), 15V (56.25W) and 20V (100W) into a resistive load. Each of the output waveforms was checked for sinusoidal characteristics and channel balance. This test was successful at all four output power levels.

Voltage Ramp Rate Test

The Voltage Ramp Rate Control is designed to provide an upper limit on the ramp rate of

Table 2. Results of Tare Power Testing at 22V, 28V and 35V.

Voltage	Condition	Input Current (Amps)	Input Power (Watts)
22	Telemetry ON	92mA (max)	2.02W (max)
22	No Telemetry	92mA (max)	2.02W (max)
22	Disabled	8.3mA (max)	0.183W (max)
28	Telemetry ON	78mA (max)	2.18W (max)
28	No Telemetry	78mA (max)	2.18W (max)
28	Disabled	9.7mA (max)	0.272W (max)
35	Telemetry ON	69mA (max)	2.42W (max)
35	No Telemetry	69mA (max)	2.42W (max)
35	Disabled	11.4mA (max)	0.399W (max)

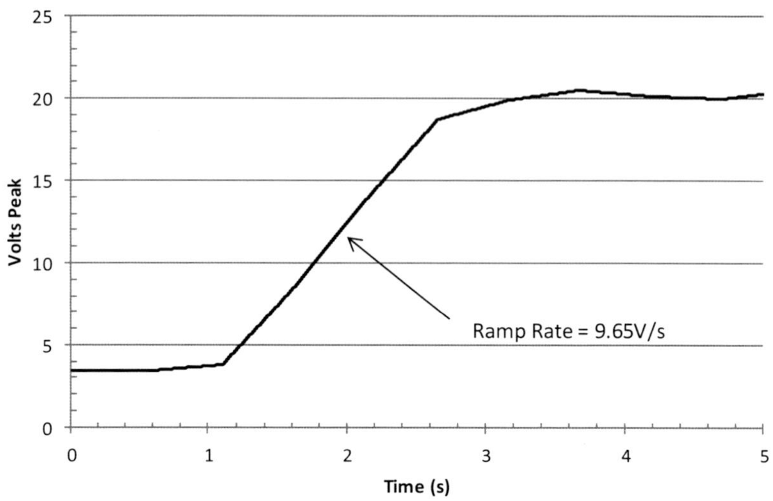

Figure 2. Ramp Rate Test Results

the peak voltage of the waveform during a single step of the of controller operation. Thus, no matter what value is set, that value cannot be exceeded. For this test the operator set the ramp rate to 10V/s. The nearest achievable ramp rate was 9.88V/s. The ramp rate is selectable in multiples of 0.309V/s. Since the Ramp Rate Control establishes an instantaneous limit for the ramp rate, the measured ramp rate will be less than the selected ramp rate. For this test the measure ramp rate was 9.65V/s, as shown in Figure 2.

Operating Frequency Test

The LCCE-2 is required to provide frequency adjustment from 40Hz to 150Hz with 0.1Hz resolution, as shown in Table 1. The LCCE-2 is designed to provide frequency steps of about 0.0013Hz steps from 24Hz to 240Hz. For this test the LCCE-2 requirement was verified. Waveforms were generated and verified for frequency accuracy for 40Hz, 44.9Hz, 45Hz, and 150Hz.

Output Efficiency Test

The LCCE-2 is required to provide 85% efficiency with output power at 100 WAC and input voltage at 28 VDC, as shown in Table 1. For this test the LCCE-2 efficiency was measured for 0W to 100W of output power at three different input voltage levels, 22V, 28V, and 35V. The test showed that at 22V input an efficiency of 83.9% was achieved, at 28V input an efficiency of 85.4% was achieved, and at 35V input an efficiency of 86.7% was achieved. The sweep results for the 28V input test are shown in Figure 3.

Reflected Ripple Current Test

The LCCE-2 is required to impart an input ripple current of less than 200 mA peak-to-peak when the LCCE-2 is providing 100W output power with an input voltage of 28 VDC. To test this requirement Iris commanded the LCCE-2 to provide 110W into a resistive load and measured the input ripple current on an oscilloscope. The resultant waveform measure 194 mA peak-to-peak and is shown in Figure 4.

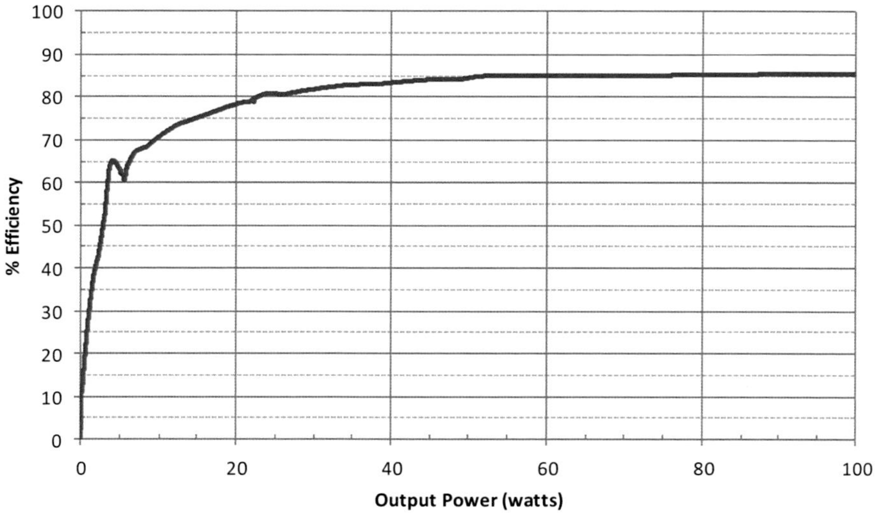

Figure 3. LCCE-2 Efficiency with 28V Input Power.

Inrush Current Test

The LCCE-2 requirement for inrush current has three parts; (1) after application of 28V power the initial inrush (charging) current due to distributed capacitance, EMI filters, etc., shall be completed in 30 µs with its peak no greater than 25 Amps, (2) the rate of change of inrush current after the initial application of +28V power shall not exceed 50 mA/µs, and (3) steady state operation shall be attained within 100 ms from turn-on or the start operating mode change, except for motors. The LCCE-2 was monitored with an oscilloscope during the initial application of power. As shown in Figure 5, for the first requirement, the initial inrush peak was 4.36A less than the 25A limit and was completed in 4.9µs, less than the 30µs limit. For the

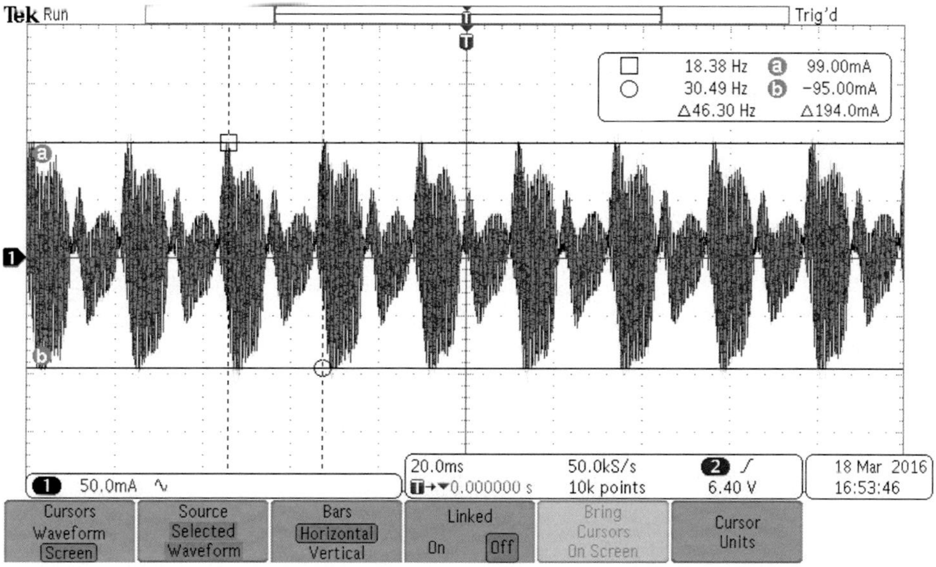

Figure 4. LCCE-2 Current Ripple Test Results.

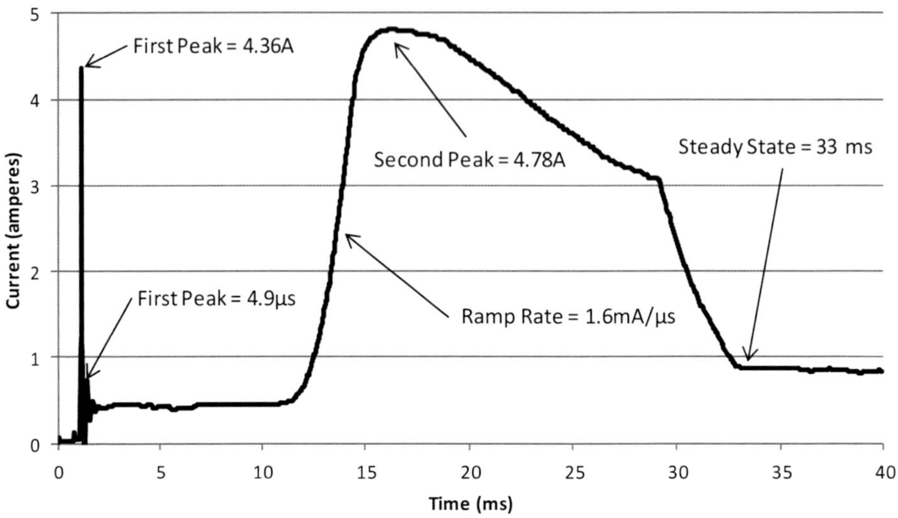

Figure 5. LCCE-2 Inrush Current Test Results.

second requirements, the ramp rate of the inrush current was 1.6 mA/μs less than the 50 mA/μs limit. For the third requirement, steady state current was achieved in 33 ms, less than the 100 ms limit.

Vibration Algorithm Test

The LCCE-2 requirement for vibration is twofold, the first requirement is to measure the cryocooler vibration at the fundamental (Fc), second harmonic (2 x Fc) and third harmonic (3 x Fc) of the drive frequency, and the second is to provide closed loop vibration control based on measured cryocooler vibration. To test these requirements, the LCCE-2 was setup to drive a Thales LPT9510 cooler in a fixture with an Endevco 7703A-1000. The LCCE-2 actually measures five harmonics so the first requirement is met, and testing showed that the five harmonic algorithm reduced the vibration measured by the LCCE-2. These results are illustrated in Figure 6. In Figure 6 time runs from left to right, and the vertical axis is non-dimensional vibration counts. When the graph in Figure 6 first starts there is no vibration as the compressor

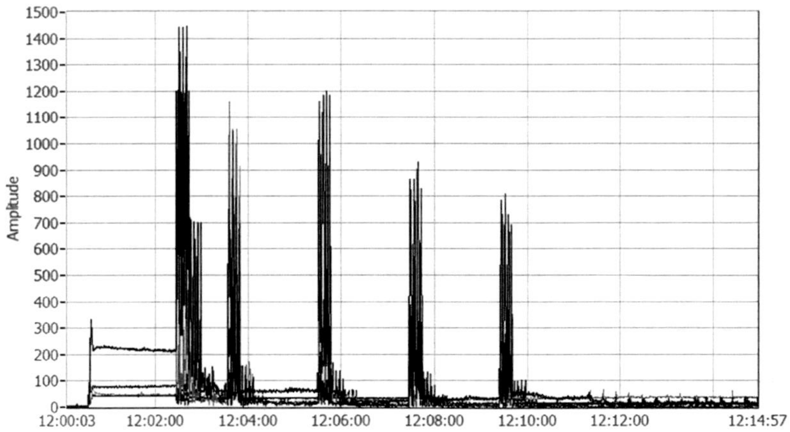

Figure 6. LCCE-2 Vibration Algorithm Test Results.

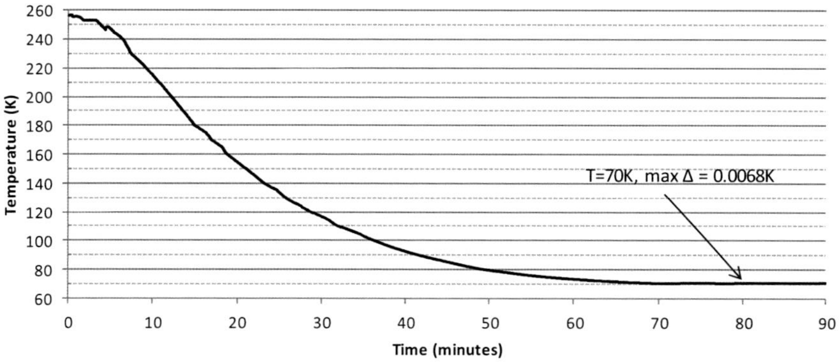

Figure 7. LCCE-2 Temperature Sensitivity Test Results.

is not on. When the compressor is turned on, we see that the vibration levels increase rapidly. The highest vibration at turn-on is the fundamental frequency, then the third harmonic, the second, fourth and fifth harmonics are all about the same at this time. As the algorithm is run, vibration extremes are seen as the algorithm adjusts the phase and amplitude of the output waveform. We can see five distinct areas, each corresponding to the five harmonics. When the algorithm is complete, we see that all of the harmonic vibrations have been reduced, and only the second harmonic has not been reduced to levels near the off conditions.

Temperature Control Test

The LCCE-2 requirements for temperature are twofold; (1) the temperature set-point is adjustable between 60K and 300K when using a Cernox CX-1080 sensor, and (2) temperature sensitivity is less than 0.1K at 70K when using a Cernox CX-1080. The first requirement was demonstrated by showing that the LCCE-2 could be adjusted to 60K and 300K. The second requirement was demonstrated with a Thales LPT9510 cooler with a Cernox-1080 sensor on the cold tip. The temperature set-point was set to 70K, and the system was allowed to cool. The system settled with a maximum deviation of 0.0068 K, much less than 0.1 K. This is illustrated in Figure 7. In addition to the tests to verify the requirements, a test was run to demonstrate system performance with a heat load. The system was run to 120 K, stabilized, and then a heat load of 0.25W was added and the system response was measured. The system settled at 120 K with a maximum deviation of 0.0055 K. This is shown in the left-hand graph in Figure 8. A heat load of 0.25 W was added, and again the system settled at 120 K with a maximum deviation of 0.0055 K; this is shown in the right-hand graph in Figure 8.

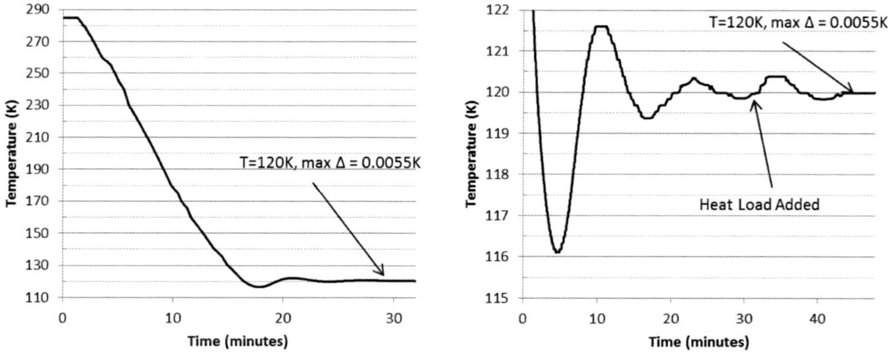

Figure 8. LCCE-2 Heat Load Test Results.

CONCLUSIONS

The Second Generation Low Cost Control Electronics (LCCE-2) and the High-Power Low Cost Control Electronics (HP-LCCE) were both successful in demonstrating the enhanced features of the LCCE product line while maintaining previous performance baselines. Some of these enhancements are listed below.

- Incorporation of an Input Ripple Filter protects the host power bus from current surges imparted when creating a cryocooler waveform, and allows the LCCE-2 and HP-LCCE to work with lower impedance cryocoolers that have higher peak voltage requirements.
- Incorporation of a firmware-only algorithm that compensates for the voltage ripple at the output of the IRF, thus reducing the total harmonic distortion in the cryocooler power waveform.
- Incorporation of the Vibration Mitigation Algorithm that reduces the vibration imparted to the structure of the spacecraft by modifying the power waveform to the cryocooler. This algorithm requires that an accelerometer be attached to the cryocooler to allow the LCCE-2 and HP-LCCE to determine the vibration components.

In conclusion, the LCCE 2 and HP-LCCE were successful in establishing and delivering a low cost, radiation hard, high performance, general purpose control electronics module to support future space missions.

REFERENCES

1. *Second Generation Low Cost Cryocooler Electronics (LCCE-2) Performance Specification*, Revision A, Iris Document Number SYS10005, dated October 22, 2015.

JT Micro Compressor Test Results

J.R. Olson, P. Champagne, E. Roth, T. Nast

Advanced Technology Center, Lockheed Martin Space Systems
Palo Alto, CA 94304, USA

ABSTRACT

We report the results of operational testing of a 200 gram closed-cycle Joule-Thomson Micro Compressor. This unique compressor is a modified version of our TRL 6 Pulse Tube Micro Compressor, using the same Oxford-style flexure bearing and clearance seal architecture common in highly reliable long life cryocoolers. The compressor is capable of delivering closed-loop dc gas flow with a high pressure ratio, suitable for driving a Joule-Thomson cryocooler.

The Joule-Thomson compressor was operated while charged with nitrogen gas at pressures ranging from 50 PSI to 150 PSI. Output pressure ratios were varied from 2:1 to as high as 7:1 at different gas flow values. In all cases, the compressor operated smoothly, without any indication of unusual mechanical or gas dynamics such as instabilities or excessive gas backflow through the check valves. Power conversion efficiency was calculated based on the theoretical power required to deliver specific pressures and gas flows vs. the measured compressor power. Conversion efficiency ranged from 15% for the highest pressure ratios, to greater than 30% for lower pressure ratios. These values are in close agreement with the expected efficiency of approximately 25%, based on predicted check valve, clearance seal, and surface pressurization losses.

INTRODUCTION

Joule-Thomson (JT) cryocoolers have only been used in space for specialized low-temperature applications, such as the 6 K cooling stage for the MIRI cryocooler [1]. The primary reasons for this are the complexity of the JT compressor and cooler, and the low JT cooler efficiency at the temperatures required by typical cryogenic applications.

The recent development [2,3] of mid-wave infrared focal planes operating at 120-150 K temperature, rather than 80 K as required by traditional InSb detectors, allows JT coolers to use optimized high-performance gas mixtures, rather than pure nitrogen, as the working fluid. JT coolers using optimized gas mixtures can potentially have higher efficiency than Stirling-cycle coolers.

Lockheed Martin reported last year [4] on the development of a closed-cycle JT micro compressor suitable for long-life cryogenic applications, including space applications. That report described the design of the JT compressor. To briefly summarize, the Pulse Tube micro compressor previously qualified to TRL 6 [5, 6] was modified for use as a JT compressor. While retaining the same flexure bearing, clearance seal, long life compressor architecture, two

Cryocoolers 19, edited by S.D. Miller and R.G. Ross, Jr.

Figure 1. Lockheed Martin Joule-Thomson micro compressor. This 200 gram compressor has piston architecture identical to the pulse tube compressor previously qualified to TRL 6, but includes check valves to rectify the ac pressure to generate the dc gas flow needed by the JT cryocooler valves to rectify the ac pressure to generate the dc gas flow needed by the JT cryocooler.

check valves were added to each compressor module in order to convert the ac pressure wave into a continuous dc flow. By operating these two modules in series, Lockheed Martin was able to achieve pressure ratios suitable for a JT cryocooler, while retaining the low exported vibration inherent in compressors with back-to-back piston modules. This compressor is shown in Figure 1.

The design of an efficient JT cryocooler using gas mixtures is a complex problem, because the gas constituents and mixing ratios of the optimum gas mixture are functions of the high and low pressures of the JT cooler, but the efficiency of the JT compressor is also a function of the high and low pressures. This optimization is more complicated than Stirling and Pulse Tube coolers because of the very high pressure ratios in the JT compressor. In order to achieve a pressure ratio of 4:1, it is necessary that each module have a pressure ratio of 2:1, whereas the pressure ratio in Stirling and Pulse Tube coolers is generally less than 1.3:1. The secondary compressor losses from surface pressurization in the compression spaces and leakage in the piston clearance seals scale with the square of the pressure ratio, causing these losses to dominate the motor i^2R coil losses, which are typically the largest loss in a Stirling or pulse tube cooler.

In order to facilitate future JT cooler optimization, extensive compressor testing was performed on the JT compressor described in [4], so that future gas mixture studies can take the compressor efficiency into account.

JT COMPRESSOR TESTING

Test Setup

The JT compressor was connected to a closed-loop gas system and instrumented with pressure sensors and a calibrated gas flow meter, as shown in Figure 2. A metering valve was

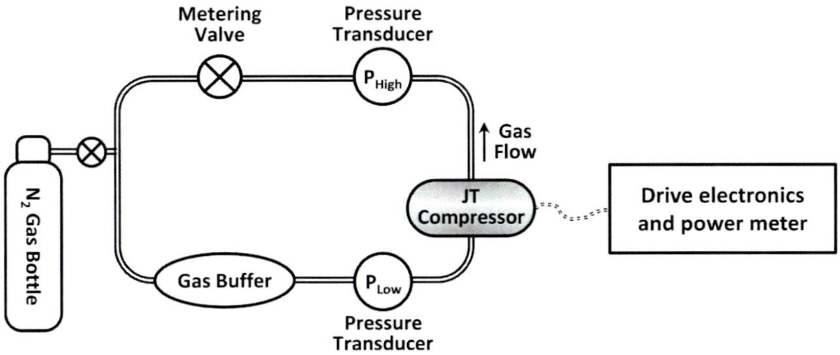

Figure 2. JT compressor test setup. The low pressure is adjustable with the nitrogen gas bottle, and the high pressure and gas flow are adjustable by varying the compressor input power and the metering valve. The compressor power, current and voltage are monitored with a wattmeter.

used to set the desired gas flow rate. The gas volume in the high pressure side was minimized, and a gas buffer was included in the low pressure side in order to decrease the response time of the system to changes in compressor operation. The compressor and gas loop were maintained at ambient temperature with air cooling. All testing was performed with nitrogen working gas.

TEST RESULTS

JT Compressor testing was performed with low pressure values of 50, 75, 100, and 150 PSIA, and pressure ratios ranging from 2:1 to as high as 7:1, with varying amounts of nitrogen gas flow. For each operating point (with a specific low pressure, pressure ratio, and gas flow rate), a frequency scan was conducted to determine the optimum operating frequency. One representative frequency scan is shown in Figure 3, for a low pressure of 100 PSIA, a pressure ratio of 4:1, and a flow rate of zero. The compressor electrical input power is shown in the figure on the left, and includes both the total electrical input power, and the individual input powers for each compressor module. It can be seen that the total compressor power has a weak minimum, in this case at 106 Hz, and that the two compressor modules have minima at slightly different frequencies, 102 and 110 Hz. The figure on the right shows the PV power, calculated by subtracting the motor i^2R loss from the electrical power. The PV power in this case, and

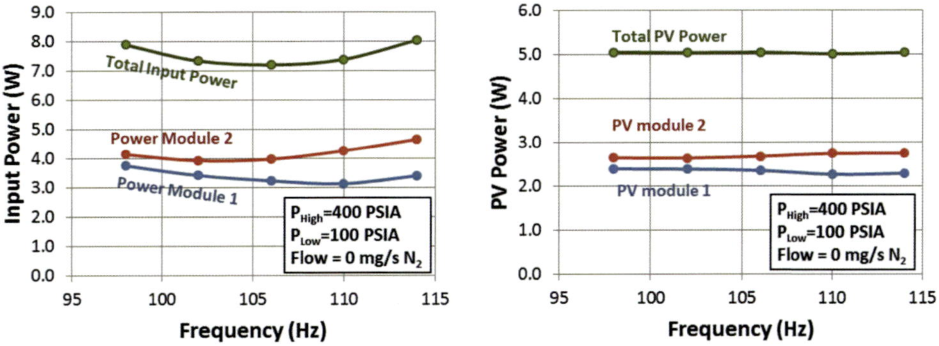

Figure 3. Measured frequency scan for JT compressor operating at 100 PSIA low pressure, 400 PSIA high pressure, with no gas flow. These representative data show that the compressor electrical input power has a weak minimum at 106 Hz (left plot), whereas the individual compressor modules have individual minima at 102 and 110 Hz. The specific frequencies corresponding to these minima vary with low and high operating pressure. The PV power (total, and for each individual module) is independent of frequency.

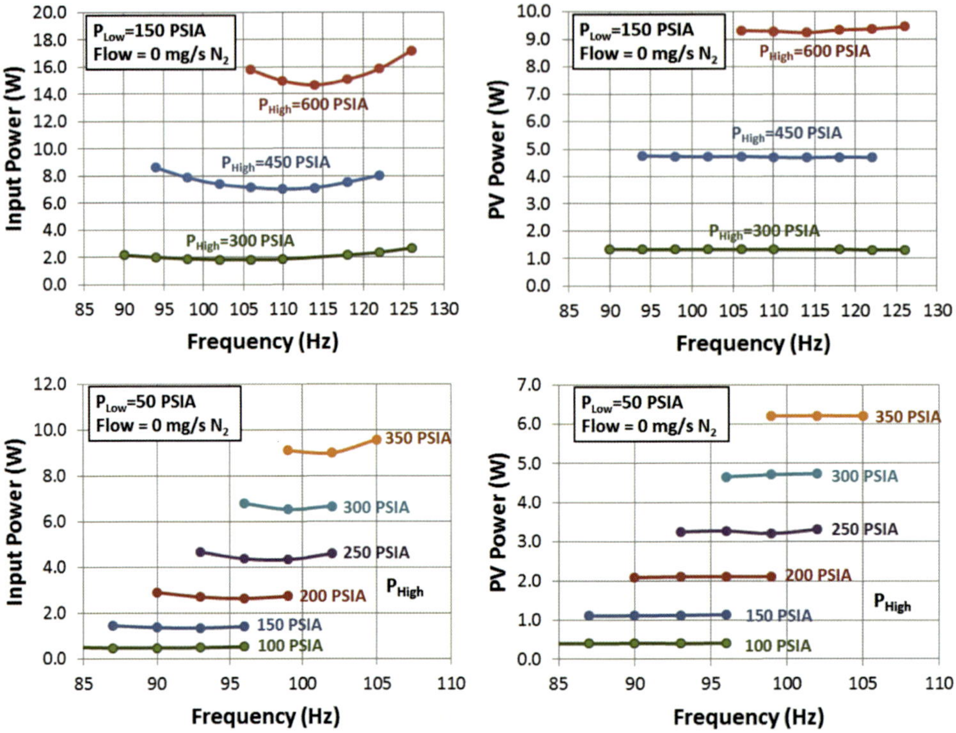

Figure 4. Other measured frequency scans for JT compressor operating at 150 PSIA low pressure (top figures) and 50 PSIA low pressure (bottom figures). The optimum frequency (minimum input power) increases as the pressure ratio increases, because of the effect of the gas spring. As in Figure 3, the PV power is independent of frequency.

indeed in every case measured in this study, is very nearly independent of frequency. This is an indication that there is no detrimental dynamics present in the check valves, which might be expected to lead to frequency-dependent backflow losses.

Several other frequency scans, all still with zero gas flow, are shown in Figure 4. The two upper figures are the total electrical input power and PV Power for 150 PSIA low pressure, with pressure ratios of 2:1, 3:1 and 4:1. The two lower figures are for 50 PSIA low pressure, and pressure ratios ranging from 2:1 to 7:1. As before, the PV power is independent of frequency. As one might expect, as the pressure ratio increases, the optimum frequency also increases, because of the increasing gas spring effect.

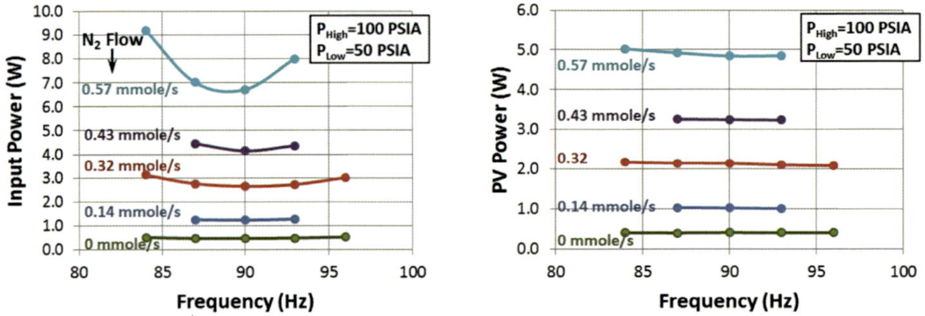

Figure 5. Effect of varying the gas flow, for one representative case (50 PSIA low pressure, 2:1 pressure ratio). As expected, more gas flow requires more input power.

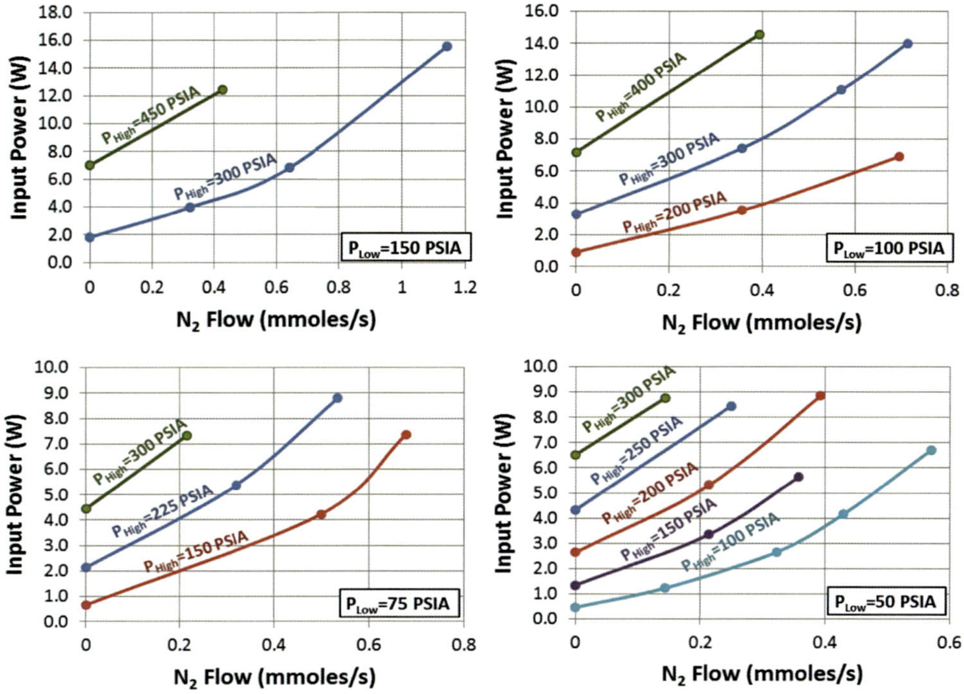

Figure 6. Optimum (minimum) compressor input power for every set of conditions tested in this study. As expected, the input power increases with increasing pressure ratio and increasing gas flow.

The effect of varying the gas flow is shown in Figure 5 for one representative case, with 50 PSIA low pressure, a pressure ratio of 2:1, and nitrogen gas flow varying from zero to 0.57 mmoles/second. As expected, as the gas flow increases while keeping the pressure ratio constant, the input power and PV power increase. As before, the PV power is nearly independent of frequency. The optimum frequency is nearly independent of flow for constant high and low pressures.

From every frequency scan such as those shown in Figure 5, we can extract the minimum input power. These optimum points are shown in Figure 6, for every set of test conditions measured during this study. Low pressures of 50, 75, 100 and 150 PSIA are shown, with nitrogen flow varying from zero up to nearly the maximum capacity of the compressor. The input power is a strong function of the pressures and flow values. For the cases of 100 and 150 PSIA low pressure, the maximum flow was limited by the compressor power we were willing to apply when fan-cooling the compressor. With good conductive cooling, we could achieve somewhat higher flow rates. For the cases of 50 and 75 PSIA low pressures, we were limited by the swept volume capability of the compressor. As expected, the input power increases as the pressure ratio and the gas flow increase.

In order to make the best use of these measurements, especially when assessing potential JT gas mixtures, it is useful to compare the measured compressor power with the ideal compressor power that one can calculate for ideal, lossless, isothermal compression. This ideal compressor power is given by

$$\text{Ideal Compressor Power} = \dot{m} C_P T_0 \left[\left(\frac{P_{High}}{P_{Low}} \right)^{1-\frac{1}{\gamma}} - 1 \right] \quad (1)$$

where $\dot{m}$ is the mass flow (g/s), C_P is the gas specific heat (J/g-K), T_0 is the ambient temperature (K), P_{High} and P_{Low} are the operating pressures, and γ is the ratio of specific heats.

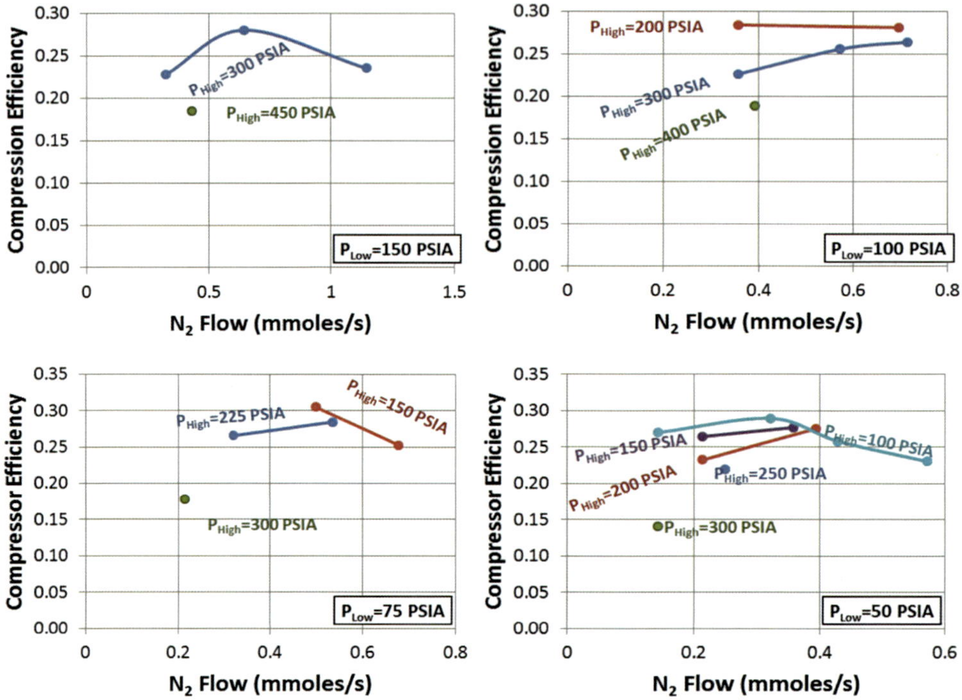

Figure 7. Measured compressor efficiency, calculated by dividing the ideal lossless isothermal compressor power by the measured compressor input power. The measured compressor efficiency of 20-30% is high enough that a JT cryocooler with an optimized gas mixture can have a cryocooler efficiency comparable to a Stirling or pulse tube cooler at 150 K cold tip temperature.

We calculate the compressor efficiency for every point in Figure 6 by dividing the ideal compressor power by the measured compressor power. The calculated compressor efficiencies are shown in Figure 7. The compression efficiency is in the range of 20-30% for most data points. This efficiency includes all the losses in the compressor, including motor i^2R coil losses, piston clearance seal and surface pressurization losses, and losses associated with the check valves and gas flow passages.

DISCUSSION AND FUTURE WORK

The data shown in Figure 7 show that the Lockheed Martin JT micro compressor is capable of achieving a compression efficiency that is 20-30% of the ideal compression efficiency over a range of pressure ratios from 2:1 to as high as 5:1, with generally higher efficiency at lower pressure ratio and lower values of the low operating pressure. According to preliminary gas mixture analyses, this compressor performance is sufficient to allow a closed-cycle JT cooler to have an efficiency comparable to a Stirling or pulse tube cryocooler when operating at 150 K. Work is ongoing to develop a high-performance, low-cost counterflow heat exchanger for a JT cooler with comparable performance to the cooler described in [5]. Future work is planned to develop the optimized gas mixture.

ACKNOWLEDGMENT

This work was supported by the United States Air Force through CU Aerospace STTR Phase II Prime Contract # FA9550-15-C-0019 "Microvascular Composites for Novel Thermal Management Devices." The authors wish to thank the CU Aerospace principal investigator Dr. Chris Mangun, and the AFRL program manager Dr. Byung-Lip "Les" Lee for their helpful interactions.

The compressor tested in this work is the property of Santa Barbara Focalplane. The authors wish to thank Vince Loung and Elna Saito for allowing the use of this hardware during this STTR program.

REFERENCES

1. Petach, M. and Michaelian, M., "Mid InfraRed Instrument (MIRI) Cooler Cold Head Assembly Acceptance Testing and Characterization," *Cryocoolers* 18, ICC Press, Boulder, Colorado (2014), pp. 11-17.

2. Ramirez, D.A. et al., "High-operating temperature MWIR unipolar barrier photodetectors based on strained layer superlattices," Proc. SPIE 945 1, Infrared Technology and Applications XLI, 945 113 (2015); doi: 10.1117/12.2176908.

3. Klipstein, P. et al., "Reducing the cooling requirements of mid-wave IR detector arrays," SPIE Newsroom, (2011), doi: 10.1117/2.1201111.003919.

4. Champagne, P., et al., "Development of a J-T Micro Compressor," IOP Conf. Series: Materials Science and Engineering 101 (2015) 012009.

5. Nast, T.C., Roth, E., Olson, J.R., Champagne, P., Frank, D., "Qualification of Lockheed Martin Micro Pulse Tube Cryocooler to TRL 6," *Cryocoolers* 18, ICC Press, Boulder, Colorado (2014), pp. 45-50.

6. Olson, J.R., Kaldas, G., Champagne, P., Roth, E. and Nast, T., "MatISSE Microcryocooler," IOP Conf. Series: Materials Science and Engineering 101 (2015) 012025.

Preliminary Experimental Study on a Precooled JT Cryocooler Working at 4 K - Open Cycle

D.L. Liu, Z.H. Gan, X. Tao, Y.F.Yao, W.K. Pan, L.M.Qiu

Institute of Refrigeration and Cryogenics, Zhejiang University
Hangzhou, China, 310027
Key Lab. of Ref. & Cryo. Technology of Zhejiang Province
Hangzhou, China, 310027

ABSTRACT

For developing a long-life precooled Joule-Thomson cryocooler (PJTC) for space use, the cold head part has been developed and tested using an open cycle. The structures and parameters of the counterflow heat exchangers and cold head heat exchanger are presented in this paper. In the testing experiment, a two-stage GM cryocooler was used as the precooler due to its sufficient and consistent cooling performance. The PJTC successfully achieved 4.4 K in the open cycle experiment. During the measurement of the cooling power, it was found that, the PJTC started to be unstable before the liquid helium was totally evaporated in the CHX. The phenomenon is described and the reasons discussed.

INTRODUCTION

The precooled Joule-Thomson cryocooler (PJTC) has become the main cooling technology at liquid helium temperatures for long-life space missions [1]. It is usually applied to directly cool the detectors or to precool other cryocoolers designed to achieve even lower temperatures. Some research institutions, companies, and space mission groups have reported their developed PJTCs [1-3]. However, few details about their designs and performance analyses have been described.

According to the analysis of Reference [4], key elements of the PJTC are the last stage recuperator, the J-T valve, and the cold head heat exchanger. These are the most important elements of the PJTC that can reflect the effects of the precooling temperature and the high pressure on cooling performance.

An open-cycle PJTC has been designed and built as part of the work reported here. For verification of the design, the PJTC was tested with a GM cryocooler as a precooler. The preliminary experimental setup is introduced in this paper. It is found from the cooling power measurement results that the PJTC went unstable before the liquid helium was totally evaporated. The reasons are discussed.

EXPERIMENT SETUP

The preliminary setup introduced here is mainly developed to verify the design of the key elements. As shown in Fig. 1, when the open cycle is operating under steadystate conditions, the helium gas discharged from the helium bottle (with reducing valve) goes through an inlet valve, a high mass flow meter (MFMh), the first counterflow heat exchanger (CFHX1), the first precooling heat exchanger (preHX1) , the second counterflow heat exchanger (CFHX2), the second precooling

Cryocoolers 19, edited by S.D. Miller and R.G. Ross, Jr.

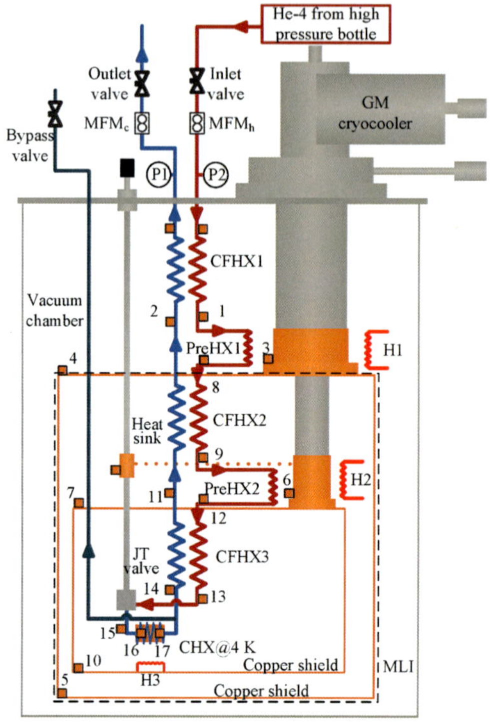

Figure 1. The open-cycle experiment setup.

heat exchanger (preHX2), and the third counterflow heat exchanger (CFHX3) before throttling to low pressure. The room temperature helium gas is precooled in these components and then goes into the two-phase region after the JT valve. The two-phase helium flow will absorb the heat load and evaporate in the cold head heat exchanger (CHX). Then the evaporated gas (or still two-phase fluid) from the CHX flows through the CFHX3, CFHX2 and CFHX1 to precool the incoming warmer helium flow. Finally it goes out to the atmosphere. The thermometers are labeled as squares numbered from 1 to 17 in Fig. 1.

For speeding up the cooling down process, a bypass line from the cold end is used. If there is no bypass, the CFHX3 and CHX will warm up the precooled gas from the preHX2 during the their cooling down process. Then the warmed gas will go back to the CFHX3 and continuously consume the precooling power provided by the precooler. It will take a extremely long time to cool down the CHX to be below the inversion temperature of Helium because the effectiveness of CFHX3 is designed with a high level and the heat conductance along its length direction is small enough to be neglected. So a bypass line is shown in Fig.1, which can let the gas from the CHX flow out to the atmosphere before the CHX approaches the temperature of point 12 (below 10 K in experiment) when the outlet of the PJTC is closed by a valve.

There are two radiation shields made of copper which are cooled by the first and second stage cold heads of the GM cryocooler, respectively. MLI is used around the outside copper shield to eliminate the heat radiation. H1 and H2 are two heaters that can be adjusted to regulate the precooling temperature. H3 is another heater which is used to measure the cooling power of PJTC. The JT valve can be controlled outside the vacuum chamber which makes it easier to have the required mass flow.

The CFHX3 is a Linde type [5] heat-exchanger, also called a tube-in-tube heat exchanger, as well as CFHX1 and CFHX2. They are all made of standard stainless tubes of which the parameters are listed in Table 1. The heat exchanger model was presented in Reference [6].

The CHX is made of 1/8 inch stainless tube, which is the same as that in Table. 1. The tube is coiled and soldered around a copper cylinder. The length of the tube available for heat exchanging is about 0.6 m. The heat load can be added on the copper cylinder by a electrical resistance.

Table 1. The parameters of the CFHXs.

label	Inner tube		Outer tube		Length (m)
	Outer diameter (inch)	Wall thickness (inch)	Outer diameter (inch)	Wall thickness (inch)	
CFHX1	1/8	0.028	3/8	0.035	0.8
CFHX2	1/8	0.028	3/8	0.035	2.0
CFHX3	1/8	0.028	1/4	0.035	3.3

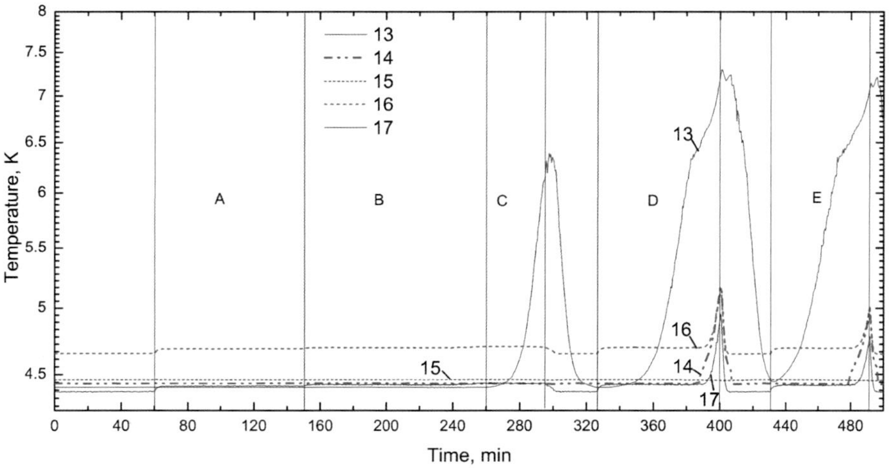

Figure 2. Cooling power

COOLING POWER MEASUREMENT

When the precooling temperature (thermometer 12) was fixed at 17.3K by controlling the H2 heater, and the high pressure was kept around 1.1 MPa, for example, the cooling power was measured. The data monitored by thermometers 13, 14, 15, 16 and 17 during cooling power measurement are plotted in Fig. 2. Thermometers 13, 14,15 and 17 are all Cernox type while 16 is a Rhodium-iron resistance type. Thermometers 13, 14 and 15 were mounted inside small copper blocks that were soldered on the connecting tubes. Thermometers 16 and 17 were mounted inside the copper cylinder of CHX. Apiezon N grease was used to keep good thermal contact between the copper walls and the thermometers.

When there was no heat load added, temperatures 13, 14, 15 and 17 were very close to each other because they were actually all cooled by liquid helium. Thermometer 16 was a little higher than the others, but it closely followed thermometer 17, probably reflecting the higher reliability of the Cernox thermometers. So thermometer 16 will not be included in the later discussion.

When the heat load was applied, the temperature of thermometer 17 increased a little bit because the heat transfer needed to be driven by the temperature difference. For region A, B, C, D and E on Fig. 2, the heat loads are listed respectively in Table 2. The other regions with no labels are for no-load conditions.

For achieving the maximum cooling power at liquid helium temperature, the heat load was added carefully. After region B, three values of heat load were tried but all of them made the cooler

Table 2. Applied heat load for each region.

Region on Fig. 2 to 5	Heat load, mW
A	39.79
B	48.94
C	58.76
D	52.94
E	49.39

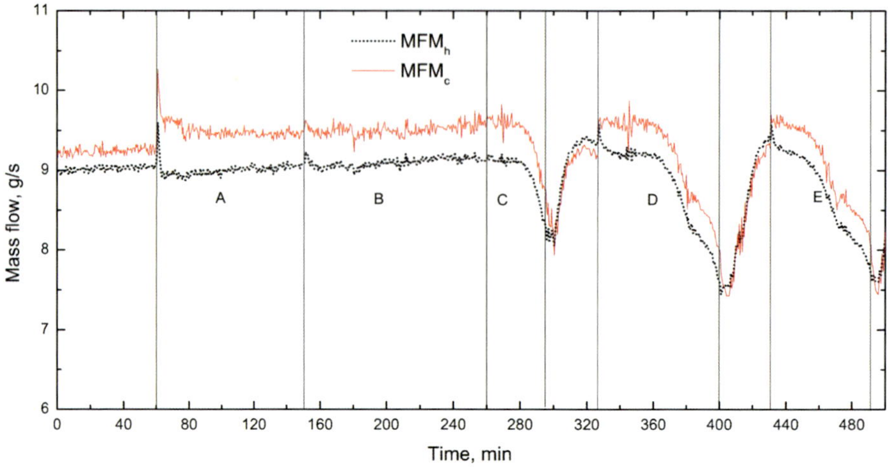

Figure 3. Mass flow

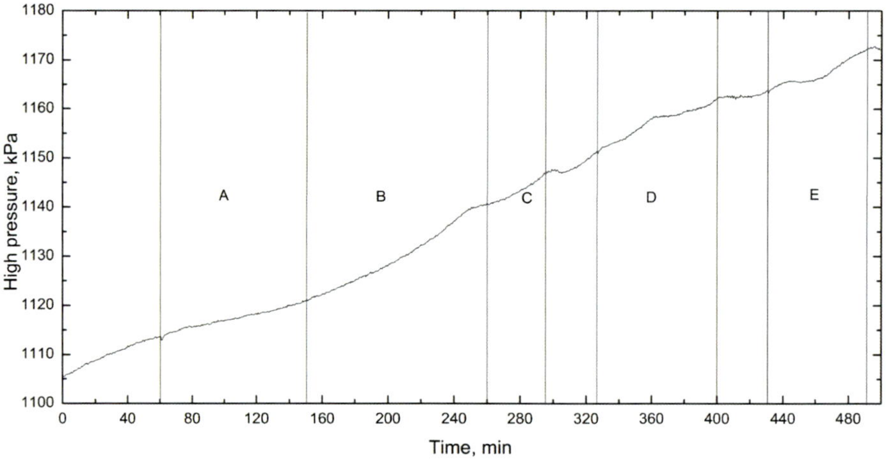

Figure 4. High pressure

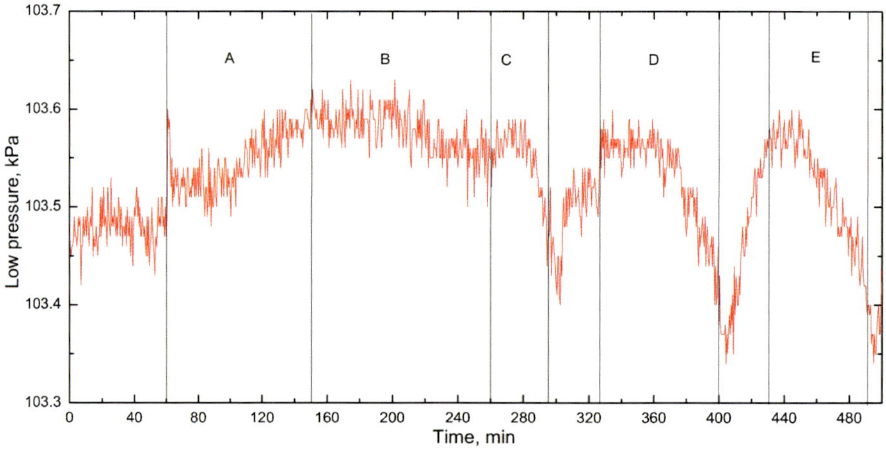

Figure 5. Low pressure

go unstable. It showed that the maximum cooling power was 48.94 mW and the unstable condition was repeatable.

The unstable condition seems related to the mass-flow drop as shown in Fig. 3. The changing of thermometer 13 and the mass flow drop happened and stopped simultaneously. The temperature increased when the mass flow decreased.

During the measurement, the high pressure continuously went higher. But the reason is not clear yet. However the total change was relatively small and growing speed is slow enough (about 8 kPa/h) to treat the measurement as accomplished with constant high pressure. The low pressure could be affected by the atmosphere pressure and the mass flow rate. The pressure changes were so small that they could be neglected. The unstable conditions shown as region C, D and E were not predicted when the PJTC was designed. It is important to find some theories or methods to explain this phenomena.

DISCUSSIONS

Intrinsic Limit of Cooling Power

The key unit of the PJTC will be analyzed in this section with a pressure-enthalpy (p-h) map similar with Reference [4]. The position of thermometers 12, 13, 14, 15 and 11 on Fig. 1 will be expressed as a, b, c, d and e ,respectively, in the p-h plot. Some assumptions have been made before the analysis:

1. The effectiveness of CFHX3 is fixed as 100%;
2. Mass flow is always constant;
3. The precooling temperature (position a) is fixed at 17.3 K;
4. The pressure drop exists only in the JT valve;
5. There is no heat transfer between the key unit and the ambient.

When there is no heat load added to the CHX, the helium flow process can be plotted as Fig. 6. As discussed in Reference [4], q labeled in Fig. 6 is the maximum cooling power achievable for these fixed precooling temperature and pressures. Thermometers b, c and d are all at liquid helium temperatures that correspond to the no-load part in Fig. 2 (before region A). The enthalpy difference between a and b should always be identical with that between e and d due to the energy balance constraint.

When the heat load is increased gradually to the maximum cooling power, the temperature of d and e will rise as shown in Fig. 7. The temperature of e reaches the precooling temperature. If a larger heat load was added, firstly d would move to right but e could not as it would be limited by the second law of thermodynamics. It means the heat transfer in the CFHX3 will decrease and the high pressure gas can not be cooled as sufficiently as before. Then b and c will move to right from the condition as Fig. 7. Because c and d are in the two-phase region but b is in the single phase region, the temperature of b is the first to rise

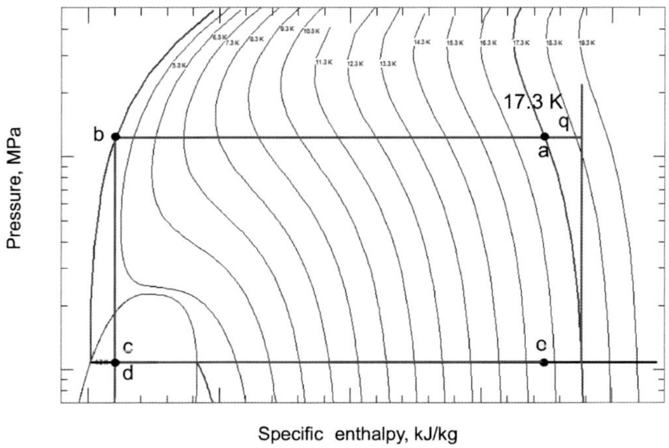

Figure 6. No-load state

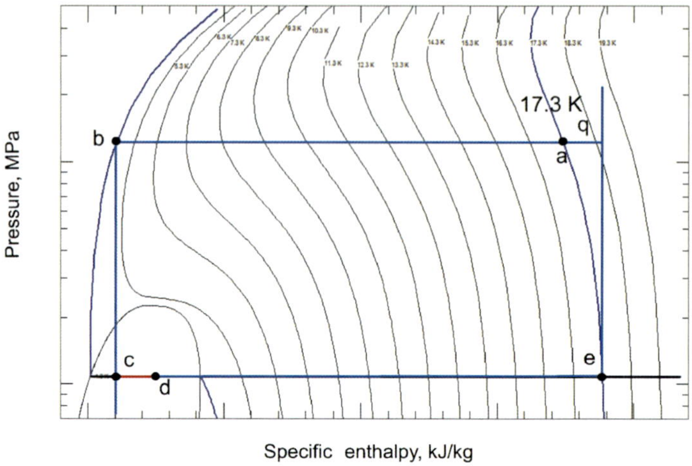

Figure 7. Maximum cooling power state

away from liquid helium temperature corresponding to the behavior of thermometer 13 in Fig. 2. The temperature of d is the second to leave the liquid helium temperature corresponding to the thermometer 14.

If the heat load was not removed, the thermometer 15 (position c) would also leave the liquid helium temperature and PJTC could never come to a steady condition. The above analysis shows how the PJTC goes unstable even without a mass flow drop. Actually the mass flow will change when the temperature before the JT valve (b) starts to rise as shown in Fig. 2 and Fig. 3. The mass flow drop will accelerate the process of the PJTC becoming unstable because the maximum cooling power will decrease.

Mass Flow Drop

Figures 2 and 3 show that when the temperature of 13 increased, the mass flow decreased dramatically. Here this phenomena will be discussed qualitatively.

Usually a PJTC working at liquid helium temperature has a pressure ratio larger than the critical pressure ratio about 2 for choked flow [5]. For an ideal gas, the molar mass flux in choked flow conditions is defined as

$$\dot{n}A = \kappa^{1/2}\left(\frac{2}{1+\kappa}\right)^{\frac{\kappa+1}{2(\kappa-1)}}\frac{p_0}{\sqrt{MRT_0}} \tag{1}$$

where A is the minimum cross section area of the passageway, κ is the adiabatic exponent, p_0 is the high pressure, T_0 is the temperature of the high pressure flow. M is the molecular mass. R is the ideal gas constant. It is shown that the mass flow is decided by the p_0 and T_0. In our experiment, the p_0 can be treated as constant. T_0 is corresponding to the temperature monitored by the thermometer 13. The tendency of the mass flow changing as the T_0 showed in Fig. 2 and Fig. 3 is consistent with Equation (1).

However, the condition in the experiment is for real fluids including subcooled liquid or supercritical state at the inlet and the two-phase condition at the outlet of the JT valve. Reference [7] measured the mass flow of helium for T_0 between 4 and 20 K when p_0 was 1 MPa. The mass flow at 4 K was more than 3 times of that at 20 K with which similar phenomena was found in our experiment.

Up to now, the characteristic of choked flow is supposed to be the main factor to cause the mass flow drop detected during our experiment. Theoretical or numerical methods may help to make it clear; this will be left to future work.

CONCLUSIONS

The open-cycle PJTC setup was built and tested. The PJTC successfully achieved the cooling power of 48.94 mW at liquid helium temperature. The reason to cause the unstable condition can be explained by the p-h map analysis. The mass flow drop accompanying the unstable condition is likely related to the characteristic of choked flow, which needs further quantitative analysis.

ACKNOWLEDGMENTS

This work is financially supported by the National Natural Science Foundation of China (No. 51176165). The authors would like to express their gratitude to A.T.A.M. de Waele from Eindhoven University of Technology.

REFERENCES

1. Gan, Z.H., Wang, B., Liu, D.L., et al, "Status and Development Trends of Space Mechanical Refrigeration System at Liquid Helium Temperature (in Chinese)," *Journal of Zhejiang University (Engineering Science)*, vol. 46, no. 12 (2012), pp. 2160-2177.

2. Liu, D.L., Wu, M., Wang, W.W., et al, "Development of Cryogenic System for James Webb Space Telescope (In Chinese)," *Cryogenics*, no. 4 (2013), pp. 5-10.

3. Gan, Z.H., Tao, X., Liu, D.L., et al, "Development Status of Space Cryogenic Technology at Liquid Helium Temperature in Japan (In Chinese)," *Journal of Zhejiang University (Engineering Science)*, vol. 49, no. 10 (2015), pp. 1821-1835.

4. Liu, D., Tao, X., Sun, X., et al, "Performance Study on ST/JT Hybrid Cryocoolers Working at Liquid Helium Temperature," *Physics Procedia*, vol. 67, (2015), pp. 468-473.

5. Maytal, B., Pfotenhauer, J.M., *Miniature Joule-Thomson Cryocooling*. ISBN 978-1-4419-8284-1, Springer Science+Business Media, New York (2013).

6. Tao, X., Liu, D.L., Wang, L.Y., et al, "A Cryogenic Heat Exchanger with Bypass and Throttling and its Thermodynamic Analysis," *Adv. in Cryogenic Engineering: Proceedings of the Cryogenic Engin. Conf. (CEC) 2015*, Vol. 101, IOP Publishing, Bristol, UK (2015), Article 012167.

7. Bradshaw, T.W., Orlowska, A.H., Jewell, C., "Life Test and Performance Testing of a 4 K Cooler for Space Applications," *Cryocoolers 10*, Kluwer Academic/Plenum Publishers, New York (1999), pp. 521 528.

Development of Cascade Non-Flammable Mixed Refrigerant Joule-Thomson Refrigerator for 100 K

C. Lee, J. Yoo, L. Jin, J. Cha, S. Jeong

Cryogenic Engineering Laboratory,
Korea Advanced Institute of Science and Technology,
Daejeon, 34141, South Korea

ABSTRACT

In this paper, a cascade type non-flammable mixed refrigerant (MR) Joule-Thomson (JT) refrigerator is presented to achieve a cooling temperature of 100 K. The working fluid utilized in the main stage (low temperature cycle) of the MR JT refrigerator is composed of nitrogen (N_2), argon (Ar), Tetrafluoromethane (R14) and Octafluoropropane (R218). An azeotropic mixture of R32 and R125 (R410A) is used for the precooling stage (high temperature cycle) refrigerant where the cooling temperature is nearly 240 K. The design specifications of the refrigerator are modeled by using a commercial software with the Peng–Robinson equation of state (EOS). The maximum COP of the optimized MR JT refrigerator is 0.24 at a temperature of 106.6 K. Also, this MR JT refrigerator is fabricated and tested to investigate the refrigeration performance in comparison with the simulated results. The compression system of the refrigeration cycle is implemented by modifying commercial air-conditioning compressors. The R410A precooling stage is simulated by a commercial Stirling cryocooler. The feasibility of the non-flammable MR in the MR JT refrigerator is validated by achieving a minimum temperature of 97.8 K, which is below the freezing point of pure R218 (125 K).

INTRODUCTION

Low temperature cooling technology, namely cryogenic engineering, draws a lot of attention from the technological area such as high temperature superconductor (HTS) applications, liquefied natural gas (LNG) and bio-preservation.

Cryogenic refrigeration technologies for large cooling capacities often adopt the Joule-Thomson (JT) refrigeration cycle. In a JT refrigeration system, a simple JT valve is installed for the expansion process instead of a mechanical expander. Thus, the reliability of JT refrigeration systems are higher than other refrigeration systems due to no moving parts at low temperatures. On the other hand, the low efficiency is an unavoidable drawback of JT refrigeration system because of the irreversible process at the JT expansion part.

Mixed refrigerants (MR) are normally applied in the JT refrigerators to boost their efficiency. Generally, MR are composed by the combination of low-boiling point refrigerants and high-boiling point refrigerants. Low-boiling point components are effective to achieve a lower cooling temperature whereas the high-boiling point components accomplish boosting the refrigeration efficiency.

Cryocoolers 19, edited by S.D. Miller and R.G. Ross, Jr.

Thus, the MR should be optimized with the target heat absorption temperature before constructing the MR JT refrigeration cycle. Since the flammable refrigerants have high JT coefficients and isothermal enthalpy differences compared to those of the non-flammable refrigerants, the flammable MR are favorable for utilization in MR JT refrigerators. Therefore, the previous studies commonly facilitated using flammable MR such as methane, ethane, propane and butane [1]. Nowadays, MR JT refrigerators are frequently applied in the natural gas liquefaction plants where the lowest temperature was approximately 110 K [2, 3]. On the other hand, flammable MR have critical drawbacks which are the flammability and the explosiveness. Non-flammable MR are attractive in this sense and start to be applied in MR JT refrigerators.

Unfortunately, there are only a few studies about non-flammable MR JT refrigerators. Boiarski et al., investigated the performance of a small-scale MR JT refrigerator by using both flammable MR and non-flammable MR in 2005 [4]. They mentioned that the freezing temperature of the non-flammable MR tested is higher than 110 K. Khatri and Boiarski discussed the development of a non-flammable single stage MR JT refrigerator where the target temperature was below 110 K, in 2008 [5]. They also mentioned that the cooling temperature of 110 K is limited by the high boiling components related to freezing temperature, and poor thermodynamic efficiency is achieved compared to using flammable MR. Thus, non-flammable MR presented a low thermodynamic efficiency to reach 100 K or below.

This paper focuses on the development of a high efficiency non-flammable MR JT refrigerator to achieve a cooling temperature of 100 K. The freezing limitation of the working fluid should be identified and therefore, the freezing point for non-flammable MR was studied in a previous study [6]. The design result for achieving the maximum Coefficient of Performance (COP) was also described in the previous studies [7]. This paper discusses the experimental verification of the design result, and the detailed description of a capillary tube as the expansion part.

DESIGN RESULT OF CASCADE MR JT REFRIGERATOR

The design of a non-flammable MR JT refrigerator is divided into two parts. First, the freezing limitation of the target non-flammable MR should be identified. Freezing points of the target non-flammable MR to achieve the 100 K cooling temperature was verified in a previous study [6]. The selected components are Nitrogen (N_2), Argon (Ar), Tetrafluoromethane (CF_4, R14) and Octafluoropropane (C_3F_8, R218). The freezing points of the mixture of Ar, R14 and R218 are experimentally verified according to the various molar compositions. N_2 is excluded in the experiments because its normal boiling temperature is lower than the target cooling temperature. Certain ternary MR (Ar, R14 and R218) have a measured freezing temperature below 77 K [6]. This freezing temperature is much lower than the triple point temperature of R218 (125 K). The constraint of the molar composition for quarternary MR (N_2, Ar, R14 and R218) is established according to a previous study.

Second, the optimal operating condition of a non-flammable MR JT refrigerator is designed by commercial software (Aspen HYSYS 8.0) with the aforementioned composition constraint [7]. The calculation was conducted for three different configurations: (1) the single stage with ternary MR (Ar, R14 and R218), (2) the single stage with quaternary MR (N_2, Ar, R14 and R218), and (3) the 2-stage (cascade type) with quaternary MR. Figure 1 (a) shows the schematic diagram of the single stage configuration and Figure 1 (b) displays the cascade type configuration [7]. Also Figure 1 (b) depicts the calculation points or instrumentation positions for this configuration.

The target cooling temperature is set at approximately 100 K, and the variables are selected as the mass flow rate of the working fluid, the discharge pressure, the suction pressure, and the molar composition. A cascade type configuration with a quaternary MR presents the highest COP of the entire calculation result. The molar composition of quaternary MR was N2:Ar:R14:R218 = 0.2:0.3:0.3:0.2 in the simulation with the maximum COP. Table 1 shows the design result of the cascade type non-flammable MR JT refrigerator with the highest efficiency. Each position of Table 1 is illustrated in Figure 1 (b), as mentioned before. The simulation assumes a two-stage compression and its isentropic efficiency is 80%. The pressure drop inside the heat exchanger is neglected.

The lowest temperature of this cascade MR JT refrigerator is calculated to be 96.3 K and the temperature after the evaporator is 106.6 K. The normalized cooling capacity is attained as 19.5 J/g,

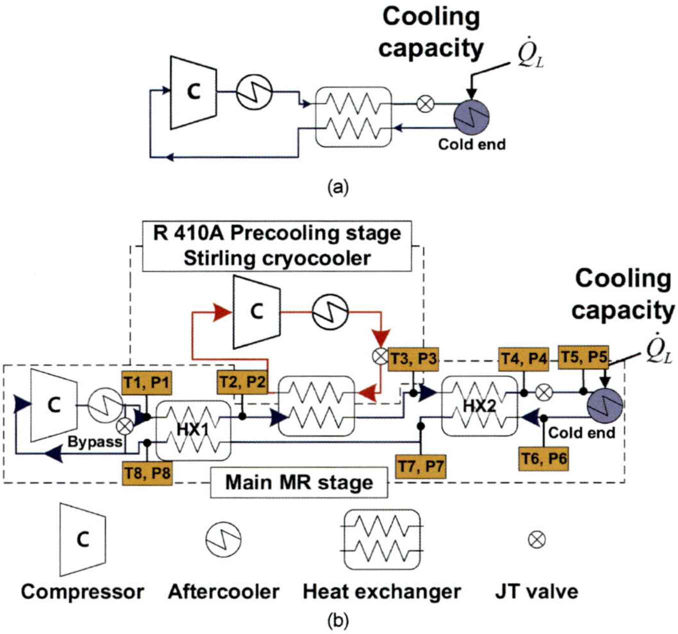

Figure 1. Schematics of (a) single stage configuration and (b) cascade type MR JT refrigerator configuration with calculation point or measurement position.

which is the cooling capacity divided by the mass flow rate of the working fluid. The normalized compression work is also obtained as 73.7 J/g, and the normalized precooling work (precooling work / mass flow rate of main MR) was calculated as 6.7 J/g. The maximum COP of this cascade type non-flammable MR JT refrigerator is calculated to be 0.24. One of the important points is that the thermodynamic state of position 4 is a subcooled liquid state. This will be discussed later in this paper.

Table 1. Design results of cascade type non-flammable MR JT refrigerator for 100 K cooling temperature

Position	Temperature [K]	Pressure [kPa]	Vapor quality (kg/kg)
1	300	1400.0	Superheated
2	254.2	1400.0	0.973
3	247.0	1400.0	0.912
4	**111.1**	**1400.0**	**Subcooled**
5	96.3	220.0	0.046
6	106.6	220.0	0.314
7	240.0	220.0	Superheated
8	297.0	220.0	Superheated

Condition	Value	Comment
Molar composition [-]	N_2:Ar:R14:R218 = 0.2:0.3:0.3:0.2	
Cooling capacity [J/g] (A)	19.5	
Compression work [J/g] (B)	73.7	
Precooling work [J/g] (C)	6.7	Cooling Temperature : 244 K Refrigerant: R410A
COP [-]	0.24	A/(B+C)

EXPERIMENTAL APPARATUS OF CASCADE MR JT REFRIGERATOR

The experimental apparatus is composed of two parts, the compressor system and the cold box.

Compression system

Figure 2 (a) shows the complete view of the developed compressor system, and Figure 2 (b) presents the assembly of each component in the compressor system. The compression system is developed with two commercial air-conditioning rotary vane compressor units (GPD420DAA and GPD240DAA, LG electronics) and an oil removal system [8]. The oil removal system utilizes two oil separators, a water-chilled aftercooler, oil mist filters, and a filter drier. A four-stage compressor system is constructed to achieve a sufficiently high compression ratio (the ratio between the discharge pressure and the suction pressure). The swept volume of the first compressor (GPD420DAA) is 40.5 cc/rev and that of the second compressor (GPD240DAA) is 23.5 cc/rev. Since the different volume flow rates of the two compressors require different operating frequencies, commercial inverters (Yaskawa cooperation, model A1000) control the operating frequencies of the two compressor units. An oil removal system is composed of oil separators (Emerson AW), oil mist filters and a filter drier, to prevent the oil and moisture in the working fluid.

Cold box

The cold box includes several heat exchangers, the evaporator, the capillary tube and the cryocooler instead of the precooling stage. The entire cold box system is attached to the top flange of the vacuum chamber. Figure 3 (a) depicts the developed cold box system and Figure 3 (b) illustrates the cold box system after the multi-layer insulation (MLI) installation.

Heat exchanger 1 (HX1) and heat exchanger 2 (HX2) are the diffusion bonded printed circuit heat exchangers (PCHEs). Stainless steel plates are applied to construct this PCHEs. The single channel is 300 μm in height and 300 μm in width. Each layer has 10 channels and is in a meandering shape. The channel layer has a height of 210 mm and a width of 77 mm. A 20-layer (10 layers for hot side and 10 layers for cold side) PCHE is fabricated for HX 1, and a 30-layer (15 layers for hot side and 15 layers for cold side) PCHE is used for HX 2. The entire layers are fabricated by using a diffusion bonding process.

A Stirling cryocooler (Sunpower Inc, Cryotel CUBE) accomplishes precooling of the MR JT system in the experiment. A machined copper block surrounded with a copper tube is attached for heat exchange. The working fluid of main MR cycle is precooled in this copper tube. The precooling temperature (position 3) is controlled by the cryocooler and the auxiliary heater at the cold tip. The temperature at position 3 is maintained between 240 K and 248 K.

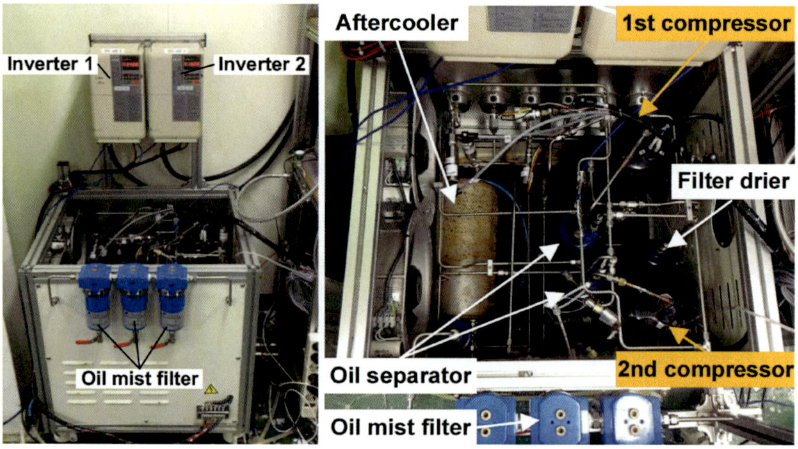

(a) (b)

Figure 2. Photographs of compression system (a) the complete view and (b) the detailed compression system.

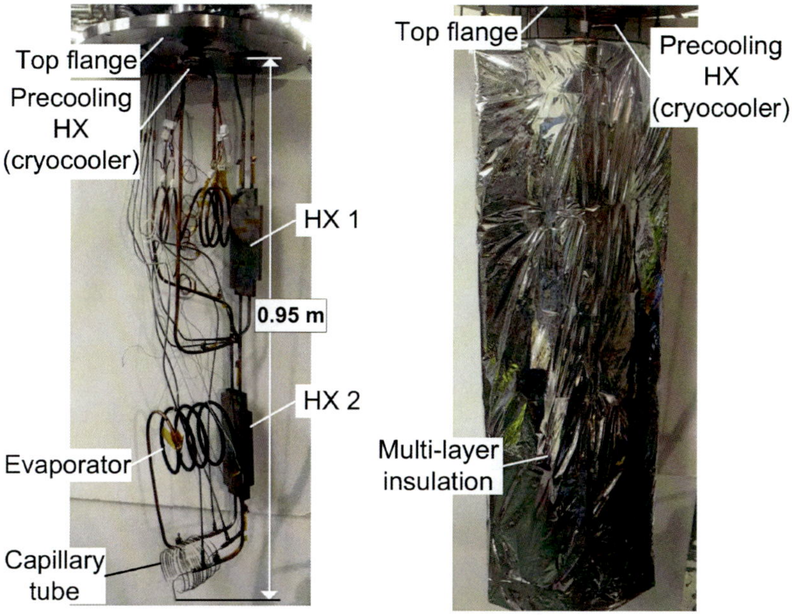

(a) (b)

Figure 3. Photographs of experimental apparatus for non-flammable cascade MR JT refrigerator; (a) the detailed view and (b) the covered view with MLI

The JT expansion part is constructed with a capillary tube to avoid a clogging problem. The capillary tube is made of a stainless steel tube, which is 1.6 mm (1/16 inch) in the outer diameter and 1.2 mm in the inner diameter. The selected lengths of capillary tubes are 1.5 m and 3.0 m. The evaporator is fabricated with a coiled 6.4 mm (1/4 inch) copper tube and the heater is covered with Stycast 2850FT epoxy.

The whole cold box system is insulated by a vacuum chamber and multi-layer insulation to prevent heat flow from the environment. Ten layers of MLI (RUAG GmbH, Coolcat 2NW) are applied to cover the entire cold box system.

In the cold box system, eight temperature sensors and eight pressure transducers were installed to measure the temperature and pressure of each location as seen in Figure 1 (b). Silicon diode type temperature sensors (Lakeshore Inc, DT-670-SD) are used, and FP 2000 pressure transducers (Honeywell Inc) are selected. To measure the mass flow rate of the working fluid, a Coriolis mass flow meter (Bronkhorst, CORI-FLOW M55) is utilized at the discharge of the compressor system. The data is collected by a DAQ system (National Instruments, SCXI-1000, SCXI-1125, SCXI-1320 and SCXI-1328) and a desktop computer. The measurement position is illustrated in Figure 1(b), as mentioned earlier.

EXPERIMENTAL RESULT AND DISCUSSION

1.5 m capillary tube experimental result

The 1.5 m long capillary tube is first applied to the experimental apparatus. Figure 4 shows the typical cool down characteristics of the non-flammable MR JT refrigerator. The composition of working fluid is the same as the design result (N_2:Ar:R14:R218 = 0.2:0.3:0.3:0.2). Figure 4 (a) shows the temperature data of the experimental results and Figure 4 (b) presents the pressure and mass flow rate data of the results. Figure 4 (c) shows the effectiveness of each heat exchanger based upon the temperature data. The lowest temperature obtained is 103 K, which is higher than the design result. The considerably higher cooling temperature is due to the relatively high pressure drop at the low pressure stream. The unexpected large pressure drop causes a pressure increase at position 5, which is higher than the design result.

Figure 4 (d) represents the quality profile of all experimental results. The quality of the working fluid is obtained by REFPROP 9.1 software [9]. For the sake of convenience, the pseudo-quality value is assumed as 1.05 and -0.05 for the superheated state and the subcooled state, respectively. The working fluid mass flow rate is drastically increased when the state at position 4 (inlet of the capillary tube) reaches a subcooled liquid state (approximately 7 hours later). The pressure at position 4 is decreased considerably when the temperature at position 4 gradually reduces.

The reason for the pressure reduction at position 4 can be explained by two effects, which are the complete liquefaction of the working fluid and the lack of pressure drop in the capillary tube. The working fluid density is greatly increased when the MR is fully liquefied. Therefore, the liquid pressure can be reduced. Furthermore, the pressure drop in the capillary tube was weaken because the state of the working fluid transforms from a two-phase state to a liquid state, at the inlet of the expansion part. The pressure drop at the capillary tube should be increased to maintain the stable operation. Thus, a 3.0 m long capillary tube is added at the expansion part in the same experimental apparatus and tested in the same operating conditions.

3.0 m capillary tube experimental result

Figure 5 displays the experimental results with the 3.0 m capillary tube expansion part. The lowest temperature in this case is maintained as 97.8 K, which is 5.2 K lower than that of the system with the 1.5 m capillary tube. When the state at position 4 enters a subcooled state, the mass flow rate significantly increases, as mentioned in the aforementioned test result. The mass flow rate is slightly reduced at the steady state therefore, the pressure drop at the low pressure stream is

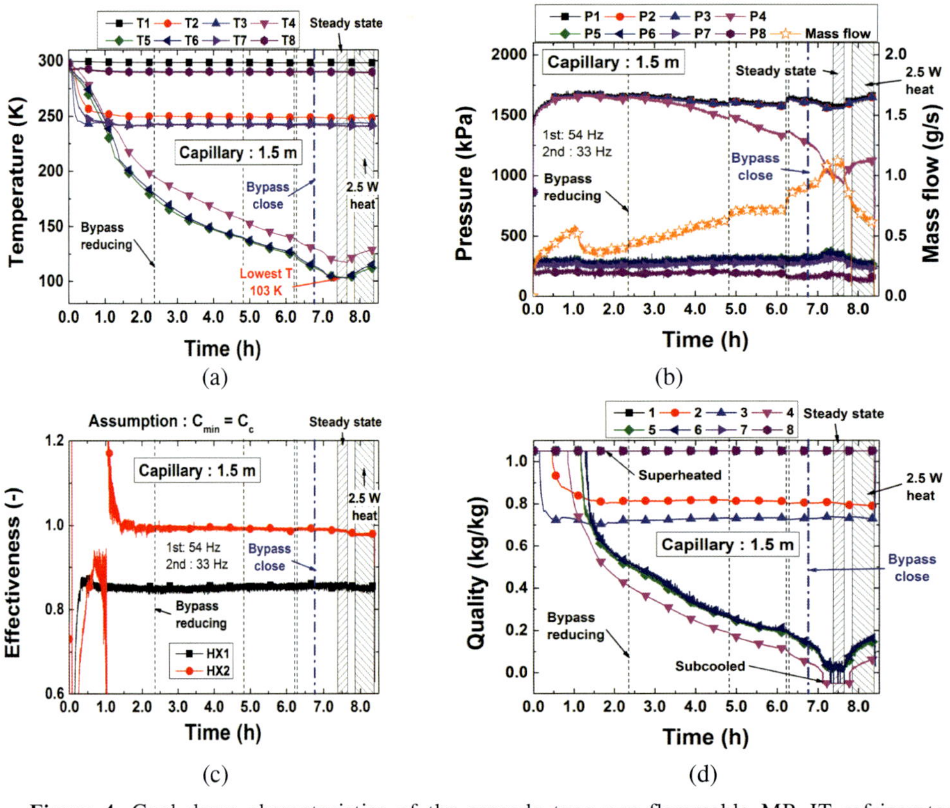

Figure 4. Cool down characteristics of the cascade type non-flammable MR JT refrigerator (N2:Ar:R14:R218 = 0.2:0.3:0.3:0.2) using the capillary tube length of 1.5 m; (a) temperature profile, (b) pressure and mass flow rate profile, (c) heat exchanger effectiveness profile and (d) quality profile at each location.

slightly reduced. Thus, the pressure at position 5 (i.e. pressure after the capillary tube) is maintained at approximately 230 kPa. Although the lowest temperature is reached as designed, the pressure at position 4 is too low, which shows a similarity to shorter length of capillary tube.

Characteristics of capillary tube

In this study, two different capillary tube lengths are reported with the same experimental apparatus. One is 1.5 m and the other is 3.0 m, each has the same diameter and cross-sectional area. The pressure drop of each capillary tube will be different at the non-choked condition. Nevertheless, the maximum mass flow rate for each result is close to 1 g/s and the pressure difference at the capillary tube is approximately 700 kPa. Figure 6 (a) and (b) plot the mass flow rate and the quality measurement of each experiment and Figure 6 (c) and (d) show the pressure profile between the inlet (P4) and the outlet of the capillary tube (P5). Although the lengths of capillary tubes are different for the two experiments, the similar pressure and mass flow rate profiles are obtained from the experiment. These results indicate that if the inlet condition reaches the subcooled liquid state, the choking phenomenon would occur in the capillary tube. While the temperature of position 4 gradually drops, the pressure and mass flow rate are gradually decreased after the choking condition. This characteristic is due to the fact that the density is increased according to the enhanced subcooling. The escalation of the mass flow rate at the choking condition leads to the pressure cutback at the position 4, and this reduced pressure causes the mass flow rate reduction in the next step. Thus, a higher degree of subcooling and controllability of the mass flow rates (i.e. by using valves) is required for achieving a stable operation of the subcooled JT expansion.

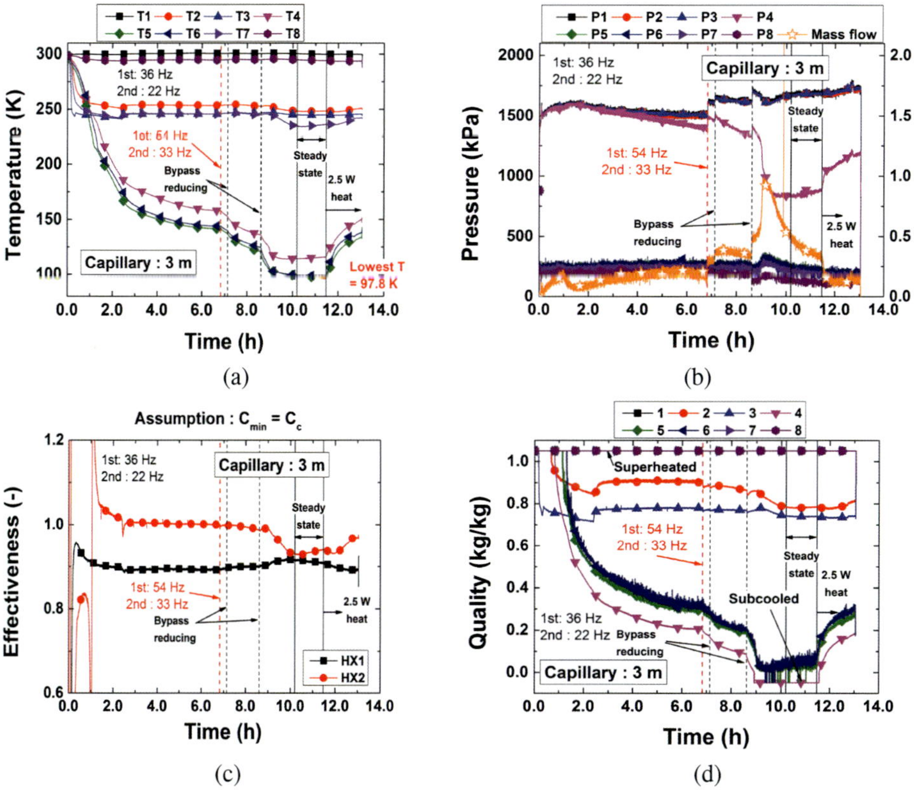

Figure 5. Cool down characteristics of the cascade type non-flammable MR JT refrigerator (N_2:Ar:R14:R218 = 0.2:0.30.3:0.2) using the capillary tube length of 3.0 m; (a) temperature profile, (b) pressure and mass flow rate profile, (c) heat exchanger effectiveness profile and (d) quality profile at each location.

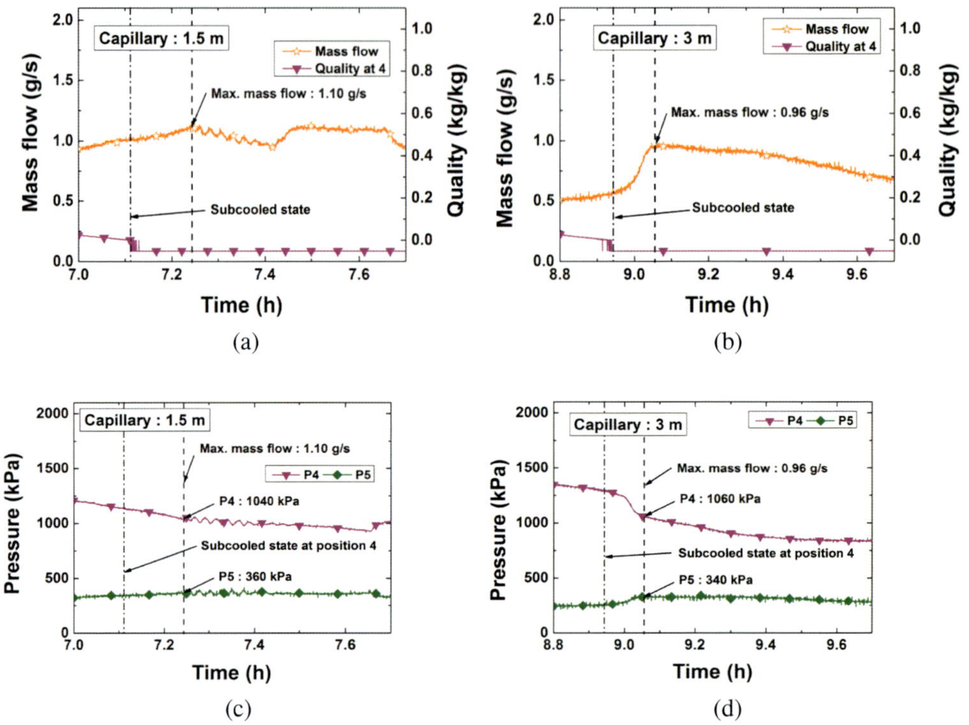

Figure 6. Mass flow rate, quality and pressure profile at the start of subcooled state before the expansion part; (a) mass flow rate and quality profile of 1.5 m capillary tube, (b) mass flow rate and quality profile of 3.0 m capillary tube, (c) pressure profile between the inlet and the outlet of the 1.5 m capillary tube and (d) pressure profile between the inlet and the outlet of the 3.0 m capillary tube (N2:Ar:R14:R218 = 0.2:0.3:0.3:0.2).

Cold end characteristics

By the design result, the target cooling capacity was calculated as 19.5 J/g at 106.6 K. Before the full loading of the design result (approximately 10 W), 2.5 W heat is applied after the steady state for each experiment. Figure 7 (a) and (b) display the temperature profiles and the quality profile of the 3.0 m capillary tube, respectively. Figure 7 (c) displays the P4 (capillary tube inlet), saturated liquid pressure calculated by T4, and the mass flow rate of the experimental result [9]. As can be seen that the MR JT refrigerator cannot endure the poor cold duty. When the tiny heat is applied in the evaporator, the mass flow rate of the working fluid is drastically reduced. This reason of the mass flow rate reduction is that the tiny heat flow changes the state at position 4, which has a tiny degree of subcooling. When the heat is applied at the evaporator, the state of the capillary inlet (location 4) becomes two-phase instead of subcooled because T4 is increased due to the increased value of T6. Then the pressure drop inside of the capillary tube is drastically raised, and the mass flow rate is decreased to maintain the certain pressure drop. As a chain reaction, the mass flow rate reduction lowers the cooling capacity at the cold end, and the lowest temperature of the cold end drastically increases. Thus, the degree of subcooling at the inlet of JT expansion part must be sufficiently developed to construct the subcooled JT expansion and the stable operation of the MR JT refrigerator.

CONCLUSION

In this paper, a cascade type non-flammable MR JT refrigerator is developed and tested according to the design result. Two types of capillary tubes are applied to construct the JT expansion part. The lowest temperature of the developed MR JT refrigerator is 97.8 K, and some interesting

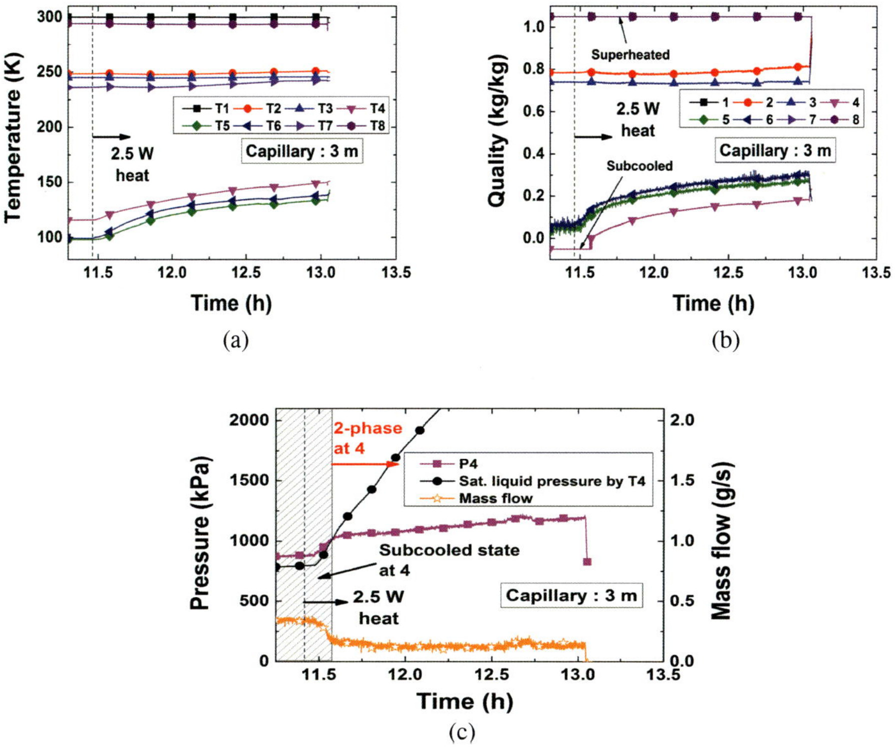

Figure 7. Temperature and quality profiles after heating for 3.0 m capillary tube; (a) temperature profile ,(b) quality profile of 3.0 m capillary tube, (c) saturated liquid pressure, capillary inlet pressure (P4) and mass flow rate profile (N2:Ar:R14:R218 = 0.2:0.3:0.3:0.2)

characteristics and limitations have been observed in association with using a capillary tube during the refrigerator operation.

- The selected non-flammable mixed refrigerant of N_2, Ar, R14 and R218 does not freeze at 97.8 K, which is much lower than the freezing temperature of R218 (125 K).
- Choking phenomenon occurs in the capillary tube. The length of capillary tube does not affect the mass flow rate of the working fluid.
- The degree of subcooling must be sufficiently secured for stable operation when the subcooled liquid JT expansion is required at the JT expansion part.
- To achieve higher degree of subcooling and controllability of the mass flow rate, a valve-control is necessary for this kind of MR JT refrigerator.

ACKNOWLEDGMENT

This work was supported by the Converging Research Center Program through the Ministry of Science, ICT and Future Planning of Korea (2014M3C1A8048836).

REFERENCES

1. Chakravarthy V. S., Shah R. K., Venkatarathnam G. "A Review of Refrigeration Methods in the Temperature Range 4–300 K," *Journal of Thermal Science and Engineering Applications*, 3, no. 2 (2011), pp. 020801.

2. Venkatarathnam, G., *Cryogenic Mixed Refrigerant Processes,* Springer, New York (2008), pp.145-153.

3. Lim, W., Choi, K., Moon, I., "Current Status and Perspectives of Liquefied Natural Gas (LNG) Plant Design". *Industrial & Engineering Chemistry Research*, 52, no. 9 (2012), pp. 3065-3088.

4. Boiarski, M., Podtcherniaev, O., Flynn, K., "Comparative Performance of Throttle Cycle Cryotiger Coolers Operating with Different Mixed Refrigerants in the 13th International Cryocooler Conference," *Cryocoolers 13*, Springer US, USA (2005), pp. 481-488.

5. Khatri, A., Boiarski, M., "Development of JT coolers operating at cryogenic temperatures with non-flammable mixed refrigerants in the Cryogenic Engineering Conference," (2008). pp. 3-10.

6. Lee, C., Yoo, J., Lee, J., Jeong, S., "Visualization of the solid–liquid equilibria for non-flammable mixed refrigerants". *Cryogenics*, 75, no. 0 (2016), pp. 26-34.

7. Lee, C., Jin, L., Jeong, S., "Design of non-flammable mixed refrigerant Joule-Thomson refrigerator for precooling stage of high temperature superconducting power cable in the 8th Asian Conference of Applied Superconductivity and Cryogenics," *Applied Superconductivity - Proceedings of ACASC 2015*, China (2015).

8. Lee, J., Oh, H., Jeong, S., "Investigation of neon–nitrogen mixed refrigerant Joule–Thomson cryocooler operating below 70 K with precooling at 100 K," *Cryogenics*. 61, no. 0 (2014), pp. 55-62.

9. Lemmon, E.W., Huber, M.L., M.O.McLinden. "NIST Standard Reference Database 23: Reference fluid thermodynamic and transport properties-REFPROP 9.1," National Institute of Standards and Technology, Standard Reference Data Program; 2013.

Investigation of the Flow Boiling Heat Transfer Characteristic for Non-Flammable Cryogenic Mixed Refrigerant

J. Yoo, L. Jin, C. Lee, S. Jeong

Cryogenic Engineering Laboratory,
Korea Advanced Institute of Science and Technology,
Daejeon, 34141, South Korea

ABSTRACT

An experimental apparatus is fabricated to measure the flow boiling heat transfer coefficients (HTC) of cryogenic mixed refrigerants (MR). The experimental system has a test section which includes a copper tube and a thin-wire heater to impose a heat flux on the experimental MR stream. The flow boiling heat transfer characteristics (HTC) are investigated for three cases of non-flammable cryogenic MR sets. R218 (Octafluoropropane, C_3F_8) and R14 (Tetrafluoromethane, CF_4) are used to constitute the experimental MRs. Three different molar compositions are selected to obtain the experimental HTC of the non-flammable mixtures (i.e. R218:R14 = 0.7:0.3 / 0.5:0.5 / 0.3:0.7). The mass flux conditions are varied as 88, 198 and 330 kg/m^{2}s during the experiment. The heat flux conditions are selected as 2800 and 9900 W/m^2. Experimental measurements are conducted for the whole MR sets at the pre-determined physical conditions. The obtained heat transfer characteristic results have 19.1% of mean uncertainty. Also, we compare three correlation sets to identify the best correlation which estimates the experimental results with the good accuracy. The most accurate correlation is Kandlikar's correlation with Fujita and Tsutsui's mixture correction factor. It predicts the experimental data within 28.2% in mean deviation.

INTRODUCTION

Recently, many efforts have been made to utilize non-flammable mixed refrigerant (MR) rather than hydrocarbon as the working fluid of an MR Joule-Thomson (JT) refrigerator because of the safety issues [1-3]. R218 (Octafluoropropane, C_3F_8) and R14 (Tetrafluoromethane, CF_4) have fairly good thermodynamic properties to be utilized as the working fluids for refrigerators from 300 K to 145 K. For this reason, R218 and R14 can be good non-flammable substitutes for propane and methane. When a MR JT refrigerator is constructed, the heat exchanger is considered as an important component which affects the size and the cost of the refrigerator. A efficient and compact heat exchanger is required to implement an excellent MR JT refrigerator. The precise estimation the heat transfer area needed is important to design of a compact heat exchanger. Since the heat exchange process in the MR JT refrigerator occurs in the form of evaporation through boiling and condensation, the flow boiling heat transfer characteristics (HTC) of the working fluid are key factors for determining the precise heat transfer area. Since little research has been conducted on the HTC of non-flammable MR, it is hard to determine the exact heat transfer area of the heat exchanger for non-flammable MR [4]. Understanding the flow boiling HTC for a specific non-flammable MR, is necessary before it is used as the working fluid of the MR JT refrigerator. Since it is difficult to identify all of the necessary flow boiling HTC by experiments, a correlation for estimating the flow boiling HTC is suggested with the acquired HTC information for the non-flammable MR.

In this paper, we examine the ensuring trustworthiness of the experimental apparatus as well as the experimental device for measuring the flow boiling HTC of a R218 and R14 mixture. The experimentally obtained HTC results are presented with their measuring uncertainties. With the experimental data, three correlations are compared to examine which correlation can describe the experimental data with minimum error. Two of the three correlations have already been suggested by previous researchers, Thome and Barraza [4, 5]. Another new one is proposed by us.

EXPERIMENTAL APPARATUS

Overall system

The schematic diagram of the overall experimental system is shown in Figure 1. In the experimental system, a refrigerant circulation system is required to create the flow of the working fluid. A compressor functions with the predetermined experimental pressure and mass flow conditions as the main component of the circulation system. The compressed working fluid, which reaches approximately 373 K, is chilled by water at the after-cooler. A throttling valve and a mass flow meter are installed next to the after-cooler. The throttling valve and the mass flow meter are used to adjust and record the mass flow rate of the working fluid, respectively. After passing through the mass flow meter, the stream is fed into a cold box.

The cold box is maintained at high vacuum condition (below 1.3×10^{-2} Pa) to avoid heat penetration from the environment. In the cold box, the working fluid is cooled down completely to the bubble point by a Stirling refrigerator (CryoTel-GT, Sunpower). The trim heater (customized-KHLV, Omega) is located at the exit stream of the refrigerator heat exchanger. It supplies heat to the liquid state refrigerant stream to precisely control the experimental temperature conditions. Then, the stream which has been finally characterized as the specific experimental temperature, pressure and mass flux condition enters the test section.

Test section

The test section is illustrated in Figure 2. The test section contains a horizontal copper tube which is 1.7 mm in the inner diameter, 3.2 mm in the outer diameter and 100 mm in length. The thin manganin wire heater (MW-32, Lakeshore) wraps the copper tube to induce heat exchange

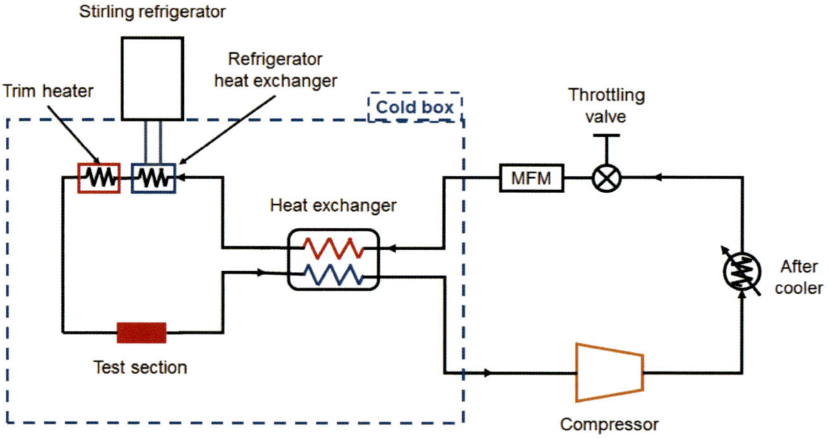

Figure 1. Schematic diagram of the overall experimental system

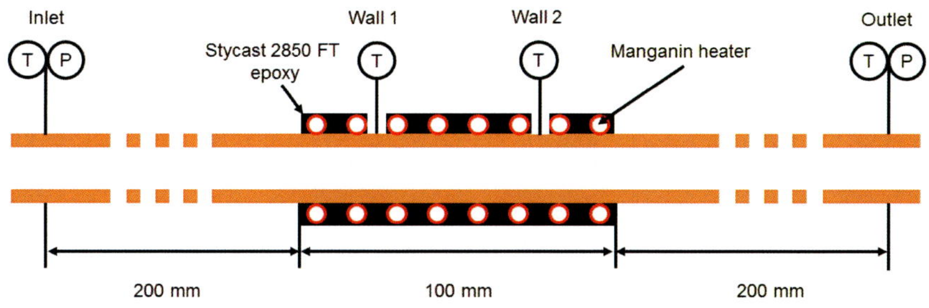

Figure 2. Schematic diagram of the test section

process between the heated wall of the test section and the MR stream through the test section. Epoxy (Stycast 2850 FT) is used to improve the thermal contact between the heater and the copper tube. The test section is instrumented by several temperature and pressure sensors. There are four temperature sensors (SD-670, Lakeshore) and two pressure sensors (FP2000, Honeywell) in the test section. The temperature sensors are attached on the inlet (T_{in}), the outlet (T_{out}) and the outer-wall of the test section ($T_{wall,1}$, $T_{wall,2}$), respectively. The pressure sensors are installed on the inlet (P_{in}) and the outlet (P_{out}) of the test section. The flow boiling HTC for the experimental MR sets are to be obtained by logarithmic mean temperature difference (LMTD) method with the collected temperature and pressure data. Equations for LMTD method are as follows [6]. All of the necessary thermodynamic properties of mixture for obtaining experimental HTC are calculated by REFPROP 9.1 [7].

$$\Delta T_{wall,in} = T_{wall} - T_{in} \tag{1}$$

$$\Delta T_{wall,out} = T_{wall} - T_{out} \tag{2}$$

$$\Delta T_{LMTD} = \frac{\Delta T_{wall,in} - \Delta T_{wall,out}}{ln\left(\frac{\Delta T_{wall,in}}{\Delta T_{wall,out}}\right)} \tag{3}$$

$$A_s = \pi D_i L \tag{4}$$

$$h_{exp} = \frac{q''}{A_s \Delta T_{LMTD}} \tag{5}$$

where T_{in}, T_{out} and T_{wall} indicate the inlet, outlet and the wall temperature of the test section, respectively. Also, A_s is the heat transfer area between the heated test section surface and the working fluid stream. D_i is the inner-diameter of the test section and L is the length of the test section. In Equation (5), h_{exp} represents the experimentally obtained flow boiling HTC value. q'' is the imposed heat flux to the test section.

Calibration of the experimental apparatus

The measurement accuracy of the experimental apparatus should be confirmed to measure the unknown HTC of refrigerants such as a mixture of R218 and R14. In this paper, the experimental apparatus is calibrated with a gas phase nitrogen which has well-known HTC estimation results from the Dittus-Boelter's correlation. It is described by Equation 6 [6].

$$Nu_D = 0.023 Re_D^{0.8} Pr^{0.4} \tag{6}$$

where Nu_D, Re_D and Pr are the Nusselt number, the Reynolds number and the Prandtl number, respectively.

The main issue for the calibration process is to determine the actual heat transfer area between the hot inner wall of the test section and the fluid in the test section. Due to the axial conduction heat transfer on the test section surface, the equivalent heat transfer length is extended during the heat exchange process. As shown in Figure 3, the actual heat transfer area should be larger than the measured area. It directly affects the experimentally obtained HTC values from Equation 5. For this reason, the actual heat transfer area should be considered to get the accurate flow boiling HTC with LMTD. In this process, the 'equivalent length' concept is applied to reflect the effect of the extended area. The equivalent length is the increased length of the test section, which can explain the effect of the extended heat transfer area. Figure 3 illustrates the 'equivalent length concept by a schematic diagram.

In the calibration process, the estimated HTC results with the Dittus-Boelter's correlation are considered as the true values of the nitrogen gas' HTC. First, the experimentally obtained HTC of the nitrogen gas is compared with the theoretically expected HTC results. For the next step, the equivalent length is deduced by minimizing the errors between the experimental and the theoretical results. Figure 4 (a) shows the experimental results with the apparent length of100 mm. They have +5~+20% errors compared to the calculation results. Figure 4 (b) represents the comparison results with 110 mm in the equivalent length. In this case, the errors are minimized as -3~+10%. As a result, we decide 110 mm as the true equivalent length for subsequent experiments.

EXPERIMENTAL RESULTS

The whole set of experimental conditions is summarized in Table 1. There are three mass flux conditions, 88, 198 and 330 kg/m^{2}s. The heat flux conditions are varied from 2800 to 9900 W/m^2.

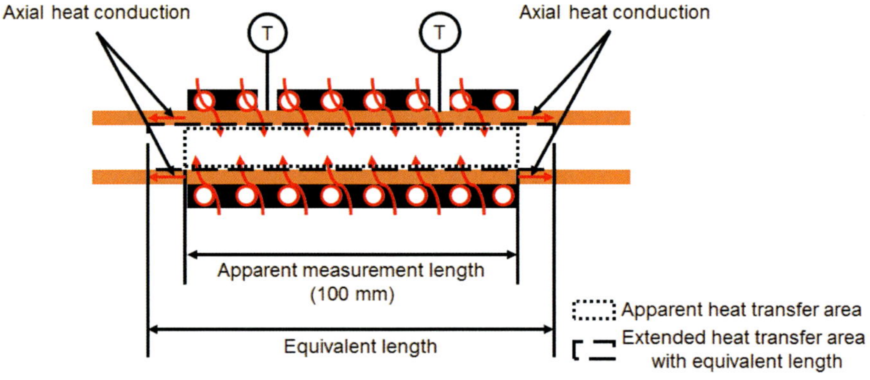

Figure 3. Picture for the equivalent length concept

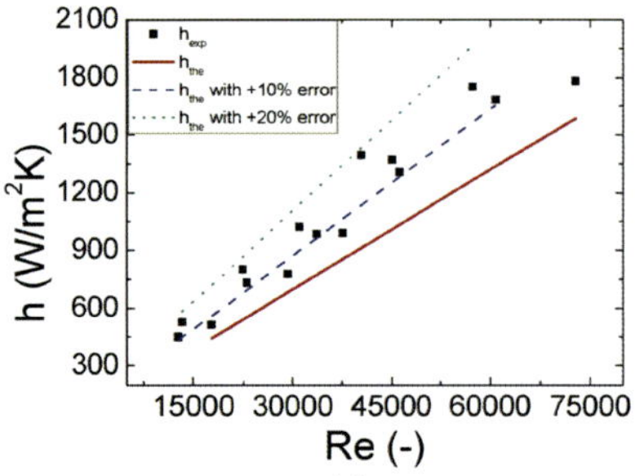

(a)

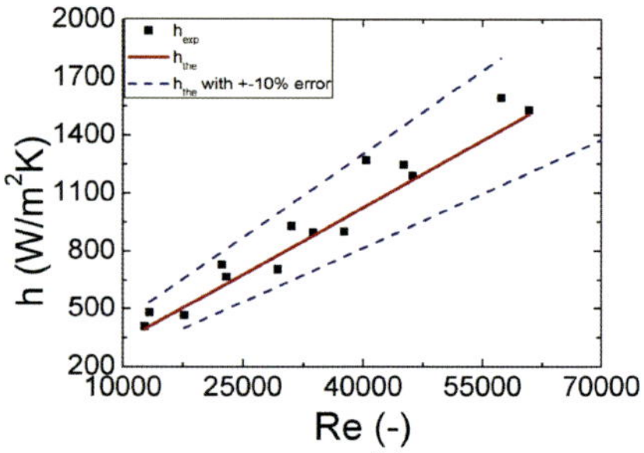

(b)

Figure 4. Comparison results between the experimental and estimated HTC results for gas phase nitrogen with (a) 100 mm in the equivalent length and (b) 110 mm in the equivalent length.

Table 1 Experimental conditions for measuring R218 and R14 mixtures

Case	Molar composition (-)	Pressure (kPa)	Heat flux (W/m^2)	Mass flux (kg/m^2s)
1	R218:R14 = 0.7:0.3	450 ± 150	2800	88.0
2			2800	198
3			2800	330
4			9900	88.0
5			9900	198
6			9900	330
7	R218:R14 = 0.5:0.5		2800	88.0
8			2800	198
9			2800	330
10			9900	88.0
11			9900	198
12			9900	330
13	R218:R14 = 0.3:0.7		2800	88.0
14			2800	198
15			2800	330
16			9900	88.0
17			9900	198
18			9900	330

The experiments are conducted for three mixtures of MR molar compositions (i.e. R218:R14 = 0.7:0.3 / 0.5:0.5 / 0.3:0.7). A total of 200 data points are experimentally obtained. The whole data sets are plotted in Figure 5. There are measurement uncertainties caused by the resolution of the temperature sensors and the measurement of the actual heat transfer area. The uncertainty of each HTC data set is analyzed by a general method [8, 9]. The uncertainties are marked in each graph by a y-axis error bar. The mean uncertainty of the experimental HTC results is 19.1%. In these experiments, the uncertainties of quality are not denoted in Figure 5 because the quality differences between the inlet and the outlet of the test section are less than 5% for the whole experimental conditions.

According to the experimental results, the flow boiling HTC of the mixtures which are composed of R218 and R14 are distributed in the range of 500 to 5000 W/m^2K. The peak HTC values exist at a quality between 0.7 and 0.9. After the maximum point, the HTC drastically decreases because of the dry-out of the residual liquid. The results generally show high flow boiling HTC values at high mass and heat flux conditions.

CORRELATION SELECTION FOR EXPERIMENTAL MR

The HTC correlation which well represents the experimental results should be formulated by using flow boiling HTC of the experimental mixture for the non-experimental conditions. There are two suggested correlations by previous researchers [4, 5]. They estimate the MR flow boiling HTC with high accuracy in the reference papers. Also, we find a fairly good correlation for our experimental results. In this paper, three different correlations are compared to find the most appropriate correlation for our experimental results. The mean deviation is used as the criterion for correlation accuracy. It means the absolute difference between the estimated and the measured HTC values as written in Equation 7. In Equation 7, N indicates the number of experimental data.

$$Mean\ deviation = \frac{1}{N}\sum_{i=1}^{i=N}\left|[h_{the} - h_{exp}] / h_{exp}\right| \times 100 \qquad (7)$$

Correlations review

There are two types of correlations to be useful for mixture's flow boiling HTC. The first type is originally developed for single-component fluid with a mixture correction factor as defined by Kandlikar [5]. The second type of the correlation is only for mixtures as defined by Little [4].

The first type correlation generally consists of addition of two terms which are convective and nucleate boiling HTC terms [5]. In order to apply the single-component correlation to the mixture, all of the required thermodynamic properties should be substituted for those of the mixtures. In addition, the mixture correction factor needs to be multiplied by the nucleate boiling term because

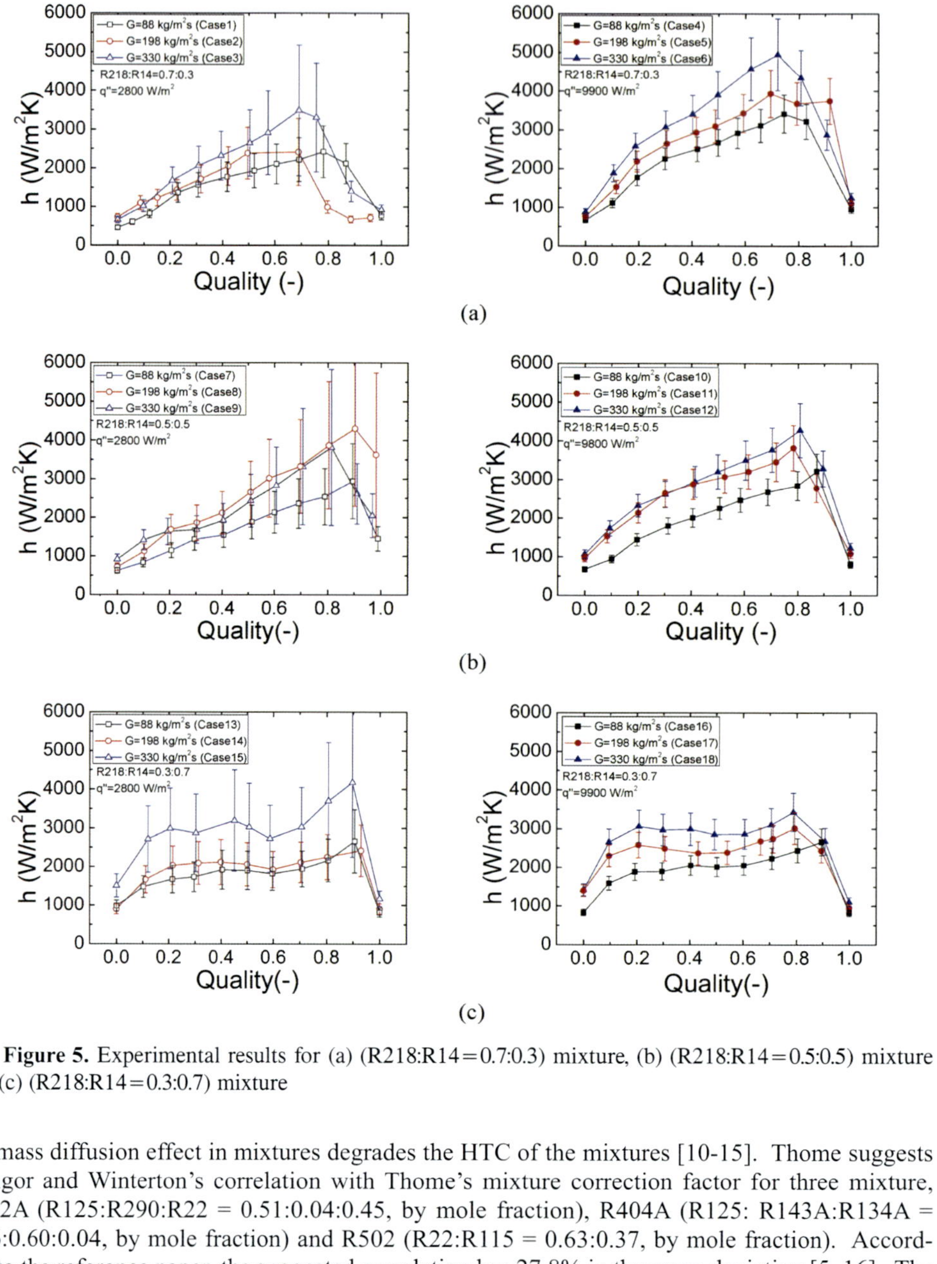

Figure 5. Experimental results for (a) (R218:R14=0.7:0.3) mixture, (b) (R218:R14=0.5:0.5) mixture and (c) (R218:R14=0.3:0.7) mixture

the mass diffusion effect in mixtures degrades the HTC of the mixtures [10-15]. Thome suggests Gungor and Winterton's correlation with Thome's mixture correction factor for three mixture, R402A (R125:R290:R22 = 0.51:0.04:0.45, by mole fraction), R404A (R125: R143A:R134A = 0.36:0.60:0.04, by mole fraction) and R502 (R22:R115 = 0.63:0.37, by mole fraction). According to the reference paper, the suggested correlation has 27.8% in the mean deviation [5, 16]. The detailed correlation and the mixture correction factor are expressed as follows;

$$h_{tp} = E\,h_L + S\,h_{nb} \times \frac{1}{1+\phi} \tag{8}$$

$$\phi = \frac{\Delta T_{bp}}{\Delta T_{id}}[1 - exp(\frac{-Bq''}{\beta_l \rho_l i_{fg}})] \tag{9}$$

where h means the HTC and ϕ indicates the mixture correction factor. The subscript tp, L and nb represent two-phase, liquid phase and nucleate boiling state, respectively.

In this paper, another first type correlation is also examined to find an accurate correlation for the experimental data. Kandlikar's correlation with Fujita and Tsutsui's mixture correction factor is adopted for this purpose. This combination of the correlation and the mixture correction factor is newly suggested in this paper. The correlation with the suggested mixture correction factor is shown in Equation 10 and Equation 11. In Equation 10, the first term at the right side signifies the convective heat transfer term and the the second term represents the nucleate boiling heat transfer term. There is an fluid dependent parameter, F_{fl}, in Kandlikar's correlation. The parameter accounts for the characteristic of the fluid. It can be determined by experimental data. Table 2 shows the fluid dependent parameters of some refrigerants [17]. As shown in Table 2, the parameters for the low boiling point fluids such as nitrogen and neon are higher than those of high boiling point fluids. The normal boiling point of R218 and R14 are 236 K and 145 K, which are relatively high boiling points compared to nitrogen and neon. The average of F_{fl} values for the high boiling point fluid (i.e. F_{fl} values of water, R11, R12, R13B1, R22, R113, R114 and R152a) in Table 2, therefore, is used as the fluid dependent parameter of the experimental mixture. For this reason, the average value 1.4 is firstly considered as the fluid dependent parameter for the experimental mixtures of R14 and R218.

$$h_{tp} = C_1 Co^{C_2} h_L + C_3 Bo^{C_4} F_{fl} h_L \times \frac{1}{1+\phi} \tag{10}$$

$$\phi = \frac{\Delta T_{bp}}{\Delta T_{id}}[1 - exp\{\frac{-60q''}{\rho_v i_{fg}}(\frac{\rho_v^2}{\tau g(\rho_l - \rho_v)})^{1/4}\}] \tag{11}$$

where Co is convection number and Bo is boiling number. F_{fl} means fluid dependent parameter which reflects the effect of each fluid.

As the second type of the correlation, in this paper, the correlation which is developed by Barraza et al., is selected. Barraza et al., modified Little's correlation to improve its accuracy for newly tested seven kinds of mixtures [18]. Their experimental mixtures include both hydrocarbon and non-flammable. The mean deviation of the suggested correlation is 18% [4]. The modified Little's correlation appeared as Equations 12 through 15.

$$\frac{1}{h_{tp}} = \frac{1}{h_{l,film}} + \frac{x^2 Cp_v^2}{\left((1-x)Cp_l + xCp_v\right)\left(\frac{\partial i}{\partial T}\right)_P} \frac{1}{h_{lv}} \tag{12}$$

$$h_{l,film} = 0.023\left(\frac{\mathrm{Re}_l}{1+\sqrt{\alpha}}\right)^{0.8} \mathrm{Pr}_l^{0.4} \frac{k_l}{ID\left(1-\sqrt{\alpha}\right)} \tag{13}$$

$$h_{lv} = 0.023\left(\frac{\mathrm{Re}_v}{\sqrt{\alpha}}\right)^{0.8} \mathrm{Pr}_v^{0.4} \frac{k_v}{ID\sqrt{\alpha}} \tag{14}$$

Table 2 Examples of the fluid dependent parameters for various fluids

Fluid (-)	Normal boiling point (K)	Fluid dependent parameter(F_{fl}) (-)
Water (H_2O)	373	1.00
R11 (CCl_3F)	297	1.30
R12 (CCl_2F_2)	243	1.50
R13B1 ($CBrF_3$)	215	1.31
R22 ($CHClF_2$)	232	2.20
R113 (CCl_2FCClF_2)	321	1.30
R114 ($CClF_2CClF_2$)	277	1.24
R152a (CHF_2CH_3)	249	1.10
Nitrogen (N_2)	77.6	4.70
Neon (Ne)	27.1	3.50

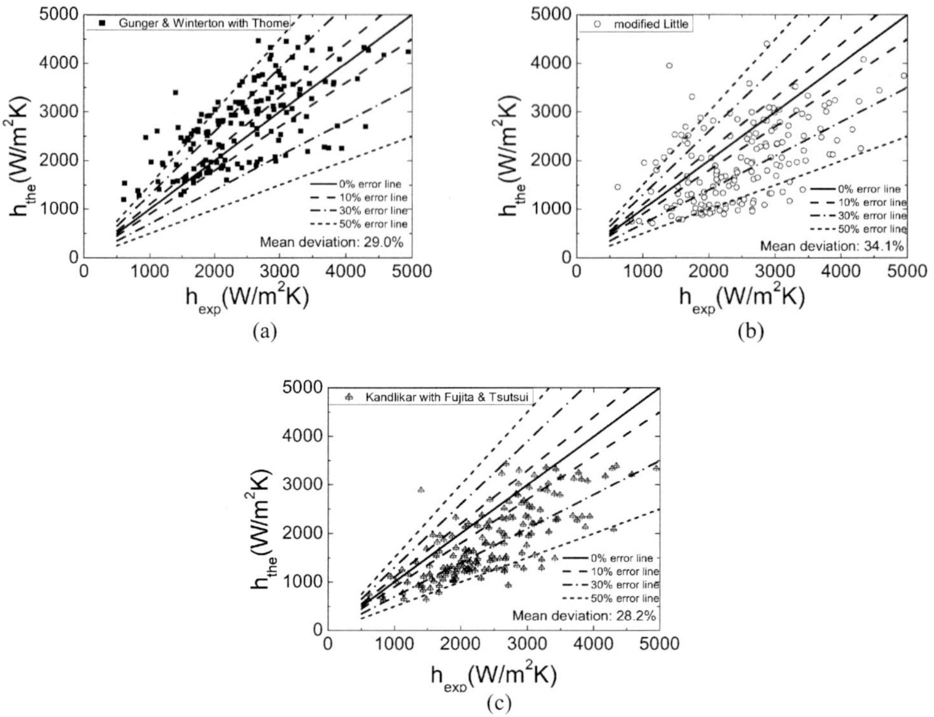

Figure 6. Comparison results of (a) Gungor and Winterton's correlation with Thome's mixture correction factor (b) modified Little's correlation (c) Kandlikar's correlation with Fujita and Tsutsui's mixture

$$\alpha = \left(1 + \frac{(1-x)\rho_v}{x\rho_l}\sqrt{1 - x\left(1 - \frac{\rho_l}{\rho_v}\right)}\right)^{-1} \tag{15}$$

Comparison results for three correlations

All the thermodynamic properties to calculate the correlations are obtained from REFPROP 9.1[7]. The correlations are tested in the quality range from 0.1 to 0.9 of the experimental data. Theoretical HTC values (h_{the}) are calculated by each correlation and are compared with the experimentally obtained HTC (h_{exp}) in Figure 6. As shown in Figure 6(c), Kandlikar's correlation with Fujita and Tsutsui's mixture correction factor has the best accuracy of 28.2% in mean deviation. The mean deviation of Gungor and Winterton's correlation with Thome's mixture correction factor is 29.0% and that of the modified Little's correlation is 34.1%.

The fluid dependent parameter is considered as a correction factor which represents the effect of each single-component fluid. The mixtures of the same components but with different molar compositions can be treated as different fluids. The fluid dependent parameters, therefore, should be varied to the molar composition of the mixture. Although the constant value of 1.4 is used in this paper for the fluid dependent parameter of the experimental mixtures, the Kandlikar's correlation with the Fujita and Tsutsui's mixture correction factor gives the best estimation result. It means that this correlation can be improved if proper fluid dependent parameters are identified for each molar composition of the experimental mixture.

CONCLUSION

This paper describes an experimental apparatus for measuring the flow boiling HTC of the MRs which consist of R218 and R14. The equivalent length concept of the heated area is used to interpret the experimental data because the heat transfer area is very sensitive parameter to obtain experimental flow boiling HTC. The experimental apparatus is calibrated by single phase nitrogen gas.

The experimental flow boiling HTC of the R218 and R14 mixtures are measured from 500 W/m^2K to 5000 W/m^2K. The mean uncertainty of the experimental results is 19.1%. Three correlations are considered to find the one with the best accuracy. For the comparison results, Kandlikar's correlation with Fujita and Tsutsui's mixture correction factor has the smallest mean deviation of 28.2%. It could be improved to become a more precise correlation when the fluid dependent parameter is appropriately defined for each molar composition of the experimental MR.

ACKNOWLEDGMENT

This research is supported by a grant from Space Core Technology Development Program of National Research Foundation of Korea (NRF-2013-042033) funded by Ministry of Science, ICT & Future planning (MSIP)

REFERENCES

1. Walimbe, N.S., K.G. Narayankhedkar, and M.D. Atrey, "Experimental investigation on mixed refrigerant Joule–Thomson cryocooler with flammable and non-flammable refrigerant mixtures," *Cryogenics*, vol. 50, no. 10 (2010), pp. 653-659.

2. Boiarski, M., O. Podtcherniaev, and K. Flynn, "Comparative Performance of Throttle Cycle Cryotiger Coolers Operating with Different Mixed Refrigerants," *Cryocoolers 13*, Springer US, (2005), pp. 481-488.

3. Lee, C., et al., "Visualization of the solid–liquid equilibria for non-flammable mixed refrigerants," *Cryogenics*, vol. 75, no. (2016), pp. 26-34.

4. Barraza, R., et al., "Description and validation of the Little correlation for boiling zeotropic mixtures in horizontal tubes from cryogenic to room temperature," *IOP Conference Series: Materials Science and Engineering*. IOP Publishing, (2015), pp. 012132.

5. Thome, J.R., "Boiling of new refrigerants: a state-of-the-art review," *International Journal of Refrigeration*, vol. 19, no. 7 (1996), pp. 435-457.

6. Bergman, T.L., F.P. Incropera, and A.S. Lavine, *Fundamentals of Heat and Mass Transfer*, John Wiley & Sons, (2011), pp. 578-593.

7. Lemmon, E., M. Huber, and M. McLinden, "NIST Standard ReferenceDatabase 23: Reference Fluid Thermodynamic and Transport Properties-REFPROP. 9.1," vol. no. (2010), pp.

8. Van Dresar, N.T., "An uncertainty analysis of the PVT gauging method applied to sub-critical cryogenic propellant tanks," *Cryogenics*, vol. 44, no. 6 (2004), pp. 515-523.

9. Nellis, G., C. Hughes, and J. Pfotenhauer, "Heat transfer coefficient measurements for mixed gas working fluids at cryogenic temperatures," *Cryogenics*, vol. 45, no. 8 (2005), pp. 546-556.

10. Jungnickel, H., P. Wassilew, and W. Kraus, "Investigations on the heat transfer of boiling binary refrigerant mixtures," *International Journal of Refrigeration*, vol. 3, no. 3 (1980), pp. 129-133.

11. Thome, J.R., "Prediction of binary mixture boiling heat transfer coefficients using only phase equilibrium data," *International Journal of Heat and Mass Transfer*, vol. 26, no. 7 (1983), pp. 965-974.

12. Fujita, Y. and M. Tsutsui, "Heat transfer in nucleate pool boiling of binary mixtures," *International Journal of Heat and Mass Transfer*, vol. 37, no. (1994), pp. 291-302.

13. Fujita, Y. and M. Tsutsui, "Heat Transfer in Nucleate Boiling of Binary Mixtures.(Development of a Heat Transfer Correlation)," *JSME International Journal Series B*, vol. 40, no. 1 (1997), pp. 134-141.

14. Inoue, T., N. Kawae, and M. Monde, "Characteristics of heat transfer coefficient during nucleate pool boiling of binary mixtures," *Heat and Mass Transfer*, vol. 33, no. 4 (1998), pp. 337-344.

15. Sun, Z., et al., "Nucleate pool boiling heat transfer coefficients of pure HFC134a, HC290, HC600a and their binary and ternary mixtures," *International Journal of Heat and Mass Transfer*, vol. 50, no. 1 (2007), pp. 94-104.

16. Gungor, K.E. and R. Winterton, "A general correlation for flow boiling in tubes and annuli," *International Journal of Heat and Mass Transfer*, vol. 29, no. 3 (1986), pp. 351-358.

17. Kandlikar, S.G., "A general correlation for saturated two-phase flow boiling heat transfer inside horizontal and vertical tubes," *Journal of Heat Transfer*, vol. 112, no. 1 (1990), pp. 219-228.

18. Little, W., "Heat transfer efficiency of Kleemenko cycle heat exchangers," *Adv. in Cryogenic Engineering*, Vol. 53, Amer. Institute of Physics, Melville, NY (2008), pp. 606-613.

Experimental Verification of a Numerical Model for Predicting the Adsorption of Mixed Gases

N. Tzabar, H.J. Holland, C.H. Vermeer, H.J.M. ter Brake

Energy Materials and Systems, Faculty of Science and Technology
University of Twente, Enschede 7500AE, the Netherlands

ABSTRACT

An ongoing effort is invested on developing a sorption compressor to drive Joule-Thomson cryocoolers with mixed-refrigerants. In previous work a numerical model is presented for determining the adsorption of mixed-gases based on the pure component adsorption characteristics. In the present paper an experimental verification of the numerical results is described. The experimental setup is suitable for simulating the sorption cycle in a compressor. The current research is focused on binary mixtures containing nitrogen and alkanes.

INTRODUCTION

Sorption Joule-Thomson (JT) cryocoolers are thermally driven and have no moving parts, both at the cold end and at the compressor (except for some check valves), which results in a high reliability, long operating life, and a vibration-free system.

Operating JT cryocoolers with Mixed-Refrigerants (MR) instead of pure refrigerants allows lower pressure ratios and higher coefficients of performance (COP). Mixed-Refrigerant Joule-Thomson (MRJT) cryocooling driven by mechanical compressors has been widely investigated with a variety of suggested MRs [1, 2].

The research presented in this paper is part of the effort to develop a sorption compressor for driving MRJTs. The basis for this study is a fundamental understanding of mixed-gas adsorption and developing the ability to predict the processes in a sorption compressor which contains mixed-gases. In a previous work [3] an analysis of a sorption compressor operating with mixed-gases is presented based on a model using only pure gas adsorption parameters. Recently, we performed a wider set of adsorption measurements and an advanced procedure for fitting the experimental results to the Sips isotherm model [4]. In the current paper we present the experimental verification of the numerical model and the pure gas adsorption measurements we performed at the University of Twente.

METHOD

Mixed gas adsorption model

A model, which has been presented elsewhere [3], predicts the adsorption of mixed-gases based on adsorption experimental data of the pure components. The model is based on the Sips isotherm and enables the prediction of the adsorbed phase composition as a function of temperature, pressure, and vapor-phase composition. The adsorption concentration of pure gases as a function of pressure and temperature for nitrogen, methane, ethane, and propane on three different adsorbents has been recently published [4], based on a modified Sips isotherm:

$$C = C_0 \frac{(ap)^{1/n}}{1+(ap)^{1/n}} \tag{1}$$

where

$$C_0 = C_A T + C_B \tag{2}$$

$$a = a_A \exp(a_B T) \tag{3}$$

$$n = n_A T + n_B \tag{4}$$

C is the adsorption concentration, T is the temperature, and p is the pressure. C_A, C_B, a_A, a_B, n_A, and n_B are fitting parameters. Based on the pure gas parameters, the adsorption of mixed gases is determined using the following relation, depending also on the vapor phase composition, y_i [3]:

$$C_i = C_{0,i} \frac{(a_i y_i p)^{\frac{1}{n_i}}}{1+\sum_{j=1}^{N}(a_j y_j p)^{\frac{1}{n_j}}} \tag{5}$$

In order to analyze sorption compressors, an algorithm for calculating the isochoric adsorption of mixed-gases is developed [3]. The sorption cell contains a known amount of adsorbent and a void volume. The composition of the vapor phase is known at a single state (a given temperature and a given pressure), which enables the calculation of the total mass for each component in the system. Then, the algorithm allows finding the equilibrium pressure, vapor phase composition, and adsorbed phase composition at any other temperature at the constant volume. The algorithm consists of the mixed-gas adsorption model, equation (5), and preserves the mass of each component in the system. The vapor phase is calculated by the Peng-Robinson equation of state (PR-EOS) with the classical van der Waals mixing rules and binary interaction parameters. A detailed description of the analysis is provided in [3]. The current research phase also includes an improvement of the numerical code to allow better convergence of the calculations.

Experimental Setup

The objective of the experimental investigation is to validate the currently used numerical model which is based on the recently measured adsorption characteristics of pure gases [4] and the latest sorption compressor model [5].

The experimental setup, depicted in Figure 1, consists of a sorption cell, a mixture vessel, a sample bag, two pressure transducers, and connections to a vacuum pump, port 1, and pure gas sources, port 2. The sorption cell contains 5.9 g of Chemviron AP 4-60 activated carbon, an outer electric heater, and two temperature sensors at the edges. The surrounding heater is wrapped with a copper sheet to obtain a uniform temperature in the sorption cell. Four valves are used for creating the mixture in the mixture vessel out of the pure gas sources, filling the sorption cell with a mixture, and sampling the mixture at different stages of the cycle. A gas chromatograph (GC) is used for determining the composition of the sampled mixture collected in the sample bag. Steady state is defined by obtaining stable temperatures at T1 and T2. A pressure transducer, p1, is connected to the mixture vessel to allow creating mixtures at low pressures. A second pressure transducer, p2, is measuring the pressure in the sorption cell.

Experimental Procedure

In order to prepare a new mixture in the experimental setup, first the mixture vessel is evacuated through port 1 when valves 1 and 2 are open. The other valves are closed (in this section valves that are not mentioned explicitly are closed). Then the source of the heaviest component is connected to port 2 (with its own valve closed), and all the tubing is evacuated, where valves 2 and 3 are open.

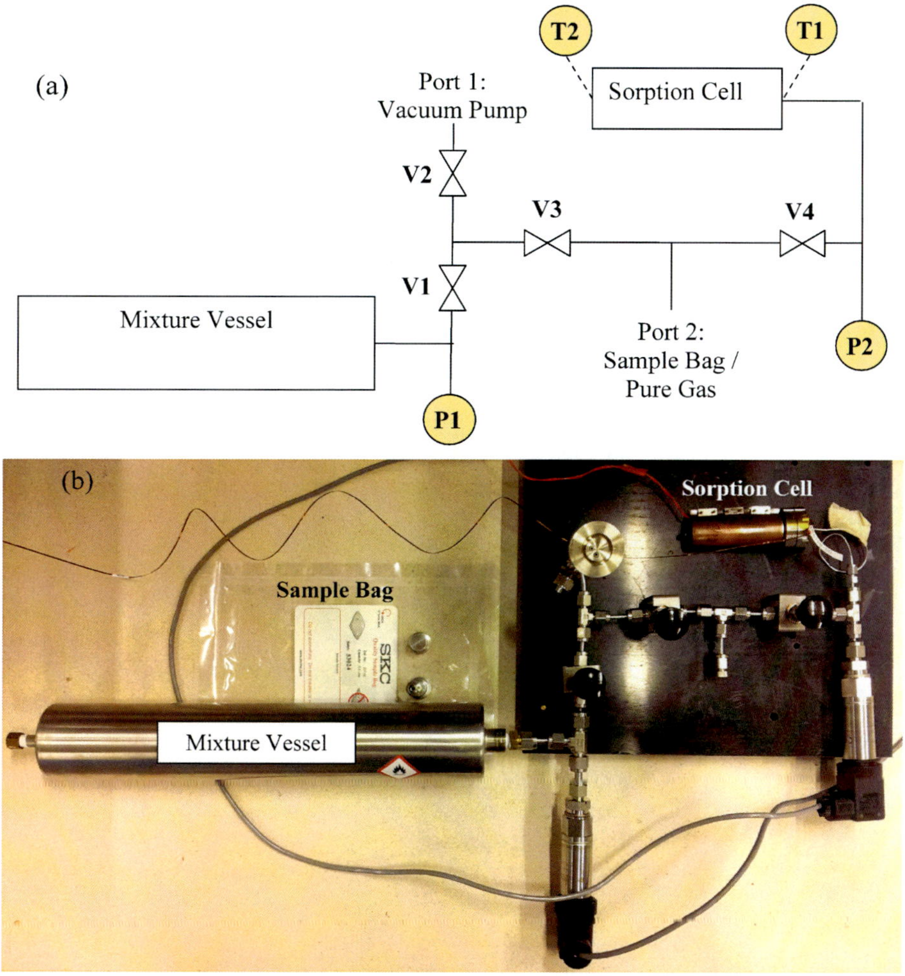

Figure 1. The experimental setup. (a) a schematic view, (b) a photo.

The gas is introduced to the mixture vessel with valves 3 and 1 open, where the pressure is monitored by p1 and controlled by valve 1. The total pressure of the mixture is designed to be of a few bars only; therefore, the pressure of the first component, divided by the total pressure, is about its molar fraction in the targeted mixture, due to the low operating pressures. When introducing the next component, it is important to maintain a relatively high pressure behind valve 1, to avoid back flow of the fluid that is already in the mixture vessel. This procedure provides a rough estimation of the mixture composition in the mixture vessel. This, however, is satisfying, since the actual composition is measured by a GC. The sample is taken with a sample bag that is connected to port 2, which is evacuated (valves 2 and 3 are open) prior to filling (valves 1 and 3 are open).

The sorption cell is vacuum baked at 400 K for at least 24 hours, with port 2 plugged and valves 2, 3, and 4 open. Then, valve 4 is closed and the heater is switched off to let the sorption cell reach room temperature. The mixture is introduced to the sorption cell by opening valves 1, 3, and 4 and waiting for the temperature sensors to reach steady state. Valve 4 is then closed and the pressure, p2, and temperatures, T1 and T2, are recorded. To enable charging the sorption cell with a pressure higher than 1 bar, the sample bag is disconnected (because its use is limited to 1 bar) and port 2 is plugged. The sorption cell is then heated to reach a higher stable temperature. When that temperature is reached, the corresponding pressure is recorded. In this way the dependency on pressure and temperature, at constant volume, is obtained.

To mimic the cycle of a sorption compressor, the closed volume is heated to obtain a desired high pressure. The temperature is further increased and the pressure is maintained constant by letting the mixture to flow out of the sorption cell into the sample bag. This flow is controlled by valve 4 which functions in a similar manner as a check valve: it is opened in a given "cracking pressure" and closed when the pressures on its two sides are equal. The sample composition is measured by the GC, and it represents the mixture composition to be circulated by a sorption compressor which operates in these conditions.

The remaining mixture in the sorption cell is then cooled back to room temperature, and thus the pressure decreases. At a certain state, the pressure equals to the initial pressure at a temperature above room temperature. This is the state where the intake valve of a compressor shall be opened.

Measurement Accuracies

In this study we measure three parameters: temperature, pressure, and mixture composition. The pressure measurements have the sensor accuracies of ± 0.1 %. The temperature sensors have accuracy of about ± 1.5 %; however, during operation we never have reached a complete uniform temperature over the sorption cell and a difference of about 5° has been obtained. Therefore, the accuracy of the temperature measurement is assumed to be ± 3 %. The accuracy of the measured volumes is ± 2.5 %, and the accuracy of the adsorbent mass is ± 0.3 %. The void volume is considered as the volume between the pellets and any other volume in the sorption cell that doesn't contain adsorbent. It does not consider the micro- and nano- pore volumes in the adsorbent. The GC that we use with the existing calibration curves has an accuracy of ± 2 %.

RESULTS AND DISCUSSION

Isochoric Adsorption of Mixed Gases

Figures 2 (a – d) shows the pressure as a function of temperature for mixtures adsorbed in a constant volume with charging vapor compositions of 0.5544 N_2 / 0.4456 C1, 0.5287 N_2 / 0.4713 C1, 0.3885 N_2 / 0.6115 C1, and 0.288 N_2 / 0.712 C1, respectively. The charging temperatures are about room temperature and the charging pressures are 1.34 bar, 3.62 bar, 4.33 bar, and 4.20 bar, respectively. The calculated and experimental results at the charging point are identical, since the calculations were fitted to simulate the experiment (the temperature, pressure, and vapor-phase composition at the charging point are the inputs to the model). The results in Figure 2 demonstrate the satisfying predictions of the pressure changes caused by temperature changes in a constant volume. The maximum deviation between the experimental and calculated results is 6.3 %.

Outflow from the Sorption Cell

When the desired pressure is obtained in simulating the compressor cycle, the temperature is further increased to provide an out flow while maintaining a constant pressure in the sorption cell. A sufficient amount of out flow is required to fill the sample bag for measuring the composition in the GC. The relatively small sorption cell in use makes it difficult to capture a valid out flow. A non valid out flow is a flow where one of the components is already fully desorbed. In the current setup the amount of mixture that leaves the sorption cell is not directly measured. However, the composition of the out flow and the temperature where the remaining mixture in the sorption cell reaches the initial pressure are measured, and these two parameters provide a complete definition of the sorption cycle.

Figure 3 shows the calculated out flow nitrogen molar fraction as a function of the flow rate for a charging mixture of 0.5287 N_2 / 0.4713 C1 at 294.4 K and 3.62 bar. In this cycle, a high pressure of 7.64 bar is obtained at 369.15 K, and the out flow is provided by further increase in temperature to 393.6 K (and maintaining the pressure equal to 7.64 bar). The dashed line in Figure 3 indicates the measured nitrogen molar fraction of the out flow as measured in the GC: 0.654 N_2 / 0.346 C1. The

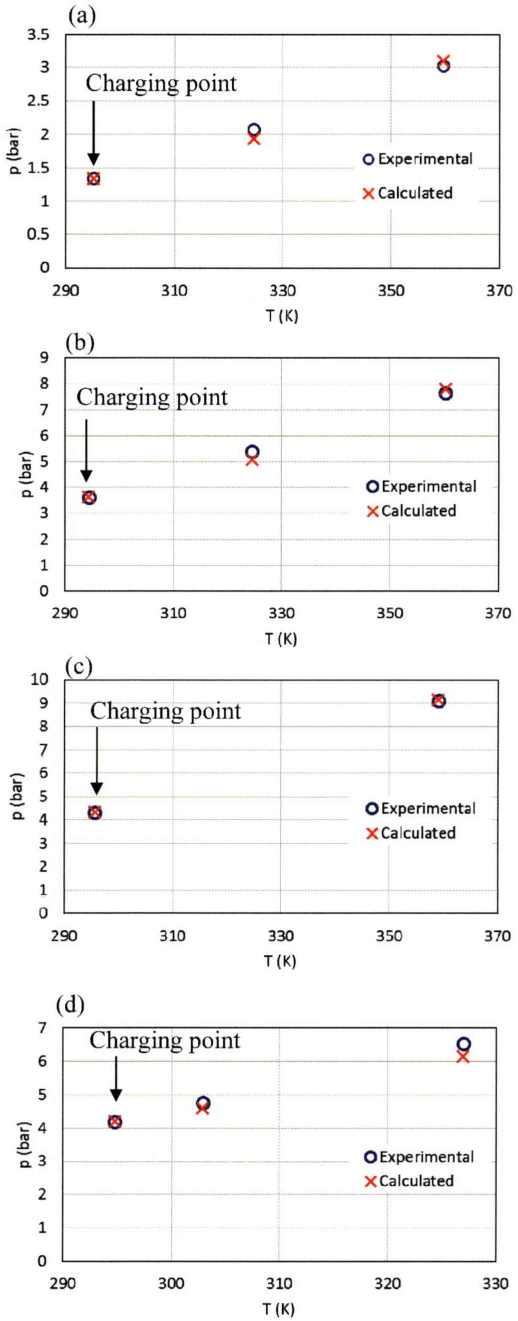

Figure 2. A comparison between experimental and calculated pressures as a function of temperature at constant volume.

(a) molar composition of charging vapor phase 0.5544 N_2 and 0.4456 C1, at 295.2 K and 1.34 bar.
(b) molar composition of charging vapor phase 0.5287 N_2 and 0.4713 C1, at 294.4 K and 3.62 bar.
(c) molar composition of charging vapor phase 0.3885 N_2 and 0.6115 C1, at 295.6 K and 4.33 bar.
(d) molar composition of charging vapor phase 0.288 N_2 and 0.712 C1, at 294.8 K and 4.20 bar.

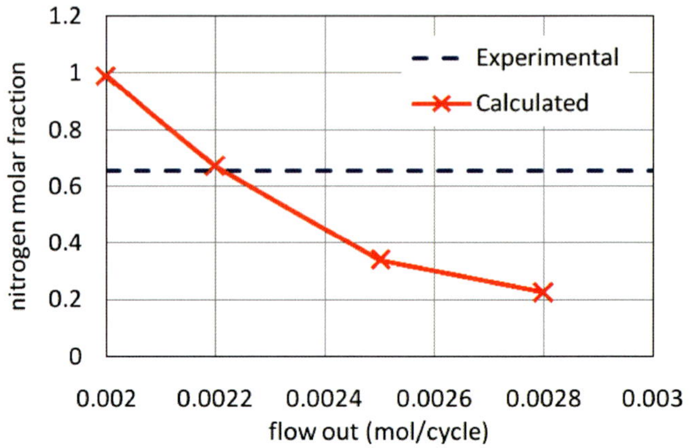

Figure 3. Determining the experimental flow rate using the measured composition of the out flow. The molar composition of the charging vapor phase is 0.5287 N_2 and 0.4713 C1, at 294.4 K and 3.62 bar. A high pressure of 7.64 bar is obtained at 360.15 bar, and the out flow is delivered up to 393.6 K

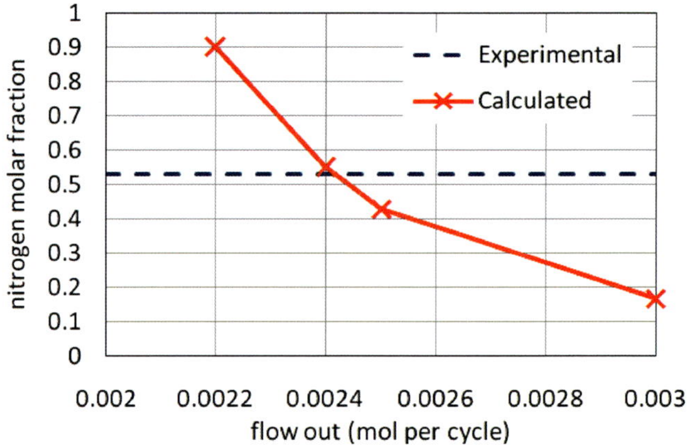

Figure 4. Determining the experimental flow rate using the measured composition of the out flow. The molar composition of the charging vapor phase is 0.3885 N_2 and 0.6115 C1, at 295.6 K and 4.327 bar. A high pressure of 9.1 bar is obtained at 358.9 bar, and the out flow is delivered up to 388.7 K.

intersect between the solid and the dashed lines shows the out flow rate in the experiments, which equals 2.21 mmol per cycle. Since the molar mass of this mixture is 23.85 g/mol and the density at 295 K and 1 bar is 0.97 mg/cc [REFPROP], the out flow is about 53 mg and occupies 54.6 cc. For this flow rate, the calculated pressure in which the remaining mixture in the sorption cell reaches the initial pressure of 3.62 bar is 344 K, whereas the experimental value is 316 K. Taking into consideration the accuracy of the experimental setup and of the numerical model, we find this correspondence reasonable.

Another example is shown in Figure 4. The charged mixture in this case is 0.3885 N_2 / 0.6115 C1 at 295.6 K and 4.327 bar. A high pressure of 9.1 bar is obtained at 358.9 K, and the out flow is provided by further increase in temperature to 388.7 K. The dashed line in Figure 7 indicates the measured nitrogen molar fraction of the out flow: 0.53 N_2 / 0.47 C1. The intersect between the solid and the dashed lines here is at 2.42 mmol per cycle. The calculated pressure in which the remaining mixture in the sorption cell reaches the initial pressure of 4.327 bar is 340 K whereas the experimental value is 301 K.

CONCLUSION

The current study is part of an ongoing research aiming at the development of sorption compressors for driving mixed-gas Joule-Thomson cryocoolers. Recently, we performed an extensive set of adsorption measurements of four pure gases on three different adsorbents. In addition, an improved procedure for fitting the experimental results to the Sips isotherm model was developed and an advanced algorithm has been developed for analyzing the mixed-gases sorption compressor. The experimental investigation presented in this paper is set up for validating this algorithm and analysis. A good agreement between the calculated results and the experimental results for the isochoric adsorption of mixtures is obtained. The experimental setup and procedure are found to be suitable for validating the complete sorption compressor model. However, further improvements are required for increasing the system sensitivity as well as the overall accuracy. Among other ideas, we plan to increase the volume of the sorption cell, to improve the temperature uniformity within this cell, and to improve the calibration of the GC.

ACKNOWLEDGEMENTS

The authors thank Songbo He and Bert Geerdink for their kind assistance with the GC measurements.

REFERENCES

1. Nayak, H.G., Venkatarathnam, G., "Performance of an auto refrigerant cascade refrigerator operating in gas refrigerant supply (GRS) mode with nitrogen–hydrocarbon and argon–hydrocarbon refrigerants," *Cryogenics*, vol. 49 (2009), pp. 350–359.

2. Walimbe, N.S., Narayankhedkar, K.G., Atrey, M.D., "Experimental investigation on mixed refrigerant Joule–Thomson cryocooler with flammable and non-flammable refrigerant mixtures," *Cryogenics*, vol. 50 (2010), pp. 653–659.

3. Tzabar, N., Grossman, G., "Analysis of an activated-carbon sorption compressor operating with gas mixtures," *Cryogenics*, vol. 52 (2012), pp. 491-499.

4. Tzabar, N., ter Brake, H.J.M., "Adsorption isotherms and Sips models of nitrogen, methane, ethane, and propane on commercial activated carbons and polyvinylidene chloride," *Adsorption*, DOI 10.1007/s10450-016-9791-9 (2016).

5. Tzabar, N., Holland, H.J., Vermeer, C.H., ter Brake, H.J.M., "Modeling the adsorption of mixed gases based on pure gas adsorption properties," IOP Conf. Series: Materials Science and Engineering 101 (2015) 012169.

Cooling Temperatures of Binary Mixed Refrigerants: Vapor-Liquid-Liquid Equilibrium versus Vapor-Liquid Equilibrium

N. Tzabar, H.J.M. ter Brake

Energy Materials and Systems
Faculty of Science and Technology
University of Twente
Enschede 7500AE, the Netherlands

ABSTRACT

Joule-Thomson cryocoolers which operate with pure gases provide stable cooling temperatures at their respective boiling points. Mixed-refrigerants have enhanced specific cooling power and they, therefore, enable operating at much lower pressures, relative to pure refrigerants. An attractive method for allowing a stable cooling temperature with binary mixed refrigerants is to choose mixtures that form vapor-liquid-liquid equilibria. In previous research, the cooling temperatures of binary mixtures were estimated by calculating their vapor-liquid equilibrium states. Recently, the vapor-liquid-liquid equilibrium of those mixtures was calculated and the current paper presents a comparison between the results of the two methods.

INTRODUCTION

Joule-Thomson (JT) cryocoolers operating with pure gases allow steady cooling temperatures without any active control, absence of fluid composition shifts, and performance independence on ambient temperature. Nevertheless, mixed refrigerants (MR) for JT cryocoolers have been widely investigated for various cooling temperatures due to two main advantages: lower pressure ratios and higher coefficient of performance (COP). Multi-component mixtures are usually in use in order to obtain enhanced performance [1-3]. Since binary mixtures are not optimal, they are rarely considered as JT refrigerants; however, they lend themselves to analytical investigation in a more convenient manner. According to the Gibbs phase rule, binary mixtures with an equilibrium of three phases behave like a pure gas in the sense that at a fixed pressure a constant temperature phase transition is obtained. In a previous study [4], the temperatures of the phase transition for several binary mixtures were estimated by Vapor-Liquid Equilibria (VLE) calculations, which were experimentally validated. Recently, we established a method for calculating the Vapor-Liquid-Liquid Equilibria (VLLE) of binary mixtures to improve the prediction of the cooling temperature and to allow extensive search for new mixed refrigerants [5]. In the present paper we calculate the VLLE of some binary MRs, using the method we developed, and compare the results with the VLE results.

METHOD

Calculating VLE

The liquid line in the phase diagram is calculated by the activity coefficient method:

$$\sum_i x_i \gamma_i(T, p, \underline{x}) p_i^{vap}(T) = p \tag{1}$$

where x_i is the liquid phase mole fraction, γ_i is the activity coefficient, p_i^{vap} is the vaporization pressure of component i, and p is the pressure. In the current research the van Laar equation is used to determine the activity coefficients of the components in binary mixtures [6]:

$$\ln \gamma_1 = \frac{\alpha}{\left[1 + \frac{\alpha}{\beta}\frac{x_1}{x_2}\right]^2} \quad \text{and} \quad \ln \gamma_2 = \frac{\beta}{\left[1 + \frac{\beta}{\alpha}\frac{x_2}{x_1}\right]^2} \tag{2}$$

where the parameters α and β are functions of the van der Waals parameters a and b of each of the components.

The complete procedure for calculating the VLE of binary mixture is described in details elsewhere [4].

Calculating VLLE

To obtain the separation of a liquid phase into two different liquid phases in equilibrium the summation of the Gibbs free energies of the two phases has to be lower than the Gibbs free energy of the single phase liquid. Therefore, it is necessary to be able to calculate the Gibbs free energy of real mixtures:

$$\underline{G} = \sum_i^N x_i \underline{G}_i + RT \sum_i^N x_i \ln x_i + \underline{G}^E \tag{3}$$

Here, $\underline{G}$ is the total molar Gibbs energy of the mixture, x_i is the molar fraction of component i, $\underline{G}_i$ is the Gibbs energy of the pure component i, R is the universal gas constant, and T is the temperature. The first term at the right side of Eq. (3) is a linear summation of the pure component properties, whereas the other two terms represent the Gibbs energy of mixing. The former of these two represents the mixing Gibbs energy for an ideal mixture whereas the latter term is the excess Gibbs energy that represents the deviation of a real mixture from an ideal one. Different methods are proposed for evaluating this excess Gibbs energy ($\underline{G}^E$), and they all rely on the activity coefficients of the components. In the current research we use the two-constant Margules expansion [6]:

$$\underline{G}^E = x_1 x_2 \left[A + B(x_1 - x_2)\right] \tag{4}$$

$$\begin{bmatrix} x_2^2 & 3x_2^2 - 4x_2^3 \\ x_1^2 & 4x_1^3 - 3x_1^2 \end{bmatrix} \times \begin{bmatrix} A \\ B \end{bmatrix} = \begin{bmatrix} RT \ln \gamma_1 \\ RT \ln \gamma_2 \end{bmatrix} \tag{5}$$

Here, the activity coefficients are calculated by the regular solution theory (RST), which assumes nearly zero excess mixing entropy. According to the RST, the solubility is defined as a function of the internal energy change of vaporization, and the volume fraction of the components is defined as a function of the mixture volume. To calculate the internal energy of vaporization we use the vaporization pressure of the pure components, and the Soave-Redlich Kwong (SRK) equation of state (EOS) is used to calculate the mixture volume. The interaction parameters are then required for the SRK EOS. As a consequence, the pure component vaporization pressures and the interaction parameters for the SRK EOS are the only input data that is required, and they are readily available. The complete procedure for calculating the VLLE of binary mixtures is described in detail elsewhere [5].

In binary mixtures, a single liquid phase is separated into two liquid phases if

$$\frac{\partial^2 G}{\partial x_1^2} > 0 \tag{6}$$

Incorporating Eq. (3) and (4), applying $x_2 = 1 - x_1$ and differentiating the result twice yields:

$$\left[\frac{\partial^2 G}{\partial x_1^2}\right]_{T,p} = \frac{RT}{x_1(1-x_1)} + 6B - 2A - 12Bx_1 \tag{7}$$

The temperature where VLLE exists is determined by:

$$f_i^{L1}(p,T,\underline{x}_1) = f_i^{L2}(p,T,\underline{x}_2) = f_i^{V}(p,T,\underline{y}) \tag{8}$$

Where the superscript *L1* and *L2* denote the two liquid phases and *V* represents the vapour phase. f_i is the fugacity of component i. p is the pressure, T is the temperature, $\underline{x}$ is the molar composition of a liquid phase, and $\underline{y}$ is the molar composition of the vapor phase.

RESULTS AND DISCUSSION

Figures 1 to 3 show comparisons between the VLE results and the VLLE results for nitrogen–ethane, nitrogen-ethylene, and nitrogen-propane mixtures, respectively. The VLE results are presented by solid lines and show the temperature of each mixture composition at a specific pressure. The VLLE results are presented by dashed lines enclosed between two big dots, which indicate the compositions of the two liquid phases. Any mixture with a composition in between (along the dashed lines) separates into these two liquid phases. The results for nitrogen-ethane mixtures, Figure 1, are given at 0.5, 1, and 1.81 MPa, the results for nitrogen-ethylene mixtures, Figure 2, are given at 1.025 and 3.51 MPa, and the results for nitrogen-propane mixtures, Figure 3, are given at 0.5, 1, and 3.5 MPa.

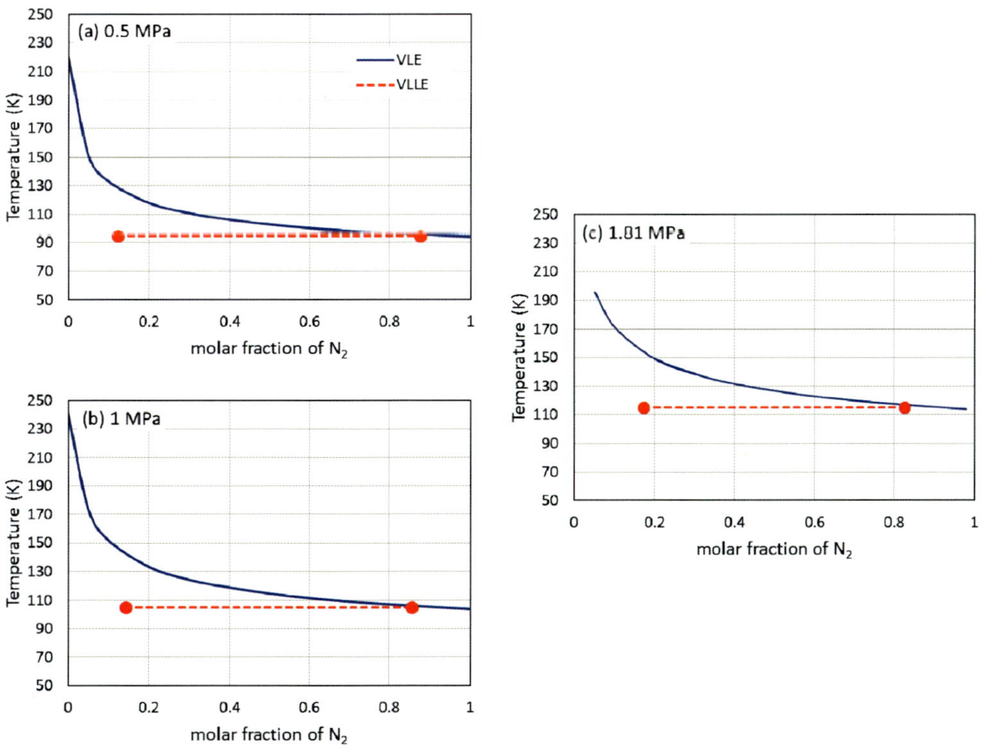

Figure 1. The liquid line of VLE calculation (solid line) versus the VLLE existence, for nitrogen-ethane mixtures at (a) 0.5 MPa, (b) 1 MPa, and (c) 1.81 MPA.

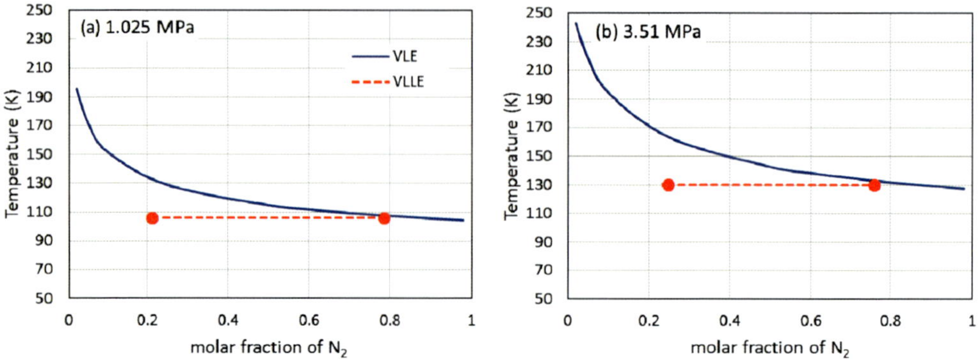

Figure 2. The liquid line of VLE calculation (solid line) versus the VLLE existence, for nitrogen-ethylene mixtures at (a) 1.025 MPa and (b) 3.51 MPa.

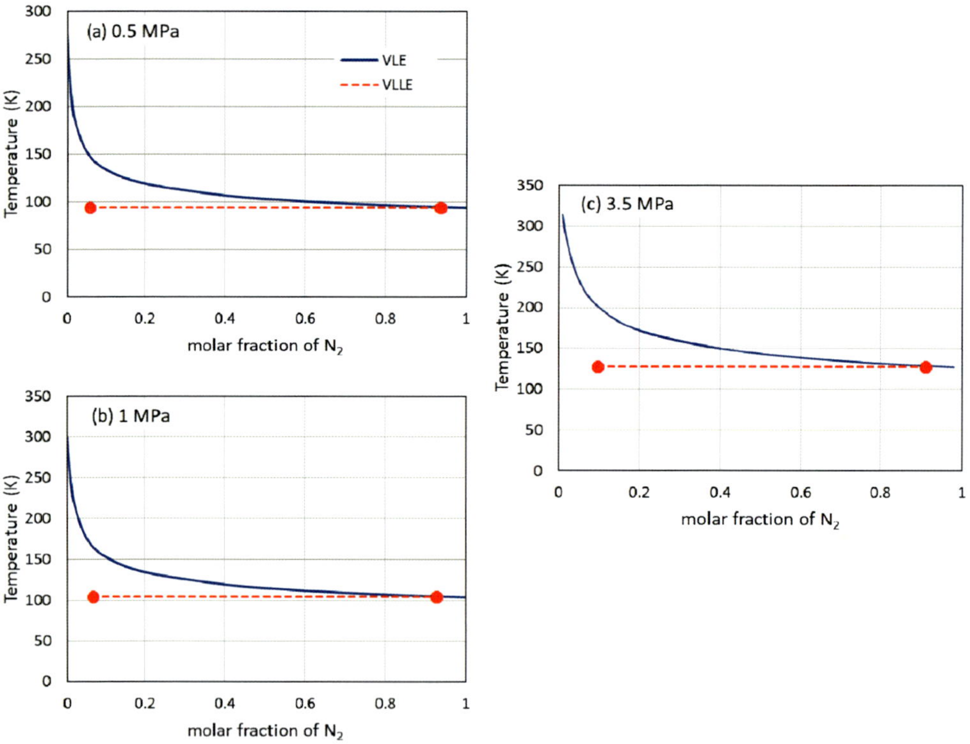

Figure 3. The liquid line of VLE calculation (solid line) versus the VLLE existence, for nitrogen-propane mixtures at (a) 0.5 MPa, (b) 1 MPa, and (c) 3.5 MPA.

One should notice that at VLLE, all the dots which refer to nitrogen enriched compositions meet the VLE curves. Moreover, all the VLLE results show lower temperatures than the VLE results, as expected, to allow a lower free Gibbs energy, which is temperature dependent. Figures 4 – 6 show these differences for nitrogen-ethane, nitrogen-ethylene, and nitrogen-propane, respectively, at different pressures. As the molar fraction of the nitrogen decreases the difference between the VLE and the VLLE results increases. It is also evident that at higher pressures the differences between the VLE and the VLLE results are larger. The absolute differences between the VLE and the VLLE results are quite similar for the three mixtures. For example, the VLE temperatures of mixtures with

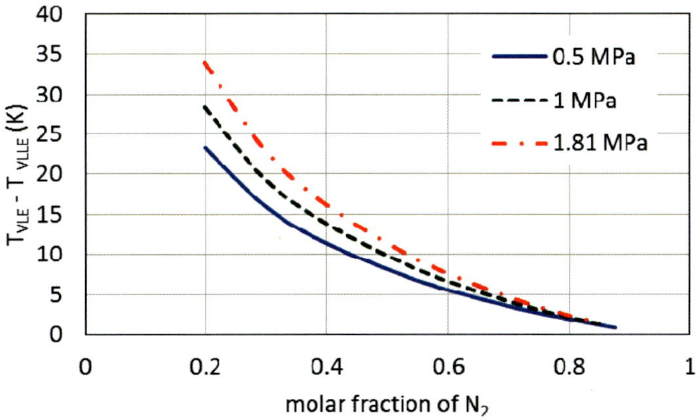

Figure 4. Temperature differences between VLE and VLLE calculations for nitrogen-ethane mixtures at 0.5, 1, and 1.81 MPa.

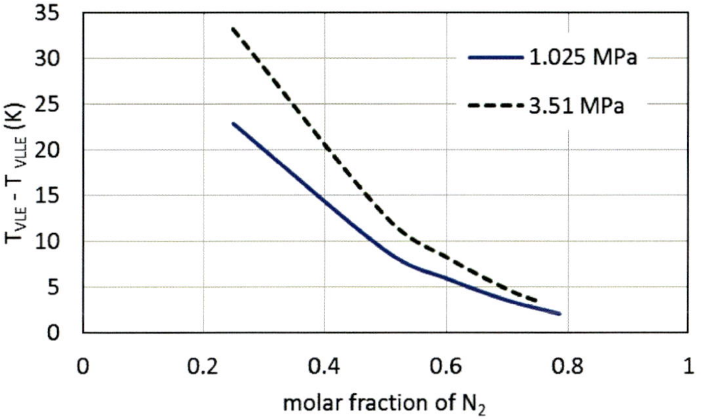

Figure 5. Temperature differences between VLE and VLLE calculations for nitrogen-ethylene mixtures at 1.025 and 3.51 MPa.

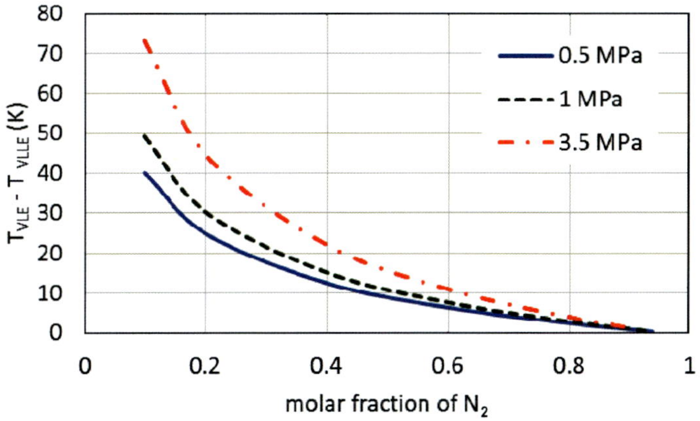

Figure 6. Temperature differences between VLE and VLLE calculations for nitrogen-propane mixtures at 0.5, 1, and 3.5 MPa.

nitrogen molar fraction of 0.3 at about 1 MPa are all about 20 K higher (19.4 K, 20 K, and 22 K for nitrogen-ethane, nitrogen-ethylene, and nitrogen-propane, respectively). Moreover, at about 1 MPa, to have a temperature difference of less than 5 K the nitrogen molar fraction has to be above 0.62, 0.64, and 0.71 for nitrogen-ethane, nitrogen-ethylene, and nitrogen-propane, respectively.

CONCLUSIONS

JT cryocoolers that operate with MRs are aimed to reach the liquid line of the phase diagram. Using VLE calculation for binary mixtures, which actually form VLLE, provides a limited estimation of the cooling temperature. For the three mixtures presented in this paper, having a large molar fraction of nitrogen in the mixture allows a relatively small deviation between the VLE and the VLLE temperatures. It is recommended to use phase diagrams with VLE prediction to roughly estimate the cooling temperatures of binary MRs; however, it is recommended to use complete phase diagrams, which incorporate the VLLE prediction, for more accurate analyses.

ACKNOWLEDGMENT

This work is supported by NanoNextNL, a micro and nanotechnology consortium of the government of the Netherlands and 130 partners.

REFERENCES

1. Gong, M.Q., Wu, J.F., and Luo E.G., "Performance of the mixed-gases Joule-Thomson refrigeration cycles for cooling fixed-temperature heat loads," *Cryogenics*, vol. 44 (2004), pp. 847-857.
2. Nayak, H.G., Venkatarathnam, G., "Performance of an auto refrigerant cascade refrigerator operating in gas refrigerant supply (GRS) mode with nitrogen–hydrocarbon and argon–hydrocarbon refrigerants," *Cryogenics*, vol. 49, no. 7 (2009), pp. 350–359.
3. Tzabar, N., Lapp, Z., "Experimental investigation on mixed-refrigerant for closed cycle Joule–Thomson cryocoolers," *Adv Cryog Eng*, vol. 55 (2010), pp. 1121–1128.
4. Tzabar, N., "Binary mixed-refrigerants for steady cooling temperatures between 80 K and 150 K with Joule-Thomson cryocoolers," *Cryogenics*, vol. 64 (2014), pp. 70-76.
5. Tzabar, N., ter Brake, H.J.M., "Modeling of vapor-liquid-liquid equilibria in binary mixtures," *Int. J. of Ref.*, JIJR3278 (2016).
6. Stanley, I.S., *Chemical, Biochemical, and Engineering Thermodynamics*, 4th edition, John Wiley & Sons Inc., Hoboken, NJ (2006).

High-Efficiency Joule-Thomson Cryocoolers Incorporating an Ejector

H.S. Cao[1], S. Vanapalli[1], H.J. Holland[1], C.H. Vermeer[2], T. Tirolien[3], H.J.M. ter Brake[1]

[1]University of Twente, 7500 AE, Enschede, The Netherlands
[2]SuperACT, 7559 AJ Hengelo, The Netherlands
[3]European Space Agency, 2200 AG, Noordwijk, The Netherlands

ABSTRACT

Joule-Thomson (JT) cryocoolers have no moving parts and therefore are vibration-free. These are attractive for cooling small optical detectors in space for earth observation missions. JT cryocoolers produce cooling by expanding high-pressure gas through a JT restriction. This, however, is a highly irreversible entropy-generating process. If work could be extracted during the expansion process, the efficiency of the cooling cycle would be significantly improved. In this paper, a JT cooling cycle with an additional ejector is proposed. The high-pressure gas, as the primary fluid of the ejector, is used to compress the low-pressure gas leaving the evaporator, thus reducing the cold-end temperature or/and the input power of the compressor. Compared to a basic JT cycle, the improvement in the COP of the cycle with an ideal ejector is analyzed. The effects of frictional and mixing losses on the real performance of an ejector are estimated through numerical simulations, which aids the understanding of ejector theory and provides information for optimizing the ejector under certain operating conditions.

INTRODUCTION

Joule-Thomson cryocoolers have been used in many applications including cooling of infrared detectors and high-electron-mobility transistor-based devices in space [1], low-noise amplifiers for radio telescopes [2], and cryosurgery [3], among others due to their special features such as compact geometry and absence of moving parts. In a normal JT expansion process, the energy of the high-pressure gas is converted into directed kinetic energy, which is not utilized as work or discharged, but instead is dissipated as heat. This cancels out a large amount of the expansion cooling potential in the gas. Only the Joule-Thomson cooling effect below the inversion temperature remains. Ejectors can use the directed kinetic energy, which is completely wasted in the normal JT expansion, to entrain and compress the low-pressure gas to an intermediate pressure.

Ejectors can be used to reduce the pressure of the evaporator and thereby lower the temperature of the evaporator. Alternatively, ejectors can be used to lift the pressure of the evaporator to a medium pressure and thus reduce the required pressure ratio of the compressor, resulting in less required work of compression. Therefore, an ejector can be used either to lower the pressure (temperature) of the evaporator of a JT cooler, or boost the outlet pressure of a JT cooler. The use of an ejector in a JT cryocooler was first proposed by Rietdijk [4] to create subatmospheric pressure in a

Cryocoolers 19, edited by S.D. Miller and R.G. Ross, Jr.

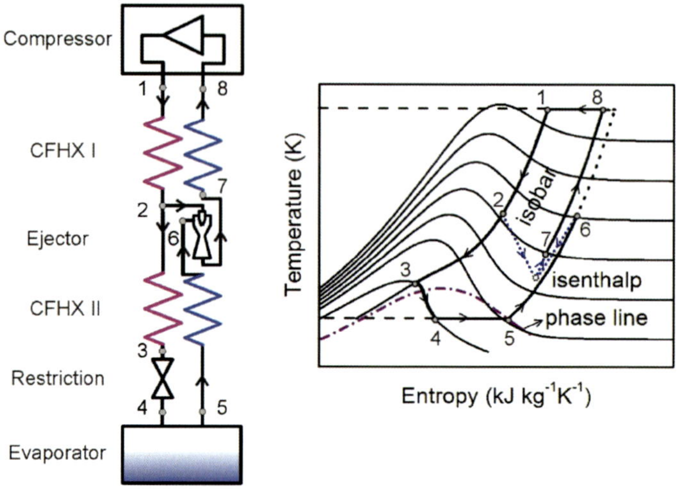

Figure 1. Schematic of Linde-Hampson cycle with an ejector (left) and corresponding temperature versus entropy diagram (right).

liquid helium evaporator. Daney et al. [5] added an ejector to a JT cryocooler with nitrogen gas as the working fluid. An assemble type ejector was used to characterize the performance of an ejector at different dimensions. Both the primary nozzle and the mixing tube-diffuser were exchangeable. There were three primary nozzles with diameters of 0.58, 0.57 and 0.46 mm and three secondary nozzles with diameters of 6.35, 7.54 and 8.73 mm. Several combinations were tested and the best ejector gave a suction pressure of 0.27 bar and an associated saturation temperature of 67.7 K. Most recent studies regarding the use of ejectors in JT cycle systems are on thermodynamic analysis [6, 7]. Very few of these were experimentally verified. It is known that the performance of an ejector not only depends on the system operating conditions (pressure and environment temperature), but also on its geometry and the working fluid. Many researchers have tried to investigate and describe the flow in an ejector in order to develop a high performance ejector [8-10]. However, not all mysteries in the operation of an ejector have been completely cleared. In this paper, the performance improvement of a JT cooler resulting from the addition of an ejector is analyzed. In order to design a JT cooler with an ejector, its performance under certain operating conditions is simulated by means of dynamic modeling. The effects of the operating conditions and the geometry on the ejector performance are discussed.

OPERATING PRINCIPLE

Figure 1 shows the schematic of a Linde-Hampson cycle with an ejector and corresponding temperature versus entropy diagram. In the cycle, the high-pressure gas flows through the counter flow heat exchanger I (CFHX I), then is split into two streams (at point 2). One stream flowing through CFHX II is expanded through the restriction (between points 3,4) and the other stream flows to the ejector as the so-called primary flow. In the ejector, the high-pressure gas (the primary fluid) expands through the nozzle to create low pressure, and then mixes with the secondary fluid from the CFHX II (at point 6) in the mixing section. The mixed gas increases in pressure in the diffuser section and flows to the CFHX I at the medium pressure (at point 7), rather than at the low pressure of the evaporator.

Figure 2a shows a schematic of a typical ejector, which consists of a nozzle, a suction chamber, a mixing section, and a diffuser. A high-pressure primary flow expands in the nozzle where its internal energy and pressure related work convert to kinetic energy. It flows out with supersonic speed to create a very low pressure region at the nozzle exit. Hence, a pressure difference between the streams at the nozzle exit plane and the secondary flow inlet is established and the secondary

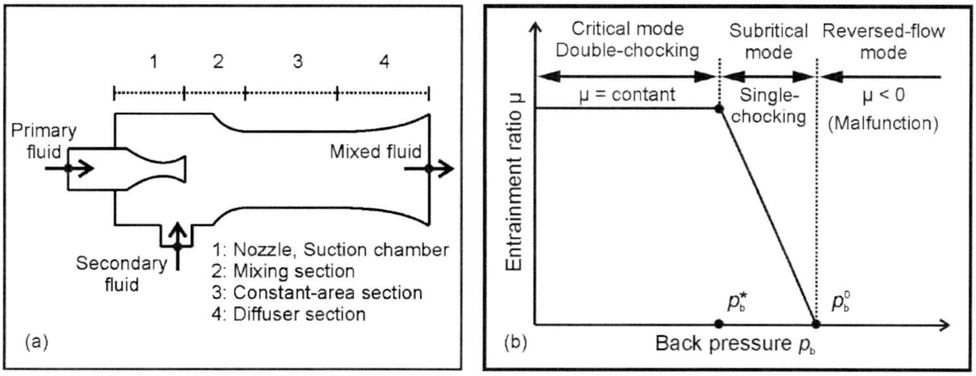

Figure 2. The schematic (a) and operational modes (b) of an ejector [7].

fluid is drawn in by the entrainment effect. In the mixing section, the velocity of the primary flow is slowed down while the secondary flow is accelerated. In the diffuser, the kinetic energy is converted into pressure related flow work.

Figure 1 shows the schematic of a Linde-Hampson cycle with an ejector and corresponding temperature versus entropy diagram. In the cycle, the high-pressure gas flows through the counter flow heat exchanger I (CFHX I), then is split into two streams (at point 2). One stream flowing through CFHX II is expanded through the restriction (between points 3,4) and the other stream flows to the ejector as the so-called primary flow. In the ejector, the high-pressure gas (the primary fluid) expands through the nozzle to create low pressure, and then mixes with the secondary fluid from the CFHX II (at point 6) in the mixing section. The mixed gas increases in pressure in the diffuser section and flows to the CFHX I at the medium pressure (at point 7), rather than at the low pressure of the evaporator.

Figure 2a shows a schematic of a typical ejector, which consists of a nozzle, a suction chamber, a mixing section, and a diffuser. A high-pressure primary flow expands in the nozzle where its internal energy and pressure related work convert to kinetic energy. It flows out with supersonic speed to create a very low pressure region at the nozzle exit. Hence, a pressure difference between the streams at the nozzle exit plane and the secondary flow inlet is established and the secondary fluid is drawn in by the entrainment effect. In the mixing section, the velocity of the primary flow is slowed down while the secondary flow is accelerated. In the diffuser, the kinetic energy is converted into pressure related flow work.

The performance of an ejector is characterized by the entrainment ratio (μ), defined as the ratio of secondary mass flow rate to primary mass flow rate ($\dot{m}_s/\dot{m}_p$), and by the pressure lift ratio (τ), defined as the ratio of the back pressure (p_b, at the exit of the ejector) to the secondary flow pressure (p_s, at the secondary-fluid inlet). Neglecting pressure drops in the CFHXs, this ratio of the inlet pressure of the compressor to the pressure of the evaporator (p_8/p_5) as shown in Figure 1. For a fixed-geometry ejector, the effect of the back pressure on the entrainment ratio is illustrated in Figure 2, in which three regions can be identified: double-choking flow, single-choking flow and reversed flow regions. If the back pressure is below the critical value (p_b^*), the primary flow and the secondary flow are both choked, causing constant entrainment ratio. However, if the back pressure is above the critical value, the entrainment ratio μ will drop sharply with increasing back pressure due to the absence of secondary flow choking. Further increase in back pressure higher than p_b^0 will lead to reversed flow.

The cold-end temperature of a JT cryocooler with an ejector is determined by the boiling temperature of the working fluid at the evaporator pressure. The cooling power $\dot{Q}$ of the cryocooler (shown in Figure 1) is defined by:

$$\dot{Q} = \dot{m}_s \cdot (h\langle T_5, p_5, x_5\rangle - h\langle T_4, p_4, x_4\rangle) \tag{1}$$

where h and x are the specific enthalpy and gas mass fraction, respectively.

Neglecting the heat losses in CFHX I, CFHX II and the ejector, the cooling power $\dot{Q}$ of the cryocooler can also be determined by the change in enthalpy of the working fluid at the warm end of the crycooler:

$$\dot{Q} = (\dot{m}_s + \dot{m}_p) \cdot (h\langle T_8, p_8\rangle - h\langle T_1, p_1\rangle) \tag{2}$$

The minimum required power ($\dot{W}$) for a reversible and isothermal compression can be calculated as the change in Gibbs free energy of the fluid during compression:

$$\dot{W} = (\dot{m}_s + \dot{m}_p) \cdot [T_1(s\langle T_1, p_1\rangle - s\langle T_8, p_8\rangle) - (h\langle T_1, p_1\rangle - h\langle T_8, p_8\rangle)] \tag{3}$$

where s is the specific entropy.

The COP of a cooler is defined as the ratio of the cooling power and the required compressor input power:

$$COP = \dot{Q}/\dot{W} \tag{4}$$

Substituting equations 2 and 3 into equation 4, and equation 4 becomes:

$$COP = (h\langle T_8, p_8\rangle - h\langle T_1, p_1\rangle)/[T_1(s\langle T_1, p_1\rangle - s\langle T_8, p_8\rangle) - (h\langle T_1, p_1\rangle - h\langle T_8, p_8\rangle)] \tag{5}$$

The cold-end temperature of a JT cryocooler depends on the evaporator pressure (p_5), and the inlet pressure of the compressor equals the product of the evaporator pressure and the pressure lift ratio: $p_5\tau$. The secondary mass-flow rate equals the product of the primary mass-flow rate and the entrainment ratio: $\dot{m}_p\mu$. The outlet pressure of the compressor and the ambient temperature are assumed to be p_h and T_h. CFHX I and CFHX II are assumed to be ideal heat exchangers, which implies that $T_8 = T_1$, and $T_6 = T_2$. Replacing p_1, T_1, p_8 *and* $\dot{m}_s$ by p_h, $T_h, p_5\tau$ *and* $\dot{m}_p\mu$, equations 2 and 5 become:

$$\dot{Q} = (\dot{m}_p\mu + \dot{m}_p) \cdot (h\langle T_h, p_5\tau\rangle - h\langle T_h, p_h\rangle) \tag{6}$$

$$COP = (h\langle T_h, p_5\tau\rangle - h\langle T_h, p_h\rangle)/[T_1(s\langle T_h, p_h\rangle - s\langle T_h, p_5\tau\rangle) - (h\langle T_h, p_h\rangle - h\langle T_h, p_5\tau\rangle)] \tag{7}$$

Equation 6 shows that the cooling power is determined by the change in enthalpy of the working fluid due to isothermal compression at ambient temperature, as is the case in a basic JT cooler without an ejector. The COP of a basic JT cooler operating under the same conditions (compressor outlet pressure p_h, compressor temperature T_h and evaporator pressure p_5) would be:

$$COP_0 = (h\langle T_h, p_5\rangle - h\langle T_h, p_h\rangle)/[T_1(s\langle T_h, p_h\rangle - s\langle T_h, p_5\rangle) - (h\langle T_h, p_h\rangle - h\langle T_h, p_5\rangle)] \tag{8}$$

Figure 3 shows the effect of the pressure lift ratio on the COP of the JT cryocooler when it is equipped with an ejector. The cryocooler is operated with nitrogen and methane at different cold-end temperatures. The outlet pressure of the compressor and the ambient temperature are assumed to be 8.0 MPa and 295 K, respectively. The COP at the pressure lift ratio of 1 is that of a basic JT cryocooler operating under the same conditions. Compared to a basic JT cryocooler, the COP of a JT cryocooler with an ejector is higher because the pressure of the evaporator (p_5) is compressed to a higher pressure ($p_5\tau$) by the ejector. Thus less input power is required to compress the working fluid to a pressure p_h.

Figure 4 shows the entrainment ratio as a function of the temperature in point 7 shown in Figure 1 at a fixed pressure lift ratio. In this analysis, nitrogen gas is considered as the working fluid and the evaporator pressure is 0.050 MPa in order to achieve a cold-end temperature of 71.8 K. The outlet and inlet pressures of the compressor are selected as 8.0 and 0.1 MPa with a fixed pressure lift ratio of 2.0. Based on energy conservation, the entrainment ratio increases as the temperature of point 7 increases from 100 to 295 K. Since a relatively low entrainment ratio is easier to realize, Figure 4 suggests that it is attractive to place the ejector at a relatively low temperature. However, at low temperature the working fluid may condense and the resulting two-phase flow may disturb the performance of the ejector. This topic is currently under investigation.

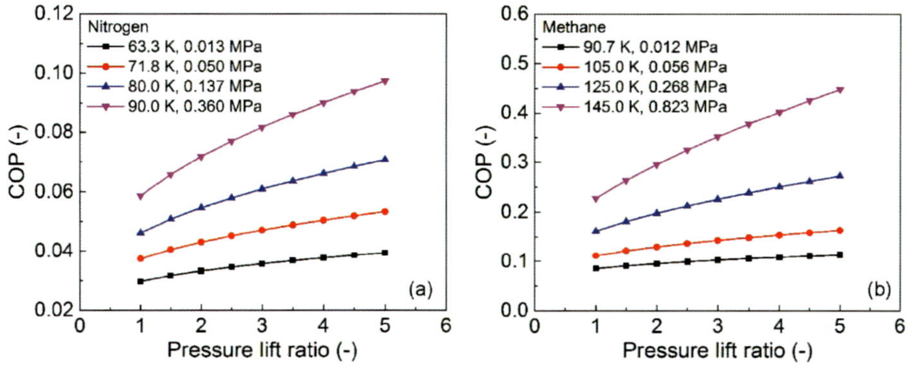

Figure 3. COP of a JT cryocooler with an ejector operating with nitrogen (a) and methane (b) at different cold-end temperatures as a function of the pressure lift ratio (p_1 =8.0 MPa and T_h = 295 K). The evaporator pressures are the saturation pressures at corresponding cold-end temperatures.

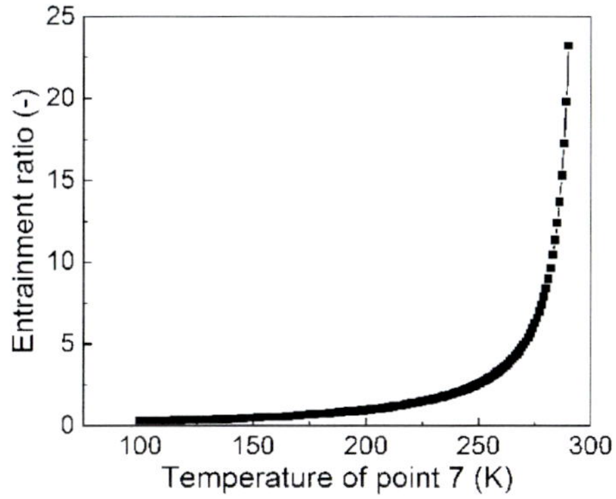

Figure 4. Entrainment ratio as a function of temperature of point 7 shown in Figure 1 (p_1 = 8.0 MPa; p_5 = 0.05 MPa, p_8 = 0.1 MPa and T_h = 295 K).

MODELLING OF AN EJECTOR

In order to design a JT cooler with an ejector, the ejector performance at different conditions needs to be investigated. There are typically two ways for simulating the flow process in an ejector: thermodynamic modeling and dynamic modeling [11]. Thermodynamic modeling uses a one-dimensional model, in which the state and operating parameters of a few sections along the ejector are obtained through steady-state governing equations. Detailed velocity, pressure, and temperature distributions along the ejector are not considered. The effects of frictional and mixing losses are taken into account by using corresponding coefficients introduced in the governing equations. However, these coefficients need to be determined experimentally or by using a dynamic model. The dynamic model is a three-dimensional model, by which detailed information such as velocity, pressure, and temperature distributions along the ejector can be obtained. Besides, the position of the shock waves can also be tracked.

In this study, dynamic models are adopted to simulate the flow process in ejectors with nitrogen gas in order to investigate the performance of ejectors at different operating conditions. Dynamic models are expressed in differential equations (continuity, momentum and total energy), which are solved by using numerical methods, such as finite difference method, finite element method, and finite volume method. In this work, the software ANSYS Fluent based on a finite volume approach is used. The flow in the ejector is assumed to be governed by the ideal gas compressible steady-state axisymmetric form. The turbulent behavior was treated using the Reynolds averaging principle (RANS). The standard version of the k-ε turbulence model was chosen in this work [12] [13] [14].

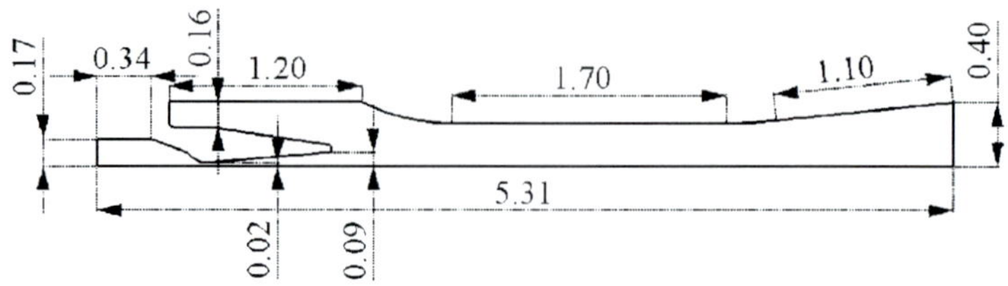

Figure 5. Geometry of the baseline ejector design used in the CFD simulations (mm).

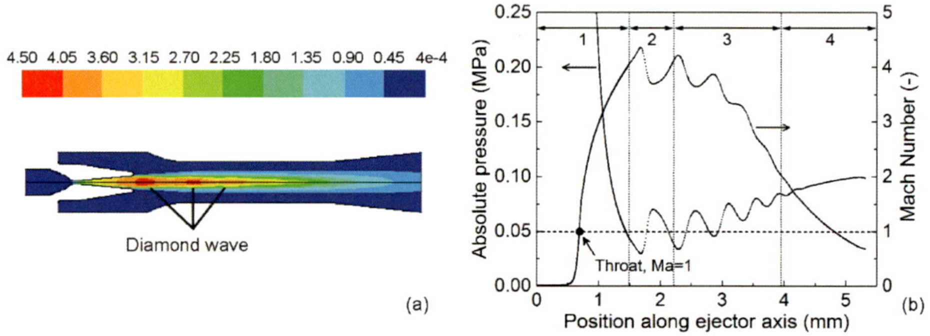

Figure 6. Mach number and absolute static pressure distributions inside the ejector under $p_p = 8.0$ MPa, $T_p = 295$ K, $p_s = 0.05$ MPa, $T_s = 295$ K and $p_b = 0.1$ MPa. (a) Mach number contour; (b) Pressure and Mach number along ejector axis. The four sections are the nozzle, the mixing section, the constant area section and diffuser section as shown in Figure 2a.

The transport equations for the turbulent kinetic energy (k) and the turbulent kinetic energy dissipation rate (ε) can be found in the ANSYS Fluent theory guide [15]. Pressure inlet boundary conditions are used to define the primary and secondary flow inlet pressure, and the outlet of the ejector is assigned to be a pressure outlet boundary condition. Temperature is applied for the inlet and outlet thermal condition, and the ejector wall is set to be adiabatic. Figure 5 shows the geometry of the baseline ejector design used in the CFD simulations. As depicted, the overall size is slightly more than 5 mm, obviously aiming at incorporation in a microcooler.

Figure 6 shows the Mach number and absolute static pressure distributions inside the ejector when the pressures and temperatures of the primary and secondary flows are set at 8.0 MPa, 0.05 MPa and 295 K, and the exit pressure is set at 0.1 MPa. The Mach number reaches 1 at the throat of the nozzle. The Mach number increases to supersonic values with the primary flow expanding to a lower pressure. The "diamond wave" phenomenon manifested by a series of oscillations of the Mach number and the static pressure along the axis of the ejector, is qualitatively consistent with observed laser tomography of Desevaux [16] and CFD investigation of Ji et al [14]. The phenomenon is caused by the large velocity difference between the supersonic primary flow and the entrained secondary flow. As the diamond wave decays the secondary fluid is accelerated and mixed with the primary flow. In the meantime, the static pressure gradually increases to the discharge value and the Mach number reduces to subsonic value.

Figure 7a shows the effect of the back pressure of the ejector on the entrainment ratio with $\dot{m}_p = 25.5$ mg s^{-1} when the pressures and temperatures of the primary and secondary flows are set at 8.0 MPa, 0.05 MPa and 295 K, respectively. Similar to the general relation between entrainment ratio and the back pressure (see Figure 2b), the entrainment ratio is independent on the back pressure when the back pressure does not exceed the critical value ($p_b^* = 0.06$ MPa). When the back pressure is higher than that critical value, the entrainment ratio decreases with increasing back pressure. At $p_b^0 = 0.108$ MPa, the entrainment ratio is zero. Further increase in back pressure will lead to reverse flow.

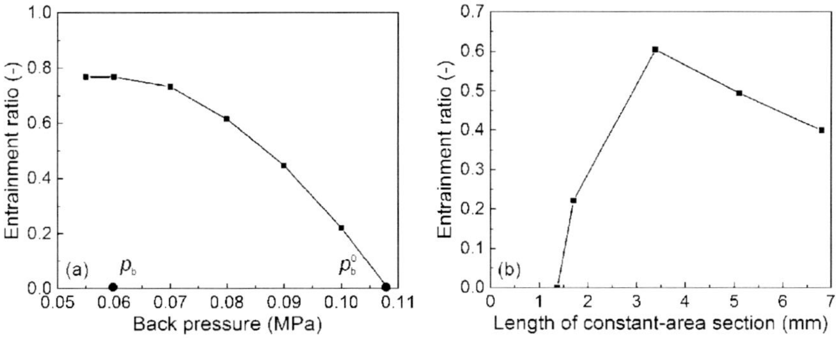

Figure 7. (a): Effect of back pressure and (b): Effect of length of the constant-area section on the entrainment ratio with p_b = 0.1 MPa. (p_p = 8.0 MPa, T_p =295 K, p_s = 0.05 MPa and T_s =295 K).

Figure 7b shows the effect of the length of the constant-area section of the ejector (see Figure 3) on the entrainment ratio with $\dot{m}_p$= 25.5 mg s^{-1} when the pressures of the primary, secondary and discharge flows are set at 8.0 MPa, 0.05 MPa and 0.1 MPa, respectively. The ejector has the dimensions shown in Figure 5 and only the length of the constant-area section is varied. The entrainment ratio increases with increasing length of the section. Figure 6a shows that the primary jet flow core leaves the nozzle exit plane with supersonic speed while the secondary flow outside the jet flow core remains at relatively low speed. The radial non-uniformity in the velocity distribution leads to momentum exchange, which is driven by the combined action of turbulent eddies and interphase drag. To achieve a complete mixing, the length of the momentum exchange (the total length of the mixing section and the constant area section shown in Figure 2) needs to exceed a certain value. However, the losses due to friction effect also increases with increasing total length of mixing section and constant-area section. As shown in Figure 7b, the optimum length under the above-mentioned operating conditions is about 3.4mm.

CONCLUSIONS

The application of an ejector in Joule-Thomson cryocoolers is discussed, and the performance improvement is explained using nitrogen and methane as working fluids. The function of the ejector is to lift the pressure of the evaporator to a medium pressure and thus to reduce the required pressure ratio of the compressor, resulting in less compression work and a higher coefficient of performance. The COP increases with increasing pressure lift ratio. With a given pressure lift ratio, the required entrainment ratio decreases as the ejector shifts from the warm end to the cold end of a cryocooler resulting from energy conservation. The optimum position of the ejector in a JT cycle depends on the ejector performance at certain operating conditions. To investigate that performance, a dynamic model is developed. The effect of the back pressure and the length of the constant area on the ejector performance are investigated at room temperature. The dependence of the entrainment ratio on the back pressure and the length of the constant area is explained. Future work will focus on the simulation of the ejector performance at low temperature with two-phase flow involved, the experimental validation thereof and the actual realization of cryocoolers equipped with ejectors.

ACKNOWLEDGMENT

This work was supported by the European Space Agency (ESA) under Contract No. 4000114493/15/NL/KML/fg.

REFERENCES

1. Collaudin, B. and Rando N., "Cryogenics in space: a review of the missions and of the technologies," *Cryogenics*, vol. 40, no. 468 (2000), pp. 797-819.

2. Cao, H.S., et al., "Cooling a low noise amplifier with a micromachined cryogenic cooler," *Review of Scientific Instruments*, vol. 84, no. 10 (2013), pp. 105102.

3. Fredrickson, K., G. Nellis, and S. Klein, "A design method for mixed gas Joule-Thomson refrigeration cryosurgical probes," *International Journal of Refrigeration*, vol. 29, no. 5 (2006), pp. 700-715.

4. Rietdijk, J.A., "The Expansion-Ejector: A New Device for Liquefaction and Refrigeration at 4 K and lower, " *Liquid Helium Technology, Proceedings of the International Institute of Refrigeration*, Pergamon Press, New York (1966) p. 241.

5. Daney, D.E., McConnell, P.M. and Strobridge. T.R. "Low-Temperature Nitrogen Ejector Performance," *Advances in Cryogenic Engineering*, vol. 18 (1973), pp. 476-485.

6. Yu, J., "Improving the performance of small Joule-Thomson cryocooler," *Cryogenics*, vol. 48, no. 9-10 (2008), pp. 426-431.

7. Rashidi, M.M., Bég, O.A. and Habibzadeh, A. "First and second law analysis of an ejector expansion Joule-Thomson cryogenic refrigeration cycle," *International Journal of Energy Research*, vol. 36, no. 2 (2012), pp. 231-240.

8. Huang, B.J., et al., "A 1-D analysis of ejector performance," *International Journal of Refrigeration*, vol. 22, no. 5 (1999), pp. 354-364.

9. Rusly, E., et al., "CFD analysis of ejector in a combined ejector cooling system," *International Journal of Refrigeration*, vol. 28, no. 7 (2005), pp. 1092-1101.

10. Pianthong, K., et al., "Investigation and improvement of ejector refrigeration system using computational fluid dynamics technique," *Energy Conversion and Management*, vol. 48, no. 9 (2007), pp. 2556-2564.

11. He, S., Y. Li, and R.Z. Wang, "Progress of mathematical modeling on ejectors," *Renewable and Sustainable Energy Reviews*, vol. 13, no. 8 (2009), pp. 1760-1780.

12. Scott, D., Z. Aidoun, and M. Ouzzane, "An experimental investigation of an ejector for validating numerical simulations," *International Journal of Refrigeration*, vol. 34, no. 7 (2011), pp. 1717-1723.

13. Hemidi, A., et al., "CFD analysis of a supersonic air ejector. Part I: Experimental validation of single-phase and two-phase operation," *Applied Thermal Engineering*, vol. 29, no. 8-9 (2009), pp. 1523-1531.

14. Ji, M., et al., "CFD investigation on the flow structure inside thermo vapor compressor," *Energy*, vol. 35, no. 6 (2010), pp. 2694-2702.

15. ANSYS Fluent Theory Guide, 2013: FLUENT Inc.

16. Desevaux, P., "A method for visualizing the mixing zone between two co-axial flows in an ejector," *Optics and Lasers in Engineering*, vol. 35, no. 5 (2001), pp. 317-323.

Design of Miniature Bellow in a Conceptual Joule-Thomson Cooler

F. Can[1], A. B. Erdogmus[1], M. Baki[1], M. K. Aktas[2], M. A. Güler[2]

[1] Department of Mechanical Design, ASELSAN INC,
MGEO Cankiri Yolu 7. Km Akyurt/Ankara/TURKEY
[2] Department of Mechanical Engineering, TOBB University
of Economics and Technology, Ankara/TURKEY

ABSTRACT

Flow control is an important issue in the solution of many engineering problems. Engineers may use different methods depending on the system size, energy consumption requirement, actuator type, etc. Since the gas source is limited in open cycle Joule-Thomson (JT) applications, the gas flow rate must be minimized to have a long operation time. When flow regulation is not considered, the flow cross sectional area in an expander must be made small in order to decrease the gas flow rate and thus extend operating time. But this situation also has some disadvantages because of the longer temperature drop time. In the literature, most of the JT coolers employ miniature bellows as a flow controller unit to maintain sustainability of the gas source. One end of these bellows, whose either inside or outside is filled with high pressure gas, is connected to a rod. The rod, which is also called as the sensing rod, is in contact with the gas liquefied by the Joule-Thomson effect. Temperature information on the liquefied gas is sensed and transmitted through the walls of the bellows to the interior of the bellow which is filled with high pressure gas. The pressure of gas inside the bellow drops while the temperature decreases. The motion of the sensing rod causes the flow cross-sectional area to decrease as a result of the temperature decrease. The length and cold side temperature of the JT cryocooler affects the bellow's surface temperature distribution, thus its internal pressure. In this study, the contraction of the bellow due to temperature decrease is investigated in a numerical way for different lengths and cold side temperatures.

INTRODUCTION

Expansion or contraction of materials is generally undesirable in engineering designs. Only tolerated displacements of materials are acceptable. However some kinds of displacements in materials may be useful when flow control with respect to flow temperature is needed. Minimizing excessive mass flow rate of cryogenic gas is the objective of a flow control unit in Joule-Thomson cryocoolers. Therefore, the flow control unit is one of the important components of advanced open cycle Joule-Thomson cryocoolers.[1] Without flow control, the cooler is designed either with a low flow rate, which is required for steady state and the temperature drop time is long, or it is designed with a high flow rate, which is required to have short temperature drop time; but the operating time of the cooler becomes shorter. The control mechanism is designed in order to use a limited gas source effectively. For this purpose, different types of control mechanisms can be used in Joule-Thomson coolers. In this application, it is observed that miniature bellows with different dimensions have been used frequently. Flow control can also be implemented utilizing bimetal materials

Cryocoolers 19, edited by S.D. Miller and R.G. Ross, Jr.

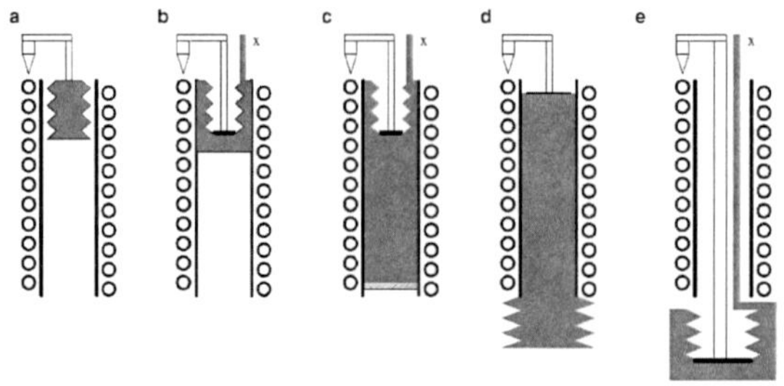

Figure 1. Different type of pressurized bellow using in JT cryocoolers[2].

that depend on the differences in expansion properties of two combined materials. These materials have a simple design. Since the spring rate of bimetals are higher, they may withstand higher drag forces and the cold end temperature stability is higher than bellows. Also, the thermal mass of bimetals is larger than bellows.[1] As a result, the temperature fluctuation is less, and the gas consumption is also lower. In spite of these advantages of bimetals, bellows are preferred in commercial products because of having faster response time. In addition, bellows are less sensitive to clogging.[2]

Bellows are generally produced by special coating techniques because of having small dimensions. This production method is known as electroforming. It is expensive, especially for very small sized bellows. It also includes a needle valve and a rod. When gas begins to liquefy, the rod attached to the bellow senses the liquefying temperature. From this time until full contraction of the bellows can be called the bellows response time. It changes between 5 to 10 seconds.[1] Bellows can be positioned at cold side (as shown in Figure 1 a, b and c) or hot side (as shown in Figure 1 d and e). Bellows can be pressurized from inside (as shown in Figure 1 a and d) or outside (as shown in Figure 1 b, c and e).[2] Since there is a sensing rod attached to the bellow, the thermal mass changes depending on the location of the rod. For a flow control unit with a long sensing rod, as long as shown in Figure 1 e, the thermal mass of the cooler will be high. It can be useful if the length of the cooler is very short.[2]

It is found that most of the studies in the literature on JT coolers do not focus on the flow control. Chien et al.[4,5] studied a Joule-Thomson cryocooler with a bellow flow control mechanism. Nitrogen gas is used as the working gas in their research, and the cryocooler includes a miniature bellows pressurized with nitrogen. Chien[4,5] not only studied the behavior of the bellow but also solved the effect of other components. In the Chien study[3], the cooling time of a cryocooler, which does not involve any flow control unit, is investigated experimentally and theoretically. In their other study[4], cooling time and the change in mass flow rate of a JT cryocooler with a flow controller is examined. It is obvious that the temperature of the gas outside the mandrel or bellow has an important effect on the determination of bellows behavior. A remarkable difference between the first[3] and second[4] studies is examination of different orifice structures. In their second study[4], the mass flow rate is decreased 95 percent with regard to its initial value. It is noted that with a cryocooler with a flow controller, the operation time can be 20 times increased after reaching a steadystate condition. In another study of Chien[5], the condensation of the gas inside the bellow is taken into consideration and a sudden pressure drop is observed at the condensation point. They determined that the amount of contraction of bellows is higher when condensation is considered. It is validated that modelling condensation has an important effect on the design of a flow control unit. In addition, the effect of the rigidity of the bellows on contraction rate is researched. The heat transfer is considered between the bellow and cold end because of contact. But the heat transfer between bellow and mandrel at the hot side is ignored. In their research[3], the temperature of the contact region between the bellow and the mandrel at the hot side is above 200 K. Therefore, a considerable amount of heat is transferred from this region. Their studies[4] demonstrate that the temperature distribution is between 83 and 90 K at the steadystate condition. It can be noted that the temperature range is wide in the real case because of the contact between bellow and mandrel at the hot side.

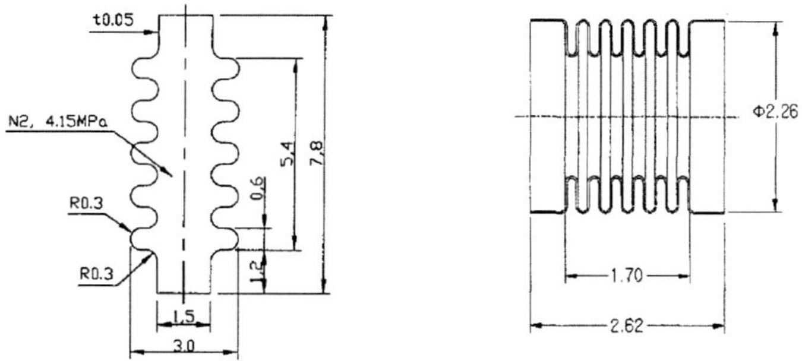

a[6] b[7,8,9]

Figure 2. Bellow dimensions in Lee's studies[6,7,8,9].

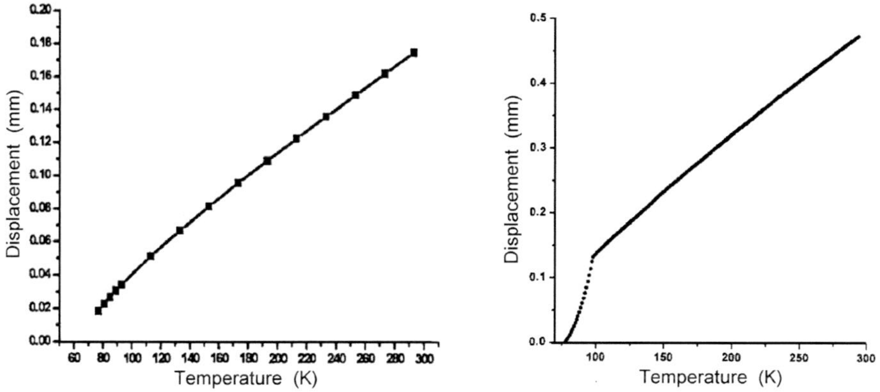

a[6] b[8]

Figure 3. Axial deformation of bellows in Lee's studies[6].

Another series of studies was performed by Lee, et al.[6,7,8,9] In those studies, only pressurized bellows were examined, and the contraction and expansion of the bellow due to different pressures and temperatures of gas were determined experimentally and numerically. The dimensions of bellows used in their studies are given in Figure 2.

The only available numerical analysis is for the bellow given in Figure 2 a. In that study[6], the pressure drop depends on the temperature decrease in nitrogen gas and is modelled by the BWR method, and then the contraction of the bellow is calculated. In their other research[7], the bellows shown in Figure 2 b is used. Different pressures are tried for the same temperature, and the amount of contraction is calculated for different points. Deformation analysis is performed by a commercial finite element code ANSYS. The bellows are modeled by 3-node axisymmetric shell elements with reduced integration. Finally, experiments were performed to validate the conducted numerical analyses with experimental values. In another study[8], the dependency of pressure drop on temperature was researched similar to their first study. The only difference stems from the gas model. The Modified Benedict Webb Rubin (MBWR) method was used in that study. When results are interpreted, it is seen that the deformation in the study[6] is nearly linear, since the BWR model does not reflect the phase change in gases as shown Figure 3 a. However, the MBWR method models the two phase region more accurately as compared to the BWR method. Therefore, nonlinear deformation is seen below the condensation temperature as shown Figure 3 b. Finally, they conducted an experimental study to support their last research.[9]

Table 1. Bellow dimensions

	Value (mm)
Inner diameter (d_{bi})	2.2
Outer diameter (d_{bo})	2.3
Thickness (t_b)	0.025
Length (L_b)	10
Number of pitch (N_b)	10

DEFORMATIONANALYSES

In this article, a bellow is studied as a flow control unit. Displacement of the bellow mainly arises from an internal pressure change. Also, a temperature decrease causes a desired movement in the bellow in addition to changing its pressure. As with the bellow, the temperature of the mandrel also decreases. The thermal expansion coefficient of the bellow and mandrel materials also affects the contraction due to the temperature drop. When the pressure drop of the gas as a result of temperature decrease is also taken into consideration, superposition of these two effects fulfills the function of the flow control unit. Flow restriction is accomplished by the rod which senses the liquefaction temperature at the exit pressure of the high pressure gas. The rod is attached to the cold side of the bellow. When liquefaction is accomplished, the needle moves with the rod towards the orifice. In this study, displacement in the bellow is modelled by numerical methods in three steps: analysis performed by in-house code, commercial finite element, and finite volume solvers (ANSYS and FloEFD).

In the first step, since three dimensional models and real gas models of Argon are complicated, the temperature distribution on the mandrel is calculated with a numerical approach. The MBWR[11] method based on NIST data is used to model the real gas. In-house code written in MATLAB is developed to solve the differential equations using a finite difference method.

Bellows with different pressures can be utilized in real applications. Therefore, the effect of inside pressure is studied to understand the behavior of the bellows. The temperature distribution and internal pressure of the gas under steadystate conditions are solved in step 2. The temperature distribution of the mandrel, which is the result of the first step, is used as a boundary condition. To define the gas inside the bellow, the default real gas model of nitrogen in FloEFD is used. When iteration times are taken into consideration, the first and second steps are completed in 30 and 150, respectively.

The calculated final pressure value is used in the next step. The bellow stiffness is calculated for the dimensions presented in Table 1. Since FloEFD is not used for structural analysis, finite element software ANSYS R 15 is preferred to calculate the displacement in the bellow. A 2D axisymmetric model with PLANE 183 as a discretization element is used. PLANE 183 has a quadratic displacement behavior and is suited for modeling irregular meshes. Pressure is defined as a surface load. PLANE 183 elements are defined by 8 nodes.

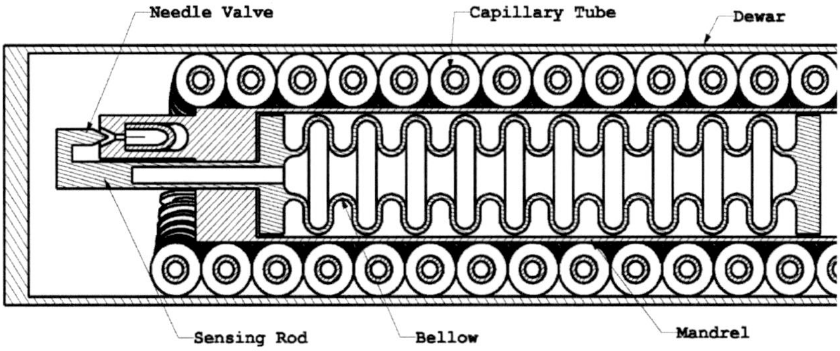

Figure 4. Three dimensional schematic model of JT

Table 2. Other dimensions

	Value (mm)
Length of JT (L)	50
Pitch of helical coil (P_c)	1
Pitch of fin (P_f)	0.3
Mean diameter of helical coil (D_{hel})	3.5
Fin diameter (d_f)	1
Fin thickness (t)	0.1

Table 3. Main dimensions

	Inner Diameter (i) (mm)	Outer Diameter (o) (mm)
Capillary tube (d_c)	0.3	0.5
Mandrel(d_m)	2.3	2.5
Shield (Dewar)(d_s)	4.5	4.8

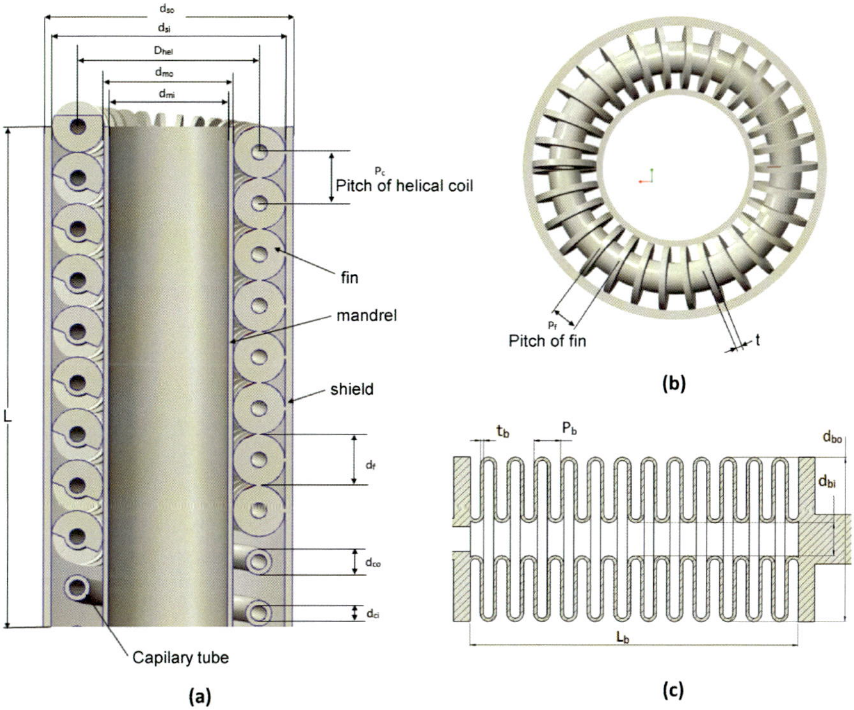

Figure 5. Schematic view of dimensions used in analysis

Models

A sectional view of whole system is demonstrated schematically in Figure 4. The cold side is located at the left side in this figure. The hot side of the bellow is assembled to the mandrel at the right side.

In Figure 5, details of the parameters used in the numerical calculations are given. Figure 5a and b show important parameters used in the first step calculations. Parameters shown in the cross-sectional view of the bellow (Figure 5 c) are input design values and used in dimensions given in the second and third steps. All dimensions used in the solution of the problem are presented in Table 1 - 3. The 3D model shown in Figure 5 is simplified to a 1D model with 500 nodes in the first step. In the second step, the three dimensional problem is solved with 570,000 cells. In the final step, the problem is solved with 130,000 elements.

Table 4. Boundary Conditions

Step	Location	Boundary Conditions
1	x=0	$T_{h,1} = T_{h,in},\ \frac{dT_s}{dx} = 0,\ \frac{dT_w}{dx} = 0, \frac{dT_m}{dx} = 0,\quad P_{h,1} = P_{h,in}$
	x=L	$T_{c,imax} = T_{c,in},\ \frac{dT_s}{dx} = 0,\ \frac{dT_w}{dx} = 0,\ \frac{dT_m}{dx} = 0,\quad P_{c,imax} = P_{c,in}$
2	x=0	$T_{m,in}$
	x=L	$T_{m,out}$
	0≤x≤L	Temperature Distribution on Mandrel
3	-	Internal Pressure

Table 5. Equations used in the calculations

Name	Equations
Continuity	$\frac{d\dot{m}}{dx} = 0$
Momentum equations	$\frac{dP}{dx} = -\rho.V\left(\frac{dV}{dx} + 2.\frac{fV}{d}\right)$
Hot fluid energy	$\dot{m}_h C_{p_h}\frac{dT_h}{dx} = h_h p_h (T_w - T_h)$
Cold fluid energy	$\dot{m}_c C_{p_c}\frac{dT_c}{dx} = h_c[p_c(T_c - T_w) + p_{si}(T_c - T_s) + p_{mo}(T_c - T_m)]$
Capillary Tube energy	$k_w A_w \frac{d^2T_h}{dx^2} = h_h p_c (T_h - T_w) + h_c p_c (T_c - T_w)$
Mandrel	$k_m A_m \frac{d^2T_s}{dx^2} = h_c p_{mo}(T_c - T_m)$
Dewar (Shield)	$k_s A_s \frac{d^2T_s}{dx^2} = h_c p_{si}(T_c - T_s) + \sigma\, p_{so}(T_a^4 - T_s^4)$

Table 6. Other equations

Fanning friction factor for Hot fluid[2]	$f = 0.046\left(1 + \frac{3.5d_{ci}}{D_{hel}}\right) Re^{-0.2}\ for\ Re > 30000$
Cold fluid Fanning friction factor[10]	$f = (0.088 + 0{,}16X_L\ (X_T - 1)^{-n})\frac{2}{Re^{0.15}}$
Hot fluid heat convection[10]	$h = 0.023C_{p_h}G\left(1 + \frac{3.5\ d_{ci}}{D_{hel}}\right) Re^{-0.2}\ \mathrm{Pr}^{-2/3}\ for\ Re > 10000$
Cold fluid heat convection[10]	$h = 0.26C_{p_c}G\ Re^{-0.2}\ \mathrm{Pr}^{-2/3}$

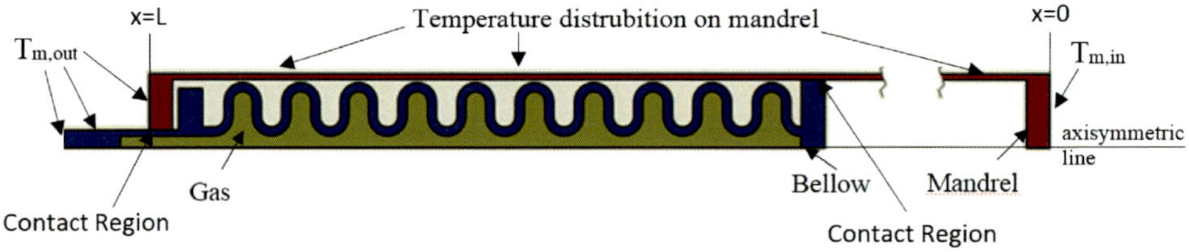

Figure 6. Boundary conditions and axisymmetric model of Step 2

Table 7. Material specification of nickel used in FEM analysis[12]

Property	Value
Young modulus (MPa)	160992
Position ratio	0.28
Yield strength (MPa)	754.8
Tensile strength (MPa)	861.8

Table 8. Comparison of numerical solution of Step 1 (L=50mm) with experimental data[13]

Pressure (bar)		Temperature(K)		Mass flow	Temperature(K) ($T_{c\ out}$)	
P_{h_in}	P_{c_in}	T_{h_in}	T_{c_in}	(SLPM)	Experiment[12]	Simulation
140.47	1.3426	291.94	90.06	10.145	284.98	285.01

Details of Analyses

The surface temperature distribution of the mandrel is found in the first step for steadystate conditions. In this step, boundary condition 1 in Table 4 is used to solve the continuity, momentum and energy equations given in Table 5. To solve the momentum and energy equations needs friction factors and heat transfer coefficients; equations for Fanning friction factors and heat transfer coefficients are required. Coefficients used frequently in the literature are given in Table 6.

Boundary conditions and the contact region between mandrel and bellows for the second step analysis are shown in Figure 6. Nickel whose material specifications[12] are detailed in Table 7 are used as the bellow material in the third step.

RESULTS AND DISCUSSION

In this section, results obtained in the three steps are summarized and some of them compared with the results published in the literature. In the first step, the problem is solved with the same boundary conditions and geometry studied by Chua et al.[13] In Table 8, the first five columns are input values for the presented numerical study. In the last two columns, experimental data for the cold fluid outlet temperature are compared with numerical results. It is seen that the two values are very close to each other. Also, this simulation is repeated with different conditions presented in the study of Chua et al.[13] Similar results are determined for other conditions. After that, the temperature distribution of the mandrel is calculated for different heat exchanger lengths and different cold end temperatures as illustrated in Figures 7 and 8. These results are used as a boundary condition for the second step in the calculation of the final pressure of the gas inside the bellow.

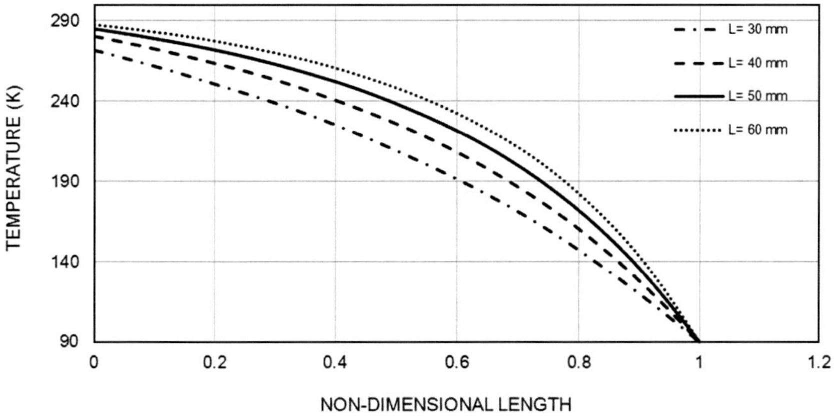

Figure 7. Temperature change along the mandrel in different heat exchanger length (Step 1)

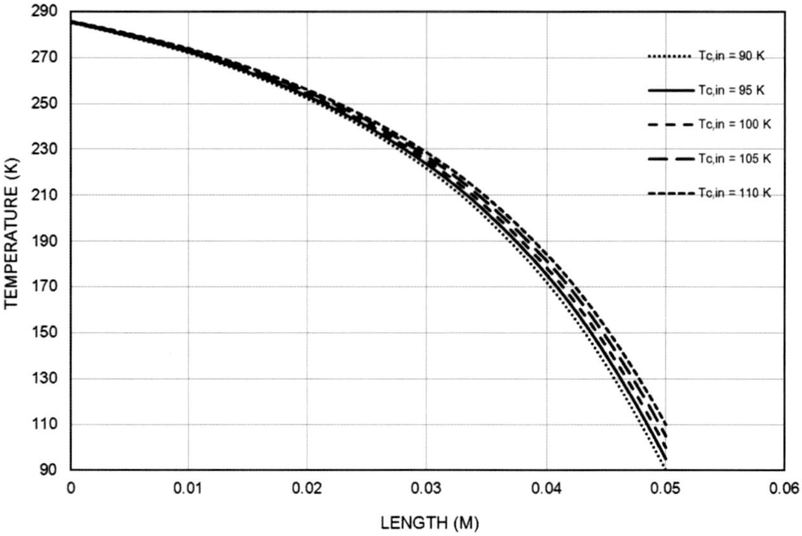

Figure 8. Temperature distribution in different cold side temperatures (L=50 mm, Step 1)

Table 9. Temperature distribution and internal pressure of gas in steady state condition for different length heat exchanger. $P_{gas,0}$=30 bar (at 295K), $T_{c,in}$=90.

L (mm)	$T_{gas,f}$(K)	$P_{gas,f}$(bar)
30	90 < T < 184.35	11.63
40	90 < T < 179.72	11.46
50	90 < T < 177.33	11.37
60	90 < T < 175.91	11.31

Table 10. Temperature interval, internal pressure of gas and deformation of bellow in steady state condition for different initial internal pressure (L= 50 mm, $T_{c,0}$=90).

$P_{gas,0}$ (bar)	$T_{gas,f}$(K)	$P_{gas,f}$(bar)	Δx (μm)
20	90 < T < 177.34	7.98	93.8
25	90 < T < 177.34	9.70	119.4
30	90 < T < 177.33	11.37	145.3
35	90 < T < 177.33	12.93	172.2

The final pressures and gas temperatures in the bellow, which is calculated in the second step, are given in the third column of Table 9. Bellow stiffness is calculated as 128.18 Bar/mm in the third step. Initial pressure of gas inside the bellow should be lower than 37.2 bar for safety operation of the bellow. This is the buckling pressure for the investigated bellow. With the calculated stiffness factor and final pressures, the deformations shown in Table 10 are calculated.

Evaluation of results are listed below.

- Temperature distribution of gas inside the bellow covers large interval when it is compared with the results given in the literature.
- The heat transfer at the hot side of bellow has important effect on inside pressure of bellow.
- The bellow should be well isolated from the hot side of the mandrel to prevent local heat transfer
- The results of this study may be taken into consideration in orifice design
- Buckling pressure should be taken into consideration in design of operating conditions. If the pressure is greater than buckling pressure, the bellow can lose its functionality.

- Needle attached to bellow moves a maximum of 172 μm with the presented bellow dimensions. This result should be taken into consideration in orifice design.

REFERENCES

1. Buelow, P.L., Ryba, E.L., Skertic, M.M., "Fast Response Joule-Thomson Cryostat", US Patent Application Publication No. US 5,913,889, filled August 20, 1996, published June 22, 1999, Hughes Electronics, Los Angeles, California.
2. Maytal, B.Z., Pfotenhauer, J.M., *Miniature Joule-Thomson Cryocooling*, Springer, New York (2013), p. 73, 220,225.
3. Chou, F.C., Pai C.F., Chien, S.B., Chen J.S., "Preliminary Experimental and Numerical Study of Transient Characteristics for a Joule-Thomson Cryocooler," *Cryogenics*, vol. 35 (1995), p. 311-316.
4. Chien, S.B., Chen, L.T., Chou, F.C., "A Study on the Transient Characteristics of a Self- Regulating Joule-Thomson Cryocooler," *Cryogenics*, vol. 36, no. 12 (1996), pp. 979-984.
5. Chou, F.C., Chien, S.B., "Two-phase coexistence analysis of the Bellows Control Mechanism for a J-T Cryocooler", *Cryogenics*, vol. 39 (1999), pp. 359-365.
6. Lee, S.E., Lee, T.W., "Deformation Analysis of Self-regulating Bellows in Joule-Thomson Cryocooler," *Journal of the Korean Society for Precision Engineering*, vol. 25, no. 4 (2008).
7. Lee, S.E., Lee, T.W., "A Study on the Structural Characteristics of Miniature Metal Bellows in Joule-Thomson Micro-Cryocooler," *Journal of the Korean Society for Precision Engineering*, vol. 25, no. 9 (2008).
8. Lee, S.E., Lee, T.W., "Deformation Analysis of Miniature Metal Bellows Charged Nitrogen for Temperature Change to Cryogenic Condition," *Journal of the Korean Society for Precision Engineering*, vol. 26, no. 10 (2009), pp. 81-88.
9. Lee, S.E., Lee, T.W., "Study on Deformation of Miniature Metal Bellows in Cryocooler Following Temperature Change of Internal Gas," *Transactions of the Korean Society of Mechanical Engineers*, vol. 39, no. 4 (2015), pp. 429-435.
10. Timmerhaus, K.D., Flynn, T.M., *Cryogenic Process Engineering*, Springer, New York (1989), p. 194-198.
11. Younglove, B.A., "Thermophysical properties of fluids 1 Argon, Ethylene, parahydrogen, nitrogen, nitrogen trifluoride, and oxygen," *Journal of physical and chemical reference data* 11(1), (1982) pp 1-8.
12. Servometer company, "Electroformed Metal Bellows Catalog," Servometer.com
13. Chua, H.T., Wang, X., Teo, H.Y., Ng, K.C., "A Numerical Study of the Hampson-Type Joule-Thomson Cooler," *Int. Refrigeration and Air Conditioning Conference* 2004, pp. R149-1.

A Turbo-Brayton Cryocooler for Future European Observation Satellite Generation

J. Tanchon[1], J. Lacapere[1], A. Molyneaux[2], M. Harris[2], S. Hill[2], S.M. Abu-Sharkh[2], T. Tirolien[3]

[1]Absolut System SAS, Seyssinet-Pariset, France
[2]Ofttech, Gloucester, United Kingdom
[3]European Space Agency, Noordwijk, The Netherlands

ABSTRACT

Several types of active cryocoolers have been developed for space and military applications in the last ten years. Performances and reliability continuously increase to follow the requirements evolution of new generations of satellite: less power consumption, more cooling capacity, increased life duration.

However, in addition to this increase of performances and reliability, the microvibrations requirement becomes critical. In fact, with the development of vibration-free technologies, classical Earth Observation cryocoolers (Stirling, Pulse Tube) will become the main source of microvibrations on-board the satellite.

A new generation cryocooler is being developed at Absolut System using very high speed turbomachines in order to avoid any generated perturbations below 1000 Hz. This development is performed in the frame of ESA Technical Research Program - 4000113495/15/NL/KML.

This paper presents the status of this development project based on Reverse Brayton cycle with very high speed turbomachines.

INTRODUCTION

The space cryogenics sector is characterized by a large number of applications (detection, imaging, sample conservation, propulsion, telecommunications etc.) that lead to various requirements (temperature range, microvibration, lifetime, power consumption etc.) that can be met by different solutions (Stirling or JT coolers, mechanical or sorption compressors etc.).

The recent years saw, in Europe, the developments of coolers to meet Earth Observation mission requirements that are capable to provide significant cooling power at an operational temperature around 50K (for IR detection). Those single stage Stirling or Pulse Tube coolers are mechanisms that involve moving parts at low frequency (in the compressor and in the Cold Finger for the Stirling) which induce microvibrations. Even if active microvibration cancellation and careful screening and manufacturing of the coolers parts can lead to low exported microvibrations (< 100 mN in all directions at all harmonics), more and more stringent system level induced microvibration requirements can lead to very complex solutions to overcome those vibrations (e.g. suspended coolers and radiators, flexible thermal link assemblies that degrade the overall thermal performances).

Cryocoolers 19, edited by S.D. Miller and R.G. Ross, Jr.

With the development of vibration-free technologies (magnetic bearing reaction wheels, microthrusters), classical Earth Observation cryocoolers (Stirling, Pulse Tube) will become the main source of microvibrations on satellites. To overcome this difficulty and cope with missions that have more and more stringent pointing performances requirements, it is necessary to develop alternative cooling solutions or to adapt current technologies to generate no mechanical disturbances.

The objectives of the activity presented in this paper are to design, manufacture and test an Elegant Breadboard Model (close to an Engineering Model) of a "Vibration-Free" cooler that can provide active cooling for temperatures in the range of 40-80 K in order to answer the needs of potential future Earth Observation IR missions.

DEFINITION OF THE PROPOSED SOLUTION

In order to meet the performances requirement and particularly the vibration-free constraint, a reverse Turbo-Brayton cooler concept has been proposed. The Turbo-Brayton cycle cryocooler uses miniature high-speed turbomachines and high-effectiveness recuperators, to provide efficient cooling with low vibration and high reliability. Gas bearings are used in the miniature machines to support the rotors, which operate at speeds of 100,000 to 600,000 rpm. The low-mass rotors are the only moving parts in the systems, and because they are precision balanced, the systems are inherently vibration-free. No supplemental vibration canceling electronics or hardware is required. The gas bearings also provide non-contact operation, so performance degradation resulting from wear or the accumulation of debris is absent. These systems are generally capable of maintenance free operating lives of 5 to 20 years.

Turbo-Brayton cryocoolers may be arranged in a number of configurations to meet a variety of cooling requirements. They are continuous-flow systems made up of appropriate numbers of compressors, expansion turbines, recuperative heat exchangers, and thermal interfaces. These components may be integrated into a compact package or distributed over fairly large areas, interconnected by lengths of tubing. Refrigeration can be delivered to multiple loads at either a single temperature or several different temperatures. The second type of delivery can be accomplished either by multistaging an integral cooler or by combining several cryocoolers at the appropriate interfaces. Cooling loads and thermal interfaces may be separated by large distances without significant effects on overall system efficiency. Thus, the turbo-Brayton cryocooler can be implemented in a variety of ways in space applications.

The Turbo-Bratyon cycle is also known to have a very good efficiency and is suitable for providing refrigeration over a broad range of temperatures. It is generally more efficient at higher loads and temperatures but operation as low as 4.2 K has been demonstrated. Decreases in performance in coolers of low capacity (less than 1 W) are primarily a function of the manufacturing technologies used in producing the small turbomachines and the recuperators. System mass and thermodynamic performances are both strongly influenced by specific design features in these components.

The in-orbit operation of these technologies during more than 10 years represents a successful demonstration of their maturity [1][2]. Today, new developments are running in the US through Creare LLC. to cover a large range of operating temperature and cold power. These new developments are also oriented to increase again the turbo-Brayton cooler efficiency by increasing the rotational speed of turbomachines up to 10,000 rev/s (600,000 rpm) [3].

The Turbo-Brayton is thus the most relevant technology to fulfill:

- a nominal operation along the required temperature range (40 K and 80 K)
- a vibration free requirement because of the the rotational speed larger than 2000 Hz which is far above the critical bandwith required for the system [0 – 1000 Hz]

The architecture of the reverse turbo-Brayton cycle proposed is presented in Figure 1.

This schematic constitutes the baseline architecture of the cooler. The centrifugal compressor is heat sunk on a 300 K radiator (the radiator surface required is below 1m^2). Then, the counter-flow heat exchanger is used to precool the gas before entering the turbine where it is expanded. The cold gas goes through a heat exchanger to be connected to the load to be cooled and goes back to the counter-flow heat exchanger up to compressor. This concept is the simplest way to build a turbo-Brayton cooler.

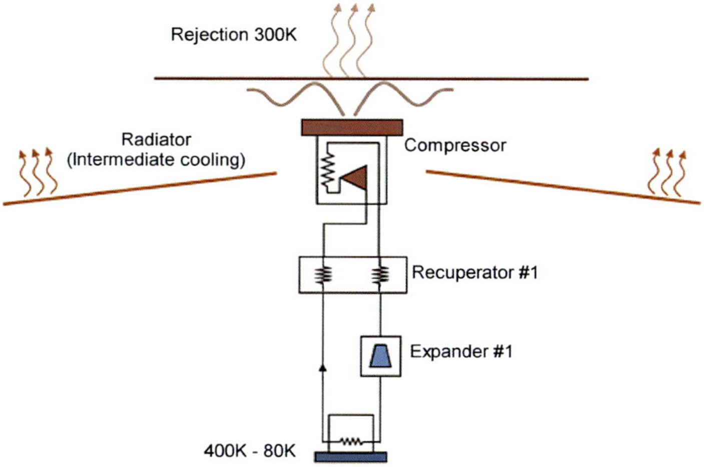

Figure 1. Baseline architecture of the 40-80K vibration-free Brayton cooler

SYSTEM TRADE-OFFS AND ANALYSIS

In order to optimize our cryocooler concept, several points have been analyzed during the first phase of the development. The different aspects confronted during the trade-off phase were focusing on the technologies to be used and the overall architecture of the system. As part of these different trade-offs, the following points have been discussed and assessed.

Number of compressor stages

The efficiency of the compressor is clearly a driver in this trade-off due to the strong impact of this parameter on the cryocooler input power. The other aspects analyzed in this trade-off are: the mass, the reliability, the modularity of the system, the manufacturing and integration issues, and the on-ground tests complexity.

Regarding the efficiency, the same isentropic efficiency can be reached by the single stage or two-stage compressor. The main difference is that the rotational speed drastically increases for the one stage compressor. An efficiency larger than 0.5 is reached at 240K rpm using two impellers. With only one impeller, rotational rate of 380K rpm is needed to reach the same efficiency of 0.5.

It has been estimated that a 2-stage compressor will operate at lower rotational speed and thus will lower the technical risks for this first development without significant impact on the compressor performance.

Compression stage arrangement

We have seen that the baseline compressor design is a 2-stage compressor. However, several options exist to arrange these 2 stages :

- The in-line arrangement is more compact but results in lower compressor efficiency due to the relatively sharp inter-stage flow distribution. As a result of these flow losses, the aerodynamic stage efficiency of an in-line two-stage compressor is lower than the single stage machine.
- In a back-to-back compressor design, each stage can be designed as a single stage machine with corresponding high stage efficiency. One of the advantages of the back-to-back configuration is obviously in term of rotor dynamics and axial load. The compressor is more balanced in this configuration. However, the flow distribution will impact the performances even if the figures should be lower than in the in-line arrangement.
- Finally the separated architecture is an arrangement with two separated compressors.

This last option (separated architecture) has been identified to be most desirable due to:

- Its modularity forease of integration.
- An optimum circulation of fluids (so optimal aerodynamic efficiency).
- Two identical designs. In our case, the impellers and the motors are strictly identical and interchangeable.
- Possibility to arrange easily inter-stage cooling (efficiency improvement).

This last arrangement concept strongly decreases the complexity of the rotor integration and will allow a more efficient management of the high precision parts.

Choice of the working fluid

Based on the operating temperature of the cooler, the working fluid needs to be optimized. In the expected operating temperature range, neon and helium can be used as a pure fluid or a mixture. Several configurations have been analyzed with pure neon as the baseline. An optimum has been found by analysis on the proportion of helium added in the pure neon. With 20 % of Helium, the cooling power increases by about 20% at 40 K but needs an increase of the compressor rotational rate up to 300K rpm (instead of 250K rpm). In order to limit the technical risks and to be conservative on the cooling power, pure neon is used for detailed design.

Recuperator technology

Different technologies have been used for the recuperator. The recuperator is one of the most critical component regarding the overall cycle efficiency. Thus, the choice of the technology to be used for our application needs to be carefully addressed. The objective of this trade-off is to confront different technologies identified as good candidates for our needs and then to evaluate their potential for the final cryocooler.

The different technologies identified are:

- The micro channels heat exchangers
- The parallel plate heat exchangers
- The metallic foam heat exchangers
- The microtube heat exchangers

For each technology, a preliminary design has been proposed and the associated performances determined. These inputs have been considered for the trade-off and the selection of the nominal technology. The baseline concept uses a microtube recuperator concept as proposed by Mezzo Technologies [4], which offers the best ratio of efficiency to mass.

SYSTEM DESCRIPTION

Following the trade-off activity run in parallel of the component design and optimization, the preliminary design of the 40-80K reverse Brayton cooler has been consolidated.

The 40-80 K vibration free cooler is composed of the following sub-systems as seen in Figure 2:

- A 2-stage centrifugal compressor.
- A microtube recuperator.
- A cryogenic expander (including a cold electrical generator).
- An external supporting structure.

2-stages centrifugal compressor

The compressor consists in a pair of identical units shown above in series formation. Each units contains a centrifugal compressor wheel optimized to be operated at 250,000 rpm, supported on gas bearings and driven by a permanent magnet motor. The neon gas flows through each compressor stage; the first is a low pressure stage and the second a high pressure stage.

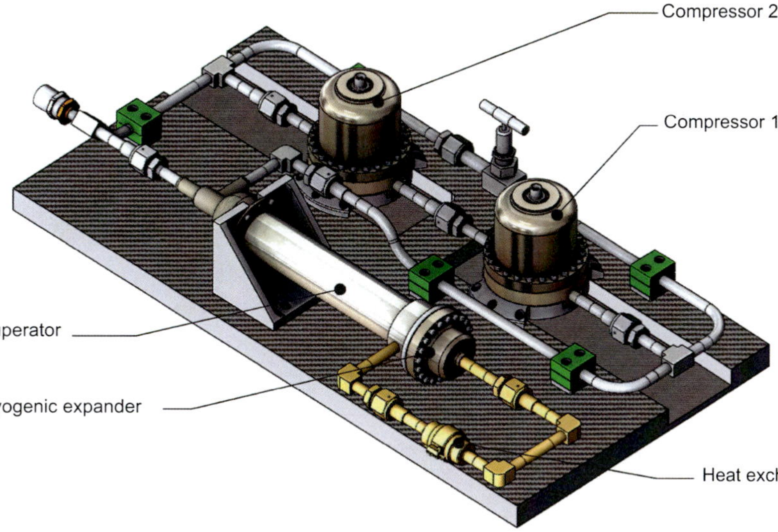

Figure 2. Overview of the 40-80 K vibration free cooler

Figure 3 details the internal structure of the centrifugal compressor. The compressor is composed of a Ti6-Al4-V cartridge where the rotor and stator components are mounted. The cartridge is then mounted in the external housing and sealed with metallic seal. Due to the relatively small change in density between stages, it was realized that one design of compressor could be applicable for both stages with minimal decrease in efficiency. This would simplify the design and manufacture process and require a smaller parts inventory. To allow this, approximately equal enthalpy rises are required on each stage, as the impeller will produce a fixed loading. Pressure ratios are then split between the two stages at about 1.40 and 1.32.

Rapid evaluation of design changes were made by using the internal 1D tools and final design was consolidated using CFX (CFD code). The design features backswept blades to achieve a high relative velocity ratio (and hence good surge margin), a vaneless space diffuser in order to also maximize the flow range, followed by a volute and conical exit diffuser, to maximize the static pressure recovery.

The full stage CFD model is shown in Figure 4, and full stage CFD predictions were carried out for both stages over a range of flows, and combined to produce an overall characteristic at 250,000 rpm. A peak overall total-static efficiency of about 57 % is predicted, while the overall pressure ratio target of 1.85 is exceeded slightly with both stages running at 250000 rpm. It is expected

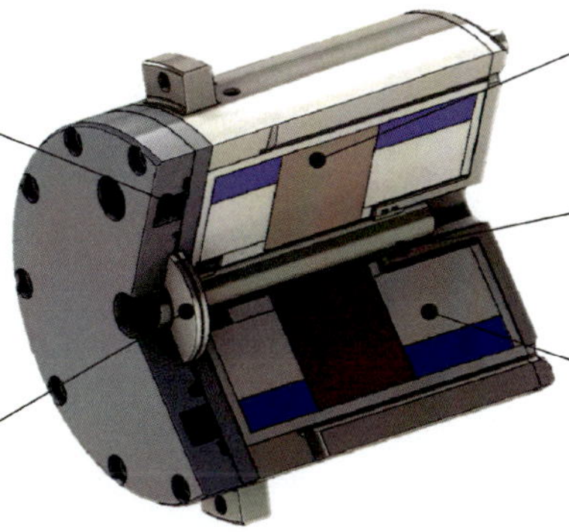

Figure 3. View of the compressor cartridge

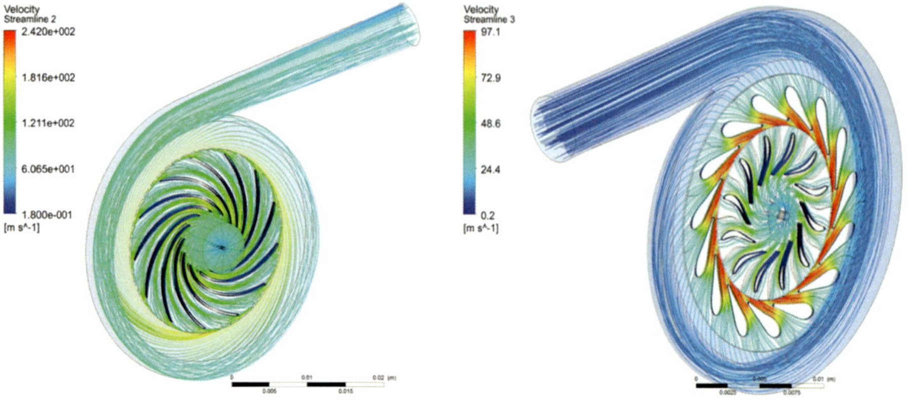

Figure 4. CFD Model (using CFX) showing streamlines (compressor on the left, expander on the right)

that further optimization at the detailed design stage may improve the efficiency slightly, and that the target of 59 % should be achieved.

The gas bearings to be used here are of the spiral groove type. These are identical in form to those used both in the MELFI project and Atlas Copco Helium expander [5], all designed using the same methods by the same experts. The journal bearings are herringbone grooved with a plain portion between symmetric patterns. These gas bearings guarantee stability and load capacity and require no adjustment or setting up once manufactured. The Figure 5 shows a typical spiral groove journal bearing.

The motor design achieves low values of electromagnetic loss in all forms (core loss, ohmic winding loss, flux harmonic and other stray sources), combined with low windage loss, which in combination are unusually low for the power rating and speed, due to careful attention to the proportions of component parts and choice of materials. The main features of the motor are:

- A cylindrical two-pole rare-earth magnet contained within a tubular shaft of material that is mechanically stiff and strong, non-magnetic, and with at least moderately high resistivity. The surface of the shaft also provides the rotating part of the gas bearings at each end of the machine.
- A multi-tooth stator core containing a 3-phase winding

Following the estimation of the different losses of the compressor, the overall efficiency has been analyzed and a global powertrain efficiency of 47% is estimated considering the different contributions regarding the electrical losses, windage losses, bearings and impeller losses.

Figure 5. Typical spiral groove journal and thrust bearing

Microtube recuperator

The recuperator is based on microtube heat exchanger technology. One advantage of this technology is that the recuperator is quite compact compared to other technologies. The recuperator has a cylindrical shape with the junctions on the two extremities. On the warm end, the recuperator is equipped with a flange which is the mechanical interface between the recuperator and the supporting structure. On the cold end, the expander is directly attached to the recuperator.

This configuration avoids any additional mechanical supports for the expander and thus lowers the thermal losses. The recuperator is shown in Figure 6. It is composed of an external shell made with stainless steel tube with two caps at the extremities where the piping is welded. Inside the shell, microtubes are welded to the end-plates.

The main characteristics of the recuperator are:

- External shell diameter: 41 mm
- Overall length (without piping): 365 mm
- Number of microtube: 2200
- Total mass: 0.81 kg
- Thermal effectiveness: 98.5 %
- Pressure loss (High pressure side): 1000 Pa
- Pressure loss (Low pressure side): 1500 Pa

Cryogenic expander

The cryogenic expander is dedicated to extract power from the fluid flow in order to decrease its temperature. This extraction is made by the gas expansion through a turbine wheel with power absorbed by a permanent magnet generator.

This cryogenic expander is composed of the following key mechanical parts:

- A rotor equipped with a wheel, a shaft and a magnet: the fluid expansion produces a mechanical work with an amount of power transmitted to the rotational shaft.
- A generator (acting like an electromagnetic brake): the shaft power produces then some electrical power in the electrical generator. This electrical power corresponds to the power extracted from the gas. It is then dissipated at the warm end of the cryocooler.
- A stator housing including the gas bearings and supporting plates, the expander housing and its electrical feedthrough, the volute with piping junction

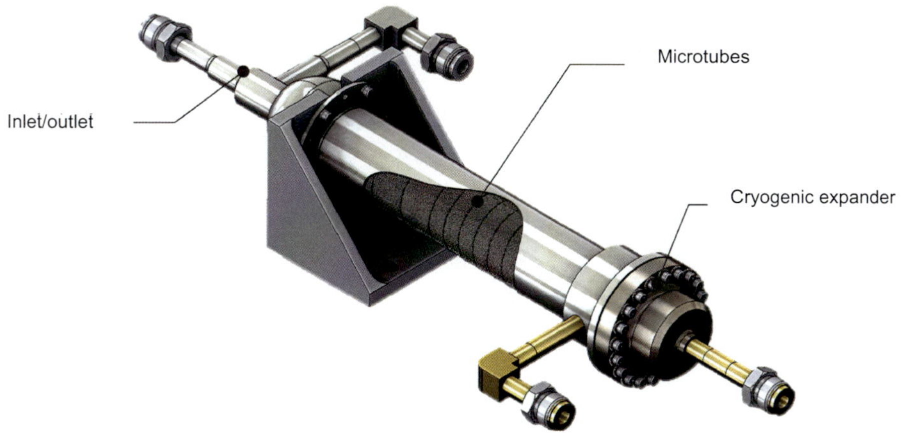

Figure 6. Recuperator design with integrated cryogenic expander

In order to reduce risks, it is intended to use similar techniques and solutions for both compressor and expander sub-systems. The conclusions of trade-offs concerning the compressor design are then directly implemented into the expander design.

The expander consists of a single turbine stage driving the electrical generator. The nominal running speed is 150,000 rpm but this is variable in order to allow control of the starting procedure and for fine tuning of the operating point. The internal architecture of the expander is detailed in the Figure 7.

As with the compressor, a full stage CFD model of the turbine has been constructed and run. The design was iterated in 1-D models and then to CFX to achieve the target flow at the design pressure ratio. The CFD results were adjusted for disc windage and leakage, and resulted in a stage total to static efficiency of 71.4 % - slightly above the target of 71%.

As per the motor of the compressor, the design details of the generator have been developed using Synthet-PM, a design program for high-speed surface-magnet machines developed over many years by OFTTECH and proven by extensive use for commercial machines over a wide range of power and speed. The optimum design is then validated through 3D time-varying finite-element study using ANSYS RMxpert [6] plus ANSYS Maxwell 3D [7], which computes power losses in all parts of the machine together with voltage-current terminal characteristics.

Following the estimation of the different losses of the turbine, the overall expansion efficiency has been analyzed and estimated in the range of 64% (electrical losses, windage losses, bearings and impeller aerodynamic losses).

BREADBOARDING ACTIVITIES

In order to mitigate the different technical risks highlighted during the preliminary design phase, different breadboard activities have been run. These activities concern different aspects of the cooler from system level analysis down to elementary component testing or manufacturing technics validation.

System level analysis validation

The objective of this activity is to perform a critical review of the system level analysis. Several assumptions in the process analysis can have a major influence on the cryocooler design. The objective herein is to perform an independent critical review of the thermodynamic cycle characteristics and trade-off on various performance features selected by Absolut System.

To do that, Creare LLC collaborates with Absolut System as reverse turbo-Brayton expert and their support has been used during the first part of the development phase in that way. During the detailed design phase, Creare LLC is involved in the review of critical design of cryocooler's components and associated manufacturing issues.

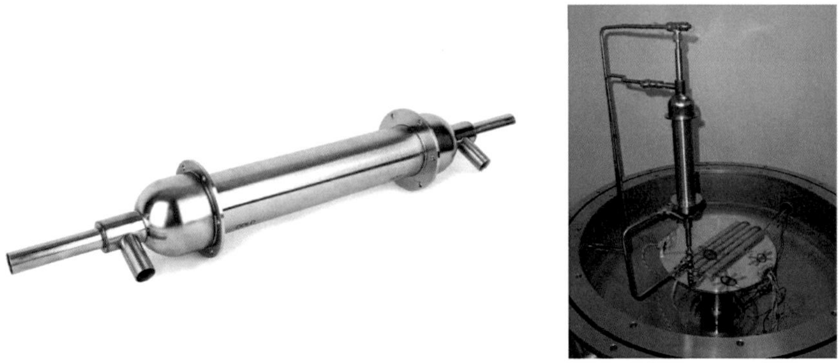

Figure 7. Picture of the recuperator prototype and test set-up in vacuum test chamber

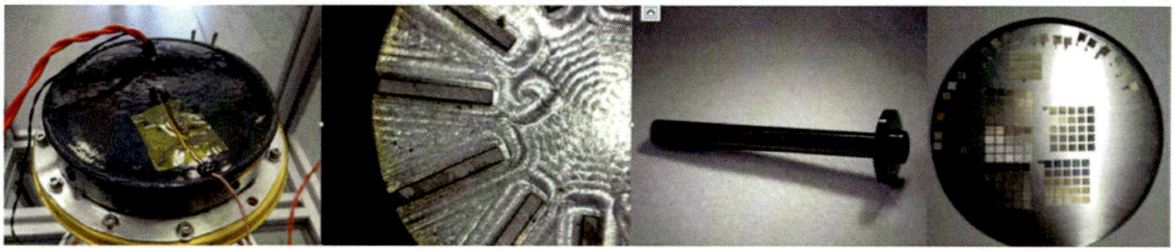

Figure 8. Breadboard samples: stator sample, dummy impeller, dummy shaft for hard coating, grooves manufacturing optimization

Recuperator performance evaluation

As detailed in the system trade-off study, several technologies could be used for the recuperator. A baseline has been selected using microtube heat exchangers and has been evaluated as promising. This technology has been prototyped in order to evaluate the real performances and to validate predicted performances (CFD models)

A prototype has been designed and manufactured to test this technology. The Figure 8 shows the prototype of recuperator manufactured by Mezzo Technologies and its implementation in Absolut System test facilities for characterization.

Stator material characterization

The efficiency of the expander is extremely critical for the cooler performances. However, it is extremely difficult to find in technical literature the electrical properties of conventional materials at cryogenic temperature range.

The evaluation of the internal losses of the generator is then linked to a good empirical understanding of the different losses and particularly permeability and loss density for stator materials. In parallel of the design of the component, a characterization test campaign has been run by Ofttech and Absolut System to measure these characteristics on different materials at different cryogenic temperatures. This yielded to several interesting datum on permeability and loss density for materials of choice, at room temperature and down to 40K (for turbine generator use).

Manufacturing breadboard tests

In complement of the design validation using breadboard tests, some manufacturing aspects like the bearing spiral grooves, the hard coatings on the rotor or the diffusion bonding of the impeller have been validated through dedicated breadboard tests.

CONCLUSIONS AND OUTLOOK

A challenging development programs is on-going in the frame of an ESA contract to provide an alternative to regenerative cryocoolers for future European Earth observation missions. A Turbo-Brayton cooler design has been developed using highly skilled engineers involved in this project with the support of US colleagues.

The detailed design of the 40-80K vibration-free cooler is bound to be finalized after an intensive breadboard test campaign used to lower technical risks and to mature as far as possible the manufacturing design of this cryocooler.

Thanks to this design, our vibration free cooler should be able to produce 1 W @ 40 K to more than 5 W @ 80 K with less than 180 W of electrical power input, satisfying ESA requirements.

ACKNOWLEDGEMENTS

This development is performed in the frame of ESA Technical Research Program - 4000113495/15/NL/KML.

The authors thank Creare LLC for their advice and their support during the development phase, MEZZO Technologies for their work performed on the high efficiency recuperator, and the intensive work performed by OFTTECH Limited on the components design.

REFERENCES

1. Nellis, G., Dolan, F., McCormick, J., Swift, W., Sixsmith, H., Gibbon, J. and Castles, "Reverse Brayton Coolers for NICMOS," *Cryocoolers 10*, Plenum Publishing Corp., New York (1999), pp 431-438.
2. Petrivelli, A., "The ESA laboratory support equipment for the ISS," *ESA Bulletin(0376-4265)*, (109), (2002), p. 35.
3. Zagarola, M.V., McCormick, J.A. and Cragin, K.J., "Demonstration of an Ultra-Miniature Turboalternator for Space-Borne Turbo-Brayton Cryocooler," *Cryocoolers 17*, ICC Press, Boulder, CO (2013), pp 453-460.
4. Becnel, C., Lagrone, J. and Kelly, K., "Micro Channel Recuperator for Reverse Brayton," *Cryocoolers 15*, ICC Press, Boulder, CO (2009) pp.397-404.
5. Molyneaux, A.K. and Leonhard, M., "The Use of Spiral Groove Gas Bearings in a 350 000 rpm Cryogenic Expander," *Tribology Transactions,* Volume 32, Issue 2 (January 1989), pp. 197-204.
6. ANSYS RMxprt, software component of Ansys/Ansoft (USA).
7. Maxwell, A.N.S.Y.S., "*3D v15 User's Guide*", Edition 6.0, ANSIS Inc., March 2012.

Performance Testing of a High Effectiveness Recuperator for High Capacity Turbo-Brayton Cryocoolers

D. Deserranno[1], M. Zagarola[1], D. Craig[2], R. Garehan[2], T. Giglio[2], J. Smith[2] J. Sanders[3], M. Day[3]

[1]Creare, Hanover, NH 03755, USA
[2]Mezzo Technologies, Baton Rouge, LA 70814, USA
[3]Edare, Lebanon, NH 03766, USA

ABSTRACT

NASA is considering multiple missions involving long-term cryogen storage in space. Liquid hydrogen and liquid oxygen are the typical cryogens as they provide the highest specific impulse for chemical propellants. The net heat load to a hydrogen tank storing 38 metric tons of liquid hydrogen is estimated to be 20 W at 20 K. To enable long-duration zero boil-off storage, this heat load must be lifted using an active refrigerator. Creare is addressing this need by developing a turbo-Brayton cryocooler that provides 20 W of refrigeration at 20 K.

The key components in the 20 K, 20 W cryocooler are the electronics, compressors, turboalternator, and recuperators. The recuperator for the 20 K, 20 W cryocooler is a five-module micro-shell-and-tube heat exchanger developed jointly by Creare, Mezzo Technologies and Edare. It has a predicted effectiveness exceeding 0.995, enabling the cryocooler to deliver 20 W of refrigeration for an input power around 1.6 kW, corresponding to a specific power of about 80 W/W. This is significantly better performance than any 20 K cryocooler existing or (to our knowledge) under development. In preparation of the cryocooler integration, a recuperator module was tested between 300 K and 20 K using helium at design flow rates. The test results indicated the recuperator exceeded its target effectiveness.

INTRODUCTION

NASA is evaluating long-term liquid hydrogen and oxygen storage in low Earth orbit to support the chemical propulsion needs of future missions. Researchers at NASA investigated a co-storage concept for these cryogenic propellants [1]. Here the liquid oxygen tank at 90 K intercepts a large fraction of the heat load to the hydrogen tank at 20 K. The shielding provided by the oxygen tank can be augmented using an actively cooled shield. As a result, the radiative heat flux to the hydrogen tank reduces from 0.25 W/m^2 to 0.02 W/m^2. The net heat load to a hydrogen tank storing 38 metric tons of liquid hydrogen is estimated to be 20 W at 20 K, including design margin. To enable long-duration zero boil-off storage, this heat load must be lifted using an active refrigerator. Unfortunately, this heat load exceeds the capacity for any space cryocooler demonstrated to date.

Creare is addressing this shortcoming by developing for NASA a turbo-Brayton cryocooler that provides 20 W of refrigeration at 20 K [2]. Turbo-Brayton cryocoolers are ideal for long-term liquid

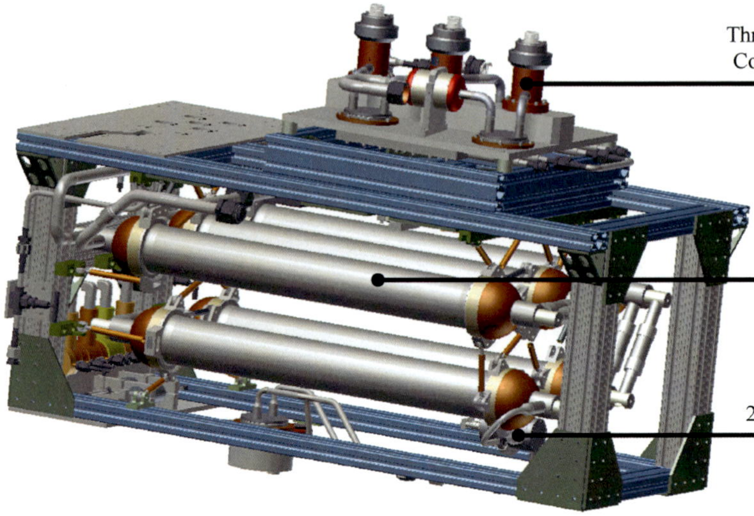

Figure 1. Creare's 20 K, 20 W cryocooler for liquid hydrogen storage.

hydrogen storage in space because of favorable mass and performance scaling to high capacities and low temperatures. In addition, the continuous flow nature of the cycle provides a means to directly interface with a Broad Area Cooling (BAC) system attached to the hydrogen tank without the mass and performance penalties associated with ancillary cryogenic heat pipes or circulation loops. The cryocooler designed for this application is shown in Figure 1. It delivers 20 W of cooling at 20 K for an input power around 1.6 kW, corresponding to a specific power of 80 W/W. The coefficient of performance of the cryocooler is 18% of the Carnot cycle. This is significantly better performance than any 20 K cryocooler existing [3] or (to our knowledge) under development.

The key mechanical components in the 20 K, 20 W cryocooler are the compressors, turboalternator, and recuperators. The compressors [4] and turboalternators [5] have been previously demonstrated at the required capacity and only minor modifications are required for use in the 20 K, 20 W cryocooler. While many recuperators have been space qualified, none have the capacity required for a 20 K 20 W cryocooler. Indeed, most turbo-Brayton cryocoolers operating near or below 20 K have a capacity below 1 W [6]. As a result, an order of magnitude increase in the cryocooler and recuperator capacity is required. Unfortunately, using many lower capacity recuperators in parallel leads to significant packaging, costs, and flow distribution concerns. A new type of recuperator is required for high capacity turbo-Brayton cryocoolers. The design, fabrication, and testing of a high capacity recuperator is the subject of this paper.

SIGNIFICANCE AND JUSTIFICATION FOR DEVELOPMENT

After an initial trade study, Creare pursued the development of a micro shell-and-tube heat exchanger to serve as the recuperator in the 20 K, 20 W cryocooler and enlisted a development team comprising Creare, Mezzo Technologies, and Edare. While conventional shell-and-tube heat exchangers are common in industry, their mass is prohibitive for space applications due to the large size tubes. The use of micro-tubes allows for significant mass savings, as a large amount of area can be packaged within a small volume. Given prior experience at Mezzo Technologies, the development team opted to pursue a unit using 0.022 inch diameter micro-tubing. The primary technical challenge is to fabricate this unit with 100% hermetic joints between the high and low pressure streams.

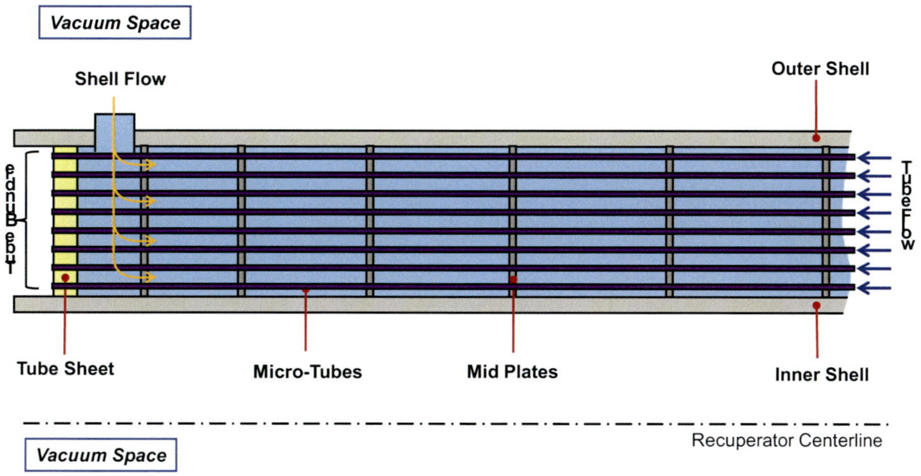

Figure 2. High capacity recuperator schematic.

RECUPERATOR DESIGN AND ANALYSIS

A recuperator was designed based on the requirements of the cryocooler, resulting in the need for over 6,600 micro-tubes having an outer diameter of 0.022 inch, each over 150 inches long. Due to the excessive length, the recuperator was divided into 5 serial modules for ease of fabrication and packaging. Each module is nominally 40 inches long, including its headers. The heat transfer zone within the module is 30 inches long. The cross section of the core is annular in nature with an inner diameter of 2 inches and an outer diameter of 4 inches (illustrated in Figure 2). Various thermal-fluid analyses were performed using computational fluid dynamics to assess the performance of the micro-tube heat exchanger and the impact of various design features on flow distribution within the core. The core overall effectiveness is assessed taking into account local flow maldistribution effects, resulting in regions of local flow capacitance imbalance. The nominal operating conditions of the recuperator are 3.6 g/s of helium; tube-side inlet pressure of 7.9 atm and inlet temperature of 300 K, and shell-side inlet pressure of 5.6 atm and inlet temperature of 20 K.

The predicted recuperator loss of the five-module recuperator is 17.7 W, including parasitics. The recuperator loss is defined as the cold end stream to stream enthalpy difference multiplied by the recuperator mass flow rate. This loss must be absorbed by the turboalternator of the cryocooler, in addition to the cryocooler heat load. The parasitics depend highly on the packaging and configuration of the cryocooler. For the configuration shown in Figure 1, the recuperator assembly is mounted to a 300 K platform and the conductive and radiative parasitics are predicted to be 5.8 W. The thermal performance of the individual modules (including conductive and radiative parasitics) is shown in Table 1. The variation in the log mean temperature difference is the result of real-gas effects of helium near 20 K and thermal parasitic. The decrease in module UA (conductance) as

Table 1. Predicted recuperator module performance with parasitic losses.

	Mod 1	Mod 2	Mod 3	Mod 4	Mod 5
Warm Temperature (K)	300.0	248.7	203.4	146.1	80.4
Cold Temperature (K)	248.4	203.0	145.6	79.7	20.0
Minimum Stream Capacity (W/K)	18.6	18.6	18.6	18.6	18.9
Log Mean Temperature Difference (K)	0.235	0.258	0.372	0.525	0.748
UA (W/K)	4082	3310	2905	2364	1521
Total Heat Transfer (W)	960	854	1081	1242	1138
Effectiveness	0.987	0.992	0.993	0.992	0.989

the recuperator temperature decreases is the result of decreasing gas thermal conductivity which decreases the heat transfer coefficient.

In addition to thermal-fluid analyses, structural analyses were conducted to assess the mechanical performance of the unit in a launch environment. Specifically, the recuperator was designed to meet the requirement of NASA's General Environmental Verification Standard (GEVS), as described in NASA TM GSFC-STD-7000A, a standardized set of environmental exposure standards for payloads, subsystems, and components. The analyses performed were (1) a hydrostatic pressure analysis and (2) a random vibration analysis. Pressure stresses were computed for a 16 atm maximum design pressure (MDP). Analysis was performed at MDP and stresses were compared against allowable values per ASME B31.3 (incorporating proof pressure conditions and design margins). Random vibration analysis was performed consistent with GEVS random vibration test levels for a 90 kg system based on the expected mass of the cryocooler. The random vibration analysis assumed a 1% dampening ratio (Q=50). The recuperators exhibited margin relative to the yield stress under all analysis conditions.

RECUPERATOR FABRICATION

The key step in the fabrication of the heat exchanger is the joining of the micro-tubes to the tubesheets. Based on analysis, it can be shown that significant performance degradation can be caused by even a few non-hermetic joints between the high and low pressure stream. For this specific design, a 99.8% weld joint reliability is required to ensure the cryocooler delivered 95% of its design cooling capacity. Laser welding was pursued to join the micro-tubes to the tubesheets. The primary advantage of laser welding is the ability to repair joints, allowing the fabrication of 100% hermetic recuperators.

The weld reliability of each joint is dependent on the laser settings as well as the geometric features of the joint. Therefore, laser welding process parameters were developed through an iterative process. Sectioning of weld joints was performed in small sample sets to identify potentially suitable conditions. Once reasonable geometric features and laser parameters were established, a reliability study was performed by welding thousands of joints with the desired settings. An iterative process was used until the resulting weld joint reliability was sufficiently high (>99.8%) such that repair options were practical.

To aid the welding process, off-axis cameras were installed to enable real-time pre- and post-inspection of the weld joint (Figure 3). This allows for the inspection of critical weld joint features such as the protrusion of the micro-tube above the tubesheets prior to completing the welding process. Furthermore, a visual inspection of the completed joint allows for the real-time identification of leaky joints. An example of pre- and post-welding images is shown in Figure 3. In the pre-weld condition, we can visualize the alignment of the laser and the weld joint (left image) and the pro-

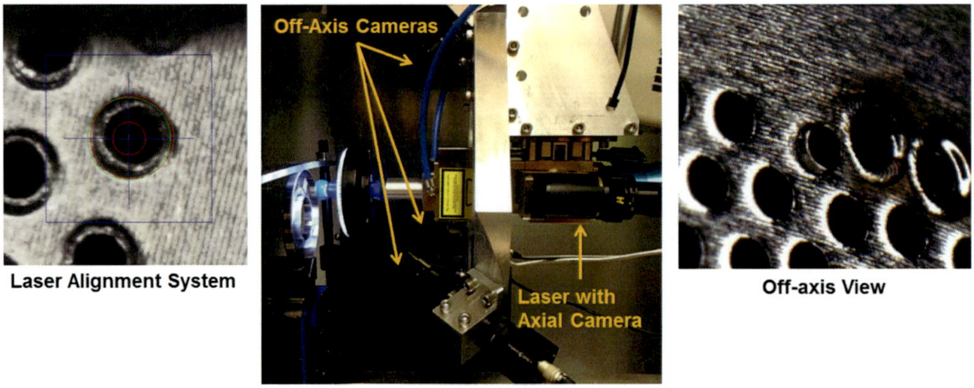

Figure 3. Micro-tube laser welding system.

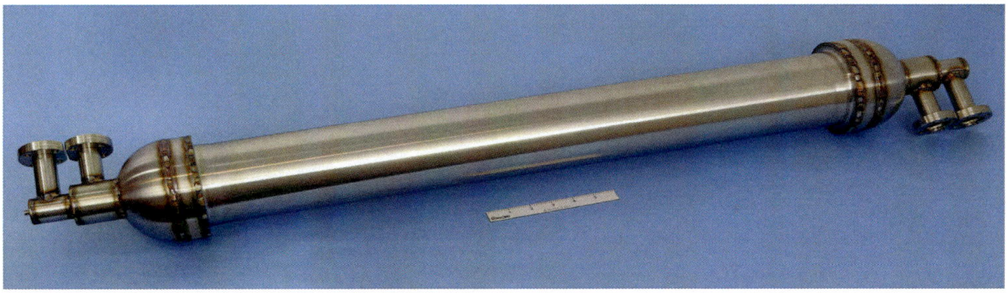

Figure 4. Completed recuperator following proof-pressure testing.

trusion of the tube above the tubesheet (right image). In the post-weld condition, we can see the completed weld joint, allowing us to identify potential flaws in the joint.

When both tubesheets are welded, a bubble-point leak check is performed to identify non-hermetic joints. These joints are subsequently repaired until a fully hermetic unit is obtained. Headers are then welded to the recuperator core, and the unit is proof-pressure tested based on the requirements of the cryocooler. The completed recuperator module is shown in Figure 4. Following fabrication, a variety of workmanship screening tests are performed, including external leak check, cross-stream leak check, and pressure versus flow tests. The unit passed all tests.

RECUPERATOR THERMAL TESTING

The recuperator was tested under cryogenic conditions to evaluate its thermal effectiveness. While the expected operation of each recuperator module spans a smaller temperature range, the resulting stream-to-stream temperature difference is too small to ensure adequate fidelity in our test due to measurement uncertainty. Therefore, the stream-to-stream temperature difference is artificially increased to allow us to more reliably characterize the recuperator performance. To accomplish this, the warm end of the recuperator was held at room temperature, while the cold end was operated between 20 K and 50 K. The test facility schematic is shown in Figure 5. Temperatures are measured at the inlets and outlets of the recuperator using redundant, in-line Platinum Resistance Thermometers (PRTs), and the flow rate is measured using a spring-loaded rotameter. The facility

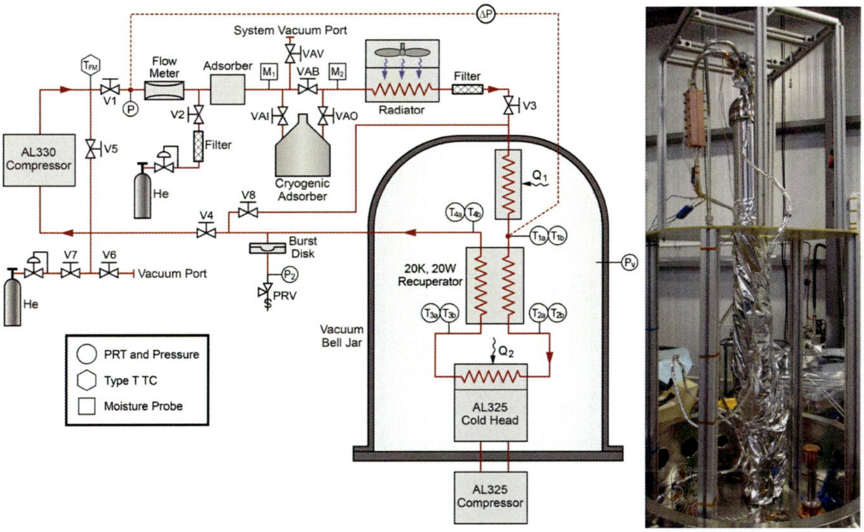

Figure 5. Cryogenic thermal test facility system.

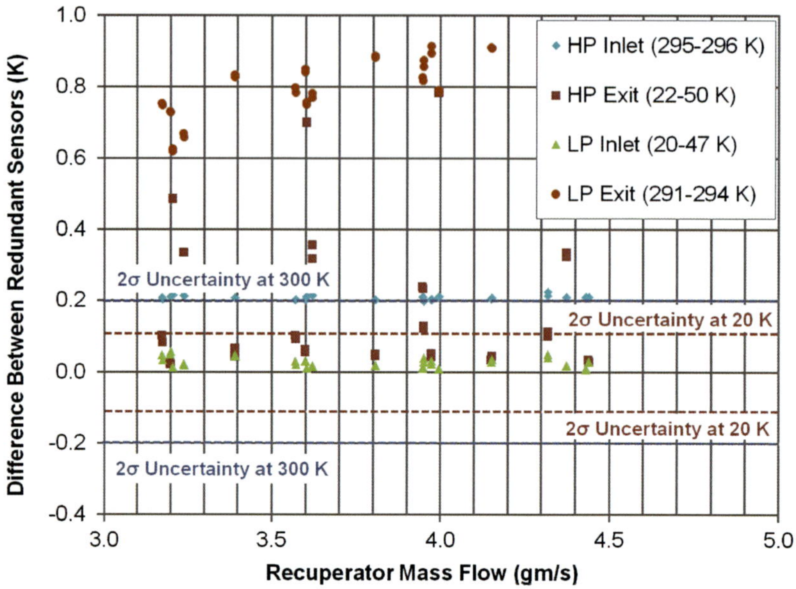

Figure 6. Agreement between redundant temperature sensors.

operates as a closed-loop, allowing us to test for extended time periods. The recuperator is covered with Multi-Layer Insulation (MLI; 20 layers at the cold end and 3 layers at the warm end) to limit parasitic losses to below 1 W, which is a small fraction of the total heat transfer rate within the unit (nominally 5 kW). Based on the instrumentation accuracy, the anticipated errors are ±3% for the ineffectiveness, and ±4% for the UA product.

A total of 19 steady-state test points were collected for a nominal warm temperature of 295 K, and a cold temperature ranging from 20 K to 50 K. The heat transfer between the two streams varied between 4.2 kW and 6.2 kW, depending on the test conditions. The heat imbalance between the two streams averaged 6.0 W or 0.11%, and was within the accuracy of the sensors. Figure 6 illustrates the agreement between the redundant temperature sensors at the inlets and exits of the recuperator. The inlet sensors have a strong agreement (within the 2σ uncertainty band) at both the warm and cold end of the recuperator. This is to be expected due to the temperature uniformity in the incoming flow. The exit sensors demonstrate some level of disagreement, likely due to stratification in the exit flow as a result of minor flow maldistribution effects.

The thermal ineffectiveness results are shown in Figure 7 as a function of mass flow rate. Throughout the development of the recuperator technology, various iterations were performed on the design and assembly process. As illustrated in Figure 7, each iteration significantly reduced thermal ineffectiveness. Ultimately, our first production unit achieved the single module target ineffectiveness of 1.2% at the design flow rate of 3.6 g/s. Ineffectiveness has a weak correlation to the recuperator cold-end temperature, but strongly correlates with the mass flow rate through the unit. As flow rate is increased, the ineffectiveness increases slightly as demonstrated by our test data and predictions.

RECUPERATOR VIBRATION TESTING

In addition to thermal performance testing, a recuperator module was vibrated in both the axial direction and one transverse direction (Figure 8). Due to symmetry of the recuperator and the fixturing, no testing was done in third direction. The recuperator was tested with the NASA GEVS spectrum, modified for a system mass of 90 kg. Testing was performed at +3 dB, 0 dB, and +3 dB relative to this spectrum. The 3 dB, 0 dB, and +3dB tests corresponds to 5.2 Grms, 7.4 Grms, and

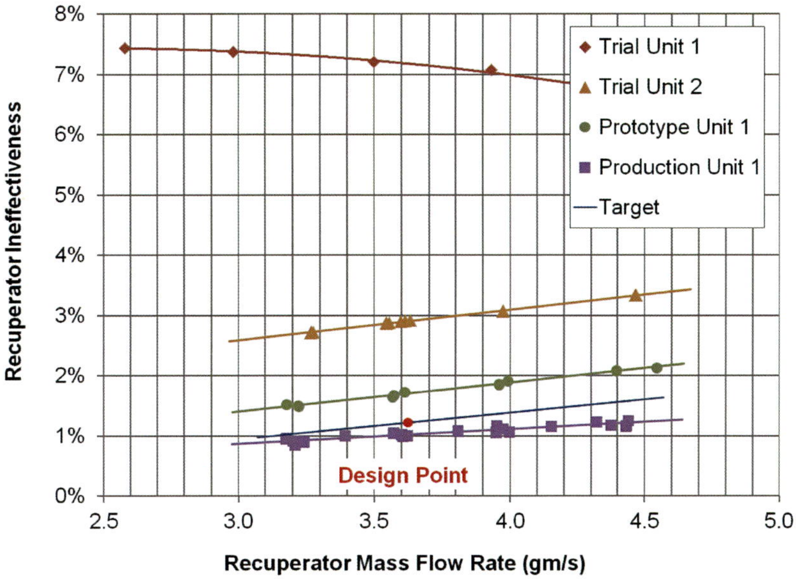

Figure 7. Thermal ineffectiveness of the recuperators throughout the development process.

10.4 Grms, respectively. Due to shaker table limitations, the +3 dB test was limited to 9 Grms. Using the modified GEVS spectrum, the following profiles were applied:

1. Axial direction, 3 dB below specification, for 3 minutes
2. Axial direction, at specification, for 3 minutes
3. Axial direction, at specification, for 27 minutes
4. Axial direction, 3 dB above specification, for 3 minutes
5. Lateral direction, 3 dB below specification, for 3 minutes
6. Lateral direction, at specification, for 3 minutes
7. Lateral direction, at specification, for 27 minutes

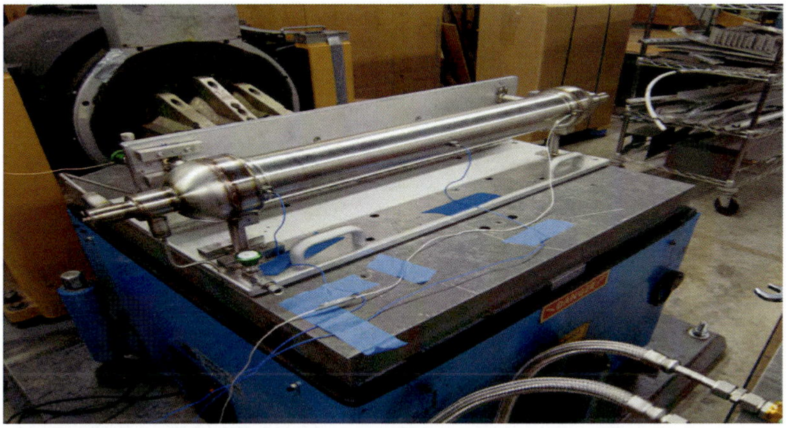

Figure 8. Vibration test configuration for the transverse tests on the prototype unit.

The 27-minute tests are substantially longer than those required by GEVS, and were accidental. Prior to testing, characterization tests were performed to determine the damping. The damping was found from the width of the response curve at the half power points. The measured quality factor in the lateral direction was Q=18 at 160 Hz resonant frequency, and in the axial direction was Q=28 at 135 Hz resonant frequency. The quality factor was considerably less than assumed in the analysis (Q=50), so notching at the resonant frequencies was not required. Before and after each vibration test, a 0.5 G sinusoidal sweep, external leakage, and cross-stream leakage tests were performed to assess recuperator failure. The recuperator module passed all random vibration testing including the extended duration testing.

CONCLUSIONS

Creare, in collaboration with Mezzo Technologies and Edare, developed a lightweight, high-capacity, high-effectiveness recuperator technology, suitable for use in space missions. This paper discussed the design, fabrication, and testing of the recuperator. A 5-module recuperator has a predicted effectiveness exceeding 0.995, enabling a 20 K cryocooler to deliver 20 W of refrigeration for an input power around 1.6 kW, corresponding to a specific power of about 80 W/W. In addition to thermal performance testing, a unit was vibration tested successfully under space launch conditions. Future work involves building the remaining recuperator modules, integrating them into a cryocoolers, and testing the cryocooler.

ACKNOWLEDGMENT

We gratefully acknowledge NASA for their support of this work (Contracts NNX13CL57P and NNC14CA15C).

REFERENCES

1. Mustafi, S., Canavan, E.R., and Boyle, R., "Co-Storage of Cryogenic Propellants for Lunar Exploration," *AIAA 2008-7800*, AIAA SPACE 2008 Conf & Expo, San Diego, CA (2008).

2. Deserranno, D., Zagarola, M.V., Li, X. and Mustafi, S., "Optimization of a Brayton Cryocooler for ZBO Liquid Hydrogen Storage in Space," *Cryogenics*, vol. 64, (2014), pp. 172–181.

3. Cha, J.S. and Yuan, S.W., "Space Cryocooler Vendor Survey Update: 2013," Aerospace Report TOR-2013(3905)-4, The Aerospace Corporation, El Segundo, CA (2013).

4. Hill, R.W., Hilderbrand, J.K. and Zagarola, M.V., "An Advanced Compressor for TurboBrayton Cryocoolers," *Cryocoolers 16*, ICC Press, Boulder, CO (2011),

5. Zagarola, M.V., Cragin, K.J. and Deserranno, D., "Demonstration of a High-Capacity Turboalternator for a 20 K, 20 W Space-Borne Brayton Cryocooler," *Adv. in Cryogenic Engineering*, Vol. 59, Amer. Institute of Physics, Melville, NY (2014), pp. 1432–1437.

6. Breedlove, J.J., Cragin, K.J. and Zagarola, M.V., "Testing of a Two-Stage 10 K Turbo-Brayton Cryocooler for Space Applications," *Cryocoolers* 18, ICC Press, Boulder, CO (2015),

Performance Testing of a High Capacity Compressor for a 20K 20W Cryocooler

K. Cragin, D. Deserranno, M. Zagarola

Creare LLC
Hanover, NH 03755

ABSTRACT

Creare is currently developing for NASA a high capacity 20 K turbo-Brayton cryocooler for long-term cryogen storage in space. The cryocooler utilizes three compressors with intercooling to compress and circulate the helium cycle gas. The compressors operate on self-acting journal bearings, are driven by brushless permanent magnet motors at speeds exceeding 6,000 rev/s, and have a nominal input power capacity of 500 W. As a result of the high speed and flow rates, the predicted aerodynamic efficiency is extremely high for all stages, resulting in net efficiencies of greater than 60% for the three compressor stages. This is significantly higher than prior permanent magnet motor compressors developed by Creare which have shown peak efficiency of around 50%. The high compressor efficiency results in a cryocooler predicted input power below 1.6 kW, corresponding to a specific power of 80 W/W. This performance is significantly better than any 20 K cryocooler existing or (to our knowledge) under development. In preparation for cryocooler integration, the compressors were tested at 300 K using helium at design flow rates. This paper describes the design, fabrication, and initial test results for these high-capacity compressors.

INTRODUCTION

To support long-duration space exploration missions, NASA must store cryogenic propellants in space for long periods of time. However, heat radiated from the sun and the Earth causes the liquid cryogen tanks to pressurize. Currently, to control the pressure inside a cryogenic storage tank, the boil-off is vented, releasing otherwise useful cryogen into space. While cryogen boil-off is acceptable for short-duration missions, long-term storage of cryogen in space requires a system to avoid loss of the propellant.[1] One solution to this problem is to employ a cryocooler to intercept the heat load on the tank and maintain the cryogen at a constant temperature. Due to the large size of the cryogen tanks, the cooling capacity is large for space cryocoolers, exceeding the current state-of-the-art by an order of magnitude.

Creare is currently developing for NASA a high capacity 20 K turbo-Brayton cryocooler (Fig. 1) that will be used in a demonstration of a zero-boil-off hydrogen system. At high cooling loads, the turbo-Brayton cycle has a clear advantage over other cycles in terms of performance, size, and mass.[2] In addition, the continuous-flow nature of the Brayton cycle is ideal for cryogen storage missions where the cycle gas can be directly interfaced with a Broad Area Cooling (BAC) system attached to the storage tank. On a recent program, Creare delivered a single-stage turbo-Brayton cryocooler to NASA/GRC to support a ground demonstration of a zero-boil-off liquid oxygen storage system and a reduced-boil-off liquid hydrogen storage tank.[3] This cryocooler provides up to 20 W of refrigeration at 90 K. The cryocooler currently under development (and subject of this paper) provides cooling at 20 K to enable zero-boil-off liquid hydrogen storage. The cryocooler utilizes three compressors with intercooling to compress and circulate the helium cycle gas. This paper reviews our progress on developing a next-generation compressor for turbo-Brayton cryocoolers. Key developments include a high-capacity 500 W motor and advanced aerodynamics for high efficiency.

COMPRESSORS FOR TURBO-BRAYTON CRYOCOOLERS

The compressor is one of the major components of a turbo-Brayton cryocooler in addition to the turboalternator, recuperative heat exchanger(s), and electronics. The refrigeration cycle uses single-phase helium gas as the working fluid, which was selected for optimal thermodynamic performance at the 20 K load temperature. The compressors provide the pressure ratio and circulate the cycle gas in turbo-Brayton cryocoolers and inter/aftercoolers remove the heat of compression from the cycle gas and compressor dissipation losses from each compression stage. The 20 K, 20 W cryocooler utilizes three compression stages with intercooling. The serial arrangement allows centrifugal compressors to obtain the desired cycle pressure ratio without exceeding material stress limits, and intercooling reduces input power.

The overall cryocooler efficiency and weight is strongly affected by the compressor performance. The compressor performance is characterized by the net efficiency, which is the ideal compressor work divided by the AC input power to the motor (Equation 1). This performance parameter accounts for losses in the motor and bearings, and aerodynamic losses in the impeller and diffuser. Therefore, the efficiency can also be represented as the product of the motor efficiency, shaft efficiency, and the aerodynamic efficiency and is given by Equation (1):

$$\eta_{net} = \frac{\dot{m}\Delta h_s}{W_{AC}} = \eta_{Motor}\eta_{Shaft}\eta_{Aero} \tag{1}$$

Figure 1. Creare's 20 K, 20 W cryocooler for liquid cryogen storage.

COMPRESSOR DESIGN

The compressor design is a modified version of Creare's 400-500 W class permanent magnet compressor shown in Fig. 2.[4] Each compressor is identical except for the aerodynamic design of the impeller and diffuser. The design is based on a prior compressor design with key upgrades to optimize the aerodynamic features of the impellers and diffusers for optimal performance at each compression stage and a new permanent magnet motor design that extends motor capacity from 400 W to 500 W. The journal bearing diameter is 6.4 mm (0.25 in.) and the impeller diameter is 19 mm (0.75 in.). Operational speed is 6,300 rev/s. The overall compressor assembly is 9.53 cm (3.75 in.) in diameter and 16 cm (6.32 in.) long and weighs 4 kg.

The compressor uses a multi-part rotor. A permanent magnet is installed within the shaft to form the electro-magnetic rotor. A precision machined impeller is mechanically fastened to the shaft. Self-acting gas bearings provide radial and axial positioning of the rotor. These bearings are supported in a bearing cartridge that also locates the motor stator. The clearances between internal components are minimized to enable heat dissipated in the motor windings and heat generated by the bearings to efficiently conduct to the outer housing. Heat is conducted through the outer housing to the base plate where it is removed by the spacecraft thermal management system.

We developed impeller and diffuser blade designs to optimize the aerodynamic efficiency of each compression stage. The performance predictions at the system operating point for each stage are shown in Table 1.

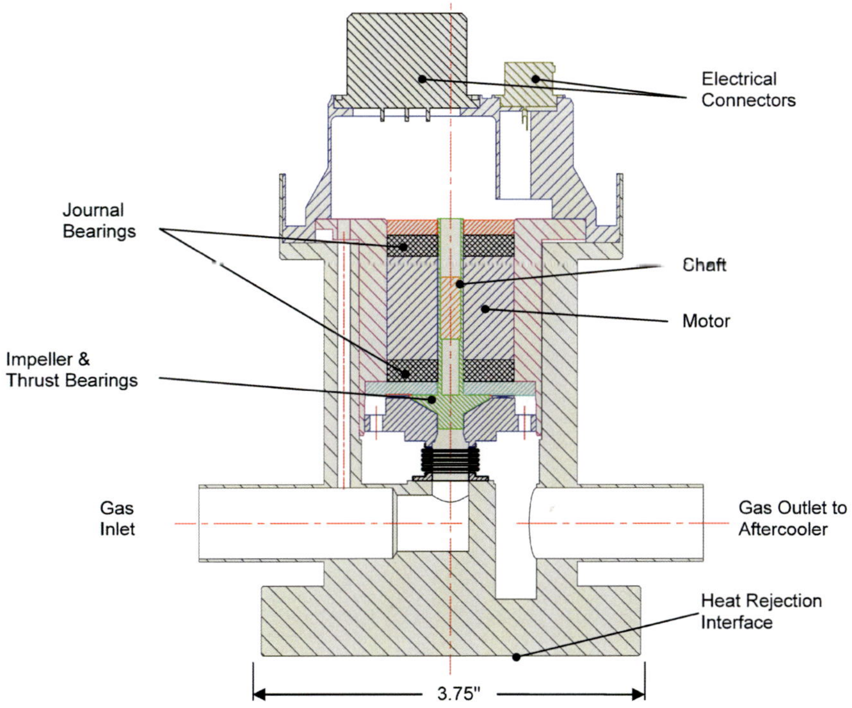

Figure 2. 400–500 W class permanent magnet motor compressor design [Hill et al. 2007].

Table 1. Compressor operating conditions and performance predictions

	Stage 1	Stage 2	Stage 3
Fluid	Helium	Helium	Helium
Inlet Pressure (atm)	4.9	5.6	6.3
Inlet Temperature (K)	300	300	300
Mass Flow Rate (g/s)	3.62	3.62	3.62
Pressure Ratio (-)	1.138	1.132	1.130
Rotational Speed (rev/s)	6300	6300	6300
Motor AC Input Power (W)	450	450	450
Head Coefficient (-)	0.58	0.56	0.55
Tip Flow Coefficient (-)	0.043	0.038	0.033
Compressor Efficiency – Predicted (%)	67	65	63

COMPRESSOR FABRICATION

Most of the compressor components are machined from bar stock. The compressor impellers are made from titanium to provide the strength needed to accommodate the large centrifugal stresses at high rotational operating speeds. Figure 3 shows a photograph of the compressor rotor after the impeller has been attached to the shaft using a threaded fastener. The shaft and impeller are coated with a hard coating for low friction and to reduce wear during start-up and coast-down when contact occurs with the bearing surfaces.

The outer housing assembly and cover with electrical passthrus for power lead wires and shaft runout probe signals is shown in Fig. 4.

The motor consists of a concentrated 3-phase winding on a 6-tooth core. The core is made from stacked laminations of Carpenter Alloy 49. The primary constituents of the alloy are nickel and iron in nearly equal proportions (about 50%). Its key feature is low core loss, compared with more commonly used silicon steel. The laminations are cut from sheet stock into the desired geometry via wire EDM. They are then coated with an oxide layer to provide electrical resistivity, stacked in a close-clearance motor housing which provides precise alignment between the laminations, and then the motor is wound with magnet wire. A photo of the completed stator assembly is shown in Fig. 5.

Figure 3. Completed compressor rotor assembly.

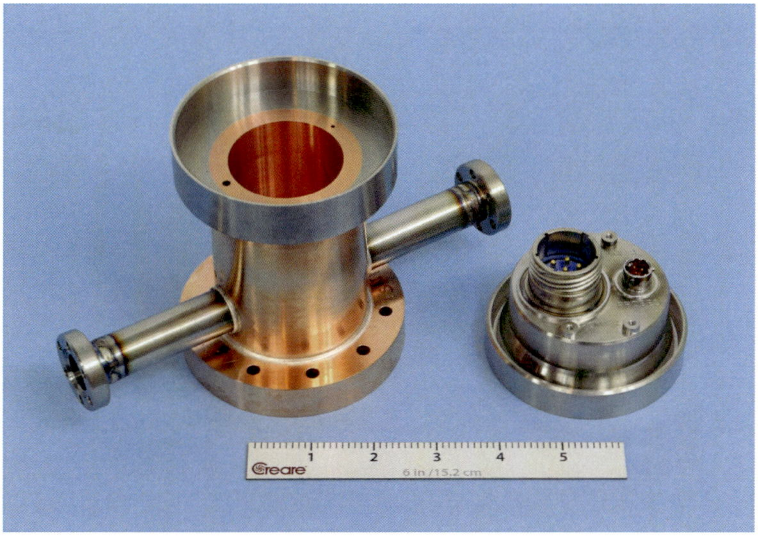

Figure 4. Compressor outer housing assembly with aft cover.

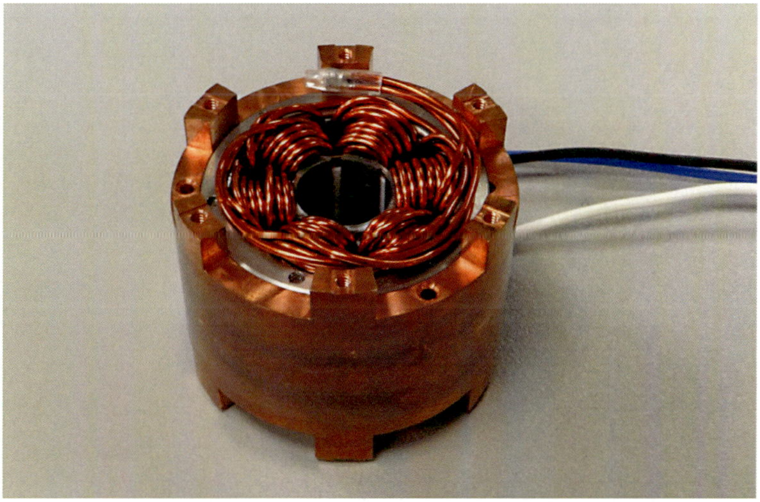

Figure 5. Compressor motor stator assembly.

The remaining compressor components are machined from copper or bronze. Figure 6 shows the three compressor assemblies ready for initial performance testing. During initial spin and performance tests, the aft covers are secured using compression clamps as shown in the photo. Once performance has been verified, a closeout weld is performed to hermetically seal the covers are welded to the housings.

Figure 6. Stage 1, 2, and 3 compressor assemblies. The aft covers are temporarily attached using clamps during initial performance screening tests.

COMPRESSOR PERFORMANCE

Compressor performance testing was performed in an existing compressor test loop. The test loop includes all the equipment—plumbing; valves; pressure, power, and temperature instrumentation; heat rejection loop; and drive inverter—that is required to execute a compressor performance test. Initial testing has been completed on the Stage 3 compressor and is presented below. Full performance characterization of the three compressors, alone and in series, is ongoing.

The compressor performance is characterized by the compressor head-flow curve and its net efficiency. These parameters are defined as:

Head Coefficient:
$$\Psi = \frac{\Delta h_S}{U^2} \tag{2}$$

Flow Coefficient:
$$\Phi = \frac{Q}{A \cdot U} = \frac{\dot{m}}{\rho \cdot \left(\frac{\pi}{4} \cdot D_{tip}^2 \right) \cdot U} \tag{3}$$

where Δh_S is the isentropic enthalpy rise, U is the impeller tip speed, Q is the volumetric flow rate, A is the flow area defined by the tip diameter, $\dot{m}$ is the mass flow rate, ρ is the inlet density, and D_{tip} is the impeller tip diameter. Compressor efficiency is defined by Equation 1.

Figure 7 shows the head-flow curve measured during initial testing on the Stage 3 compressor along with the predicted performance from CFD analysis. In this test the compressor was operated at 5,500 rev/s instead of the 6,300 rev/s design speed. The compressor bearings currently require further adjustment to attain the design operating speed and a few iterations where the bearing clearances are systematically reduced are typical in the development of these gas bearing machines. The head-flow performance is close to predictions, but the peak in head coefficient occurs at a lower than expected flow coefficient. The difference is modest, and can be ameliorated by increasing the system pressure slightly.

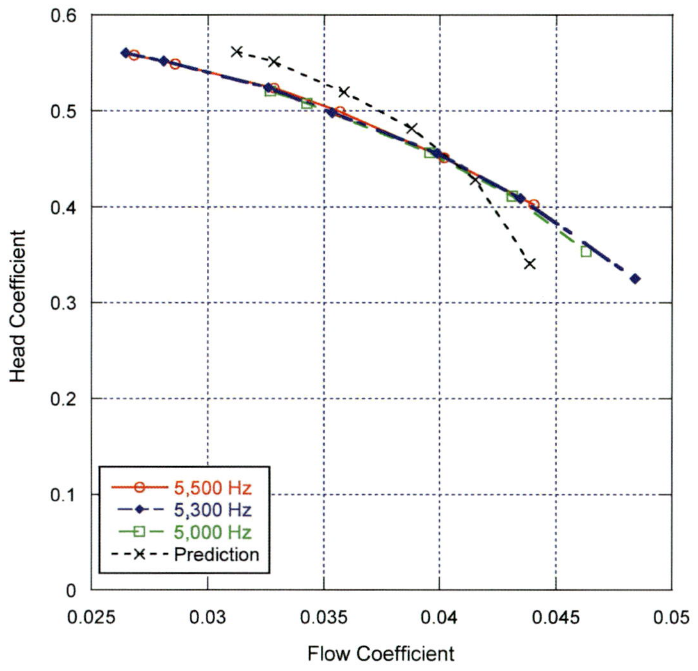

Figure 7. Preliminary head-flow curve from Stage 3 compressor testing.

The head-flow behavior is independent of operating speed, so data taken at lower operating speeds (Fig. 7) are representative of the compressor performance. However, the compressor net efficiency measured during the test was 57%, lower than the design target of 63%. The efficiency was reduced in part to the lower operating speed and further testing at the design speed is needed to quantify the difference from predictions, if any. Nevertheless, this net efficiency represents a considerable improvement over the efficiency of induction motor compressors of similar size (less than 40%), and corresponds to a reduction in input power of up to 30% as compared to prior Creare induction-motor compressors at this power level.

CONCLUSIONS AND FUTURE WORK

The design and fabrication of a 500 W class PMM compressor for a 20 K turbo-Brayton cryocooler is complete, and initial testing has commenced. This paper discussed the design, fabrication and initial test results for three such compressors for a 20 K cryocooler. The compressors have predicted efficiencies exceeding 60%, enabling the 20 K cryocooler to deliver 20 W of refrigeration with an input power below 1.6 kW, corresponding to a specific power of about 80 W/W. The compressor efficiency significantly exceeds that of prior induction motor compressors (less than 40%), resulting in a substantial reduction of input power to the cryocooler. Future work involves complete performance characterization for each compressor stage, integrating them into a cryocoolers, and testing the cryocooler.

ACKNOWLEDGMENT

We gratefully acknowledge NASA for their support of this work (Contract NNC14CA15C).

REFERENCES

1. Hastings, L.J., Plachta, D.W., Salerno, L. and Kittel, P., "An overview of NASA efforts on zero boiloff storage of cryogenic propellants," *Cryogenics*, vol. 41(11–12) (2002), pp. 833–839.
2. Hill, R.W., Hilderbrand, J.K. and Zagarola, M.V., "An Advanced Compressor for Turbo Brayton Cryocoolers," *Cryocoolers 16*, ICC Press, Boulder, CO (2011), pp. 391–396.
3. Zagarola, M.V. and McCormick, J.A, "High-capacity turbo-Brayton cryocoolers for space applications," *Cryogenics,* vol. 46; (2006), pp. 169–175.
4. Zagarola, M.V., Breedlove, J.J. and Cragin, K.J., "Demonstration of a high-capacity cryocooler for zero boil-off cryogen storage in space," *International Cryocooler Conference 17* (2012).

Modified Brayton Refrigeration Cycles for Liquid Hydrogen in Spallation Neutron Source Moderator

H.-M. Chang[1,2], S.G. Kim[1], J.G. Weisend II[2], H. Quack[3]

[1] Hong Ik University, Seoul, 121-791, Korea
[2] European Spallation Source, SE-221 00 Lund, Sweden
[3] TU Dresden, D-01062 Dresden, Germany

ABSTRACT

A thermodynamic study of standard and modified Brayton refrigeration cycles is carried out to cool liquid hydrogen in spallation neutron facilities. The target moderators under development at ESS require a refrigeration capacity of 30 kW at 20 K to remove the dissipated energy from sub-cooled liquid hydrogen. A standard Brayton refrigeration cycle to perform this refrigeration is designed with helium as refrigerant. A variety of modified processes are examined for improved efficiency and suitability. The modification includes the comparison of LP (low-pressure) and HP (high-pressure) cooling, depending on the location of cryogenic expander in the cycle, and two-stage and dual-turbine cycles, depending on the manner of combining the two expanders. Based on existing and proposed designs, eight different cycles are selected and analyzed fully with the assumed models of the components. The results are presented in terms of the figure of merit (*FOM*) as an index of the thermodynamic performance, and the detailed exergy expenditure is also investigated. As a result of this analysis, a modified dual-turbine cycle is recommended for large-capacity refrigeration at 20 K.

INTRODUCTION

Cryogenic refrigeration at 20 K is required for liquid-hydrogen moderators in spallation neutron source facilities[1-5]. As shown in Figure 1, the dissipated energy of neutrons is removed at the moderators by a circulating flow of sub-cooled liquid hydrogen at temperatures below 20 K, and then delivered to a refrigerator via a helium-hydrogen (He-H_2) heat exchanger (HX). The closed loop of liquid hydrogen can be plotted as a counterclockwise triangular cycle on the phase diagram, as shown in Figure 1. The cooling flow through the moderators is represented by a descending line according to the magnitude of the pressure drop and the temperature rise, which are restored by a pump and by the refrigerator, respectively. In order to avoid any vaporization, liquid hydrogen is maintained at a sub-cooled temperature (at 17~20 K) and at the approximate critical pressure (1.32 MPa). The refrigeration load is 1.5~6 kW for the existing systems under operation or nearing completion[1-3]. A much larger cryogenic moderator system has been recently designed and will be completed by 2019 for the European Spallation Source (ESS)[4,5], where the maximum heat load is estimated to be over 30 kW during the operation with a full level of beam power.

Cryocoolers 19, edited by S.D. Miller and R.G. Ross, Jr.

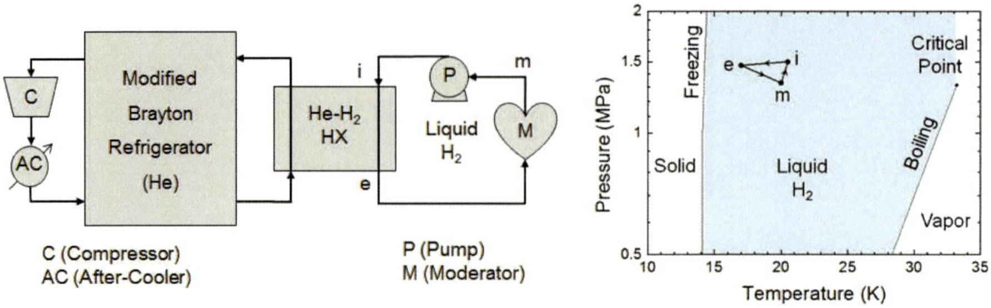

Figure 1. Cryogenic refrigeration system and liquid-hydrogen cycle on phase diagram.

The cryogenic refrigerators to cover 1.5~6 kW at 20 K have been designed as a standard reverse-Brayton cycle[1-3]. Helium is the only possible gas refrigerant for this application, as a cryogenic turbine should be used at temperatures below 20 K. The standard Brayton cycle could be modified in various ways to increase the thermodynamic efficiency and refrigeration capacity, or to satisfy other design requirements. The thermodynamic design of the 30 kW refrigerator for ESS is a new and exciting challenge at 20 K, since it requires not only the largest capacity ever, but also a set of thermo-hydraulic and geometric constraints[5]. This study investigates the structures of standard or modified cycles in a systematic way, and provides a thermodynamic basis for the development of large-scale refrigerators at 20 K.

STANDARD AND MODIFIED CYCLES

Eight different refrigeration cycles are selected and shown in Figure 2. Cycle (I) is the standard Brayton cycle that has one recuperative heat exchanger (HX1) and one expander (E) or turbine. This cycle is called "Standard cycle with LP (low-pressure) cooling", since the cold helium (state 3) enters HX2 after expansion for the thermal contact with LH_2. Cycle (II) is another simple cycle with one He-He HX and one expander, but is called "Standard cycle with HP (high-pressure) cooling", since the cold helium (state 2) enters HX2 before the expansion. The efficiency of these standard cycles has a certain limit (no matter how highly effective HX1 may be), because there is a mismatch in the specific heat of He between the HP and the LP streams.

The limit of standard cycles can be overcome by employing two expanders[6]. The first way is to arrange two expanders in series such that the first (warm) expander is used to compensate the mismatch of specific heat and the second (cold) expander plays the main role of cryogenic refrigeration. Cycles (III) and (IV) are such examples, which are called "Two-stage cycle with LP cooling" and "Two-stage cycle with HP cooling", respectively. In these two-stage cycles, the flow rate is the same for the two expanders, but the operating pressure is different.

Another way is to arrange two expanders in parallel so that they work at the same pressure level. Among a variety of possible structures, Cycles (V) and (VI) are selected, as called "Dual-turbine cycle with LP cooling" and "Dual-turbine cycle with HP cooling", respectively. At a location of the HP stream (state 2), a small fraction of gas is diverted, expanded through the first (warm) expander, and reunited with the LP stream below HX2. The main HP stream continues through HX2 and HX3, and is finally expanded through the second (cold) expander to the coldest temperature. The structure of these cycles is similar to the Claude cycle, except that the JT valve is replaced by a cold expander. In HX2, the LP stream has a higher flow rate than the HP stream in order to compensate for the mismatch in specific heat.

The cold end of the dual-turbine cycles could be further modified by adding a warm-up flow to the HP stream before entering the cold turbine, as shown in Cycles (VII) and (VIII), and called "Modified dual-turbine cycle with LP cooling" and "Modified dual-turbine cycle with HP cooling", respectively. In these cycles, the number of He-He HX's is five, and HX4 is a triple-stream HX, having two cold streams and one warm stream. The warm-up flow could be effective in reducing the temperature difference at the cold end as described later.

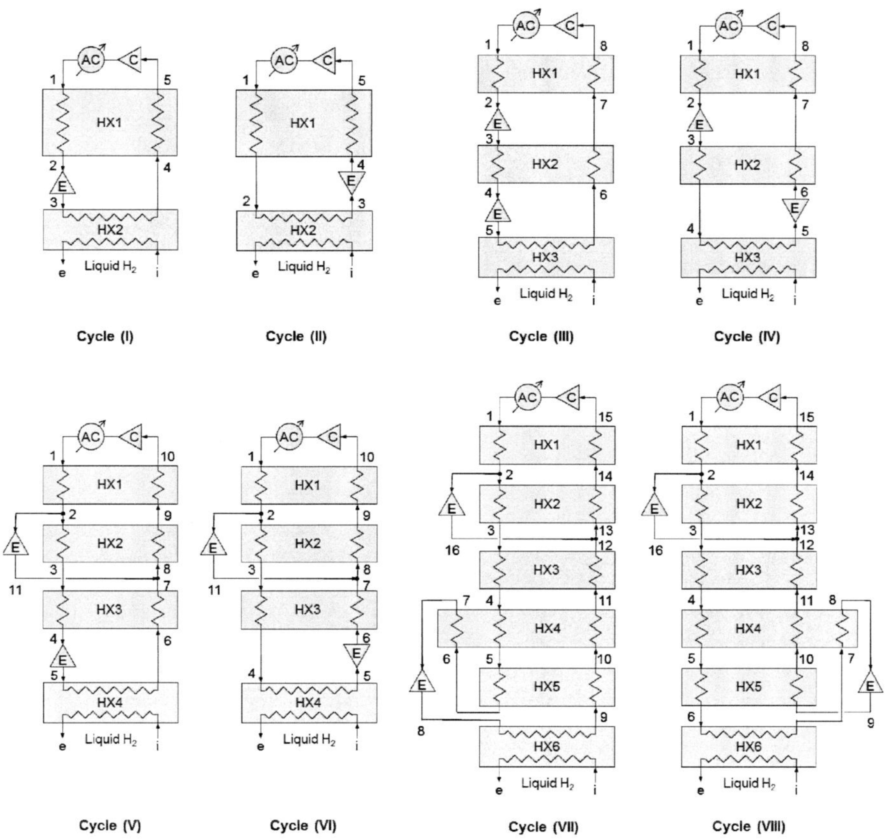

Figure 2. Standard or modified Brayton cycles for sub-cooled liquid hydrogen. (AC: after-cooler, C: compressor, HX: heat exchanger, E: expander)

CYCLE ANALYSIS

The following assumptions are made to analyze and compare the selected cycles.

- The ambient temperature (T_0) is 298 K.
- The refrigeration load by liquid hydrogen is 30 kW.

$$\dot{m}_{H_2}\left(h_i - h_e\right)_{H_2} = 30\ \text{kW} \qquad (1)$$

- At the He-H_2 HX, the inlet and exit temperatures of He are 15 K and 20 K, respectively, and the pressure drop of He flow is 0.1 MPa. The minimum temperature difference between H_2 and He is 0.5 K.
- The high pressure of all cycles is 2 MPa.
- The pressure drop in all He-He HX's is zero.
- The minimum temperature difference between hot and cold streams is 3 K (or 4.5 K) for HX1, and 1% (or 1.5 %) of the absolute temperature at HP stream for other HX's.
- The adiabatic efficiency of all compressors and turbines is 80%.

The second and third assumptions are specified by the design requirements of the target moderator system at ESS[7]. The sixth assumption is made to investigate the size effect of HX's on each cycle with two different cases (ΔT_{min} = 3 K and 4.5 K), which represents relatively high and low effectiveness, respectively. The temperature difference given as 1% (or 1.5%) of absolute temperature means, for example, that ΔT_{min} = 1 K (or 1.5 K) at T_{HP} = 100 K and ΔT_{min} = 0.4 K (or 0.6 K) at T_{HP} = 40 K. Since the temperature of the HP stream is determined as a result of the cycle analysis, this assumption will be imposed by a few iterative calculations. In calculating the input power to the compressors, a single-stage compression is assumed, because the pressure ratio is between 3 and 7 and screw compressors could be used in practice. The only exception is Cycle (II), where a two-stage compression with inter-cooling is assumed for the pressure ratios exceeding 10.

A general-purpose process simulator (Aspen HYSYS®) with the thermodynamic properties of helium and para-hydrogen is used for the cycle analysis. Under the specified conditions, Cycles (I) and (II) are uniquely determined, as the number of unknowns is the same as the number of given equations. For Cycles (III) through (VIII), however, the number of unknowns is more than the number of given equations by one. The cycle analysis is repeated over one variable, until the input power can be minimized under the assumptions.

The thermodynamic performance of a refrigeration cycle is evaluated with a dimensionless index, *FOM* (the figure of merit)[8], which is defined as the ratio of minimum to the actual work.

$$FOM = \frac{\dot{W}_{\min}}{\dot{W}_C - \dot{W}_E} \tag{2}$$

The minimum work is the thermodynamic limit for a reversible cycle, which can be expressed as the increase of exergy or flow availability of liquid hydrogen.

$$\dot{W}_{\min} = \dot{m}_{H_2}\left[\left(h_e - h_i\right)_{H_2} - T_0\left(s_e - s_i\right)_{H_2}\right] = 447.0 \text{ kW} \tag{3}$$

where h and s are specific enthalpy and entropy, respectively. The subscripts e and i denote the exit and inlet of He-H_2 HX, respectively, as shown in Figure 1 and 2, and T_0 is the ambient temperature at which heat is rejected by the refrigerator. In practice, the power output of the expanders (W_E) may be used to drive the compressors or may be simply dissipated, but the "net" input (W_C - W_E) is counted in the denominator of Eq. (2).

For a better understanding of the thermodynamic nature of cycles, an exergy (or second-law) analysis is presented as well. Combining the energy and entropy balance equations[9], the exergy balance can be written as:

$$\dot{W}_C - \dot{W}_E = \dot{W}_{\min} + T_0\left[\left(\dot{S}_{gen}\right)_C + \left(\dot{S}_{gen}\right)_{AC} + \left(\dot{S}_{gen}\right)_E + \sum_{HX}\left(\dot{S}_{gen}\right)_{HX}\right] \tag{4}$$

where the left-handed side is the exergy input for refrigeration, and the right-handed side shows so-called the exergy expenditure. As defined in Eq. (2), the fraction of the first term is the *FOM*, and the remaining terms are the irreversibility or the entropy generation rate multiplied by ambient temperature. The total irreversibility can be itemized for the components in the system, including compressors (C), after-coolers (AC), expanders (E), and heat exchangers (HX). In case of Cycles (IV) through (VIII), the mixing of two streams is another source of entropy generation, but is not included here, since the magnitude is negligibly small for an optimized cycle[10].

RESULTS AND DISCUSSION

The selected eight cycles are fully analyzed with the assumptions described above, and the results are plotted as temperature-entropy diagram in Figure 3 and exergy expenditure in Figure 4. The cycles with ΔT_{min} = 3 K (1%) are presented here, and will be discussed later together with ΔT_{min} = 4.5 K (1.5%). As mentioned above, Cycles (III) through (VIII) have been optimized for the maximum *FOM*.

In the case of the standard cycles, Cycle (I) with LP cooling is more efficient than Cycle (II) with HP cooling. The main reason is that the HP cooling needs a larger pressure ratio to perform the refrigeration. The pressure ratio is 6.9 and 19.7 in Cycles (I) and (II), respectively. In a sense, the HP cooling is regarded as "indirect" cooling, because the coldest LP stream cools the HP stream, and then the HP stream cools the LH_2 flow. It may be stated in standard cycles that the LP cooling is essentially superior in efficiency to the HP cooling. On the other hand, the HP cooling could have merits in other aspects, such as a smaller diameter of transfer lines for cold helium. The HP cooling could also be safer against the freeze-out of liquid hydrogen, because the coldest helium (at the exit of turbine) indirectly affects the LH_2 flow, when the temperature temporarily drops according to a scheduled or unscheduled decrease of thermal load.

Two-stage cycles, Cycles (III) and (IV), have considerably higher *FOM's* than the standard cycles. In Figure 4, the irreversibility in HX's of Cycles (III) and (IV) is smaller than that of Cycles

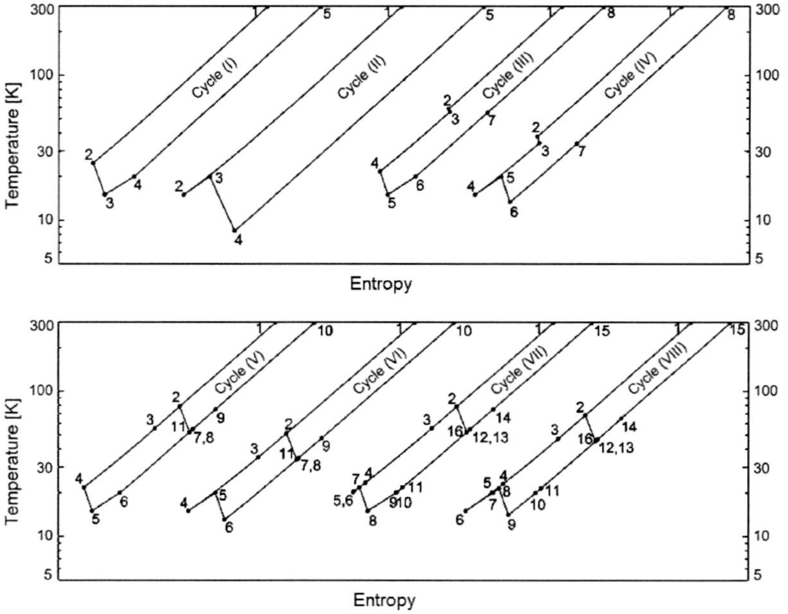

Figure 3. Temperature-entropy diagram of eight cycles at optimized condition with ΔT_{min} = 3 K (1%).

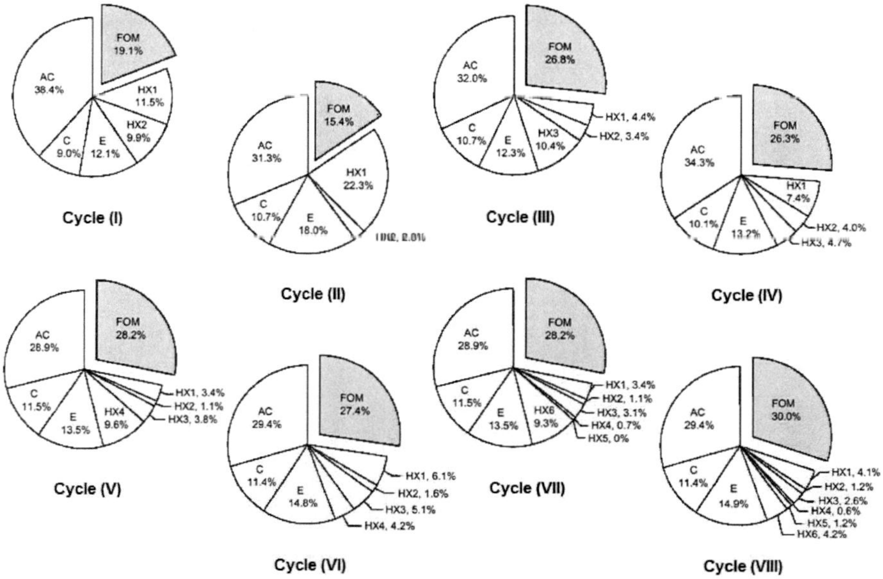

Figure 4. Exergy expenditure of eight cycles at optimized condition with ΔT_{min} = 3 K (1%). (FOM: figure of merit, AC: after-cooler, C: compressor, E: expander, HX: heat exchanger)

(I) and (II), as the temperature difference in HX's is reduced by the addition of warm expander. A key point here is that the LP cooling is still more efficient than the HP cooling, but their difference in *FOM* is not so large. It is interesting to notice in Figure 4 that Cycle (IV) has a larger irreversibility in HX1+HX2 than Cycle (III), but a smaller irreversibility in HX3. This means that the HP cooling still requires a larger pressure ratio (for a larger temperature difference), but the penalty of pressure drop (0.1 MPa) in HX3 is less severe in HP cooling. The pressure ratio is 4.7 and 5.3 for Cycles (III) and (IV), respectively.

The dual-turbine cycles, Cycles (V) and (VI), have an even higher *FOM's* than the two-stage cycles. The branching flow through the warm turbine is effective in reducing the temperature difference in recuperative HX's, and the pressure ratio is small for both cycles. Like typical Claude cycles, the main design parameter is the ratio of the flow rate through the warm turbine to the total flow rate through the compressor[8]. The optimal condition is determined such that the exit temperature of the warm turbine (state 11) matches closely with the LP temperature at the point of mixing. The flow ratio is 0.114 and 0.210 for Cycles (V) and (VI), respectively, and the pressure ratio is 3.9 and 4.0 for Cycles (V) and (VI), respectively.

The modification of dual-turbine cycle with a warm-up flow is not as effective in improving the *FOM* for LP cooling, but is very effective for HP cooling. Cycles (VII) has nearly the same *FOM* as Cycle (V), even though the detailed cycles are slightly different from each other. On the other hand, Cycle (VIII) has a significantly higher *FOM* than Cycle (VI). Two additional HX's at the cold end and the warm-up flow make the operating temperature of the warm turbine higher and accordingly the overall temperature difference in the HXs is smaller. The flow ratio to the warm turbine is 0.114 and 0.136 for Cycles (VII) and (VIII), respectively, and the pressure ratio is 3.9 and 4.0 for Cycles (VII) and (VIII), respectively.

In order to examine the size effect of the HXs, Figure 5 compares the *FOM* of eight cycles with the minimum temperature difference of 3 K (1%) and 4.5 K (1.5%). The standard cycles, Cycles (I) and (II), are not only poor in efficiency itself, but also more sensitive to the size of HXs than the modified cycles. Among the modified cycles, dual-turbine cycles are superior to two-stage cycles, if highly effective HXs are used. The penalty of small HX size is least for Cycle (VIII). This means that an efficient refrigeration with HP cooling can be achieved by the warm-up flow without an extremely large size of HXs.

In summary, two cycles are suggested from a thermodynamic point of view. First, the dual-turbine cycle with LP cooling or Cycle (V) is a preferred cycle for both simplicity and efficiency. This cycle is especially recommendable when highly effective HXs can be used. Alternatively, the modified dual-turbine cycle with HP cooling or Cycle (VIII) has an advantage in efficiency, even when the HXs are not so highly effective. On the other hand, this cycle is complex in structure and could yield more pressure drops in the HXs. Table 1 lists the detailed thermodynamic data, including temperature, pressure and flow rate at each state for Cycle (V) with $\Delta T_{min} = 3$ K (1%) and Cycle (VIII) with $\Delta T_{min} = 4.5$ K (1.5%).

In determining a refrigeration cycle, there are a number of practical factors to take into account as well as efficiency. Although the details are beyond the scope of this thermodynamic study, a few comments are mentioned for discussion. First, the availability and performance of cryogenic turbines should be considered for the selection between the two-stage cycle and the dual-turbine

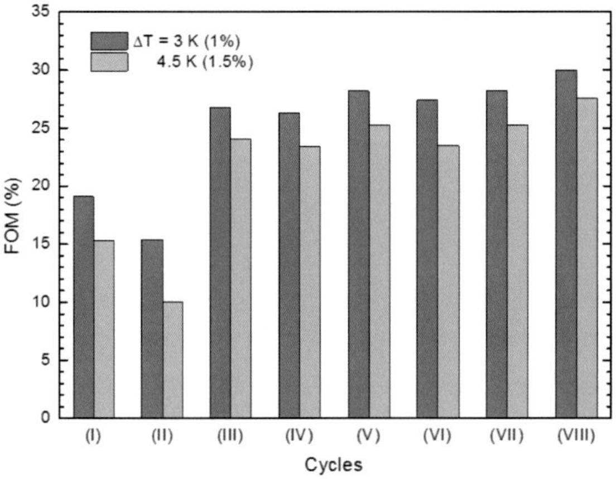

Figure 5. Figure of merit (FOM) for eight cycles with $\Delta T_{min} = 3$ K (1%) and 4.5 K (1.5%).

Table 1. Temperature, pressure and flow rate of two suggested cycle

		1	2	3	4	5	6	7	8	9	10	11	12	13	14	15	16
Cycle (V)	T [K]	298.0	78.5	55.4	21.8	15.0	20.0	54.9	54.6	75.3	295.0	52.3					
	P [MPa]	2.00				0.61	0.51										
	m [kg/s]	1.18		1.04					1.18			0.14					
Cycle (VIII)	T [K]	298.0	85.7	57.5	23.9	20.3	15.0	20.0	21.9	14.0	20.0	21.9	56.7	56.5	81.0	293.5	55.2
	P [MPa]	2.00						1.90		0.50							
	m [kg/s]	1.09		0.94										1.09			0.15

cycle. In two-stage cycles, the warm turbine requires a small pressure ratio with a large flow rate. In dual-turbine cycles, on the contrary, the warm turbine requires a large pressure ratio with a small flow rate.

Detailed HX design could be important for a thorough and fair comparison of the cycles. It is recalled that the results presented are based on a simplified model of minimum temperature difference and an assumption of no pressure drop in the HXs. The actual size and thermo-hydraulic characteristics of the HXs could affect the selection to an extent. Finally, an optimization theory is cited here with regards to the effect of finite HX sizes, especially when a cycle is composed of multi-stages of HXs. It is always true that as any HX area increases, the temperature difference between hot and cold streams in the HX decreases. It is an important design strategy, however, to allocate the HX area to each stage, if the total sum of HX area is fixed. This is a well-known optimization problem subject to a constraint, which has been solved by the method of Lagrange multiplier. The results show that the best thermodynamic performance is achieved when the temperature difference is proportional to the absolute temperature[9,11].

$$\left(\frac{\Delta T}{T_{HP}}\right)_{\text{opt}} \approx \text{constant} \tag{5}$$

This condition has been already incorporated as an assumption of the cycle analysis.

CONCLUSIONS

Standard and modified Brayton refrigeration cycles are investigated, with an aim at the efficient refrigeration of 30 kW for sub-cooled liquid hydrogen at 20 K. Based on existing and proposed systems, a variety of modifications are tried in a systematic way, and their efficiency is calculated in terms of *FOM* (figure of merit) through a full thermodynamic analysis. The modifications include the LP (low-pressure) or HP (high-pressure) cooling, depending on the location of the cryogenic expander in a cycle, and the two-stage or dual-turbine cycles, depending on the combination of the two expanders. Among the eight cycles under consideration, two cycles are suggested along with the required size of HXs and other constraints. A dual-turbine cycle with LP cooling or Cycle (V) is preferred for simplicity and efficiency at the same time, if highly effective HXs can be used. A modified dual-turbine cycle with HP cooling or Cycle (VIII) has an advantage in efficiency, even if less effective HXs are used. The results of this thermodynamic study are immediately applicable to the neutron source moderators at ESS, and should be useful in large-scale refrigeration at 20 K as well.

ACKNOWLEDGMENT

This work was partially supported by 2016 Hong Ik University Research Fund for sabbatical leave in a foreign institute.

REFERENCES

1. Aso, T., Tatsumoto, H., Hasegawa, S., Ushijima, I., Kato, K., Ohtsu T., Ikeda, Y., "Design result of the Cryogenic Hydrogen Circulation System for 1 MW Pulse Spallation Neutron Source (JSNS) in J-PARC," *Adv. in Cryogenic Engineering*, Vol. 51, Amer. Institute of Physics, Melville, NY (2006), pp. 763-770.

2. Lee, M., Choi, H.Y., Han, J.S., Cho, S.H., Kim, M.S., Hur, S.O., Son, W.J., Ahn, G.H., Lim, I.C., "Operational characteristics of the helium refrigeration system at HANARO-CNS," *Adv. in Cryogenic Engineering*, Vol. 57, Amer. Institute of Physics, Melville, NY (2012), pp. 361-367.

3. Wang, G.P., Zhang, Y., Xiao, J., He, C.C., Ding, M.Y., Wang, Y.Q., Li, N., He, K., "Design progress of cryogenic hydrogen system for China Spallation Neutron Source," *Adv. in Cryogenic Engineering,* Vol. 59, Amer. Institute of Physics, Melville, NY (2014), pp. 233-238.

4. Arnold, P., Hees, W., Jurns, J., Su, X.T., Wang, X.L., Weisend, J.G., "ESS cryogenic system process design," *IOP Conference Series: Materials Science and Engineering* Vol. 101, No. 1 (2015), IOP Publishing, p. 012011

5. Jurns, J., Ringnér, J., Quack, H., Arnold, P., Weisend, J.G, Lyngh, D., "Spallation target cryogenic cooling design challenges at the European Spallation Source," *IOP Conference Series: Materials Science and Engineering* Vol. 101, No. 1 (2015), IOP Publishing, p. 012082.

6. Chang, H.M., Park, J.H., Cha, K.S., Lee, S., Choe, K.H., "Modified Reverse-Brayton Cycles for Efficient Liquefaction of Natural Gas," Cryocoolers 17, ICC Press, Boulder, CO (2013),, pp. 435-442.

7. European Spallation Source, *Annex 4 Technical Specification*, Document Number ESS-0034501 (2015), pp. 41-42.

8. Barron, R. F. *Cryogenic systems*, Clarendon Press, 1985, pp.237-261.

9. Bejan, A., *Advanced Engineering Thermodynamics*, 3rd ed., John Wiley & Sons, New Jersey (2006), pp. 101-142.

10. Chang, H.M., Lim, H.S., Choe, K.H., "Thermodynamic design of natural gas liquefaction cycles for offshore application," *Cryogenics*, 63 (2014), pp. 114-121.

11. Chang, H.M., Chung, M.J., Lee, S., Choe, K.H., "An efficient multi-stage Brayton-JT cycle for liquefaction of natural gas," *Cryogenics*, 51, no. 6 (2011), pp. 278-286.

Characterization of Emitted Vibration from Turbo-Brayton Cryocoolers

K. Cragin, K. Rule, M. Zagarola

Creare LLC
Hanover, NH 03755

ABSTRACT

Turbo-Brayton cryocoolers are known for low emitted vibrations due to the lack of reciprocating parts. The only moving parts are the miniature turbomachine rotors which are precisely balanced and operate at extremely high rotational speeds (3000 to 10,000 rev/s are typical). The high operating speed, low rotor mass and low rotor imbalance results in physical displacements of cryocooler components that are on the order of nanometers. The first space implementation of a turbo-Brayton cryocooler was for cooling the NICMOS instrument on the Hubble Space Telescope (HST). On-orbit operations indicated that the cryocooler could not be detected by the super-precise instruments on the HST. Future missions may have more restrictive vibration requirements or different integration approaches than HST and a quantitative assessment of emitted vibration from a turbo-Brayton cryocooler is needed. During the development of the NICMOS cryocooler, several tests were conducted to characterize the emitted vibrations from the cryocooler, but these tests did not produce comprehensive information due to either high background vibrations or non-prototypical operation of the cryocooler. Recent work at Creare has focused on quantifying the emitted vibration of turbo-Brayton cryocooler technology using computational fluid dynamics/structural simulations of representative turbo-Brayton components at prototypical operating conditions. This paper describes the analysis work.

INTRODUCTION

Creare has developed and demonstrated a unique class of mechanical cryocoolers, turbomachine-based reverse-Brayton cryocoolers, which has an inherent advantage over competing active cryocoolers when it comes to vibration emittance. Turbo-Brayton cryocoolers utilize miniature turbomachines to compress and expand the working gas as part of a reverse-Brayton thermodynamic cycle. The only moving parts in these cryocoolers are miniature turbomachine rotors which have masses of 1 to 10 grams. The rotors have diameters of a few millimeters and operate at speeds of several thousand revolutions per second. These small sizes and high speeds are required for high thermodynamic efficiency at the low power levels associated with space cryocooling applications. The rotors are supported on non-contacting gas bearings, and the turbomachines use clearance seals so there is no mechanical contact or wear mechanism during normal operation. Contact between the rotor and bearings occurs briefly during start-up and shutdown operations, but the extremely low contact forces prevent any

Cryocoolers 19, edited by S.D. Miller and R.G. Ross, Jr.

appreciable wear during thousands of start-up and shutdown cycles. Numerous demonstrations support this claim (Breedlove et al. 2001). In addition, the rotors are precisely balanced for high-speed operation, and the cycle gas flow is continuous. The result is that emitted vibrations are extremely small and occur at high frequencies where the predicted mechanical displacement caused by vibrations is on the order of nanometers.

The initial space demonstration of a turbo-Brayton cryocooler occurred during NASA's NICMOS Cooling System (NCS) Program (Swift et al. 2008). The NCS consists of a single-stage turbo-Brayton cryocooler coupled with a cryogenic circulator and capillary pump loop. The system provides 7 W of cooling at 70 K to the NICMOS instrument on the Hubble Space Telescope and replaced the solid nitrogen cryogen that had depleted prematurely after less than two years in space. The NCS was deployed and integrated with NICMOS by astronauts during STS–109 (Space Shuttle Columbia) in March 2002 and operated nearly continuously for more than 6.5 years before a problem with an external cooling loop caused the system to be deactivated.

During the NCS program, several tests were conducted to characterize the emitted vibrations from the cryocooler. The Vibration Emittance Test (VET) described in Clapp et al. (2002) showed continuous performance below the derived maximum Root-Sum Square acceleration allocated to the center of mass for the NCS (4.1 micro-g RMS over the frequency range of 0.001 to 50 Hz). This test was conducted with the compressor in surge and considered by Clapp, et al., to be an upper bound on vibrations expected during steady operation. Unfortunately, no testing was performed at design operating conditions. Even at this off-design condition, measured vibrations (3.5 mas jitter) were well below the required budget for the Hubble Space Telescope of 7.7 mas jitter. Additional test data were obtained during the servicing mission to the Hubble Space Telescope responsible for the installation of the NCS. During that mission (Hubble Servicing Mission 3B), measured jitter was recorded during cooldown (surge) as 0.9 mas. It is worth noting that at steady-state temperatures, Clapp (2003) reports no measureable jitter from the NCS.

Future missions may have more restrictive vibration requirements or different integration approaches than HST and a quantitative assessment of emitted vibration from a turbo-Brayton cryocooler is needed. Recent work at Creare has focused on quantifying the emitted vibration of turbo-Brayton cryocooler technology using two approaches. The first approach utilizes computational fluid dynamics/structural simulations, and will be the subject of this paper. The second approach involves testing the vibration emitted and will be the subject of a future paper.

VIBRATION SOURCES

Unlike Stirling and pulse tube cryocooler systems, the fundamental operating frequencies for turbo-Brayton cryocoolers are well above vibration harmonics that excite spacecraft resonant frequencies (Ross 2003). The high operating frequencies make turbo-Brayton cryocoolers ideal for space telescopes requiring high pointing accuracy. The primary vibration sources of concern for a turbo-Brayton cryocooler are described below.

Rotor Mass Imbalance

Emitted vibration from rotor motion at steady state is caused by synchronous imbalance of the rotor and sub-synchronous rotor whirl. Synchronous imbalance is a lateral motion of the rotor within the hydrodynamic bearing clearance that occurs as a result of mass imbalance in the shaft/impeller. In this case, emitted vibrations occur at the same frequency as the shaft rotational speed. These resulting forces are emitted at frequencies that are expected to be well above structural resonances of most payloads and should not affect pointing accuracy. Rotor whirl occurs at resonant frequencies for the rotor and bearings. By design, these frequencies are below the rotational speed of the shaft. We have observed rotor whirl to be greater than 1 kHz for rotor operating speeds of greater than 3,000 rev/s.

Turbomachine-Fluid Interaction

The primary flow interaction between the turbomachines and the balance of the fluid volume that has the potential to emit forces is compressor surge. Surge occurs when there is a mismatch between the compressor head-flow characteristics and the flow resistance of the cryocooler flow loop, which is dominated by the flow resistance of the turbine. High loop flow resistance causes separation and reversed flow inside the compressor impeller resulting in unsteady flow and significant pressure fluctuations that produce detectible vibration. For our turbo-Brayton cryocoolers, surge only occurs during cooldown when the compressor flow rate is low due to high flow resistance at the turbine. Decreasing cold-end temperature during cooldown reduces turbine flow resistance and relieves surge. Prior testing of the NCS has characterized the measured vibration levels during surge to be in the 10's of mN at frequencies of 0.1 to 10 Hz. Because surge will not be present during steady-state operation of the cryocooler, no further analysis is presented here.

Flow Induced Forces

Lower frequency vibration in the cryocooler system can be caused by flow-induced pressure fluctuations in the cryocooler tubing, load heat exchangers, and auxiliary components. Two mechanisms contribute to flow induced vibration—vortex shedding and flow turbulence. Vortex shedding is an oscillating flow condition that can occur in a flow when the fluid passes over a bluff body. Abrupt changes in passage geometry can also result in vortex shedding, such as flow discharge into a plenum or through an internal restriction. The vortex shedding frequency can be approximated via an appropriate Strouhal number, defined as:

$$St = \frac{fL_{char}}{V_{avg}} \qquad (1)$$

where f is the vortex shedding frequency (Hz), L_{char} is the characteristic length of an obstruction, and V_{avg} is the average flow velocity. The second mechanism, flow turbulence, is characterized by a much broader frequency range than is typical of vortex shedding. These fluctuations are random in nature and can produce a net force through RMS velocity fluctuations accompanied by pressure fluctuations in the flow.

We utilized Fluent® to assess the turbulent eddies and vortices inside representative flow passage geometries for the cryocooler. We can bound the exported RMS force magnitude by evaluating the integrated pressure force along the tube wall. The turbulence was not modelled at all length scales, but rather, we utilized a Large Eddy Simulation (LES) (Pope 2000; Sagaut 2006; Leonard 1974), which accurately and directly computes the energy containing turbulent flow scales and incorporates a turbulence model to include the effect of the small flow scales that have negligible contribution to the overall turbulent kinetic energy. This approach strikes a balance between accuracy and computational time.

COMPRESSOR AND TURBOALTERNATOR VIBRATION ANALYSIS

The two primary factors associated with emitted vibration in the compressor and turboalternator are synchronous imbalance of the rotating shaft and sub-synchronous whirl. The characteristic frequency of synchronous imbalance is equal to the rotational frequency of the system (3000 to 10,000 rev/s are typical). The shaft and bearing system can be modeled as a simple mass-spring system. Based on observed runout, typical rotor masses and typical speeds, the forces transmitted to the turbomachine housings are on the order of 1 to 2 N and occur at frequencies greater than 3000 Hz, which will have negligible impact on jitter due to the high frequency.

In addition to synchronous imbalance, rotor induced vibration is possible at lower rotational frequencies as a result of rotor whirl. During component-level testing of each turbomachine, operation is verified and bearing clearances are adjusted to operate without detectible whirl, so this source of vibration is not expected to be present in the cryocooler.

FLOW-INDUCED VIBRATION ANALYSIS

To determine the magnitude of emitted vibration from flow disturbances, we utilized the turbulence modeling capabilities in Fluent. We analyzed representative components in the cryocooler to obtain a bounding estimate of flow-induced vibration, and assumed that the effects of flow coupling between subsections are negligible relative to the magnitude of flow induced noise within each analysis region. The three items assessed were: 1) abrupt changes in flow passage geometry of interconnecting tubing between components representing a weld penetration or step; 2) small radius bends in the interconnecting tubing; and 3) flow entering and exiting the load heat exchangers. The predicted RMS force magnitude for each geometry and flow condition analyzed is summarized in Table 1. The details of the analysis are given below.

Modeling Approach

The LES turbulence simulations used to produce the results in the subsequent sections were found to be very computationally intensive. One of the major downsides of LES results from the presence of boundaries where the dynamically important turbulent eddies are very small. The size of the turbulent eddies scales with flow Reynolds number. As the Reynolds number increases, the size of the viscous sublayer decreases, along with the near wall turbulence length scales. To accurately model the boundary layer in LES, one must use a very fine mesh resolution in the near wall region.

One option to overcome the difficulties with mesh size at high flow Reynolds numbers is to utilize a suitable eddy-viscosity model to separately model the inner part of the logarithmic boundary layer (Davidson 2004). This was accomplished in Fluent by utilizing an algebraic Wall Modeled Large Eddy Simulation (WMLES). Here, the effects of small energy containing eddies near the wall are captured using a Reynolds Averaged Navier Stokes (RANS) model. The outer portion of the boundary layer, which contains the largest eddies, is directly modeled using LES.

To assess the impact of the WMLES approach on model accuracy, we can examine the energy content associated with the small eddies (which are simulated with RANS) versus the large scale eddies (which are directly modeled). Figure 1 illustrates the distribution of kinetic energy as a function of turbulent eddy size. We can see that the majority of energy is contained within the largest scale eddies. The energy content decreases with eddy size down to the approximate limit of the smallest eddy size in the flow (the Kolmogorov microscale, η).

While WMLES is largely Reynolds number independent for channel and pipe flows, the computational cost increase relative to a RANS models is still high (Menter 2012). Progressing solutions out to reasonable flow periods for these analyses necessitated the use of dedicated computing resources. Furthermore, a comprehensive time step and mesh resolution study was required to ensure that proper results were being generated.

Table 1. Flow induced vibration analysis summary.

Geometry	Analysis Case	Temp. (K)	Pressure (kPa)	Sample Period (s)	RMS Emitted Force (mN)	Frequency for Maximum Force (Hz)
Weld-in-tube	1	320	209.6	0.4	0.012	20.0
	2	85	324.7	0.4	0.013	52.2
90 Degree Tube Bend	3	320	209.6	0.4	0.173	22.5
	4	85	324.7	0.4	0.030	145.0
IFHX Inlet Section	5	85	324.7	0.35	1.285	74.3
IFHX Outlet Section	6	85	324.7	0.4	1.758	75.0

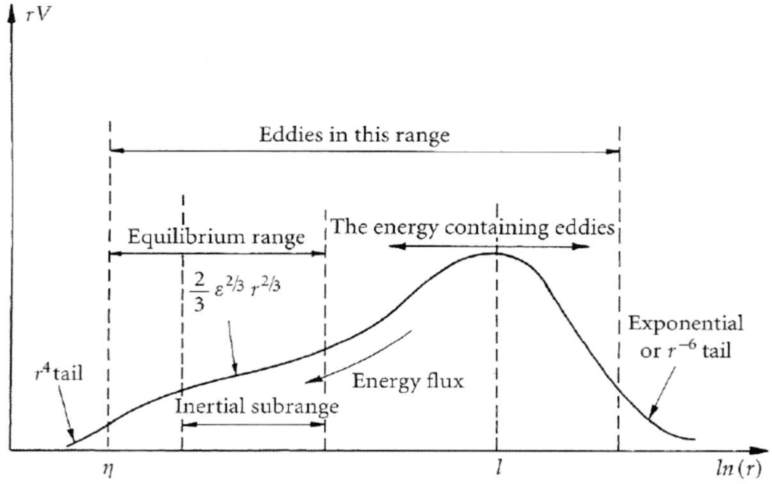

Figure 1. Distribution of turbulent kinetic energy as a function of eddy size (General Shapes per Davidson 2004). Here l is the size of the large, energy-containing eddies in the flow and η represents the approximate lower limit on eddy size.

Vibration Analysis Setup and Results

As a baseline analysis, we considered a weld projection in the ID of the cold- and warm-end tubing. To estimate the ID at the weld location, we examined welded tube joints with the same tube OD (0.5 in.) as typically used in our cryocoolers. We used a worst-case weld penetration thickness of 0.16 in. for this analysis, restricting the available flow area by 13%.

We analyzed two gas flow conditions as bounding cases for this investigation. The first condition corresponds to an upper limit on the flow velocity in the cryocooler, with the fluid at the warm-end temperature and compressor inlet pressure (Case 1). For the second analysis, (Case 2) we roughly doubled the flow's Reynolds number, consistent with flow conditions at the cold end of the cryocooler. We selected a tube length upstream and downstream of the internal weld based on an analytical correlation of the hydrodynamic development length for turbulent flow provided in Nellis and Klein (2009). This value also set the analysis region for flow induced vibration downstream of the weld. A visual representation of the flow field downstream of the weld feature is shown in Fig. 2. We can clearly see an area of flow separation resulting from the step change in passage diameter after the weld. Coherent flow structures appear in the near wall region (colored in blue) and dissipate as we move further downstream.

We calculated the integrated radial force downstream of the weld feature at each time step in Fluent by summing the dot product of the pressure force vector on the interior wall face with a unit vector in the y and z directions (Fig. 2). Figure 3 shows the force for Case 1 for a flow period of 0.4 s. We can combine the results from these two orthogonal directions at each time step to assess the magnitude of the greatest wall-normal force. We utilized post-processing code in MATLAB® to resolve spectral content over the analysis period (Fig. 4). The falloff in magnitude at high frequencies should follow a power law with a negative exponent according to turbulence theory (Pope 2000). We see that the forces predicted by the LES simulation are dominated by broadband turbulence. Furthermore, we do not observe any significant contribution to emitted force from vortex shedding, which would appear as a large increase in force magnitude at a single frequency. The emitted force predicted in ANSYS® is very small, on the order of 0.012 mN RMS.

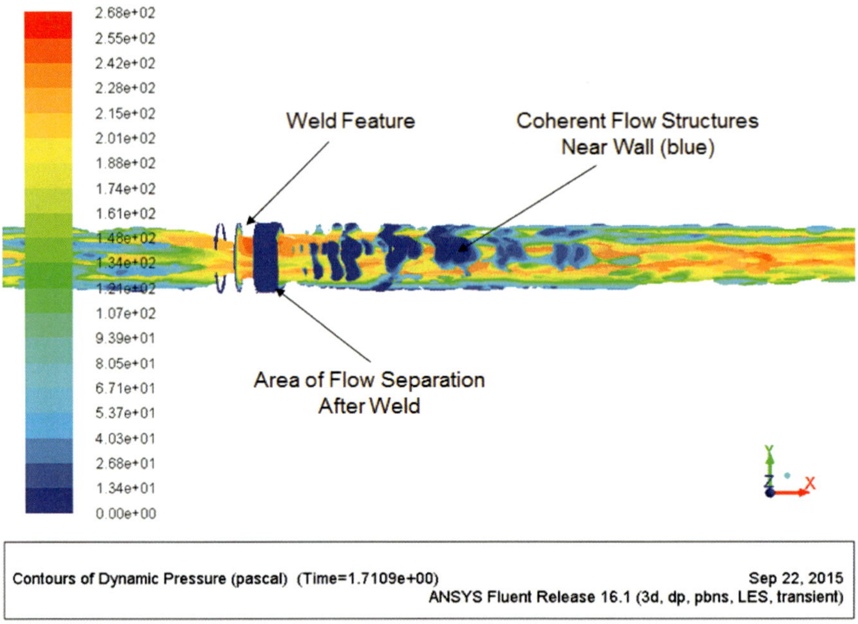

Figure 2. 3-D Flow field downstream of the weld feature. Results are shown for analysis Case 1.

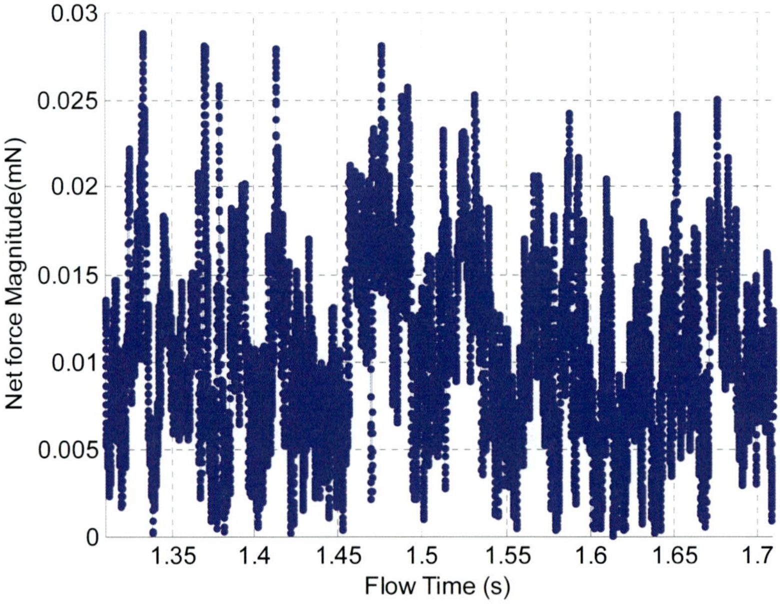

Figure 3. Net force magnitude on the interior tube wall downstream of the weld feature as a function of flow time. Results are shown for a flow period of 0.4 s for analysis Case 1.

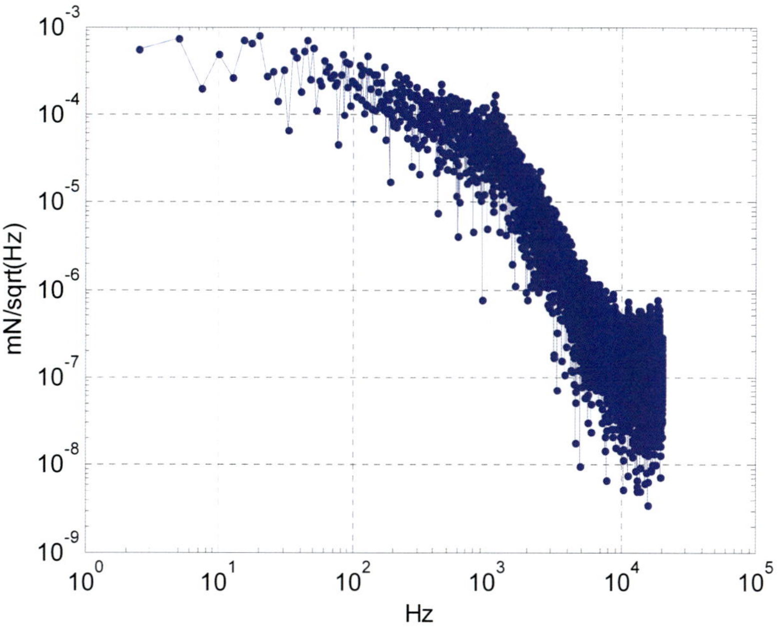

Figure 4. Spectral analysis of wall-normal force magnitude for the weld-in-tube flow geometry. Results are shown for analysis Case 1.

Results Summary

We repeated the vibration analysis for the additional five cases given in Table 1. Each analysis case required several weeks of computational time on a high-performance parallel computer cluster. The spectral results for each case are plotted in Fig. 5. We see that the IFHX remains the dominant contributor at frequencies up to 1 kHz. Flow induced vibrations are dominated by broadband-turbulence for every case considered.

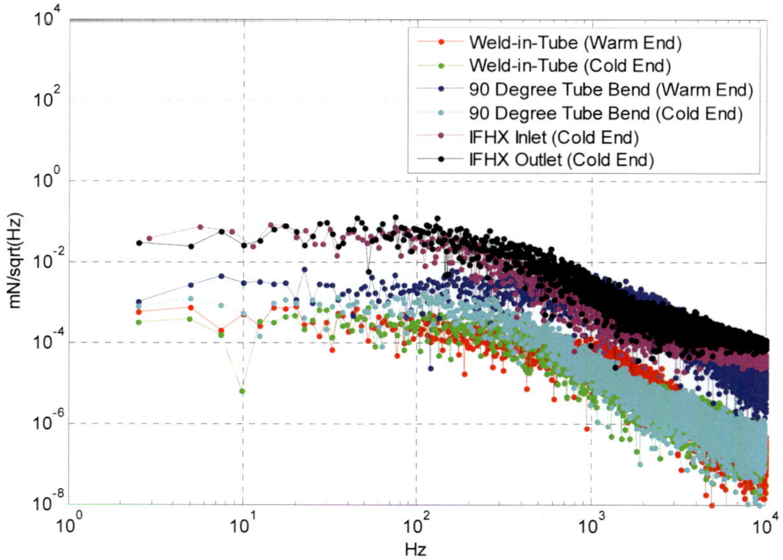

Figure 5. Power spectrum for simulated flow geometries.

CONCLUSIONS AND FUTURE WORK

The primary objective of this analysis is to quantify limits for emitted vibration from a turbo-Brayton cryocooler. For the compressor and turboalternator, we have shown a conservative emitted disturbance on the order of 1 N due to mass imbalance at frequencies greater than 3 kHz. Forces resulting from sub-synchronous rotor whirl will be of the same order of magnitude and have been shown to occur experimentally at frequencies 25 to 35% of the rotating speed at steady state (greater than 1 kHz for operating speeds above 3,000 Hz). The high frequency of these forces results in imperceptible vibrations of the turbomachines.

Using Fluent, we demonstrated an approach to assess turbulence and flow noise in the interconnecting tubing and load heat exchangers and performed an analysis to quantify the magnitude of flow induced vibrations on interior surfaces of the cryocooler flow loop. We examined three key flow geometries for the cryocooler assembly including a straight section of tubing with an over-penetrated weld joint, a small radius 90 degree tube bend, and the inlet and outlet sections of the load heat exchanger. We found that vibrations produced by broadband turbulence dominate in these sections of the cryocooler assembly. Results from these analyses can be imported into a structural harmonic analysis to predict the actual emitted disturbance for a specific payload.

The outlet section of the IFHX remains the largest contributor of flow induced vibrations. We predicted a worst-case RMS force of 1.76 mN for this geometry. No attempt was made to redesign the load heat exchanger to reduce the emitted forces at this location.

Future work involves measuring the emitted vibration of representative turbo-Brayton cryocooler components at prototypical operating conditions. This work is planned to be performed at the Cryocooler Vibration Output Test Facility that was developed and is operated by the Aerospace Corporation.

ACKNOWLEDGEMENTS

We gratefully acknowledge NASA for their support of this work on Contract NNX14CG57C.

REFERENCES

1. Breedlove, J.J., Zagarola, M.V., Nellis, G.F., Gibbon, J.A., Dolan, F.X. and Swift, W.L., "Life and Reliability Characteristics of Turbo-Brayton Coolers," *Cryocoolers 11*, edited by R.G. Ross, Jr., New York, NY, Kluwer Academic/Plenum Publishers (2001), pp. 489-498.

2. Clapp, B.R., "Vehicle Disturbance Test Report for SMOV-3A/B," Mission Operations, System Engineering & Software Program Engineering Memorandum, EM MOSES 1226, Mar 31, 2003.

3. Clapp, B.R., Stills, J.W. and Voorhees, C.R., "Hubble Space Telescope Pointing Performance Due to Micro Dynamic Disturbances from the NICMOS Cryogenic Cooler," AIAA 2002-1249 (2002).

4. Davidson, P.A., *Turbulence: An Introduction for Scientists and Engineers*, New Yord: Oxford University Press (2004).

5. Leonard, A., "Energy Cascade in Large-Eddy Simulations of Turbulent Fluid Flows," *Advances in Geophysics A*, vol. 18 (1974), pp. 237-248.

6. Menter, F.R., *Best Practice: Scale-Resolving Simulations in ANSYS CFD*, ANSYS, Inc. (2012).

7. Nellis, G. and Klein, S., *Heat Transfer*, Cambridge University Press (2009).

8. Pope, S., *Turbulent Flows*, Cambridge University Press (2000).

9. Ross, R. G., "Vibration Suppression of Advanced Space Cryocoolers – an Overview," Presented at the International Society of Optical Engineering (SPIE), Mar 2-6, 2003.

10. Sagaut, P., *Large Eddy Simulation for Incompressible Flows*, Springer (2006).

11. Swift, W.L., Dolan, F.X. and Zagarola, M.V., "The NICMOS Cooling System – 5 Years of Successful On Orbit Operation," *Trans. of the Cryogenic Engineering Conference—CEC*, vol. 53, edited by J.G. Weisend II (2008), pp. 799-806.

Development of a 3-Stage ADR for Space including CCA 50 mK Characterization

D. A. Paixao Brasiliano[1], J-M. Duval[1], N. Luchier[2], S. D'Escrivan[3], J. André[3]

[1]Univ. Grenoble Alpes, INAC-SBT, F-38000 Grenoble, France
[2]CEA, INAC-SBT, F-38000 Grenoble, France
[3]CNES, 18, Avenue Edouard Belin, 31401 Toulouse Cedex 9, France

ABSTRACT

Highly sensitive detectors required for astrophysics missions, such as ATHENA and SPICA, require temperatures as low as 50mK to operates with the targeted sensitivity. Adiabatic demagnetization refrigerator (ADR) systems are a suitable solution for sub-kelvin space applications. We present a design of a 3 stage ADR for the range 4 K – 50 mK with 0.45 μW of cooling power at 50 mK. Results of our preliminary study and design of the main components such as the heat switches and paramagnetic materials are discussed. We also report a sub kelvin characterization of the magneto caloric properties of chromic caesium alum (CCA) and compare its performance as a magnetic refrigerant to chromic potassium alum (CPA).

INTRODUCTION

The next generation of astronomy missions requires cooling down to 50 mK for highly sensitive detectors such as used in SPICA/SAFARI[1] for far IR observations and ATHENA/X-IFU[2] for X-ray spectroscopy. ADRs are a suitable solution for sub kelvin space applications for a broad range of starting temperature[3]. For instance, the last stage of the cryogenic chain of the SXS instrument on Astro-H uses a three stage ADR, combined with an intermediary 1.2 K liquid helium cryostat or a 4.5 K in degraded mode, to produce 50 mK for cooling down the X-ray detector array[4].

We hereby report our study of a one-shot multi-stage ADR for the 4 K – 50 mK temperature range having 0.45 μW of cooling power at 50 mK for an autonomy of 24 hours. Our pre-study indicated that a system composed of three stages represents a good compromise between the overall mass and the system complexity. We present the design that was chosen and its leading criteria. We also report on the impact of using CCA as an alternative magnetic refrigerant for CPA for the low temperature stage.

DESIGN

The design of our three-stage ADR is shown in Figure 1. Each stage is composed of a solid state refrigerant (a paramagnetic material), a superconducting magnet and a gas-gap heat switch (HS). Similar to the design proposed by Shirron et. al.[5], the stages are set in series, which allows the multi-stage ADR to cover the large range of temperature that is required. For convenience, the stages are called S1, S2 and S3, from the high temperature stage to the low temperature; and HS1,

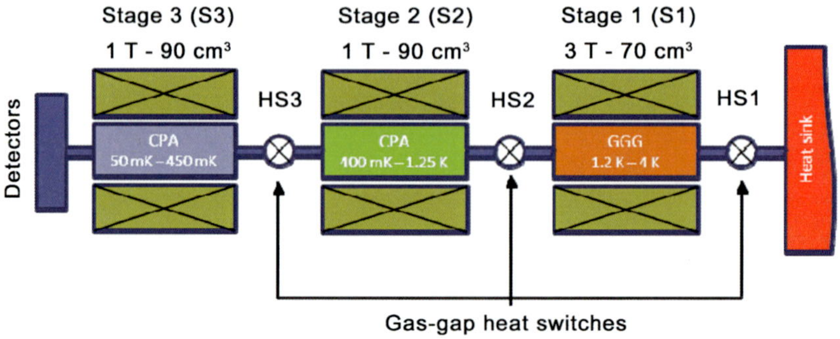

Figure 1. Schematic of the three-stage ADR. The magnetic fields indicated are averaged over the salt pill volume.

HS2 and HS3, their respective heat-switches. Some of the main elements that drive the design, namely the magnetic refrigerants and heat-switches, are discussed.

The paramagnetic materials chosen for this initial design are GGG and CPA for the high and low temperature stages, respectively. These materials are typical used for their respective temperature range and are readily available at the required amounts. In parallel, work has been done to identify, grow and characterize more efficient materials that can replace GGG and CPA in future designs, and will reduce the overall system mass.

The role of the heat switches is to provide a high thermal conductance in the closed or ON position, for magnetizing the stages and to otherwise isolate them in the open or OFF position. A high ON conductance shortens the recycling time and/or decreases the temperature gradient through the HS, and increases the heat-transfer efficiency. A low OFF conductance reduces the parasitic heat load on each stage.

^{3}He gas gap heat-switches are suitable for all stages. The coldest temperature at which the HS3 needs to be ON is 400 mK. At this temperature, care must be taken to prevent the formation of liquid by limiting the amount of helium inside the switch. Condensation of liquid would deteriorate the operation and the switching time. The HS design adopted is similar to the heat-switches developed for the Herschel space telescope program[6] in order to benefit from its heritage.

The choice of the number of stages is a trade-off between the overall complexity of the system, which increases with the number of stages, and the total mass. The former is related to the management of a higher number of elements and their associated electronics. The latter can be reduced by increasing the number of stages through reduction in the parasitic heat load and the magnetic field requirements[5]. Considering the one-shot operation, the use of classical magnetic refrigerants such as GGG and CPA and the chosen thermal budget, the gain from 3 to 4 stages is marginal in terms of mass. The conclusion reached was that a 3 stages system represents a good compromise for the design.

The primary criterion for sizing the stages is that each one must withstand 24 h at its low operating temperature under nominal heat load conditions, the starting point being its high operating temperature and maximum magnetic field as shown in Figure 1. The second criterion is that the upper stage must be able to recharge the lower stage in one cycle in order to minimize the recycling time.

In principle, only the extreme operating temperatures of the system are fixed (4 K and 50 mK). The intermediate temperatures are chosen in order to minimize the global mass. This choice depends mainly on the magnetic refrigerant properties and on the HS OFF conductance. Combined, those two factors imply that the highest stage temperature ratios rT (i.e. ratio between higher and lower stage temperatures) are found for the coldest stages in order to minimize the required magnetic field and parasitic heat load. Figure 1 shows that rT is close to 3.3 for S1, 3.1 for S2 and 9 for S3. The estimated parasitic heat load through the heat-switches are 20.9 μW for HS1, 1.1 μW for HS2 and about 70 nW for HS3.

The total heat losses of each stage during the "cold phase" is composed of the parasitic losses due to the heat switch, Kevlar suspensions and attached to the heat sink at 4 K, the latter being neg-

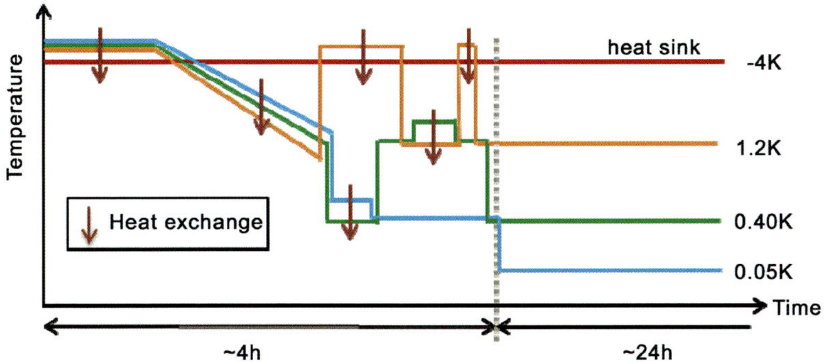

Figure 2. Schematic of the warm start recycling strategy

ligible. In order to reduce the heat load of S2 and S3, S1 is used to intercept heat from the Kevlar suspensions of S2 and S3, as well as S2 is used to intercept heat from S3. In this configuration, heat losses due to the Kevlar suspensions are estimated to be 18 μW for S1, 1.9 μW for S2 and 0.05 μW for S3, respectively. The net cooling power available at 1.2 K, 400 mK and 50 mK are estimated to be 15 μW, 1.2 μW and 0.45 μW, respectively. The power of S1 and S3 are typical external heat loads for the 1.2 K and 50 mK temperatures (e.g. precooling of wires reaching S3 and sensors load). The cooling power of S2 is a consequence of the recycling strategy we have chosen (i.e. S2 must be able to recharge S3 in one cycle) and may be used as well for precooling the sensor wires.

RECYCLING STRATEGIES

Two recycling strategies that will be employed in function of the starting condition of the multi-stage ADR is reported in this section. For this, two extreme conditions are considered: warm and cold start. The former is characterized by all stages being at the sink temperature. The latter is characterized by all stages being at their nominal cold temperature and correspond to the end of the previous cycle. At the beginning of both conditions, the stages are fully discharged i.e. in zero magnetic field.

The warm start procedure is represented in Figure 2. The cycle starts with the 3 ADR stages recharging their magnetic field and rejecting the heat load to the sink at 4 K. The second step takes advantage of the great cooling capacity of S1 at high temperatures for precooling the other stages: with HS1 OFF and HS2 and HS3 ON, S1 is slowly demagnetized in order to efficiently cool S2 and S3 down to about 1 K. This procedure allows the reduction of the recycling time compared with the alternative of independently demagnetizing the 3 stages for a few reasons: the heat exchanges for recycling stages 2 and 3 happen at high temperature, where the thermal conductance of the heat switches is enhanced. For the same magnetic field, the entropy variation at high temperature is significantly smaller than at low temperature; after the precooling, S2 is already fully recharged and can then carry on the final step to recharge of S3. The third step completes the recycling procedure: S3 is recharged by S2 at 400 mK while S1 is recharged at 4 K. Once finished, S2 is recharged by S1 at 1.2 K and then S1 is recharged at 4 K. Now that all stages are fully charged, they are demagnetized to attain their nominal temperatures. A proportional–integral–derivative (PID) algorithm is used to maintaining them for about 24 h under nominal heat load.

The cold start procedure is represented in Figure 3. The stages are first magnetized to their nominal hot temperature, which allows S1 to be recharged and the reduction of heat losses over S2 and S3. After being recharged, S1 recharges S2, which can then recharge S3. S1 afterwards recharges S2 and is finally recharged at 4 K. After demagnetization, the 3 stages are ready to deliver their cold temperature for about 24 h under nominal heat load.

CHARACTERIZATION OF CCA

The viability of using CCA for space ADR was first reported by Hagmann et al[7]. CCA has the same chemical composition of CPA with potassium replaced by caesium. Furthermore, they have

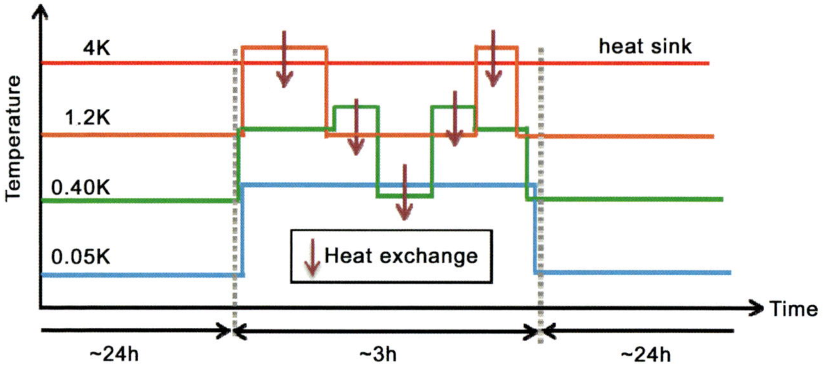

Figure 3. Schematic of the cold start recycling strategy

the same magnetic ion and many close properties such as effective angular momentum J, Landé factor g and molar volume of magnetic ions. It has been reported that CCA has greater resistance to thermal degradation than CPA, withstanding temperatures up to 50 °C in a sealed pill with no observable degradation, while the CPA was found to be degraded in the same conditions[7]. The degradation temperature is an important concern if the satellite is baked out.

CPA has been characterized during the characterization of an ESA engineering model of a hybrid 50 mK ADR/sorption cooler previously developed in our laboratory[8]. The CPA stage operates in the 450 mK – 50 mK range and is fully magnetized at 0.8 T using a heat sink at about 300 mK. CCA has been chosen to be characterize in similar conditions and using the same hybrid cooler.

The CPA pill was replaced by a CCA pill containing 102 g of the salt (0.178 mol) grown onto copper wires as has been described by Timbie et al[9]. The pill filling rate is close to 82% according to the density reported by Hagmann et al[7] and may be improved with further development of the growth process. The impact of the relatively low filling rate may affect the thermal resistance of the bus but it's unlikely to significantly affect the magnetocaloric properties of the salt.

The CCA pill was equipped with a 50 mK Lakeshore Ruthenium Oxide thermometer and a heater.

Specific Capacity, Thermal Gradient and Parasitic Heat Load Measurement Methods

Figure 4 summarizes how the specific heat, thermal gradient and parasitic heat load are measured: the evolution of the pill temperature is measured before, during and after a heat pulse is applied to the pill.

When the pill is heated, a thermal gradient appears between the extremity of the thermal bus where the thermometer is mounted and the salt. It is assumed that the salt temperature is homogeneous and that the majority of the pill heat capacity is composed by the salt. The total gradient, which includes the gradient in the copper bus and at the interfaces, can then be measured from $\Delta T_{decoupling}$ and the applied power (see Figure 4a).

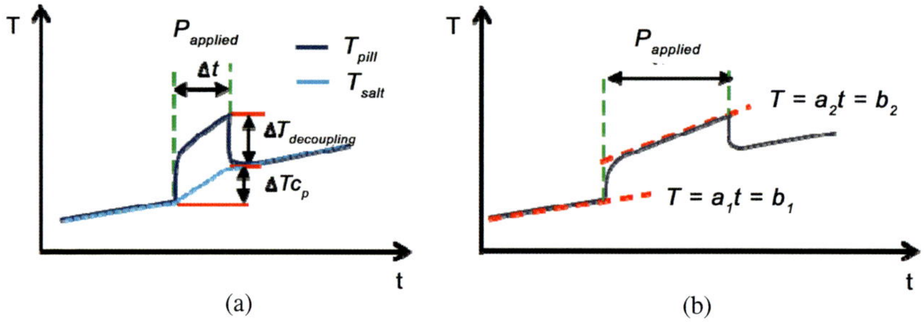

Figure 4. Method of measurement of specific heat, thermal conductance (left) and parasitic heat load (right).

Measurements of the pill thermal gradient showed that it could be predicted by the power law $\Delta T_{decoupling} = 3926 \cdot PT^{-1.92}$, with the applied power P in μW and the mean temperature T of the salt pill in mK. The thermal decoupling of the CCA pill is 2.1 mK/μW at 50 mK as compared to 1.1 mK/μW for the CPA salt pill[8].

Specific capacity can be measured from the applied power and its duration. If the parasitic heat load can be neglected compared to the applied power, the amount of energy transferred to the pill is PΔt. Thus, the pill heat capacity is given by $C(T, B) = P\Delta t/\Delta T$, where B is the magnetic field applied by the superconducting coil. If the parasite heat load is not negligible, P must be replaced by P + Pparasitic.

The parasitic heat load can be estimated by comparing the temperature slopes in the presence and absence of a constant applied power. If the temperature variation is small enough, the pill specific heat can be considered constant and the parasitic heat load can be calculated as function of the applied power and the ratio $r = a_1/ a_2$ (see Figure 4b) as follow: Pparasitic = Papplied*r/(1-r). To maximize the sensitivity of these measurements, the applied power is chosen to be of the same order of magnitude as the estimated heat load.

A parasitic heat load of 0.12 μW at 50 mK has been measured, which is in agreement with the calculated load from parasitic conductivity.

Specific Heat

The power and duration used in this measurement ranges from 2 μW to 50 μW with a duration from 5 min to 20 min depending on the temperature and magnetic field. The parasitic heat load is limited to 5% of the applied power in the worst case scenario.

As verified by Luchier et al[8] for a similar pill design, the components of the pill other than CCA have a negligible contribution to the pill specific capacity in the temperature range covered by our measurements. Thus, the specific heat of CCA is given by $C_{CCA} = P\Delta t/\Delta T/n$, where n = 0.178 mol of CCA.

The results are shown in Figure 5, as well as our measurements of CPA specific heat done in similar experimental conditions. The similar behavior of both pills suggests that the magnetocaloric properties are close. Therefore, it is expected that CCA could replace CPA in an ADR stage without a major change in performance.

Available Energy

In these measurements, the salt is magnetized at the 300 mK heat sink and then is thermally isolated by turning the HS OFF. At this point, a power is applied by the heater and the salt is demagnetized. The demagnetization is controlled by a PID algorithm so the pill temperature is kept constant until the magnetic field is depleted. The available energy at a temperature T is given by the integral over time of the total load reaching the pill. With no power applied, the pill was able to reach 28 mK when completely demagnetized from 325 mK and 0.8 T.

A measurement done at 56 mK with Papplied = 1.57 μW is shown in Figure 6. The measured parasitic heat load is less than 10% of the applied power. The temperature of the salt is estimated to be 52 mK due to the pill thermal decoupling. For the measurements at 100 mK and 200 mK, the parasitic heat load is less than 5% of the applied power.

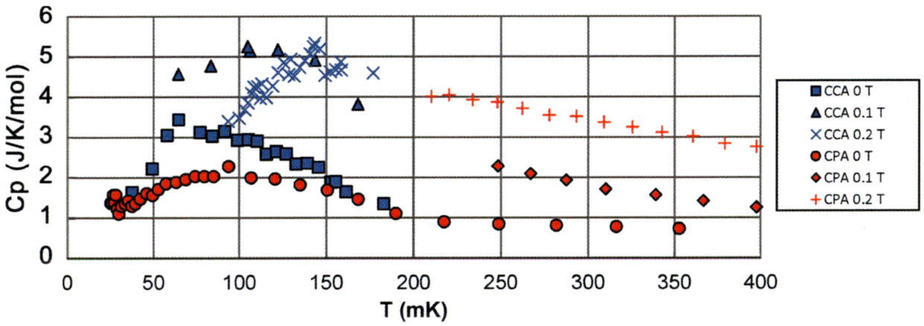

Figure 5. Specific heat measurements of CCA and CPA in different magnetic fields.

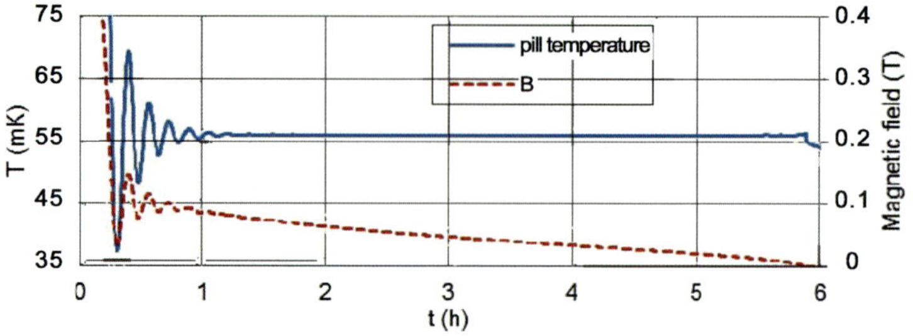

Figure 6. Temperature regulation of CCA salt pill at 56 mK with 1.57 ∝W applied load

Table 1. Available energy in function of the working temperature. The predicted energy was calculated using the theoretical entropy diagram published by Hagmman et al[7].

T0 (mK)	B0 (T)	T (mK)	P (μW)	Autonomy (h)	Predicted energy (J/mol)
315	0.8	52	1.7	5.6	0.19
302	0.8	100	5.4	6	0.65
305	0.8	200	10.6	6	1.28

Table 1 summarizes the results of these measurements. The starting magnetic field is 0.8 T for all cases but the starting temperatures are slightly different, varying between 302 mK and 315 mK. Direct and precise comparison of available energy is limited by experimental uncertainties in these types of measurements, such as the thermometers precision at high and low temperatures, magnetic field homogeneity over the salt pill volume and temperature gradient through the thermal bus. The estimation of each contribution is difficult but we think it can explain the variations of Table 1. Nevertheless, we validated that CCA can be considered a replacement of CPA for producing 50 mK from temperatures up to about 400 mK without major impact on the stage design considering our experimental precision.

CONCLUSION

A one-shot 3 stage ADR for space working on the 4 K – 50 mK temperature range with 24 h autonomy with 0.45 μW at 50 mK has been designed. The prototype is currently in its final stage of fabrication and will be characterized this fall.

The CCA in the 50 mK – 400 mK temperature range has also characterized using a maximum magnetic field of 0.8 T. This material gives the possibility of a higher temperature bake-out. This first characterization shows that the CCA has similar magneto-caloric properties to CPA within the precision of experimental measurement. Further characterization will be needed for precisely quantifying the differences between both salts and for extending the temperature and magnetic field ranges.

CURRENT DEVELOPMENTS

In parallel to this work, alternative materials with better predicted performances are currently being developed such as YbGG ($Yb_3Ga_5O_{12}$). The difficulty for their availability in the short term and large quantities prevents us from implementation in this prototype. This work will be presented in a future publication.

ACKNOWLEDGMENT

This work is financially supported by CNES and CEA. We wish to thank Lionel Dubois and Johann Breitenstein from CEA/SYMMES for their work on the development of CCA growth process. Thanks also to Jean-Louis Durand for the experimental work and his expertise in the design and set up. Thanks to Florian Bancel for the design of the complete model.

REFERENCES

1. Duband, L., Duval, J.M., Luchier, N., D'Escrivan, S., "SPICA/SAFARI sub-Kelvin cryogenic chain," *Cryogenics*, Vol. 52, Issues 4-6 (April-June 2012), pp. 145-151.

2. Ravera, L., et al. "The X-ray integral field unit (X-IFU) for Athena," *Proc. SPIE 9144 Space Telescopes and Instrumentation* 2014: Ultraviolet to Gamma Ray, 91442L (July 24, 2014)

3. Shirron, P. J., Canavan, E.R., DiPirro, M.J., Tuttle, J.G., Yeager, C.J., "A multi-stage continuous-duty adiabatic demagnetization refrigerator," *Adv. in Cryogenic Engineering*, Vol. 45B, Springer US, pp. 1629-1638.

4. Shirron, P.J., Kimball, M.O., Wegel, D.C., Canavan, E.R., DiPirro, M.J., "Design of a 3-stage ADR for the soft x-ray spectrometer instrument on the ASTRO-H mission," *Proc. SPIE 7732,* Space Telescopes and Instrumentation 2010: Ultraviolet to Gamma Ray, 773212 (July 29, 2010)

5. Shirron, P.J. "Applications of the magnetocaloric effect in single-stage, multi-stage and continuous adiabatic demagnetization refrigerators,"*Cryogenics, Vol.* 62 (July-August 2014), pp. 130-139.

6. Duband, L. "A thermal switch for use at liquid helium temperature in space-borne cryogenic systems," *Cryocoolers 8*, Springer US, pp. 731-741.

7. Hagmann, C., Benford, D. J., Richards, P. L., "Paramagnetic salt pill design for magnetic refrigerators used in space applications," *Cryogenics,* Vol. 34, Issue 3 (March 1994), pp. 213-219.

8. Luchier, N., Duval, J.M., Duband, L., Tirolien, T., "Performances of the 50mK ADR/sorption cooler," *Cryogenics,* Vol. 52 Issues 4 -6 (April-June 2012), pp. 152-157.

9. Timbie, P.T., Bernstein, G.M., Richards, P.L., "Development of an adiabatic demagnetization refrigerator for SIRTF," *Cryogenics,* Vol. 30 Issue 3 (March 1990), pp. 271-275.

Qualification Program of the Engineering Model of the 50 mK Hybrid Sorption-ADR Cooler

L. Duband and J.M. Duval

University Grenoble Alpes, CEA INAC-SBT
F-38000 Grenoble, France

ABSTRACT

For several years, we have been working on a sub-kelvin hybrid cooler concept based on the combination of a 300 mK sorption stage and a small adiabatic demagnetization stage. A first prototype of this hybrid 50 mK cooler has been developed with ESA support to demonstrate its feasibility for the IXO (International X-ray Observatory) mission, which is now the European-led mission called ATHENA. Based on this successful development and upon the knowledge gained during this work and funding support from CNES, an advanced hybrid cooler has been designed and built to fulfill the requirements of the SAFARI instrument on-board the SPICA mission as of 2014. This compact 5 kg hybrid cooler provides has a net heat lift of 0.4 µW and 14 µW, respectively at 50 mK and 300 mK, for an overall cycle duration of 48 hours and a measured duty cycle of over 78%. This result has been obtained with limited resources, 5 and 10 mW, respectively at 1.8 K and 4.9 K. The cooler successfully went through its qualification program, which includes thermal and vibration tests, and has now reached a TRL 6 level.

INTRODUCTION

The need for sub-kelvin cooling has been voiced for several years and several solutions have been successfully developed and flown. Based on our heritage on sorption coolers for several projects[1,2] we have developed a so-called hybrid cooler, which is a combination of a 300 mK helium three sorption stage and a miniature adiabatic demagnetization stage. Several prototypes and models have been built and tested[3,4]. We focus here on the engineering model (EM) specifically developed for the SAFARI instrument onboard the SPICA satellite. The definition of this mission is currently being reworked and will be submitted at the M5 selection of the European Space Agency. During the years 2011-2016, extensive work on the sub-kelvin cooler has been carried out and a first EM unit was designed, built and tested[5]. This EM has been sized to operate from two interface temperatures, 1.8 K and 4.9 K, and most importantly using very limited resources, respectively 5 and 10 mW. In fact during the course of the project the 5 mW specification was reduced to 3.5 mW at 1.8 K.

This one shot system provides simultaneously net heat lift of 0.4 µW at 50 mK and 14 µW at 300 mK, with a duty cycle of 80% and an overall cycle duration of 48 hours. The whole cooler has a low weight of 5.1 kg and is 312 x 182 x 156 mm in size. A schematic along with 2 photos are displayed in Figure 1.

Cryocoolers 19, edited by S.D. Miller and R.G. Ross, Jr.

QUALIFICATION PROGRAM

The experimental qualification campaign of the cooler comprises a test sequence, which contributes, together with analysis and other verification methods, to qualification. This test program covers the functional performance and environmental verifications required to provide confidence in the ability of the cooler to meet the specified requirements. Performance measurements have been made before and after the environmental tests, designed to detect changes in performance parameters, which may indicate a potential failure. Prior to this campaign, the heat switches undergo their own selection process.

The cooler features 3 gas gap heat switches, a proven technology already flown on several systems but has also been identified as single point failure. The redundancy solution has been extensively discussed in the past and shows that a complete redundancy architecture against the ON and OFF position of the switches would require a set of 4 heat switches for any single switch. Thus these items are not redundant, a general conclusion made for ultra low temperature coolers. Yet in order to gain confidence on the performance of the EM, additional switches are built and undergo a dedicated test campaign featuring geometrical, thermal and vibration tests. A tool has been developed to check whether or not the interlocked copper parts in the switch are centered. The tool comprises two sensors, which measure respectively the force and the displacement on a moving part. This moving part is used to push gently on the switch end in all four directions (perpendicular to the switch axis). From the recording of the force versus displacement, the position of the inner copper part is obtained and the switches are ranked.

The switches are then thermally and mechanically tested. An example of the results is reported in Figure 2. The best set is then integrated in the cooler. In practice, the vibration tests very rarely affect the performance of the switches; most of the time, degraded switches, if any, are identified during the first thermal tests and the cause is, in general, due to the brazing process (misalignment of the internal copper parts).

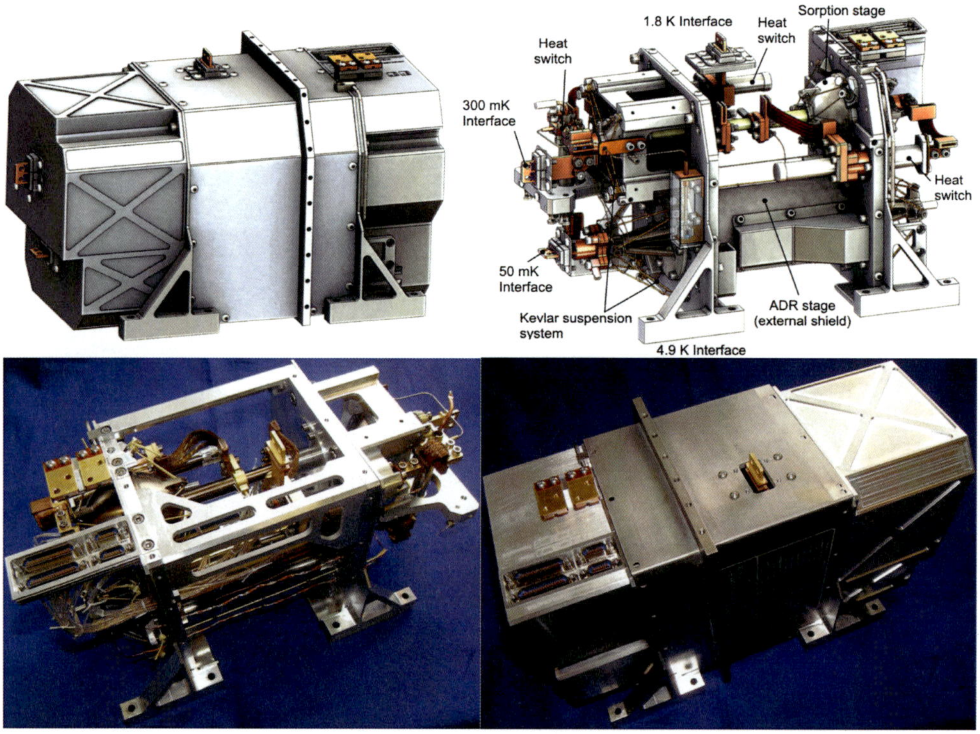

Figure 1. Schematic and pictures of the SPICA SAFARI engineering model

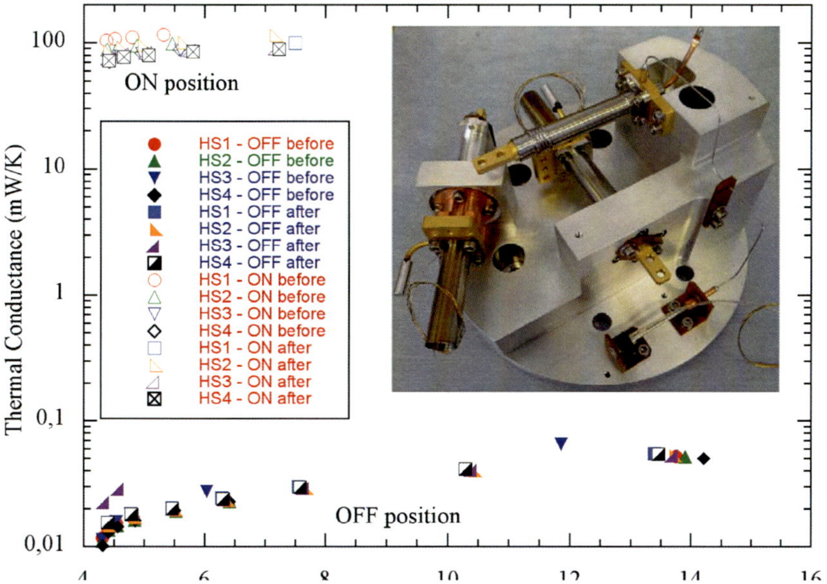

Figure 2. Heat switches thermal conductance before and after vibration tests. The picture depicts the set-up used for the vibration tests

The sorption stage is also thermally tested prior to integration, mostly to verify that the amount of helium-3 gas matches the expected performance. The cooler is charged under a cold condition. The pressure at room temperature is close to 8.4 MPa. By essence this step is straightforward assuming the sizing has been done properly. To qualify as a leak before burst, the pumping line of the cooler is designed with a calculated burst pressure of 35 MPa. This is the weakest point with respect to the internal pressure. The system was proof pressure tested at 18 MPa with ^4He (and kept at this level of pressure for 2 minutes). During this test, the recorded leak level was 2.6×10^{-11} Pa m^3 s^{-1}.

Integration in the test cryostat

A specific test cryostat featuring a nitrogen tank and two helium reservoirs has been designed to provide an extended autonomy of about 10 days at a temperature as low as 1.2 K. In this arrangement most heat inputs to the last reservoir are heat intercepted by the 4.2 K tank. Thus with a limited liquid helium volume and once pumped down, the coldest reservoir can last for one to two weeks without the need for refilling which would disturb the cooler operation. On the other hand, the 77 K and 4.2 K reservoirs, both under atmospheric pressure, can be refilled as necessary. The cryostat is fitted with two interfaces that can be temperature regulated respectively at 1.8 K and 4.9 K for the present campaign. The objective of this system is to measure (and control) the heat flows while the cooler is recycled. Thus each thermal interface is thermally connected to the helium cryostat cold plate by a calibrated copper link, which acts as a fluxmeter. Pictures of the EM cooler mounted in the test cryostat are displayed in Figure 3.

Initial thermal tests campaign

The fully assembled cooler is thermally tested under the nominal conditions, i.e., with the interfaces regulated at 1.8 K and 4.9 K. A typical result is provided in Figure 4. A duty cycle of 80% is reached, slightly in excess of the predicted performance (78%). Several scenarios are possible for the recycling of the cooler, and in the case reported a double magnetization (see Figure 5) was used. The ADR stage is first magnetized at about 2 K while the sorption stage is recycled. Then as the sorption stage is cooled down by evaporative process to 300 mK, the ADR is demagnetized so that the cost of cooling this stage is not supported by the ^3He stage. Once at 400 mK, the remaining

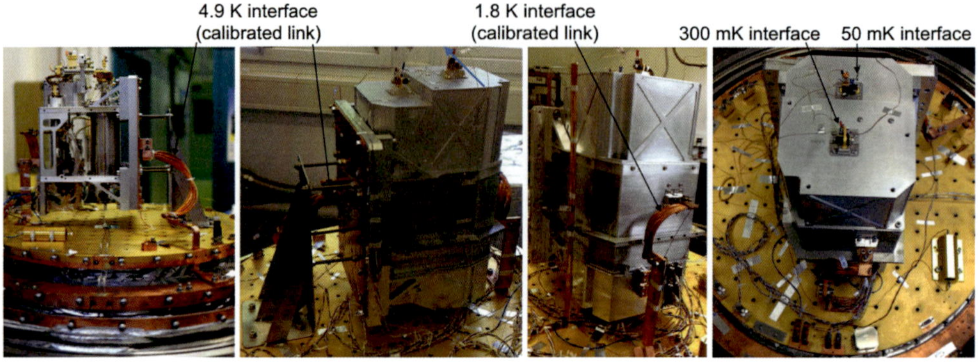

Figure 3. EM cooler mounted in the test cryostat

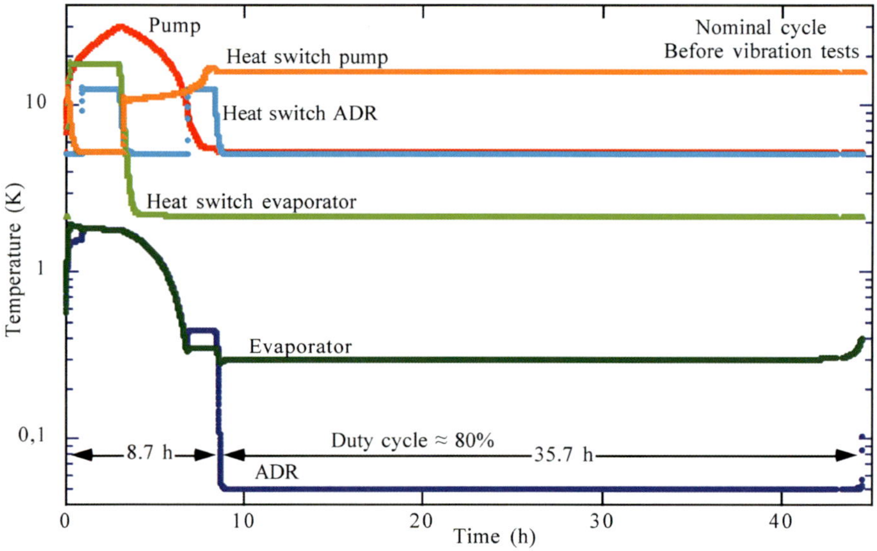

Figure 4. Typical cooler cycle, with 0.4 ∝W @ 50 mK and 14 ∝W @ 300 mK.

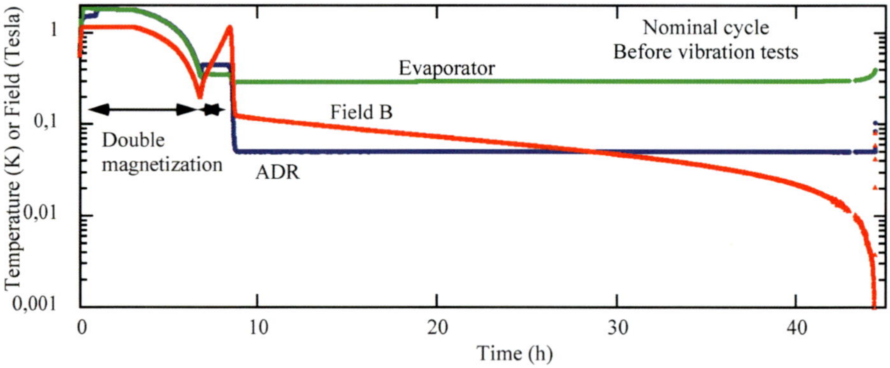

Figure 5. Typical cooler cycle: time evolution of the magnetic field

field is ramped up to about 1.1 T and the ADR is then demagnetized again to bring the salt pill to 50 mK, at which point the field is slowly reduced to keep 50 mK until the salt runs out of magnetic entropy. A few percent of the duty cycle are gained due to this recycling mode.

During the operation of the cooler, the dissipation on the two interfaces (1.8 K and 4.9 K) is monitored and controlled using a Labview® based homemade software. As mentioned the dissipations are measured in real time and the heat switch conductances are tuned accordingly. This method is very efficient but can only be used if the heat flows are measured. An alternative, previously tested on a single sorption cooler, is to produce a "thermal map" of the dissipation for heat switch conductance and sorption pump temperature. This information can then be used as an input by the drive electronic. This technique works equally well as long as the performance of the various thermal links remains the same.

As seen in Fig. 6, the dissipated powers never exceed the specified limits, i.e., 5 mW at 1.8 K and 10 mW at 4.9 K during the recycling phase, and 0.5 and 1.5 mW, respectively, during the operating phase. Additional work on the software and control electronics is currently being carried out to smooth out the behavior and to suppress the slight overshoot on the 1.8 K interface. In fact on this interface, the average dissipated power during recycling falls below the 5 mW limit.

Vibration tests campaign

The EM cooler was mounted on a vibration pot (Figure 7) and tested for each axis following:

- low sine sweep 0.5 g, 20 to 2000 Hz, 2 Oct./min
- high sine 20 g, 30 to 100 Hz, 2 Oct./min
- low sine sweep 0.5 g, 20 to 2000 Hz, 2 Oct./min
- random 13.1 g rms, 20 to 2000 Hz, 2 mn
- low sine sweep 0.5 g, 20 to 2000 Hz, 2 Oct./min

The vibration tests were successfully conducted and can be seen in Figure 8. The signatures recorded before and after each test remain fairly identical. The slight shift seen on the 50 mK interface is attributed to the accelerometer which had to be glued back. In any case no visual defect was spotted and the coil resistance of the ADR superconducting magnet was unaffected. Prior to the

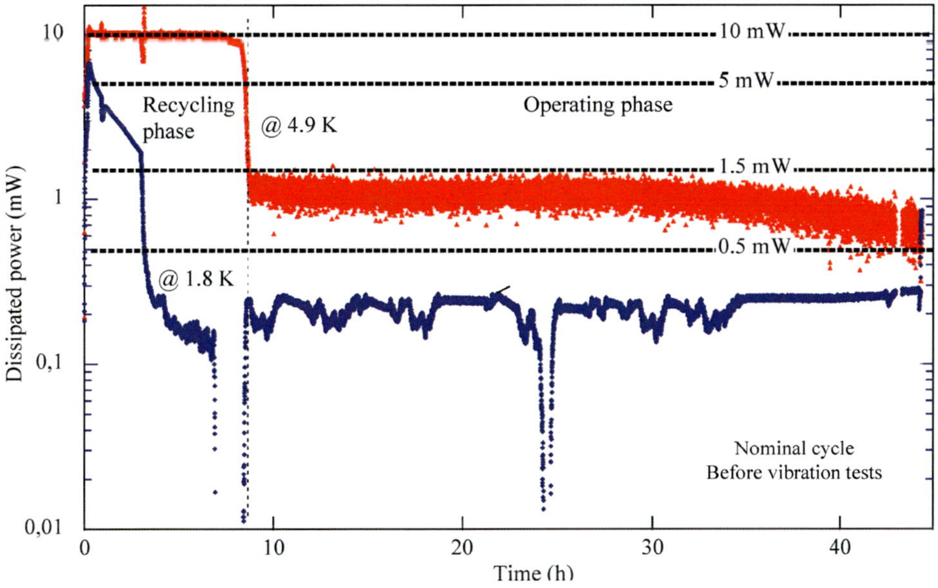

Figure 6. Dissipation of the cooler during operation

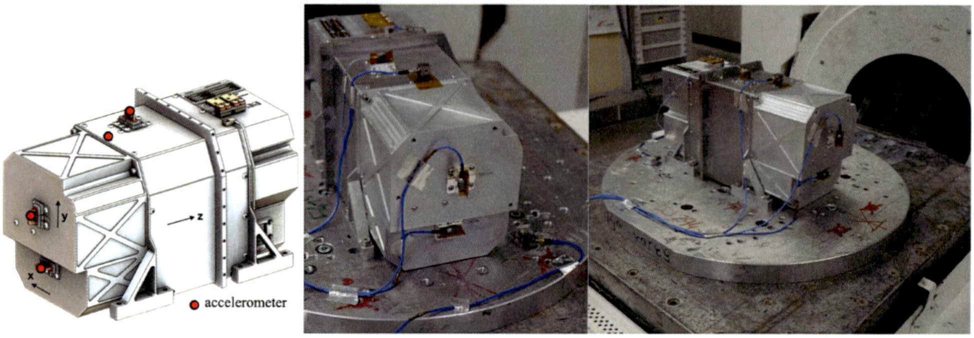

Figure 7. EM cooler ready for the vibration tests

tests, several photographic records of the interfaces and particularly of the way they were centered in their tight opening of the cover allowed us to verify that indeed the suspended masses (sorption stage and salt pill) and the thermal interfaces did not change in position. The cooler was successfully leak checked before integration in the test cryostat.

Final thermal tests campaign

In order to verify the performance of the cooler and check for any internal damage, the unit was tested again under the same operating conditions. The results did not reveal any significant differences between each set of tests.

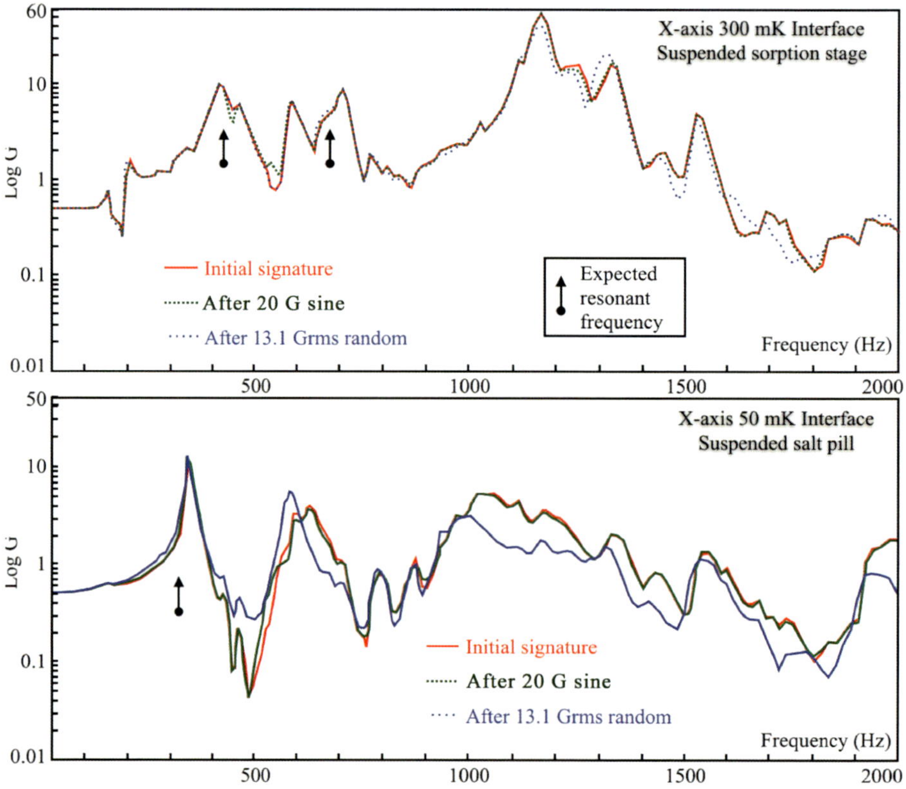

Figure 8. Example of mechanical signatures along the various vibration tests carried out

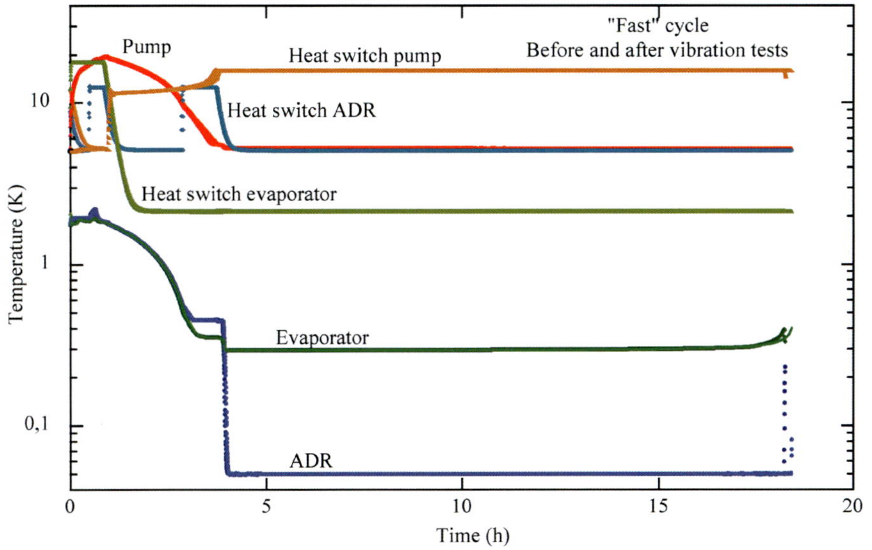

Figure 9. "Fast" cycle. Results of tests before and after the vibration tests; almost perfect superimposition

In Figure 9, we have reported what we called the fast cycle. Rather than doing a nominal test which requires 48 hours (see Figure 4), we set up a faster test which could be done overnight. In fact the cooler is flexible and can be operated over a wide range of conditions. In this case the main sorption pump is heated to 20 K instead of 30 K, and the magnetic field is ramped up to 0.6 T for the second magnetization. Note that the nominal test was done also and lead to identical results. In the results reported in Figure 9, no differences can be distinguished between the cycles obtained before and after the vibration tests. They perfectly superimpose leading to the conclusion that the vibration tests did not affect the performance of the cooler.

The cooler is currently tested under various conditions for which the interface temperatures and available resources are varied. These results will be reported in a subsequent publication. One of the goals is to improve our modeling of the cooler and to be able to better predict its performance. The current model is already reliable to better than 90%, and it can be perfected to provide a better predictions.

CONCLUSION

We have designed, built and tested an engineering model hybrid cooler comprising a 300 mK stage and a 50 mK miniature adiabatic demagnetization. This cooler was originally designed for the SPICA SAFARI instrument to provide cooling powers of 0.4 μW and 14 μW, respectively at 50 and 300 mK. This mission is under redefinition and will presented at the ESA M5 selection. The cooler went through its qualification program and has now reached TRL 6 level.

In the meantime, the X-Rays mission ATHENA (ESA Class L) has been confirmed and the sub-kelvin requirements in terms of temperature match those of SAFARI. The net heat lifts to be provided are not clearly defined at this stage but the SAFARI EM cooler is currently identified as a baseline. Due to its flexibility of operation, it can cover an extended range of operating conditions. If needed, the same design could be scaled up or scaled down to produce and qualify the new cooler to a high TRL within limited time.

ACKNOWLEDGMENT

We thank the French Space Agency (CNES) for the funding of these works. We thank Florian Bancel, Laurent Clerc and Jean Louis Durand for their expert technical work on this project.

REFERENCES

1. Duband, L, Clerc, L., Ercolani, E., Guillemet, L. and Vallcorba, R., "Herschel flight models sorption coolers," *Cryogenics*, vol. 48 (2008), pp. 95-105.

2. Ercolani, E., Relland, J., Clerc, L., Duband, L., Jourdan, T., Talvard, M., Le Pennec, J., Martignac, J. and Visticot, F., "Cryogenic system for the ArTeMiS large sub millimeter camera" *SPIE Astronomical Telescopes+ Instrumentation*, International Society for Optics and Photonics (July 2014), pp. 915324-915324.

3. Duband, L., Duval, JM. and Luchier, N., "Development status of 50 mK hybrid coolers for space applications at CEA-SBT" *Cryocoolers 17*, ICC Press, Boulder, CO (2013),

4. Prouvé, T., Duband, L., Hodis, J., Bock, JJ., Bradford, CM. and Holmes, W., "SPICA/BLISS cryo-chain demonstrator" *Cryogenics,* vol. 70 (2015), pp. 70-75

5. Duband, L., Duval, J.M. and Luchier, N., "SAFARI Engineering Model 50 mK Cooler" *Cryogenics*, vol. 64 (2014), pp. 213-219.

Performance Analysis of an Active Magnetic Regenerative Refrigerator for 20 K

I. Park, S. Jeong

KAIST, Daejeon, Republic of Korea

ABSTRACT

An active magnetic regenerative refrigerator (AMRR) with the conduction cooled high temperature superconducting (HTS) magnet is fabricated and experimentally investigated. Key components of the AMRR are a two-stage multi-layered active magnetic regenerator (AMR), a conduction cooled HTS magnet, and the helium gas flow system. Since magnetic refrigerants are only effective in the limited temperature range near their Curie temperatures, the layered structure with four magnetic materials ($GdNi_2$, $Gd_{0.1}Dy_{0.9}Ni_2$, $Dy_{0.85}Er_{0.15}Al_2$, $Dy_{0.5}Er_{0.5}Al_2$) is employed on the AMR to increase its temperature span. The AMR is designed to operate with the temperature range between 80 K and 20 K. The HTS magnet, being cooled down by a two-stage Gifford-McMahon (GM) cryocooler, produces maximum magnetic field of 3 T for the AMR. DC power supply, a solenoid current switch and an external dump resistor are used for a continuous ramping operation of the HTS magnet. Oscillating helium flow in the AMR is induced and controlled by the helium compressor and four solenoid valves. This paper presents the experimental results of the AMR system operating between 80.8 and 21.8 K and discusses technical issues for the results.

INTRODUCTION

An AMRR is a type of magnetic refrigerator which utilizes magneto-caloric effect (MCE) for refrigeration. Early magnetic refrigerator such as an adiabatic demagnetization refrigerator (ADR) has been only employed to obtain sub-Kelvin temperature [1]. However, adoption of an active magnetic regenerator concept has enabled researchers to use MCE for refrigeration with a large temperature span. An AMR system can achieve much larger temperature lift than that of ADR by effectively utilizing a magnetic refrigerant as a regenerator and as an active magnetic component. Due to its advantage, an enormous number of AMRRs operating at room temperature have been reported [2-4]. However, only a few AMRRs working below liquid nitrogen temperature have been developed and investigated over the past several decades [5 - 11].

Since the energy interaction of MCE is an inherently reversible process, AMRR has high thermodynamic potential to improve the efficiency of hydrogen liquefaction. In order to achieve the wide temperature span of an AMR, large magnetic entropy change over broad temperature range is indispensable [12]. Although an AMR can operate at a larger temperature span than the adiabatic temperature rise of a magnetic refrigerant, the temperature span is still limited because the sufficiently large MCE of a typical magnetic refrigerant appears over a limited temperature range. For instance,

Cryocoolers 19, edited by S.D. Miller and R.G. Ross, Jr.

Numazawa et al. [8] experimentally investigated single layered AMRs with three different kinds of magnetic refrigerants (GGIG, $HoAl_2$, and $DyAl_2$) and the maximum temperature span of 12 K was achieved at the maximum magnetic field of 1.8 T. Moreover, they proposed a three stage AMR to cover the temperature span between 77 K and 20 K. Limitation of a single AMR can be overcome by using a multi-layered AMR. A two stage, two-layered AMR [10, 11] is reported in the previous research with an operating temperature span of 50 K.

An AMR requires magnetic field variation during operation. A superconducting magnet is an appropriate candidate as a magnetic field generator for a cryogenic device such as a hydrogen liquefier. Previous research groups [8, 9] used direct current (DC) superconducting magnet to apply the necessary magnetic field to the AMR. The magnetic field variation of the AMR was possible by relative displacement between the AMR and the magnet. However, continuous movement of an experimental device in a cryogenic environment often undermines its reliability. For a static AMRR, an alternating current (AC) low temperature superconducting (LTS) magnet was selected by previous researchers [10, 11] to create magnetic field variation. During the AMR experiment, a cryostat was filled with expensive liquid helium to remove heat from the LTS magnet caused by AC loss. Using liquid helium is neither an attractive nor efficient way for creating 20 K which enables a potential refrigeration system to liquefy hydrogen. In order to fabricate a cryogen-free system, a conduction cooled high temperature superconducting magnet must be prepared for operation above 20 K. This paper describes an AMRR operating between 80 K and 20 K with the conduction cooled HTS magnet. The AMR system including the AMR, the HTS magnet and the helium oscillating flow system is presented and the experimental results are discussed.

EXPERIMENTAL SETUP

Conduction cooled HTS magnet system

The HTS magnet system employed in this paper is a modified version of the previously studied magnet [13]. Figure 1 shows the overall HTS magnet system. Ramping current generated by a DC power supply (KLN 40-76, Kepco) is applied to the magnet by passing through the conduction cooled hybrid current leads. The hybrid current lead consists of HTS and copper current leads. Since

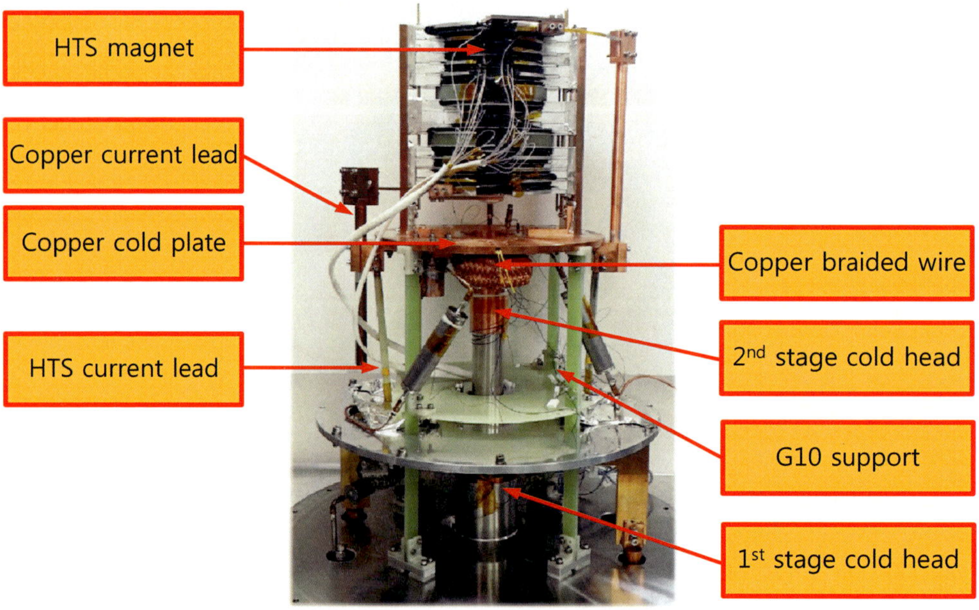

Figure 1. A photo of the HTS magnet system

AC flows through the magnet, heat generation due to AC loss has to be effectively minimized and dumped to the environment. The maximum heat generated by the magnet in the experiment is 12 W with the operating frequency of 0.1 Hz and the magnet is conductively cooled to 20 K by a two-stage GM cryocooler (RDK-415D, Sumitomo). Thermally conductive components such as copper braided wires, a copper plate and copper supports enable the cryocooler to cool the magnet with negligible temperature difference. The magnet is composed of twelve double pancake coils which are connected to each other by superconducting joints. Each double pancake coil was wound by GdBCO conductor insulated with polyimide tape. The HTS coils are thermally connected to aluminum plates and aluminum wings by Stycast 2850FT in order to effectively remove heat generated from the coils. The maximum central magnetic field of 3 T is produced when the current of 150 A is provided for the magnet whose diameter is 26 mm.

Active magnetic regenerator

The two stage AMR is composed of four different kinds of rare-earth compounds and shown in Fig. 2. The compounds are filled in the thin walled stainless steel tube with the inner diameter of 21.6 mm. The top end of the stainless steel tube is welded with a stainless steel flange. The stainless steel flange is mechanically fixed to an aluminum radiation shield by screws. Each compound is allocated by its primary transition temperature. $GdNi_2$ and $Dy_{0.85}Er_{0.25}Al_2$ which have relatively high transition temperatures are stacked in AMR1 while $Dy_{0.5}Er_{0.5}Al_2$ and $Gd_{0.1}Dy_{0.}9Ni_2$ are filled in AMR2 to cover low temperature region. A heater made by NiCr wire is installed in order to maintain the warm end of the AMR1 at 80 K. The filling ratio of the AMR is determined by a one-dimensional time dependent numerical model and the numerical results are indicated in Fig. 3. Each operation temperature of AMR1 or AMR2 is set to 80 - 45 K or 45 - 20 K by numerical simulation. Precise numerical calculation method is explained thoroughly in previous study [14]. Because the heat capacity of AMR1 is much higher than AMR2, the optimum shuttle mass for AMR1 is larger than AMR2. A bypass line made of 1/16 inch stainless steel tube is installed between AMR1 and AMR2 to control the amount of the shuttle mass for each AMR. Four calibrated temperature sensors (Cernox, Lakeshore) are installed on the external surface of the AMR to measure the temperature variation. The experimental apparatus including the AMR and the HTS magnet in the cryostat is evacuated below 10^{-4} torr during the experiment.

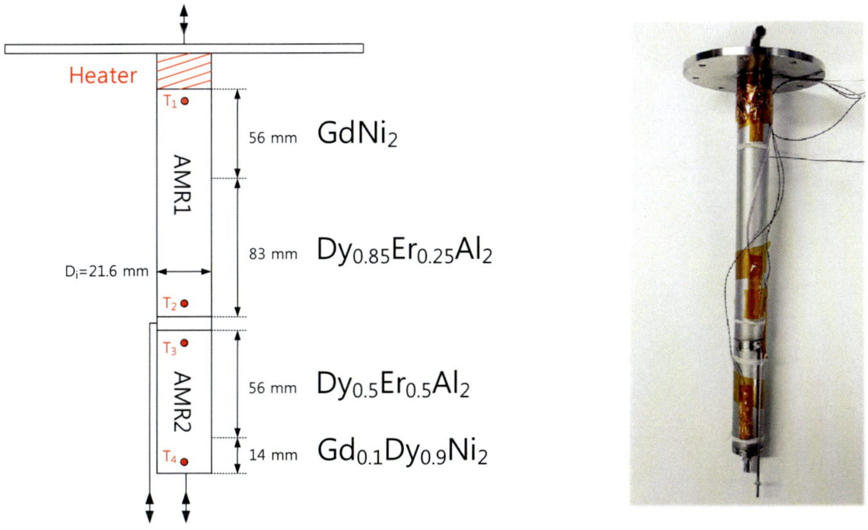

(a) (b)

Figure 2. (a) Schematic diagram and (b) a photo of the AMR

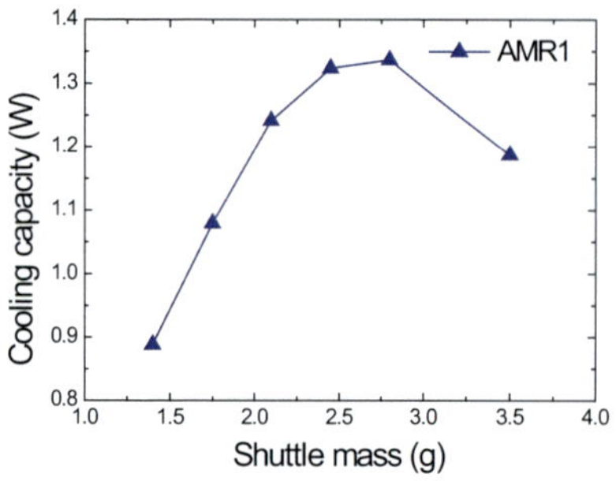

(a)

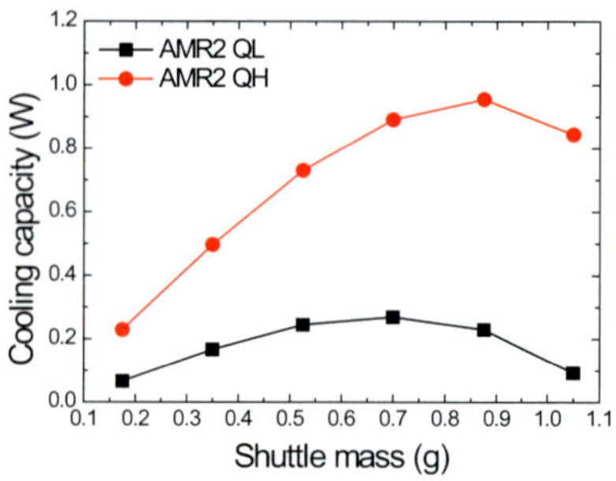

(b)

Figure 3. Cooling capacity of (a) first (AMR1) and (b) second stage (AMR2) of the AMR calculated by numerical simulation.

Helium oscillation flow system

The helium gas oscillation flow system is illustrated in Fig. 4. The oscillating flow is provided by a helium compressor (M125, Austin Scientific) and two solenoid valves. Helium compressor generates a pressure difference and the inlet pressure (P1) changes by opening two solenoid valves alternatively. When periodic inlet pressure variation occurs, helium gas flows through the AMR and its shuttle mass is determined by the size of the buffer volumes, V1 and V2. Figure 5 shows the fabricated buffer tanks to create the desirable amount of oscillating flow. Since the helium gas with ambient temperature enters into the cryostat, the gas is cooled by the first stage of GM cryocooler. Moreover, passive regenerators (R1, R2, R3) are added at the oscillating helium flow line to reduce the cooling load of the cryocooler and AMR1.

EXPERIMENTAL CONDITIONS AND RESULTS

Experimental process and operating conditions

The experimental procedure is as follows:

1. Evacuate the AMR system for good thermal insulation.
2. Cool down the AMR system including the HTS magnet and the AMR by a cryocooler.
3. Proceed with HTS magnet test : continuous AC operation.
4. Oscillate helium gas flow.
5. Conduct the AMR experiment until the cyclic steady state is obtained.
6. Change the operating conditions.

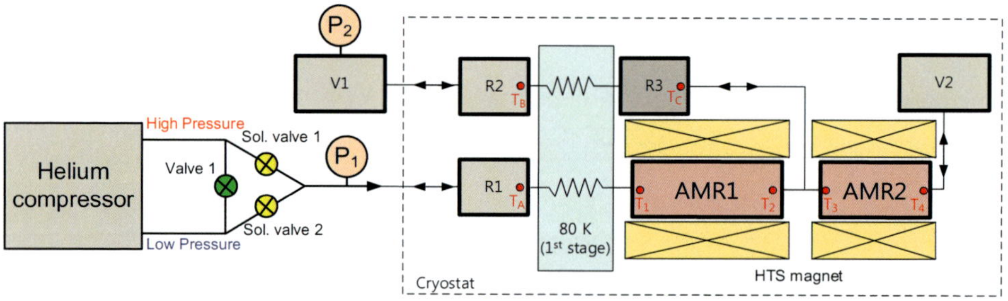

Figure 4. Helium gas oscillation flow system

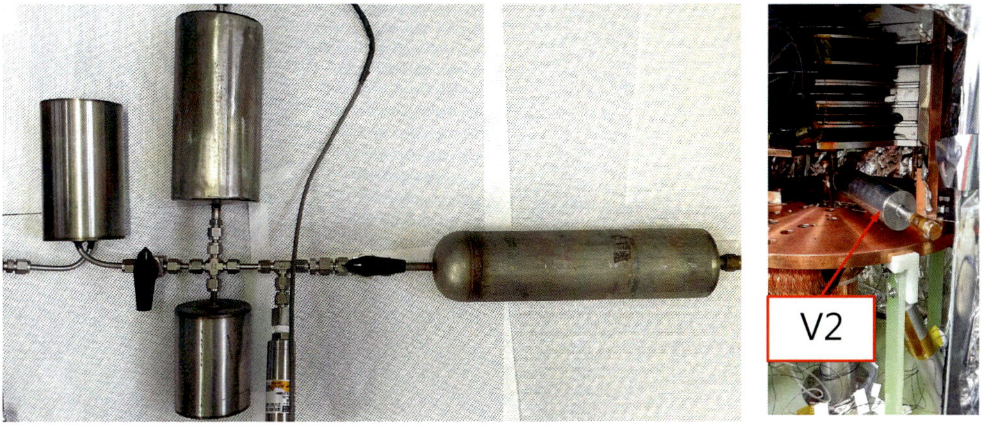

(a) (b)

Figure 5. Photo of the buffer volumes : (a) V1 composed of four stainless steel reservoirs and (b) V2 connected to the cold end of the AMR2

The helium gas flow and the magnet field variation during one cycle are illustrated in Fig. 6. The AMR is pseudo-adiabatically magnetized without helium flow at first. When the magnetic field reaches its maximum value, the up flow is induced so that the heat generated during the magnetization process is rejected through the top of the AMR. Next, the AMR is demagnetized in a pseudo-adiabatic manner to create low temperature. The down flow after demagnetization produces a cooling effect at the bottom of the AMR. Three experimental cases are reported in this paper. Ramping up or down time of the magnetic field is set to 3 seconds in all cases while the constant magnetic field times or the flow period of the three cases are specified as 7, 12, 17 seconds, respectively. The oscillating helium flow is generated during the constant magnetic field period to simulate magnetic reverse Brayton cycle. The maximum amplitude of buffer volumes' pressure swing (P2) which is in proportion to the amount of the shuttle mass is controlled by the helium flow time. The pressure swing amplitude increases as the gas flow time during half cycle lasts longer. The temperatures of the AMR, the pressures of the inlet and the buffer volume (V1) and the electric current supplied to the magnet are recorded at a rate of 10 samples per second using the Labview® program during the experiment. A summary of the operating conditions for three cases is presented in Table. 1.

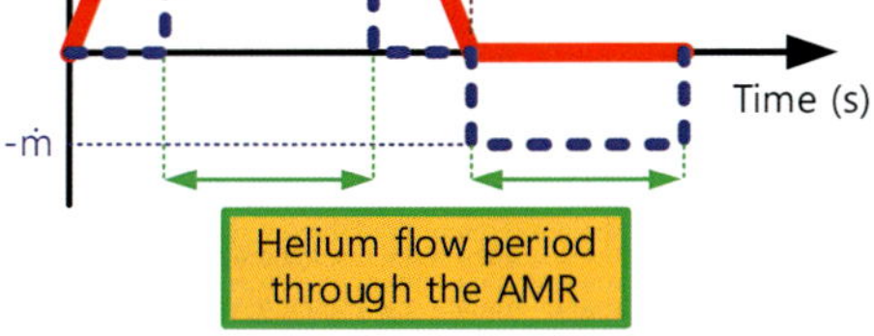

Figure 6. Magnetic field and mass flow rate variation during one cycle (the minus sign means down flow of the helium gas)

Table 1. Operating conditions of the AMR system

			Case 1	Case 2	Case 3
Cycle period (s)			20	30	40
Duration of mass flow during half cycle (s)			7	12	17
magnetization or demagnetization period (s)			3		
Buffer volume (m^3)		**V1**	2.79×10^{-3}		
		V2	4.83×10^{-5}		
Regenerator	**Length (m)**	**R1, R2**	0.2		
		R3	0.15		
	Diameter (m)	**R1, R2, R3**	0.248		
	Porosity (-)	**R1, R2, R3**	0.65		
P_1 (MPa)		**Max.**	1.3		
		Min.	0.3		
P_2 (MPa)		**Max.**	1.09	1.13	1.17
		Min.	0.76	0.67	0.58
Shuttle mass calculated by P_2 during half cycle (g)		**AMR1**	1.81	2.53	3.25
		AMR2	0.30	0.42	0.54

Experimental results and discussions

Figure 7 shows the temperature history of the AMR for Case 1. The warm end of the AMR1 (AMR1H) is cooled to 60 K by the first stage of the GM cryocooler at the beginning of the operation. During of the AMR operation, we can observe that the AMR temperature slowly decreases as shown in Fig. 7(a). A heater installed at the warm end of the AMR is turned on after the thermal cyclic steady state of the AMR is reached. The temperature of the AMR1H is held near 80 K by the heater as shown in Fig. 7(b). The no-load temperature spans and the lowest temperature of the AMR are 50.2 K and 28.3 K, respectively in Case 1. If the AMR does not experience magnetic field variation, the overall temperature of the AMR slowly rises and it can be observed in Fig. 7(b) after 18,000 seconds. The temperature variation during two cycles at the cyclic steady state is shown in Fig. 8(a). When the AMR is magnetized, the AMR temperature increases due to MCE. The cold ends of the AMR1(AMR1L) and AMR2(AMR2L) are cooled down by the upflow of the helium gas after magnetization. In contrast, the temperature of AMR1H slightly increases in upflow period. This

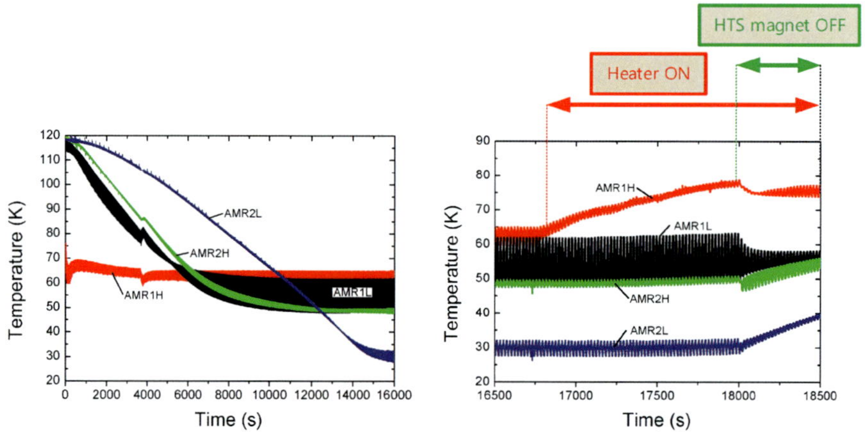

(a) (b)

Figure 7. Temperature history of the AMR of case 1 (a) without heating (b) with heating of the AMR1H.

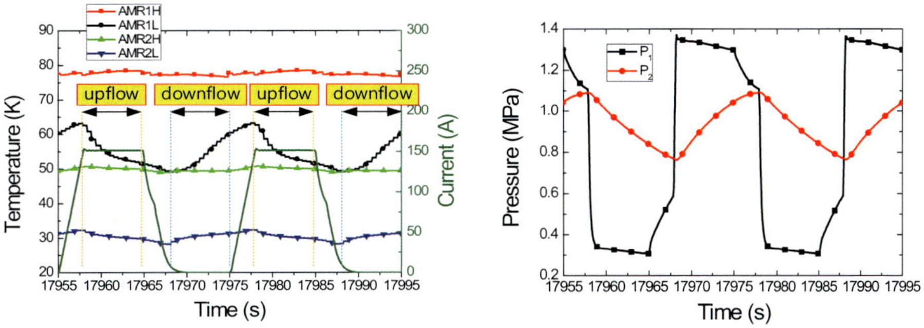

(a) (b)

Figure 8. (a) Temperature variation of the AMR and (b) pressure variation of the P1 and P2 during two cycles at case 1.

means that the AMR dumps heat at the warm end by the up flow of helium gas which exchanges heat with the magnetic refrigerant. The AMR is demagnetized after the up flow of the helium gas and its temperature decreases. Finally, the temperature of the AMR increases in the down flow period because the heated helium gas from the warm end flows through the AMR and exchanges heat with the magnetic refrigerants. The pressure swing of the inlet of the AMR (P_1) and V1 (P_2) is shown in Fig. 8(b). The pressure variation of P_1 is identical in all cases. However, the amplitude of P_2 depends on the flow duration. It is interesting to note that the temperature span of the AMR increases in accordance with the amount of shuttle mass, even though the operating frequency of the AMR system decreases. The largest temperature span of 59 K is achieved in case 3 and the result is shown in Fig. 9(b). The shuttle mass of AMR1 in case 2 was estimated as the amount for optimal performance of the AMR1 in the numerical results as shown in Fig. 3. However, the largest temperature span of the AMR1 is achieved in the case 3. This discrepancy can stem from the inefficiency of R3. Since R3 is not an ideal regenerator, the cold end of R3 connected with those of AMR1 deprives cooling capacity of AMR1 to maintain low temperature. The parasitic heat load by R3 is influenced by the operating frequency, the shuttle mass, and temperature span of R3. If V1 is installed in the cryostat by removing R2 and R3, we can directly compare the experimental results with the numerical results. The pressure swing of helium gas in V2 is not directly measured in this paper. If we assume the pressure swing in V2 is identical to that in V1, the shuttle mass of AMR2 can be estimated and the results are included in Table 1. Because the calculated shuttle masses in three cases are smaller than the optimal value (0.7 g) which is predicted by the numerical results of Fig. 3, the temperature span of AMR2 is expected to be further increased by transferring more shuttle mass of helium gas.

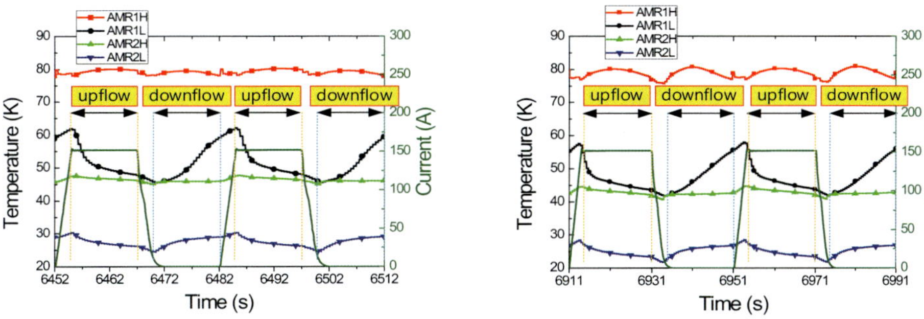

(a) (b)

Figure 9. Temperature variation of the AMR at (a) case 2 and (b) case 3

CONCLUSIONS

The multi-layered AMR operating between liquid nitrogen and hydrogen temperatures is designed, fabricated and tested. Four different kinds of magnetic refrigerants are arranged in the AMR bed. The cold end of the AMR reaches no-load temperature of 21.8 K and the maximum temperature span of 59 K is achieved. The performance of the AMR is investigated with various shuttle masses of helium gas. The temperature span of the AMR is increased with the increased the shuttle mass of helium. The numerical results predict that the performance of the AMR should decrease if the amount of the shuttle mass exceeds a certain value. In order to observe the performance tendency of the AMR, the buffer volumes (V1, V2) should be enlarged to increase the shuttle mass. Furthermore, the ineffectiveness of the auxiliary passive regenerator (R3) deteriorates the performance of the AMR1. Installation of the buffer volume, V1 in the cryostat will minimize heat leak at the cold end of the AMR1.

ACKNOWLEDGMENT

This research was supported by the Converging Research Center Program through the Ministry of Science, ICT and Future Planning, Korea (2014M3C1A8048836).

REFERENCES

1. Giauque, W.F.. MacDougall, D.P., "Attainment of temperatures below 1 K absolute by demagnetization of $Gd_2(SO_4)_3 \cdot 8H_2O$," *Physics Review*, Vol. 43 (1933), pp. 768.

2. Gschneidner, K.A., Pecharsky, V.K., "Thirty years of near room temperature magnetic cooling: Where we are today and future prospects,," *International Journal of Refrigeration*, Vol.31 (2008), pp. 945-961.

3. Scarpa, F., Tagliafico, G., Tagliafico, L. A., "A classification methodology applied to existing room temperature magnetic refrigerators up to the year 2014," *Renewable and Sustainable Energy Reviews*, vol .50 (2015), pp. 497-503.

4. Yu, B., Liu, M., Egolf, P. W., Kitanovski A., "A review of magnetic refrigerator and heat pump prototypes built before the year 2010" *International Journal of Refrigeration*, Vol.33 (2010), pp. 1029-1060.

5. Jeong, S., Smith Jr, J. L., Iwasa Y., "Tandem magnetic refrigerator for 1.8 K," *Cryogenics*, Vol. 34 (1994), pp.263-269.

6. Cogswell, F.J., Iwasa, Y., Smith Jr, J. L., "A regenerative magnetic refrigerator operating between liquid helium and liquid hydrogen reservoirs," *IEEE Transactions on magnetics*, Vol. 24 (1988), pp. 1011-1014.

7. Seyfert, P., Bredy, P., Claudet, G., "Construction and testing of a magnetic refrigeration device for the temperature range of 5–15 K," *Proceedings of International Cryogenic Engineering Conference* (1988), pp. 607–611.

8. Numazawa, T., Kamiya, K., Utaki, T., Matsumoto, K., "Magnetic refrigerator for hydrogen liquefaction," *Cryogenics* , Vol. 62 (2014), pp.185-192.

9. Hirayama, Y., Okada, H., Nakagawa, T., Yamamoto, T., Kusunose, T., Numazawa, T., Mastumoto, K., Irie, T., Nakamura, E., "Experimental study of active magnetic regenerator (AMR) composed of spherical GdN," *Cryocoolers*, Vol.16 (2011), pp.531.

10. Kim, Y., Park, I., Jeong, S., "Experimental investigation of two-stage active magnetic regenerative refrigerator operating between 77 K and 20 K," *Cryogenics*, Vol. 57 (2013), pp. 113-121.

11. Park, I., Jeong, S., "Experimental investigation of 20 K two-stage layered active magnetic regenerative refrigerator," *IOP Conference Series: Materials Science and Engineering*, Vol. 101 (2015), pp.012106.

12. Jeong, S., "AMR (Active Magnetic Regenerative) refrigeration for low temperature," *Cryogenics*, Vol. 62 (2014), pp. 193-201

13. Park, I., Lee, C., Park, J., Kim, S., Jeong, S., "Ramping operation of the conduction-cooled high-temperature superconducting magnet for an active magnetic regenerator system," *IEEE Transactions on applied superconductivity*, Vol. 26 (2016), pp.4600505.

14. Park, I., Kim, Y., Park, J., Jeong, S., "Design method of the layered active magnetic regenerator (AMR) for hydrogen liquefaction by numerical simulation," *Cryogenics*, Vol. 70 (2015), pp.57-64.

Improved Power Efficiency for Cryogenics at the Very Large Array

D. Urbain, W. Grammer, G. Peck, J. Jackson, S. Durand

National Radio Astronomy Observatory
Socorro, NM 87801

ABSTRACT

The National Radio Astronomy Observatory (NRAO) recently completed a major upgrade to the electronics in the Karl Jansky Very Large Array (VLA) telescope, including all eight cryogenic receivers. The focus of development has now shifted toward improving efficiency and reliability to reduce maintenance downtime and power consumption. The cryogenic system, in particular, requires frequent maintenance by skilled labor, and consumes a considerable fraction of the electrical power supplied to the array. Careful evaluation of the current system and preliminary tests of new techniques have demonstrated significant power savings are possible, along with an overall reduction in maintenance. This paper will describe the application of a Variable Frequency Drive (VFD) for powering the cryogenic refrigerators, to tailor the flow requirement for an individual receiver and extend the cold-head life. Additionally, the benefits of Multilayer Insulation (MLI) for the receivers are described and quantified. Combined, these improvements should allow up to a one-third reduction in electrical power consumption by the elimination of one helium compressor per antenna. Predictions on the improvements in reliability and maintenance interval are also given.

INTRODUCTION

The VLA is a radio telescope array with eight receivers on each of 27 antennas that provide continuous frequency coverage from 1 to 50 GHz. The front end electronics in each receiver are cryogenically cooled by a Gifford-McMahon (GM) refrigerator to minimize thermal noise for maximum receiver sensitivity. Three helium compressors are currently used on each antenna, which supply gas at ~280 psi for 2 to 3 cold-heads each. All 81 compressors run continuously, the combined 18 kW load per antenna costs the observatory almost half million dollars annually in electrical power. In addition, these compressors along with the 216 cold-heads in the array require regular maintenance by a team of experienced technicians.

The NRAO has begun an initiative to reduce power consumption and to improve reliability, called the Green Antenna Project (GAP). Our research into more efficient helium compressors and more reliable refrigerators led us to consider variable-speed operation of both of these. As VFD helium compressors and cold-heads are becoming commercially available[1], we examined our current system to see how it could be retrofitted with this new technology.

CURRENT VLA CRYOGENIC SYSTEM EVALUATION

GM Refrigerators

All VLA cold-heads are 2-stage units, with a first stage at ~50 Kelvin, and a second stage at ~15 Kelvin. The smallest cold head is designed to run at 200 rpm, while all the other ones run at 72 rpm, for the nominal 60 Hz line frequency. Regardless of the refrigerator size, the electric motor that drives the displacer is an AC-synchronous type with a rotor and stator that have no physical contact. It requires two quadrature-phased AC voltage inputs, which runs the motor at a rate directly proportional to the input frequency. However, the winding impedance varies with the frequency, so the supply voltage amplitude needs to be adjusted as the drive frequency is changed, to maintain a constant torque and keep the current below the rated maximum.

Helium Compressors

The VLA helium compressors are the reciprocating (piston) type, and are based on an older model CTI 450 air-cooled compressor. They use the same motor assembly, but have been highly modified over the years to reduce construction cost and ease maintenance. During the upgrade, 60 more compressors had to be built to accommodate the increased number of cold-heads. Varying the speed of a reciprocating compressor poses challenges, such as torque pulsation and lubrication problems, which forced us to look at other options.

RECEIVER TEMPERATURE CHARACTERIZATION AT VARIABLE SPEED

VLA receivers span a range of physical size and cooling requirements, although all except the largest and smallest use exactly the same refrigerator. This suggests that many if not most of the receiver types have cold-heads that are oversized, at least when run at 60 Hz. To quantify this, laboratory testing of each receiver type was done using a commercial variable-frequency inverter driving our standard refrigerator drive unit, and the cold stage temperatures were recorded[2]. The measurements were then repeated after the receiver under test was retrofitted with MLI on the first stage radiation shield, and the improvement over the first test recorded.

Test #1: Characterization of VLA Receivers at Variable Motor Speeds

After the cold-head motors were verified to be compatible with VFD operation, one of each type of receiver was randomly selected, and run with the inverter over a range of frequencies, from 60 Hz down to 30 Hz. The receiver under test was in a laboratory environment at ambient temperature; the helium compressor supply pressure was frequently monitored and adjusted to keep it constant. Temperatures on both cold stages were recorded, after allowing enough time between frequency steps for the temperatures to re-stabilize. The temperature data for each stage versus frequency is plotted in Figure 1 and Figure 2, respectively, for the 8 receiver types tested.

The plotted results are revealing. For both the L-band and S-band receivers, it's clear that they can be run down to two-thirds the nominal speed (40 Hz), with a nearly negligible temperature rise on the critical second stage where the low-noise amplifiers are mounted. The first stage temperature was also nearly flat for both of them, and for the Ka-band receiverthe second stage temperature actually decreased slightly at 45 Hz, then returned to the original temperature at 40 Hz. The C-band receiver shows as well a flat temperature response on both stages down to 40Hz.

Conversely, the X-band, Q-band and Ku-band receivers all show a significant proportional increase in temperature on both stages as the drive frequency drops from 60 to 30 Hz, indicating the cold head thermal loading is closer to the rated capacity. The Q-band receiver does have the small cold head, which could account for the rise; likewise, the first stage radiation shield area on the X-band and Ku-band receivers are relatively large compared to the remaining three receivers.

Overall, the relative increases on the first and second stages for all receivers appear to be correlated. This very likely indicates radiative loading on the first stage as the dominant contributing factor, which the next test is intended to confirm.

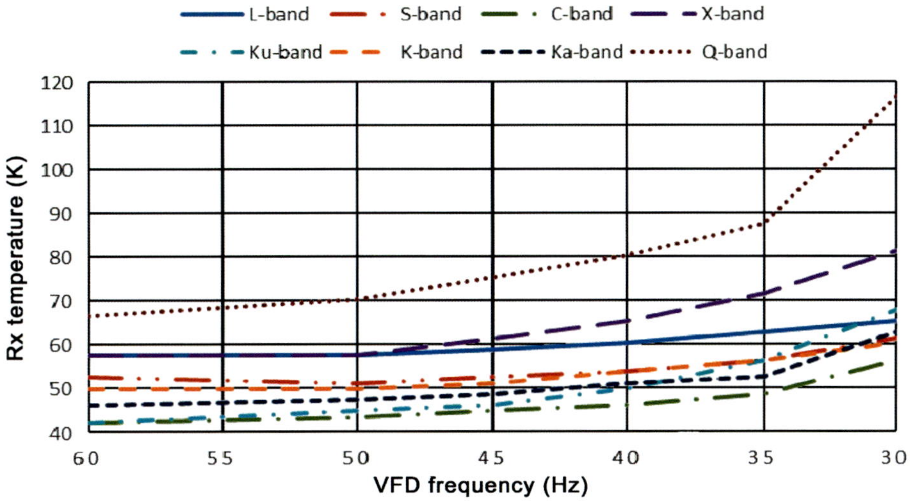

Figure 1. VLA receivers 1st stage temperatures when run with VFD

Test #2: Addition of MLI to the First Stage (~50K) Radiation Shield

Of the eight VLA receiver types, only two (L-band and S-band) currently incorporate MLI on the first stage radiation shield. In this phase, MLI was added over the radiation shields of the remaining 6 receiver types in the test group, and the testing above was repeated on these receivers to quantify the effect of thermal loading on the refrigerator.

A brief summary of the results is given in the Table 1 below, showing the temperature reduction after MLI installation on each temperature stage for the 6 bands, at the nominal 60 Hz drive frequency. The addition of MLI caused a drop in the temperature of the first stage by several degrees, with a corresponding smaller drop on the second stage. The 10K reduction seen on the Ku-band second stage was questionable, so another Ku-band receiver was tested to confirm the results. The second receiver initially had a warmer first stage and colder second stage compared to the first receiver. After adding MLI , the final temperatures for both receivers were almost identical.

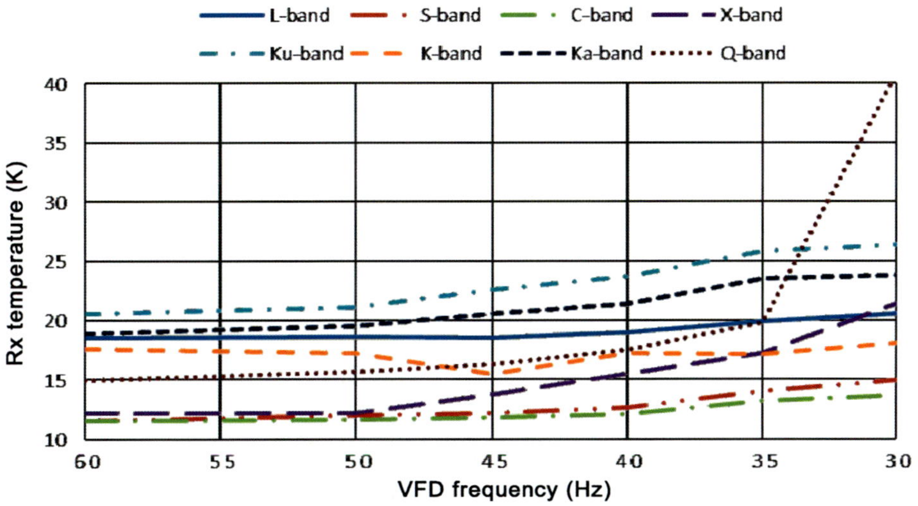

Figure 2. VLA receivers 2nd stage temperatures when run with VFD

Table 1. Temperature Reduction with MLI, at 60 Hz drive

	C Band	X Band	Ku Band	K Band	Ka Band	Q Band
ΔT 1st stage	5.2K	14.1K	6.5K/33.2K	3.8K	7.8K	23K
ΔT 2nd stage	0.4K	0.9K	10K/4.9K	0.5K	0.9K	1.8K

Table 2 shows the temperature increase on both stages relative to the 60 Hz result, as the drive frequency is lowered in steps to 30 Hz. Again, the data was recorded after sufficient time had passed for the temperatures to re-stabilize, when the drive frequency was changed.

The most striking result from the test is that all the receivers can be run at 40 Hz, and still maintain lower temperatures with MLI than at 60 Hz without MLI. This assumes the receivers tested are representative of the majority in the VLA. Retrofitting these receivers with MLI is relatively inexpensive and straightforward, but the refrigerator drives currently in the VLA are fixed at 60 Hz. Without a VFD, the benefits of reduced power consumption and longer life cannot be achieved. This will be the topic of the next two sections.

IMPLEMENTATION OF VFD OPERATION FOR POWER REDUCTION

Change in Compressor Configuration

A cold-head that runs at a lower speed requires less helium flow from the compressor, effectively allowing a smaller unit to be used. A compressor capable of variable-speed operation would be ideal, as it can be adjusted to precisely match the required flow. Unfortunately, the VLA compressors are fixed-speed, and cannot be easily modified for variable-speed operation.

However, another solution is possible, if all three compressors on an antenna can be adapted to feed a single supply manifold common to all 8 receivers, rather than having separate circuits. This could allow selective remote shutdown of a compressor, when the total demand is low enough to allow it. In steady-state operation (all receivers cold), running the cold-heads at 40 Hz should reduce the required flow by a third, which would allow one compressor to be shut off. When more flow is required, such as during receiver cool-downs, the third compressor could be brought back online. While such an arrangement isn't as flexible as varying the compressor speed, it could still allow a substantial power savings when the extra capacity is be needed for a very small percentage of the time.

Implementation of New Compressor Configuration

Combining multiple compressors on a single circuit present risks, and a few precautions must be taken. In order for the flow from one compressor not to enter or interfere with that of another, a one-way valve must be placed at the output of each. If a compressor is shut down, its supply and return pressures must equalize, and it must be isolated from the rest of the circuit. When a compressor is turned off, the static pressure is ~200 psi, so the one-way valve is adjusted to allow flow at a slightly higher pressure (~230 psi). Therefore, helium will only flow from a compressor into the supply input manifold if the compressor is actually running. In a like manner, a second

Table 2. Temperature Rise with Decreasing Drive Frequency, MLI Installed

	C Band		X Band		Ku Band		K Band		Ka Band		Q Band	
Frequency	Δ1st in K	Δ2nd in K	Δ1st in K	Δ2nd in K	Δ1st in K	Δ2nd in K	Δ1st in K	Δ2nd in K	Δ1st in K	Δ2nd in K	Δ1st in K	Δ2nd in K
50 Hz	0.0	0.1	0.0	0.2	0.0	0.6	1.2	0.1	1.3	0.1	2.6	0.4
45 Hz	1.3	0.2	1.3	0.6	1.3	0.8	2.5	0.2	3.9	0.6	7.7	0.7
40 Hz	2.6	0.3	2.6	0.8	1.3	1.2	3.8	0.4	5.2	1.2	12.8	1.8
35 Hz	5.2	1.2	5.1	2.0	2.6	3.1	5.1	0.6	3.9	4.4	17.9	2.6
30 Hz	7.8	1.3	9.0	2.5	3.9	5.6	9.0	1.1	7.8	6.0	33.0	7.7

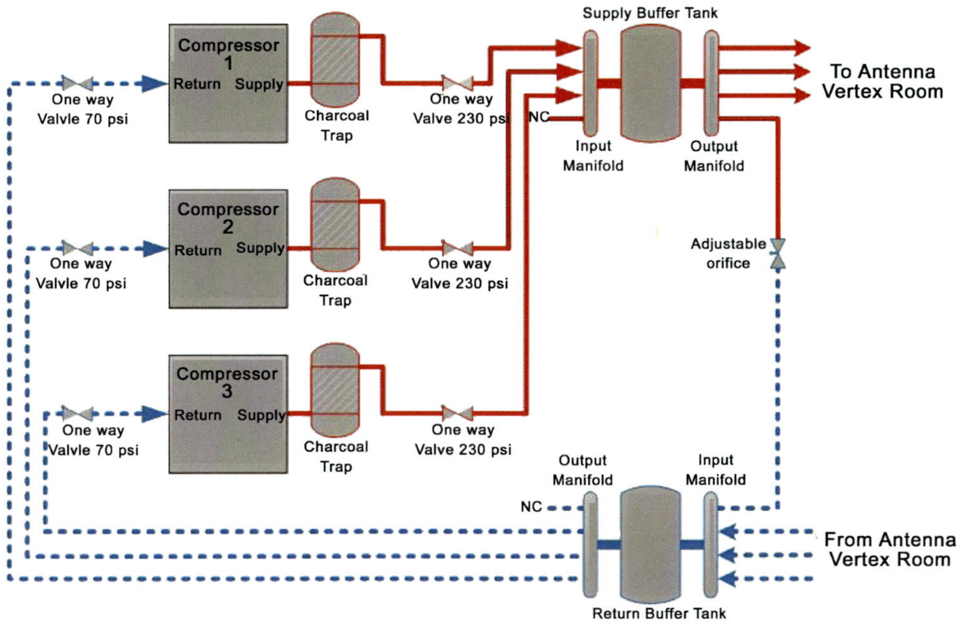

Figure 3. Green Antenna cryogenics block diagram

one-way valve set to ~70 psi is placed on each compressor's return port, to prevent back flow and set a minimum return pressure.

The output from all compressors feed into a 20-liter stainless-steel buffer tank through an input manifold. The tank is equipped with a relief valve for safety, and a transducer to measure the pressure. An identical buffer tank and manifold arrangement are used on the return side, to combine the return flows from the cold-heads and on to the individual compressors.

Finally, an adjustable orifice is added to connect the two buffer tanks, providing a bypass for excess flow. The orifice is set to provide a return pressure of 75 psi when all the cold-heads are not running.

Figure 3 shows a diagram of the overall system, with helium supply paths as solid lines, and the return paths as dashed lines.

CURRENT DEVELOPMENT STATUS AND FUTURE PLANS

Design of Custom VFD Electronics

The variable-speed tests on receivers in the lab used a commercially-supplied line frequency inverter to drive a standard VLA refrigerator power supply, for simplicity and expedience. However, in the implementation for the antenna a custom inverter will be designed. The new inverter will generate two quadrature-phase supply voltages, without the bulky and expensive transformers currently used. Output amplitudes will be adjusted automatically with changing frequency, to keep the motor torque constant and limit the current.

Figure 4 shows a block diagram of the VFD electronics for a single cold-head. An FPGA is utilized to digitally synthesize the quadrature-phase waveforms. A 16,384-point representation of a sine wave is stored in a dual-port ROM, and sampled at 31.25 kHz with an incrementing address input. The frequency f (Hz) of the sinusoid is changed by adjusting the address increment N used to step through the table, where

$$f = (31{,}250 / 2^{14}) \times N \approx 1.90735 \times N \quad ; N < 8192 \tag{1}$$

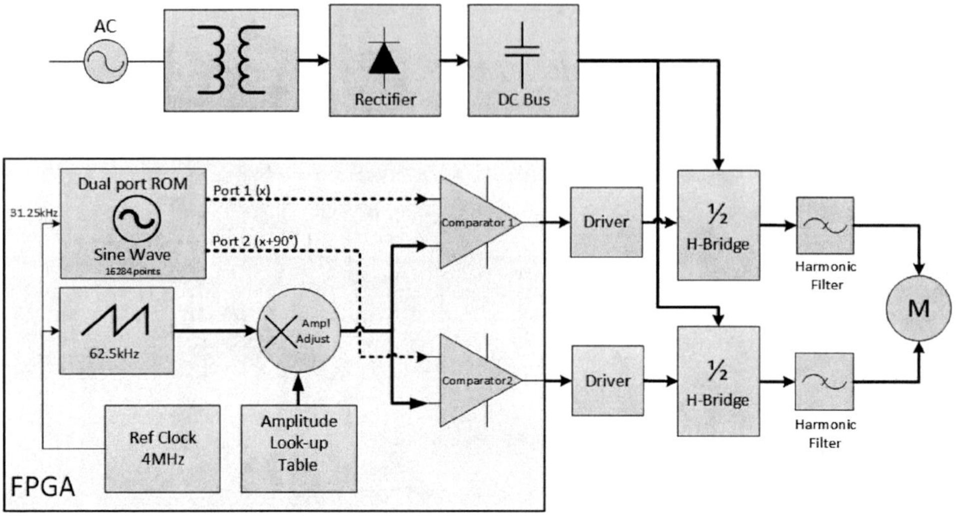

Figure 4. VFD block diagram

Hence for ~40 Hz output, N would be set to 21. The sawtooth waveform generator and comparators are used to generate pulse-width modulated outputs at 62.5 kHz, each with a duty factor (ratio of pulse width to period) directly proportional to the sampled output values from the dual-port ROM. A second lookup table adjusts the sawtooth amplitude with the selected drive frequency, in order to maintain a constant motor torque. These pulse train outputs are amplified and filtered to produce two analog sinusoidal voltages for the direct motor drive.

A bench prototype has been thoroughly tested with a cold-head motor, validating the design approach. We are currently focused on harmonic mitigation, to minimize potential RFI and to insure smooth motor operation throughout the drive frequency range. Future work will be design of an advanced prototype unit suitable for installation and test in an operational VLA antenna.

Buffer Tank Assembly

Two 20-liter stainless steel buffer tanks have been fabricated as well as the input and output manifolds. The one way valves were adjusted to the respective pressure before being mounted to the manifold. Before the completion of the assembly, both tanks and the four manifolds will have to pass the hydrostatic pressure test. Once completed, both assemblies will be leak-tested before being charged with high-purity helium for system testing with the actual compressors.

Prediction on Improvements in Reliability and Maintenance Interval

Analysis of maintenance history for the VLA cryogenic equipment shows the service interval for a cold head is directly proportional to the speed of the displacer. For example, a model 22 that runs at 200 rpm has to be serviced almost three times more often than a model 350 that runs at 72 rpm. Based on this information, we can expect an increase in the maintenance interval proportional to the decrease in cold-head speed. The collected data show that most of the compressor problems are due to cooling fan failure. When the fan stops working, the compressor heats up and shuts down requiring human intervention to connect the spare compressor to the proper helium circuit. Having three compressors on a single circuit will allow us to shut down a failing compressor while starting up the standby unit, entirely remotely and without interruption.

CONCLUSION

We have demonstrated that adding MLI to the radiation shield of our receivers allows a 33% reduction in running speed on the cold-heads, with essentially no increase in cold stage temperatures. Receiver sensitivity will be unaffected, while allowing a ~30% overall power savings from the compressors. The slower movement of the cold-head displacers should result in a lower wear rate and longer maintenance interval, with attendant savings in labor and material costs.

The Green Antenna Project is a test bed for more energy-efficient technologies and techniques that can be implemented on the present VLA, but also a pathfinder for advanced cryogenic systems ia needed for a next-generation VLA telescope.

ACKNOWLEDGMENTS

The authors would like to thank Raul Lower, Pat Madigan and his team in the NRAO machine shop for the fabrication of the buffer tanks. Technical expertise was provided by Jim Gregg and Michael Zamora in the VLA Cryogenics group.

REFERENCES

1. G. Jakob, J.L. Lizon, "Advanced high-cooling power 2-stage Gifford-McMahon refrigerator system", *Proc. SPIE 7739*, Modern Technologies in Space- and Ground-based Telescopes and Instrumentation, 773931 (July 22, 2010)

2. D. Urbain, W. Grammer, "Results of the Variable Frequency Drive experiment on the EVLA cryogenic receivers", Internal NRAO engineering meeting presentation

Integrated Refrigeration and Storage for Advanced Liquid Hydrogen Operations

A. M. Swanger[1], W. U. Notardonato[2], W. L. Johnson[3], T. M. Tomsik[3]

[1] NASA Kennedy Space Center, Cryogenics Test Laboratory, NE-M5, KSC, FL 32899 USA

[2] NASA Kennedy Space Center, Cryogenics Test Laboratory, UB-R1, KSC, FL 32899 USA

[3] NASA Glenn Research Center, Fluid and Cryogenic Systems Branch, LTF, Cleveland, OH 44135 USA

ABSTRACT

NASA has used liquefied hydrogen (LH_2) as a fuel on a large scale since the beginning of the space program for the Centaur and Apollo upper stages, and more recently for the three Space Shuttle main engines. The LH_2 systems currently in place at the Kennedy Space Center (KSC) launch pads are aging and inefficient as compared to the state-of-the-art. Therefore, the need exists to explore advanced technologies and operations that can reduce commodity costs, and provide increased capabilities. The Ground Operations Demonstration Unit for Liquid Hydrogen (GODU-LH2) was developed at KSC to pursue these goals by demonstrating active thermal control of the propellant state by direct heat removal using a cryocooler. The project has multiple objectives including zero loss storage and transfer, liquefaction of gaseous hydrogen, and densification of liquid hydrogen. The key technology challenge was efficiently integrating the cryogenic refrigerator into the LH_2 storage tank. A Linde LR1620 Brayton cycle refrigerator is used to produce up to 900 W cooling at 20K, circulating approximately 22 g/s gaseous helium through the hydrogen via approximately 300 m of heat exchanger tubing.

The GODU-LH2 system is fully operational, and is currently under test. This paper will discuss the design features of the refrigerator and storage system, as well as the current test results.

INTRODUCTION

In the 1960s NASA drove the large scale liquid hydrogen technology development in preparation for the Apollo moon missions, and continued to be one of the largest consumers of LH_2 through the end of the Space Shuttle program in 2011. In the 50 years since this early period of innovation, the industry has taken the lead in developing state-of-the-art in cryogenic systems, and NASA has found itself utilizing inefficient equipment. Over the duration of the Space Shuttle program, NASA lost approximately 50% of the hydrogen purchased due to continuous heat leaks into ground and flight vessels, transient cool-down of warm cryogenic equipment, boil-off during transport, liquid bleeds, and vent losses [1]. Since production of liquid hydrogen is an energy intensive process [2], these losses constituted a large, unrecoverable cost on the order of several million dollars per year. This cost prompted NASA to invest in next-generation cryogenic storage and transfer systems, and drove the development of the Ground Operations Demonstration Unit for Liquid Hydrogen (GODU-LH2) at KSC. Figure 1 shows a ground level view of GODU-LH2 test site.

At the heart of the GODU-LH2 system is the concept of Integrated Refrigeration and Storage (IRAS), controlling the state of the fluid inside the storage tank via direct removal of energy from the liquid using an integrated heat exchanger coupled with a cryogenic refrigerator. The IRAS concept pursues four capabilities: (1) Zero loss LH_2 tanker offloads: tankers can be offloaded into the cold IRAS tank with only minimal purging required, and then the refrigerator is used to decrease the pressure that builds up during liquid transfer as opposed to venting the ullage; (2) Zero Boil-Off (ZBO): if the refrigerator lift and intrinsic tank heat-leak are balanced ($Q_{lift}/Q_{tank} = 1$) and no mass is being added or removed then the pressure and liquid level will remain constant indefinitely; (3) Liquefaction: gaseous hydrogen can be introduced into the IRAS tank and liquefied in-situ to fill the vessel as opposed to using liquid tankers; and (4) Densification: if the refrigeration lift is greater than the tank heat-leak ($Q_{lift}/Q_{tank} > 1$) and no mass transfer exists then the LH_2 can be cooled below its normal boiling point (NBP), becoming denser as a result.

Capabilities 1 through 3 have economic benefits for launch pad architectures, the advantages of densification however, are a little more abstract. Cooling LH_2 from its NBP of 20.4 K to just above the triple-point of 13.8 K yields a density increase of roughly 9%, yet requires a hefty 3.8 kJ of energy removal per liter of liquid. The required cooling becomes increasingly difficult as the temperature decreases, resulting in an additional Carnot penalty of 33% that must be paid in order to reach the triple-point. There are substantial benefits in propellant densification as it pertains to space launch vehicles [3-5]. It has been reported that the positive impact on payload mass fraction can be quite significant, ranging from 4.9% to 17.5% for low Earth orbit depending on launch platform and densification temperatures, and up to 26% if slush hydrogen is employed [6]. Other advantages can be derived from the lower vapor pressures and increased sensible heat associated with densifying propellants [7]. Lower operating pressure can have a profound effect on storage and flight tank designs, potentially decreasing weight and/or increasing life cycle; and gaining any amount of sensible heat can ease pumping and chill-down of equipment, decrease overall loading time, and may have a positive effect on unfavorable phenomena such as geysering.

The IRAS concept was first demonstrated on a small scale (180 L) with hydrogen in 2004 by Notardonato et.al. [8]. This system employed a Cryomech model AL330 Gifford-McMahon cryocooler, heat-pipe, and braided copper heat exchanger integrated to a dewar. Lift capacity of the AL330 was 40 W at 20 K, and the heat leak of the vessel was roughly 8.7 W; this yielded a refrig-

Figure 1. GODU-LH2 Test Site

eration ratio (Q_{lift}/Q_{tank}) of 4.6. During testing the vessel was successfully filled via liquefaction of GH_2, and the liquid was densified to approximately 15.7 K. Demonstration on a more significant scale was necessary however, in order to validate the use of IRAS on a system as large as the LH_2 facility at the KSC launch pads (3,200 m^3 or more).

In 2010 a larger scale (400 L) IRAS system was tested at KSC using a liquid nitrogen heat exchanger to densify liquid oxygen. Testing was conducted with the heat exchanger completely submerged, and the data were used to validate an associated mathematical model. The results were reported by Partrideg et.al, and found that that a submerged cooling source is a viable method to obtain densified propellants [9]. These findings guided the development of the internal heat exchanger for the GODU-LH2 IRAS tank.

To-date, ZBO, liquefaction, and densification of liquid hydrogen has been successfully demonstrated with the GODU-LH2 system at the 30% fill level. Testing at the 60% level is underway, and 90% is scheduled for late 2016.

GODU-LH2 SYSTEM OVERVIEW

The GODU-LH2 site is located in an open field within a controlled area at KSC, and is comprised of four primary subsystems (1) IRAS tank, (2) Refrigeration system, (3) IRAS tank vent system, and (4) Pneumatics (gaseous nitrogen, helium and hydrogen). Ancillary systems include a water chiller unit, electrical power distribution (480 V and 120 V), hazgas detection for hydrogen leaks, command and control (C&C), and liquid nitrogen (LN_2) for refrigeration precooling. Figure 2 shows and aerial view of the entire site.

IRAS Tank

GODU-LH2 repurposed a 125 m^3 cylindrical LH_2 storage tank originally built by Minnesota Valley Engineering in 1991 to support the Titan-Centaur rocket program at Cape Canaveral Air Force Station. It is a high performance vessel—vacuum-jacketed with 80 layers of multi-layer insulation (MLI)—has original pressure and temperature ranges of 0 to 655 kPa (gauge) and 20 K to 311 K respectively, an inner diameter of 2.9 m and length of 21.8 m.

Five primary tank modifications were required to facilitate GODU-LH2 testing: (1) construction of the internal heat exchanger, (2) incorporating internal stiffening rings to accommodate the sub-atmospheric pressure associated with densification testing, (3) an updated man-way feed-through

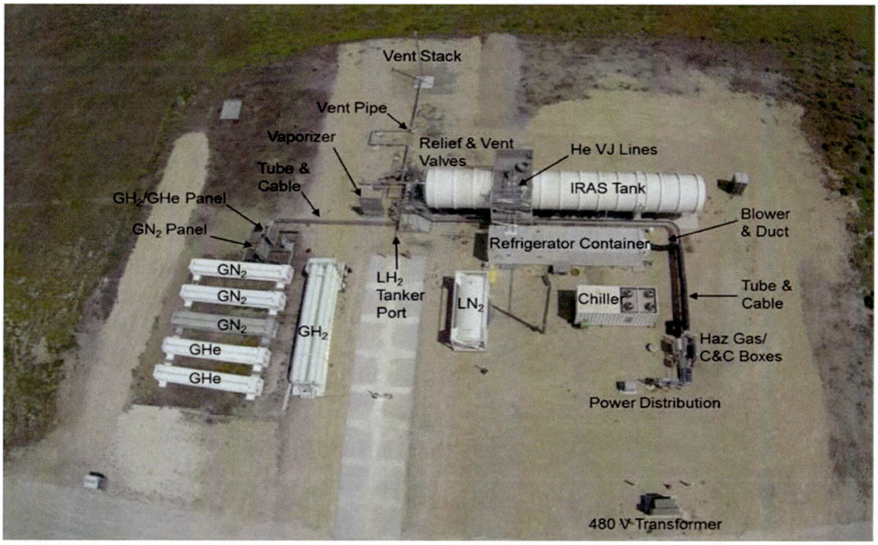

Figure 2: Aerial View of the GODU-LH2 Test Site

plug to incorporate instrumentation and refrigerant penetrations, (4) recertification of the tank per the American Society of Mechanical Engineers (ASME) Boiler and Pressure Vessel Code (BPVC) to the new operating conditions, and (5) temperature rakes to map the vertical, axial and radial tank temperature profile (24 silicon diodes in total).

Installation of all internal components can best be described as a "ship in a bottle" approach, since all hardware, tools, personnel, etc. had to fit through the 53 cm diameter man-way port at the top of the tank. This required that the stiffening rings, heat exchanger and rakes be modular—fabricated in pieces capable of being transported into the tank, and then assembled by hand. All modifications were performed by KSC personnel in 2013, and were covered extensively by Swanger et al., [10,11].

Design of the internal heat exchanger is highly three-dimensional in order to effectively distribute the cold helium refrigerant throughout the tank volume, and provide some structural rigidity. It consists of supply and return manifolds (total of 37 m of 25 mm OD tubing, and located at the 25% and 75% fill levels respectively), 40 cooling coils (total of 244 m of 6.4 mm OD tubing), and has a total heat transfer area of approximately 11 m^2 (including flex hoses and fittings). Refrigerant flows were balanced end-to-end using precision orifice fittings as previously reported by Fesmire et al., [12]. Two 25 mm diameter by 3 m long flexible lines connect to the man-way feedthrough to each manifold. Each of these tubing sections had to be interfaced using fittings with ultra-low leak rates and high reliability. Swagelok VCR type units with silver-plated nickel gaskets and retainer rings were chosen for this task. Figure 3 shows the as-built internal IRAS configuration with heat exchanger, stiffening rings, and temperature rakes.

Refrigeration System

The cryocooler selected for the GODU-LH2 system is a Linde LR1620 model with an RSX helium compressor. The LR1620 is a reverse-Brayton cycle refrigerator that employs two parallel piston expanders with 4-stages of recuperation. Engine speed is governed by an inertial flywheel and variable speed motor to a maximum of 230 RPM (normal operation is between 225 and 228 RPM). Flow into and out of the expanders is controlled by spring-loaded stainless steel valve assemblies that are actuated by cams mounted to the flywheel. Also integrated into the vacuum-jacketed cold-box assembly are activated charcoal adsorber beds for purification, LN_2 boiler for optional precooling, and numerous valves used for different operational modes. Manual valves are employed at the helium supply and return bayonet connections to isolate the IRAS tank from

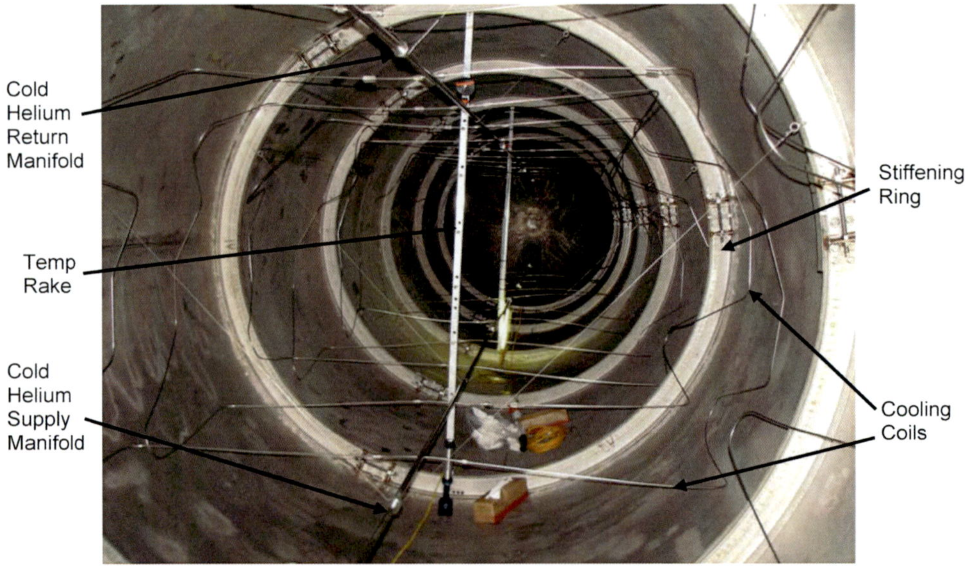

Figure 3. Internal IRAS Tank Modifications

the refrigerator. Physical dimensions of the unit are 1.8 m x 1.5 m x 1.1 m (H x W x D), and the weight is approximately 1000 kg.

The RSX helium compressor consists of a hermetically sealed vertical screw compressor pump, multiple coalescers for oil separation, water after-cooler assembly to cool the compressed helium and compressor oil, and an activated charcoal adsorber bed. The compressor supplies the LR1620 unit with ambient temperature helium at roughly 1.7 MPa, and ingests the return gas at around 2.5 kPa. It has two modes of operation, loaded and unloaded—both discharge the same pressure, however, the latter produces less mass flow and is only used during refrigerator cool-down, while the former yields the maximum flow rate and is used during normal operation. The unit is driven by 480 V 3-phase power at 60 Hz, its physical dimensions are 1.5 m x 1.4 m x 1.5 m (H x W x D), and the weight is approximately 1135 kg.

As the system cools to operational temperatures the ultra-high purity (UHP) helium refrigerant pressure drops dramatically. In order to maintain an adequate pressure within the closed system, automatic solenoid valves control flow into and out of a 450 L accumulator tank rated to 2.1 MPa. This vessel is filled with UHP helium from k-bottles periodically to a 1.0 MPa working pressure by GODU-LH2 personnel.

The system can officially provide 400 W lift at 20 K, or 850 W with LN_2 precooling at a maximum mass flow rate of 22 g/s; however, initial performance testing yielded elevated capacities of roughly 500 W and 900 W with and without precooling respectively. Cold helium from the refrigerator is routed to and from the IRAS tank through 8.5 m (each way) of 25.4 mm x 63.5 mm vacuum-jacketed (VJ) hard piping.

During system conception it was estimated that the intrinsic heat leak of the retrofitted LH_2 dewar was around 350 W. This yielded a refrigeration ratio of 1.1 and 2.4 at the rated lift capacities, with and without precooling respectively; and 1.4 and 2.6 using the actual system performance numbers. As previously noted, a ratio greater than 1.0 was necessary to achieve all the test objectives; however, a large positive margin is desirable, as it affords even greater test flexibility and control by increasing system response.

Housing the refrigeration system—which utilizes equipment that does not comply with the National Fire Protection Association (NFPA) requirements for operation in explosive gas environments—in close proximity to the IRAS tank was achieved using a standard 12 m ISO shipping container constantly purged with fresh air from outside the NFPA Class I, Division II area via a blower unit. Pressure inside the semi-sealed container is kept at roughly 0.5 in H_2O above atmospheric at all times when hydrogen was present in the tank to ensure that, in the case of a leak, no GH_2 could infiltrate the purged environment. Integrating the entire refrigeration system into the shipping container also allowed for the possibility of transporting it to a future location in an assembled, serviced state. Only the refrigerant supply and return lines, cooling water lines, and power to the system must be de-mated prior to transporting.

Rejecting the heat of compression and cooling the compressor oil is accomplished via a custom built chilled water unit that complies with NFPA regulations to operate within a Class I, Division II zone. Similar to the refrigerator and compressor modules, this unit was built into an ISO shipping container (6 m instead of 12 m), and can be easily transported. 25 mm diameter PVC lines interface the chiller to the RSX compressor, and provides a constant supply of 16°C water to the system.

Liquid nitrogen for helium precooling is supplied from a 20,000 L US Department of Transportation (DOT) approved transportable dewar placed adjacent to the refrigeration container. It interfaces to the refrigerator module through 11 m of 13 mm x 51 mm VJ hard piping. Pressure is allowed to build in the LN_2 tank from natural boil-off, and is maintained around 103 kPa by periodically venting the ullage space. While running in precooling mode the internal Linde control software maintains a boiling LN_2 pool inside the refrigerator using a bang-bang control scheme. Incoming GHe from the compressor is partially routed through the LN_2, heat exchanger, dropping its temperature close to the LN_2 boiling point before being fed back into the recuperators. Boil-off vapor is vented to atmosphere out the side of the refrigeration container. Use of LN_2 in this manner accounts for the 400 W increase in refrigeration capacity over the standard closed-loop cycle.

Control of the lift capacity is accomplished by adding heat to the cold GHe exiting the piston expanders via an in-line resistive heater. An exit or return temperature set point can be entered into the standard Linde software, which in turn adds heat as required. Custom software was also developed for GODU-LH2 that interfaces with the Linde package and controls the heater to achieve a desired IRAS tank pressure. Figure 4 shows the refrigerator and compressor modules inside the shipping container, and the VJ interface piping.

IRAS Vent System & Pneumatics

The IRAS tank vent system was designed to meet the unique test requirements of the GODU-LH2 project, as well as the mandatory ones established in the ASME BPVC. Two vent paths can be utilized depending on the mode of operation: A 150 mm line and 7 m tall vent stack accommodates large flow rates, and routes flow from the relief devices in case of an over pressurization event; and a 25 mm path directs normal boil-off flow to a 3 m tall vent stack during IRAS tank heat leak testing, and when the refrigeration system is down for repair or maintenance. Each of these legs ties into the 100 mm primary IRAS tank vent, and are activated remotely from the command and control trailer. A vertical, open-air vaporizer is also employed in the system to provide tank pressurization for possible future LH_2 transfer testing. Since an over arching theme of GODU-LH2 is to minimize, or completely eliminate, any hydrogen losses, under normal operation the IRAS tank is isolated from the vent system, and the refrigerator is used to control the pressure. During densification testing, when the tank is allowed to go sub-atmospheric, any interface to the ullage space (flanges, valves, fittings, etc.,) is "bagged" and purged with gaseous helium. This is done as a precaution, in case a leak occurs helium instead of air will be drawn into the tank.

To reduce project cost, several pneumatic panels and gas storage bottles were reused from the Space Shuttle program to provide gaseous helium, hydrogen and nitrogen to the system. Two 16.5 MPa mobile storage units (MSU) provide a 2,265 m^3 helium storage capability, and three additional 16.5 MPa MSUs store up to 3,400 m^3 of gaseous nitrogen.. The facility nitrogen panel reduces the 16.5 MPa down to several 5 MPa sources as well as up to twenty 700 kPa sources used for valve actuation, panel inerting, and purging of lines. Gaseous hydrogen is supplied to the site by compressed gas trailers, is regulated from 25.5 MPa down to 1,030 kPa and used for liquefaction supply as well as purge and pressurization of lines and the IRAS tank.

GODU-LH2 TESTING RESULTS

The GODU-LH2 test matrix consists of three primary objectives (zero-loss offloading and storage, liquefaction, and densification) conducted at three different IRAS tank fill levels (30%, 60%, and 90%). Measurable differences in system efficiency and response are expected at the different fill levels as the internal heat exchanger becomes increasingly submerged in LH_2.

Figure 4. Refrigeration Container Showing the Refrigerator Module (foreground) and RSX Compressor (background) (left); GHe VJ Piping Interfaced to IRAS Tank (right)

Initial system start-up and testing began in March 2015 with an LN_2 cold-shock of the IRAS tank, followed by evacuation and GH_2 purge cycles until the proper purity was achieved. The refrigerator was then used to bring the inner tank wall temperature down to roughly 55 K in order to facilitate a zero-loss tanker offload of approximately 45,000 L of LH_2 (i.e. no hydrogen was lost during the offload process), and afterward to re-condense the ullage vapor that accumulated during the transfer [13].

This first tanker offload effectively brought the fill level to around 30%, and once the pressure stabilized close to the NBP, testing to determine the intrinsic tank heat leak commenced. Over a three week period the steady-state boil-off was measured using a Brooks Instruments mass flow meter, and was determined to be 257 SLPM on average, at a tank pressure of 104.8 kPa (absolute). This equates to a heat leak of 292 W, and increases the refrigeration ratios to 1.7 and 3.1 with and without LN_2 precooling respectively. Once the heat leak was known the refrigerator was started, IRAS tank brought online, and ZBO, liquefaction and densification testing began.

Zero Boil-Off/Pressure Control

In order to demonstrate full control of the tank pressure with the GODU-LH2 software, it was desirable to test the response of the system when a set-point is chosen at both higher and lower values than the tank pressure at that time to converge to the desired pressure from "above" and "below." From above, the control logic has to allow the refrigerator to run at maximum capacity in order to drop the pressure as fast as possible, and then introduce heat until the refrigerator lift balances the tank heat leak (i.e. a refrigeration ratio of 1). Whereas, from below, the software needs to create a refrigeration ratio less than 1 by introducing excessive heat into the tank in order to drive the pressure up to the set-point. This test sequence was performed over a 9 day period, using four different set-points, and without LN_2 precooling for the refrigerator. Figure 5 shows the results of the ZBO testing at four different set-points.

Figure 5 shows that the system is able to control the tank pressure very precisely. It is apparent from the transient periods that the system responds faster when approaching the set-point pressure from below rather than above. From above, the average slope is roughly 0.5 kPa/hr, and from below it is 1.4 kPa/hr. Another notable feature is the negative lift values when approaching the set point from below. This is due to the refrigerator supply temperature being higher than the return when the cryocooler is actually acting as a heater to drive the tank pressure up. On average the tank

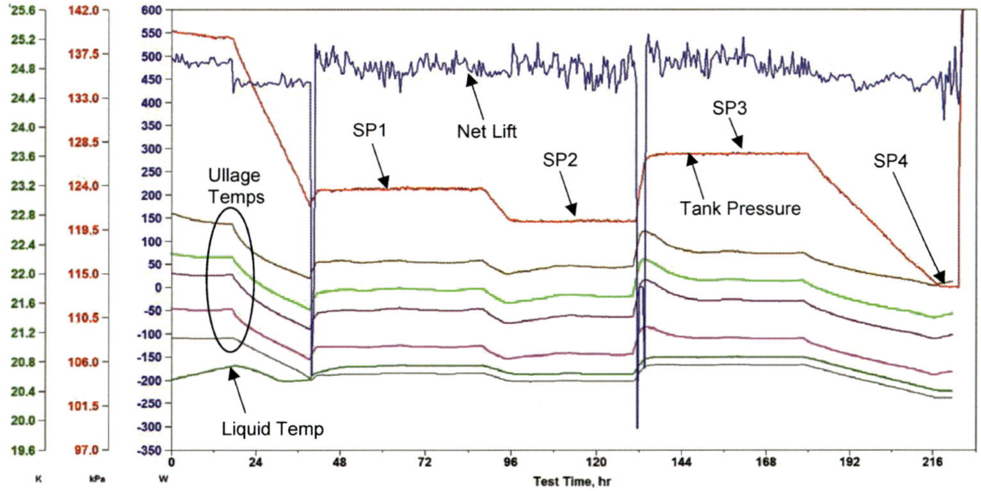

Figure 5. ZBO Test Results at the 30% Fill Level
(ZBO Set Points: SP1 = 124.1 kPa, SP2 = 120.7 kPa, SP3 = 127.6 kPa, SP4 = 113.8 kPa)

pressure was 0.35 kPa below the set-point, attributed to PID loop tuning, and the average standard deviation over the duration of time for set points 1, 2 and 3 was only 0.07 kPa. Average positive net refrigerator lift was 474 W at $T_r = 23$ K (where T_r is return temp from the IRAS heat exchanger). As expected, the tank temperatures trend with the pressure, and show a small but obvious amount of ullage stratification, save the inversion of the bottom two readings seen at roughly 40 hours. This anomaly is most likely due to the proximity of the liquid-vapor interface to the bottom manifold of the heat exchanger. The top most ullage temperature line in Figure 5 is located 2 m from the inner tank floor, just below the top manifold, while the lowest line (labeled as the "liquid temperature") is 0.6 m from the floor, just below the bottom manifold, and is completely submerged at the reported fill level. The location of the other four temperature readings are spaced evenly between these two.

Hydrogen Liquefaction

Liquefaction testing at the 30% fill level was conducted over a day and a half period and consisted of feeding gaseous hydrogen from a mobile compressed gas trailer to a mass flow controller, which fed directly into the IRAS tank. The refrigerator was operated at full capacity, and the control logic managed the mass flow in order to achieve a constant tank pressure.

It must be stated that due to budget constraints the GODU-LH2 system in its current configuration is not optimized to conduct hydrogen liquefaction. Feeding ambient temperature GH_2 directly into the tank is not the preferred method due the non-trivial ortho-to-para energy penalty that must be paid. As a result, the GODU-LH2 liquefaction testing is aimed more at validating the ability of IRAS to liquefy on a large scale, and testing the software control schemes, rather than maximizing the liquid yield. To these ends 30% testing was successful. The control logic maintained a constant tank pressure using the mass flow controller, and the increase in liquid volume was captured by the liquid level sensor via a ΔP reading.

Figure 6 shows the results of the liquefaction testing. Between hour 26 and 58 the average tank pressure was 127.6 kPa, with a standard deviation of 0.2 kPa. Average net refrigerator lift was 1284 W at $T_r = 28$ K, GH_2 mass flow rate was 113.6 SLPM, and the head pressure increase (ΔP reading) was 0.02 inH_2O. In all, roughly 295 L of liquid was produced during the test.

Hydrogen Densification

Densification testing officially began when the tank pressure went sub-atmospheric and the liquid hydrogen dropped below its normal boiling point of 20.4 K. The refrigerator was run at full power, and successfully densified approximately 49,000 L of LH_2 for roughly 14 days prior to a

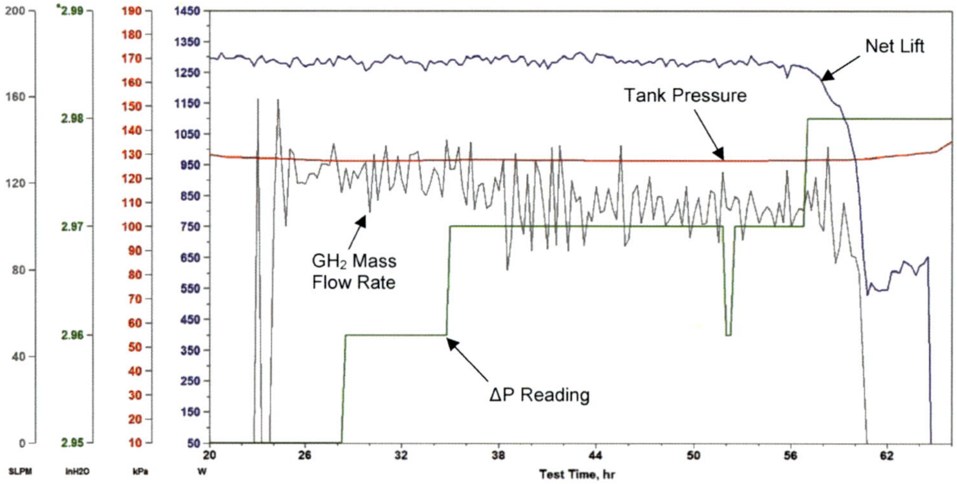

Figure 6. Liquefaction Test Results at the 30% Fill Level

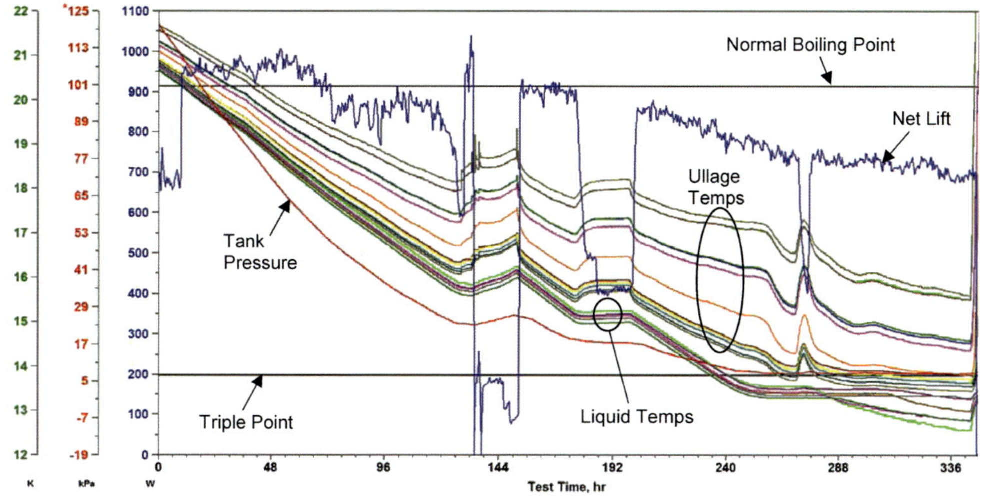

Figure 7. Densification Test Results at the 30% Fill Level

component failure and subsequent shutdown. Figure 7 shows the temperature and pressure data from the 30% densification testing.

Many notable features can be seen in figure 7, most obvious is the minimum temperature reported by many of the diodes. Values well below the triple point were seen not only in the bulk liquid, but also in the vapor space a significant distance from the liquid-vapor interface, the farthest being approximately 0.6 m away. This result is not completely understood at this time. It implies that not only was a significant amount of solid hydrogen being formed inside the IRAS tank, but that it was somehow also being cooled below its freezing point. Preliminary calculations estimate that roughly 1,134 kg of hydrogen ice was produced during the test, and it is believed that it existed as a homogenous bulk slush mixture as opposed to solid formations clinging to the cold heat exchanger surfaces. The minimum temperature witnessed during the test was 12.6 K—a full degree below the triple point value of 13.8 K.

Another interesting feature is the sharp dip in ullage temperatures seen around hour 260, at the same time the tank pressure appears to reach the triple point. This is not believed to be a coincidence. Instead, when solid hydrogen began to form—which would take place mostly at the bottom heat exchanger manifold submerged in the liquid, and coincide with the tank pressure reaching the triple point—the bulk of the cooling power transitioned from the bottom manifold to the cooling coils running up the tank wall. This resulted in a swift decline in ullage temperatures until the LN_2 precooling was turned off briefly. Densification testing at the 60% and 90% fill levels will hopefully shed more light on the unique behavior seen here since there will be greater resolution into the bulk liquid (or slush) temperatures. Dips in the refrigerator lift seen at hours 144 and 192 are attributed to expander valve adjustments, and at 276 hours is due to cycling of the LN_2 precooling. Maximum refrigerator lift was 1007 W at $T_r = 22.5$ K, and the minimum was 680 W at $T_r = 14.6$ K.

CONCLUSION

The GODU-LH2 system has successfully demonstrated zero-boil-off, liquefaction, and densification of hydrogen in a 125 m^3 tank at approximately 30% full using Integrated Refrigeration and Storage (IRAS). Two un-vented tanker off-loads were also performed with and without liquid hydrogen in the tank; the latter being the worst case scenario. Further demonstrations will repeat these at LH_2 fill levels of approximately 60% and 90%. Performance of the Linde LR1620 reverse-Brayton refrigerator exceeded rated capacities by up to 25% during the testing, but also presented numerous maintenance challenges operating outdoors in a remote area of costal Florida.

Initial results indicate that this type of IRAS system could be readily scaled to support space launch vehicle and pad operations by lowering the cost of cryogenic propellant maintenance, affording greater flexibility for propellant handling, and providing increased mission capability.

REFERENCES

1. Partridge, J. K., "Fractional consumption of liquid hydrogen and liquid oxygen during the space shuttle program," *Adv. in Cryogenic Engineering*, Vol. 57, Amer. Institute of Physics, Melville, NY (2012), pp.1765-1770.
2. Fukano, T., Fitzi, U., Lohlein, K., and Vinage, I., "Efficiency of Hydrogen Liquefaction Plants," available at: http://www.lindekryotechnik.ch/public/fachberichte/efficiency_of_hydrogen_liquefaction_plants.pdf
3. Friedlander, M., Zubrin, R., and Hardy, T., "Benefits of Slush Hydrogen for Space Missions," NASA TM 104503, October 1991.
4. Fazah, M., "STS Propellant Densification Feasibility Study Data Book," NASA TM 108467, September 1994.
5. LeMoyne, R., "Fundamental Analysis of Cryogenic Propellant Densification," 15th AIAA International Space Planes and Hypersonic Systems and Technologies Conference, Dayton OH, (2008), AIAA 2008-2500
6. Haberbusch, M. S., "Study Task for Determining the Effects of Boost-Phase Environments on Densified Propellants Thermal Conditions for Expendable Launch Vehicles," NASA/CR—2002-210808, (2002)
7. Mustafi, S., Johnson, W., Kashani, A., Jurns, J., Kutter, B., Daniel, K., and Shull, J., "Subcooling for Long Duration In-Space Cryogenic Propellant Storage," *Proceedings of the AIAA SPACE 2010 Conference & Exposition*, California, Anaheim, (2010), AIAA 2010-8869
8. Notardonato, W.U., Baik, J.H., McIntosh, G.E., "Operational testing of densified hydrogen using G-M refrigeration," *Adv. in Cryogenic Engineering*, Vol. 51, Amer. Institute of Physics, Melville, NY (2006), pp. 64-74, DOI: 10.1063/1.1774668,
9. Partridge, J.K., Tuttle, J.W., Notardonato, W.U., and Johnson, W.L., "Mathematical model and experimental results for cryogenic densification and sub-cooling using a submerged cooling source," *Cryogenics*, Vol. 52, Issues 4-6, (2012), pp. 262-267
10. Swanger, A. M., Jumper, K. M., Fesmire, J. E., and Notardonato, W. U., "Modification of a Liquid Hydrogen Tank for Integrated Refrigeration and Storage," *Advances in Cryogenic Engineering*, IOP Conference Series: Materials Science and Engineering 101 (2015) 012080 doi:10.1088/1757-899X/101/1/012080
11. Swanger, A. M., Notardonato, W. U., and Jumper, K. M., "ASME Section VIII Recertification of a 33,000 Gallon Vacuum-Jacketed LH2 Storage Vessel for Densified Hydrogen Testing at NASA Kennedy Space Center," *Proceedings of the ASME Pressure Vessels and Piping Conference*, Massachusetts, Boston, pp. V003T03A055 (2015), doi:10.1115/PVP2015-45625
12. Fesmire J E, Tomsik T M, Bonner T, Oliveira J M, Conyers H J, Johnson W L and Notardonato W U, "Integrated heat exchanger design for a cryogenic storage tank," *Advances in Cryogenic Engineering*, AIP Conference. Proceedings, 1573, 1365-1372. (2014)
13. Notardonato, W. U., Johnson, W. L., Swanger, A. M., and Tomsik, T., "Ground Operations Demonstration Unit for Liquid Hydrogen Initial Test Results", *Advances in Cryogenic Engineering*, IOP Conference Series: Materials Science and Engineering 101 (2015) 012081 doi:10.1088/1757-899X/101/1/012081

Cryocooler-Based Helium Recovery Plant for Applications Requiring Gas or Liquid with Extreme Purity

M. Gabal, C. Rillo, J. Sesé[1], J. Diederichs[2], G. Rayner[2], S. Spagna[2]

ICMA (CSIC – Universidad de Zaragoza), Zaragoza, Spain
1 INA, Universidad de Zaragoza, Zaragoza, Spain
2 Quantum Design Inc., San Diego, CA, USA

ABSTRACT

A cryocooler based helium recovery plant designed to store, compress and purify helium gas collected from one or more helium-using instruments, as well as to liquefy and redistribute the purified gas within a closed system, is presented. The recovery plant is composed to match the purification and liquefaction rate of the system with the consumption rate of the coupled instruments in an attempt to achieve zero loss. In addition, extreme purity helium (gas and liquid) is obtained by the combination of cryo-condensation of impurities and further chemical adsorption by a getter. Problems like clogging of low temperature impedances by impurities normally present in helium have been definitely solved using this plant concept.

INTRODUCTION

Although helium (He) is the second most abundant element in the universe, on earth it is scarce and only extracted with difficulty. Helium is found underground, in a gaseous state, as a byproduct of natural radioactive disintegrations.

Liquid-phase helium is generally obtained from high power industrial liquefaction plants, referred to herein as "Class XL" liquefiers, which produce liquefied helium in quantities greater than 1000 liters/hour, and require more than 1000 kW of power. These Class XL plants generally yield a liquefaction efficiency of around one liter/hour/kW. Within these Class XL liquefiers, the helium gas undergoes one or more cyclical thermodynamic processes and the gas is then cooled until it reaches its liquefaction temperature. The technology of these liquefaction plants dates from the last century, and can be referred to as "Collins Technology" [1].

Moreover, in view of the Class XL liquefiers specified above, various other liquefaction classes may be referred to herein. For example, Class L liquefiers will refer to those liquefiers which produce greater than 100 liters/hour; and Class M liquefiers will refer to those liquefiers which produce greater than 20 liters/hour. All of these large-scale liquefiers typically achieve a liquefaction efficiency of about 10 liters/day/kilowatt of input power.

The scientific and industrial applications of helium gas are numerous. In a gas-phase, helium under ambient temperature is useful for welding and as a means for floating balloons; at the highest purity levels is used as a carrier gas in gas chromatography. Moreover, in a liquid-phase, helium at atmospheric pressure is at or near -269 ^{0}C, and thus is commonly used for refrigeration

Cryocoolers 19, edited by S.D. Miller and R.G. Ross, Jr.

of medical and scientific equipment, among other things. Helium is therefore considered to be a valuable resource, lending interest to advancements in helium gas recovery and reuse, especially such recovery and reuse as might be accomplished with negligible or zero loss.

In modern recovery plants, helium gas is generally processed throughout multiple stages, that is, stage 1: recovery; stage 2: storage under pressure; stage 3: purification; stage 4: liquefaction; and stage 5: distribution. These modern plants are generally known to suffer losses in each and every stage as outlined above. Furthermore, even where the loss is very small at one or more stages, when aggregated together the total loss can be significant and often exceeds 10% per cycle. Furthermore, these plants require complex facilities for the storage of vast volumes of highly pressurized gas, regardless of the liquid consumption rate at the particular facility, since the liquefaction rate generally cannot be coordinated, regulated, or adapted for consumption. Finally, without the capability to adjust the liquefaction rate, the liquefied helium is produced in volumes that exceed consumption, which necessitates the use of very large storage dewars, and consequently requires smaller transportation dewars to distribute the liquid to end users of liquid helium.

With the advent of closed-cycle refrigerators capable of achieving temperatures below that of liquid helium, such as those based on known Gifford McMahon (GM) and pulse tube technologies, liquefiers having lower liquefaction rates and lower maintenance requirements have been developed [2-3]. In such liquefiers, the gas to be liquefied does not undergo complex thermodynamic cycles, but rather condenses by convection and direct thermal exchange with the different stages of the cryogenic refrigerator and is subsequently stored in a dewar. However, at present there has yet to be developed such a helium recovery plant based on GM or pulse tube technologies which is designed to yield liquefaction efficiencies comparable to the class XL liquefaction plants described above; that is, one liter/hour/kW.

In an attempt to solve the problem for an individual medical or scientific instrument, that is, providing liquefied helium as needed, liquefaction systems have been developed which incorporate a closed-cycle refrigerator adapted to collect and re-condense helium that is evaporated by the medical or scientific instrument using helium. However, these systems are constructed to use one refrigerator per instrument, and thus underutilize the refrigerator's capacity. For installations in which the direct installation of a refrigerator is technically not feasible, these closed-cycle refrigerator systems do not solve the problem of providing helium as and when required. Moreover, when a large number of instruments require refrigeration, the acquisition and maintenance costs associated with the corresponding number of refrigerators make this solution impractical.

Accordingly, helium-gas recovery and purification plants, based on closed-cycle refrigerator technologies, are of immediate and significant interest. With such plants, helium gas being employed as a trace gas in leak detection processes, or as a cooling medium, can be recovered and reutilized over several cycles with little or no loss, thereby reducing of the need to acquire virgin helium gas. Moreover, the recovery of helium would provide an economic advantage for processes that require pressurized helium gas. During the last years, in order to implement efficient solutions, we have developed a new technology for helium liquefiers and purifiers based on cryocoolers (ATL 'Advanced Technology Liquefiers', ATP 'Advanced Technology Purifiers' [4-7]).

Below its critical temperature (T_c = 5.2 K) any unwanted substance present in the liquid phase, i.e. any impurity, will be in solid form, resulting in mist, snow, suspensions or particulates. The vapor pressure of these solid impurities will be in general negligible small ($\ll 10^{-9}$ Pa), except for the case of the hydrogen isotopes and their molecular combinations [8] for which this is of the order of 10^{-2} Pa and 10^{-5} Pa, at 5.2 K and 4.2 K, respectively. The solid impurities are usually charged and can be easily eliminated by electrostatic precipitation using Petryanov filters to obtain "optically clean" liquid, as demonstrated by Abrikosova and Shal'nikov [8]. But, even "optically clean" filtered liquid helium may contain a relevant quantity of non-solid hydrogen, i.e. molecular hydrogen traces. These small quantities are eventually the cause for blockages in small capillaries used in instruments to produce temperatures below

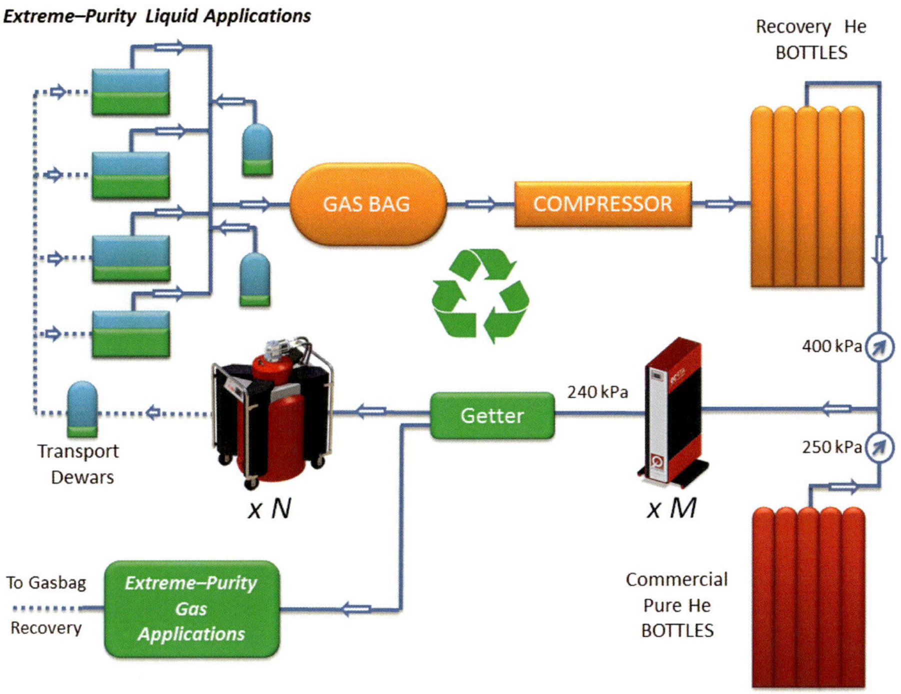

Figure 1. Schematic view of a helium recovery plant for the production of helium with extreme purity.

4.2 K. Therefore, the complete elimination of hydrogen has to be ensured in a helium recovery plant to produce high extreme purity helium either in liquid or gas form. This is a concept plant that we called ¨Green Helium¨ recovery plant.

CRYOCOOLER-BASED HELIUM RECOVERY PLANT

A closed-cycle, high-efficiency, automated helium recovery plant is herein described. This high-efficiency helium recovery plant is designed for computerized control and switching between a liquefaction mode and a standby mode, and is further adapted to collect, re-liquefy, and redistribute recovered helium amongst a plurality of medical or scientific instruments or other equipment. As helium gas is utilized among several instruments within the closed system, it is recovered and subsequently liquefied before being re-introduced to the equipment. As a finite supply of helium is continuously recycled, with no or minimal loss within this closed system, thus reducing the cost associated with resupply of virgin helium gas.

The helium recovery plant presented here covers a liquefaction range of zero liters/hour when operated in standby mode, and 1.25 x N liters/hour when operating in liquefaction mode. In this regard, this helium recovery plant yields liquefaction rates generally comparable to the prior art Class M plants. Additionally, the performance efficiency of the helium liquefaction is around 4 liters/day/kW, nearly matching the production and performance attributes associated with Class L plants, but with simplified operating and maintenance procedures.

The recovery plant shown in figure 1 comprises several parts:

(i) a recovery module for collecting and storing helium gas from one or more helium using instruments;

(ii) a gas collection and storage module adapted to collect gas under atmospheric pressure in a balloon or a container, filter the gas through one or more filters, pressurize the gas to an absolute pressure above 200 bar by means of a purge-free compressor, and store the gas in a gas storage container at the compressor output pressure;

(iii) a purification module ATP based on a cryocooler, which allows the removal of low vapor pressure impurities such as water, nitrogen, oxygen;

(iv) a non-evaporable getter purifier dedicated to the elimination of hydrogen;

(v) a liquefaction module ATL by means of a closed-cycle based refrigerator, the liquefaction module being adapted to vary the liquefaction rate to closely match the gas recovery rate, and therefore to match the consumption of liquefied gas of the connected equipment;

Maximum efficiency in the liquefaction of helium gas within this helium recovery plant is accomplished with an electronic control adapted to provide precise regulation of the vapor pressure within the dewar. Each pressure value "P" at which liquefaction is performed within the helium recovery plant has a corresponding liquefaction rate "T1," expressed in liters/hour, whereas "T1" is an increasing function of "P" [6].

This helium recovery plant is adapted to adjust the liquefaction rate of helium gas and thus minimizes the storage time lapse of the evaporated gas and therefore reduces the acquired impurities therein. The volume of the stored gas prior to liquefaction is also minimized which simplifies the plant. Furthermore, the liquefier allows permanent storage of the produced liquid within its own thermally insulated dewar, during which the system is operated in a standby mode with a zero liters/hour liquefaction rate and a loss of 0%, thus providing a reserve of liquefied helium for its immediate on-demand use.

This helium recovery plant is scalable by increasing the number of liquefaction (N) and purification (M) units (with $M \approx N/2$ because one ATP can handle enough gas flow to feed two ATLs), resulting in a simplified procedure. Moreover, as the available power of closed-cycle refrigerators continues to increase, fewer refrigerators will be required in each liquefaction unit within the plant.

HELIUM WITH EXTREME PURITY

The purity of helium is important to be maintained at the highest level, either in liquid phase to avoid blockage produced by impurities in small capillaries, and in gas phase for applications like gas chromatography where any impurity will jeopardize the ultimate sensitivity of the analysis.

In order to obtain helium with extreme purity with the above recovery plant, the gas is purified by passing it through the purifier ATP. The ATP cools down the helium to $T<30$ K, and thus all the relevant impurities having very low vapor pressures at 30 K (nitrogen, oxygen) are eliminated by cryo-condensation, except hydrogen.

The main impurities in the recovered helium, i.e. nitrogen and oxygen, come from the atmosphere by gas diffusion processes through polymeric materials (e.g. gasbag and/or polymeric pipes) [9]. Meanwhile, hydrogen traces can come from the helium sources, since molecular H_2 is naturally present in helium gas as obtained from natural gas sources [10], or from oil degradation in the recovered helium compression stage.

The elimination of hydrogen is performed afterwards by passing the helium through a non-evaporable, thermally activatable getter (e.g. Zr-V-Fe) where the hydrogen becomes trapped. It diffuses easily into the bulk because it dissociates on the getter surface into atomic hydrogen. The hydrogen atoms slip into the atomic lattice of the metal grains. The process is represented by the following reaction [11]:

$$H_2(gas) + M \leftrightarrow 2H(adsorbed) + M \leftrightarrow 2H(Mbulk) \qquad (1)$$

Where M represents the metal constituents of the getter alloy.

This is a reversible process that allows regeneration of the getter material by thermal activation. Regeneration can be performed every 20,000 hours, taking profit of cryocooler maintenance, to minimize the impact on the operation of the plant.

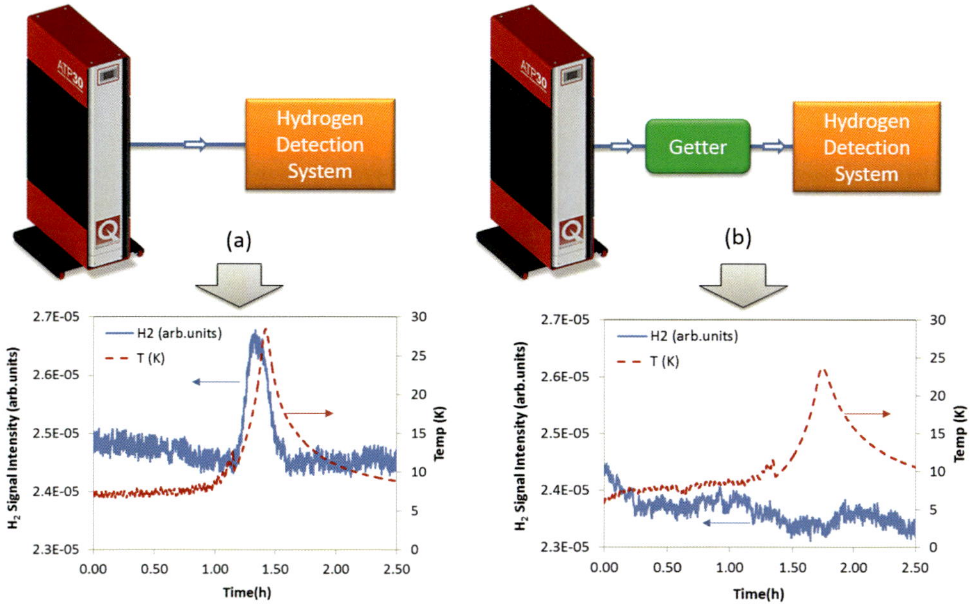

Figure 2. Experimental verification for the elimination of hydrogen with a getter. (a) After purification of 100,000 sL of He gas, the ATP is slowly warmed up and a peak of hydrogen is detected (blue continuous line) when the temperature (red dashed line) is close to 14 K. (b) No peak of hydrogen is detected after purification of the same amount of He gas when a getter is connected at the output.

The validity of the "Green Helium" plant concept is demonstrated by the fact that impedance blockages in small capillaries have been completely eliminated for more than one year, when the plant configuration was implemented in the Cryogenic Liquids Service at the University of Zaragoza and in Quantum Design Inc. Furthermore, the efficiency of the thermally activated getter media in trapping H_2, was verified by the continuous on line H_2 monitoring described below. The O_2 content was also verified to be well below 0.1 ppm using a Servomex Multiexact 5400 O_2 trace analyzer. Since the N_2 content is known to be of the same order of magnitude than the O_2 content, the gas purity will be > 99.99999 %. The graphs in Figures 2a and 2b show the filter temperature of the ATP as a function of time. During the test this temperature is varied from the base temperature of the purifier up to about 25 K, and back to the base temperature, in about 2.5 hrs. The hydrogen concentration in the helium purified carrier gas is monitored using a mass spectrometer (helium leak detector QualyTest HLT260 from Pfeiffer Vacuum with hydrogen mass selected). During the test, the ATP collected solid H_2 sublimates. The getter effectivity capturing H_2 molecules is very well demonstrated by placing the H_2 detector to analyze the helium gas at the output of the ATP (Figure 2a) and at the output of the heated getter (Figure 2b), respectively.

CONCLUSIONS

A helium recovery plant has been presented that utilizes liquefiers and purifiers based on cryocoolers. Special care has been given to the purity of the helium produced by the plant. The purification of helium is performed by a combination of the cryocondensation of all the relevant impurities (except hydrogen) at T<30 K, and the subsequent elimination of hydrogen with a (heated) getter. With this plant concept we obtain extreme purity helium (purity > 99.99999 %) liquid, from recovered helium with impurity levels up to 1%, that has completely eliminated the occurrence blockages of liquid helium small flow impedances.

ACKNOWLEDGMENT

The authors express their gratitude for the financial support from the Spanish Ministry of Economy and Competitiveness through project MAT2015-64083-R.

REFERENCES

1. S. C. Collins, "Helium Refrigerator and Liquefier," *Adv. in Cryogenic Engineering,* vol. 11 (1966), pp. 11–15.
2. C. Wang, "Small scale helium liquefaction systems," *J. Phys*. Conf. Ser. 150, 012053 (2009).
3. "Series 4x10 QUANTUMCOOLER, manufactured by Quantum Technology Corp., Squamish, Canada." 2016.
4. "ATL160 and ATP30, manufactured by Quantum Design, Inc., San Diego, CA, USA.," 2016.
5. C. Chialvo, T. Sayles, J. Diederichs, S. Spagna, M. Gabal, J. Sesé, and C. Rillo, "Smart Energy Compressors for Helium Liquefiers," *Cryocoolers 18*, ICC Press, Boulder, CO (2014), pp. 551-556.
6. C. Rillo, M. Gabal, M. P. Lozano, J. Sesé, S. Spagna, J. Diederichs, R. Sager, C. Chialvo, J. Terry, et al., "Enhancement of the Liquefaction Rate in Small-Scale Helium Liquefiers Working Near and Above the Critical Point," *Physical Review Applied* 3, 051001 (2015).
7. C. Rillo and L. Tocado, "HELIUM-RECOVERY PLANT," PCT/ES2010/070632, US patent 8,973,397 B2.
8. I. I. Abrikosova and A. I. Shal'nikov, "The purity of liquid helium," *Cryogenics* (Guildf). 11 (1971), pp. 137–138.
9. M. Gabal, M. P. Lozano, A. Oca, M. P. Pina, J. Sesé, and C. Rillo, "Helium Contamination through Polymeric Walls," *Phys. Procedia*, vol. 67 (2015), pp. 141–146.
10. D. F. McFarland, "Composition of Gas from a Well at Dexter, Kan.," *Trans. Kansas Acad. Sci.* 19, (1903), pp. 60–62.
11. R. Rameshan, "Evaluation of Non-Evaporable Getters for High Vacuum Hermetic Packages," NEPP Progr. Doc., pp. 1–31.

Design of Cryogenic Helium Gas Circulation with Multiple Cryocoolers for Superconducting Power Devices

N.G. Suttell[2], C.H. Kim[1], J. Ordonez[2], S. Pamidi[2]

[1]Center for Advanced Power Systems, Tallahassee, FL 32310
[2]FAMU-FSU College of Engineering, Florida State University, Tallahassee, FL 32310

ABSTRACT

Finite element analysis using COMSOL MultiPhysics® is used to study the relative performance of the various potential configurations for multiple units of cryogenic helium circulation systems. The design of large cryogenic systems needed for superconducting power devices require large mass flow rates with significant pressure drops. The models were used to identify the best configuration for an application that operates at around 50 K and requires a helium gas mass flow rate of 12 g/s with a pressure drop of 20 kPa. The configuration with two sets of two parallel circulators is shown to be the best option. The utility of the modeling techniques in the design of a large cryogenic helium circulation systems is demonstrated.

INTRODUCTION

Many High Temperature Superconducting (HTS) power devices have been successfully demonstrated to validate their capabilities and utility in the electric power grid [1]-[16]. The cryogenic systems used in the demonstrations are complex and expensive. Simpler, rugged, and operationally less expensive versions of the cryogenic systems are necessary for realizing the dream of safe, secure, and environmentally acceptable electrical power network that can be achieved with widespread applications of HTS power devices [17]. The cryogenic cooling capacity required for many HTS applications is rather large for a single cryocooler. As a consequence, multiple cryocoolers with heat exchangers and circulation pumps need to be integrated into a single system [18]. Cryocoolers have been used for HTS power applications. Most HTS power devices are cooled with liquid nitrogen and cryocoolers have been used to sub-cool liquid nitrogen that is circulated through the application. Many applications are designed to operate at temperatures below the liquid nitrogen temperature range (65 – 77 K) and require gaseous helium or neon in the circulation systems [19] [20]. One of the challenges with cryogenic gaseous circulation systems is that the pressure drop across the heat exchangers attached to the cryocoolers and the superconducting applications are high and the circulation pumps cannot handle the pressure drops at the required flow rates without losing extensive cooling capacity. Use of multiple cryocoolers and circulators is necessary for cryogenic helium circulation systems needed for many applications. Combining multiple units requires a thorough analysis to enable optimal designs [21]. This paper describes modelling efforts that would support the design of a versatile cryogenic helium circulation systems and the validation of the design parameters.

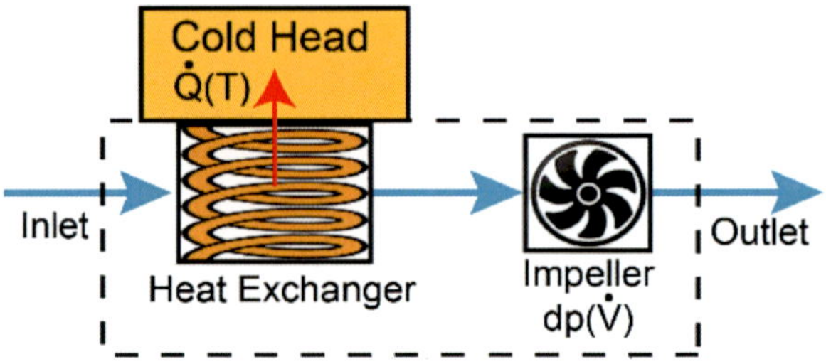

Figure 1. Schematic for one CHCS unit.

MODEL DESCRIPTION

Finite element method was used to analyze the cryogenic helium circulation system (CHCS) because it can handle complicated geometry in less time and with less effort than models developed using Fortran, Matlab, or Simulink. COMSOL Multiphysics® was used to develop the model. The built in Heat Transfer in Fluids physics module was coupled with the Low Reynolds Number k-epsilon Turbulent Flow physics where "Low Reynolds Number" refers to the fact that it resolves the boundary layer where the Reynolds number is significantly lower than the rest of the helium gas domain. The Reynolds number ranges from 50,000 to 125,000 and is within the turbulent regime. Only the helium domain is defined in the model, and the viscous dissipation is neglected. This domain consists of a heat exchanger followed by an internal impeller. Figure 1 shows a schematic of one CHCS which consists of a cryocooler, a heat exchanger (HX) and an impeller. The impeller is defined at an internal boundary that is perpendicular to the fluid flow. In addition, the impeller is characterized by a static pressure performance curve that is a function of the volumetric flow rate of the gaseous helium as shown in Figure 2 (a). This curve was defined within COMSOL in tabular form with a piece-wise cubic interpolation function. As the volumetric flow rate increases, the static pressure due to the impeller decreases.

The heat exchanger attached to the cold head in the gas flow stream has a smaller diameter than the rest of the CHCS resulting in a larger pressure drop due to friction. Comparable pressure drop is observed in the current heat exchangers tested at the Center for Advanced Power Systems (CAPS). The pressure drop through the heat exchanger on average is $\Delta p_{HX} = 4$ kPa. The boundary condition for the heat exchanger is characterized by the cooling power capacity curve of an AL325 Cryorefrigerator in which the cooling capacity is a function of the temperature of the cold head as shown in Fig 2 (b). A trend line with a second degree polynomial was generated from the power capacity curve, and it was applied to the interface between the cryocooler and heat exchanger in

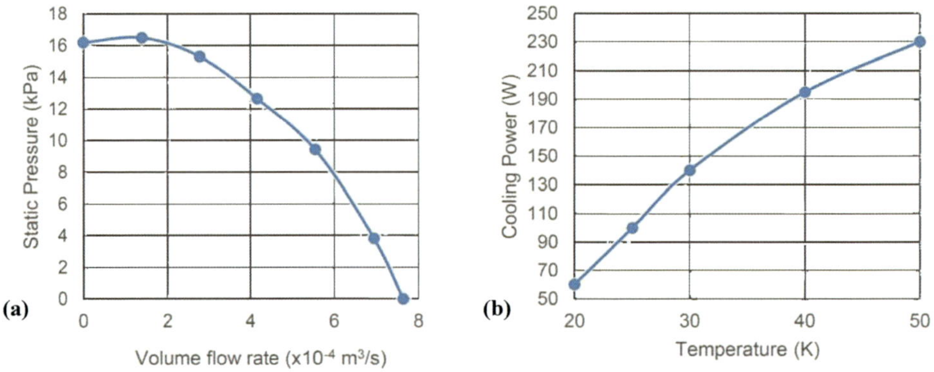

Figure 2. (a) Static pressure performance curve for CryoZone Noordwind CryoFan operating at 21,000 RPM. **(b)** Capacity curve for Cryomech AL325 cryocooler..

the CHCS. The boundary conditions for the entire model are as follows: inlet BC specifying mass flow rate, outlet BC specifying pressure, outward heat flux on the heat exchanger boundary, and all other boundaries are insulated. The heat capacity, viscosity, and thermal conductivity of helium are all functions of temperature, and the gas is considered ideal and compressible in order to calculate the density at each node in the mesh. The mesh of one CHCS unit contains 142,000 elements. The model was solved on a PC with an Intel i7-6700K processor and 32 GB of RAM running Windows 10. The former model description comprises one CHCS unit, and multiple units were added in series and/or parallel to analyze the relative performance of different configurations.

Figure 3 shows a few example configurations that were analyzed in this paper. The configurations denoted by parenthesis ( ) of interest are the following: two fans in series followed by two fans in parallel (1-1-2), one fan followed by two fans in parallel followed by one more fan (1-2-1), 2 fans in parallel followed by 2 fans in series (2-1-1), 2 fans in parallel followed by two more fans in parallel (2-2), and four fans in series (1-1-1-1). A parametric sweep of the mass flow rate between 6 g/s and 14 g/s was conducted while keeping the outlet pressure constant at 2 MPa for each configuration.

MODEL VERIFICATION

Before analyzing the various configurations of the 4 impeller system, the model had to be verified. This was done by running simulations on just one CHCS with the heat flux on the heat exchanger initially turned off. This provided the ability to test the behavior of the impeller and to determine if the gas can be assumed to be incompressible across the impeller. A further verification was made by adding another unit of CHCS in series and/or in parallel to see if these modifications behave similarly to that of a truly incompressible fluid. A final verification was made by turning on the heat flux and observing the temperature difference and density variation. Some of these verifications were also re-confirmed by manual calculations. All of these observations on the basic verifications are presented next.

For one case with a mass flow rate m. = 6 g/s at the temperature of 50 K, and the pressure before the impeller of p_i = 2 MPa, the model resulted in an increase in static pressure across the impeller of $\Delta p = p_o - p_i$ = 14.6 kPa, where subscripts i and o denote the property before and after the impeller, respectively. This result could easily be verified by manually calculating the volumetric flow rate and referring to the performance curve for the impeller from Figure 2(a) to find Δp. Equation (1) shows how the density before the first impeller was calculated.

$$\rho_i = \frac{p_i}{RT} \tag{1}$$

where R is the gas constant of helium, and T is the absolute temperature of the helium gas. The density before the impeller was used to calculate the volumetric flow rate shown in Equation (2).

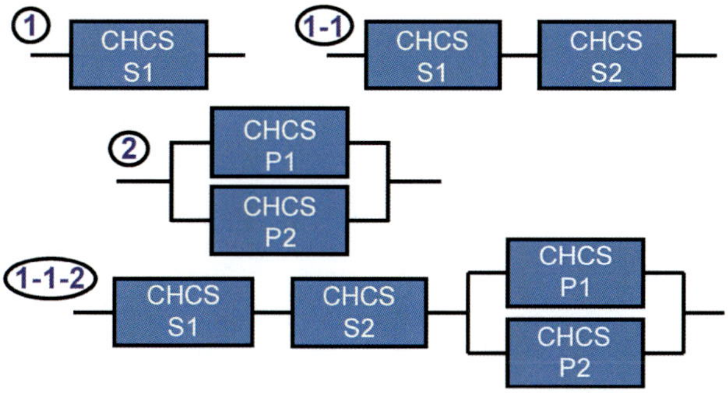

Figure 3. Examples of different configurations analyzed.

$$\dot{V}_i = \frac{\dot{m}}{\rho_i} \tag{2}$$

where $\dot{m}$ is the helium mass flow rate that is defined as the boundary condition at the inlet, the density before the impeller is $\rho_i \approx 19.26$ kg/m^3. This makes the volumetric flow rate before the impeller to be $\dot{V}_i \approx 3.12\text{x}10^{-4}$ m^3/s. By referencing the performance curve for the impeller, this volumetric flow rate corresponds to an increase in static pressure of $\Delta p \approx 14.7$ kPa. The error was within 0.7 %.

Another verification that was made was on the compressibility of the helium gas in our operational range. The density after the impeller is found using the following relationship.

$$\rho_o = \frac{p_o}{RT} \tag{3}$$

For the previous operational conditions, this results in a density after the impeller of $\rho_o \approx 19.4$ kg/m^3 which is an increase of only 0.7 %, and it verifies our assumption that helium gas can be assumed to be an incompressible fluid across the impeller. This is because the operational pressure is relatively high (2 MPa), and an increase in static pressure of only 15 kPa has almost no effect on the density. In addition, the process across the impeller is assumed to be adiabatic. Therefore, without cooling or heating, gaseous helium at high pressure may be treated as an incompressible fluid.

It is known that, for incompressible fluids, having two impellers in series will double the static pressure increase compared to that by one impeller. Additionally, having two impellers in parallel will double the mass flow rate and increase the static pressure by the same amount compared to that by one impeller. To ensure that the model will also provide the same result and for further validation, two additional cases were generated, one with two CHCS units in series (1-1) configuration and the other with two units of CHCS in parallel (2) configuration, both with heat transfer turned off. These results are shown in Figure 4, and they are as expected.

Finally, heat transfer was included in the model to verify the effects on the temperature of the helium. The heat flux that was applied to the outer boundary of the heat exchanger was initially set at a constant value of $\dot{Q}$= -200 W to simplify the verification. The model result shows a temperature difference through the heat exchanger of ΔT_{HX} = -6.3K. This can be verified by using a form of the first law of thermodynamics seen in Eq. (4).

$$\Delta T_{HX} = \frac{\dot{Q}}{\dot{m}C_p} \tag{4}$$

With $\dot{m}$ = 6 g/s and C_p = 5254 J/(kg K), Equation (4) gives ΔT_{HX} = -6.3K This verifies that the heat transfer physics within the model is accurate.

RESULTS

A parametric sweep was performed on the model for the mass flow rate from 6 – 14 g/s. To start, Figure 5 (a) shows how including nominal friction affects the increase in static pressure through

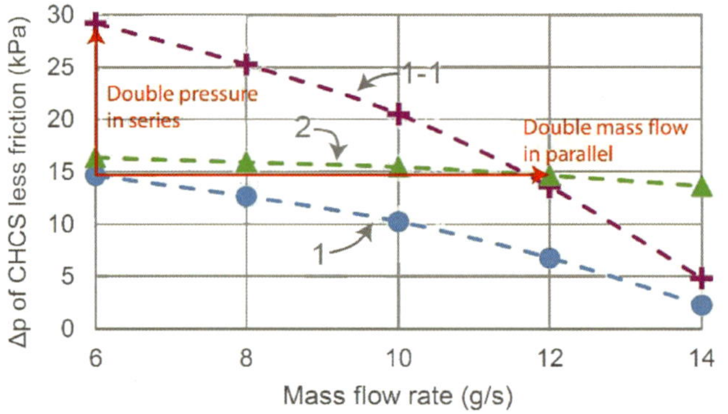

Figure 4. Increase in static pressure due to the impellers neglecting friction through the pipe.

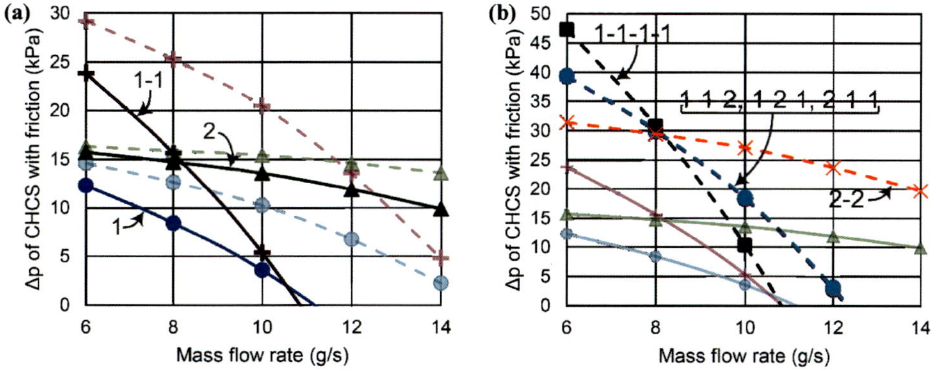

Figure 5. (a) Increase in static pressure vs. mass flow rate through cooling system with friction of the (1), (2), and (1-1) configurations (solid lines) compared to that without friction from Fig. 4 (transparent, dashed lines). **(b)** Increase in static pressure vs. mass flow rate through the cooling system for the four impeller configurations with friction (dashed lines) compared to the configurations with friction from Fig 5 (a) (transparent, solid lines). Note that this data is excluding the effects of heat transfer..

the cooling system. This is a necessary step in understanding the effects that friction has on each configuration and to provide clarity when designing a CHCS. The lengths of the tube channels were identical in all configurations. By including friction, the increase in static pressure of the (1) and (1-1) configurations dropped considerably more than that of the (2) configuration, and they can only provide flow rates up to 11 g/s. This is an example of a major drawback of the (1) and the (1-1) configurations. Many heat load applications require a flow rate of more than 8 g/s and the (2) configuration is the optimal choice. This shift occurs because having 2 impellers in parallel doubles the cross-sectional area of the flow channel providing less constriction and reduces the mass flow rate by half in each path. As the mass flow rate increases as in Figure 5 (a), the difference in the pressure drop between the (1-1) and (2) configurations increases exponentially. This is because the pressure drop due to friction is proportional to the square of the velocity.

To add even more complexity and versatility to the study, Figure 5 (b) shows the increase in static pressure of each of the four impeller configurations. The lines of the 4 different impeller configurations intersect at a mass flow rate of 8 g/s and a static pressure increase of 30 kPa. Another interesting result is that the combinations of two CHCS in series and two CHCS in parallel (1-1-2, 1-2-1, and 2-1-1) are all lined up. This implies that, with no cooling present, the total pressure drop is the sum of the individual pressure drops of each sub-scale CHCS configuration regardless of their order.

The following section presents the results that include heat transfer and friction for the various configurations with four impellers. Figure 6 (a) shows the increase in static pressure through the total cooling system for the four configurations with cryocoolers turned on (solid lines), and they are compared to the same configurations with cryocoolers turned off (transparent, dashed lines). The cryocoolers, when turned on, are removing heat from the helium gas and lower the temperature, which increases its density, and therefore the impellers perform better (refer to Figure 2). The configurations that benefit the most by adding heat transfer are (1-1-1-1) and (2-1-1). One of the most interesting observations of this analysis is how the relative performance of the various configurations, that have a combination of CHCS units in series and parallel, changes after including heat transfer and which configurations outperform the others. The (2-1-1) configuration performs slightly better than (1-2-1) and (1-1-2). This happens because the helium gas travels through the parallel section first where its density is increased more than that by a single CHSC, and this allows the following two CHCS units in series to produce higher static pressure. On the contrary, total temperature drop by the cooling system does not change with the configuration of CHCS units as long as the mass flow rate and the number of CHCS units are the same as shown in Figure 6 (b). In conclusion, the impeller performs better by decreasing the temperature first, but the ultimate temperature achieved by the cryogenic system remains unchanged regardless of the configuration of CHCS units.

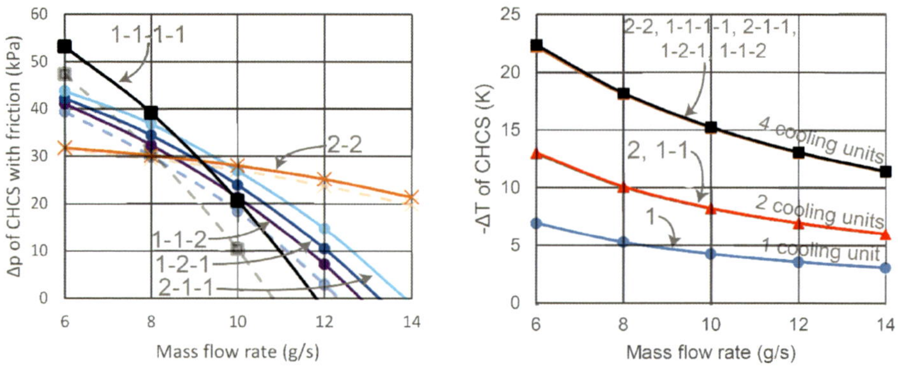

Figure 6. (a) Increase in static pressure vs. mass flow rate through cooling system for the four CHCS configurations including heat transfer through the heat exchanger (solid lines) compared to those without heat transfer (transparent, dashed lines). **(b)** Change in temperature of helium gas through the cooling system vs. mass flow rate for different CHCS unit configurations.

DISCUSSION

The modeling efforts described in this paper are necessary, educational, and beneficial for the development of versatile and optimized helium gas circulation systems. During the development of the model, it has been proven that gaseous helium may be assumed to be incompressible across the impeller, pipe friction affects the configurations of CHCS units that have more in series than in parallel, and that the orientation of the CHCS sub-scale configurations have no effect on the temperature differential through the cooling system (Figure 6 (b)).

The following discussion will present the process of choosing a CHCS configuration for an example set of specifications as a case study of the utility of the modelling. These specifications include a mass flow rate of $\dot{m}$ = 12 g/s required to cool an HTS power cable that has a pressure drop of Δp = 20 kPa. Another requirement is that the cable system remain at or below T = 50 K, and the system load experiences a temperature rise of less than ΔT = 7.5 K. Looking at Figure 7, it appears that, in order to get a temperature difference of ΔT = 7.5 K and a mass flow rate of $\dot{m}$ = 12 g/s, it is required to have more than two CHCS units. It is possible to use just three CHCS units in this case, but these configurations were omitted in this analysis, and one of the configurations with four CHCS units will be chosen. Figure 6 will help in selecting one of the specific configurations among the possible four unit options. For a pressure drop of Δp = 20 kPa and mass flow rate of $\dot{m}$ = 12 g/s, only the (2-2) configuration is capable of satisfying the required operational specifications. This is a generic design, but the (2-2) configuration appears to be the optimal configuration of CHCS units in parallel and series for most large-scale HTS applications.

Another important factor to consider when designing a set of multiple CHCS units is the friction. Here, pipe length and diameter are important parameters. It has been observed in previous experimental results that parallel channels with different pipe lengths produce undesirable effects such as backflow. Also, having parallel channels converge to a single channel with the same diameter creates flow constriction.

This model will be used in the future design of an experimental cryogenic helium circulation system where the different configurations of four CHCS units will be tested and the performances compared to that of the model results. Figure 7 shows an example of a (2) configuration setup that will be tested and compared to the FEM model. The lessons of pipe friction and performance of different configurations will be applied to the experiment to aid in designing a versatile gaseous helium cryogenic systems needed for supporting many different types of HTS applications.

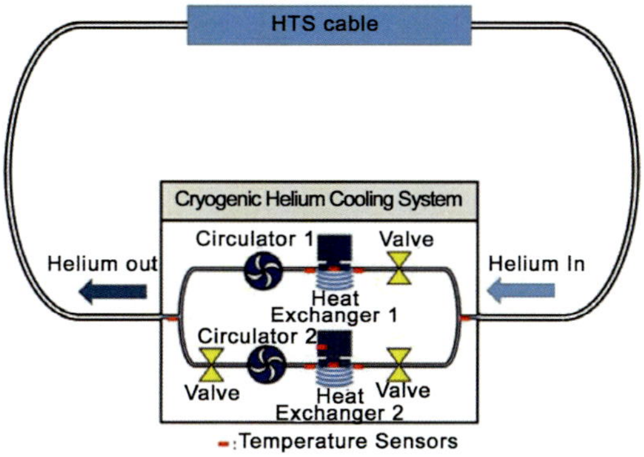

Figure 7. Physical system schematic used for experimental validation of CHCS FEM model.

ACKNOWLEDGMENTS

This work was sponsored by the US Office of Naval Research under contract N000141410198.

REFERENCES

1. Grant, P. M., Starr, C., and Overbye, T. J., "A power grid for the hydrogen economy," *Scientific American*, 295(1), pp. 76-83.
2. Trevisani, L., Fabbri, M., and Negrini, F., "Long-term scenarios for energy and environment: Energy from the desert with very large solar plants using liquid hydrogen and superconducting technologies," *Fuel processing technology* 87, no. 2 (2006): 157-161.
3. Haugan, T. J., Long, J. D., Hampton, L. A., & Barnes, P. N. "Design of compact, lightweight power transmission devices for specialized high power applications," *SAE International Journal of Aerospace* 1, no. 2008-01-2930 (2008): 1088-1094.
4. Ballarino, A., "Design of an MgB2 feeder system to connect groups of superconducting magnets to remote power converters," *Journal of Physics: Conference Series*, vol. 234, no. 3, p. 032003. IOP Publishing, 2010.
5. Rey, C. M., Duckworth, R. C., Demko, J. A., Ellis, A., James, D.R., Gouge, M. J., "Test results for a 25 meter prototype fault current limiting HTS Cable for project Hydra," *Adv. in Cryogenic Engineering*, AIP Publishing, 2010: 453-460.
6. Ferrara, P. J., Uva, M. A., and Nowlin, J., "Naval ship-to-shore high temperature superconducting power," *IEEE Transactions on Applied Superconductivity* 21, no. 3 (2011): 984-987.
7. Kalsi, S. S., *Applications of high temperature superconductors to electric power equipment*, John Wiley & Sons, 2011.
8. van der Laan, D. C., Lu, X. F., & Goodrich, L. F., "Compact GdBa2Cu307-o coated conductor cables for electric power transmission and magnet applications Contribution of NIST, not subject to US copyright," *Superconductor Science and Technology* 24, no. 4 (2011): 042001.
9. Maguire, J. F., Yuan, J., Romanosky, W., Schmidt, F., Soika, R., Bratt, S., "Progress and status of a 2G HTS power cable to be installed in the long island power authority (LIPA) grid," IEEE Transactions on Applied Superconductivity 21, no. 3 (2011): 961-966.
10. Gamble, B., Snitchler, G., & MacDonald, T., "Full power test of as 36.5 MW HTS ship propulsion motor," *IEEE Transactions on Applied Superconductivity* 21, no. 3 (2011): 1083-1088.
11. Ohsaki, H., Lv, Z., Sekino, M., & Tomita, M., "Application of superconducting power cables to DC electric railway systems," *Physics Procedia* 36 (2012): 908-913.
12. Ryu, C., Jang, H., Choi, C., Kim, Y., Kim, H. "Current status of demonstration and commercialization of HTS Cable system in grid in Korea," *Applied Superconductivity and Electromagnetic Devices (ASEMD), 2013 IEEE International Conference on*, pp. 539-542. IEEE, 2013.
13. Xiao, L., Dai, S., Lin, L., Zhang, Z., & Zhan, J., "HTS power technology for future DC power grid," *IEEE Transactions on Applied Superconductivity* 23, no. 3 (2013): 5401506-5401506.

14. Dai, S., Xiao, L., Zhang, H., Teng, Y., Liang, X., Song, N., "Testing and demonstration of a 10-kA HTS DC power cable," *IEEE Transactions on Applied Superconductivity* 24, no. 2 (2014): 99-102.
15. Stemmle, M., Merschel, F., Noe, M., and Hobl, A., "AmpaCity - Advanced superconducting medium voltage system for urban area power supply," In *2014 IEEE PES T&D Conference and Exposition*, pp. 1-5. IEEE, 2014.
16. Morandi, A., "HTS dc transmission and distribution: concepts, applications and benefits," *Superconductor Science and Technology* 28, no. 12 (2015): 123001.
17. Radebaugh, R., "Cryocoolers: the state of the art and recent developments," *Journal of Physics: Condensed Matter* 21, no. 16 (2009): 164219.
18. Pamidi, S. V., Kim, C. H., Kim, J-h., Crook, D., & Dale, S., "Cryogenic helium gas circulation system for advanced characterization of superconducting cables and other devices," *Cryogenics* 52, no. 4 (2012): 315-320.
19. Saji, N., Asakura, H., Yoshinaga, S., Ishizawa, T., Miyake, A, Obata, M., "Design of oil-free simple turbo type 65 K/6 kW helium and neon mixture gas refrigerator for high temperature superconducting power cable cooling," *Adv. in Cryogenic Engineering: Proceedings of the Cryogenic Engineering Conference-CEC*, vol. 613, no. 1, pp. 893-902, AIP Publishing, 2002.
20. Zhang, T., Haran, K., Laskaris, E.T., Bray, J. W., "Design and test of a simplified and reliable cryogenic system for high speed superconducting generator applications," *Cryogenics* 51, no. 7 (2011): 380-383.
21. Suttell, N., Kim, C. H., Ordonez, J., Kephart, J., Graber, L., Pamidi, S., "Three-dimensional finite-element analysis of terminations for gaseous-helium-cooled high-temperature superconducting power cables," *IEEE Transactions on Applied Superconductivity* 26, no. 4 (2016): 1-5.

Connecting Coolers to Superconducting Magnets with a Thermal-Siphon Cooling Loop

M. A. Green[1,2]

[1]FRIB Michigan State University, East Lansing, MI, USA 48824
[2]Lawrence Berkeley Laboratory, Berkeley, CA, USA 94720

ABSTRACT

Coolers are used to cool-down and cool cryogen-free superconducting magnets and other devices that operate at temperatures from 4 K to 40 K. Coolers are also used to re-condense helium boil-off gas in superconducting magnet cryostats, such as for MRI and NMR magnets. There is an in-between case where a natural-convection thermal-siphon cooling loop is used to connect the cooler cold heads to the load being cooled and cooled down [1]-[3]. This case works well when the magnet is indirectly cooled by the coolant flowing through tubes attached to the magnet cold mass. A thermal-siphon loop fluid can work with fluids other than helium, which permits a magnet to be liquid-cooled (over the two-phase temperature range for the fluid) and thus ensure that the temperature variation within the magnet is minimized. For a flammable gas like hydrogen or a rare gas like neon, the loop must be prefilled and sealed at room temperature. A thermal-siphon loop will work when the coolers are remotely located from the magnet provided that the cold head heat exchanger is above the highest point in the magnet cryostat. This type of cooling system is used to cool and cool-down the 4 K cyclotron gas-stopper magnet at Michigan State University [4], [5]. Calculated cool-down times will be compared for a cooling loop between a cold mass of 1250 kg and a single Cryomech PT415 pulse tube cooler for the cases when the cooling loop is filled with helium, hydrogen, and neon for pressures of 0.2 MPa, and 0.8 MPa.

INTRODUCTION

In a typical cryogen free magnet, heat from the load is conducted to the cooler's or coolers' cold head through a copper or aluminum strap from the part of the magnet closest to the cooler. During the cool down and during normal operation there is a temperature drop ΔT_1 between the part of the magnet closest to the cooler cold head, and the cooler cold head. In addition, there is a second temperature drop ΔT_2 between the points on the magnet that is farthest from the cooler connection point to the cooler connection point. For a large magnet, ΔT_2 can be large [6]. The total $\Delta T = \Delta T_1 + \Delta T_2$ between the magnet and the cooler is directly proportional to the heat that can be taken into the cooler cold head and the conduction path length. The ΔT is inversely proportional to the thermal conductivity of the conduction path and the cross-section area of the path carrying the heat. For large 4 K magnets, this ΔT may cause the magnet to fail to reach its design goal.

When one cools a cryogen free magnet by conduction, the heat is conducted through a pure metal strap with a high thermal conductivity. The strap must be flexible to allow for thermal

Cryocoolers 19, edited by S.D. Miller and R.G. Ross, Jr.

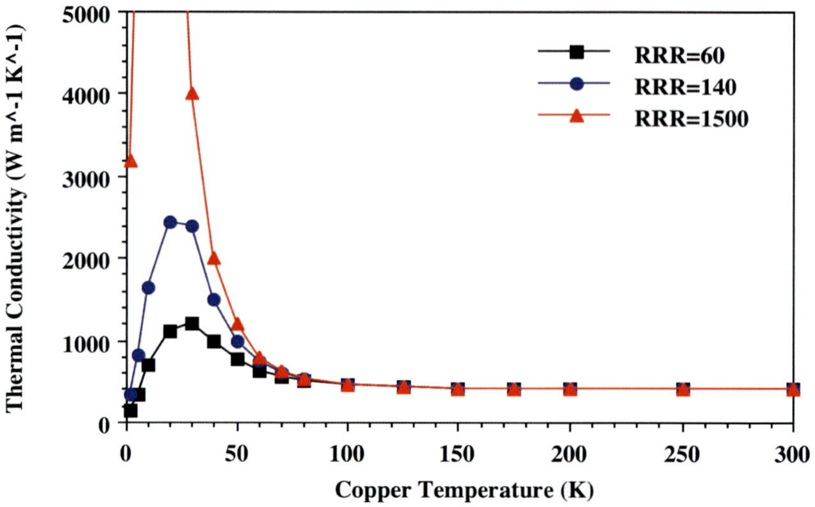

Figure 1. The thermal conductivity of three different annealed copper samples as a function of temperature and copper purity. RRR = 60 copper is 0.9995 pure. RRR = 140 copper is 0.9998 pure and annealed. RRR = 1500 copper is 0.99999. All three copper samples are very soft. Cold working will reduce the copper RRR and decrease the thermal conductivity.

contraction of the magnet being cooled relative to to the cold head of the cooler, which is usually fixed with respect to the cryostat vacuum vessel. To allow for movement, the pure metal strap is long and has a small relative cross-section. In general, the strap is made from a metal-like copper with a residual resistance ratio RRR greater than about 30. The RRR of a metal is dependent on the metal purity and the state of cold work. Figure 1 shows a plot of copper thermal conductivity as a function of temperature for three copper materials with an RRR = ~60, an RRR = ~140, and an RRR= ~ 1500 [7]. At temperatures above 80 K, the three materials have nearly the same thermal conductivity. Below 80 K, the thermal conductivity is different for the materials.

If one wants to calculate the heat flow Q through a strap of length L with a constant cross-section area A_c from temperature T_1 to temperature T_2 one can use the following expression;

$$Q = K_{1-2} \frac{A_c}{L}, \tag{1}$$

where an expression of K_{1-2} is as follows;

$$K_{1-2} = \int_{T2}^{T1} k(T)\,dT . \tag{2}$$

and $k(T)$ is the thermal conductivity of the strap material as a function of temperature T.

$K_{1-2,}$ the thermal conductivity integral (with units W m^{-1}), is an interesting function. For all of the copper materials between 80 K and 300 K, K_{1-2} is nearly the same. This is the region where a magnet has ninety percent of its heat capacity between zero and 300 K. For example, the RRR=60 copper has a K_{1-2} of 91,000 W between 80 K and 300 K, whereas the RRR=1500 copper has a K_{1-2} of 91,300 W m^{-1} over the same temperature range. In the low heat capacity temperature range from 5 K to 80 K, the RRR=60 copper has a K_{1-2} of ~63,000 W m^{-1}, but the RRR=1500 copper has a K_{1-2} of ~300,000 W m^{-1} over the same temperature range. For a copper with a high RRR at temperatures <80 K, one has a large thermal conductivity integral at low temperatures, but the heat removed is relatively small compared to the total heat that must be removed during the magnet cool-down. The thermal conductivity of a high RRR copper will degrade with the cold working of the metal over time. One should use a metal with a RRR>25 for magnet shields that are designed to operate at temperatures ~ 40 K, because the thermal conductivity of the metal is greater than a factor of two higher at 40 K than it is at 80 K [8], [9].

During a cool-down, this ΔT between the magnet and the cooler cold head determines the cooler cold head temperature, which in turn determines the amount of refrigeration that can be applied to the load being cooled down. The cool-down of a magnet with the cooling applied at a single point can be estimated with a spreadsheet. Iteration is required because the heat being removed from the magnet must be matched with capacity of the cooler as a function of the cold head temperature. With a two-stage cooler, the first stage cold head temperature has an effect on the cooling capacity of the second stage.

Even after the magnet is cooled-down, the temperature drop between the warmest place in the magnet and the cold head can have an effect on the operating temperature margin of the superconducting magnet. When the heat flow to the cooler is too large for a magnet with a low operating temperature, the magnet may not reach its design goal [6]. For this and other reasons, it is often desirable to use a two-phase natural-convection thermal-siphon cooling loop.

TWO-PHASE THERMAL-SIPHON COOLING LOOP

A two-phase thermal-siphon natural free-convection cooling loop spreads the cooling around the magnet when there is liquid cryogen in the tube. The temperature drop from the inner surface of the tube to the cold head of the cooler is very low for helium. If the heat transfer to the condenser is < 40 W per square meter, and the heat transfer boiling heat transfer of the tube inner surface is < 6 W m^{-2}, the ΔT between the magnet cooling tube is < 0.1 K for liquid helium at 4.2 K for flow circuits <5 m in length. For a liquid hydrogen system, the total ΔT is <0.2 K for circuits <5 meters in length. For circuits with neon or nitrogen, the ΔT is <0.5 K for circuit lengths <5 m. The ΔT between the cooled surface and the cold head is much lower that it would be for a RRR = 100 copper strap of the same length and cross-sectional area [10].

There are a number of other potential advantages of using a thermal-siphon cooling loop in place of a cooler solidly connected to a magnet as is the case in the typical cryogen-free cooling system. These advantages include: 1) There can be good vibration isolation between the magnet and the cooler. 2) One can use drop in cooler [11], which permits magnet shipment without the coolers and the coolers can be easily changed for maintenance while the magnet is cold. 3) For gases like hydrogen and neon, all of the needed gas can be contained in a sealed system for very low gas loss and enhanced (hydrogen) safety [12], [13] (a closed system rather than a system that is potentially open to the air). 4) Coolers can liquefy room temperature gas provided the gas is pre-cooled to remove as much of it's sensible heat as possible before the gas goes through final cooling and condensation [14], [15], [5]. Figures 2a and 2b show an open free-convection cooling loop and a closed free-convection cooling loop. Both types of cooling loops can be used to cool-down a magnet as well as keep the magnet at its desired operating temperature. A closed loop can be used with any cryogen, but it is essential to use a closed loop for hydrogen and neon. The open cooling loop was used for the cyclotron gas-stopper magnet at Michigan State University (MSU). The liquid helium volume in each cryostat is about 12 liters. In order to make the closed sealed system viable, it is useful to limit the amount of liquid cryogen in the loop to the 3 to 5 liters. The surge tank volume in Figure 2b is a function of the system liquid cryogen volume and the maximum working pressure of the surge tank. Figure 2b shows the surge tank inside if the cryostat vacuum vessel and the shield. This tank can be located outside of the cryostat vacuum vessel at any reasonable distance from the cryostat. The surge tank should be designed and fabricated in accordance with the applicable pressure vessel codes.

Experience with the MSU magnet showed the following [15], [16]: 1) The cool-down is faster if the magnet has open channels. The mass flow per unit area should be as low as possible for a rapid cool-down of a magnet. 2) When multiple coolers are used, the flow channels to and from each condenser heat exchanger must be identical, so that each cooler contributes the same amount of cooling to the cooling loop [15]. 3) A cooling loop system should be designed to have a working

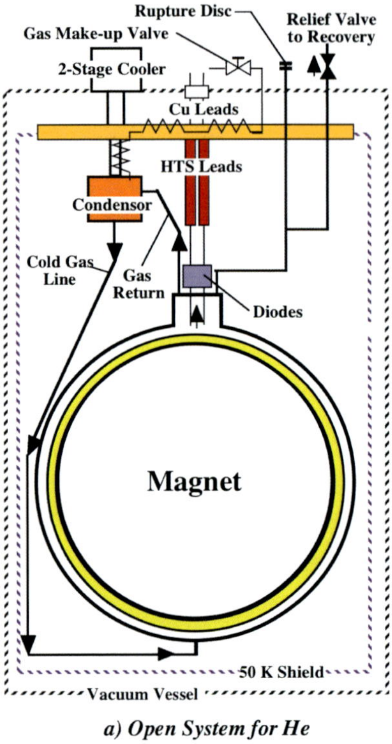

Figure 2. Open and closed two-phase, free-convection thermal-siphon cooling loops for use with cryogens such as helium, hydrogen and neon. Helium can be used in either an open loop or a closed loop. Hydrogen, a highly flammable gas is best used in a closed loop with the loop pre-charged and sealed. The tank and piping must be designed for a flammable gas. Neon is a very expensive gas, so it is best put into a closed sealed circuit to avoid loss of the gas.

pressure of at least 0.4 MPa. The bellows design pressure is important for achieving high cooling circuit operating pressures. High cryostat pressures during the cool-down will result is a shorter cool-down time, especially when the channel area is small [5], [16].

THE COOLER AND ITS PERFORMANCE AT ELEVATED TEMPERATURES

The Cryomech PT-415 pulse tube cooler (see Figure 3) was selected for the study of cooling loops using liquid helium from 4.2 to 4.7 K, liquid hydrogen from 15 K to 28 K and liquid neon from 28 K to 35 K. A PT415 cooler produces ~1.5 W of cooling on the second-stage with ~40 W of cooling on the first-stage. This cooler has been used for cooling superconducting magnets operating at 4.2 to 4.7 K. MSU is using six of the remote motor version of this cooler on the cyclotron gas stopper magnet. Six coolers are used to cool-down the two coils, but only four coolers are needed to keep the magnet at 4.3 K. The PT415 has been tested over a range of first stage and second stage temperatures [17], [18], [19]. This cooler will produce ~20 W at 20 K and ~40 W at 35 K with 40 W going into first-stage at ~52 K (see Figure 4).

COOL-DOWN CALCUATION FOR A MAGNET COOLING CIRCUIT

The author calculated the cool-down time using a spreadsheet program for a loop that has four channels from the bottom to the top. Each channel attached to the cold mass of 1250 kg is 3.8-meters long. The channels are round with an inside diameter of 12.7 mm. A magnet coil cross-section

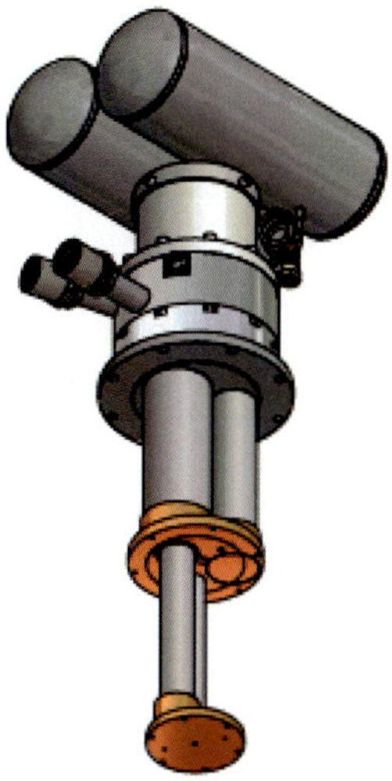

Figure 3. PT415 cold head [17]

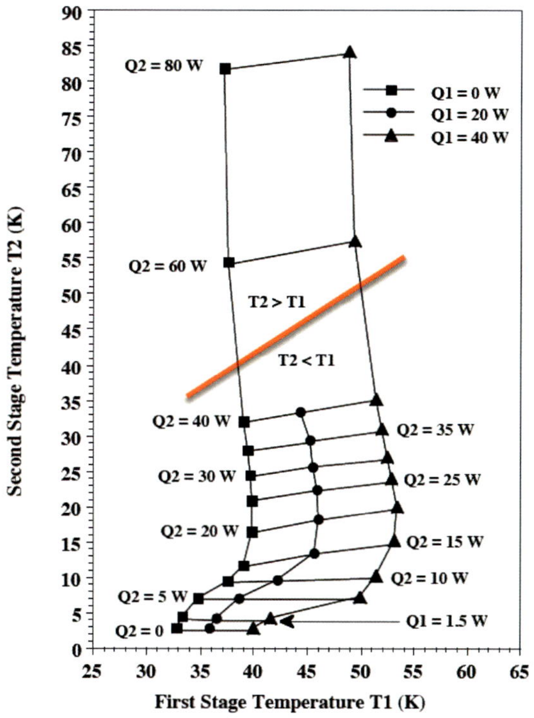

Figure 4. Performance curves for a PT415 cooler with zero to 80 W on the second stage and zero to 40 W on the first stage [19].

is shown in Figure 5. A single tube 4.0-meters long with an ID of 12.7 mm carries the fluid from the condenser heat exchanger to the bottom of the magnet. The tube leaving the top of the magnet and returning to the condenser heat exchanger has an ID of 25.4 mm. The total liquid volume in the magnet tubes and the tube from the condenser is about 2.5 liters, which is under the 3 to 5 liquid liters suggested for a sealed closed circuit system. If 0.5 m of the tube between the top of the magnet and the condenser contains liquid cryogen, the total cryogen in the system is ~2.8 liters. The heat transfer area within the condenser heat exchanger is 0.184 m^2. The surface

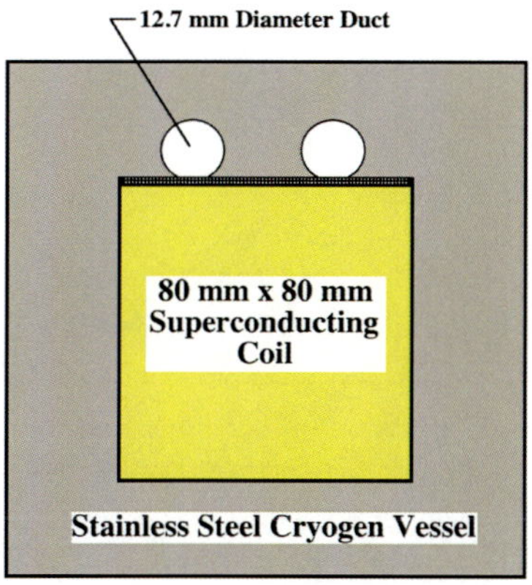

Figure 5. The superconducting coil cross-section with two 12.7 mm ID cryogen cooling-tubes. The flow area is 253.4 mm^2 and the inner surface area is ~0.62 m^2 [16].

area of the cooling tubes that are part of the cold mass is ~0.62 m^2. Cross-section area of the tubes in the magnet is 0.0002354 m^2. The tube from the condenser has a flow area of 0.00005885 m^2, and the cross-section are of the tube back to the condenser is the same as the four tubes carrying cryogen in the magnet coil.

The circuit is a free-convection circuit to get the circuit to flow the gas within the condenser durin cool down by the cooler cold head. The temperature of the cold head decreases rapidly until the cold dense gas begins to flow down the tube to the bottom of the magnet. Once the flow circuit starts to flow, the cold head temperature goes up because there is warm gas entering the condenser heat exchanger from the top of the magnet. This effect was clearly seen during the cool-down of the MSU cyclotron gas-stopper magnet coils [15]. Once the circuits start to flow the circuit continues to operate until the magnet is cooled-down and the cooler starts to liquefy the gas.

The rate of liquefaction is dependent on the excess refrigeration available. The MSU gas stopper magnet coolers were able to make about 6 liters a day of helium with about 1.2 W of refrigeration available to liquefy the helium. Each of the two coils in the cyclotron gas stopper magnet was cool-down using three PT415-RM coolers that can develop 1.35 W at 4.2 while delivering 36 W on the first stage at 40 K. Once the helium was liquefied in the magnet cryostat, one of the three coolers was turned of. Each coil was kept cold with two operating coolers. The MSU cyclotron magnet coils have nearly the same parameters as the coils in this study.

The magnet cool-down was modeled using a single PT415 cooler to cool-down 1250 kg of cold mass that consists of stainless steel, copper and about ten percent organic material. The total energy removed from the coils during the coil down from 300 K to 4.2 K (with helium gas) was 104 MJ. Slightly less energy was removed when cooling the cold mass to 20 K (with hydrogen gas) and 25 K (with neon gas). The cool-down calculations were done with a gas pressure of 0.2 MPa (about two atmospheres absolute) and 0.8 MPa (about eight atmospheres absolute). The object of the calculation was to determine the cool-down time as a function of the gas and a function of the gas pressure.

The cool-down time for a single PT415 cooler to cool-down a 1250 kg coil as a function of temperature is shown in Figure 6. From Figure 6, one can see that the cool-down is fastest when when hydrogen is used as a cooling loop fluid. This is true both at 0.2 MPa and 0.8 MPa. The cool-down was slowest when neon was the gas circulating within the loop.

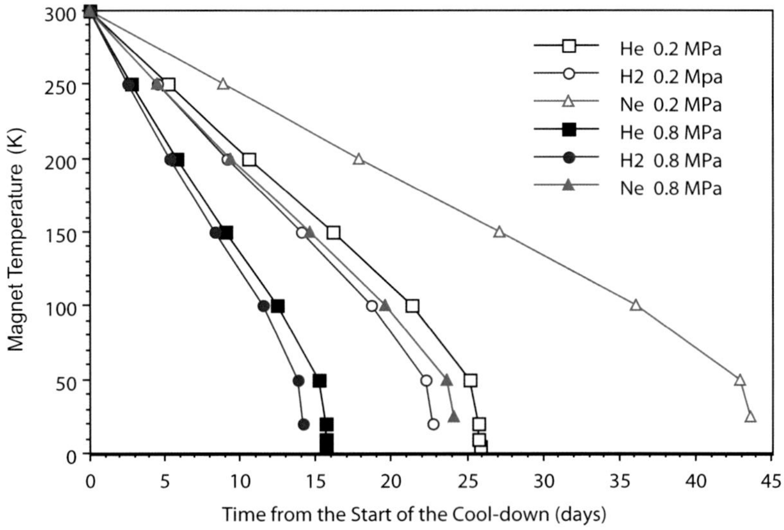

Figure 6. This figure shows cool-down time using a single PT415 cooler as a Function of temperature and loop Pressure for a single Loop with a mass of 1250 Kg using He, H_2 and Ne in the cooling loop. Helium is a black square; hydrogen is a blue circle and neon is a green triangle. Open data points are for the cases with a loop pressure of 0.2 MPa. Filled data points are for the cases with a loop pressure of 0.8 MPa.

Table 1. Properties of Helium, Hydrogen, and Neon [20]

Parameter	Cryogen		
	He	H_2	Ne
Molecular Weight	4	2	20.2
Gas Density @ STP (kg m^{-3})	0.1786	0.0893	0.902
Triple Point T (K)	2.17	13.81	24.6
Triple Point P (MPa)	0.0051	0.007	0.0423
Boiling T $T_{b\ at}$ 0.1013 MPa (K)	4.22	20.4	27.2
Liquid Density at T_b (kg m^{-3})	125	70.8	1212
Critical T (K)	5.19	32.3	44.4
Critical P (MPa)	0.221	1.292	2.710
Heat of Vaporization (J g^{-1})	20.9	442	86.0
C_p Liquid at T_b (J g^{-1} K^{-1})	~2.5	~9.8	~0.44
C_p Gas at T > $2T_b$ (J g^{-1} K^{-1})	~5.2	~14.2	~1.03
k_f Liquid at T_b (W m^{-1} K^{-1})	0.027	0.119	0.04
k_f Gas at T_b (W m^{-1} K^{-1})	0.011	0.021	0.014
μ liquid at T_b (kg m^{-1} s^{-1})	3.5×10^{-6}	1.3×10^{-5}	~7×10^{-5}
μ gas at T_b (kg m^{-1} s^{-1})	0.9×10^{-6}	1.1×10^{-6}	~4.3×10^{-6}

As seen in Table 1, hydrogen is a better working fluid than helium. The reason that hydrogen is better than helium is that its specific heat is 2.7 times the specific heat of helium. For a given amount of cooling. the mass flow of hydrogen is 36 percent of that of helium. A free convection loop is a balance between mass flow pressure drop and the driving force, which is proportional to density. Helium has twice the density of hydrogen, but this does not make for the advantage hydrogen has with respect to the mass flow.

Neon on the other hand has a specific heat that is a factor over a factor of five lower than helium. As a result, the mass flow must be at least a factor of five higher than for helium. The density difference between neon and helium is about a factor of five favoring neon. The momentum pressure drop is proportional the mass flow squared. The driving force is proportional to the density. While at low Reynolds number the friction factor can approach 1, but overall the pressure drop cut the flow of neon enough to make the rate of cooling about sixty-one percent of the rate of cooling for helium.

Increasing the pressure increases the rate of cooling almost a factor of two for all of the gasses. The loop pressure effect has been demonstrated experimentally at MSU [5]. Whenever a system is designed for free-convection cool-down, it is desirable to increase the pressure. For an open helium-cooling loop like the one shown in Figure 2a, it is desirable to set the maximum pressure to a least 0.4 MPa. For a closed cooling loop like that shown in Figure 2b, one should design the loop for higher pressure to keep the surge tank volume down to a reasonable size.

CONCLUDING COMMENTS

The cooling system for the MSU cyclotron gas-stopper magnet has shown that a low-temperature superconducting magnet can be designed to be cooled-down and be kept cold using small coolers. The liquid cryogen can be liquefied into the magnet cooling system by the cooler provided the warm gas entering the magnet cryostat is pre-cooled with liquid nitrogen and/or the first stage of the cooler, if the unit is a two stage cooler.

This paper shows that a magnet can be cooled-down and be kept cold using a cooling loop filled with helium, hydrogen or neon. For a rapid cool-down of the cold mass, it is best to use more than one cooler, such as the PT415. For cooling circuits containing hydrogen that can operate at 15 to 20 K or neon at temperatures from 26 to 30 K, one can use multiple PT415 coolers and use the first stages to cool the leads and shields. One can also use a single stage cooler such as the Cryomech AL-325 [21], [16] single stage GM cooler in the temperature range from 15 to 30 K to cool-down an MgB_2 or HTS magnet and liquefy hydrogen or neon in a closed loop. Both a PT415 cooler and an AL-325 cooler will go to temperatures below the triple point of para hydrogen and neon. A cold head of either of these coolers must be prevented from going down to the triple point temperature of either gas. If one uses an AL325 cooler, the leads and shields must be liquid nitrogen cooled. A single stage cooler can also cool and cool-down an MgB_2 magnet or an HTS magnet with helium gas in the loop. The maximum temperature in the cooling loop must be low enough to provide adequate temperature margin for the magnet.

Both hydrogen and neon must use a closed sealed loop. A hydrogen loop must be sealed to prevent the migration of hydrogen into air. A hydrogen loop must be designed with the appropriate safety systems and engineering design margins commensurate with the flammable-gas pressure vessel code. Closed neon loops must use a larger surge tank for a given liquid volume because its expansion ratio is over 40 percent higher than for either helium or hydrogen. A neon loop must be leak tight to avoid losing an expensive gas. In the view of this author, there is little to be gained from using a neon loop as compared to a hydrogen or helium loop. The cool down time for a neon loop is nearly double that of a hydrogen loop for a given mass being cooled down and a given cooler. Using helium in a cooling loop extends the cool-down time about ten percent over hydrogen. If an MgB_2 magnet can be operated in a helium gas cooled loop (with no liquid helium), there is almost no need for a cooled loop filled with hydrogen.

ACKNOWLEDGMENT

This work had partial support by the DOE Office of Science, under contract DESC0000661. This work was supported in part by NSF grants PHY-09-58726 and PHY-11-02511.

REFERENCES

1. Taylor, C. E., Abbott, C. R., and Leitner, D., "An Efficient Cooling Loop for Connecting a Cryocooler to a Helium Reservoir," *Adv. in Cryogenic Engineering,* Vol. 45B, Springer US (2000), pp 635.

2. Green, M. A. et al, "Design Issues for Cryogenic Cooling of Short Period Superconducting Undulators," *Adv. in Cryogenic Engineering,* Vol. 49B, Amer. Institute of Physics, Melville, NY (2004), pp 783-790.

3. Green, M. A., "Cooling MICE Magnets using Small Cryogenic Coolers," MICE Note 109, www.mice.iit.edu/notes/, (2004)

4. Green, M. A., Chouhan, S. S., and Zeller, A. F., "Design Limitations on a Thermal Siphon 4 K Helium Loop for Cooling-down the Cyclotron Gas Stopper Magnet Coils," *Adv. in Cryogenic Engineering,* Vol. 59, Amer. Institute of Physics, Melville, NY (2014), pp 1543-1550.

5. Green M. A., et al, "Cooling and Current Testing of the MSU Cyclotron Gas-stopper Magnet," *IEEE Transactions on Applied Superconductivity* **26**, No 4, p (2016), DOI-10.1109/TASC.2016.2525769.

6. Green, M. A. and Yang, S. Q., "Heat Transfer within the 4.4 K Region and the 40 K Shields of the MICE Focusing and Coupling Magnets," MICE Note 101, www.mice.iit.edu/notes/, (2004).

7. Victor J. Johnson, Editor, *A Compendium of the Properties of Materials at Cryogenic Temperatures, Volume 2 Properties of Solids*, Pergamon Press, New York, (1961).

8. Green, M. A., Pan, H., and Preece, R. M., "Changes Made on a 2.7-m Long Superconducting Solenoid Magnet Cryogenic System that allowed the Magnet to be kept cold using 4 K Pulse Tube Coolers," *Adv. in Cryogenic Engineering,* Vol. 59, Amer. Institute of Physics, Melville, NY (2014), pp 1551-1558.

9. Green, M. A., "Lessons Learned Concerning the use of 4 K Coolers to Cool LTS Magnets," *Cryocoolers 18*, ICC Press, Boulder, CO (2015), pp 567-576.

10. Green, M. A., "The Integration of Liquid Cryogen Cooling and Cryocoolers with Superconducting Electronics Systems," *Superconducting Science and Technology*, Volume 16, No. 12, p 1349-1359, (2003).

11. Green, M. A. and Wang S. T. "Tests of Four PT-415 Coolers Installed in the Drop-in Mode," *Proceedings ICEC-22* Seoul Korea, 21 to 25 July 2008, pp 105-110, (2008).

12. Green, M. A., "Hydrogen Safety Issues Compared to Safety Issues with Methane and Propane," A*dv. in Cryogenic Engineering*, Vol. 51, Amer. Institute of Physics, Melville, NY (2006), pp 319-326.

13. Green, M. A., "Safety Issues for an Absorber that is Designed for Liquid Helium Operation as well as Liquid Hydrogen Operation," MICE Note 113, www.mice.iit.edu/notes/, (2005).

14. Green, M. A., "Re-condensation and Liquefaction of Helium and Hydrogen using Coolers," *Adv. in Cryogenic Engineering*, Vol. 55, Amer. Institute of Physics, Melville, NY (2010).

15 Green, M. A., et al, "Lessons learned from the Cool-down of a Superconducting Magnet using a Thermal-siphon Cooling-loop," submitted to *Advances in Cryogenic Engineering* **61**, *IOP Conference Series* **101**, article 012150, (2015)

16. Green, M. A., "A Case Study of MgB_2 or HTS magnets being cooled and Cooled-down using a Hydrogen Thermal-siphon Cooling-loop with Coolers," *Physics Procedia,* International Symposium on Superconductivity ISS2015, Tokyo Japan, 16 to 18 November 2015.

17. Data sheets on the PT415 cooler from Cryomech Inc. at 113 Falso Drive, Syracuse NY 13211, USA. Data sheets can be found www.cryomech.com.

18. Choi, Y.S., Painter, T.A., Kim, D.L., Lee, B.S., Yang, H.S., Weijers, H.W., Miller, G.E. and Miller, J.R., "Helium-liquefaction by Cryocoolers for High Field Magnet Cooling," *Cryocoolers 14*, ICC Press, Boulder, CO (2007).

19. Green, M. A., Chouhan, S. S., Wang, C., and Zeller, A. F., "Second-stage Cooling from a PT415 Cooler at Second-stage Temperatures up to 300 K with Cooling on the First-stage from 0 to 250 W," *Advances in Cryogenic Engineering* **61**, *IOP Conference Series* **101**, article 012002, (2015).

20 Johnson V. J., Editor, *A Compendium of the Properties of Materials at Cryogenic Temperatures, Volume 1 Properties of Fluids*, Pergamon Press, New York, (1961).

21. Cryomech data sheets on the AL-325 GM cooler, Cryomech Inc. 113 Falso Drive, Syracuse NY 13211, USA. See the Cryomech website at www.cryomech,com.

20 K Cryogenic Helium Forced Flow Circulation Loop

T. Trollier, A. Ravex, J. Tanchon, J. Lacapere

Absolut System SAS
Seyssinet-Pariset, France

ABSTRACT

MgB2 based superconducting electro-technical applications under development require distributed high cooling power in the 20 K temperature range. For such applications, Absolut System designed a supercritical (20 bar) helium forced flow circulation loop with a net cooling power in the range of 70-100 W at 20 K.

The circulation loop makes use of two Cryomech AL325 mono stage Gifford-MacMahon type cryocoolers as the cold source. In order to enhance the efficiency and to reduce the foot print, a Noordenwind type cryogenic circulator from Cryozone is used for the forced flow instead of a traditional room temperature compressor and counterflow heat exchangers. Thermal shielding is provided with lost LN_2 open loop forced flow.

The design optimizations will be reported in this paper, as well as outcomes of manufacturing and preliminary performance testing.

INTRODUCTION

Absolut System has developed and produced [1, 2] several remote helium cooling loop systems ranging from 30 W @ 50 K to 80 W @ 30 K using helium room temperature compressor packages as circulators to flow the required helium mass in a closed loop to the recuperator heat exchangers, the cold head heat exchanger, the transfer line and the application heat exchanger to absorb the heat load.

Potential application is to cool down and operate a MgB2 based superconducting cable system at 20 K. The objective helium mass flow should be in the range of 10 g/s.

Commercial compression units providing such a mass flow require a large foot print, a high electrical input power and are expensive. Such a mass flow does not cope with the use of cold check valves to convert a small portion of the oscillating flow from the GM compression unit to circulating flow for the loop as proposed by Cryomech [3].

The conductor 20 K cooling will be provided by a supercritical (20 bar) helium forced flow closed loop implemented by a mono stage GM type cryorefrigerators and a cold pump.

The Piping and Instrumentation Diagram (P&ID) of the 20 K cryorefrigeration closed loop is found in Figure 1.

DESCRIPTION OF THE SYSTEM

The cryogenic helium circulation loop gathers the following components identified on the P&ID:

- One Cryostat interfacing with two cryorefrigerators cold heads, I/Fs with the cryofan, Johnston coupling K13 female Input/Output cold helium ports, Input/Output LN2 ports, GHe Inlet port, ISO K100 vacuum port mated with a gate valve and three KF40 ports.

Cryocoolers 19, edited by S.D. Miller and R.G. Ross, Jr.

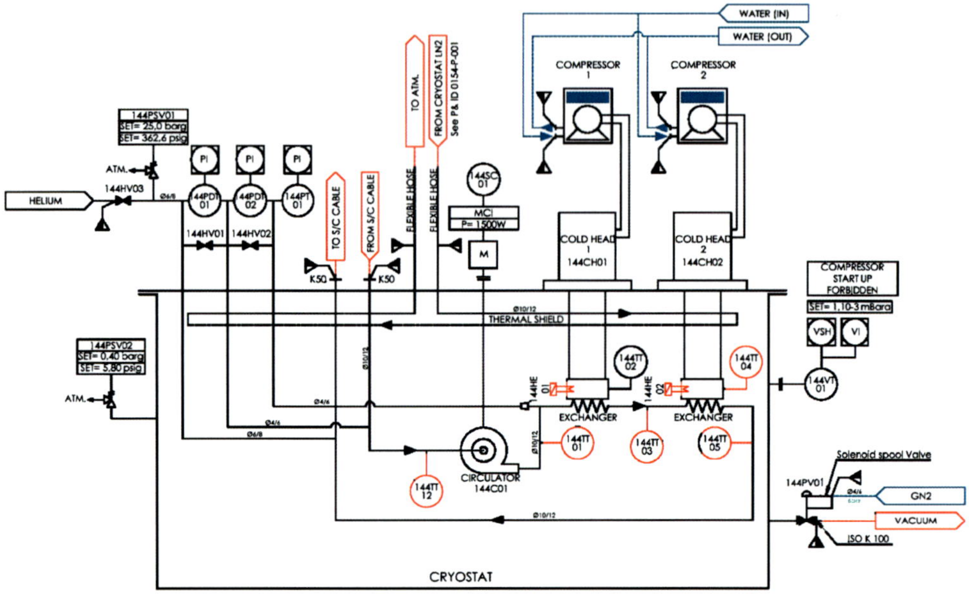

Figure 1. Piping and Instrumentation Diagram of the 20 K cryorefrigeration closed loop.

- Two Cryomech AL325 type cryo-refrigerator systems providing cooling power at 20 K and consisting of:
    - Two Cryomech AL325 cold heads (CH01 and CH02) mated to a cryostat upper flange,
    - Two Cryomech CP1110 helium Compressor Packages connected to the cold heads via two (LP and HP) helium flex lines 3 meters long and a cold head motor cord.
- One Cryofan cold pump (144C01) Noordenwind type from Cryozone for a supercritical helium forced flow at 20 bar.
- One pair of Cryoflex transfer lines type DN12/DN32 terminations K13 He male, 4.4 meters long, for Input/output to the S/C cable with cold helium flow at 20 bar.
- One pneumatic single-acting vacuum gate valve (144PV01) for 20 K cryostat vacuum insulation.
- One vacuum gauge (144VT01).
- One vacuum safety relief valve (144PSV02).
- Six calibrated Lakeshore Cernox type CX-1070-CU-HT-4L-QT temperature sensors (144TT01 to 144TT05 & 144TT12) to control the temperature at the cold heads and at the inlet and outlet of the heat exchangers and the cryofan circulator.
- Two heaters 144HE01 & 144HE02 (1 at each cold head exchanger).
- Two differential pressure transmitters (144PDT01 and 144PDT02).
- One absolute pressure transmitter (144PT01).
- One high pressure safety relief valve (144PSV01).
- One GHe Inlet manual valve (144HV03), straight ball type valve.
- Two manual valves (144HV01 & 144HV02) for the by-pass of the pressure differential transmitters, angle pattern ball valve type.

The two AL325 cold heads have been specifically adapted by Cryomech with an efficient integrated cold heat exchanger equipped with 1/2" VCR type fittings as shown on the Figure 2. The copper heat exchanger is directly silver brazed around the expansion chamber of the GM cryocooler to reduce the thermal gradient.

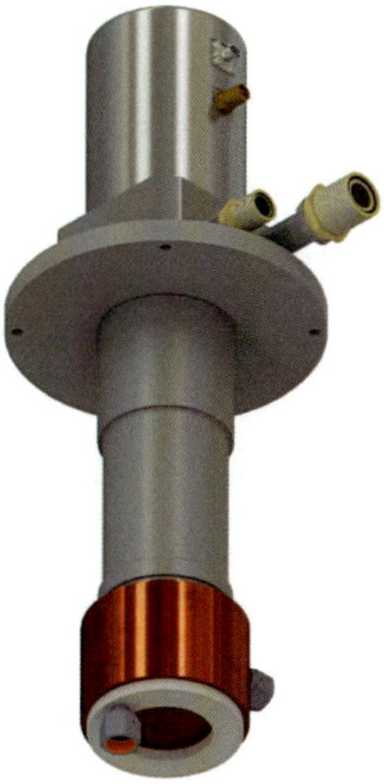

Figure 2. Artist view of the Cryomech AL325 cold head with integrated cold heat exchanger.

The Noordenwind type cryofan from Cryozone used as the circulator in the forced flow loop is illustrated in the Figure 3. The cryofan consists of a high speed motor under helium pressure terminated with an elongated stainless steel shaft and a 31 mm impeller. It can operate between 6,000 and 21,000 rpm and sustain a maximum static system pressure of 30 bar. Pump housing has also been equipped with 1/2" VCR type fittings. A picture of the cryorefrigerator assembly is seen in Figure 4.

A copper thermal screen is used for the thermal shielding of the 20 K cold parts as shown on the Figure 5. This copper thermal shield is also composed of two flanges and a cylindrical shell. The upper flange of the copper shield is mechanically anchored to the upper flange of the cryostat thanks to low conductive struts made of G10 glass fiber. The upper flange of the thermal shield comprises a LN2 loop with In/Out fittings mated on the cryostat upper flange. The three parts of the copper thermal shield are separately wrapped into 20 layers of MLI.

Figure 3. Picture of the Cryozone Noordenwind type cryofan used as the circulator.

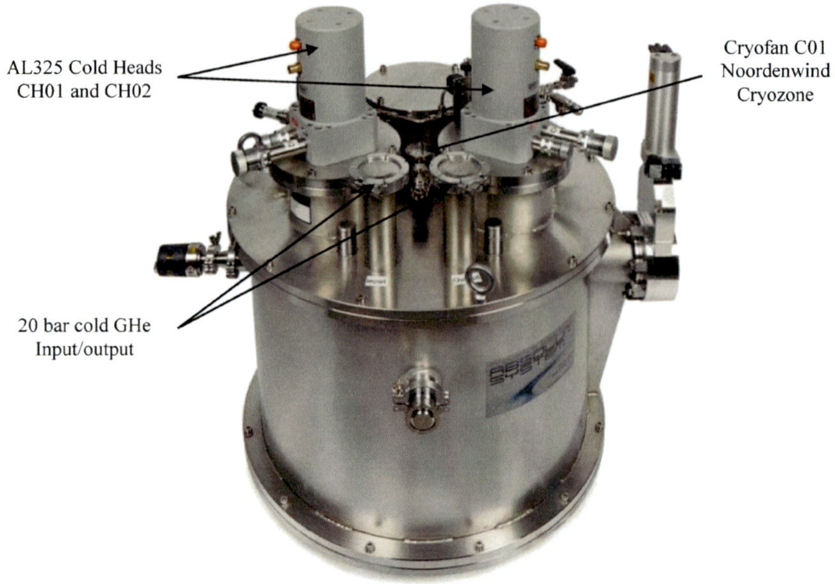

Figure 4. Picture of the 20 K cryorefrigeration closed loop overall assembly.

PERFORMANCE TESTS

Performances at cryocooler integration

The cryogenic performances of both AL325 cold heads have been measured with the cold heads integrated in the cryostat with the final Multi-Layer Thermal Insulation (MLI) and connected to the stainless steel piping. The circuitry is kept under vacuum to avoid any thermal loss via helium gas. The LN2 shield was cooled during the performance characterization tests. Thermal sensors 144TT02 & 144TT04 and heaters 144HE01 & 144HE02 have been used.

The performance is reported in Figure 6. Measured performance at Absolut System is slightly better that the one reported by Cryomech due to the use of a LN2 cold thermal screen at the 20 K cryostat.

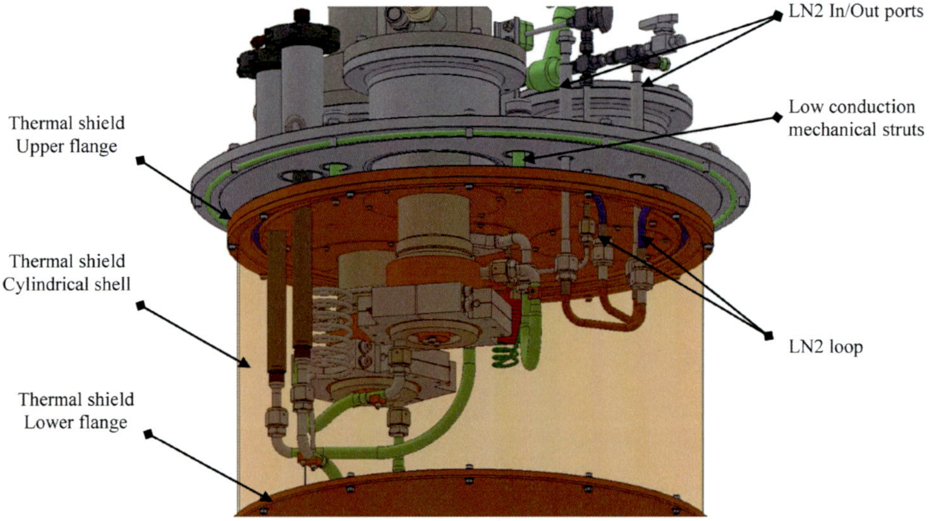

Figure 5. CAD view of the copper thermal shield.

20 K cryostat performance tests using a test cryostat

To simulate a future superconducting cable, a test cryostat has been manufactured and connected to the helium loop for the test campaign. Two different heat exchanger designs have been implemented to determine the effect on the 20 K refrigeration loop performances and estimate future performance expected with a superconductor cable. The test cryostat has been equipped with two temperature sensors and a differential pressure transducer to estimate the heat exchanger pressure drop and to discriminate the flexible helium lines pressure drops.

With this test configuration we are able to get sufficient information to measure and analyze the thermal losses and pressure drops of the main components of the helium cooling loop:

- Cold circulator,
- Cold heads heat exchangers,
- Helium flex lines,
- Test cryostat,
- Cable simulating heat exchanger.

A specific test has been performed with the helium loop by-passed through a single helium flex line to determine the helium flexible line pressure drop and thermal losses. A total of about 65 runs with different parameters set up have been performed and analyzed, covering a large range of parameters:

- Helium loop pressure: 10, 15, 17.5 and 20 bar (nominal).
- Cold pump rotation speed: 12.000, 12.500, 15.000, 18.000 and 21.000 rpm.
- Future superconductor cable heat losses: 25, 50, 75, 87.5, 100 and 125 W.
- Applied heat load on cryorefrigerators cold heads in the by-passed helium loop configuration: 75 W and 100 W per cold head.

Performance tests analysis

The helium mass flow rate is determined via a thermal balance at the cold head heat exchangers:

$$(Q_{\mathrm{CH01}} - Q_{\mathrm{CH02}}) = \dot{m}_{He}(h_1 - h_5) \quad (1)$$

Where,

Q_{CHO1} or Q_{CHO2} are respectively the thermal loads on cold heads 1 & 2 (watt).

$\dot{m}_{He}$ is the helium mass flow rate (g/s).

h_1 or h_5 are respectively the helium enthalpy at cold head 1 heat exchanger inlet and at cold head 2 heat exchanger exit (J/g).

Q_{CHO1} and Q_{CHO2} are determined from the cold head load curve measurements as reported in the Figure 4, using measured cold head temperatures (TT02 and TT04). From the above-calculated

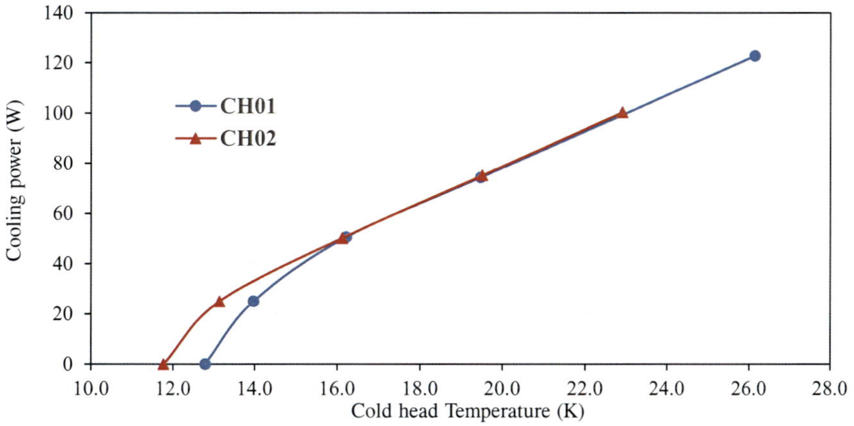

Figure 6. Performance characterization of two AL325 cold heads. Helium circuit pumped and LN2 thermal shield active.

helium mass flow rate $\dot{m}_{He}$ (Eq. 1), the the helium volumetric flow rate V_{He} (m³/h) is determined, using the helium density ρ_{12} (kg/m³) at cold pump inlet:

$$V_{He} = \dot{m}_{He}/\rho_{12} \tag{2}$$

From the measured differential pressure ΔP_2 (Pa) between cold pump inlet and outlet, the pressure head ΔH (m) delivered by the cold pump is calculated using the helium density ρ_{12} (kg/m³) at cryogenic circulator inlet:

$$\Delta H = \Delta P_2/(\rho_{12} * g) \tag{3}$$

It is now possible to represent the cold pump hydraulic characteristics in a universal graph ΔH versus V_{He}.

In Figure 7, a graph to report the values obtained from all experimental tests performed is found: the points have been determined at different helium pressure, cold pump rotational speeds and thermal loads.

Three different sets of points can be observed corresponding to the three different test families: 20 K refrigeration system with by-passed test cryostat by connecting In/out of the loop with a single helium flex line (1 flexline only, by-pass) and two sets with helium circulation loop connected with a pair on In/Out flexlines to the test cryostat equipped with heat exchanger 1 (In/out flexlines + Exchanger 1) or equipped with heat exchanger 2 (In/Out flexlines + Exchanger 2).

Both heat exchangers are simulating the future superconductor cable. These three sets of points have been correlated with quadratic laws, the usual way to take into account linear pressure drops (in piping) and singular pressure drops (for cold head's heat exchangers and for different flow singularities such as elbows, restrictions or expansions in piping). We have also implemented "guides for eyes" on this graph corresponding to the different rotational speeds of the cold pump. The data is in quite good agreement with the curve based on measurements with air at 20°C and atmospheric pressure provided by Cryozone for the cold pump.

We also evaluated the hydraulic efficiency of the cold pump η_P. To do so we first calculated the "gross" heat dissipated by the cold pump:

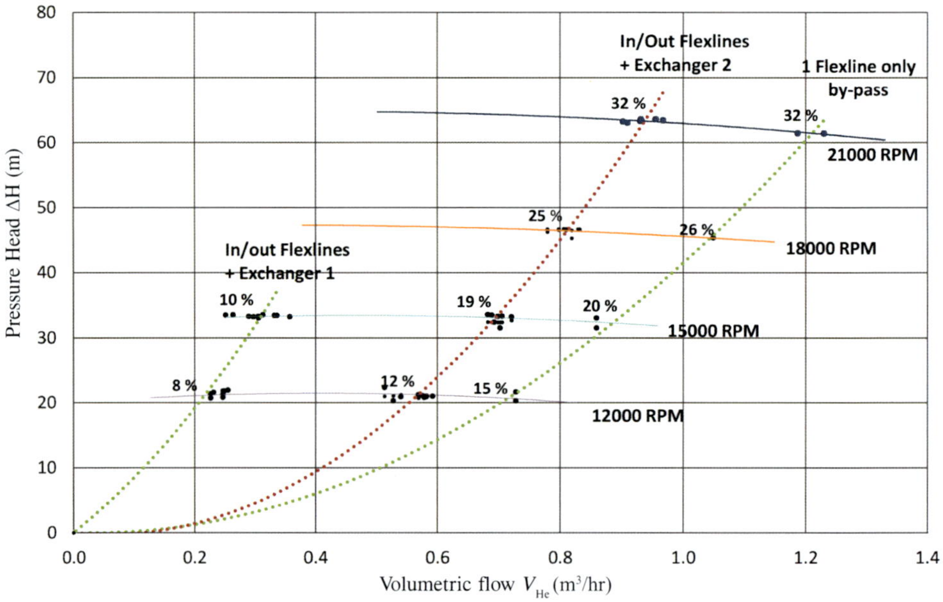

Figure 7. Measured hydraulic characteristics of cold pump and helium loop for various configurations and rotational speeds.

$$Q_{cold\,pump} = \dot{m}_{He} \cdot (h_{12} - h_1) \quad (4)$$

This "gross" heat dissipation includes both static losses by thermal conduction through the cold pump shaft from room temperature and the actual dynamic work of compression. We used the static losses value (6 Watt @ 20 K) given by Cryozone in the technical documentation delivered with the cold pump and subtracted it from the "gross" heat to get the actual dynamic work of compression.

$$W_{comp} = Q_{cold\,pump} - 6 \quad (5)$$

The theoretical work of compression can be determined knowing the pressure drops on the pump sides (ΔP_2) and the calculated volume flow (Eq. 2).

$$W_{comp\,theo} = V_{He} * \Delta P_2 \quad (6)$$

Comparing the calculated actual work (Eq. 5) to the theoretical work of compression (Eq. 6), we can evaluate the cold pump hydraulic efficiency:

$$\eta_P = \frac{W_{comp\,theo}}{W_{comp}} = \frac{V_{He} * \Delta P_2}{Q_{cold\,pump} - 6} \quad (7)$$

The results are also reported in Figure 7. They are logical in terms of relative variation. In terms of absolute values these calculated efficiencies are smaller than those generally reported by Cryozone on their commercial brochures. In practice there is no efficiency data given nor guaranteed by Cryozone in the data package delivered with cold pump. Only the performance curves (pressure head versus volumetric flow for different rotational speeds) of the delivered cold pump are given. They are measured with air at atmospheric temperature and pressure. An explanation could also be the underestimated static losses.

Pressure drops analysis

From the differential pressure drop transducers (ΔP_1: 20 K cryostat inlet/outlet, ΔP_2: cold pump inlet/outlet, ΔP_3: test cryostat heat exchanger inlet/outlet) installed on the helium loop, we can determine the pressure drops at different subsystems level:

- Pressure drop in cold head heat exchangers: $\Delta P_2 - \Delta P_1$
- Pressure drop in helium flexible lines: $\Delta P_3 - \Delta P_1$

The contribution of the main helium loop components (cold head heat exchangers, helium flex lines, test cryostat heat exchanger) for different cold pump rotational speed is reported in Fig. 8 in terms of pressure drop (mbar). The values reported correspond to the tests performed with the second version of the test cryostat heat exchanger, which is the closest in terms of pressure drops of a superconductor cable.

It appears that the helium flex lines pressure drop contribution is quite significant (in the range of 100 to 150 mbar at 15,.000 to 18,000 rpm cold pump rotational speed and 10 g/s or 0.8/0.95 m^3/hr helium flow rate). A significant improvement can be done on these components.

The pressure drop in the heat exchanger simulating a superconductor cable (about 45 to 60 mbar at 15.000 to 18.000 rpm cold pump rotational speed and at 10 g/s or 0.8/0.95 m^3/hr helium flow rate) is significantly larger than that expected in a superconductor cable. This still allows for some margin in cold pump operation and cooling performances for future operation.

The pressure drop in the cold head heat exchanger is in agreement with our expectations for an efficient heat transfer as requested to minimize temperature gradients between cold head and helium in the cooling loop.

Thermal performances analysis

The foreseen operation pressure of the helium-cooling loop is 20 bar. We made some tests with lower pressure (17.5 bar, 15 bar and 10 bar). Performances are effectively better at 20 bar as

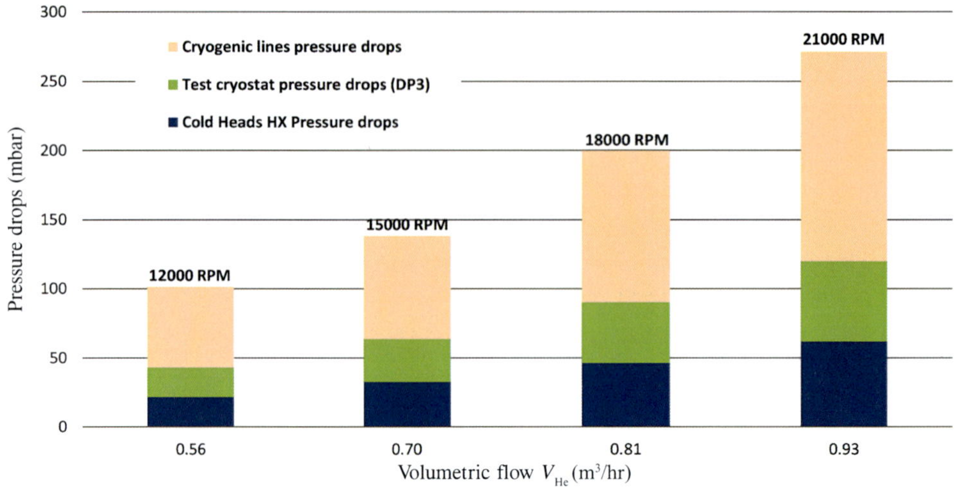

Figure 8. Relative contributions of cold head heat exchangers, helium flex lines and test cryostat heat exchanger to helium loop total pressure drop for different cold pump speeds (for test heat exchanger #2).

expected. We did not performed tests at higher pressure which would be even better since a typical superconductor cable burst disc generally set in the range of 25 bar.

To test and qualify the helium cooling loop thermal performances with the 20 bar nominal helium loop pressure, we applied progressively different thermal loads (25 W, 50 W, 75 W, 87.5 W and 100 W) on the heat exchanger simulating a superconducting cable via a resistive heater.

We report in Table 1 temperature measurements at different points of the helium cooling loop for test configurations representative of a superconductor cable operation: helium pressure 20 bar, cold pump rotational speed 15,000 rpm / 18,000 rpm and applied heat load of 87.5 W / 100 W.

With these operation parameters, the temperature at cable inlet (18.4 K to 19.2 K) and outlet (20.5 K to 21.4 K) are in good agreement with the requirement to have a cooling loop helium average temperature of 20 K.

An even lower temperature can be expected in a future cable operation due to the fact that a larger mass flow rate may be achievable with a lower pressure drop in the cable than in the test heat exchanger.

Table 1. Temperatures at different helium loop locations for representative heat loads and cold pump rotational speed (measurements performed at 20 bar).

Applied load (W) Cold Pump speed (rpm)	**100 15.000**	**100 18.000**	**87.5 15.000**	**87.5 18.000**
Cooling power at cable estimation applied load + test cryostat losses	~ 114 W	~ 113 W	~ 102 W	~ 100.5 W
TT_{12} Cold Pump inlet / (~ Cable outlet)	21.41 K	21.29 K	20.64 K	20.54.K
TT_{01} Cold Head #1 HX inlet	21.76 K	21.66 K	21.07 K	20.99 K
TT_{02} Cold Head #1	20.04 K	20.18 K	19.45 K	19.50K
TT_{03} Cold Head #2 HX inlet	20.15 K	20.29 K	19.48 K	19.64 K
TT_{04} Cold Head #2	18.90 K	18.95 K	18.38 K	18.50 K
TT_{05} Cold Head #2 HX outlet / (~ Cable inlet)	18.85 K	19.14 K	18.37 K	18.59 K

Table 2. Thermal loads or losses of different helium loop components for representative heat loads and cold pump rotational speeds (measurements performed at 20 bar).

Applied load (W) **Cold Pump speed (rpm)**	**100** **15.000**	**100** **18.000**	**87.5** **15.000**	**87.5** **18.000**
Cooling power at cable estimation applied load + test cryostat losses (W)	~ 114	~ 113	~ 102	~ 100.5
Mass flow rate (g/s) calculated	8.55	9.98	8.66	9.84
Volume flow rate (m^3/h) calculated	0.699	0.806	0.693	0.779
Cold pump heat dissipation (W) [Eq. 4]	17.4	21.4	21.8	25.9
2 x flex lines heat dissipation (W) [Eq. 8]	17.0	15.8	17.0	15.8
Test cryostat losses (W) [Eq. 10]	14.2	13.4	14.6	13.1

We can determine from the above temperature measurements and associated helium enthalpy the heat dissipation at some specific subsystems:

- Heat dissipated by cold pump: from previous Eq. (4)
- Thermal losses of a pair of helium flex lines, determined in a specific test with a single flex line connected to the cooling loop (by-pass test cryostat conditions):

$$Q_{flex\ lines} = 2 \cdot \dot{m}_{He} \cdot (h_{12} - h_5) \tag{8}$$

- Heat dissipated out of the 20 K cooling system cryostat, including 2 flex lines + test cryostat losses + test heat exchanger applied heat load:

$$Q_{out} = \dot{m}_{He} \cdot (h_{12} - h_5) \tag{9}$$

- Test cryostat losses:

$$Test\ Cryostat\ losses = Q_{out} - Q_{flex\ lines} - applied\ heat\ load \tag{10}$$

These thermal loads or losses are reported in the Table 2.

Measured cold pump heat dissipations are significantly larger than that evaluated. This can be explained by the measured lower efficiency of the pump than expected from Cryozone commercial documentation and also the larger overall loop pressure drop mainly in helium flex lines and test cryostat heat exchanger. The pressure drop in a superconductor cable will be lower than the one in test cryostat heat exchanger, resulting in less dissipation by cold pump.

Thermal losses measured for the pair of flex lines (about 16 to 17 W) are larger than those estimated. This can be explained by the fact that insulation performances are probably a bit worth than estimated.

The calculated test cryostat thermal losses are in the range of 13.1 W to 14.6 W. This corresponds to a future cooling power margin for a cable cooling (it is a reasonable value for a test cryostat without LN2 cooled thermal shielding).

CONCLUSIONS

A cryogenic helium forced flow circulation loop has been designed, manufactured and tested by Absolut System. It makes use of two Cryomech AL325 GM coolers connected in series with a Noordenwind type cryofan from Cryozone and a pair of 4.2 meters of cryoflex lines. A remote cooling performance of 113 W at 20 K has been demonstrated with 20 bar helium loop pressure and 18.000 rpm cryofan rotational speed.

Performances improvements are expected by reducing the pressure drop budget on the pair of flex lines and while connected to a superconducting cable demonstrator.

REFERENCES

1. T. Trollier, J. Tanchon, Y. Icart and A. Ravex, "Remote Helium Cooling Loops for Laboratory Application", *Cryocoolers 17*, ICC Press, Boulder, CO (2013), pp. 503-510.

2. T. Trollier, Y. Icart, J. Tanchon and A. Ravex, "High capacity 30 K remote Helium cooling loop", *Adv. in Cryogenic Engineering,* Vol 59B, American Institute of Physics, Melville, NY (2014), pp. 1461-1466.

3. C. Wang and E. Brown, "A GM Cryocooler with Cold Helium Circulation for Remote Cooling", *Adv. in Cryogenic Engineering,* Vol 59B, American Institute of Physics, Melville, NY (2014), pp. 1149-1156.

In-Space Cryogenic Propellant Storage Applications for a 20 W at 20 K Cryocooler

D. W. Plachta[1], J. R. Feller[2], M. C. Guzik[1]

[1]NASA Glenn Research Center, Cleveland, OH 44135
[2]NASA Ames Research Center, Moffett Field, CA 94035

ABSTRACT

The National Aeronautics and Space Administration is currently developing technologies for future space exploration missions and scientific discovery missions by addressing the need to raise the technology readiness level of cryogenic fluid management storage technologies. Cryogenic propellants are part of exploration plans due to their inherent high specific impulse, yet their inherently low boiling points cause high heat loads that can cause substantial boil-off losses in the long term. Recent developments to reduce boil-off and scale up the concept for large tanks provides important information on the benefit of an active cooling system applied to liquid hydrogen propellant storage. Findings point to zero-boil off technology as a means of reducing mass in liquid hydrogen storage systems when low Earth orbit loiter periods are greater than 2 months. These findings have spurred the development of the 20 Watt (W) at 20 Kelvin (K) flight representative cryocooler. The analysis presented explores the propellant tank sizes and heat loads within the range of this cryocooler, to achieve zero-boil-off propellant storage. This analysis is performed both with and without the integration of a 90K reduced boil-off cryocooler. Larger tanks, such as those considered in the Mars Reference Mission, are considered. Results are highly dependent on the cryogenic upper stage structural heat, however, a thermally optimized structure considered herein combined with self-supporting MLI shows that 20 W at 20K cooling is an appropriate size to eliminate boil-off for three of the four liquid hydrogen propellant storage tank sizes considered. Two 20 W coolers are needed, along with a 90K cooler, for the largest tank considered in the trade space.

INTRODUCTION

To achieve zero boil-off of liquid hydrogen (LH_2), it is critical that the heat loads into the tank, particularly the tank structure, are minimized. Also, it is necessary to advance the state-of-the-art in LH_2 temperature cryocoolers for flight, which are at a relatively low state-of-the-art for propellant storage application. Because of this, the National Aeronautics and Space Administration (NASA) is advancing the most promising cryocooler technology for LH_2 Zero Boil-Off (ZBO) storage, the reverse turbo-Brayton cycle cryocooler. This advancement is important in that the present state-of-the-art 20K cryocooler technology has a lift capacity of less than 1 W[1], much less than the projected tank heat leak rates, while offering no ability to distribute the cooling. The purpose of this paper is to clarify the usefulness of a 20 W capacity with missions that NASA may pursue, with the goal of minimizing power and mass. The issues of power availability and the expense of power systems

Cryocoolers 19, edited by S.D. Miller and R.G. Ross, Jr.

limits the consideration of active cooling. Mass is also critical as it increases the size and cost of the launch vehicle and reduces the potential payload mass.

One method of reducing power and mass is to optimize the active cooling system for propellant storage. Since 2007, several developments[2,3] and tests[4] have been conducted by NASA with the goal of increasing the technology readiness level of Cryogenic Boil-Off Reduction Systems (CBRS) and Zero Boil-Off (ZBO). The focus to date has been on cryocooler integration into the large surface areas of cryogenic propellant tanks, to understand the feasibility and losses involved. In addition to that, an effort was started to develop a higher capacity liquid hydrogen cryocooler.

With respect to the application of the 20 W at 20K cryocooler, however, an analysis into tank heat leaks and methods to achieve two circuits of cooling is required. Given the relatively high power and costs associated with this cryocooler, the goal is to study methods to achieve ZBO with low power and low mass. To accomplish this, a second cryocooler is considered. The second stage, a 90K cryocooler system, was found to reduce tank heat by 60%, as measured by the heat reduction of the components cooled by the cryocooler, in Reduced Boil-Off (RBO) testing at NASA Glenn[4] (see Figure 1). A schematic representation of the two circuits of cooling envisioned, using both the 90K cooling system and a 20K cryocooler system, is shown in Figure 2.

Previous modeling efforts[5] determined that 90K cryocoolers, as part of an RBO system, begin to reduce mass for liquid hydrogen loiter periods in space when compared to passive only storage, that is with multi-layer insulation (MLI) only and no active cooling, with its constant boil-off, after several weeks, and ZBO storage reduces mass after two months. This follow-on effort uses two-stage cooling with both the 90K and 20K cryocooler systems, to determine potential mass and power reductions. As part of that, an explicit approach to size the 90K cryocooler is developed as an improvement on the previous model that indirectly optimized mass. This analysis incorporates RBO testing heat exchanger data. As done in the previous scaling study,[5] this analysis was applied to a range of four tank sizes, as shown in Figure 3.

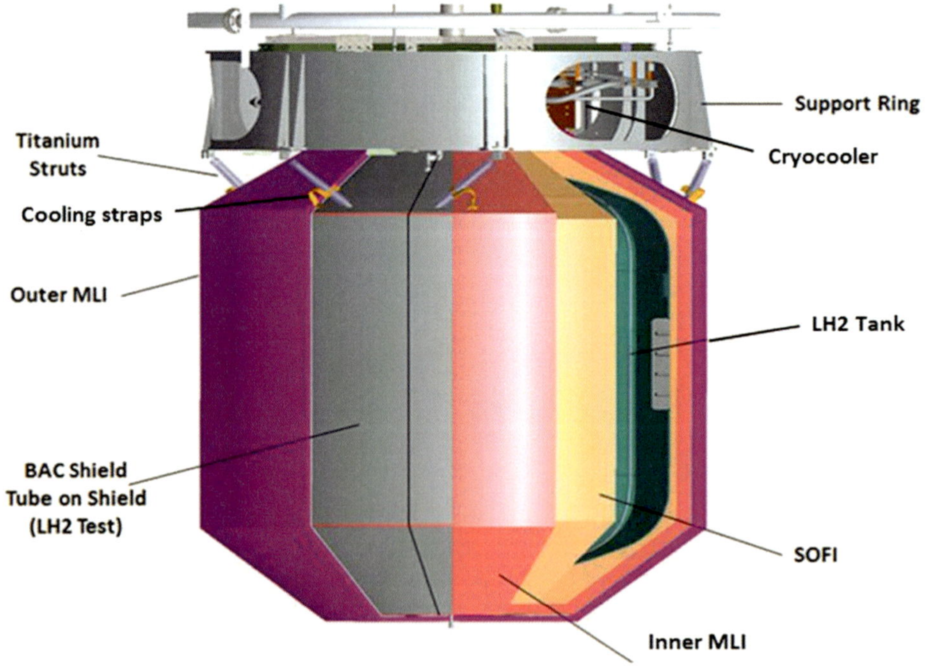

Figure 1. Cut-a-way image of the Cryogenic Boil-Off Reduction System test article, designed and tested at NASA GRC. This test article serves at the basis for component scaling.

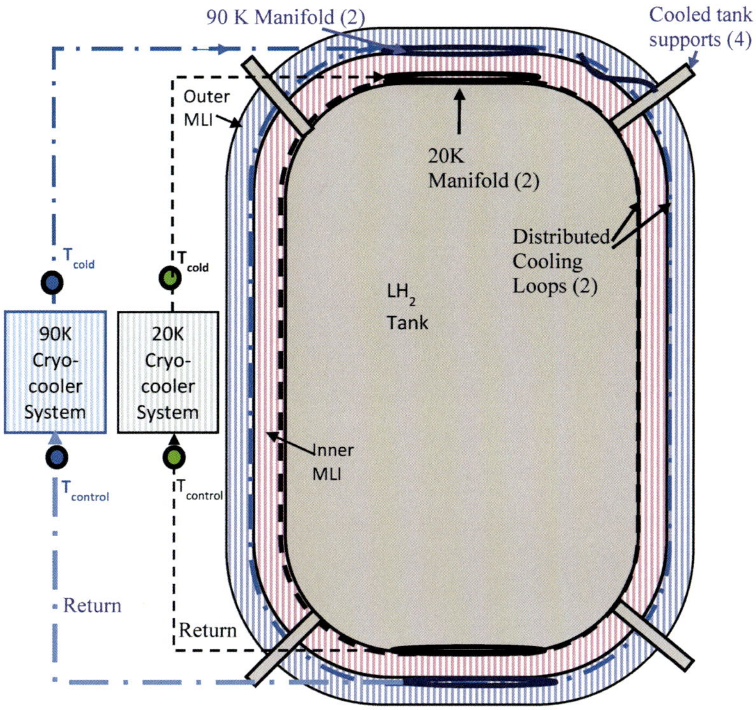

Figure 2. Two cryocooler system with 20 and 90K cooling

THERMAL BALANCE APPROACH

The two circuits of cooling, as configured in Figure 2, integrates a nominally 90K broad area cooled (BAC) shield to intercept heat on the structural supports, the fluid lines, and within the MLI. The heat load of the 90K shield is determined by optimizing the location of the heat intercept, knowing the thermal performance of the cooling straps from the RBO testing and developing the thermal balance relations for a penetration with a heat intercept, via a cooling strap. Also, the MLI is modeled outside the shield and between the shield and the tank.

Note that in these discussions and analysis, 20K and 90K are stated as the cooling points on the cryocooler but these are just approximations and refer to a 20K class cryocooler and an a 90K

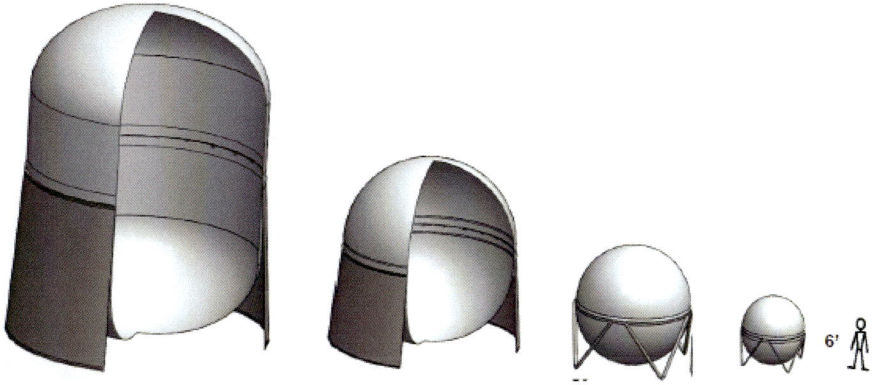

Figure 3. Notional LH2 tanks analyzed in this study. From left to right, the Earth Departure Stage, 8.3 m dia with a Ti skirt, a 6 m dia spherical tank with a Ti skirt, a 4 m dia sphere with 8 Ti struts, and a 2.2 m dia tank with 8 Ti struts. Show in the right is a 1.8 m tall person.

class cryocooler and BAC system. Liquid hydrogen will more likely be stored between 21.5 and 23K, its saturation temperatures at the likely tank pressure range required by propulsion systems. Also, the 90K cooling stage is assumed because previous analysis[5] showed that there is little sensitivity to mass and power if this stage is between 80 and 110K; 90K is within that range. The nodes at each of these temperatures and of the environment appear in subscript and they are defined as: 0 is at the 20K tank, 1 is at the 90K broad area cooling (BAC) shield, and 2 is at a representative low Earth orbit (LEO) environment, which is assumed to be at 220K.

Heat Load on 20K Tank

The total heat load on the LH2 tank is:

$$Q_{tank} = Q_{mli,0} + \Sigma_i Q_{pen,0,i} + Q_{const} - Q_{bac} \tag{1}$$

$Q_{mli,0} = Q_{mli,0}(T_{shield}, T_0)$, is the heat leak from the shield to the tank wall through the inner MLI blanket. $Q_{pen,0,i} = Q_{pen,0,i}(T_{1,i}, T_0)$ is the solid conduction through the ith penetration, from the thermal intercept (near 90K, but specifically at temperature) to the tank wall. Q_{const} is the sum of all approximately constant heat leaks to the tank (e.g., through instrumentation leads). Q_{bac} is the heat removed to the 20K cryocooler via the tube-on-tank BAC heat exchangers. This does not include heat removed via thermal straps to the 20K heat distribution manifolds.

Heat Load on the 20K Cryocooler

The total heat load on the nominally 20K cryocooler is:

$$Q_{20} = Q_{bac} + Q_{man20,supply} + Q_{man20,return} + Q_{supply20} + Q_{return20} \tag{2}$$

$$Q_{man20,supply} = \Sigma_i Q_{strap20,i} + Q_{man20,supply,parasitic} \tag{3}$$

Eqn. (3) is the total heat load on the inlet manifold. $Q_{strap20,i} = Q_{strap20,i}(T_{1,i}, T_{man20})$ is the heat conducted by the ith strap from the ith penetration to the supply manifold. $Q_{man20,supply,parasitic}$ is the total parasitic heat load on the supply manifold. Parasitic heat is the portion of the cryocooler system heat that is not useable by the system. Also, for simplicity, it is assumed that all heat removed from the penetrations is conducted to the two (20K and 90K) supply manifolds. $Q_{supply20}$ and $Q_{return20}$ are the parasitic heat loads on the supply and return lines running from the cryocooler to their respective manifolds.

Heat Load on the 90K Shield

The 90K shield is embedded between the inner and outer MLI blankets. The net heat load on it is simply:

$$Q_{shield} = Q_{mli,1} - Q_{mli,0} \tag{4}$$

$Q_{mli,1} = Q_{mli,1}(T_{env}, T_{shield})$, with T_{env} equal to the temperature of the thermal environment, is the heat leak to the shield through the outer MLI blanket and $Q_{mli,0} = Q_{mli,0}(T_{shield}, T_0)$ is the heat leak from shield to tank wall through the inner MLI blanket.

Heat Load on the 90K Cryocooler

Total heat load on the 90K cryocooler is:

$$Q_{90} = Q_{shield} + Q_{man90,supply} + Q_{man90,return} + Q_{supply90} + Q_{return90} \tag{5}$$

$$Q_{man90,supply} = \Sigma_i Q_{strap90,i} + Q_{man90,supply,parasitic} \tag{6}$$

Eqn. (6) is the total heat load on the supply manifold, where $Q_{strap90,i} = Q_{strap90,i}(T_{1,i}, T_{man90})$ is the heat conducted by the ith strap from the ith penetration to the inlet manifold, and $Q_{man90,supply,parasitic}$ is the total parasitic heat load on the supply manifold. $Q_{man90,return}$, $Q_{supply90}$, $Q_{return90}$ and are 90K counterparts of $Q_{man20,return}$, $Q_{supply20}$, and $Q_{return20}$.

Thermal Balance Relations for a Penetration with One Intercept

For the penetration shown in Fig. 4 the thermal balance at the intercept node is:

$$Q_{strap} = Q_1 - Q_0 = (T_1 - T_{man})/R_{strap} \tag{7}$$

where $T_0 = T_{tank}$ and T_2 are the known boundary temperatures and R_{strap} is the thermal resistance of the strap, including the contact resistances at each strap end. Generally Eqn. 7 must be solved numerically. However, once it is solved and Q_0, Q_1, and Q_{strap} are known for a given T_{man}, then if T_{man} is increased or decreased slightly ($T_{man} \rightarrow T_{man} + \delta T_{man}$), the resulting heat leaks can be calculated approximately. They are:

$$Q'_0 = Q_0 + \left(\tfrac{A}{L}\right)_0 \kappa(T_0, T_1, \delta T)\,(T_1 + \delta T + T_0) \tag{8}$$

$$Q'_1 = Q_1 - \left(\tfrac{A}{L}\right)_1 \kappa(T_1, \delta T, T_2)\,(T_2 - (T_0 + \delta T)) \tag{9}$$

and

$$Q'_{strap} = Q'_1 - Q'_0 \tag{10}$$

It is assumed that R_{strap} = 5 K/W, which was the lowest strap resistance measured in the LH2 RBO testing performed at GRC.

ASSUMPTIONS

MLI

The assumed multi-layer insulation (MLI) in the study was load-bearing MLI designed by Quest[6] and tested by NASA in RBO testing. This insulation was selected because of its excellent performance and because it can support the weight of a broad area cooling shield. A 40 layer blanket was assumed, with 25 layers outside of the shield and 15 layers underneath the shield; this location was found optimal in the previous scaling study effort. A scale factor of 3 was applied to the predicted heat load, which was calculated by a layer-by-layer thermal model,[7] which was developed as a function within Excel using Visual Basic for Applications. The scale factor, which has been used in other studies[8], to account for the historically higher heat loads of tank applied MLI, was applied to the outer and inner layers. Additionally, since the RBO testing found a reduced scale factor for the outer MLI and an increased scale factor for the inner MLI, an indication of higher temperature sensitivity than the model predicts, a second analysis was performed with the RBO test scale factors of 2 on the outer MLI and 5 on the inner MLI. Traditionally constructed MLI was not considered in this study.

It should be noted that the RBO testing demonstrated a 55% heat reduction by integrating the broad area cooled shield into the MLI, significantly less than that predicted by the model. Post-test inspections showed no inherent issue with the MLI and no explicit reason for the less than expected performance. New work at NASA on MLI performance between 20K and 90K is in progress. However, the same model predicting high boil-off reduction is used in this analysis.

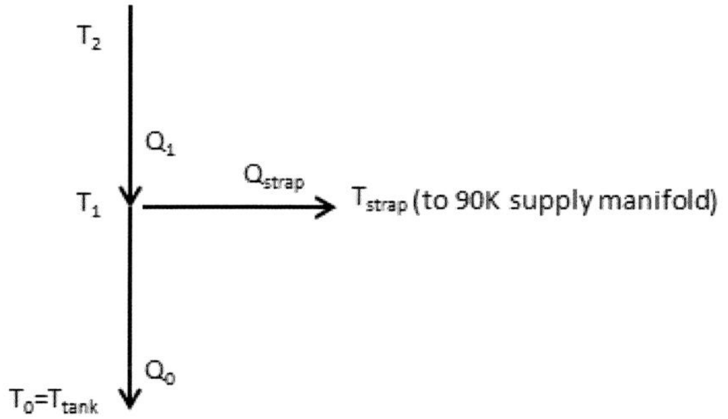

Figure 4. Heat flow between the environment at temperature T2 (node 2), the T1 ~ 90K cooling stage (node 1), and the tank at temperature $T_0 = T_{tank}$ (node 0).

Structure and Plumbing

The same structural relations shown in the scaling study were used with the notional tanks considered in this study. Titanium skirts were assumed for the 8.3 and 6 m diameter tanks, and 8 titanium struts were assumed for the 4 and 2.2 m diameter tanks, as notionally shown in Figure 3. Note that this analysis,[9] performed by Orbitec's partner Lockheed Martin Space Systems Corp. Advanced Technology Center, was thermally and structurally optimized. Lockheed met the natural frequency and factors of safety for stress and buckling while minimizing the heat load through the support structure. It is important to note that for the larger tanks, analysis revealed that the torsional vibration mode is a dominant driver and would have resulted in a relatively massive support structure. However, Lockheed assessed that this vibration mode is not a major concern for the launch vibration environment and can be ignored for the 8.3 and 6 m diameter tanks. Another assumption that could have a substantial effect on results is, given that no detailed model of the temperatures along the structure was performed, no structural insulation was assumed and therefore no associated structural insulation heat load was calculated. Similarly, no insulation on the plumbing was assumed and no heat load calculated. The vent and fill lines could possibly be under the tank insulation for some of the path. Solid conduction for an assumed length and diameter of the vent, fill, and pressurization lines, as well as for instrumentation was estimated. However, unlike the scaling study, no feed line heat leak was included. The lack of engine detail and the knowledge of the feed line valve location prevents any definition of this heat leak. Also, it, too, could be adjacent to the tank and could follow the perimeter of the oxidizer tank down to the engine, adding length which decreases heat load.

Strap Cooling

While the tank and MLI are cooled by the tube-on-tank or tube-on-shield concept, the tank penetrations are not. To reduce the tank penetration heat, copper cooling straps were included to transfer heat from the penetrations to the broad area cooled shield, with an assumed 2.5 cm wide strap for each strut and for each 25 cm of tank skirt circumferential length. The straps in the test reduced penetration boil-off by 68% for the tank struts and 64% for the fill and drain line. The resultant contact resistance of 5 K/W from the test was integrated into the model. The cooling strap location was optimized, by changing the location and summing the thermal system mass, including the active thermal control system with power and heat rejection. The optimal location was the same for each tank size, which was to intercept heat closer to the warm end, at 80% of the length of the structure being cooled. Radiative effects were ignored in that optimization.

Cryocooler

The cryocooler power and mass were determined from the results of a contracted study with Creare, Inc. (contract number NNG12LN29P). This was done for several point designs between 20 and 80 W at 20K, and a curve was created to predict mass and power between these points. Similarly, point designs were performed between 100 and 500 W, a curve was created, and the mass and power were predicted for the heat loads. Note that this effort does not consider standard compressor or turbo-machinery sizes; rather they are floating notional sizes to match the required LH_2 tank heating rate, with margin. Included with the selected cryocooler mass and power is the integration losses to the LH2 tank, which includes the parasitic loss and thermal ineffectiveness predictions of the NASA design. The assumed losses are based on RBO testing with a 15 W 77K reverse turbo-Brayton cycle cryocooler. A listing of the major assumptions in the study is shown in Table 1.

A listing of the major assumptions in the study is shown in Table 2.

RESULTS

Heat Loads

The predicted heat loads are shown in Table 3.

The thermal control system mass was determined by summing the MLI, foam, and boil-off mass for the passive case and summing the MLI, foam, and active cooling system mass including

Table 1. The assumed value and associated basis for the major parameters used in the study.

Parameter	Assumed Value	Basis
Tank Pressure	172 kPa	Pump fed engine
Thermal Margin	50% applied to all thermal loads	JPL Design Standards
Tank and Structure Mass	Approx. 1.2x propellant mass	Orbitec contract (LM sub)[9]
Insulation Mass	0.08 kg/m^2 per layer self-supporting MLI	CBRS test
Spray on foam mass	0.9 kg/m^2	CBRS test
Broad area cooled shield mass	1.2 kg/m^2	CBRS test, 5 mil Al foil
Radiator mass	0.021 kg/watt heat rejected; 10% margin added for structure	CBRS, including cooler interface plate
Radiator surface area	0.0013 m^2/watt; 10% margin added	CBRS test
Solar array mass/area	2.3 kg/m^2; 30% contingency added to surface area	The New Space Mission Analysis and Design[10]
Solar array specific power	48 W/kg; 30% contingency added to mass	
Tank Ullage	3% of tank volume	Prevent tank rupture
Tank Residual	2% of tank volume	Inaccessible propellant
Environmental Temperature	220 K	Representative temp. of Earth and Sun oriented low Earth orbits
Heat Rejection Temperature	T=300 K	Discussions with cryocooler vendor, Creare

Table 2. Predicted heat loads.

Tank Diameter	8.3	6	4	2.2	m
MLI	87	30	14	4	W
Structure	21	12	3	.4	W
Plumbing	1	.4	.4	.4	W
Passive Heat Load	108	43	17	5	W
Heat Load with 50% Margin	163	64	26	7	W

Table 3. Mass, power, and 20K lift comparison for MLI scale factor (SF) changes.

	8.3 m	6 m	4 m	2.2 m
2-Stage Mass, kg, SF 3	1362	664	418	285
2-Stage Mass, kg, SF 5, 2	1350	660	414	282
2-Stage Power, W, SF 3	6900	2815	1335	643
2-Stage Power, W SF 5, 2	6126	2550	1210	600
20K Lift, W, SF 3	32.3	13.4	5.5	1.8
20K Lift, W, SF 5, 2	47.9	18.8	7.9	2.5

the cryocooler, radiator, and solar array mass for the active zero boil-off cases, both with a single stage of cooling and two circuits of cooling. Figure 5 shows the results of this analysis for the four tank sizes considered. The point where curves cross, the equal mass points, are when the passive system mass, including boil-off, equals the active system mass. At loiter periods that exceed these intersect points, active cooling should be considered. Note that these intersect points decrease with size, non-linearly, indicating the efficiencies gained with turbo-machinery as the cryocooler size

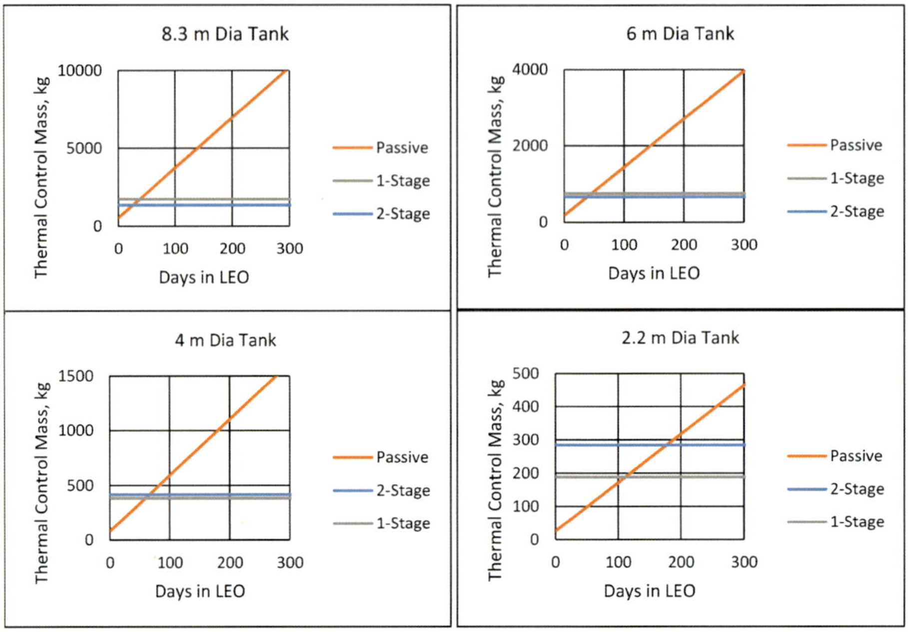

Figure 5. Thermal control system mass is shown for the four tank sizes considered in the trade space.

increases. Also, the two stage concept weighs less than the one stage cooling for the larger tanks. For the two smaller tank cases, the opposite is true—a one stage design is lighter in weight. Weight is not the only factor, however. Input power availability is critical--this is shown in Figure 6 plotted versus tank heat load. It is clear that for the larger tank sizes, the two cryocooler configuration reduces power by thousands of Watts but that this advantage is not realized for the lower heat loads (smaller tanks). Figure 7 shows the effect that the two circuits of cooling has on the 20K cryocooler lift. Minimizing the 20K lift requirement is important, as new technology is part of the development effort in the 20 W at 20K cryocooler presently being manufactured by Creare. For the largest tank with highest tank heat of 182 W, the 2 circuits of cooling concept decreases the 20K lift to just 32 W. The 90K cryocooler lift varies from 535 W for the 8.3 m diameter tank to 33 W for the 2.2 m tank.

MLI Scale Factor Variation

As stated in the assumptions, the scale factor used in determining the MLI heat loads was 3. However, this scale factor was not realized during RBO testing, which found that under the shield,

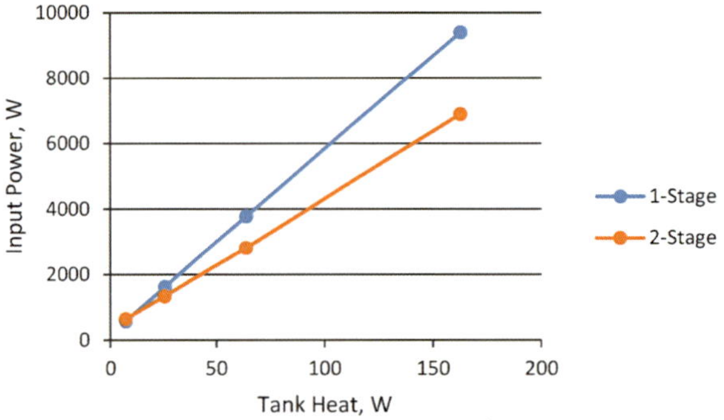

Figure 6. Input power vs. tank heat load.

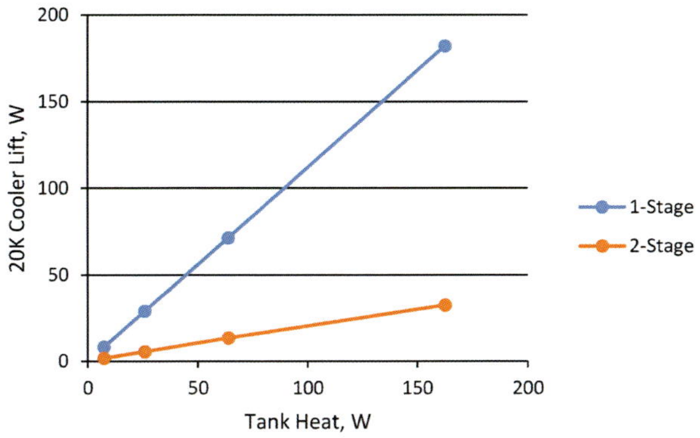

Figure 7. 20K cryocooler lift versus tank heat for the 1-stage and 2-stage cooling concepts.

where MLI was between 20 and 90 K, the MLI heat leak was higher than expected for both traditional and self-supporting MLI. The reasons for less than expected performance is unclear but is presently being investigated by NASA. A second analysis was done that increased this scale factor from 3 to 5, the scale factor found during RBO testing. Also, the MLI over the shield performed better than expected, so that scale factor was adjusted downward from 3 to 2. The results of this trade increased the 20K lift significantly but did not change the overall mass significantly (see Table 3).

The total power actually decreased as the increase in the 20K cryocooler power was more than offset by the decrease in the 90K cryocooler power.

Boil-Off Reduction

With the MLI scale factor set at 3 for inner and outer MLI, the percentage of boil-off reduction with the broad area cooled shield is 83% for all tank sizes. This is in contrast to a boil-off reduction of 73% when the scale factor of 5 is applied to the inner MLI and 2 for the outer. Note that both of these predictions are theoretical for an idealized shield. These reductions remain much greater than the RBO tested shield, which reduced boil-off by 55%.

SUMMARY

There were two major assumptions in this analysis with the first being that the insulation around the tank structure had no heat leak. This was done because of the lack of configuration information of the structure. The second major assumption was that the torsional vibration mode can be ignored in the tank structure sizing. Note that this analysis does not represent a current stage design; in fact, the current upper stage design shows a higher tank structural heat load of several orders of magnitude.

With those assumptions in mind, an analysis was performed to determine the size and application of 20K cryocoolers to a range of propellant tank sizes with and without the inclusion of a second circuit of cooling. The focus of this sizing effort was on the direct sizing of the 90K cryocooler system, based on the equations presented and the RBO test program results. Results show that the 20K cryocooler size and power is reduced by the inclusion of the 90K cooling stage for the two large tank cases considered. The largest system considered, the 8.3 m diameter tank, only requires 32 W of lift at 20 K, a substantial reduction from the single stage concept, which required a 182 W at 20K cryocooler. However, for the small tanks the second stage of cooling does not reduce mass.

ACKNOWLEDGMENT

This effort is sponsored by the Space Technology Mission Directorate's Small Business Innovative Research (SBIR) and Game Changing Development Programs at NASA.

REFERENCES

1. Northrop Grumman., "Cryocoolers Data Sheet," Model HCC at 20K, 2014, www.northropgrumman.com/aerospacesystems

2. Plachta, D. W., et. al. "Cryogenic Boil-Off Reduction System, *Adv. in Cryogenic Engineering*, Vol. 53, Amer. Institute of Physics, Melville, NY (2008), p. 1457.

3. Feller, J. R., et. al., "Analysis of Continuous Heat Exchangers for Cryogenic Boil-Off Reduction," *Adv. in Cryogenic Engineering,* Vol. 53, Amer. Institute of Physics, Melville, NY (2008), p. 401.

4. Plachta, D. W., Johnson, W. L., Feller, J. R., "Cryogenic Boil-Off Reduction System Testing," NASA TM 2015-0000134, AIAA/ASME/SAE/ASEE Joint Propulsion Conference; 50th; 28-30 Jul. 2014; Cleveland, OH

5. Plachta, D, W., Guzik, M. C., "Cryogenic Boil-Off Reduction System Scaling Study," *Cryogenics* Vol. 60 (2014), pp. 62–67, www.elsevier.com/locate/cryogenics

6. Johnson, W. L., et. al., "Tank Applied Testing of Load Bearing MLI," AIAA paper 2014-3581, presented at the Propulsion and Energy Forum, Cleveland, OH, 2014.

7. Johnson, W. L., et. al., "Thermal Coupon Testing of Load-Bearing Multilayer Insulation," *Adv. in Cryogenic Engineering*, Vol. 59, Amer. Institute of Physics, Melville, NY (2014), http://dx.doi.org/10.1063/1.4860775

8. Knoll, R., et. al., 1991, "A Review of Candidate Multi-layer Insulation Systems for Potential Use on Wet-Launched LH2 Tankage for the Space Exploration Initiative Lunar Missions," *AIAA 91-2176*

9. Orbital Technologies Co., "Cryogenic Propellant Tank Structural/Thermal Database," Contract NNC10BA11B, 2012, Report No. OTC-GS-0261.1-FR-12-01

10. Wertz James R, et al., *Space Mission Engineering: the New SMAD, Vol. 28.* Space Technology Library, Microcosm Press, 2011.

Integrated Cryocooler Assemblies for Miniature Satellite Applications

Andy Burkic, Reginald Little

Iris Technology Corporation
Irvine, CA 92614

ABSTRACT

The Microsat Cryocooler System (MCS) and Cubesat Cryocooler System (CCS) are radiation-hard, space-qualified integrated cryocooler assemblies (ICA) for miniature satellite platforms. The MCS and CCS have been developed on AFRL and NASA SBIRs, respectively, and are comprised of a high-reliability miniature cryocooler, a set of miniature Low Cost Control Electronics (mLCCE), and supporting thermal management components.

The mLCCE supports any of a wide range of linear cryocoolers in its design output power range, nominally 25W, and with minor adaptation, can accommodate rotary coolers as well. This paper highlights the evolution of the mLCCE supported by test results from the flight module phase of the more mature MCS program. For the MCS, a space-grade mLCCE rated for 55 krad has been built and integration tested with the AIM SX030. The test results for the system are discussed herein, and the overall cubesat-compatible mechanical subsystem design is also presented, including a description of the thermal management approach.

More broadly, the Iris Technology mLCCE represents a software-based general purpose, radiation hardened platform for the control of a variety of device types such as motors, actuators, solar power / battery charging systems, and optical bench thermal stabilization.

INTRODUCTION

The dramatic increase in small satellite use for commercial and government applications can be linked to their growing technological capabilities. As requirements for remote sensing and science data gathering missions grow more demanding, the need to mature small cryocooler systems that enable high performance mid-wave infrared (MWIR) and shortwave infrared (SWIR) sensors becomes more evident. To satisfy these needs, the MCS and CCS have been developed and designed with the purpose of simplifying integration in various microsat and cubesat applications, without the need for costly customized drive electronics and systems.

To demonstrate the versatility of the mLCCE, several different cryocoolers have been driven and controlled successfully: a Thales LPT 9510 pulse-tube cooler, an AIM SX030 single piston linear cooler, and a Lockheed Martin Microcryocooler. To further explore compatibility with different temperature sensors needed for precise temperature control, the MCS utilizes a 1 mA diode sensor compatible with the AIM SX030 while the CCS utilizes a 100 μA Cernox sensor compatible with the Lockheed Martin Microcryocooler. Both systems can be adjusted to accommodate either sensor with a simple change to the component assembly of the board.

MINIATURE LCCE EVOLUTION

Iris Technology initially leveraged its higher power LCCE designs under IRAD funding to produce a miniature COTS version of a proof-of-concept board that demonstrated full drive capability and temperature control of the AIM SX030 cryocooler. The effort evolved through the support of an AFRL SBIR which allowed the initial design to be expanded to generate an appropriate path for flight electronics. The new MCS met this criterion and has become a mature flight system while CCS, developed through a NASA SBIR, for now remains a COTS system with mLCCE efficiency improvements and flexible temperature sensor implementation for diode and cernox sensors. These improvements have also been included in the flight version of the MCS mLCCE.

Gen I for MCS

The MCS, comprised of a space-grade mLCCE and the SX030 cryocooler, has reached flight maturity through AFRL's Phase II SBIR award and is currently being evaluated for NASA's Lunar IceCube mission to the moon. The mLCCE specifications and test results will be discussed in the following sections. Further funding is expected to qualify the complete system to TRL-6, including TVAC, vibration, and EMI testing as well as triple mode redundancy (TMR) functionality.

Gen II for CCS

The CCS developed under NASA's Phase I SBIR uses a COTS mLCCE fully compatible with Lockheed Martin's mature TRL-6 Microcryocooler. The cooler was successfully tested with the mLCCE using a Cernox temperature sensor at Lockheed Martin's facility in Palo Alto, CA. The CCS has served as another platform to show how quickly a new ICA prototype can be built up and tested using the mLCCE and a cryocooler other than the SX030.

MINIATURE LCCE SPECIFICATIONS

Table 1 provides a summary of critical performance criteria and design performance specifications of the mLCCE.

Table 1. mLCCE Performance Specifications

Parameter	Characteristic
Operating Input Voltage	9 to 35 VDC
Number of Motor Drives	1
Total Output Power	Continuous operation @ 25 WAC maximum at 22 VDC
Max Output Voltage @ 27°C	> 39 Vpp @ 22 VDC; > 16 Vpp @ 9 VDC
Efficiency	> 85%
Tare Power	< 1.5W
Output Voltage Sinusoidal Symmetry	Sine wave output, with less than 0.3V DC offset and < 3% THD including harmonics up to 5 kHz
Operating Frequency Control	Output drive frequency shall be adjustable from 40 to 200Hz with 0.1 Hz resolution
Number of Temp Sensors	1 (configurable for RTD (e.g. Cernox, PRT) or diode)
Temperature Sensor Bias	- 1 mA ±0.05 mA bias current to the temperature sensor (Diode configuration) - 100 μA ±5 μA bias current to the temperature sensor (Cernox configuration)
Temperature Sensor Set-point	- Adjustable in range 0 V to 3.2 V with 0.8 mV resolution (Diode configuration) - Adjustable in range 0 V to 3.2 V with 49.0 μV resolution (Cernox configuration)
Temperature Sensor Sensitivity	- Detectable in range 0.73 V to 1.13 V with 99 μV resolution (Diode configuration) - Detectable in range 0 V to 2.00 V with 3.06 μV resolution (Cernox configuration)
Temperature Response Time	Response to a detected temperature change shall be asserted in drive voltage within 36 ms

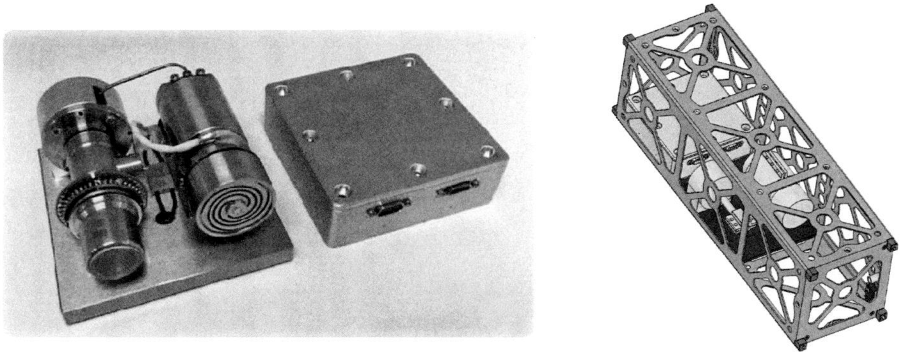

Figure 1. MCS SX030 Cryocooler with mLCCE (left) with implementation in a 3U skeleton (right)

SYSTEM SOLUTIONS

Since the MCS is a flight mature system, it will be the primary point of discussion in the following sections which will provide test data collected in a clean room environment, appropriate for radiation-hardened, high-reliability electronics and hardware.

MCS with AIM SX030

The MCS and its mock-up implementation within a 3U size cubesat skeleton are both shown in Figure 1. The AIM SX030 cryocooler, shown with a test dewar, as well as the flight mLCCE each fit inside of a 1U frame, streamlining full system integration for cubesats. The maximum input power to the cooler is 12 W at 65 Hz and its cooling capacity is shown in Figure 2. At this power level, the efficiency of the mLCCE is approximately 88% at 28 V but depending on the input voltage, can reach as high as 90%.

CCS with Microcryocooler

The CCS Microcryocooler and a mock-up implementation of the whole system within a 3U cubesat skeleton are shown in Figure 3. The maximum input power to the cooler is 20 W at 90-105 Hz, allowing a higher cooling capacity up to 2 W. The mLCCE is compatible with the Microcryocooler as well and was able to drive it at full power with 28 V applied to its input.

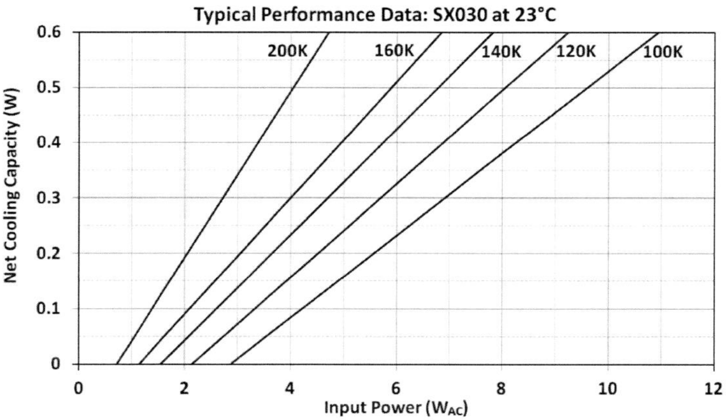

Figure 2. AIM SX0303 Cooling Capacity vs. Input Power

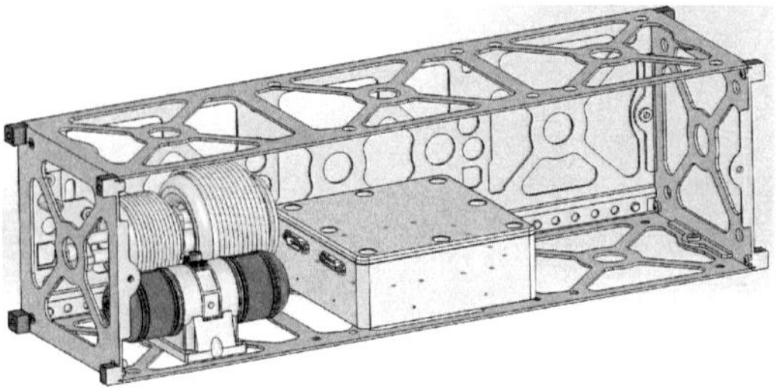

Figure 3. CCS Microcryocooler (left) with mLCCE implementation in a 3U skeleton (right)

MINIATURE LCCE AND MCS TEST RESULTS

The following subsections present the test results obtained for the flight version of the mLCCE as well as the whole system using the AIM SX030 cryocooler.

Tare Power

The mLCCE tare power testing was performed by applying four different input voltages: 9V, 12 V, 28 V and 35 V. During the test, the AC output of the mLCCE was disabled and only the power consumption of the board was measured. The results are summarized in Table 2. Note that a higher input voltage translates to a higher tare power due to increased switching losses in the onboard power supplies.

Output Power and Efficiency

The mLCCE efficiency was first measured with a resistive load to demonstrate its output power capability at power levels beyond what the SX030 can handle. Using a 2.65 Ω resistive load, the mLCCE is able to produce 20 W of output power at efficiencies between 87-89% with either 12 V, 28 V or 35 V applied to its input. Furthermore, it was driven up to 25 W and as high as 35 W at 28 V, reaching efficiencies of 92%.

Next, the efficiency was measured by applying 12 V, 28 V and 35 V and driving the SX030. Since the mLCCE is not capable of boosting its input voltage, it cannot output a peak voltage higher than the input voltage. For example, if 12 V is applied to the mLCCE, its maximum sinusoidal peak output voltage is approximately 11 V, or 22 V peak-to-peak due to some losses in the board and limitations of the duty cycle of the sine wave converter. With an input of 28 V, the maximum output peak voltage is 27 V, or 54 V peak-to-peak. Considerations should be taken when choosing the right input voltage based on the impedance of the cooler that needs to be driven and its maximum allowed voltage rating.

An efficiency of 90% was achieved with a 12 V input to the mLCCE, however, the impedance of the SX030 and the output voltage limitation of the mLCCE only allowed the cooler to be driven up to 8 W. A minimum input voltage of 15 V is recommended for MCS in order to drive the cooler to its full rated power of 12 W Alternative coolers with lower impedances can be driven at higher power since the mLCCE can handle input currents up to 3 A. Similarly, applying a 28 V input to the mLCCE allowed the SX030 to be driven at 12 W with 88% efficiency while 35 V produced an efficiency of 85%. The efficiency plots summarizing the results discussed are shown in Figure 4.

Table 2. mLCCE Tare Power

Input Voltage (V)	Input Current (mA)	Tare Power (W)
9.006	60.75	0.547
12.011	46.40	0.557
28.041	32.00	0.897
35.047	31.40	1.100

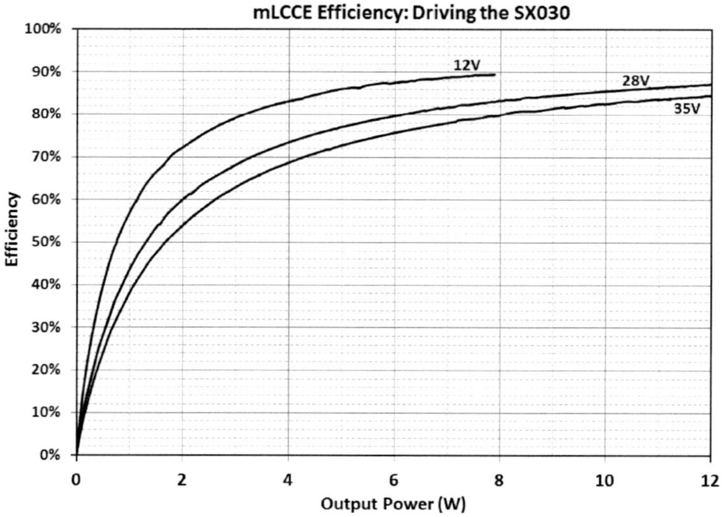

Figure 4. mLCCE Efficiency vs. Output Power

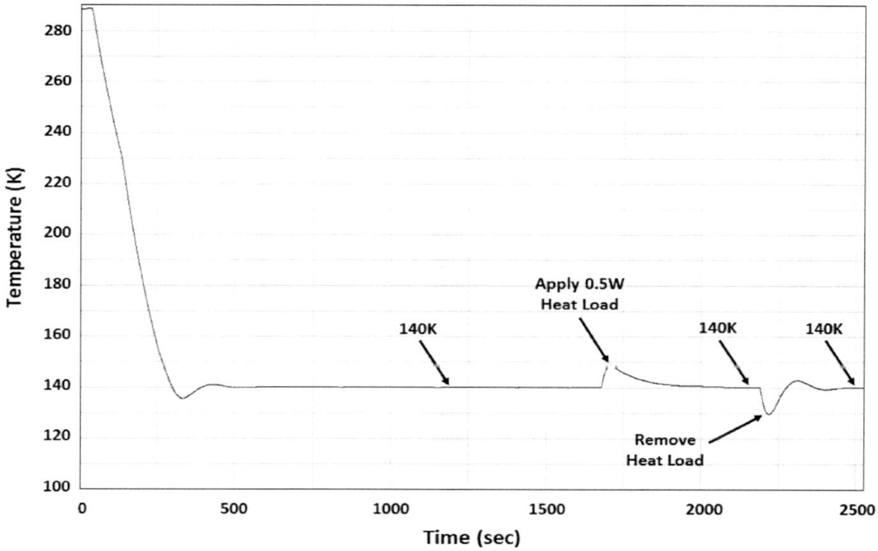

Figure 5. SX030 Cold Tip Temperature with Applied Heat Load

Temperature and Voltage Control

The temperature vs. time plot shown in Figure 5 illustrates the mLCCE operation with the SX030 in temperature mode. Using the Iris communications protocol GUI via RS422, the user can manually control the mLCCE output voltage and set limits for the drive voltage and cooler temperature, at the same time collecting telemetry while plotting various data curves.

Starting at room temperature, as shown at the beginning of the plot in Figure 5, the Iris GUI was set to Temperature Control Mode and a control temperature of 140 K. During the cool down period, safe voltage limits need to be set to protect the cooler pistons from being overdriven at higher temperatures. More power can be applied once the cooler reaches 200 K or below. Once the cold tip of the SX030 reached 140 K, a 0.5 W heat load mounted to the cold finger was applied. The cold tip temperature again reached its 140 K setpoint and stabilized with the mLCCE running continuously at a higher nominal output power. For the final step, the 0.5 W heat load was removed and the mLCCE readjusted its drive level to the cooler to re-stabilize at 140 K. The Iris GUI allows the user to set the PI values for the PI loop of the mLCCE which can be optimized for a desired operating speed and reaction time during cool downs, application of heat loads, and other transitions in the operation of the system.

The Iris GUI also allows the user to run the mLCCE in continuous Voltage Control Mode. The output peak voltage can be set manually for constant voltage outputs needed to test with resistive loads and other control applications independent of temperature sensing.

Total Harmonic Distortion (THD)

For the THD measurement of the mLCCE, an input of 28 V was used to drive a 2.65Ω resistive load at 65 Hz and 12.8 Vpk, providing the maximum rated power output of 25 W as specified in the requirements. At this worstcase condition, the highest THD level measured by the Yokogawa power meter was 0.768%, well below the 3% requirement. Additional testing with the SX030 operating at its maximum rated power of 12 W, revealed maximum THD measurements no higher than 0.72%.

Frequency Control

The mLCCE is designed to support a variety of cryocoolers and is capable of providing drive output frequencies between 40-200 Hz. The frequency of its AC output waveform can be finely adjusted in 100 mHz increments and has been verified and tested for the entire specified range. Scaling the drive frequency by such small increments allows complete drive optimization between the mLCCE and the SX030, as well as any other cryocooler able to operate within this range.

MCS THERMAL MANAGEMENT

The mLCCE chassis draws heat from the PCB via two internal and thermally conductive chassis planes fabricated in the middle of its layer stackup. The copper on the bottom side of the PCB, containing the majority of the power components, is exposed around the perimeter of the board and thermally connected to the internal chassis layers with arrays of vias. The heat generated in the components dissipates through the board to the outer ENIG edges. The chassis is connected to these exposed copper edges of the board with gold foil thermal gaskets to decrease interface resistance. Screws make up the final connection to bring the heat away from the components and the PCB to the exterior of the chassis and a base plate connected to the skeleton of the aircraft.

Figure 6 shows a complete MCS implementation using a common radiator plate connected to both the mLCCE and the SX030. A thermal strap connecting the compressor of the cooler to the radiator plate is also shown. The arrows indicated the thermal flow from the compressor to the thermal strap and finally to the radiator plate. The reason for using the thermal strap rather than securing the compressor rigidly to a plate is to allow for movement. Isolation through flexures from the compressor to the radiator plate further expands flexibility for movement, allowing the mitigation of vibrations escaping from the passive balancer of the cooler.

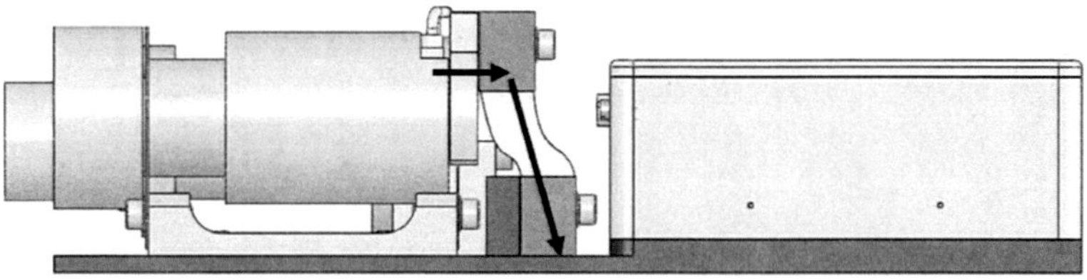

Figure 6. Thermal Heat Strap for SX030 Compressor

CONCLUSIONS

The flight-ready MCS has met all of its requirements and the mLCCE is able to support various different cryocoolers with its wide input voltage range, flexible drive frequency, and support for different temperature sensors. The CCS SBIR serves as an example of how quickly a new ICA can be developed together with the mLCCE for missions requiring a different cryocooler. A novel MCS thermal approach completes the system package and allows for immediate integration into a larger system.

We expect the radiation hardened, software-based mLCCE to soon find many applications in general purpose control systems used aboard the smallest satellite platforms.

ACKNOWLEDGMENT

Iris Technology Corporation would like to thank Dr. Ingo Ruelich at AIM Infrarot-Module GmbH and Dr. Jeff Olson at Lockheed Martin for providing the cryocooler TMUs used in these SBIR programs. We also would like to thank AFRL and NASA for supporting the development of these cubesat and microsat ICAs and working together with us to send the MCS to the moon through the Lunar IceCube mission.

REFERENCES

1. J.R. Olson et al., "MatISSE Microcryocooler," *Adv. in Cryogenic Engineering: Proceedings of the Cryogenic Engin. Conf. (CEC) 2015*, Vol. 101, IOP Publishing, Bristol, UK (2015), Article 012025.
2. Olson, J.R. et al., "Coaxial Pulse Tube Microcryocooler," *Cryocoolers 18*, ICC Press, Boulder, CO (2014), pp. 51-57.
3. Nast, T.C et al., "Qualification of Lockheed Martin Micro Pulse Tube Cryocooler to TRL6," *Cryocoolers 18*, ICC Press, Boulder, CO (2014), pp. 45-50.

Design of a Flight Like Cryostat for 30-50 K Two-Stage Pulse Tube Cooler Integration

J. Tanchon[1], T. Trollier[1], P. Renaud[1], J. Mullié[2], H. Leenders[2], T. Prouvé[3], I. Charles[3], T. Tirolien[4]

[1]Absolut System SAS, 38170 Seyssinet Pariset, France
[2]Thales Cryogenics BV, 5626 DC, Eindhoven, NL
[3]University Grenoble Alpes, CEA INAC-SBT, 38000 Grenoble, France
[4]European Space Agency ESA-ESTEC, Noordwijk, NL

ABSTRACT

An ESA Technical Research Program (TRP - 4000109933/14/NL/RA) is run by a TCBV/CEA/AS consortium for the development, optimization and testing of a cryostat actively cooled by a 2-Stage High Reliability Pulse Tube cryocooler (presented in a companion paper). The interest of this concept is to allow the detector operation – for example Quantum Well IR Photodetector (QWIP) or Mercury Cadmium Telluride (MCT) detectors - at lower temperature, in the range of 40K-45K, for an overall input power similar to the one required by current Earth Observation programs.

Several cryostat concepts have been traded to optimize mechanical, thermal and electrical interfaces between the 2-stage cryocooler, the detector assembly and the external structure while minimizing the impact of 2 stage cooling on the Detector's Assembly Integration and Test phases.

For this project, an innovative supporting structure concept has been developed by Absolut System to offer a good compromise between thermal and mechanical performance. High performance flexible Thermal Links Assemblies (TLA) have been also optimized using our background developed for European on-going space projects (MTG, IASI-NG and others, presented in a companion paper).

This paper will present the conceptual study output, the detailed design phase including thermal and mechanical justification and the final design description of the cryostat. The outcomes of current manufacturing will be also presented.

INTRODUCTION

The space cryogenics sector is characterized by a large number of applications (detection, imaging, sample conservation, propulsion, telecommunications) that lead to various requirements (temperature range, microvibration, lifetime, consumption) that can be met by different solutions (Stirling or JT coolers, mechanical or sorption compressors). Recent years in Europe saw the development of coolers to meet Earth Observation mission requirements that can provide more than ~3 W of cooling power at an operational temperature around 50K [1][2]. Those single stage Stirling or Pulse Tube coolers have a no load temperature in the range of 30K with limited cooling power at 40K with reasonable rejection temperatures.

On the other end, the trend of focal plane cooling requirements for Earth Observation is to aim either at higher cooling power at current temperature levels (around 50K) due to the use of bigger detectors, or to lower the ultimate temperature to limit dark current or to use new detector types. For instance, the QWIP detectors seem promising but need to operate at 35 - 45K, which is not reasonably obtainable with the currently available coolers developed for Earth Observation.

In addition to detector cooling, this cooler can also be used in a multistage cooling system (e.g., Shield cooler). To satisfy this trend, one solution consists of using the current Pulse Tube and Stirling technologies in a double stage configuration to intercept parasitics at a higher temperature and to make cooling power available at lower temperature. This solution makes the most use of the existing technology building blocks (e.g., compressor) but requires mastery of the design/manufacturing/integration constraints linked to double stage cryocoolers used in cryostats.

In parallel to the development and the optimization of a 30-50K two-stage pulse tube cooler performed by TCBV and CEA/INAC [3], Absolut Ssystem focused on the system level analysis and the best way to improve the cryostat concept taking benefits of the two-stage cryocooler capability. This paper reports the activities performed on the target Cryostat from the system level trade-off to the manufacturing and integration of the final hardware going through justification analysis.

SYSTEM LEVEL TRADE-OFFS AND CONCLUSIONS FOR DETAILED DESIGN

Several aspects have been analyzed during the first phase of the development. The objective was to propose different cryostat concepts and to evaluate their performance for comparative study. The main aspects analyzed are: cryostat integration and accessibility (in particular detector's alignment constraints), thermal performance, mechanical robustness and stability regarding optical constraints, and orientation of the cold finger regarding gravity for ground testing.

After intensive trade-off, Concept 4 surrounded by the border in the Figure 1, has been selected. This concept offers a good compromise between the thermal performance, mechanical aspects and integration of the components. The main advantages of this configuration are:

- Very compact Optical Cold bench. The optical cryostat is split from the cold head cryostat. Both sub-assemblies are thus optimized in term of volume, and mass.
- Design with cold fingers downward. The parasitic losses of the cold finger will be representative of flight conditions.
- Distance required between the 2 sub-assemblies limited due to the direct coupling between cold head cryostat and optical bench (limited thermal links length).
- Radiative and conductive loads intercepted by cryocooler 1st stage.
- Cold shield implemented after mounting of all internal components (better accessibility)

The main design driver identified is the mechanical behavior of the supporting structure which could be critical and required detailed analysis. At a lower level, some trade-offs have been conducted regarding the component selection like the "Thermal link concepts and material selection", the "Radiative insulation selection" or the "Supporting structure concept".

Thermal link concepts and material selection

The thermal links required for this project are critical for thermal performance, mass, stiffness and mechanical loads. To answer these requirements with the best compromise, different materials have been studied. The material to be selected for the thermal links shall combine:

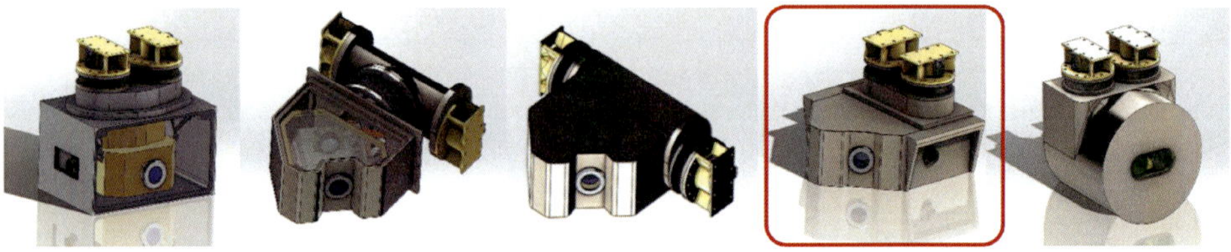

Figure 1. Different cryostat concepts considered

- an excellent thermal conductivity at ambient temperature.
- a low density.
- a low Young modulus (low stiffness).

To evaluate the best candidate for the thermal links, we utilize the ratio between the thermal conductivity and the density of the different interesting materials in our operating temperature range. As can be seen in the Figure 2, the Pyrolytic Oriented Graphite (POG) is the best candidate in the 75K-160K temperature range while the 5N aluminium is the best candidate in the 40K-75K temperature range. Both materials offer a very low Young modulus.

The aluminium will be used for the 2nd stage thermal links operating in the 30-50K temperature range and the POG will be used for the 1st stage thermal link operating in the 100-160K temperature range.

The two thermal links are made with thin aluminium and POG foils staked together and thermally and mechanically connected on the extremities to 1080 aluminium grade end-fittings (see Figure 3). The specific coupling processes have been developed by Absolut System and qualified for European space programs. The typical foils thickness is in the range of 50 μm to 70 μm.

Radiative insulation selection

The radiative loads on the thermal shield represent a significant part of the parasitic heat losses for the complete cryostat. For most of the cryostat, low emissivity coatings are used to limit the

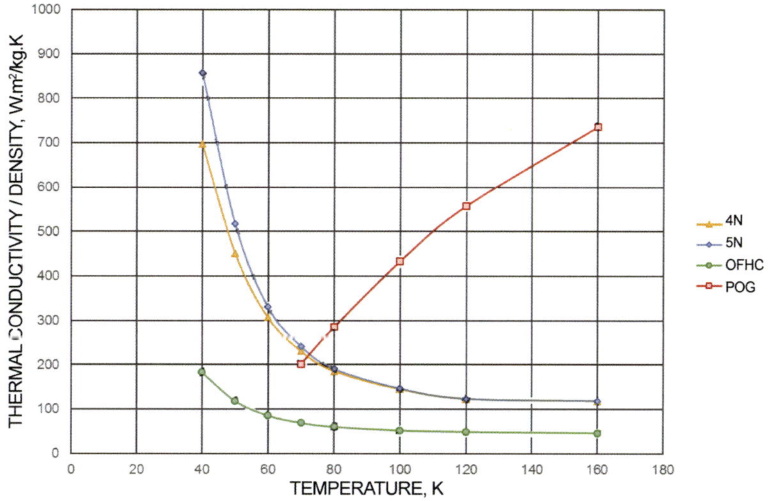

Figure 2. Ratio between thermal conductivity and density for pure aluminium, OFHC copper and POG

Figure 3. CAD views of the high purity aluminium thermal link on the left and POG thermal links on the right

radiative coupling between the warm side and the cold side with very good results. However, the low emissivity coatings are not as easy to produce and the final emissivity is not always expected.

The best candidates for the low emissivity coatings are aluminium, gold and silver with a high purity grade metallized coating. If we consider the surface contamination and manufacturing discrepancy, the "typical" emissivity of the coatings need to be corrected for the worst case analysis. For gold coating, the emissivity range is between 0.03 and 0.05 which increases the uncertainty of thermal analysis by a factor 2.5.

The low emissivity coatings are a very efficient solution to reduce radiative coupling between the surfaces but are very sensitive to surface contamination and in particularly to ice contamination. A thin layer of ice on the thermal shield will induce a direct degradation of the emissivity of the coating because ice tends to absorb in the IR wavelengths and scatters in visible wavelengths.

As detailed by R. G. Ross [4], very thin deposited films of contaminant gases can significantly influence the emissivity of coated surfaces such as polished gold or aluminum. It has also been demonstrated that the water-ice film has a very strong impact on the sample emissivity. The emissivity of the polished metal surface increases initially by a factor 2 per micrometer of water-ice thickness and finish by a factor close to 1 for water-ice thickness greater than 20 μm. For a polished gold surface, a water-ice film thickness of approximately 1.0 μm would be expected to increase the emissivity up to 0.06-0.1 and thus greatly increase the radiative loads.

Several aspects of R. G. Ross [4] are interesting. For metallic surface with a temperature below 250K, the "Surface Residence Time" of water molecule is several years, and the surface film of water is immobile when temperature reaches 100K. Thus, once water is trapped on a cryogenic surface, it is very difficult for it to leave and migrate to another surface.

The second interesting aspect of this paper is that it suggests that using multi-layer insulation (MLI) is much better than a single low-emissivity surface in controlling the effects of contamination. The MLI will be less sensitive to water-ice contamination due to the number of layers which act like a stacked series of radiation shields operating from environment temperature down to cryogenic temperature. The "equivalent" emissivity of MLI blanket is in the range of 0.001 to 0.01 for a space qualified MLI which will then provide, in worst case conditions, 6 to 10 times better thermal performance than a gold coating. However, MLI is generally not recommended in optical cryostats which are sensitive to surface contamination. Polyimides used in MLI blanket are a source of water vapor. This material is known to retain water and have a saturation temperature of 152K. As a result, this material should not be used where it has a view factor with optical elements below 152K.

In our case, the optical configuration is very favorable. The thermal shields will cryopump most of the water vapor and then decrease its partial pressure. When the water is trapped onto a cold surface (temperature below 150K), the molecules are fixed and will not migrate to other surfaces. The emissivity degradation mainly impacts the intermediate cold shield but it is far less critical with a two-stage cryocooler configuration compared to a single stage cooler which is directly impacted by the emissivity degradation. We concluded that the MLI is a very good candidate to limit the contamination sensitivity on the cryostat while reducing the radiative losses.

Supporting structure concept

One of the most critical components highlighted in the study is the supporting structure. The overall design of the system and its optimization is strongly driven by this component. The primary function of the mechanical supporting structure is to ensure a robust positioning of the optical components against mechanical and thermo-mechanical loads including the high stability of the structure during cool-down, thermal cycling, and mounting/dismounting.

The other aspect in the design of the supporting structure is to reduce the conductive losses through the supports. Ti6Al4V Titanium alloy and composite materials like Glass-Fiber Reinforced Plastic (GFRP) are the best candidates for this function. However both don't offer the same level of performances. GFRP is a good candidate considering its mechanical and thermal performance. GFRP is 10 times better in terms of thermal conductivity, 2 times better in terms of density and 3 times better in term of Yield limit (in fibers directions) than Ti6Al4V. However, this material is difficult to implement because it is ortho-tropic, and thus the mechanical properties are significantly different with the direction of the load. When the load is applied perpendicular to the fibers, the Yield limit drops down to 100 MPa which is considerably low. Furthermore, the material requires specific design due to the complex manufacturing process. It is then difficult to get the optimal shape, fibers orientation and thickness due to the manufacturing constraints.

To obtain a robust supporting structure while minimizing the thermal conductive heat losses, an innovative solution has been proposed using 3-D lattices in titanium alloy (Ti6Al4V) produced with an Additive Manufacturing (AM) process. The combined use of Titanium alloy and the particular structure of the lattice (see Figure 4) makes the supportive structure very stiff and with a low thermal conductivity.

Following the mechanical analysis, the optimal shape has been selected for the lattice structure, and a good mechanical performance has been reached. As a first comparison, with the same conditions, the first Eigen mode of the concept with lattice design is above 300 Hz with a thermal conductance 45% lower than with composite blades.

To validate this concept, some breadboard samples have been designed and produced with several dimensions to evaluate the robustness of the manufacturing process. A baseline design has been selected for the lattice, and the mechanical and thermal analyses have been updated using these inputs.

DESIGN DESCRIPTION OF THE EQUIPPED CRYOSTAT

After optimization of the design regarding thermal and mechanical analysis, the detailed design of the target cryostat has been validated by the project team during the Critical Design Review Process. The manufacturing and assembling processes started following this design review.

The core of the target cryostat is composed with a Cold Box system where dichroic optics are mounted to split the entrance beam towards the two infrared detectors attached on the cold box (see Figure 5 and 6). With the innovative design of the supporting structure, the cold box is directly part of the supporting structure including warm interface and struts. Thus, no screwed interfaces are implemented on the optical/mechanical path between the mechanical reference and the detectors. A single monolithic metallic part ensures this function.

On the back of each detector, the high purity aluminium thermal link is mounted. This thermal link shape has been adapted to optimize the temperature homogeneity while reducing the temperature gradient and the overall mass. With the specific manufacturing process developed and qualified by Absolut System for thermal links, the thermal link's shape is free of constraints and can be designed with extremely complex routing, shape or interfaces. The other extremity of the thermal link is attached directly on the second stage of the cold tip operating at 40 K.

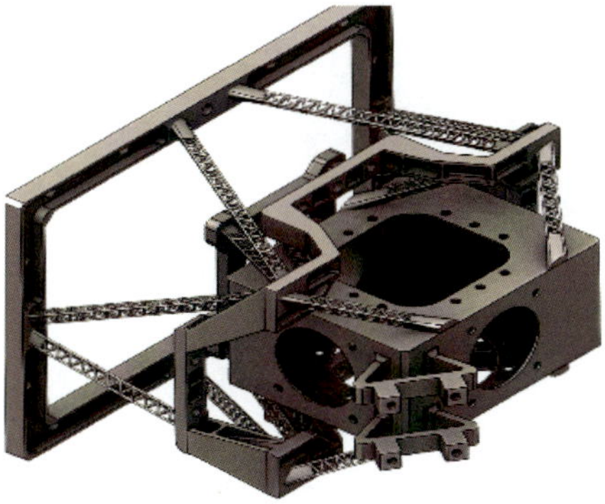

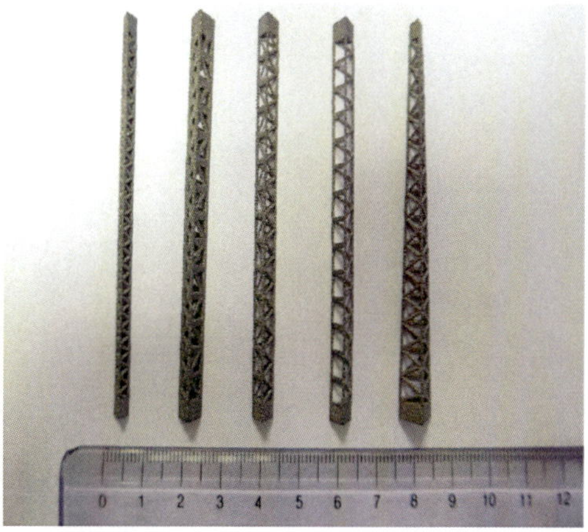

Figure 4. Overview of the monolithic supporting structure (left) and Ti6Al4V lattice breadboards manufactured using AM process (right)

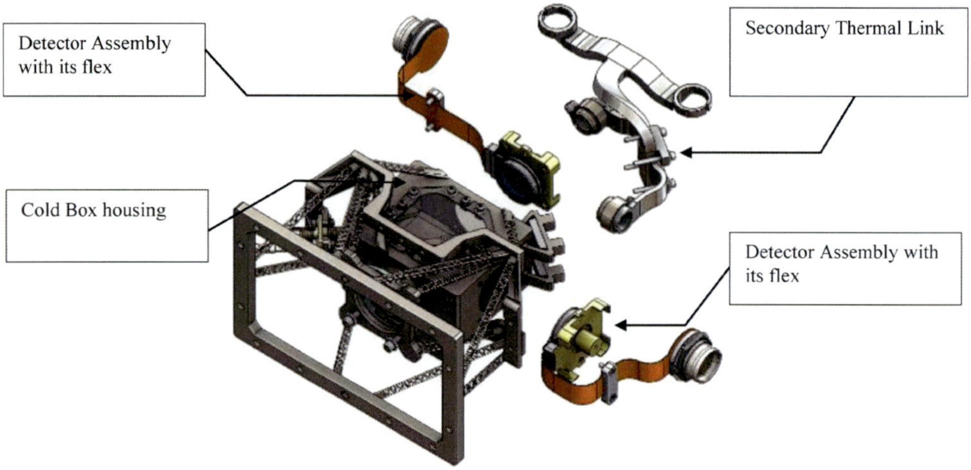

Figure 5. Exploded view of the cold box assembly

In order to minimize the parasitic heat fluxes on the detectors and on the cold box, a baffle assembly is required. The different parts of the baffle are made using aluminium alloy to offer the best compromise between mass, thermal conductivity and mechanical performance. The external surfaces will be covered with MLI blanket and the internal surface black coated. Following thermal optimization, the cold window is mounted on the baffle at the intermediate temperature (in the range of 100 K). The baffle will be black coated on its internal face. The external face will be covered with MLI blanket as per the cold shield.

The cold box is surrounded by an aluminium thermal shield to intercept radiative heat losses. The thermal shield is separated in 4 main parts in order to implement the thermal shield after final

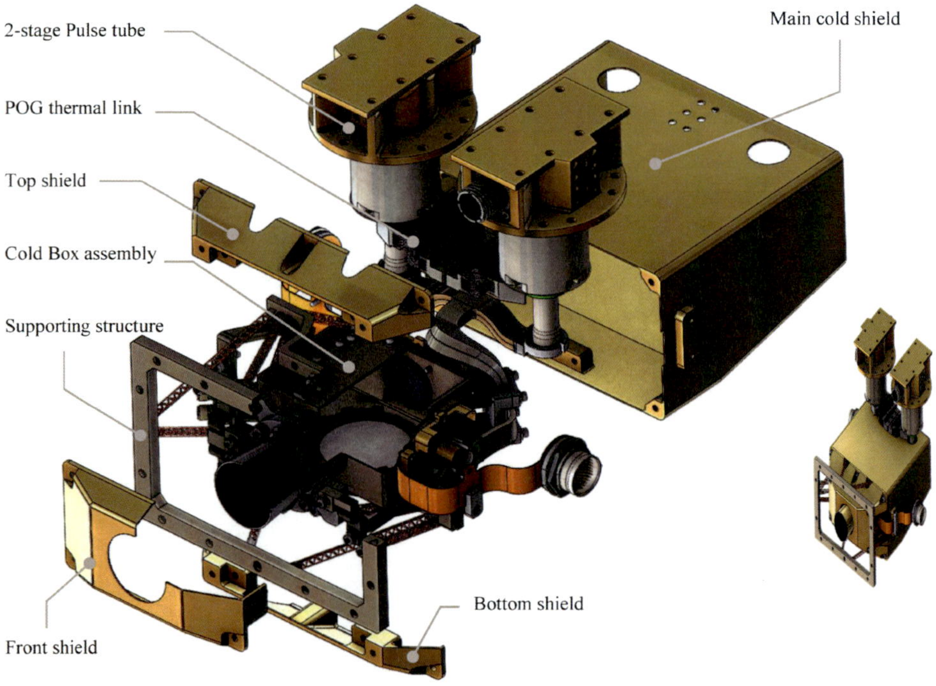

Figure 6. Exploded view of the optical bench assembly and assembled view in miniature

integration and alignment of the detectors. The thermal shield is attached mechanically onto the supporting structure intermediate interface. This intermediate interface is implemented between the warm flange at room temperature and the Cold Box in the 40 K temperature range. The back of the thermal shield is thermally coupled to the second stage of the Pulse tube cold head through POG thermal links. Around this thermal shield, MLI blankets are implemented to lower radiative heat losses.

The cryostat housing is made with several components. The main part is the housing of the cryostat and the other components are mainly the covers which ensure the vacuum tightness after assembly of all internal components. The cryostat housing is a massive part made with aluminium alloy. The use of titanium alloy is possible to improve the stiffness and to lower the mass. On the front side of the cryostat, the front cover is equipped with the entrance window. This cover is mounted on the same mechanical interface as the supporting structure to have the same mechanical reference for optical chain alignment (see Figure 6).

On the top and on the back, two covers are screwed after assembling of the sub-system inside the housing. The sides of the cryostat are used for hermetic feedthroughs. 4 hermetic feedthroughs have been implemented: 2 for the video harnesses and 2 for sensors and heaters. The 4 connectors are mounted on the cryostat using intermediate plates which are easier to change or adapt during development process than the re-machining of the cryostat.

THERMAL AND MECHANICAL ANALYSIS

To predict the thermal and mechanical performance of the target cryostat FEM, modelling has been run supported by an elementary breadboard tests. Two models have been developed to perform the validation of the design. The first is dedicated to the thermal modeling, including radiative modeling and the second one, more detailed, to check the mechanical behavior of the cryostats and particularly, the supporting structure (see Figures 7 and 8).

Using these models, the thermal analysis of the cryostat has been studied for different operating condition with the objectives to evaluate the parasitic heat losses and to validate the thermal balance with cryocooler capability.

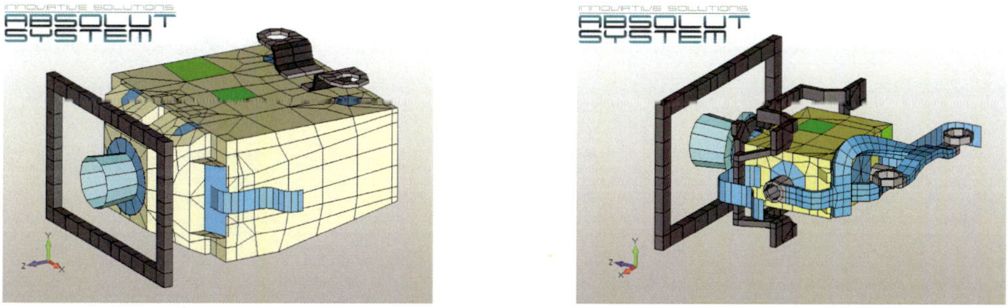

Figure 7. Thermal model of the cryostat

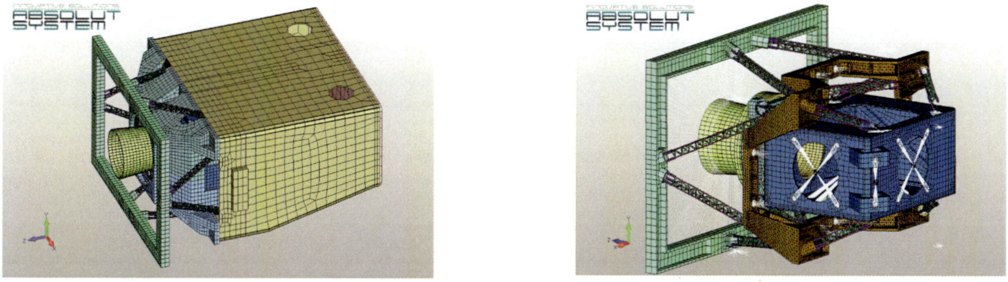

Figure 8. Overview of the meshing of the complete cryostat used for FEM analysis

In parallel, the 3-D lattice has been tested using breadboard test to evaluation the thermal conductance of the elementary structure. This test was important to decrease the uncertainty related to the contribution of the supporting structure on the conductive heat losses.

After several iterations, the thermal losses of the complete cryostat have been evaluated and are reported in the Table 1. We can see that a significant portion of the losses are from the redundant cold head. The other significant contribution is coming from the radiative load on the cold shield. Considering these outputs and the project margin philosophy, the final cooling capacity required on the cryocooler is: 1095 mW@ 42K and 2648 mW@126K.

Regarding the mechanical analysis, the cryostat and associated components has been modelled with a very refined model including a representative modelling of the 3-D lattice supporting structure. The objective was to evaluate the stress in the structure when the cryostat is submitted to mechanical environmental tests (sine and random loads) (see Figure 9). One critical parameter in the dynamic response prediction of structural components is the hypothesis made on the amplification factor. To give more consistency in our analysis, the supporting structure has been produced and submitted to sinus sweep at low level to refine our initial hypothesis (initial Qfactor = 50 and measured at 40). Following these considerations, the stress analysis in sinus and random reported positive margins on all the components. The stress in the elementary beams of the lattice contains a comfortable Margin of Safety.

MANUFACTURING AND ASSEMBLING PROCESS

The final configuration of the test cryostat has been simplified compared to the flight-like design. The aim of the tests is to validate the different performances and for this reason, the redundant cooler will not be implemented. Other simplifications have been made on the optical chain where dummy detectors equipped with temperature sensors and heaters will be used for the thermal test campaign or the entrance window which is replaced by a plain aluminium disk black coated. With the exception of these simplifications, we kept the design of the optical bench as close as possible to the flight-like design. The supporting structure, the thermal links, the cold shield and other components critical for performance validation are fully representative. The MLI blanket will be made with conventional MLI using spacer. This MLI should be replaced for flight with a blanket made with embossed foils far less critical in term of particulate contamination. Figure 10 shows the different components during their integration in the ISO5 environment.

Table 1. Summary of Thermal Losses Analysis

Heat fluxes in mW	**Intermediate stage 126 K**	**Cold stage 42 K**
Internal dissipation		
IR sensors	0	220.0
Conductive heat fluxes		
Supporting structure	147.0	99.6
Wiring - heaters	117.3	47.6
Wiring - sensors	43.3	4.9
Video harness	183.4	183.4
Redundant cold finger	1425.1	294.3
Radiative heat fluxes		
External radiation on the cold shield	434	-
Internal radiation with the cold shield	-	18
Radiation on the baffle	436	-
Radiation from the cold windows	-	15.6
Radiation on the Video Harness	30	30
Total of parasitic heat losses on each stage	**2816.1**	**693.4**
Total net heat load on each stage	**2122.7**	**913.4**

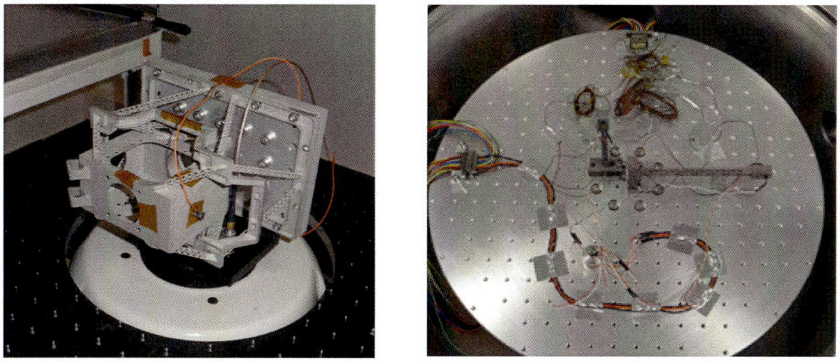

Figure 9. Elementary breadboard test: structural test of the supporting structure on the left and thermal conductance characterization of the elementary struts on the right

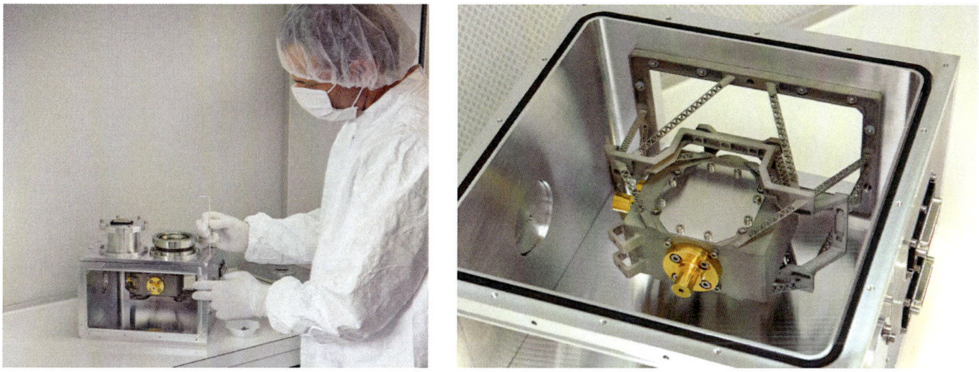

Figure 10. Picture of the optical bench during integration and integrated into the cryostat housing in Absolut System premises

CONCLUSIONS AND OUTLOOK

A flight-like cryostat has been designed by Absolut System and is under assembly prior testing. This cryostat is designed to take benefits of a high efficiency two-stage coaxial pulse tube cooler to operate infrared detectors in the 40 K temperature range with the same input power budget than conventional single stage coolers. An innovative supporting structure has been developed, manufactured and tested in order to offer a robust and efficient structure. A monolithic design with high thermal and mechanical performance has been designed which simplify the mechanical path and reduce drastically the number of parts and screwed interfaces. Most of the parts of this demonstration cryostat have been manufactured and are under assembly. The final test will run this summer after integration of the two-stage pulse tube cooler.

ACKNOWLEDGEMENTS

This development is performed in the frame of ESA Technical Research Program - 4000109933/14/NL/RA.

The authors thank ESA for their funding and technical supports during this development, and Thales Alenia Space for their technical supports regarding cryostat system engineering and peer review.

REFERENCES

1. T. Trollier, J. Tanchon, J. Buquet et al. « Design of a Large Heat Lift 40 to 80 K Pulse Tube Cryocooler for Space Applications," *Cryocoolers 14*, ICC Press, Boulder, CO (2007), pp. 75-82

2. S. Watson, "Design and development of a high power stirling cooler," *European Space Mechanisms & Tribology Symposium*, ESMATS 2013

3. P. Bruins, T. Prouvé, I. Charles, T. Trollier, J. Tanchon, J. Mullié, T. Benschop, « A 30-50 K Dual-Stage Pulse Tube Space Cooler," *Cryocoolers 18*, ICC Press, Boulder, CO (2015), pp. 33-37

4. Ross, R.G., Jr., "Cryocooler Load Increase due to External Contamination of Low-E Cryogenic Surfaces," *Cryocoolers 12*, Springer US (2003), pp. 727-736.

Design of Superconducting Gravity Gradiometer Cryogenic System for Mars Mission

X. Li[1], F. G. Lemoine[1], H. J. Paik[2], M. Zagarola[3], P. J. Shirron[1], C. E. Griggs[2], M. V. Moody[2], S. C. Han[4]

[1]NASA Goddard Space Flight Center, Greenbelt, MD 20771, USA
[2]University of Maryland, College Park, MD 20742, USA
[3]Creare LLC, Hanover, NH 03755, USA
[4]University of Newcastle, NSW 2308, Australia

ABSTRACT

Measurement of a planet's gravity field provides fundamental information about the planet's mass properties. The static gravity field reveals information about the internal structure of the planet, including crustal density variations that provide information on the planet's geological history and evolution. The time variations of gravity result from the movement of mass inside the planet, on the surface, and in the atmosphere. NASA is interested in a Superconducting Gravity Gradiometer (SGG) which will measure the gravity field of a planet from orbit. An SGG instrument is under development with the NASA Planetary Instrument Concepts for the Advancement of Solar System Observations (PICASSO) program, which will be able to resolve the Mars static gravity field to degree 200 in spherical harmonics, and the time-varying field on a monthly basis to degree 20 from a 255 x 320 km orbit. The SGG has a precision of two orders of magnitude better than the electrostatic gravity gradiometer that was used on the European Space Agency (ESA) Gravity field and steady-state Ocean Circulation Explorer (GOCE) mission. The SGG operates at the superconducting temperature lower than 6 K. The study presented in this paper designed a cryogenic thermal system to maintain the SGG at the design temperature in the above-mentioned Mars orbit. The system includes fixed radiation shields, a low thermal conductivity support structure, and a two-stage cryocooler. The fixed radiation shields use double aluminized polyimide to emit heat from the warm spacecraft into deep space. The support structure uses carbon fiber reinforced plastic, which has low thermal conductivity and very high strength at cryogenic temperature. The low vibration cryocooler has two stages, of which the high temperature stage operates at 65 K and the low temperature stage works at 6 K, and the heat rejection radiator works at 300 K. The study also designed a second option with a 4 K adiabatic demagnetization refrigerator (ADR) and two-stage 10 K cryocooler.

PLANETARY GRAVITY MAPPING MISSIONS

A planet's gravity field reflects the distribution of mass in its interior and across its surface. Thus a study of a planet's gravity field provides insights into the interior structure of the planet, allows

Cryocoolers 19, edited by S.D. Miller and R.G. Ross, Jr.

us to elucidate the thickness of its crust, and provides a means to better understand the geophysical processes that produced many features visible in the surface geology. A planet's gravity field also bears the imprint of the transport of mass across the surface and within the planet. In the case of Mars, the study of time-variation in the gravity field tells us about how a portion of the planet's mass is redistributed as the Martian polar caps form and dissipate seasonally.

There are three ways to measure a planet's gravity field using the spacecraft. The traditional method uses Doppler tracking of planetary orbiters via NASA's Deep Space Network (DSN), measuring changes in the spacecraft velocity as it flies over mass anomalies, such as mass concentrations or mass deficits[1,2]. The spacecraft is tracked at X band (~8 GHz) with a typical precision of 0.1 mm/s.

The second method uses satellite-to-satellite tracking of two spacecraft that follow each other in the same orbit by measuring the change in range between the two spacecraft as they orbit the planet. With these data, it is possible to estimate the characteristics of the gravity field underneath the spacecraft and how it changes. The U.S./German Gravity Recovery and Climate Experiment (GRACE)[3], and the NASA Gravity Recovery and Interior Laboratory (GRAIL)[4] spacecraft used highly accurate microwave ranging systems to measure the change in range between two identical spacecraft flying in a low polar orbit of the Earth and the Moon, respectively. Data from GRACE were used to construct a detailed map of the Earth's gravity anomalies so that scientists were able to detect changes in the distribution of water and ice across the planet, estimate changes in ocean bottom pressure, and detect changes in the gravity field due to other geophysical processes. The main objective of GRAIL was to understand the internal structure of the Moon and map the lunar crust and lithosphere[5]. The GRAIL data had a precision of 0.05 – 0.07 μm/s[4].

The third method to map a planetary gravity field is with a gravity gradiometer. A gravity gradiometer measures gravity changes across a short baseline. Gradiometers are used in terrestrial (airborne) gravity surveys, because the differential measurement means spurious accelerations are removed. Launched in 2009, the European Space Agency (ESA) Gravity Field and Steady-State Ocean Circulation Explorer (GOCE) mission included a highly accurate gradiometer to map the Earth's gravity field. GOCE used an electrostatic gravity gradiometer (EGG), which had a precision of 10^{-3} E/Hz$^{1/2}$, where 1 E (Eötvos) is a unit of 10^{-9} s^{-2}, which represents a unit of acceleration divided by distance that is used to describe gravity gradients. The GOCE data have been used in combination with a mean ocean surface defined from 20 years of ocean radar altimetry to compute the mean dynamic ocean topography, serving studies of the Earth's ocean circulation and transport. The data have also been used in studies of the Earth's lithosphere and mantle, as well as to map the gravity field of the Earth, where previously surface gravity measurements were hard to obtain[6]. The EGG measured the gravity gradients in three directions with a set of six three-axis accelerometers mounted in a diamond configuration. The on-board GPS receiver was also used as satellite-to-satellite tracking to supplement the EGG measurement. The EGG was about 100 times more accurate than any gravity measurement instrument in space.

SUPERCONDUCTING GRAVITY GRADIOMETER

NASA has been interested in Superconducting Gravity Gradiometer (SGG) since the 1980s. Unlike an EGG, an SGG uses a set of six two-axis accelerometers, and the SGG also uses a Superconducting Quantum Interference Device (SQUID) to measure extremely weak magnetic fields. Superconducting circuits offer high stability and sensitivity. Since levitation and sensing coils replace mechanical springs, the restoring force can be tuned to adjust the sensitivity of the instrument. The instrument can be tuned to provide sensitivity to either the short-wavelength gravity field (measuring the gravity expression of surface features at high resolution), or to provide sensitivity to the variations in the long-wavelength gravity field that are due to the effects of mass transport. For the high resolution field, a sensitivity of 10^{-4} E/Hz$^{1/2}$ is envisaged, whereas for the long-wavelength time-variable field, sensitivity better than 10^{-5} E/Hz$^{1/2}$ is expected. Thus, a single spacecraft and mission can accomplish two separate goals for mapping the gravity field of Mars. The SGG instrument is under development at the University of Maryland under a three-year NASA program to develop a prototype instrument that can be used on a mission to map the gravity field of Mars at high resolution[7]. Figure 1 shows the configuration of the SGG instrument.

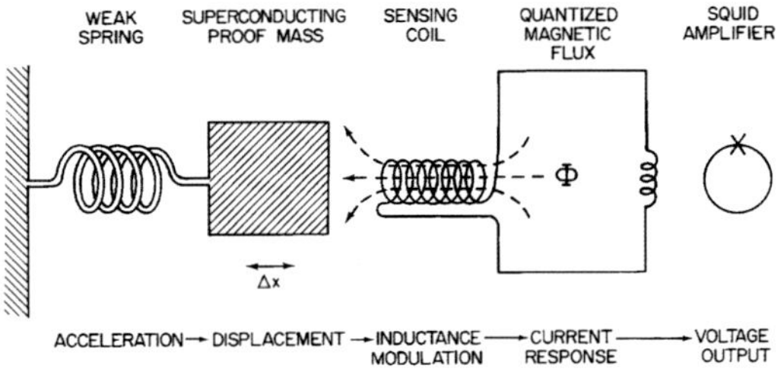

(a) Principle of superconducting accelerometer

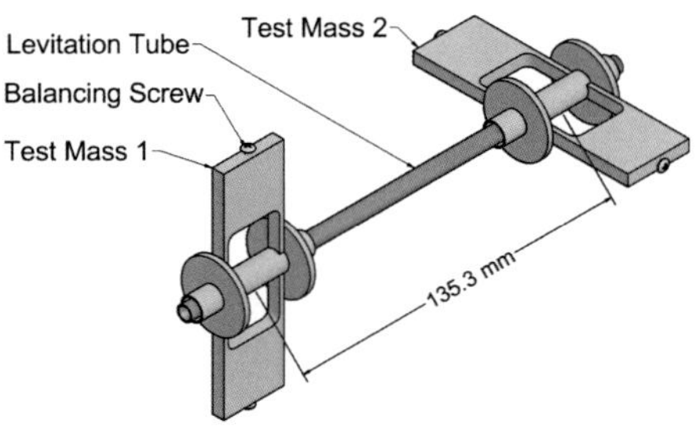

(b) Two test masses levitated around a tube

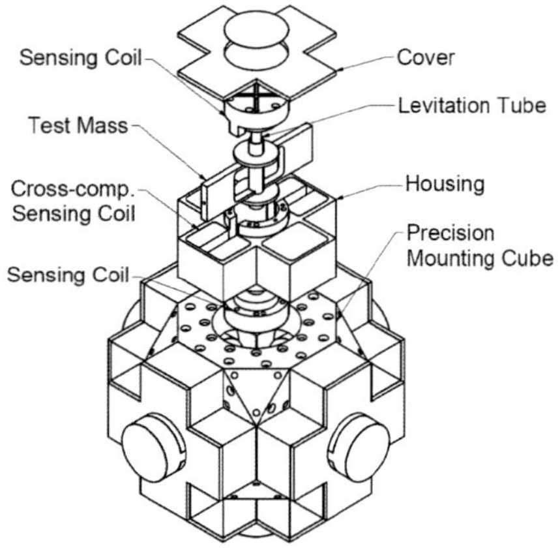

(c) Six test masses form a tensor gradiometer

Figure 1. SGG instrument.

The advantage of using an SGG is that a single spacecraft, flying at an altitude of about 200-225 km, will be able to double the resolution of the static gravity field at the least over that determined by using the standard method of Doppler tracking for an orbiting spacecraft[8]. In addition, the Mars time-variable field can be mapped on a monthly basis with a resolution of ~1500 km, whereas today with Doppler tracking, only the variations in the zonal spherical harmonic coefficients of the Mars gravity field, C_{20} and C_{30}, are resolvable[8].

This paper focuses on a preliminary design of a cryogenic system which can maintain the SGG instrument at the required temperature in Mars orbit. The orbit design, environment, and spacecraft configuration are also considered prior to the thermal study. Finally, design options, mass, and power budgets are provided for future mission design activities.

INSTRUMENT REQUIREMENTS

The SGG instrument, as shown in Figure 1, is about 20 cm in each direction and has a mass of ~10 kg. The driving requirements for this design are listed below.

1. The SGG operates with the highest efficiency at ≤ 4 K, and it operates with reduced accuracy at any temperature below 8 K.
2. In general, it takes 6-8 months for the spacecraft to travel from Earth to Mars, and another 5-6 months for insertion into the low-altitude mapping orbit using aerobraking. For example, this was the experience of the Mars Reconnaissance Orbiter (MRO).
3. The SGG should be precisely located at the center of gravity of the whole spacecraft.
4. The SGG instrument is sensitive to vibration. Linear acceleration from the spacecraft shall be less than 10^{-6} m/s^2/Hz$^{1/2}$, and angular acceleration shall be less than 10^{-5} rad/s^2/Hz$^{1/2}$ during the operation. The instrument measures at frequencies below 10 Hz.

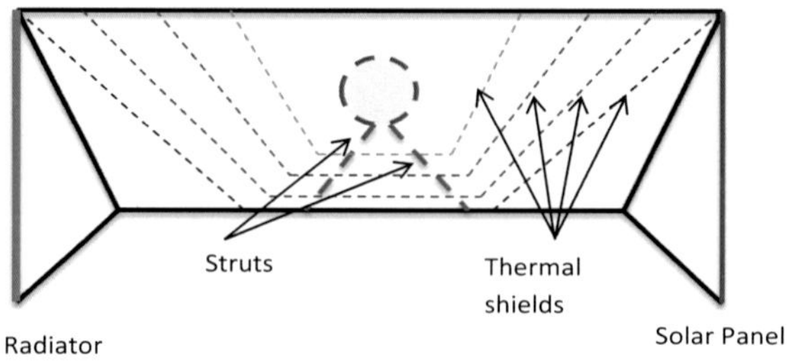

Figure 2. SGG spacecraft cross-section.

5. The operating orbit shall be sun-synchronous, facing the sun to minimize thermal variations, and the mean altitude is assumed to be 200 km.
6. The mass of spacecraft shall be less than 500 kg, and the power shall be less than 300 W.

SPACECRAFT CONFIGURATION

We used the thermal shield design for the Primordial Inflation Explorer (PIXIE)[9] astronomy mission as the basis for the SGG[10] thermal design. The SGG instrument and thermal shields are embedded inside the spacecraft to satisfy requirement 3. The subsystems of the spacecraft will be positioned to meet this requirement.

The cross section of the spacecraft is narrow in the flight direction to minimize the drag force, as shown in Figure 2. Although the Mars atmosphere is more tenuous than that of Earth, it can still exert a drag force on the orbiting spacecraft. Orbits lower than 200 km degrade quickly, especially near the Mars perihelion, when the upper atmosphere density increases due to the melting of the Mars ice caps just prior to the southern hemisphere summer. The shape of the cross section is designed to accommodate the shape of radiation shields, which will be discussed in the following section. All the shields face deep space away from the sun, since the spacecraft is assumed to be in a near dawn-dusk orbit.

The solar panels have been sized to generate 300 W of power at Mars. A radiator is attached to the opposite side of the solar panel. The radiator removes the heat generated by the electronics and cryocooler at close to room temperature, 300 K. The bottom of the spacecraft, which faces Mars, uses high thermal conductivity material to transfer the heat from solar panel to the radiator. The top of the spacecraft will be made of lightweight carbon fiber material, and thermally isolated from the spacecraft. The carbon fiber material can be used as a higher temperature radiator to further reduce the parasitic heat to the instrument.

CRYOGENIC THERMAL SYSTEM

The objective of the cryogenic thermal system is to maintain the instrument at operating temperatures. In this study, we studied both 6 K and 4 K cryogenic system options for the instrument. The cryogenic system includes thermal shields, low thermal conductivity support structure (struts) and cryocooler(s).

Thermal model

Thermal Desktop® was used to perform a thermal analysis for the cryogenic thermal system design and optimization[11]. Thermal Desktop® discretizes the surfaces into surface elements in preparation for the use of a finite difference method. RADCAD, which is one of the add-on programs, uses the Monte Carlo method to calculate the thermal radiation between surface elements. Thermal conduction between elements is calculated using SINDA. The thermal analysis calculates the radiation and the conduction heat load on the instrument. The thermal model includes the spacecraft body, solar panels, radiator, wiring, struts, radiation shields, and radiator, as shown in Figure 3.

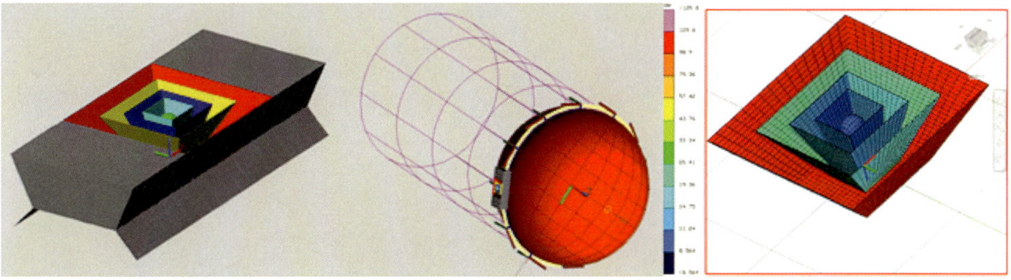

(a) Spacecraft configuration (b) On orbit (c) Thermal shield

Figure 3. SGG spacecraft.

Parametric studies were performed with different spacecraft configurations, strut sizes and materials, and heat interception points, at which locations the cryocooler removes heat. It was found that the heat transfer by thermal conduction through the wiring and struts is much higher than the heat transfer by radiation. The materials, number of wires, and conductive heat loads are listed in Table 1. More analysis has been done on how to minimize the radiation heat load.

Instrument shield

The instrument, as shown in Figure 1, has a large surface area. A spherical aluminum shield, as shown in Figure 3, is used to reduce the surface area and thermal radiation from the environment. The shield can also protect the instrument from environmental hazards throughout all phases of the mission. The shield is actively cooled by the cryogenic thermal system. The aluminum shield conducts heat, and it creates a uniform temperature boundary for the instrument. The instrument shield is covered by 40 layer multi-layer insulation (MLI) to reduce exposure to radiation from space and thermal shields.

Thermal shielding

There are four layers of thermal shields, which protect the instrument from solar radiation, Mars infrared (IR) radiation, as well the spacecraft IR radiation. All shields face deep space, which is assumed to have a temperature of 7 K. Each layer of the shields, as shown in Figure 3, has four sides to accommodate the spacecraft configuration. It can be modified to six sides or circular in future studies. The shields are constructed by using double aluminized Kapton (DAK). Lab tests of the radiation properties and shield design were performed by NASA GSFC[12]. The angle to the vertical of layer one is 45°, and reduced 5° for each layer to 30° for shield four. Layer one is mounted on the spacecraft and MLI is used to further reduce the heat loads from the spacecraft bus (assumed to be at room temperature) and IR radiation from Mars. Layer two, three and four are mounted on the hexapod struts. The shields will be launched as shown in the design, and use T10 to stabilize all edges. Additional T10 support tubes can be used locally, if any area of the shields cannot withstand the launch load after a detailed stress analysis. The thickness of the DAK is about 25 microns. Therefore, the thermal conduction can be neglected, and thermal radiation dominates the shield design. The sides of layer two facing deep space are painted black, which is used as a radiator to remove heat at 150 K from struts. Thermal Desktop® is used to predict the thermal performance of the shields. The temperatures of the different layers of shields are listed in Table 2.

Table 1. Wiring material and heat load.

Material	Number of wires	Diameter (AWG)	Heat load	Heat load	
			6 K only (mW)	68 K (mW)	6 K (mW)
Manganin	144	36	8.0	463.0	0.7
Phosphor bronze (PhBr)	50	36	53.7	193.0	5.1
	23	28			
	10	22			
Stainless steel	576	36	22.1	70.0	7.7

Table 2. Temperature of the shields (thermal model results)

Shields	Temperature (K)
Shield 4 (outer layer)	171
Shield 3	70
Shield 2	28
Shield 1 (inner layer)	10

Struts

The hexapod struts use carbon fiber reinforced plastics (CFRP). CFRP is used on the JWST MIRI instrument, which is at 7 K[13]. It has higher yield strength and modulus of elasticity than the other support materials used for cryogenic system design, but has low thermal conductivity at cryogenic temperature, which significantly reduces the thermal conduction to the low temperature instrument. The thickness, length and the angle between pod symmetry, as shown in Figure 4, are designed based on a simple stress analysis. The minimum natural frequency of the struts is higher than 100 Hz, and the safety factor is greater than two for a launch load of 10 *g*.

Active cooling

There are a couple of ways of maintaining the instrument operating temperature. A helium cryostat can be used for cooling the SGG. However, the Mars mission lasts significantly longer than an Earth mission, because of the distance to Mars and the time needed for orbit insertion. Helium venting will also affect of the spacecraft center of gravity, which fails to meet requirement 3. The other way to cool the instrument is to use cryocoolers. Most of the mechanical cryocoolers generate vibration, which cannot meet requirement four.

Low-frequency vibration creates noise to the instrument, which causes errors in the measurement. Turbo-Brayton cooler is considered to be a vibration-free cooler[14] for space applications. This type of cooler was used to replace NICMOS dewar for the Hubble Space Telescope (HST)[15]. The NICMOS cooler operated at 70 K, and provided continuous cooling of 7 W to HST. Low temperature turbo-Brayton coolers can provide hundreds of ~0.1 W of cooling at 6 K[16]. A schematic of the Brayton cycle cooler is shown in Figure 5. The compressors and turboalternators operate at hundreds of thousand rpm; therefore, the vibrations generated by these components are at the frequency of thousands of hertz, which is much higher than the frequency band of the SGG instrument. Flow turbulence in the working fluid (low pressure helium gas) can cause pressure fluctuations at frequencies below 10 Hz, which can also cause mechanical vibrations in the transfer tubes between components. The most recent study[17] shows that the mechanical vibrations in the transfer tubes can be managed to satisfy the SGG vibration requirement.

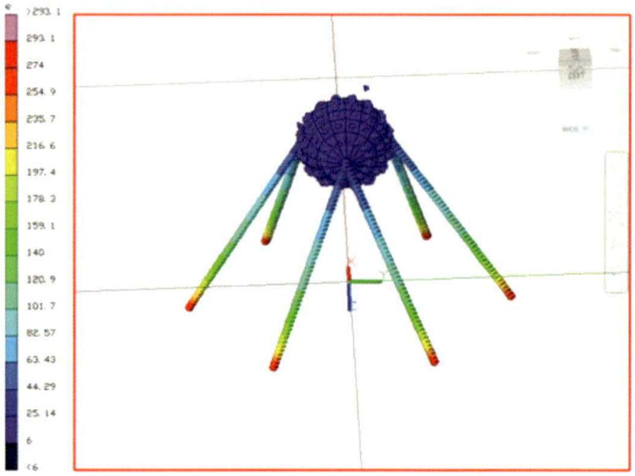

Figure 4. Temperature of the SGG support structure (K).

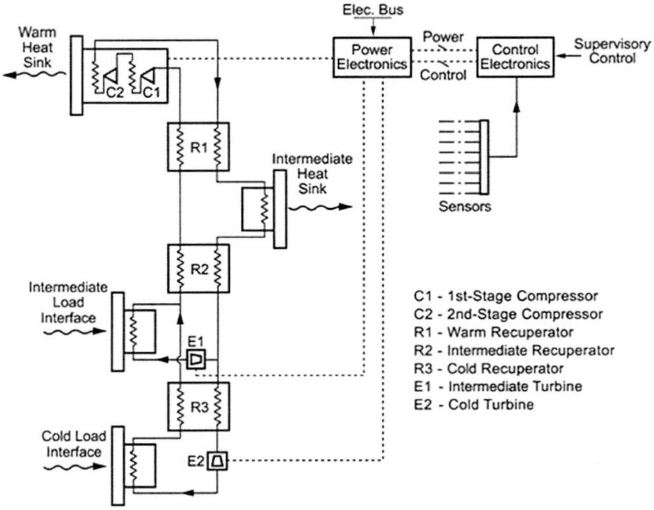

Figure 5. Reverse Brayton cycle cooler.

An adiabatic demagnetization refrigerator (ADR) can cool an instrument to superconducting temperature without vibration. ADR uses a material's magnetocaloric effect, in which a temperature change in the material is caused by exposing the material to a changing magnetic field, to remove heat with an adiabatic thermal boundary condition. ADR works more efficiently than mechanical coolers at temperatures below 10 K. NASA GSFC conducted a study of 10 K ADR[18], which uses a 10 K heat sink to cool the instrument to 4 K. Figure 6 shows an example of a 4 K ADR[19].

In this study, two options for cryocoolers are considered. One option is to use a two-stage turbo-Brayton cryocooler to cool the instrument to 6 K. The higher temperature stage removes conductive heat from struts at 68 K, and the lower temperature stage maintains the SGG instrument at 6 K. The second design is to use a two-stage 10 K Brayton cycle cooler with a 68 K upper stage and an ADR to maintain the instrument at 4 K with a 10 K heat sink.

The higher temperature stage at 68 K intercepts heat on struts at a cooling point. The location of the cooling point was calculated using the 1-D thermal conduction equation, and then verified using the Thermal Desktop® model. The 68 K cooler also removes heat from wiring to reduce thermal

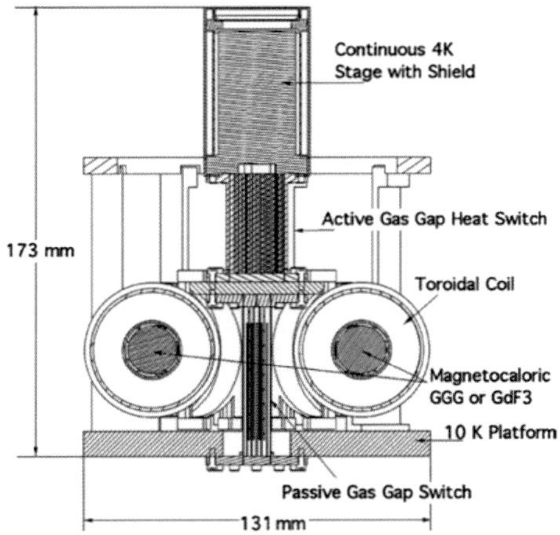

Figure 6. 10 K to 4 K ADR

Table 3. Comparison of 6 K and 4 K cryocoolers.

	Option 1		Option 2	
Instrument temperature	6 K		4 K	
Cooler(s)	Reversed Brayton 68 K/6 K two-stage cryocooler		Reversed Brayton 68 K/10 K two-stage cryocooler with ADR from 10 K to 4 K	
Cooling load	68 K	1500 mW	68 K	2888 mW
	6 K	100 mW	10 K	115 mW
			4 K	10 mW
Power	68 K 6 K	120 W	68 K 10 K	193 W
			4 K	50 W
Mass	26 kg		36 kg	

conduction through the wiring. Specifications of Brayton cycle cryocoolers and ADR for options one and two are listed in Table 3. The cooling loads are estimated based on the thermal analysis. A margin of 100% is applied to both options.

SUMMARY

This paper presents a preliminary thermal design of the cryogenic thermal system for an SGG Mars mission. The system includes thermal shields, a low thermal conductivity support structure, and cryocoolers. The design also defines the size of the thermal shields and the size of struts. Two options were given for the active cooling. One uses a 6 K/68 K two-stage Brayton cycle cryocooler, and another uses a 10 K to 4 K ADR and 10 K/68 K two-stage Brayton cycle cryocooler.

REFERENCES

1. Lemoine, F. G., Smith, D. E., Rowlands, D. D. *et al.*, "An improved solution of the gravity field of Mars (GMM-2B) from Mars Global Surveyor," *J. Geophys. Res.* 106 (E10) 23359-23378, 2001, doi: 10.1029/2000JE001426.

2. Mazarico, E., Genova, A., Goossens, S., Lemoine, F.G. et al, "The gravity field, orientation and ephemeris of Mercury from MESSENGER observations after three years in orbit," *J. Geophys. Res.* 119(12), 2417–2438, 2014, doi:10.1002/2014JE004675.

3. Tapley, B. D., Bettadpur, S. Ries, J. C., Thompson, P. F., and Watkins, M. M., "GRACE measurements of mass variability in the Earth system," *Science* 305, 503-505, 2004, doi:10.1126/science.1099192.

4. Asmar, S. W., Konopliv, A. S, Watkins, M. M. et al., "The scientific measurement system of the Gravity Recovery and Interior Laboratory (GRAIL) mission," *Space Sci. Rev.* 178(1), 25-55, 2013, doi: *:10.1007/s11214-013-9962-0.*

5. Zuber, M. T., Smith, D. E., Lehman, D. H. et al., "Gravity Recovery and Interior Laboratory (GRAIL): Mapping the lunar interior from crust to core," *Space Sci. Rev.* 174*, 3-24, 2013, doi:10.1007/s11214-012-9952-7.*

6. Floberghagen, R., Fehringer, M., Lamarre, D., Muzi, D., Frommknecht, B., Steiger, C., Piñeiro, J. and Da Costa, A., "Mission design, operation and exploitation of the gravity field and steady-state ocean circulation explorer mission," *J. Geodesy* 85, 749-758, 2011, doi: 10.1007/s00190-011-0498-3.

7. Griggs CE, Paik HJ, Moody MV, Han SC, Rowlands DD, Lemoine FG, Shirron PJ, Li X., "Tunable superconducting gravity gradiometer for mars climate, atmosphere, and gravity field investigation," InLunar and Planetary Science Conference 2015 Mar (Vol. 46, p. 1735). (http://www.hou.usra.edu/meetings/lpsc2015/pdf/1735.pdf).

8. Genova, A., Goossens, S., Lemoine, F. G. et al. (2016), "Seasonal and static gravity field of Mars from MGS, Mars Odyssey and MRO radio science," *Icarus* 272, 228-245, 2016, doi:10.1016/j.icarus.2016.02.050.

9. DiPirro, M., Fixsen, D., Kogut, A., Li, X., Marquardt, J., Shirron, P., "Design of the PIXIE cryogenic system," *Cryogenics* 52 (2012), 134-139.

10. KKogut, A., Chuss, D.T., Dotson, J.L., Dwek, E., Fixsen, D.J., Halpern, M., Hinshaw, G.F., Meyer, S.S., Moseley, S.H., Seiffert, M.D. and Spergel, D.N., "The Primordial Inflation Explorer (PIXIE)," *SPIE Astronomical Telescopes and Instrumentation Conference,* June 22-27, 2014, Montreal, Canada.

11. Panczak, T.D., Ring, SG Welch, MJ Johnson, DA Cullimore, BA and Bell, DP *Thermal Desktop® User's Manual, A CAD Based System for Thermal Analysis and Design, Version 5.6*, C&R Technologies, Boulder, CO (2013).

12. DiPirro, M, Tuttle, J., Ollendorf, S., Mattern, A., Leisawitz, D., Jacksonk, M., Francis, J., Hait, T., Cleveland, P., Muheim, D., Mastropietor, A. J. "High fidelity cryothermal test of a subscale large space telescope," *Proc. SPIE*, vol. 6692, 669202.1-669202.9, 2007

13. Makowski, K. P., Larson, M. E., Loc, A. S., Zhang, B. X., Leland, R. S., Hayashi, B., "Development of a low heat leak CFRP stand for MIRI cooler JT heat exchanger stage," *AIP Conf. Proc.* **1218**, 10-15, 2010, doi: 10.1063/1.3422260.

14. Neillis, G., McCormick, J., Sixsmith, H., Swift, W., "Turbo-Brayton cooler for a range of loads and temperatures," *Proc. SPIE,* vol. 3436, 1998, pp. 813-823

15. Swift, W. L., McCormack, J. A., Zagarola, M. V., Dolan, F. X., Sixsmith, H., "The NICMOS turbo-Brayton cryocooler - two years in orbit," *Cryocoolers 13*, 633-639, 2004.

16. Zagarola, M. V., Breedlove, J. J., McCormick, J. A., Swift, W. L., "Turbo-Brayton cryocooler technology for low temperature space applications," *Proc. SPIE IR Space Telescopes and Instruments,* vol. 4850, 2003, pp. 1029-1037.

17. Cragin, K., Rule, K., Willbur, J., McCormick, J., Mellinger, S. Zagarola, M., "Characterization of Emitted Vibration from Turbo-Brayton Cryocoolers," *Cryocoolers 16*, ICC Press, Boulder, CO (2011),

18. Tuttle, J., Pourrahimi, S., Shirron, P., Canavan, E., DiPirro, M., Riall, S., "A 10-K magnet for spaceflight ADRS," *Cryogenics 44* (2004), pp. 383-388.

19. DiPirro, M., Canavan, E., Shirron, P., Tuttle, J., "Continuous cooling from 10 to 4 K using a toroidal ADR," *Cryogenics 44* (2004), pp. 559-564.

Flexible Thermal Link Assembly Solutions for Space Applications

T. Trollier, J. Tanchon, J. Lacapere, P. Renaud, J.C. Rey and A. Ravex

Absolut System SAS
Seyssinet-Pariset, France

ABSTRACT

Absolut System is designing and producing Thermal Link Assemblies (TLA) to be used on space observation programs including CNES IASI-NG and MTG ESA programs.

TLA have the following main functions: to ensure a high conductive coupling between both cryocoolers' cold tips (nominal and redundant) and detectors or cold optical bench, to have a reduced stiffness allowing misalignment and relative dynamic displacement between cold tips and detectors, and to filter micro-vibration coming from cryocooler cold tips to the detectors. The TLA should also be compliant with stringent constraints which are common for space products as follows: to have a reduced mass, to stay inside the static and dynamic Interface Requirement Document (IRD) reduced volume, to be compliant with the cleanliness requirements imposed by the detector proximity, and to survive without performance degradation from the launch loads and thermal cycling.

This paper will present different technical trade-offs performed on the material candidates and production constraints. Current thermal, mechanical, and cleanliness performance of TLA FMs (Flight Model) made of 5N (99.999) high purity aluminum foils and OFHC copper foils. Several on-going TLA designs and performance will be presented, including a TLA made of Pyrolytic Oriented Graphite (POG) foils developed for a 2-stage cryostat (presented in a companion paper [1])

INTRODUCTION

Absolut System is producing cryogenic Thermal Link Assemblies (TLA) for European space observation missions flying pulse tube cryocoolers. TLA are critical equipment in the overall cryostat performance as they shall provide the main following functions:

- Ensure a high conductive coupling between both cryocoolers cold tips (nominal and redundant) and detectors or cold optical bench.
- Have a reduced mass.
- Allow misalignment and relative dynamical displacement between cold tips and detector.
- Stay inside the static and dynamic volumes of the IRD.
- Filter micro-vibration coming from cryocooler cold tips toward detectors.
- Be clean.
- Survive without performance degradation to the launch loads and to the thermal cycling.

	Thermal conductivity (W/m.K)				
	Aluminum grade				Copper
T (K)	1100	1050	4N	5N	OFHC
50	369	425	948	1087	1173
60	338	389	645	695	816
70	308	354	484	507	646
80	283	326	390	401	558
Purity %	99	99.50	99.990	99.999	99.99

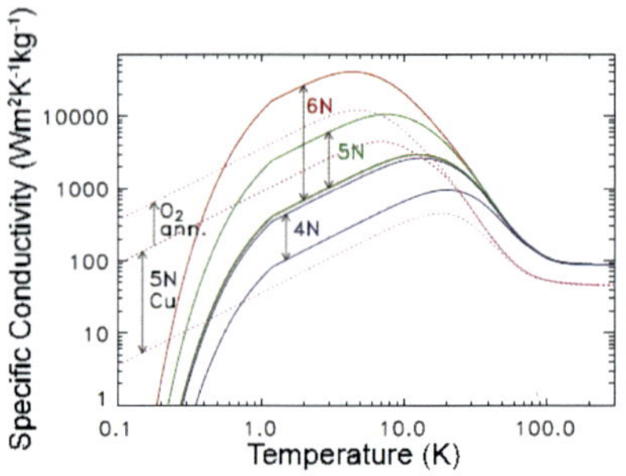

Figure 1. Thermal conductivities (left) and thermal conductivity versus density ratio (right) of OFHC copper and various grades of aluminum as function of the temperature.

MATERIAL TRADE-OFF FOR 50-80 K CRYOGENIC APPLICATIONS

For cryogenic applications of TLA below 80 K, two materials can be selected: high purity aluminum or OFHC copper. As shown in the Figure 1 [2], in the 50-80 K temperature range of interest, OFHC copper offers the highest thermal conductivity.

However, looking at the ratio of thermal conductivity to density, pure aluminum gives a considerable advantage over copper material for on-board applications. High purity aluminum appears the best candidate for TLA attached to the cryocooler cold tip where the mass constraint is very stringent.

Furthermore, in the 50-80 K temperature range, the thermal conductivity of aluminum is less sensitive to the purity, and thus, Al4N or Al5N grades can be selected (Al6N being less common and quite expensive).

Thin foils of Al5N can be sourced with thickness ranging from 50 to 100 μm. The procurement can be made with several hundred meters long foil wounded in coil format.

Since aluminum is more difficult to assemble than copper, high vacuum Electron Beam (EB) welding process is selected (Table 1) because it induces a perfect mechanical and thermal continuity between the foil material and the end-fittings without the use of external filling material, that are generally a source of additional thermal resistances.

Table 1. Feasibility study of assembling processes according to the foils and fittings materials.

Processes available for foil assembly	Aluminum foils and aluminum fittings	Copper foils and copper fittings	Copper foils and aluminum fittings
TIG/MIG	**Good** but not critical on foils	**Good**	**Not feasible**
Press-welding	**Not feasible**	**Very good** no thermal resistance between foils	**Not feasible**
Diffusion welding	**Feasible** with force applied during process	**Good** no deformation and good performances	**Feasible** with force applied during process
Laser welding	**Feasible** but spot welding only	**Feasible** but spot welding only	**Not feasible**
EB welding	**Very good** no thermal resistance between foils	**Very good** no thermal resistance between foils	**Not feasible**
Brazing	**Good** with additional thermal resistance	**Good** with additional thermal resistance	**Good** with additional thermal resistance
Swaging	**Good** with additional contact resistance between foils	**Good** with additional contact resistance between foils	**Good** with additional contact resistance between foils

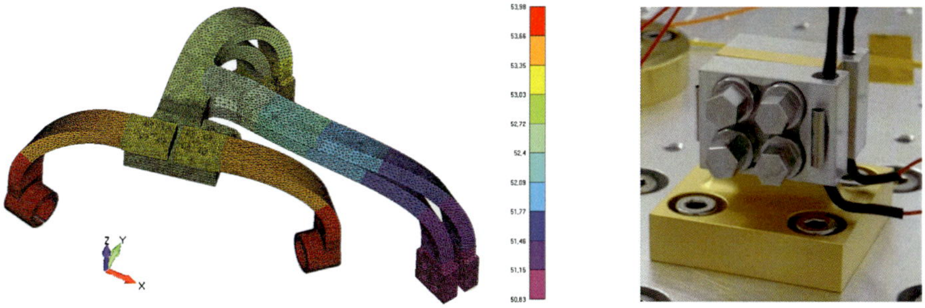

Figure 2. Illustrations of a FEM thermal model (left) and thermal contact conductance characterization at cold tip interface (right).

THERMAL AND MECHANICAL ANALYSES DURING DEVELOPMENT PHASE

Thermal and mechanical performances of the TLA always result in a trade-off of competing requirements. Thus some compromises shall be performed between the thermal performance (primary aim of the TLA) and other mechanical requirements such as stiffness, mass, and robustness to the loads which all depend on the routing of the TLA foils.

Thermal Analyses

An iterative thermal modelling phase is performed taking into account the thermal conductance and AIT requirements for various routed designs. For the thermal analyses, we use FEMAP software with TMG thermal module (MAYA). Thermal analyses are coupled with prototyping mainly performed on the thermal contact conductance measurements at cold tips and detector dismountable bolted interfaces (Figure 2). Thermal contact conductances are then characterized as function of the cryogenic temperature, screws tension, fillers (such as gold foils), mounting and dismounting and thermal cycling impacts.

Mechanical Analyses

The support of FEM analysis is limited during the development phase due to the difficulty to model the mechanical behavior of the flexible TLA composed of a stack of hundreds of foils. For the stiffness optimization, we use prototyping for various routings of the foils stack as illustrated in the Figure 3.

Dynamic displacement tests are also performed to complement the stiffness characterization. During this test, one extremity of the thermal link is fixed while the others are submitted to dynamic displacement in the three axes, from few tenths of a millimeter up to several millimeters. This dynamic test is quite easy and is performed in house due to the low shaker capacity required. This allows for fast reactivity during the design phase.

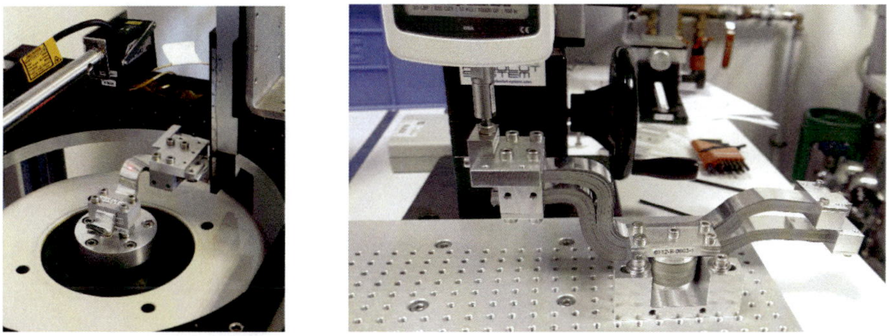

Figure 3. Illustrations of a cumulated dynamic relative displacements tests (left) and stiffness characterization (right) on prototyped routing of foils stack.

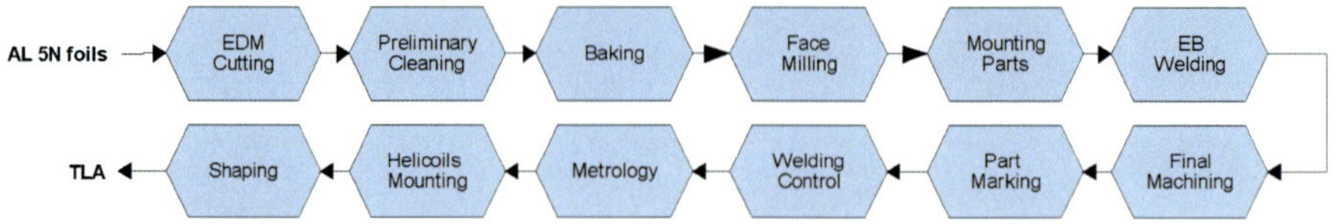

Figure 4. Manufacturing flow chart of aluminum TLA

TLA PRODUCTION

Aluminum TLA Manufacturing

The manufacturing of an aluminum TLA consists of a succession of several processes as presented in the schematic manufacturing flow chart attached in the Figure 4. A brief summary of the processes is given as follows:

- Aluminum 5N foils are cut to the dedicated shape by Electron-Discharge Machining (EDM) process. To do so, a stack of foils is pressed between two masks made of thick stainless steel. The masks are reused for the mass production. The cutting is performed by an external supplier with an extra length at the end fitting sides.
- After the cutting, the foils are cleaned one-by-one in a bath with isopropanol alcohol.
- The cleaned foils are baked at 140 °C in a furnace under vacuum environment.
- Al5N cut foils with extra length are mounted in dedicated U shaped Al1080 support and the stacking is ensured by a dedicated bolted assembly. In-house high speed face milling process is used to remove the extra length of the Al5N foils so that a perfectly planed surface is produced in the weld joint.
- Then perfect contact is ensured between the machined face of the Al5N foil stack and the massive Al1080 block prior to the welding step.
- An external qualified EB welding process is used to weld the Al5N foil stack to the Al1080 blocks.
- The final end-fitting interfaces (cold tips, detector assembly or optical bench interfaces) are machined into the massive Al1080 blocks.
- All the parts are marked for identification. The marking is a machining process performed by the supplier insuring also the final machining.
- After the production tests described below, helicoils are implemented, if required, on dedicated machined end-fittings.
- Finally, the produced TLA is shaped by Absolut System according to dedicated manufacturing drawings (example of shaping is attached in Figure 5).

Copper TLA Manufacturing

The manufacturing of an OFHC copper thermal link is performed with a similar process to the aluminum material thermal link. The technology is the same because thin copper foils are stacked together, cut to shape and welded at end-fittings interfaces.

The main difference is that the press-welding process is used instead of EB welding. With aluminum, the foils are welded on a massive end-fitting made with Al1080 while with press-welding process, the end-fittings are constituted directly with the foils themselves.

The foils are pressed and a current is applied into the foils in the contact area. With the resistivity of the copper, the current heats up the material near the fusion point (in the range of 965°C) and the foils are welded together without contact resistance or additional filling materials. This process is performed under an inert gas environment to prevent oxidation of the copper. Figure 6 shows an overview of the press-welding process during operation. The press-welding process is very good for the thermal performance (no thermal resistance between the foils) and for mechanical homogeneity (conservation of the material properties in the junction area) due to molecular connection reached after press-welded process. In our case, after the cutting of the foils to the required shape, the foils are stacked and placed between two OFHC copper plates increasing the thickness of the end-fittings. This stack is press-welded and the final shape of end-fitting is realized after the final machining.

Figure 5. Examples of Al5N and OFHC cooper TLAs produced by Absolut System.

Production Testing

For the EB welding control a nondestructive inspection (NDI) tomography process is used. This control consists of an X-ray source to visualize the internal material characteristics of the end-fittings at the weld location. This control is used to validate the integrity of the welding. The definition is in the range of 100μm, and it is possible to estimate with this method a maximal porosity of the welding. In order to maximize the definition, this process is performed after the final machining of the welded end-fittings.

The metrology control is a standard metrology process used in order to ensure the tolerances of the machined interfaces according to the manufacturing drawings are respected. Depending on the manufacturing flow chart, some protective tubes tool are used after this step to protect the machined interfaces to accomplish the following steps.

Qualification and Acceptance Testing

A typical flow chart of an acceptance test is attached in the Figure 7. Cleanliness control is measured during the test sequence. Physical properties (such as mass), electrical resistivity, mechanical stiffness, and static/dynamic displacements measurements are performed under ISO 5 class laminar hood at Absolut System premises.

For the thermal conductivity measurement, TLA specimens are mounted and encapsulated in metallic housing under ISO 5 environment (Figure 8). Protective housing set-up are implemented prior to integration in the thermal test bench and are removed after test also under ISO 5 clean hood. The same test bench and precautions apply for the thermal cycling tests.

For the sine and random (and shock if applicable) tests, the TLA specimens are mounted on JIGs under ISO 5 clean hood and encapsulated in polyester housing closed by Kapton tape. Those mechanical tests are performed externally to Absolut System. Protective housing are removed after mechanical tests under ISO 5 clean hood.

Particularly for the TLA Qualification Models, thermal conductivity and stiffness measurements are performed before and reproduced after the application of stresses (sinus, random, accumulated displacements, thermal cycling) to track any change in performance.

Figure 6. Example of press-welding process.

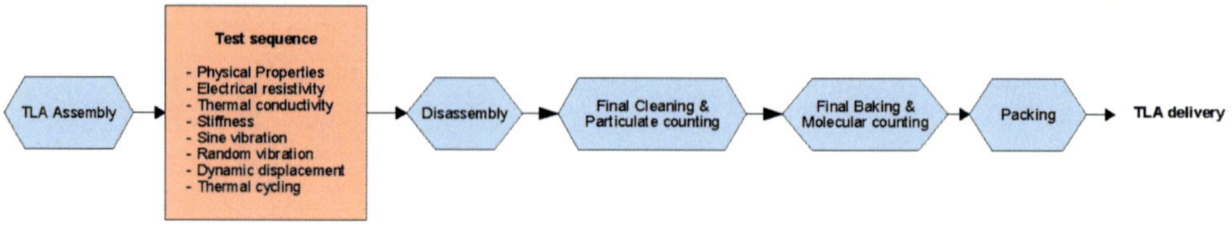

Figure 7. Typical acceptance flow chart of TLA.

The final cleaning process is performed after the completion of the complete test sequence, before the contamination measurement. This process is performed in ISO 5 environment by Absolut System. This process consists of:

- Ultra-Sonic (US) bath with Isopropanol (IPA),
- US bath with de-ionized water,
- Cleaning with particulate counting with de-ionized water.

After final cleaning, a sample extraction is made by running the TLA specimen entirely in a filtered bath of IPA with manual agitation. This bath is then analyzed internally by a specific HIAC 8011+ type analyser from Beckman Coulter able to count particles from class 5 (> 100 μm) down to class 1 (5 μm – 15 μm).

After the final cleaning with US bath using IPA and de-ionized water, a bake-out procedure is applied in order to remove all volatile components present on the specimen's surface. This bake-out process is done in a specific furnace able to heat the system up to the required value during a long period (min. 72 hours) in vacuum (typically < 10^{-5} mbar).

For the molecular counting, the analyses are performed with a Gas Phase Chromatography coupled with mass spectrometry (GC / MS). The sample is extracted by wiping process (according to ECSS-Q-ST-70-05C). Tissue is wetted by methanol and wiping is performed on the TLA surfaces. This sample extraction is done at Absolut System premises under ISO 5 environment and then sent to external laboratory for molecular analysis.

For the packing, a double bag approach with nitrogen flushing is used. This double packing is performed under class ISO 5 and flushed with dry Nitrogen. Desiccant is used in the second bag. Along the MAIT process, components are stored in a dry cabinet (humidity controlled to a value lower than 1 %).

TYPICAL TLA PERFORMANCES

Each project flying TLA comes with its specific set of user requirements, so only typical performances can be listed here. As primary function of the TLA, the thermal conductance shall be maximized giving the overall dimensions and mass constraints. Typical conductance of aluminum

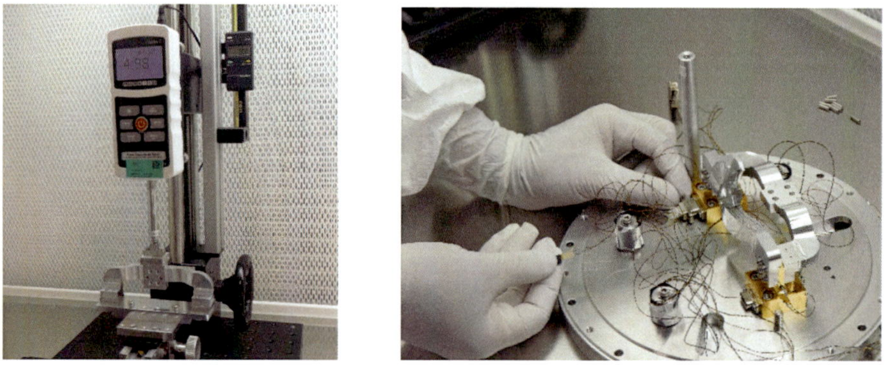

Figure 8. Picture of TLA during stiffness measurement and TLA installation on dedicated cold plate for thermal conductance characterization.

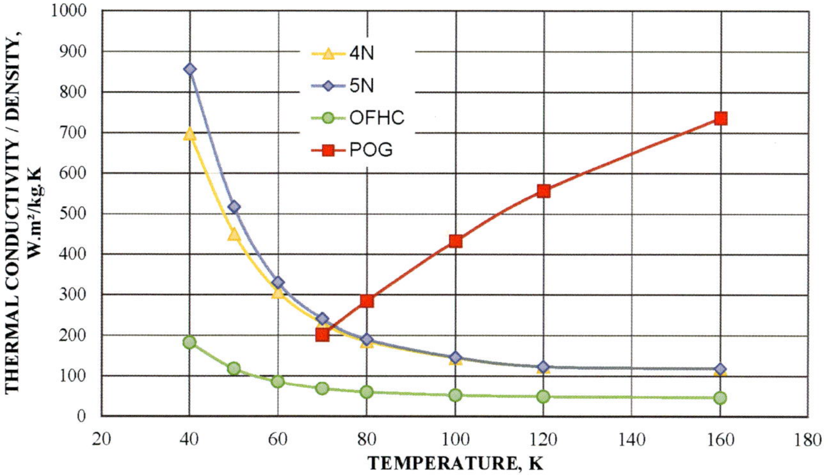

Figure 9. Ratio between thermal conductivity and density for pure aluminum, OFHC copper and POG

FM TLA, including contact interfaces, are in the range of 0.5 to 0.9 W/K in the 50-70K temperature range. Thermal performance achieved are now predictable within less than 10%. Typical overall mass budget allowed is in the range of 200 to 400 grams. The more stringent budget specified for the mass is linked to the limitation of the mass suspended to the cold finger cold tip of less than 75 grams. For the mechanical environment, typical figures are 20-50 g for quasi-static and launch sine vibrations qualification levels in the [5-100 Hz] bandwidth and 0.5 to 1 g^2/Hz power spectral density in the [100-500 Hz] bandwidth for the random vibration qualification levels. Depending on the mechanical launch vibration levels and TLA shape (imposed by the stiffness and routing requirements), the flexible foils stack of TLA can be wrapped in a PET polyester tube of 100 μm thick which is heat shrunk around the foils with a dedicated process. The PET shrink tube allows the TLA to support high mechanical levels without delamination of the stacked foils and with very limited impact of the overall stiffness of the TLA. With respect to flexibility, typical stiffness requirement is in the range of few N/mm at cryogenic temperature. This requirement applies for displacements up to 5 millimeters to accommodate the misalignments of detector and cold finger and relative displacements under mechanical and/or thermal load. The stiffness measured at room temperature is corrected by an amplification factor to take into account the Young modulus variation of the foil material. Finally, the TLA cleanliness is in accordance with standard $5x(10)^{-8}$ g/cm² and 100 ppm respectively for molecular and particulate cleanliness requirements.

TLA UNDER DEVELOPMENT

For flight applications, the mass and thermal conductance need to be improved. OFHC copper thermal link can be used in application not sensitive to mass, but in most of the applications, the volume of the thermal link and its perfosrmance are critical for instrument performance and thus need to be optimized. As we can see on Figure 9, the Pyrolytic Oriented Graphite (POG) is the best candidate in the 75-160 K temperature range while the 5N aluminum is the best candidate in the 40-75 K temperature range. Both materials offer very low Young modulus which will make those the optimized choice for our application.

For this reason we adapted the thin foils thermal link technology to POG material which represents an excellent candidate for performance optimization. The POG thermal link is made with thin foils of POG which are stacked and connected at extremities on Al1080 end-fittings. The foils are swaged on the end-fittings to insure the thermal contact between POG and Al1080. On paper, the performance of such a thermal link is very high due to the high POG thermal conductivity which exceeds the OFHC copper with a factor of 3 at room temperature. However, several issues need to be solved to be able to exploit the material capability.

Figure 10. Pictures of a breadboard thermal link used to validate the FEM modelling. Before swaging on the top and after swaging on the bottom.

Management of the Contact Conductance on End-Fittings

The contact conductance is a key characteristic of the POG thermal links. To benefit from the high POG thermal conductivity, the contact resistance between foils and aluminum end-fitting needs to be optimized as well. Furthermore, the POG has the attribute to offer a very high thermal conductivity in the plane but a poor thermal conductivity out of the plane (thermal conductivity 50 times lower out of the plane than in the plane).

To optimize this interface, FEM modelling has been performed, supported by breadboard tests' (Figure 10) to determine the most efficient design for this junction. If the thermal link is made with a single stack, the contact conductance across the POG foil will be very low due to the poor thermal conductivity of the foils out of the plan. So the number of slots in the aluminum block and their height has been optimized to lower the thermeal conductance through the complete end-fitting. After several iterations and correlation with tests, we defined an optimal configuration for this interface with specific dimensions for the slots for an optimal elementary stack thickness.

Contamination Due to Particles Release

The contamination is a driver in the production of our flight cryogenic thermal links. Most of the time, these thermal links are integrated between the detector and the cryocooler, and thus, the thermal links need to offer a perfectly controlled contamination. With aluminum and copper, this issue is managed with the implementation of several cleaning steps along manufacturing process and the performance tests are performed in clean environment. However with POG, the problem is to avoid release of foil material particles during its life. It is particularly critical during integration where particles can scratch the foils. To solve this issue, a thin membrane will encapsulate the foils. The membrane will surround the complete flexible part of the thermal link in order to trap particles. To do so, a PET membrane is used. This membrane has been qualified in the frame of other projects down to 50 K. The membrane is very thin (about 20 μm) and thus doesn't impact the stiffness of the foils.

Example of POG Thermal Link

Following this development, several POG thermal links are under production. The one presented in Figure 11 is made to connect the first stage of a Pulse tube cryocooler to an optical bench. This thermal link is able to provide a thermal conductance of 1.1W/K @ 110K for a mass of less than 50 g.

Figure 11. Overview of POG thermal link

CONCLUSIONS

Following the different developments and qualification phases performed by Absolut System for European projects, we are now able to provide flight qualified high performance thermal links using aluminum and OFHC copper. Flight hardware has been delivered in Europe and several other components are under production. In complement, the thin foils thermal link technology is tailored to POG material to offer thermal and mechanical architects an efficient alternative to aluminum and copper between 75K and room temperature.

REFERENCES

1. J. Tanchon, T. Trollier, P. Renaud, J. Mullié, H. Leenders1, T. Prouvé, I.Charles, T. Tirolien, "Design of a Flight Like Cryostat for 30-50K Two-Stage Pulse Tube Cooler Integration," *Cryocoolers 19*, ICC Press, Boulder, CO (2016), (this proceedings).

2. A. L. Woodcraft, "Recommended values for the thermal conductivity of aluminum of different purities in the cryogenic to room temperature range, and a comparison with copper," *Cryogenics,* Volume 45, (2005), pp. 626-636.

The Measurement of RRR and Resistance for Aluminum Alloy Based on a Cryocooler

C.H. Yin, Z.H. Gan, S.H. Wang, D.L. Liu, L.M.Qiu

Institute of Refrigeration and Cryogenics, Zhejiang University
Hangzhou 310027, P.R.China
Key Laboratory of Refrigeration and Cryogenic Technology of Zhejiang Province
Hangzhou 310027, P.R.China

ABSTRACT

RRR (Residual Resistivity Ratio) is a common index for measuring the quality of a metal. This paper reviewed the three main methods of measuring alloy RRR, and compared the respective advantages and disadvantages. A new cryocooler based measurement method is proposed. Theoretical analysis shows this method can measure the resistance at various temperatures. There is no need to reach the liquid helium temperature (4.20K), as achieving temperatures below 38 K is sufficient to measure RRR. The analysis is later verified by experimental results, and the results differ little from PPMS (Physical Property Measurement System) measured results.

INTRODUCTION

Residual Resistivity Ratio or Residual Resistance Ratio (RRR) is defined as the ratio of a metal's electrical resistance at room temperature to that at 0 K, which is usually replaced by liquid helium temperature (4.20K). It is the one of the common indexes for characterizing the quality of a metal, including purity and structural characteristics. RRR can be used to evaluate operational stability of commercial devices made of superconductive materials.[1] Moreover, the RRR of a metal strongly controls its cryogenic conductivity and emissivity, including the conductivity of a superconductive material cavity and the performance of a superconducting resonator cavity.[2] Thus, efficient and accurate measurements of RRR are particularly useful in the cryogenics field.

There are three mainstream conventional methods for RRR measurement: the eddy current decay method, the curve method, and the fixed-point method. Developed in the 1960s, the eddy current decay method is a mature technology with established international standards.[3] This method is based on the relationship between the magnitude and decay of eddy currents and the resistance of a metal sample with a certain size and shape. The eddy current is introduced by a changing magnetic field, and the resistance at liquid helium temperature is then determined according to the time constant of the eddy current decay. The sample is placed in a liquid helium dewar. For instance, when the metal sample is of a cylindrical form, its resistance can be calculated via Eq. (1):

$$\rho = 2.17r^2 \times 10^{-11} / \tau \qquad (1)$$

where, r is the radius of sample in m; τ is the time constant of eddy current decay time in s; and ρ is the resistivity. It is applicable for pure metal samples with a diameter between 5 mm and 20 mm, a length-to-diameter ratio larger than eight, and a resistance ranging from $10^{-12}\,\Omega\cdot$m to $10^{-8}\,\Omega\cdot$m; otherwise a large measurement error may occur.

Table 1. Comparisons of conventional RRR measurement methods

Method	Measuring principles	Advantages	Disadvantages
Eddy current method	Induced eddy currents are generated when the metal sample is in a changing magnetic field; eddy current decay time constant is measured to determine the resistivity.[4]	Non-destructive to the samples; No welding of electric potential or current wire lead to samples; Impurity distribution can be determined.	Strict requirements of shape and size of samples; Only discontinuous points can be measured; No measurement of resistance varying with time.
Fixed-point method	Constant temperature points achieved by helium dewar and ice-water bath are used to measure a sample's resistance at certain temperature are used to calculate RRR.[1]	Relatively short measuring time; low cost; Suitable for batch measurement.[4]	Only discontinuous points can be obtained, not resistance versus temperature curve.
Curve method	Helium dewar is used for cooling and achieve low temperature condition while heating units are used to control temperature.[1]	Can obtain sample resistance curve changing with temperature.	Complex system setup; Long measuring time and high cost.

The Fixed-point method calculates the RRR based on the measurement of material resistance at fixed temperature points. Room temperature (293.0 K) or ice point (273.15 K) and helium temperature (4.20 K) are typical pairs of fixed points. Chinese and some international manufacturers mainly employ the fixed-point method to measure the residual resistance at low temperature. One characteristic of the fixed-point method is that the residual resistance measurement is conducted at a thermal equilibrium state.[1]

The Curve method employs measurements of a series of temperature points before numerically fitting the curve. The RRR is calculated using the resistance read from the curve.[1] Typically, a liquid helium dewar is used to provide the required temperature (4.20 K).

The advantages and disadvantages of the methods are summarized in Table 1. As can be seen, all three methods require a helium dewar, and thus there exist inherent limitations: the resistance can only be measured at limited temperature points, and the cost is high. Therefore, this paper proposes a new cryocooler-based method for RRR measurement. The temperature of a cryocooler can be continuously adjusted simply while reaching 4.20 K as well. The proposed method is capable of accurate RRR measurement. The method is also applicable to studying the variations of resistance with temperature. Moreover, the accuracy of the measurement depends less on the sizes and shapes of samples, it is more versatile for industrial applications.

BASIC PRINCIPLES OF MEASUREMENT

Principle Analysis

The resistance of metals can be described by the Mathieson rule:

$$\rho = \rho_l + \rho_r \tag{2}$$

where: ρ_l originates from the thermal motion of lattice points and is also called phonon or ideal resistivity; it is related to temperature. The residual resistance ρ_r, which originates from impurities and defects in the metal lattice, is independent of temperature.[5-6] There is no phonon excitation at absolute zero, which means ρ_l=0. Increasing temperature causes faster movement of the lattice points, thus interaction between electrons and phonon will increase ρ_l with increasing temperature.

The relationship between ρ_l and temperature can be expressed by the Bloch-Greisen equation:

$$\rho_l = \frac{AT^5}{M\Theta^6}\int_0^{\Theta/T}\frac{z^5}{(e^z-1)(1-e^{-z})}dz = \frac{AT^5}{M\Theta^6}J_5\left(\frac{\Theta}{T}\right) \tag{3}$$

where

A is constant,

M is the molar mass of the substance,

Θ the characteristic temperature of the metal element and close to the Debye temperature Θ_D.[7]

If the temperature is relatively high or low, the equation can be simplied as follows when $T > \Theta/2$:

$$\rho_l \approx \frac{AT}{4M\Theta^2} \propto T \tag{4}$$

ρ_l is linear function of temperature.

When T>Θ/10, the integral upper limit of Equation (3) can be set as infinity, thus:

$$\rho_l \approx \frac{124.4AT^5}{M\Theta^6} \propto T^5 \tag{5}$$

The residual resistance ρ_r depends only on impurities and the types and quantities of defects.

The total resistance can be expressed as $\rho = \rho_l + \rho_r$. At high temperature, $\rho_l >> \rho_r$, so $\rho \approx \rho_l$; at very low temperature, for instance T=4.2 K, $\rho_l << \rho_r$,, so $\rho \approx \rho_r$ which means the total resistance is mostly independent of phonon resistance. An alloy can be regarded as a highly impure metal.[8]

RRR is expressed as the ratio of the resistance of the material at ice water bath temperature (273.15K) to that at a low temperature (4.20K):

$$RRR = \frac{\rho_{T=273.15K}}{\rho_{T=4.20K}} \tag{6}$$

The relation between resistance and resistivity can be expressed as:

$$R = \frac{\rho L}{A} \tag{7}$$

where

R is the resistance in Ω;

L is the length in m;

A is the cross-sectional area in m^2.

Since L and A can be considered as constant through the temperature range, the problem of measuring resistivity can be simplified as resistance measurement, which can be expressed as:

$$RRR = \frac{R_{T=273.15K}}{R_{T=4.20K}} \tag{8}$$

The basic principle of resistance measurement follows Ohm's law:

$$R = \frac{U}{I} \tag{9}$$

where

U is the voltage in volts;

I is the current in amps.

The terminal voltage of the sample can be measured when constant current is applied at both ends; therefore the RRR can also be expressed as the ratio of the voltages measured at the two different temperatures:

$$RRR = \frac{U_{T=273.15K}}{U_{T=4.20K}} \tag{10}$$

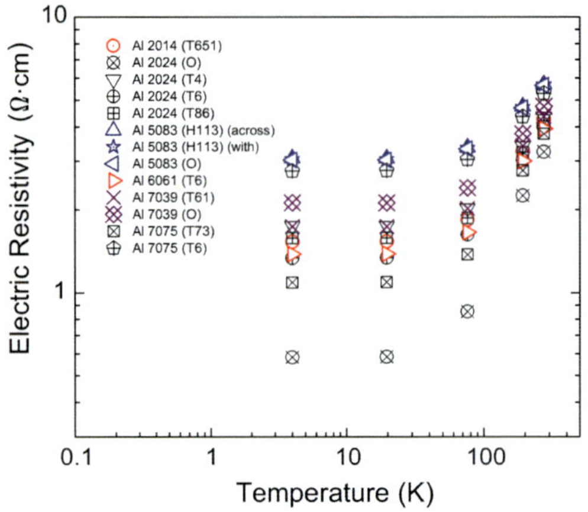

Figure 1. Resistance changes at varied temperature for different types of aluminum alloy.

Aluminum Alloy Resistance Analysis

The Debye temperature of aluminum is 380 K, which means that its resistance is nearly constant under 38 K. The resistance of aluminum alloys at low temperature is over 1000 times greater than that of pure aluminum.[9-10] For different types of aluminum alloys. the resistivity changes as a function of temperature is shown in Figure 1, where resistivity varies between 0.5 and 6 μΩ·cm. Figure 1 shows that aluminum alloy resistivity changes little under 77 K. Especially when the temperature is under 20 K, the resistance remains nearly constant when the temperature decreases. If the sample size (width × thickness × length) is 5.0 ×0.5 ×90 mm, its resistance at low temperature will be (0.18-2.16) $\times 10^{-3}\Omega$. With 100 mA excitation current applied, the terminal voltage will be 18-216 μV.

Measuring Strategy

According to the range of signal induced, a six-and-a-half-digit high precision digital multimeter (2700 type) was used in this experiment with a measuring range of 100 mV and measuring precision of 3 μV. A Lake Shore 120 Current Source was used for 100 mA current excitation, which provides constant current within 11 V with precision of ±0.1%.

A Cernox thermometer from Lake Shore Inc. was used, which is applicable from 1.4 to 325 K. The measuring precision is 0.014 K at 4.2 K, 0.009 K at 20 K and 0.046 K at 77 K, which satisfies the requirements for low temperature measurements.

CRYOCOOLER MEASUREMENT PLATFORM

As shown in Figure 2, the cryocooler measurement platform consists of the cryocooler, the data acquisition system, and the sample setup. A self-developed two-stage pulse tube cryocooler was used to achieve the low temperature environment.[11] Its lowest cooling temperature is 2.2 K, thus meeting the measurement requirement. In the experiment, the constant current source for exciting the aluminum alloy sample was attached to the cold head at the second stage of the cryocooler and to measure the voltage at both ends of the sample. By adjusting the temperature of the second stage cold head through the rotary valve and minor orifice valve, the respective voltage of both ends can be measured when the temperature varies.

Four wires were used for the measurements. Two copper wires were welded to both ends of the sample respectively. As shown in Figure 3, to the outer end was welded the current leads (I+,I-), and the inner end was attached to the voltage leads (U+,U-). The samples were wrapped in electrical tape in order to insulate them electrically. Outside the lead end tape was wrapped another highly adhesive aluminum tape to avoid the influence of cryocooler vibration. Samples were placed on the side wall of the copper block at the second stage cold head. To improve the sample attachment to the

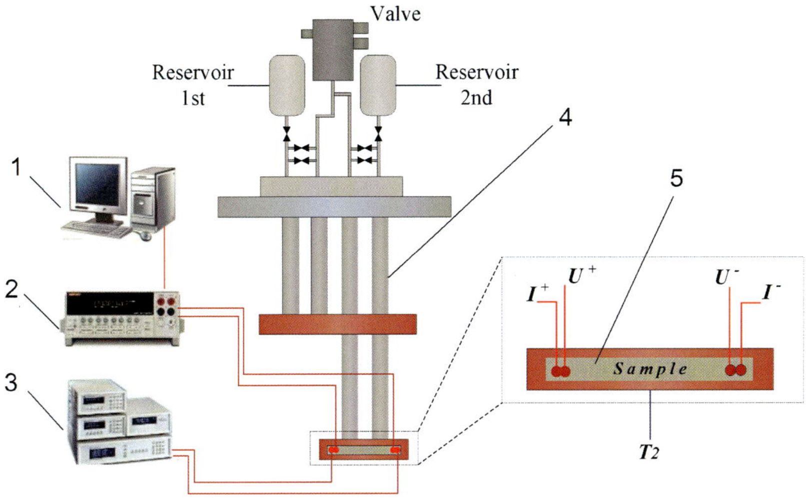

1:data acquisition 2:Keithley2700 3:Lakeshore120 4:cryocooler 5:sample

Figure 2. Diagram of cryocooler measurement platform

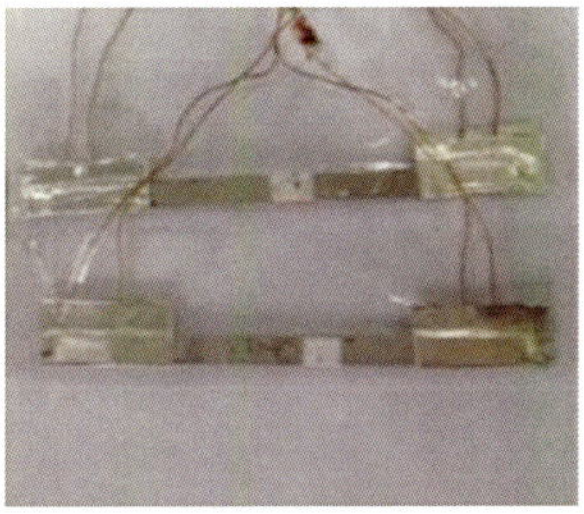

Figure 3. 90 mm samples and four-wire connection method

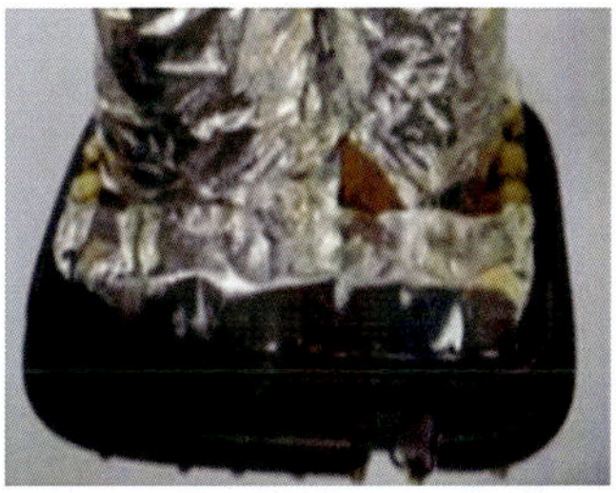

Figure 4 Spring fastening method on outside of test samples

second stage cold head, a spring fastening method was adopted in addition to the traditional aluminum tape, as shown in Figure 4. This method ensured that the sample was closely attached to the cryocooler cold head at all times during the measuring process, preventing the sample from falling off from unstable adhesion or cryocooler vibration, i.e. it ensured that the temperature of the sample was equal to that of the cryocooler cold head.

In order to eliminate contact potential and effects caused by zero drift, separate measurements were made with currents flowing in opposite directions to obtain two voltages, one positive and one negative. The average voltage value was obtained by subtracting these two values and dividing the result by 2.

$$\bar{U} = \frac{1}{2}(|U^+| + |U^-|) \tag{11}$$

RESULTS AND DISCUSSION

The experimental results are shown in Figure 5, which shows resistance versus temperature curves for samples A and B. Below 38 K, the curves remain almost horizontal and the resistance changes little with temperature. Above 90 K, the aluminum alloy resistance is mostly linearly proportional to temperature. Furthermore, the ratio between resistance at 273.15 K and other temperatures can be obtained as shown in Figure 6, where the resistance ratio is almost constant below 38 K. To further analyze the accuracy of the test results from the cryocooler measurement platform, the

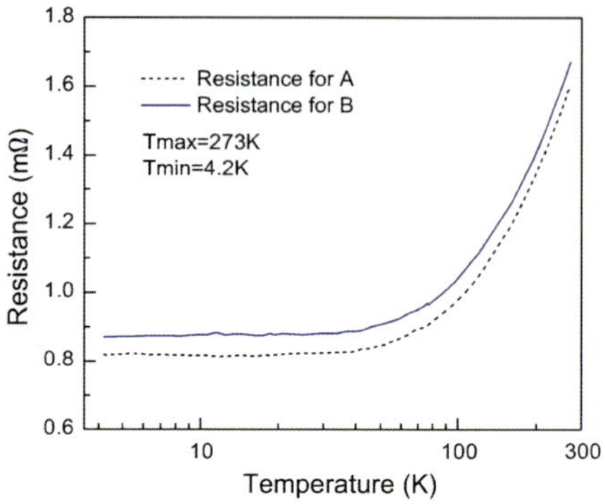

Figure 5. Resistance vs. temperature of the cryocooler measurement platform

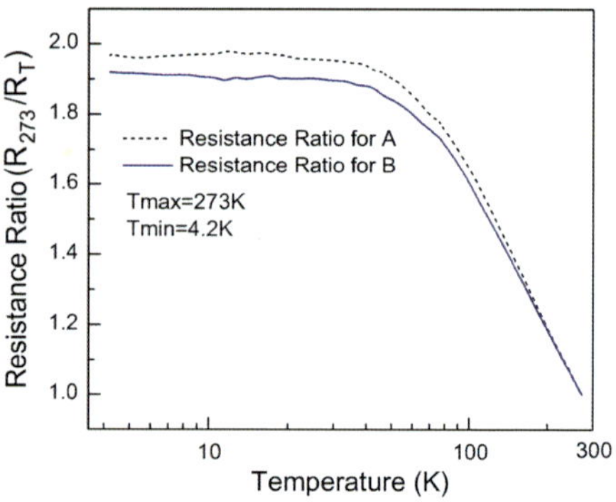

Figure 6. Resistance ratio vs. temperature of the cryocooler measurement platform

Figure 7. Schematic diagram for sample A and B on the sample holder of PPMS.

electric transport measurement option of PPMS (Physical Property Measurement System) was used to measure the sample resistance. Both samples were 5.0 ×0.5 ×2.5 mm in size and were placed as shown in Figure 7. Gold wires with high ductility were used for four-wire connection method and they were adhered by elargol.

Measurement results using PPMS for the two aluminum alloy samples are shown in Figure 8. The ratio between the resistance at 273.15 K and at other temperatures is shown in Figure 9, where the resistance ratio is also shown to change little below 38 K.

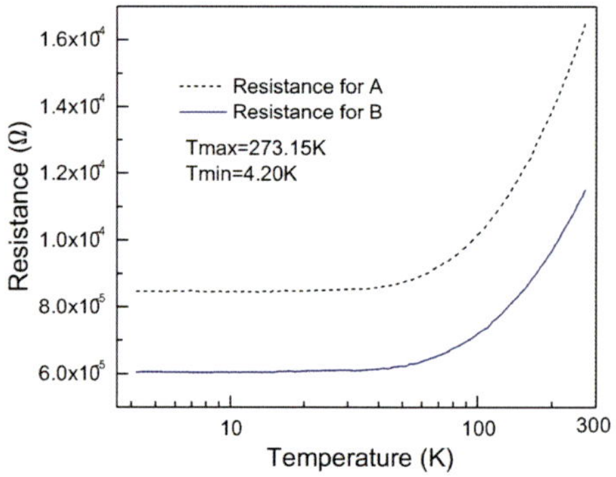

Figure 8. Resistance changes at varied temperature on PPMS

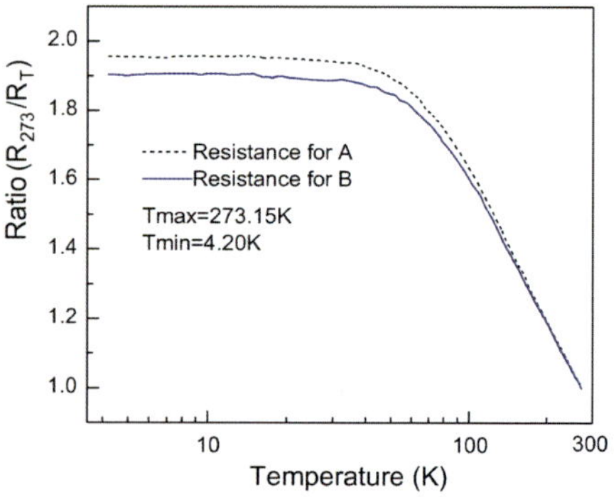

Figure 9. Resistance ratio changes at varied temperature on PPMS

Table 2. Resistance ratio of samples for 273.15K/4.20K (/20K/38K).

Sample	Sample A (4.20K)	Sample A (20K)	Sample A (38K)	Sample B (4.20K)	Sample B (20K)	Sample B (38K)
PPMS	1.957	1.953	1.944	1.905	1.896	1.891
Cryocooler	1.966	1.960	1.952	1.918	1.903	1.892

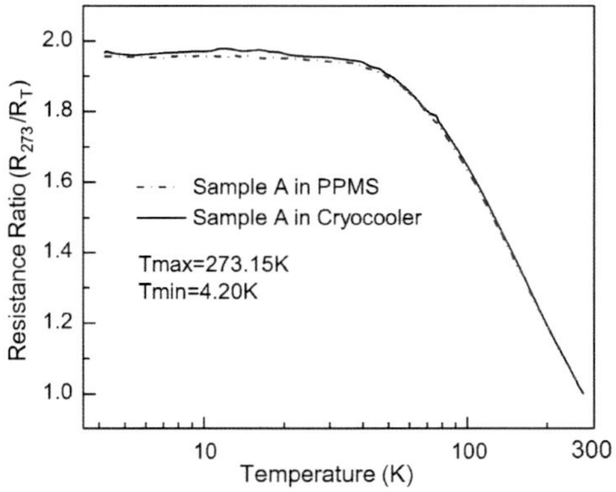

Figure 10. Comparison of Resistance Ratio vs Temperature curve, sample A.

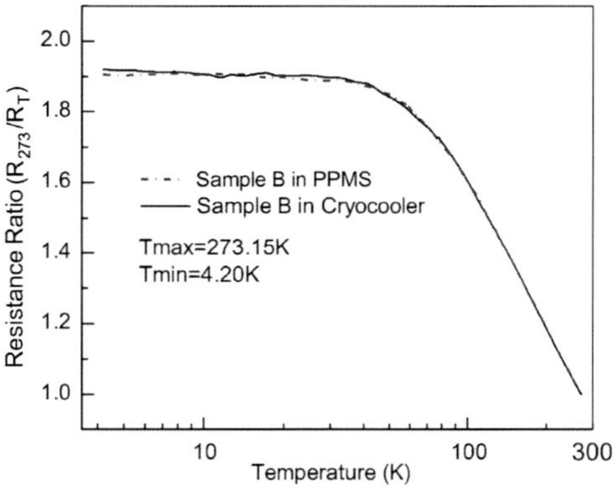

Figure 11. Comparison of Resistance Ratio vs. Temperature curve, sample B

Comparing the results from the cryocooler measurement platform and the PPMS system, the resistance value at 273.15K is divided by the resistance at 4.20K, 20K and 38 K to obtain the resistance ratio, as shown in Table 2. The deviation for samples A and B between measurement results from the cryocooler and PPMS systems is 0.46% and 0.68% at 20 K, respectively. This proves the measurement results of the cryocooler platform to be accurate and reliable.

Note also that the difference in the resistance ratio obtained from cryocooler measurements made at 4.20 K and at 38 K is only 0.66% and 0.73%, respectively, for samples A and B.

For both samples, Figures 10 and 11 compare the resistance ratio versus temperature curves from the two methods. The curves of the PPMS platform agree well with those of the cryocooler platform.

CONCLUSION

Several methods for measuring the RRR of alloy materials have been compared. The respective advantages and limitations are analyzed. The cryocooler measurement method is proposed and verified experimentally. The deviation of this new method is within 1%. On the other hand, the theoretical analysis and experiment results have shown that the resistance remains nearly the same between 38 K and 4.20 K for the alloy examined. If 4.2 K cannot be reached due to limits of the test setup, the resistance value of materials with relatively low RRR values (say RRR< 10) should not be significantly affected by using at any temperature between 38 K and 4.20 K.

ACKNOWLEDGMENTS

This work is supported by the National Magnetic Confinement Fusion Science Program (No. 2015GB121001).

REFERENCES

1. Hua, C.Y, Deng, T.Z, "Discussion about several issues were in the international standards for testing residual resistivity ratio of the compositing superconducting wire NbTi," Journal of low temperature physics. 1997(06), pp. 21-27. (in Chinese)

2. Wen, M.H, "the RRR testing of niobium material in Superconducting resonator cavity through eddy current nondestructive method," Journal of low temperature physics. 1998(03), pp. 54-58. (in Chinese)

3. Hua, C.Y, "Determining the content and distribution of impurities in high purity metal through testing residual resistivity ratio," Rare metal. 1987(02), pp. 126-129. (in Chinese)

4. Ni, Y.M, "The resistivity measurement for metal samples based on eddy current method," Low temperature physics, 1981(02), pp. 139-143. (in Chinese)

5. Rosenberg, H.M, "Low Temperature Solid State Physics," Oxford: Clarendon Press, 1963.

6. Fickett, F.R, "Aluminum—A review of resistive mechanisms in aluminum," Cryogenics. 1971, 11(5): pp. 349-367.

7. Yan, S.S, "Principle and method of physical experiment in low temperature," Beijing: Science Press, 1985. (in Chinese)

8. Reed, R.P, Clark, A.F, "Materials at Low Temperatures," Metals Park, Ohio 44073: American Society for Metals, 1983.

9. Clark, A.F, Childs, G.E, Wallace, G.H, "Electrical resistivity of some engineering alloys at low temperatures," Cryogenics (1970), pp. 295-305.

10. Qiu, L.M, Zhang, K.H, Dong, W.Q, et al, "Study of mass flow distribution between stages in a two-stage pulse tube cryocooler capable of 1.1 W at 4.2 K," International Journal of Refrigeration. 2012, 35(8), pp. 2332-2338.

Lifetime Test and Heritage On-Orbit of SHI Coolers for Space Use

K. Narasaki, S. Tsunematsu, K. Otsuka, K. Kanao, A. Okabayashi, S. Yoshida, H. Sugita[1], Y. Sato[1], K. Mitsuda[2], T. Nakagawa[2], T. Nishibori[1]

Sumitomo Heavy Industries, Ltd. Niihama, Ehime, 799-13, Japan
[1]Japan Aerospace Exploration Agency (JAXA),
Tsukuba, Ibaraki 305-8505, Japan
[2]Institute of Space and Astronautical Science, JAXA,
Sagamihara, Kanagawa 252-5210, Japan

ABSTRACT

Since 1987, Sumitomo Heavy Industries, Ltd. (SHI) has developed small Stirling coolers of two types and two Joule–Thomson (JT) coolers with operating temperatures ranging from 80 K to 1 K for space use. Ground lifetime tests of four coolers were conducted to demonstrate their long life and reliability. A single-stage Stirling cooler was tested for 113,640 hr. Two single-stage Stirling coolers were tested for 108,370 and 104,771 hr. Also, a two-stage Stirling cooler was tested for 72,906 hr. An advanced 4 K cooler with ^{4}He was tested for over 36,140 hr. Also, a 1 K-class cooler with ^{3}He was tested for 6721 hr from last year. All of these coolers, except for the 1 K-class cooler, have shown good results on orbit. Three single-stage Stirling coolers were carried on the "SUZAKU" X-ray astronomical satellite, the "KAGUYA" Japanese lunar polar orbiter, and the "AKATSUKI" Japanese Venus Climate Orbiter. Two units of a two-stage Stirling cooler were carried on the "AKARI" infrared astronomical satellite. Moreover, a 4 K cooler was carried on the Superconducting Submillimeter-Wave Limb-Emission Sounder (SMILES) aboard the Japanese Experiment Module (JEM) of the International Space Station (ISS). A Japanese X-ray astronomy satellite "ASTRO-H (Hitomi)" including a 4 K cooler, two sets of two-stage Stirling coolers, and two sets of single-stage coolers was launched on February 17, 2016. This report describes the updated results and operating status from an earlier report[1] of ground lifetime testing and achievements on orbit of SHI coolers for space use.

INTRODUCTION

Since 1987, Sumitomo Heavy Industries, Ltd. has been developing small cryocoolers of four types for space use, with temperatures ranging from 80 K to 1 K. The single-stage Stirling cooler used for cooling to temperatures of 80 K was developed in 1987 for low-noise amplifier cooling in space by modifying a commercial cooler. Subsequently, its development has continued for space use.[2] In all, five coolers were loaded on SUZAKU, KAGUYA, AKATSUKI, and ASTRO-H for cooling detectors, thermal shields, etc. The first, SUZAKU, which launched in July 2005, is the fifth in a series of Japanese astronomy satellites designed for observations of celestial X-ray sources. A

Cryocoolers 19, edited by S.D. Miller and R.G. Ross, Jr.

single-stage Stirling cooler cools an outer vapor cooled shield to extend the observation lifetime. The second, KAGUYA, a Japanese lunar polar orbiter, was launched in July 2007. The gamma-ray spectrometer (GRS) in KAGUYA uses a 250 cc high-purity germanium (Ge) detector cooled to 80 K by a single-stage Stirling cooler. The third, AKATSUKI, is a Japanese Venus Climate Orbiter launched in July 2010. After five years, it entered an elliptical orbit around Venus on December 9, 2015. The mid-wave infrared camera (IR2: 2-μm camera) of AKATSUKI uses a CCD camera cooled to around 65 K by a single-stage Stirling cooler. The fourth, ASTRO-H, is the sixth in a series of Japanese astronomy satellites devoted to observations of celestial X-ray sources. It was launched into low earth orbit with an altitude of 550 km on February 17, 2016. The CCD camera in a Soft X-ray Imager (SXI) of ASTRO-H is cooled to around 150 K by two reduced vibration models of single-stage Stirling coolers.

A two-stage Stirling cooler for space application with temperatures below 20 K began development for AKARI in 1991.[3] AKARI, the second infrared astronomy mission of the JAXA/ISAS, was launched into a sun-synchronous polar orbit at an altitude of 750 km for an infrared sky survey in February 2006. Two units of the cooler were used to cool the inner vapor cooled shield and extend the lifetime of the liquid helium of AKARI.

A 4 K cooler, which is a combination of a two-stage Stirling cooler and a JT cooler with ^{4}He, began development in 1993 to cool the submillimeter detectors for the SMILES mission.[4] SMILES, which was launched in September 2009, was aimed at highly sensitive observations probing chemical processes related to ozone depletion and used submillimeter technologies such as superconductor–insulator–superconductor (SIS) mixers and the 4 K mechanical coolers.

The soft X-ray spectrometer (SXS) of ASTRO-H is a high-resolution spectrometer using an X-ray microcalorimeter array. To achieve high energy resolution of 7 eV at 0.5–10 keV, the detectors must be kept extremely cold (50 mK), requiring a complex cryogenic system that includes a three-stage adiabatic demagnetization refrigerator (ADR), liquid helium (30 L), two sets of advanced two-stage Stirling coolers and an advanced 4 K-class cooler.[5, 6, 7]

The 1 K-class coolers with ^{3}He were developed, starting in 2001, for the Space Infrared Telescope for Cosmology and Astrophysics (SPICA) mission.[8] Lifetime testing started this past year.[9]

HIGH-LEVEL DESCRIPTION OF COOLERS

Single-Stage Stirling cooler

For cooling at 80 K temperature levels, SHI has a split type single-stage Stirling cooler with the specifications shown in Table 1. A photograph of the original cooler is presented in Figure 1(a). The cooler, which consists of a cold head unit with a single-stage displacer, a compressor, and a gas feed connecting tube, has a moving cylinder, a clearance seal by diaphragm springs, a twin pole magnet system, and a pneumatically driven displacer. The compressor has two oppositely moving cylinders to reduce the vibration level. For reducing the induced vibration in the drive axis of the cold head, the cooler was improved by adding an active counter balancer mechanism to the cold head of the original model. A photograph of the reduced vibration model is shown in Figure 1(b). The specifications are shown in Table 1. The mechanical vibration forces are reduced below 0.1 N at the operating frequency of 52 Hz.

Two-Stage Stirling cooler

Table 1 shows the cooling temperature specifications for the 20 K split two-stage Stirling cooler. The cooler consists of a cold head unit with a two-stage displacer, a compressor, and a gas feed connecting tube. The compressor has two opposed pistons mounted on a drive shaft to reduce the vibration level. The drive shaft is supported by two sets of linear ball bearings. This linear-ball-bearing supporting system achieves the piston clearance seal, long piston stroke operation, and low-frequency operation. These features present benefits for a long useful life, compactness, light weight, and high efficiency of the second regenerator at low temperatures.

Table 1. Specifications of single-stage and two-stage Stirling coolers.

Item	Single-stage Stirling Cooler (1ST)	1ST (Reduced vibration model)	Two-stage Stirling Cooler (2ST)
Cooling Capacity	2 W at 80 K	2 W at 80 K	200 mW at 20 K 1000 mW at 100 K
Environment Temp.	200 K – 300 K	200 K – 300 K	200 K – 300 K
Power Consumption	50 W	50 W	90 W
Size	Compressor: ϕ 98× 230 L (mm) Cold Head: ϕ78×181 L (mm)	Compressor: ϕ 98× 230 L (mm) Cold Head: ϕ82×236 L (mm)	Compressor: ϕ 108× 400 L (mm) Cold Head: ϕ81×329 L (mm)
Mass	4.2 kg	5.0 kg	9.5 kg
Drive Frequency	52 Hz	52 Hz	15 Hz

(a) Original model (b) Reduced vibration model

Figure 1. Single-stage Stirling cooler on test bench.

The advanced two-stage Stirling cooler has been improved from the original cooler for AKARI as described below.[5]

- Change from contact seal to clearance seal at displacer of cold head for reducing abrasion
- Re-selection of materials with lower outgassing and reduced adhesive amount
- Improvement for managed processes of baking and methods of purification

Additionally, it is important to reduce the total time during cooler assembly under atmosphere condition for reducing outgassing. Because the main gas impurities that degrade cooling performance are CO_2 and H_2O, the allowable concentrations of CO_2 and H_2O contamination are set respectively as 500 ppmv during operation. The newest model of two-stage Stirling coolers is used for the ASTRO-H project. A photograph of a flight model cooler is presented in Figure 2.

Figure 2. Two-stage Stirling cooler on the cryostat of ASTRO-H/SXS.

Table 2. Specifications of Advanced 4 K Cooler and 1 K Class Cooler.

Item	Advanced 4 K Cooler	1K-class Cooler
Cooling Capacity	40 mW at 4.5 K	10 mW at 1.7 K
2-stage Stirling Cooler Cooling Capacity Power Consumption	 0.2 W at 20 K, 1 W at 100 K 90 W	 0.15 W at 15 K, 1 W at 100 K 90 W
JT Compressor Pressure Mass Flow Rate Power Consumption	 Supply: 2 MPa, Return: 0.1 MPa 2.0 NL/min (=6 mg/s) of ^{4}He 90 W	 Supply: 0.7 MPa, Return: 7 kPa 1.0 NL/min (=2.23 mg/s) of ^{3}He 90 W
Heat Exchanger Efficiency Pressure Drop	 97% 0.03 MPa on the low-pressure side	 97% 3 kPa on the low-pressure side
JT Orifice Diameter	20–30 μm	20–30 μm
Mass	30 kg (excluded cryostat)	30 kg (excluded cryostat)

Figure 3. 4 K cooler for ASTRO-H.

4 K Cooler from SMILES to ASTRO-H

The 4 K cooler consists of a JT cooler with ^{4}He as the working fluid and a two-stage Stirling cooler as a precooler. Major components of the JT cooler are two JT compressors, three coaxial-double-tube heat exchangers, an orifice type JT valve, and a bypass route to shorten the cooling time in the precooling process. The JT compressor system comprises two units connected in tandem. The 4 K cooler was developed originally to achieve a cooling capacity of 20 mW at 4.5 K for SMILES.[4] The 4 K cooler for SMILES used linear ball bearings as the piston support in the JT compressors, which is the same as the two-stage Stirling cooler. To improve the cooling power and long life operation, the JT compressors were changed, in terms of the piston supporting mechanism, from linear ball bearings to flexure bearings; this reduced mechanical abrasion. Moreover, low-outgassing materials were adopted to avoid cooling performance degradation caused by impurities. In addition, a "getter" was mounted upstream of the JT closed cycle to trap impurity molecules.[6] Managed baking processes, gas purification, and limited time during cooler assembly under atmospheric pressure were are also incorporated to further reduce internal contaminants in the JT coolers.

An advanced type of two-stage Stirling cooler is used as the precooler for the advanced 4 K cooler. Design specifications of the advanced 4 K cooler are presented in Table 2. A photograph of the advanced 4 K cooler for ASTRO-H is presented in Figure 3. The 4 K cooler for ASTRO-H consist of a JT cooler and two units of advanced two-stage Stirling coolers as the precooler (PC-A and PC-B).

1 K-Class Cooler

The 1 K-class cooler comprises a two-stage Stirling cooler and a JT cooler with ^{3}He as the working fluid. The 1 K-class cooler is an improved type of 4 K cooler. ^{3}He is used in the JT cooler

Table 3. Summary of Operating Hours.

Test Model or Project		Temp. Level	Operating hours			Remark
			Earth	Orbit	Total	
Single-Stage Stirling Cooler						
1 EM		80 K	113640	N/A	113640	1999 Nov.-2015 Jan.: Stopped
2 PM-1		80 K	108370	N/A	108370	2001 Nov: Operation Start
3 PM-2		80 K	104771	N/A	104771	2002 May: Operation Start
4 SUZAKU/XRS		150 K	2000	51400	53400	2009 June: Mission Completed
5 KAGUYA/GRS		80 K	9200	12600	21800	2009 June: Mission Completed
6 AKATSUKI/IR2		65 K	500	3570	4070	2010 July: Operation Start
7 ASTRO-H/SXI - A&B		150 K	443	456	899	2016 March : Operation Start
Two-stage Stirling Cooler						
8 EM for AKARI		20 K	72906	N/A	72906	1999 March – 2008 August
9 AKARI-A		20 K	2098	40350	42448	2010 May: Mission Completed
10 AKARI-B		20 K	2121	13800	15921	2007 October : Operation Stop
11 PM&PC for SMILES		20 K	22283	N/A	22283	2007 May – 2010 Sep.: Stopped
12 PC for SMILES	Part-1	20 K	2250	6035	8285	2009 Sep. – 2010 June
	Part-2		2250	16124	18374	2009 Sep. – 2014 March
13 EM for ASTRO-H		20 K	36140	N/A	36140	2010 January : Operation Start
14 ASTRO-H/SXS SC-A&B		30 K	5746	902	6648	2016 Feb.: Operation Start
15 ASTRO-H/SXS PC-A&B		15 K	5746	902	6648	2016 Feb.: Operation Start
16 PC for 1K-level Cooler		20K	6721	N/A	6721	2015 May: Start
JT Cooler with ^{4}He for 4 K level						
17 EM for SMILES		4.5 K	9315	N/A	9315	1998 Sep. – 1999 Dec.: Stopped
18 PM for SMILES		4.5 K	22283	N/A	22283	2007 May – 2010 Sep.: Stopped
19 SMILES	Part-1	4.5 K	1172	6010	7782	2009 Sep. – 2010 June
	Part-2	20 K	1172	30428	32200	2009 Sep. – 2014 March
20 EM for ASTRO-H		4.5 K	36140	N/A	36140	2010 January: Operation Start
21 ASTRO-H/SXS		4.5 K	5746	863	6609	2016 Feb.: Operation Start
JT Cooler with ^{3}He for 1 K level						
22 EM for SPICA		1.7 K	6721	N/A	6721	2015 May: Start

Note: EM, Engineering Model; PM, Prototype Model; SC, Shield Cooler; PC, Pre-Cooler

instead of ^{4}He. The JT compressor capacity and pressure drop in the heat exchangers on the low-pressure side are redesigned to reach temperatures below 2 K. The selection of ^{3}He as the working fluid was made based on much higher vapor pressure than ^{4}He by about one order of magnitude at 1.7 K. Design specifications of the 1K-class cooler were described in an earlier report.[8] The updated model[6] was designed and tested to improve its performance and reliability.[9]

LIFETIME TEST AND OPERATION RESULTS ON ORBIT

Table 3 presents the number of operating hours of coolers until April 29, 2016. Specifically, the table presents the operating hours of lifetime tests of coolers under the development phase, along with the operating hours in ground tests and operating hours on-orbit for launched projects. The 4 K and 1 K-class coolers are divided into the JT cooler and the precooler in Table 3.

Lifetime test results and operation results on orbit of Nos. 1, 4, 5, 8, 9, 10, 11, 12 (Part-1), 17, 18 and 19 (part-1) in Table 3 were presented in an earlier report.[1]

Single-Stage Stirling Cooler

Three lifetime tests of single-stage Stirling coolers using an engineering model (EM) cooler and two prototype model (PM) coolers demonstrate reliability of greater than 100,000 hr. Figure 4 depicts lifetime test results obtained for two PM coolers (Nos. 2 and 3 in Table 3). Figure 4 shows the temperature of the cold stage and the environment temperature. The test conditions are input

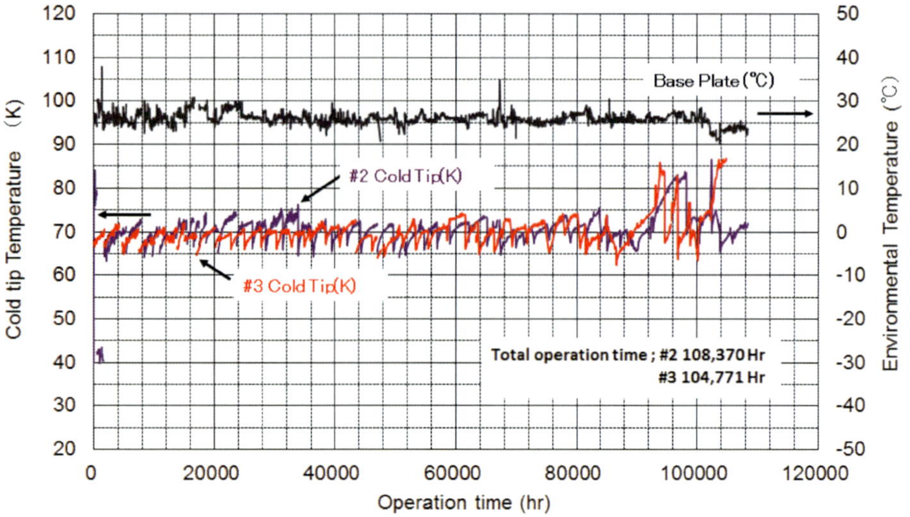

Figure 4. Status of ground lifetime tests of two single-stage Stirling coolers (PM).

power of 40 W (nominal), heat load of 1 W on the cold tip, and normal environment temperature of 295±5 K. Slow and repeatable temperature degradation was observed, with recovery following restart of operation after power interruption. The cooler is still functioning, providing good cooling performance. Therefore, we infer that a mechanical fault is unlikely. At around 90,000 hr and later, disturbing performance degradation was observed, although the temperature maintained the 80 K level. The behavior might be mostly attributable to internally cooled condensation in the displacer because these units were not baked out sufficiently. The internal gas was insufficiently purified. The flight model coolers (Nos. 4 to 7 in Table 3) have been built with improved and managed processes of baking and methods of purification to reduce outgassing.

Although the mid-wave infrared camera (IR2) of AKATSUKI operated minimally during more than five years of space transit, the cooler (No. 6 in Table 3) started to operate after entry into the ellipsoidal orbit around Venus. During operation, the CCD camera is cooled to around 65 K using the cooler with an input power of 45 W. From now on, we look forward to obtaining useful observation results for Venus.

Two-Stage Stirling cooler

The lifetime test on the ground of two advanced models of the two-stage Stirling coolers (Nos.13 and 16 in Table 3) have been conducted as a precooler for the JT cooler of the advanced 4 K cooler and the 1 K-class cooler reported respectively hereinafter.

4 K Cooler

After 9 months (6,480 hr) from the operation start, SMILES stopped operation because of a disorder related to the water cooling system for JEM. The cryogenic system returned to room temperature level. By that time, the cryogenic system for SMILES had demonstrated sufficiently good cooling performance for the submillimeter atmospheric observation. The sum of real operating hours including the ground operation was 8,285 hr for the precooler and 7,782 hr for the JT compressors (Part-1 of Nos.12 and 19 in Table 3). After restoration of the water cooling system, the SMILES cryogenic system was restarted for cooling in July 2010. However, it did not achieve a temperature of the 4 K level. When the temperature of the 100 K-stage reached around 150 K, the lower stage temperatures were still warm, and the gas flow in the JT cooler seemed to have stopped. The cessation was indicated by a change in suction pressure at the lower side of the JT compressor and by the bypass valve temperature. We believe the cause to be blockage by internal solidification of a

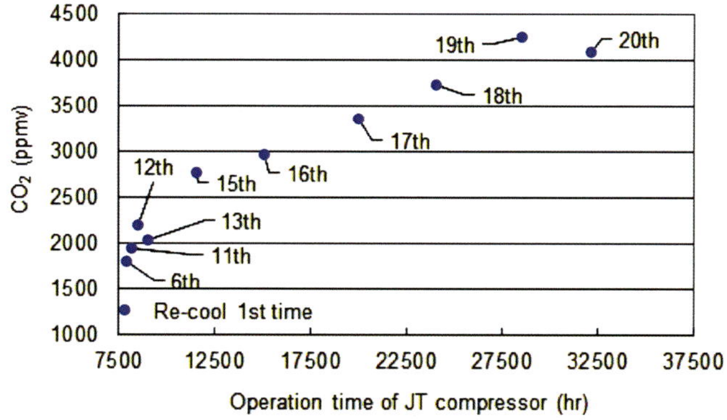

Figure 5. Status of CO_2 contamination concentration in the 4 K cooler of SMILES on orbit.

contaminant gas in the narrow high-pressure piping around the 100 K stage. The main gas impurity was CO_2, which becomes solid at such temperatures and its saturation pressure region. The CO_2 contamination concentration was estimated as about 1264 ppmv from the 100 K stage temperature of 147.2 K, which is an unexpected amount. Because the cooling capability of the precooler and the performance of the JT compressors showed no degradation, re-cooling for SMILES has been attempted by changing various procedures of the cooling process. However, it did not achieve a temperature of the 4 K level again. Subsequently, the 4 K cooler of SMILES has been investigated to analyze the CO_2 contamination concentration depending on the operation time and to confirm the mechanical reliability and performance for the future of cooler development through March 2014 (Part-2 of Nos.12 and 19 in Table 3). Figure 5 shows that the CO_2 contamination concentration increases with the operation time of the JT compressors. The maximum value is estimated as about 4251 ppmv from 100 K stage temperature of 156.0 K at the 19th investigation period. It is apparently saturated between 19th and 20th. The total operation duration of the two-stage Stirling cooler was 18,374 hr. That of the JT compressors was 32,200 hr without remarkable degradation of performance such as pressure, cooling temperature, or current.

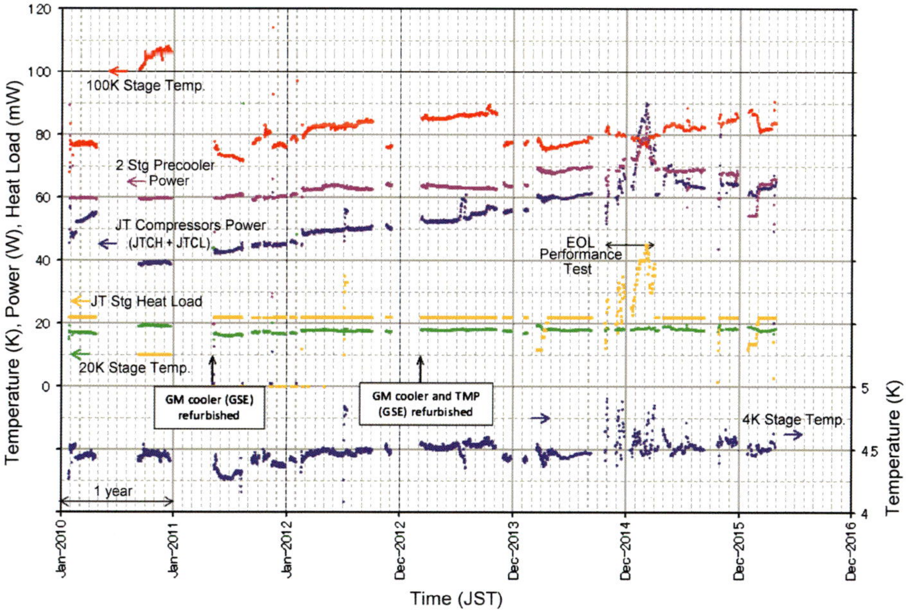

Figure 6. Lifetime test progress of an advanced 4 K Cooler (EM).

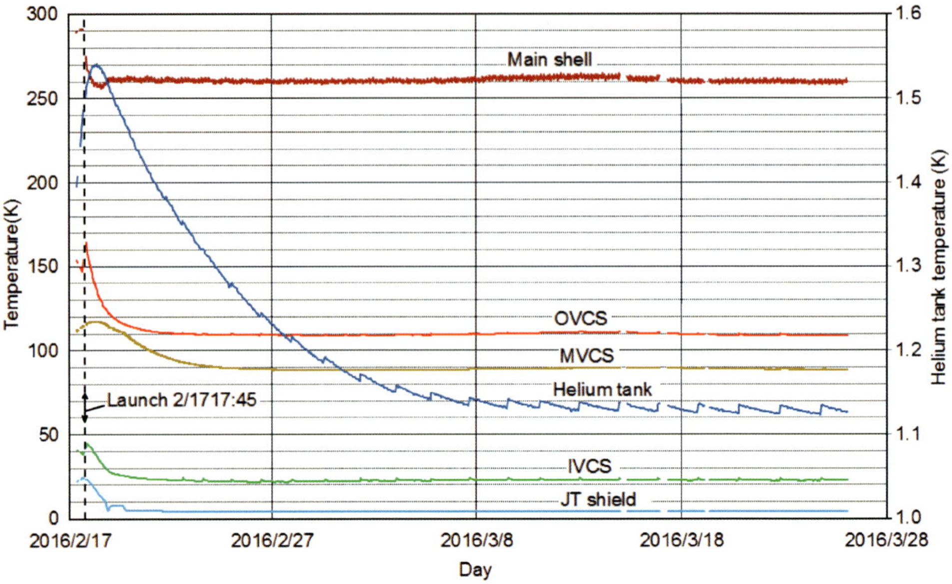

Figure 7. Status of early stages of cooling operation in Cryogenic system of STRO-H/SXS on orbit.

An engineering model of the advanced 4 K cooler with ^{4}He (Nos. 13 and 20 in Table 3) has been tested under a room temperature environment to verify its cooling performance and lifetime. Figure 6 shows the lifetime test progress of the advanced 4 K cooler. The figure shows temperatures of 100 K, 20 K, and 4 K-stage, in addition to the input power to the JT compressors and precooler and heat load to the 4 K-stage. As of April 29, 2016, the total running time was over 36,140 hr, demonstrating three years of operation, which is the required design mission life of ASTRO-H. Furthermore, the design requirement for the cooling performance of 40 mW at 4.5 K at the end of life was also verified to have an input power of 80 W for the precooler and 85 W for the JT compressors. In the lifetime test, the input power to the EM cooler needed to increase from the beginning state to 10 W for the precooler and 20 W for the JT compressors to maintain constant cooling capacity at a temperature of the 4-K level.

The cooldown and initial cooling performance of the cryogenic system of the soft X-ray spectrometer (SXS) of ASTRO-H were completed successfully after launch. Figure 7 presents the status of early stages of cooling operation in the cryogenic system for the ASTRO-H/SXS. (Nos. 14, 15, 16 and 21 in Table 3) The figure presents temperatures of the main shell, three vapor cooled shields (OVCS, MVCS and IVCS), the JT shield, and the helium tank of the cryogenic system. The OVCS and IVCS are cooled by two units of advanced two-stage Stirling coolers (SC-A and SC-B) with the input powers of 46.6 W and 49.1 W. The JT shield is cooled by the 4 K cooler. The input power of the two pre-coolers A, B and the JT compressors are 42.5 W, 50.6 W and 15.2 W, respectively. Finally, the SXS detector was cooled to 50 mK by the two-stage ADR. The initial cooling results of the cryogenic system of the SXS of ASTRO-H are matched to the thermal design and demonstrated to fulfill the requirement of mission lifetime. Initial observations were performed successfully. However, ASTRO-H lost all its solar panels by an attitude rotation anomaly caused by an attitude anomaly on March 26, 2016. Consequently, with deepest regret, observations by ASTRO-H were given up on April 28, 2016.

1K-Class Cooler

A lifetime test of a 1K-class cooler with ^{3}He using an engineering model (Nos. 16 and 22 in Table 3) has continued since May 2015. Figure 8 shows a photograph of the engineering model of the 1K-class cooler. Figure 9 presents the lifetime test progress of the EM cooler, which passed 6721 hr with no degradation. The figure shows the cooler temperatures of the 100 K, 20 K, and 1 K-stage and

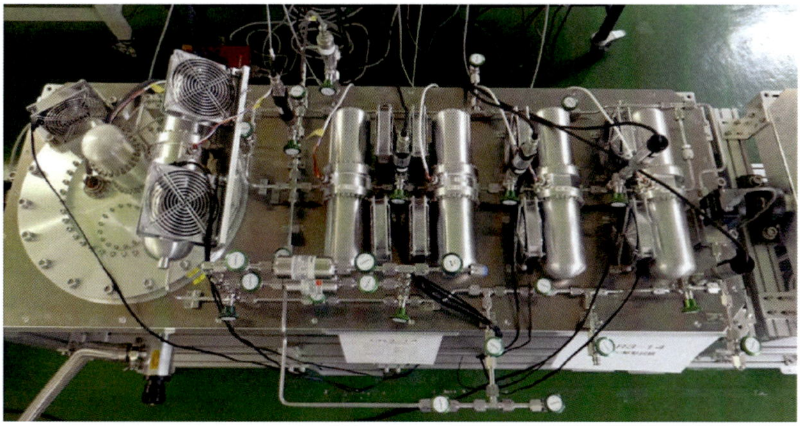

Figure 8. A photograph of the engineering model of the 1-K class cooler.

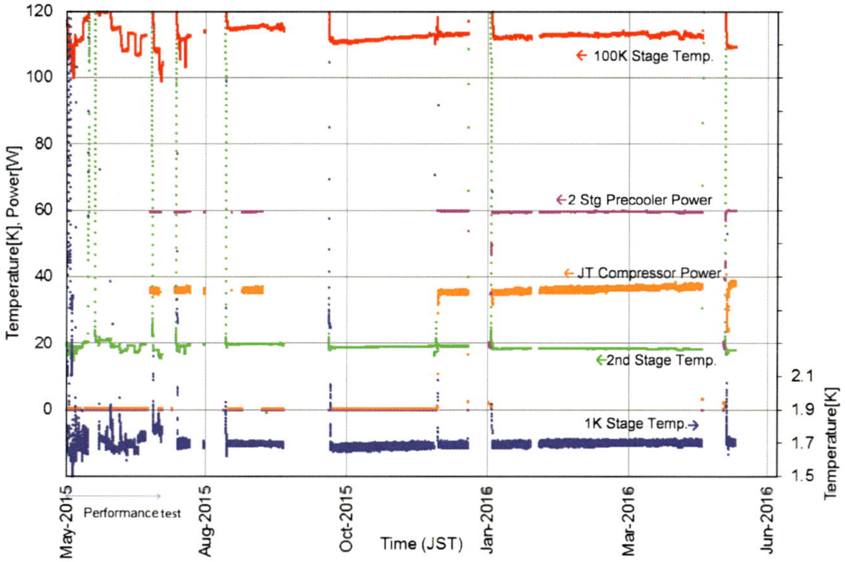

Figure 9. Lifetime test progress of a 1-K Class Cooler (EM).

the power consumption. The EM cooler, controlled by laboratory electronics, was tested under laboratory conditions with a local vacuum chamber around the cold stages. The environment temperature was 295 ± 5 K. The heat load applied to the 1 K-stage was 5 mW. When the lifetime test started, the power consumption of the two-stage Stirling cooler and the JT compressors was 60 W (nominal) and 30 W (nominal), respectively. The operating voltage of the two-stage Stirling cooler has not changed during the test.

CONCLUSIONS

Ground lifetime tests were conducted of coolers of three types: a single-stage Stirling cooler, two-stage Stirling cooler, and JT cooler. Three single-stage Stirling coolers have been tested for over 113,640 hr, 108,370 hr and 104,771 hr. Advanced 4 K coolers with ^4He have been tested for 36,140 hr. Results show proof of long life and reliability for space use. A lifetime test of 1-K class cooler with ^3He has begun for future missions. Five single-stage Stirling coolers were carried on each of SUZAKU, KAGUYA, AKATSUKI, and ASTRO-H/SXI. Four sets of two-stage Stirling coolers were carried on AKARI and ASTRO-H/SXS. Two 4 K coolers were carried on SMILES and ASTRO-H/SXS. These coolers produced useful results for future projects on orbit. From investigation results related to cooling performance degradation, we have learned that filled gas contamination, especially by CO_2 and H_2O, is crucially important. Advanced models of a two-stage Stirling cooler and JT cooler have been improved in terms of abrasion, and development and investigation are continuing to decrease filled gas contamination in the coolers.

ACKNOWLEDGMENTS

We wish to thank all staff of SUZAKU/XRS, KAGUYA/GRS, AKARI, SMILES, AKATSUKI, and ASTRO-H project teams, and the cryogenic group at SHI Niihama for their support of this work.

REFERENCES

1. Narasaki, K., Tsunematsu, S., Ootsuka, K., Kanao, K., Okabayashi, A., Mitsuda, K., Murakami, H., Nakagawa, T., Kikuchi, K., Sato, R., Sugita, H., Sato, Y., Murakami, M., Kobayashi, M., "Lifetime Test and Heritage on Orbit of Coolers for Space Use," *Cryogenics* Vol. 52, (2012), pp. 18-195.

2. Narasaki, K., Tsunematsu, S., Kanao, K., Otsuka, K., Hoshika, S., Fujioka, K., Tsurumi, K., Hirabayashi, M., "Development of Single-stage Stirling Cooler for Space Use," *Adv. in Cryogenic Engin.* Vol. 51B (2006), pp. 1505-12.

3. Narasaki, K., Tsunematsu, S., Ootsuka, K., Kyoya, M., Matsumoto, T., Murakami, M., Nakagawa, T., "Development of Two-stage Small Stirling Cooler for ASTRO-F," *Adv. in Cryogenic Engin.* Vol. 49B (2004), p. 1428.

4. Narasaki, K., Tsunematsu, S., Yajima, S., Okabayashi, A., Inatani, J., Kikuchi, K., Satoh, R., Manabe, T., Seta, M., "Development of Cryogenic System for SMILES," *Adv. in Cryogenic Engin.* Vol. 49B, (2004), pp. 1785-1796.

5. Sato, Y., Sugita, H., Komatsu, K., Shimizu, R., Uchida, H., Nakagawa, T. et al., "Development of Advanced Two-stage Stirling Cryocooler for Next Space Missions," *Cryocooler* Vol. 15, (2008), pp. 13-20.

6. Sato, Y., Shinozaki, K., Sugita, H., Mitsuda, K., Yamasaki,Y. N, Takei, Y., Nakagawa, T., Fujimoto, R., Murakami, M., Tsunematsu, S., Otsuka, K., Yoshida, S., Kanao, K., Narasaki, K., ASTRO-H SXS team, "Development of mechanical cryocoolers for the cooling system of the Soft X-ray Spectrometer onboard Astro-H", *Cryogenics* Vol. 52, (2012), pp. 158-164.

7. Yoshida, S., Miyaoka, M., Kanao, K., Tsunematsu, S., Otsuka, K., Hoshika, S., Mitsuda, K., Yamasaki, N., Takei, Y., Fujimoto, R., Sato, Y., DiPirro, M., Shirron, P., "Flight model performance test results of a helium dewar for the soft X-ray spectrometer onboard ASTRO-H," *Cryogenics* Vol. 74, (2016), pp. 10-16.

8. Narasaki, K., Tsunematsu, S., Ootsuka, K., Watanabe, N., Matsumoto, T., Murakami, M., Nakagawa, T., Sugita, H., Murakami, M., Awazu, H., "Development of 1 K-class Mechanical Cooler for SPICA," *Cryogenics* Vol. 22, (2004), pp. 375-381.

9. Sato, Y., Sawada, K., Shinozaki, K., Sugita, H., Mitsuda, K., Yamasaki, Y., Nakagawa, T., Tsunematsu, S., Ootsuka, K., Narasaki, K., "Development of 1K-Class Joule-Thomson Cryocooler for the Next-Generation Astronomical Mission," *Cryogenics* Vol. 74, (2016), pp. 182-188.

Results from a Year of Running a 12L Per Hour Pulse Tube Liquefier in an Industrial Application

A. Caughley[1], N. Emery[1], M. Nation[1], A. Kimber[1], H. Reynolds[2], C. Boyle[2], J. Meier[2], J, Tanchon[3]

[1]Callaghan Innovation, Christchurch, New Zealand
[2]Fabrum Solutions, Christchurch, New Zealand
[3]Absolut Systems SAS, Seyssinet-Pariset, France

ABSTRACT

Callaghan Innovation and Fabrum Solutions, in collaboration with Absolut System, have produced a range of large pulse tube cryocoolers. The cryocoolers are based on Callaghan Innovation's metal diaphragm pressure wave generator technology (DPWG). The metal diaphragms in a DPWG separate the clean cryocooler working gas from the oil lubricated reciprocating mechanism. DPWG technology has matured over the last 10 years to become a viable option for providing acoustic power to large pulse tube cryocoolers. The largest cryocooler consists of three in-line pulse tubes working in parallel on a 1000 cc swept volume DPWG. It has demonstrated 1280 W of refrigeration at 77 K, from 24 kW of input power and was subsequently incorporated into a liquefaction plant to produce liquid nitrogen for an industrial customer. The liquefier has now completed one year of operation, producing 11 liters of liquid per hour. Development of the pulse tube has continued with a single in-line pulse tube directly mounted to a 330 cc DPWG. The pulse tubes on the large cryocooler each produced 450 W of refrigeration at 77 K. Further optimization of the single unit has increased the cooling power to 500 W at 77 K, with no change in input power. This paper presents the experience obtained from running the large liquefier for a year in an industrial setting and the results of the optimization exercise on the smaller cryocooler.

INTRODUCTION

Callaghan Innovation and Fabrum Solutions have been developing cryocoolers based on metal diaphragm pressure wave generators (DPWG) for pulse tube[1–3] and Stirling [4] cryocoolers.

A range of industrially robust cryocoolers have been developed that are suited to cooling High Temperature Superconductor (HTS) applications such as transformers, power cables and fault current limiters, or for on-site production of industrial liquid nitrogen. The DPWG uses metal diaphragms to separate the pulse tube or Stirling cold-head's clean working gas from an ambient pressure, oil lubricated, motor-crank reciprocating mechanism. The result is an environmentally robust long-life cryocooler that is easily maintained.

The PTC1000 is the largest pulse tube refrigerator made to date and it has demonstrated 1280 W of refrigeration at 77 K from 24 kW of input power [2]. The PTC1000 consists of three in-line pulse

Cryocoolers 19, edited by S.D. Miller and R.G. Ross, Jr.

tubes sharing the pressure wave from a single 1000 cc swept volume DPWG[5]. The pulse tubes' cold heat exchangers were fitted with condensers to allow liquefaction of nitrogen. The system produced an average 12.1 litres of nitrogen per hour during initial testing[3].

The liquefier has been installed and operated at a gas company in Christchurch, New Zealand. The gas company primarily supplies bottled gas, which is either imported or produced from its own air separation column, and liquid nitrogen to a small local market.

The development of the triple pulse tube cryocooler required development of a single pulse tube working on a 330 cc swept volume DPWG. The single pulse tube refrigerator, the PTC330, is a viable cryocooler in its own right and development of it has continued with results applicable to the large cryocooler. Areas for development were simplification of the pulse tube construction to reduce manufacturing cost, experimental optimization for increased refrigeration power and condenser design for efficiency.

A PTC1000 based liquefier incorporating the alpha-prototype 1000 cc DPWG was installed in July 2015 and ran for 11 months. The alpha-prototype DPWG was replaced with a production DPWG in June 2016 which, at the time of writing, is still in service. This paper presents the performance and experiences of running the liquefier on site and presents results from the single pulse tube's performance optimization.

THE LIQUEFIER SYSTEM

Figure 1 shows a schematic of the liquefier. The cryocooler's pressure wave generator (1) was powered by a standard 3 phase induction motor (2), which is controlled by a variable speed drive (3). A Nitrogen generator (4) supplied gas to the system. A solenoid valve (5) controlled by a pressure switch controlled the pressure in a buffer tank (6). Nitrogen gas was fed from the buffer tank to the cryocooler's condenser via gas line (7). The condensers were incorporated into the pulse tubes' cold heat exchangers inside the cryostat (8). Condensed nitrogen from each pulse tube was combined in the cryostat and drained via a vacuum insulated line (9) into the buffer Dewar (10). When the buffer Dewar became full, a solenoid valve (11) on the relief of the tank (12) to be filled was opened allowing pressure to drop and liquid to flow into the storage Dewar. An industrial PLC controller monitors the buffer Dewar level, controlled the output solenoid and the cryocooler motor (for example turning the cryocooler off when Dewars were full). Heat rejection for the cryocooler was achieved with a closed water circuit exchanging heat with the ambient air via a radiator.

THE INDUSTRIAL SITE

Figure 2 shows the liquefier installed in the main factory building. The industrial building is not insulated and has an iron roof and concrete walls. The liquefier was positioned close to a steel

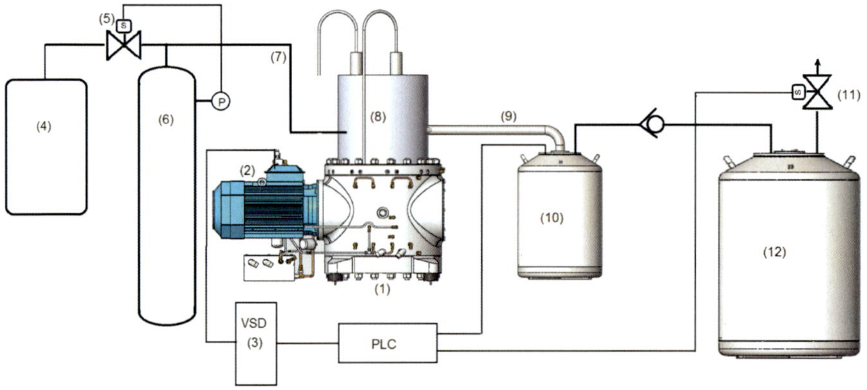

Figure 1. Schematic of the liquefier system.

Figure 2. The liquefier installed in the factory.

overhead door which could be opened on hot days to allow air to circulate and provides little insulation in cold weather.

The liquefier was installed in July, which is mid-winter in New Zealand. The winter temperatures in July had a maximum of 20 °C and minimum of -5 °C with an average daily maximum of 11 °C; contrasting to summer when the December maximum was 36 °C and minimum 2 °C with average daily maximum of 22 °C. The general machine and oil temperatures followed the ambient temperatures as would be expected from an air-cooled system. The general machine temperature was typically 15 degrees warmer than the ambient air in the factory and the oil temperature 30 degrees warmer than ambient.

DATA LOGGING

Running a machine in the field rather than in the development laboratory requires a degree of separation between operation and engineer. A data logging system was set up on the liquefier to allow the development team to monitor and record the liquefier's operation. The data logging system was purposely separate from the machine control. If a failure of the logging occurred, the machine control would continue to operate the liquefier in a safe manner.

The data logger measured machine parameters such as temperatures, motor power, gas and oil pressures, and Dewar level. The data logger hosted a web page and an FTP site on Callaghan Innovation's local area network via a cellular network router. Once the connection was made, the data logger could be accessed from the office network and allowed the development team to remotely view and control the data logger. A script running on a Linux computer was used to automatically upload the data from the data logger to a database file on the office network; where it could be analyzed by the development team. Most of the issues encountered during the field testing exercise were centered on getting the different components of the data logging system working smoothly. The system needed to automatically restart and send information in situations such as network connection loss or power-off by the customer. By the end of the 11 month trial we had encountered and accounted for most situations, so the monitoring was working reliably.

RESULTS

Figure 3 shows the running history of the cryocooler. Running time was 3338 hours with a duty cycle of 42%, over 30,000 liters of liquid nitrogen was produced for sale. There were 67 start-stop cycles with periods of running up to 271 hours at a time. One of the realities of an industrial site was a variable duty cycle driven by customer demand for the liquid nitrogen produced. This was caused by surplus liquid nitrogen production and shortages of Dewar capacity, where the empty Dewars would be at customers' premises. For the period from late August to the start of February, the liquefier produced significantly more liquid nitrogen than could be sold, so the liquefier was stopped for significant periods while waiting for Dewar capacity before the cryocooler was able to be restarted. At the beginning of February, a new order that required the full capacity of the liquefier was secured by the gas company and it was run full-time. By this time, the cryocooler liquefier had demonstrated that it was significantly cheaper to run than the gas company's air separation plant. Even so, there were short periods when the storage Dewars were full and the cryocooler was stopped.

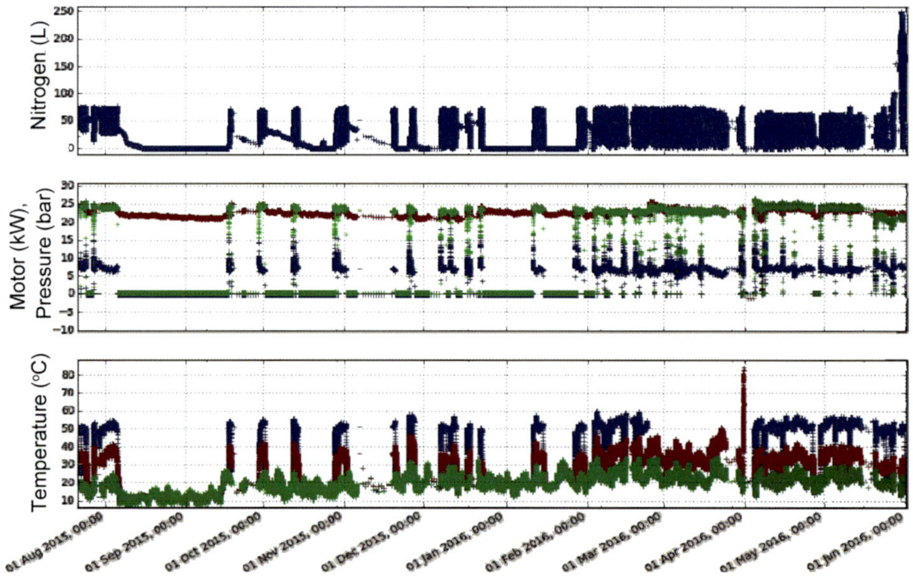

Figure 3: Running history of cryocooler over the 11 months of use. The top plot is the quantity of liquid in the buffer Dewar. The middle plot shows the oil pressure (blue), mean gas pressure (red) and motor power (green). The lower plot shows the oil temperature (blue), machine temperature (red) and ambient temperature (green).

Liquefaction Rates, Condenser Performance, Pulse Tube Performance

Figure 4 shows a typical production for two days of operation with the liquefaction rate approximately 11 liters per hour. The water temperature shows that the fans and radiators of the cooling circuit maintained the cooling water at approximately 4 degrees above the ambient temperature. The motor power shows a slight variation synchronized with the ambient temperature. The machine temperature affects the working gas pressure and therefore the acoustic power input.

The three pulse tubes consistently held temperatures within 4 degrees of each other. The thermal link between the pulse tubes had a conduction capacity of 0.5 W/K [3] which means that the pulse tube performances were very similar, transferring only 2 W between them.

The motor power input was steady at 24 kW when warmed up. The cooling system's circulation pump and radiator fans consumed 2.5 kW of electricity. The energy cost for the nitrogen liquefied, at a rate of 11 litres per hour was therefore 2.5 kWh/litre.

Liquefier System

Over the course of the year, optimizing the condensing and liquid handling system was carried out with many variations of the Dewar pressure, and control scenarios tried. There were combinations that produced stable liquefaction and others that could produce run-away freezing due to liquid build-up in the condenser. Run-away freezing of the nitrogen occurs when the liquid nitrogen film on the condensing surface becomes too thick, causing a temperature gradient through the liquid. As the film thickness increases, the temperature gradient increases, and the cold head temperature decreases, until the cold head reaches 64K and freezing occurs. The frozen layer then quickly thickens and blocks the condenser, at which point the cold head temperature plummets to its no-load temperature of approximately 40 K. Once the cryocooler was stopped and warmed up above the freezing temperature of 64 K, the liquid drained away and normal operation could be resumed.

Event

After initial bug fixing during installation and commissioning, the cryocooler settled down to steady operation. There was only one event in the 11 months of operation. On the 30th of March, a fitting started to leak cooling water into the cryostat. Cooling water filled the cryostat and the cooling

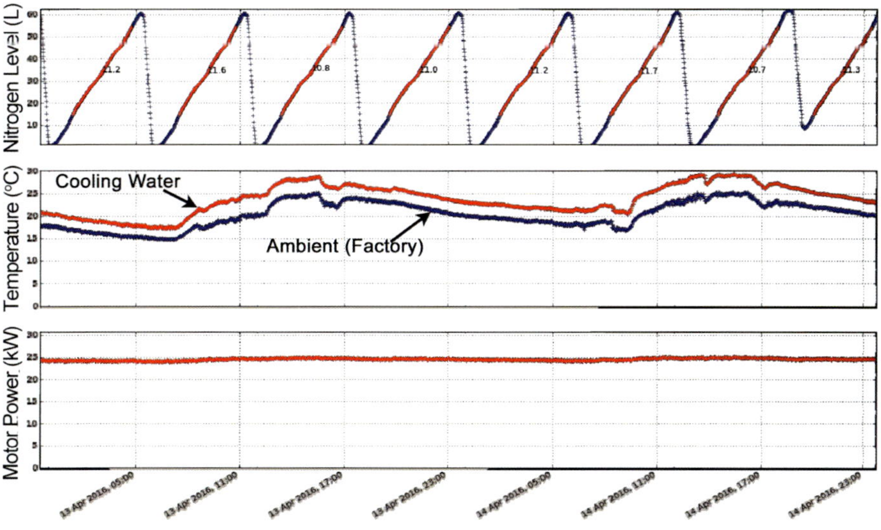

Figure 4. Two typical days of production. The top plot is the liquid nitrogen level in the buffer Dewar, with the average production rate for each filling cycle. The middle plot shows the ambient temperature in the factory (blue) and the cooling water temperature (red). The lower plot shows the motor power.

water reservoir eventually ran out. The loss of cooling water caused the temperature to rise (Figure 3 shows a temperature spike just before 1st of April) and the machine's safety interlocks shut it down.

The data logging and reporting back to base told the story as shown in Figure 5. The whole sequence took approximately 20 minutes. Firstly a small rise in condenser temperature at 07:48 indicated that the vacuum in the cryostat was degrading (when the water entered the cryostat), next the outlet water temperature became unstable when the water supply started getting low and produced an unsteady flow. The machine temperature started to rise and the sudden rise in the condenser temperature at 8:07 occurred when water reached the cold parts of the liquefier. Soon after, one of the safety switches was tripped and it stopped the machine. The pool of water in the cryostat protected the pulse tube aftercoolers from overheating. All the other machine signs such as gas pressure, pressure ratio, motor power, looked normal.

We were able to remotely diagnose the above scenario and made the decision to bring the cryocooler back to base. The cryocooler was removed from the liquefier on-site and returned to base on a Thursday, the water leak was repaired, cryostat cleaned out and the wet MLI replaced on the Friday. As a precaution, we replaced any sensitive parts that may have been affected by the temporary thermal load. The cryostat and gas spaces were vacuumed over weekend to remove left over moisture from the leak. It was refilled with helium, delivered, fitted into the cryocooler and back in service on Monday. The water fitting that caused the sequence of events has been removed from the production version to prevent the event from recurring.

PULSE TUBE DEVELOPMENTS ON PTC330

The testing and development of the single pulse tube cryocooler, the PTC330, was focused on confirming manufacturing improvements and management schemes for steady nitrogen condensation. The original pulse tube prototype was constructed with bolted joints between the regenerator, cold head and pulse tube to allow experimental optimization of the geometry. This design, whilst required for experimentation, is not a viable production solution. Once the regenerator proportions had been finalized, a design incorporating brazed joints was tested.

Another improvement feature was the pattern of the slots in the aftercooler and warm heat exchanger. The original design used radial slots with a pin filling the hole in the centre where the slots intersected, similar to the heat exchangers used by Ercolani et al [6]. In the original pulse tube the pin loosened over time. An alternative heat exchanger pattern was devised which allowed cutting of the slots without a central hole, thus eliminating the pin.

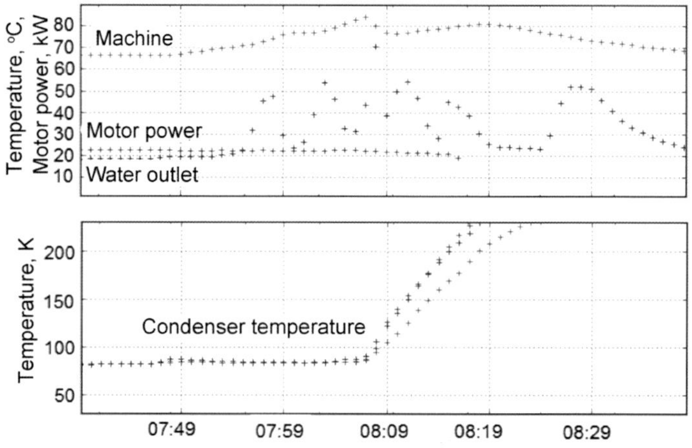

Figure 5. Data logger record of event showing (from top to bottom at start of record): machine temperature monitor (green), motor power (blue), water outlet temperature (red) and on the lower plot the three pulse tube condenser temperatures.

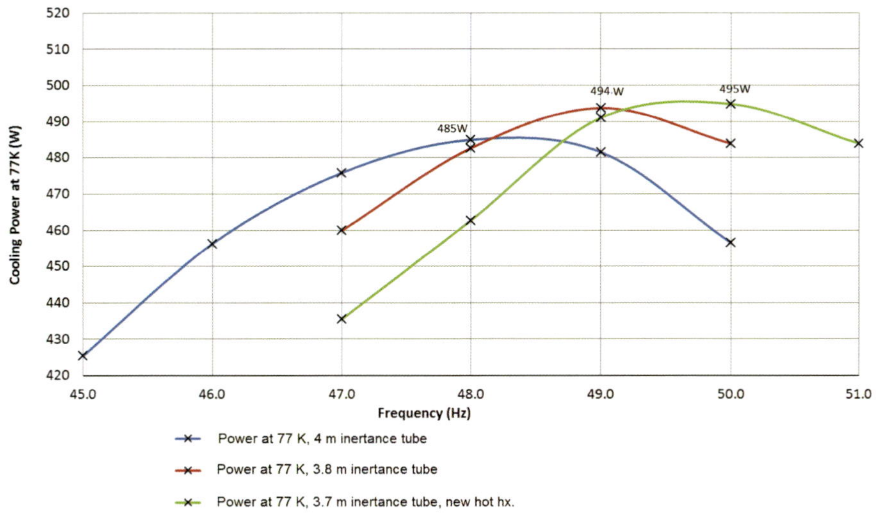

Figure 6. Frequency sweeps for the production design PTC330's cooling power at 77 K; tuning for 50 Hz running and the new hot heat exchanger design.

Figure 6 shows that tuning of the inertance tube length and incorporation of the new hot heat exchanger design has increased the operating frequency from 48 Hz to 50 Hz, with an increase in cooling at 77 K from 485 W to 495 W. The ability to tune for a given operating frequency with minimal effect on the refrigeration power or efficiency is a function of the non-resonant DPWG. It is very useful if the larger system that the cryocooler is installed into has a particular operating frequency to avoid.

The PTC330 has been configured as a liquefier and was installed for endurance testing at the gas company alongside the PTC1000. It has been producing liquid nitrogen at a rate of 4.7 liters per hour.

CONCLUSIONS

Fabrum Solutions and Callaghan Innovation now have an excellent endurance test site for cryocoolers. The gas company provides an industrial environment, real duty cycles, and a commercially viable use for the liquid nitrogen produced. The close location of the site to the development team aids monitoring but there is enough separation to provide the operation of the liquefier by the gas company and therefore provides a real-environment test.

The DPWG has proved to be a reliable pressure wave generator for the pulse tubes and ran for 3338 hours over the 11 months of testing. It ran in a dusty, un-insulated, environment throughout a winter and a summer, with large daily temperature swings. This is seen as an exceptional result as the DPWG used in the test was the first prototype of the 1000 cc DPWG. The DPWG technology has improved considerably since its design and construction and the alpha-prototype DPWG in the liquefier has now been replaced with a production version. The next step for the alpha-prototype DPWG is to strip and inspect so that learnings from its running can be incorporated into future machines.

The pulse tubes produced consistent performance liquefying at approximately 11 l/hr. Condenser design and management are continuing to be developed. The radiator based cooling system has proven itself over a wide temperature range.

Performance improvements and production refinements on the PTC330, single pulse tube cryocooler have resulted in a liquefier producing 4.7 liters per hour.

The next steps are to continue running the PTC1000 liquefier alongside the PTC330 based liquefier at the gas company. The reduced cost of liquid nitrogen production, made possible by the pulse tube based liquefiers, has allowed the gas company to secure more customers and has increased liquid nitrogen demand. These two liquefiers will be joined by three more in four months' time, making a total of five units producing commercial liquid nitrogen.

ACKNOWLEDGMENT

The authors acknowledge Callaghan Innovation, New Zealand, for support of this work, New Zealand's MBIE for funding, and Fabrum Solutions Ltd for supporting the commercialization of the technology.

REFERENCES

1. Emery, N., Caughley, A., Nation, M., Tanchon, J., "Large Pulse Tube Developments," *Cryocoolers 18*, ICC Press, Boulder, CO (2015), pp. 180–186.

2. Caughley, A., Emery, N., Nation, M., Allpress, N., Kimber, A., Branje, P., Reynolds, H., Boyle, C., Meier, J. and Tanchon, J., "Commercialisation of Pulse Tube cryocoolers to produce 330 W and 1000 W at 77 K for liquefaction," In *IOP Conference Series: Materials Science and Engineering*, vol. 101, no. 1, p. 012060. IOP Publishing, 2015.

3. Caughley, A., Emery, N., Nation, M., Allpress, N., Kimber, A., Branje, P., Reynolds, H., Boyle, C., Meier, J. and Tanchon, J., "Commercial pulse tube cryocoolers producing 330 W and 1000 W at 77 K for liquefaction," *IEEE Transactions on Applied Superconductivity 26, no. 3 (2016): 1-4.*

4. Caughley, A., Tucker, A., Gschwendtner, M. and Sellier, M., "Novel diaphragm based Stirling cryocooler," *Adv. in Cryogenic Engineering*, Vol. 57, Amer. Institute of Physics, Melville, NY (2012), 667–674.

5. Caughley, A., Branje, P. and Klok, T., "30 kW Metal Diaphragm Pressure Wave Generator," *Adv. in Cryogenic Engineering*, Vol. 59, Amer. Institute of Physics, Melville, NY (2014).

6. Ercolani, E., Poncet, J.M., Charles, I., Duband, L., Tanchon, J., Trollier, T. and Ravex, A., "Design and prototyping of a large capacity high frequency pulse tube," *Cryogenics* 48, no. 9 (2008)*: 439-447.*

Quantifying Cryocooler Contamination

P. S. Spoor

Chart Industries
Troy, NY 12180

ABSTRACT

Contamination of the cryocooler working gas (usually helium) with other species results in performance degradation if those species freeze at or below the cooler's intended process temperature. Even relatively small amounts of contaminant can result in harmful blockage of the fine fluid passages in the cold zones[1,2]. Fortunately, avoiding contamination is fairly straightforward: use clean parts and low outgassing materials, and vacuum-bake (or hot dwell-and-purge[1]) the assemblies to remove the remaining volatiles. Verifying cleanliness, however, may require many months of run-time to establish temperature stability. For low-cost mass production and rapid R&D cycles, it is very desirable to have a method for actually quantifying the level of contamination in a cryocooler against a pass/fail specification at the time of manufacture. This paper proposes a very simplified model of outgassing that leads to a quantitative estimate of available contaminants in milligrams, based on the accumulated vapor-pressure and the RGA measurements on evacuated units. This estimate can be compared to a figure-of-merit derived from contamination studies in the literature, hence yielding a pass-fail test for cleanliness.

INTRODUCTION

The issue of contamination of working gas in cryocoolers is widely recognized[1,2], as are many of the best practices for minimizing it. The impact of specific levels of contamination has been studied, as well[1]. The measurement of contamination level in cryocoolers is not as well understood, however. In the cited studies, the contamination is deliberately introduced, so its level is at least approximately known before the start of the test. On the other hand, contamination coming from within the cooler itself is challenging to measure, because it is stored not in the working gas but on surfaces, in non-metallic or porous solids, and in any trapped volumes. Analyzing a sample of the helium from a cryocooler does not tell us the actual contaminant level of interest, because:

1. the cryocooler acts as its own cryopump, so as long as the cooler is running, it continually freezes out any contaminants in the working gas.
2. when the cryocooler is not running, the partial vapor pressure of any contaminant is limited by the balance of release vs. reabsorption.

What we really want to know is the quantity of contamination that could potentially outgas over a long period of time, e.g., how much would outgas if we could continually keep the vapor pressure low.

THEORY

To begin, we start by making some simplifying assumptions. By far the most common and stubborn contaminant is water; occasionally there are other volatiles such as methanol from cur-

Cryocoolers 19, edited by S.D. Miller and R.G. Ross, Jr.

ing epoxies, but we can take the analysis pretty far by assuming water as the only contaminant of interest (for now). There may be several simultaneous outgassing processes, each with its own characteristics; nonetheless, it is generally true that the release of contaminants slows down as the amount available is reduced. If we make the very simplifying assumption that the overall outgas rate is proportional to the amount of (unfrozen) contaminant in a cooler, we can write:

$$m'(t) = m_0(1 - e^{-t/\tau}) \quad (1)$$

where

$m'(t)$ = the mass of contaminant frozen in the cold zones at time t;

m_o = the total mass of contaminant in the cryocooler available for outgassing at time $t = 0$; and

τ = a time constant, characteristic of a given size and type of cryocooler.

In other words, the outgassing process is assumed to obey a relaxing exponential, with an asymptote at m_0, and where $m'(t)$ reaches 63% of m_0 at time $t = \tau$. The buildup of pressure in an initially evacuated vessel due to outgassing is fundamentally similar, except that the pressure saturates not because the available contaminant has been exhausted, but because the release and reabsorption of contaminant are in equilibrium. Nonetheless, there is an important connection between them: if (vapor) pressure is uniform in a cryocooler during an outgas test (a big if, we should note) then the initial rate of pressure buildup roughly corresponds to the rate of frozen mass accumulation that the cryocooler would experience if running. In other words, if the cryocooler is evacuated, then isolated from vacuum, the pressure will rise according to

$$p(t) = p(\infty)(1 - e^{-bt}) \quad (2)$$

where the relationship between m_0 from Eq. (1) and $p(t)$ is

$$m_0 \simeq \tau \left.\frac{dp}{dt}\right|_{t=0} \cdot V \frac{M}{RT_H} \quad (3)$$

Here V is the cryocooler internal gas volume, R is the universal gas constant, T_H is the temperature of the outgas test (which is chosen to equal the expected, or, to be conservative, the hottest expected operating temperature of the warm parts of the cryocooler), and M is the molecular weight of the dominant vapor (presumed to be water in most of our analysis).

Finding m_0 and τ

Without knowing τ, we can't get m_0 directly from Eq. (3). However, τ can be deduced from a sequence of two outgas tests, where the crycooler is vacuum baked at the same temperature T_H for a time Δt in-between the two tests. In this case, the two outgas tests are only run long enough to establish an initial slope, for a much shorter time than Δt. Using the familiar dot notation for time-derivative, and the subscript i for "initial", we can say that

$$\frac{\dot{p}_{i,2}}{\dot{p}_{i,1}} \simeq e^{-\Delta t/\tau} \quad (4)$$

In principle, the initial rate of increase of vapor pressure and the ultimate saturation pressure are both proportional to the source mass, so either one can be used to find the parameters of interest. This redundancy permits independent calculations for the same value, which can be compared for consistency. Disagreement between the calculations can lead to insights and ways to improve the method beyond the very simple version presented here (for instance, how to handle multiple outgassing processes with different time constants).

For example, if there exists a bakeout temperature $T_B > T_H$ that can accelerate the bakeout process (in other words, if the cryocooler can be baked out at a higher temperature than its expected operating temperature) then m_0 can be deduced from a sequence of two outgas tests at T_H bracketing a dwell (where the vapor pressure is allowed to build) at T_B. Knowing the volume, the temperature, and the gas species reveals how much mass Δm was removed between outgas tests; the higher bakeout temperature ensures that Δm is much greater than what is removed during the outgas tests at T_H (recall that vapor pressure tends to rise exponentially with temperature, so even a modestly higher bakeout temperature will have a big effect). Supposing we use the saturation pressures in the outgas tests before and after the dwell; then

$$\frac{p(\infty)_2}{p(\infty)_1} \simeq \frac{m_0 - \Delta m}{m_0} \tag{5}$$

In a similar fashion, if a vacuum bake (or other process, such as a hot pressurized dwell-and-purge[2]) is inserted between two outgas tests, and run for a controlled time, the reduction in outgas saturation pressure or initial outgas rate can be used to predict the time required to reduce the contamination level below a given threshold.

Threshold Criteria for Contamination

Equations (1) through (5), combined with the proper series of outgas tests, can determine the amount of contaminant in a given cooler and how quickly it can be eliminated. It remains to establish how much contamination can be tolerated. For that we turn to the literature, and subsequently, to our own experiments running coolers with contamination that has been quantified. There are numerous studies published on the effects on contamination on cryocoolers; one of the most thorough is a 1998 study[1] by Hall, Ross *et al*, detailing the effects of known amounts of deliberately introduced contamination on a pulse-tube cryocooler's ultimate (no-load) temperature. This study concluded that amounts of water vapor (the most prevalent and most stubborn contaminant) below about 50 ppm would keep the temperature from rising more than 1K. Our own experiments, on the other hand, seem to indicate that the threshold is closer to 150 ppm. One intuitively appealing way to reconcile these results is to state the contamination level a different way. Because the assumed mechanism by which contamination harms performance is by freezing and blocking flow passages, we can argue that the important statistic is not ppm level in the working fluid (or theoretical ppm level that might exist if all the available contaminant could be suspended as vapor in the working fluid at one time) but the actual mass per unit area of heat exchanger or regenerator area.

On that basis, our results and those of Hall & Ross are fairly consistent, and suggest that a contaminant load of ~0.01 mg per mm^2 of regenerator area will cause a ~1K rise in no-load temperature (H&R observe how astounding it is that such a small amount, covering the RGR cross-section with a layer a only few molecules deep, could have any measurable effect). Hence:

$$T_c(t) \sim T_{c,i} + \frac{m'(t)}{A_{RG} \cdot 0.01\frac{\text{mg}}{\text{mm}^2}} \tag{6}$$

If Equation (1) is used for $m'(t)$, then the empirical knowledge from the literature is combined with the simple theory expressed by Equations (1) through (5) to yield a prediction that can be tested against actual cryocooler behavior.

RESULTS

The method outlined here was used to verify that the contamination load in a particular cryocooler design was more than sufficient to explain its warming trend, and to identify which components were the major sources of contaminants Not surprisingly, the single biggest source turned out to be the stator VPI compound, which had been recommended to us by the manufacturer as "low outgassing" but was actually uncurable by any bakeout process we could discover.

We decided to set our warming target to $\Delta T_c \leq 1$K (i.e. the cryocooler should not warm up more than 1K due to contamination); in this particular cryocooler design, with a capacity of ~30W at 77K, the regenerator area is ~1460 mm^2, so Eq. (6) can be rewritten as

$$\Delta T_c(t) \sim 0.07\frac{\text{K}}{\text{mg}} m'(t) \tag{7}$$

Eqn. (7) suggests that for this cryocooler, to keep $\Delta T_c \leq 1$K, m_0 must be ≤ 15 mg. We performed series of outgas tests on all the cryocooler components, to obtain a 'contamination inventory'.

Figure 1 shows the results for a pair of motor stators; a sequence of two outgas tests were done at 60C (the expected max temperature of operation) with a 'dwell' at 90C in-between (the solid lines are exponential fits to the data; note here that our DAQ system was still getting debugged and there are frequent dropouts).

From the volume of the vacuum vessel used in these tests, the species of contaminant removed (identified by RGA) and the temperature, we can tell how much mass was released during the dwell (and subsequently pumped out before the second 60 °C outgas test). Because the vapor pressure was so much higher during the 90 °C dwell (~9000 mTorr. maximum) we can neglect the small amount evolved during the 60 °C tests. The modest change in slope between the two tests tells us that only a small fraction of the total contaminant load was removed. Overall, the tests suggest that the stators start out with over 90 mg of contaminant, and that it would take over 20 cycles of hot dwell and pump-out to reduce the level down to single digits (or equivalently, over a month of constant vacuum bakeout at 90 °C). The stators were found to be the biggest single source of contaminant, and the cleanliness of the units was vastly improved by changing VPI compounds.

This is not *prima facie* a surprising result, that the stator potting compound would be a significant source of contamination for a cryocooler with motors housed inside the pressure vessel; but quantifying the contamination and the minimum processing time to remove it helped to justify the cost of discarding existing inventory and buying new parts.

Contaminated vs. clean cryocooler tests

Our outgas tests measured the effects of various bakeouts and dwells, and hastened refinements to our manufacturing and processing, resulting in coolers with an estimated contamination load at or below 15 mg. Not only was the stator VPI compound changed, but we also began baking out coldheads separately from compressors (they can stand a higher temperature), among other process improvements. The original, heavily contaminated prototypes provided a good means of contrasting the 'before' and 'after' cryocooler temperature stability. In particular, we could use Eq. (7) to predict the temperature rise of each cryocooler, and compare it to actual measurements as a means of evaluating the validity of the overall approach to understanding contamination.

Figure 2 shows the comparison for the first week or so running a cryocooler with a total estimated contaminant load of 160 mg. Two things are apparent from this graph; one, the agreement appears excellent, and two, the warmup rate at this level of contamination is easily detectable within a few days, since the fastest warmup occurs in the beginning (if we believe the functional form of Eq (1) and its cousins).

However, tests on 'clean' cryocoolers provide a more nuanced view. One challenge that became apparent to us in the process of setting up a fleet of cryocoolers to run long-term is that the running conditions have to be tightly controlled in order to verify that a change of less than 1K in coldtip

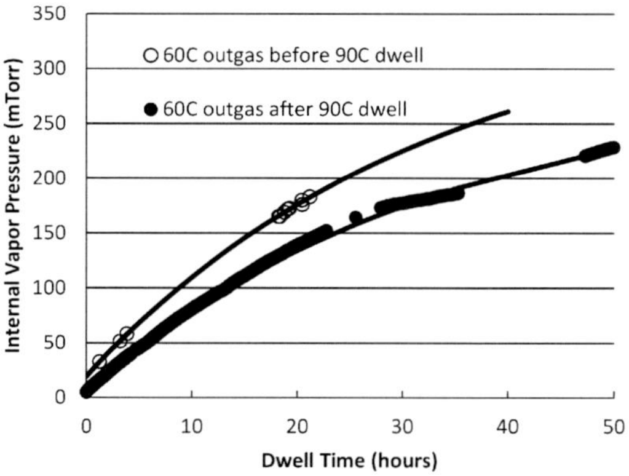

Figure 1. Vapor pressure versus time for two 60°C outgas tests, one before and one after a "dwell" at 90C, of two linear motor stators in a 3.5-liter vacuum vessel.

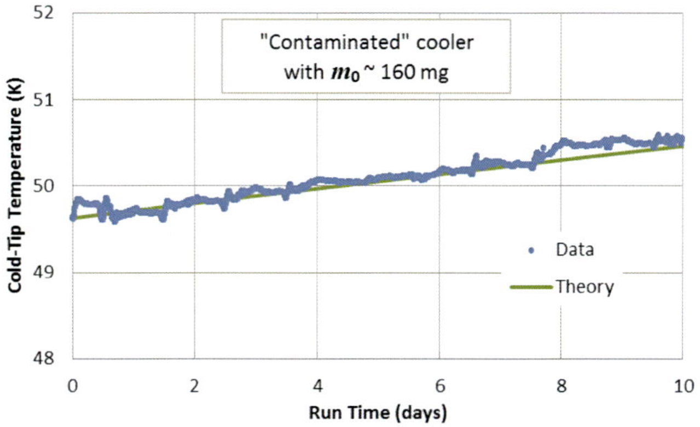

Figure 2. Contaminated cooler, for its first few days of running, compared to theory.

temperature is indeed due to contamination. The cryocooler of Figure 2 was water-cooled, with a dedicated controlled-temperature bath; but for a fleet of many coolers this would not be practical (and the fundamental design for these assumes air-cooling). Changes in room temperature cause changes in internal charge pressure, which result in power consumption fluctuations as well as variations in reject temperature. Mapping the performance of each cooler over various conditions enables some correction for these variations, but these corrections are imprecise. Furthermore, there are many variables that can affect the no-load temperature of a cryocooler, especially during its early burn-in period, which are hard to control or correct for (slight wear of clearance seals that have incidental contact, for instance).

Finally, as the contamination load is reduced to fairly low levels, we find that the estimates based on initial outgas rate and the estimates based on saturation vapor pressure don't agree particularly well. This is due to a variety of factors; for instance, the vapors outgassed in one of these coolers include quite a bit of hydrogen from the stainless-steel. We could try to de-gas the SS, but the hydrogen will not affect the performance of these coolers. It does, however, mean that the evolution of the 'light' components has to be somehow separated from the 'heavy' components, using RGA spectra, etc., and this makes resolving the initial slope of the heavy-component vapor-pressure curve particularly challenging.

Figure 3 shows a comparison of theory and experiment for a relatively 'clean' cooler, over its first few days of running. Here, the contamination load is given as a range of values, since the two methods of measuring contamination do not agree especially well. This cooler does not have a particularly steady temperature, due to fluctuating conditions; it is difficult to draw a conclusion about

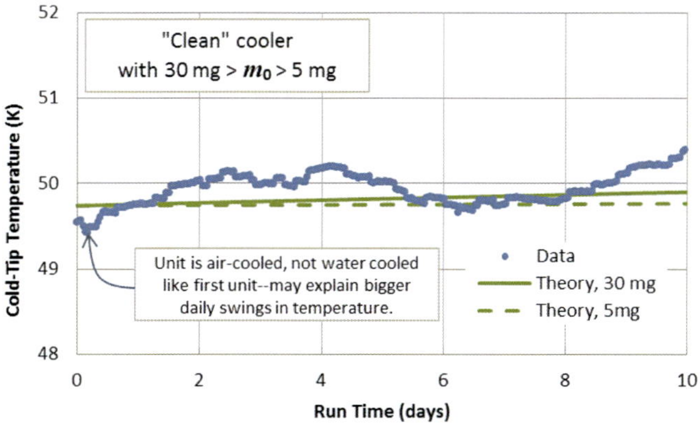

Figure 3. "Clean" cooler over 10 days.

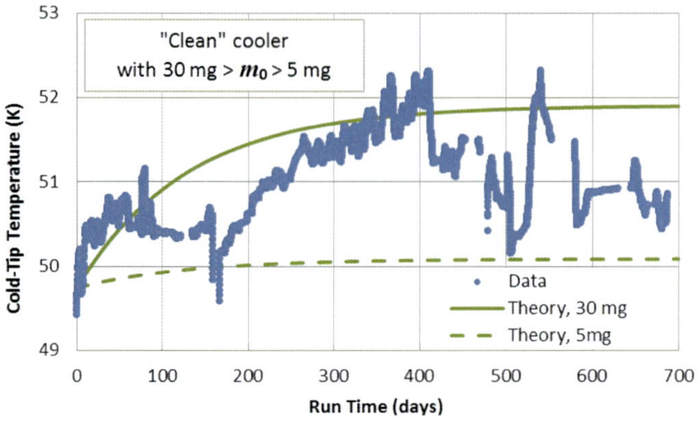

Figure 4. "Clean" cooler over two years.

the comparison of theory to experiment in this case (indeed, the amount of expected temperature rise could hardly be resolved in this length of time).

As it happens, this paper is written roughly two years after the bulk of the work described here was accomplished; and so we have a much longer record for this and other cryocoolers in this study. Figure 4 shows the cooler of Figure 3 but after almost 700 days of running. During that time there were observed some rather significant fluctuations in cold-tip temperature, not all of them understood. One likely cause of some (perhaps all) of these fluctuations is variation in vacuum level in the vessel enclosing the coldhead. In our run-in racks, several coolers' vacuum vessels are served by a common vacuum pump; whenever a cooler is added or removed from the rack, the individual coolers' vessels must be valved off from the common manifold. Occasionally one would be left valved off for some time, resulting in steady degradation of the vacuum surrounding the coldhead. The data over time clearly show these trends, yet quantifying the effects of vacuum loss on coldhead performance have proved challenging, and the inexpensive (Pirani) gauges we typically use have shown inconsistencies as well, so the data cannot be rigorously corrected for changes in vacuum level.

The overall trend for two years, however, seems to land squarely between the contamination load estimates. In a sense, this justifies our reliance on the contamination measurement as a leading indicator of cryocooler health, before we had enough long-term data to smooth out the short-term changes.

Further refinements to this method (in analysis and measurement techniques) may enable confident predictions regarding cryocoolers or components with very low levels of contamination. Its most obvious utility at this point is in estimating performance penalties or processing times in systems with much higher contamination levels.

CONCLUSIONS

Based on the results in the previous section, it would appear that the method we have outlined for quantifying cryocooler contamination is reasonably accurate and can reliably predict the long-term warming trend of a cryocooler. However, this appearance must be at least somewhat coincidental. There are a myriad of simplifying approximations that have been made in order to arrive at a method that is amenable to simple measurements and rules of thumb. Some of these are discussed in the beginning of the "Theory" section, such as the assumption that there is one dominant outgassing process, with one overall time constant. Others involve how the tests are performed.

For one, we use Pirani gauges for all the vapor-pressure measurements described here; but these are not especially accurate above 1000 mTorr and are especially inaccurate for water vapor, the main species of interest. For another, we find that the evolution of vapor pressure in a device or vacuum vessel containing a component often does not conform to the simple relaxing-exponential form of Equation (2); sometimes there are very clearly two relaxing exponentials (in other words, two superimposed exponentials provide a much better fit to the data), and sometimes there is a linear component that must be subtracted out, as described in "Results". This may be due to a small leak, or in the case of a cryocooler or compressor, it may be hydrogen outgassing, which comes to dominate the pressure trace (strictly speaking, the hydrogen outgassing rate is slowing down with time, but the curvature is not always detectable on the time scale of our experiments). An exponential component is not always visible against this background. For another, as a device or component becomes clean, it may well be cleaner than the vacuum lines used to connect it with a gauge or RGA, especially if these are the same lines used for bakeout. Notwithstanding attempts to maintain the vacuum lines themselves at an elevated temperature, it is often the case that water-vapor outgassing from a fully processed cryocooler can't be detected above the background in the vacuum lines.

Rather than a definitive study, this work should be viewed as an introduction to what is possible, a frame on which refinements can be added. Good baseline measurements with capacitive manometers in place of Pirani gauges, and a more sophisticated view of outgassing mechanisms will perhaps take this method to a point where it is more generally useful.

ACKNOWLEDGMENT

This work benefited greatly from discussions with colleagues knowledgeable about vacuum systems and cryocoolers. Thanks in particular are due to Keith Gustafson, of Chart Industries, and Willy Gully (formerly of Ball Aerospace, now retired).

REFERENCES

1. Hall, J. L. and Ross, R. G. Jr. "Gas Contamination Effects on Pulse Tube Performance," *Cryocoolers 10,* Springer US (1999), pp. 343-350.

2. Castles, S., Price, J. D., Glaister, D. S., Gully, W., Reilly, J., Nast, T., Kotsubo, V., "Space Cryocooler Contamination Lessons Learned and Recommended Control Procedures," *Cryocoolers 11,* Springer US (2001), pp. 649-657.

Reliability Evaluation of Stirling Cryocooler for an Electric Vehicle High Temperature Superconducting Motor System

K. Yumoto, K. Nakano and Y. Hiratsuka

Sumitomo Heavy Industries, Ltd.
Nishitokyo-city, Tokyo, Japan

ABSTRACT

Due to the performance improvement of high temperature superconducting wires, the development of superconducting motors has been actively developed. To demonstrate the efficiency improvement of an electric vehicle driven by a superconducting motor, development of superconducting motors and cooling systems has been performed by the joint research of Sumitomo Heavy Industries, Ltd. (SHI) and Sumitomo Electric Industries, Ltd. (SEI). The results of the actual vehicle tests and simulations showed that it is possible to improve the efficiency by about 10% compared to a conventional electric motor. In this project, SHI developed a high-efficiency Stirling cryocooler with a cooling capacity of 151 W at 70K and a compressor input power of 2.15 kW with a corresponding COP of 0.07.

More compact, efficient and reliable cryocoolers are required for the commercial use of a superconducting bus. In this paper, the investigation of the reliability of the cryocooler is introduced. The operation system, the safety measures, and the long-term operation and environmental performance test results will be presented.

INTRODUCTION

High Temperature Superconducting Motor System Development Project

After the high temperature superconductivity (HTS) was discovered, the application of HTS devices has been studied in many fields. The application of HTS equipment is possible for commercial use. In addition, superconducting technology has been attracting attention as a means to resolve environmental issues such as energy conservation and carbon dioxide emissions reduction. In the field of transportation, efforts have been made to improve the efficiency of an electric vehicle. Because of its high efficiency in at low speeds, a superconducting motor is a potential choice for a bus, which requires frequent stops.

Recently, the development of a superconducting motor and cooling system has been completed by the joint research of Sumitomo Heavy Industries, Ltd. (SHI) and Sumitomo Electric Industries, Ltd. (SEI). This project aims to improve the overall electric efficiency of the motor by 10% relative to a conventional electric motor. The motor's superconducting state is maintained by circulating liquid nitrogen which is cooled by a cryocooler. To improve the total efficiency of a superconducting motor system, it is vital to improve cryocooler performance. Thus SHI has been developing a high-efficiency Stirling cryocooler for such superconducting motor systems.

Cryocoolers 19, edited by S.D. Miller and R.G. Ross, Jr.

Stirling Cryocooler for Superconducting Motor System

Owing to its high cooling capacity and efficiency, a Stirling cryocooler has been developed for cooling a superconducting motor. It is possible to achieve a high efficiency at liquid nitrogen temperature range since there is no valve loss in a Stirling cryocooler and low temperature space P-V work can be recovered. Figure 1 shows a photograph of the cryocooler under discussion. The schematic cross-section diagram of a Stirling cryocooler is shown in Figure 2. A split-type Stirling cryocooler was selected because the compressor and expander can be arranged independently in the motor room. Helium gas is charged in the cryocooler and the initial gas pressure is 1.7 MPa. The compressor consists of a moving-magnet type motor and two opposed pistons which are driven by the linear motor. The moving cylinders are guided by flexure bearings, which can maintain clearance of several micro meters between the pistons and cylinder. A water-cooled heat exchanger is built in the outer body of the compressor to transfer heat generated by the motor. The expander consists of a cold-head, a regenerator, a heat exchanger and a free-piston type displacer. The displacer piston and the regenerator are coaxially arranged. The displacer piston is also guided by flexure bearings. The regenerator is packed with thousands of stainless-steel screens. The heat exchanger in the expander is a shell and tube type and is also water-cooled. To suppress vibration from the displacer, a vibration absorber is attached.

As to the cryocooler performance, a cooling capacity of 151 W at 70K with a compressor input power of 2.15 kW and cooling water temperature of 30 ℃, corresponding COP of 0.07, and a no-load temperature of 33 K, has been achieved[1].

Figure 1. Photograph of Stirling cryocooler developed.

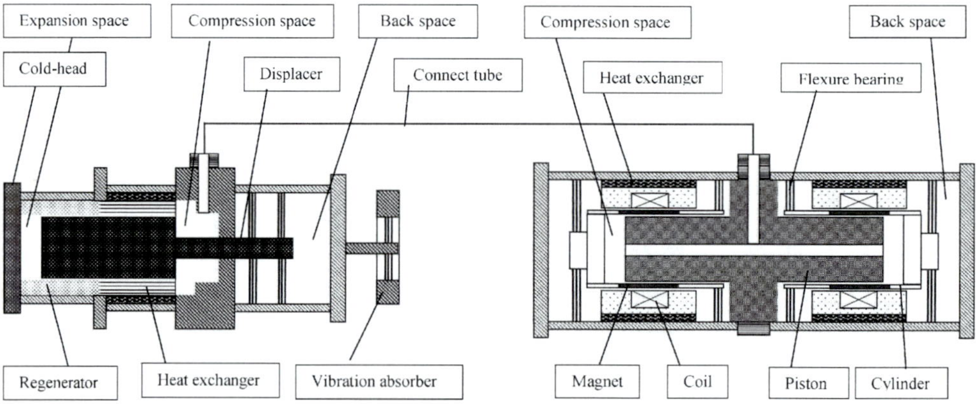

Figure 2. Schematic cross-section diagram of the expander and compressor.

Actual Vehicle Test

After the cryocooler performance was tested, the cryocooler and superconducting motor unit were mounted in an electric bus, and preliminary driving tests were conducted. The cryocooler was driven by an inverter power source and the input power was controlled by a proportional–integral–derivative (PID) algorithm to maintain a constant cold-head temperature. For safe operation of the superconducting motor system, the inverter power source is designed to shut-down automatically if any abnormality is detected. For example, the voltage and current levels, the cryocooler outer wall temperature and the displacer piston stroke are monitored by a control system. Cooling water for the heat exchanger is supplied from the electric bus radiator. Actual bus running test and simulation of the running pattern were carried out. As a result, it was found that the efficiency can be increased by about 10% compared to a conventional motor if the weight of a superconducting motor is the same as that of a conventional motor[2].

RELIABILITY EVALUATION

For practical use of a superconducting motor, high reliability and safety is also required. Some experiments on the reliability evaluation have been conducted with the afore-mentioned cryocooler. In this paper, results of long term operation and vibration tests will be introduced.

Long Term Operation Test

In the practical use of a superconducting motor, the cryocooler will be continuously operated for several years in order to suppress the liquid nitrogen boil-off. Therefore, the cryocooler is required to maintain its cooling capacity over thousands of hours. After a short-term vacuum baking, the cryocooler has been operated at a lowest temperature of 33K for about 5,000 hours to detect potential initial failures. Initially, the cryocooler was operated with a cooling water of 30℃. In order to investigate the performance under severe conditions, the cooling water temperature was changed to 50℃ after 2,500 hours operation. The results are shown in Figure 3. As a result, the cooling capacity decreased 4.5% compared to that at the initial status and no load temperature rose about 3.5K. Mechanical factors and contamination are considered as causes of the performance degradation. Measures and isolation of performance degradation factors is under considering at the moment and will be solved in near future.

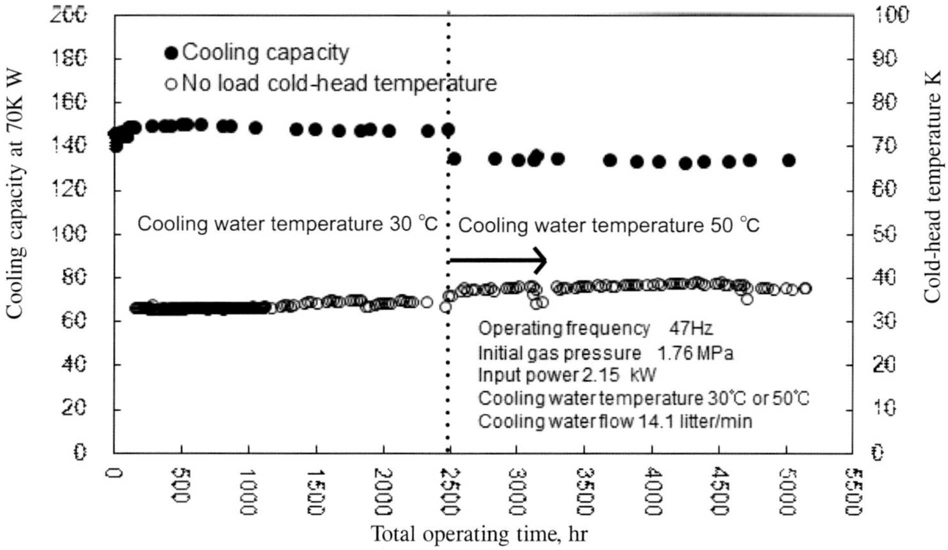

Figure 3. Experimental results of long term operation test.

Vibration Test

As an automotive cryocooler, it is important that the impact from road and cyclic vibration from the motor be considered in the reliability evaluation. In order to make the cryocooler more robust against vibration, it is necessary to measure the impact pattern, simulate the vibration mode and conduct a vibration test. Some simple vibration tests were conducted to measure the basic anti-vibration performance and vibration failure modes of the cryocooler. Figure 4 shows a photograph of the cryocooler unit mounted on a vibration exciter. The cryocooler was mounted as close to the same orientation as the actual vehicle test condition as was possible. Figure 5 explains the definition of the vibration direction and position of the acceleration sensors. The acceleration sensors were attached to the cryocooler body at four points. The vibration of the moving parts of the cryocooler (cylinder in the compressor, the displacer in the expander and the vibration absorber) was also measured through view ports by laser vibration detectors.

First, a random vibration test was performed to investigate the natural frequency and response magnification of the cryocooler body. A random vibration tests includes frequency components of

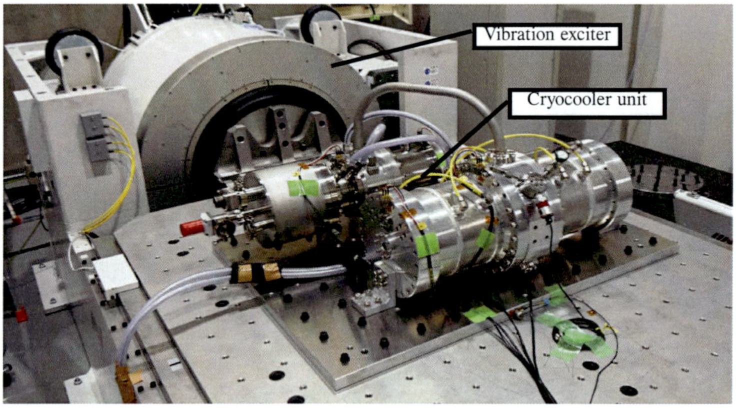

Figure 4. Photograph of cryocooler unit mounted vibration exciter.

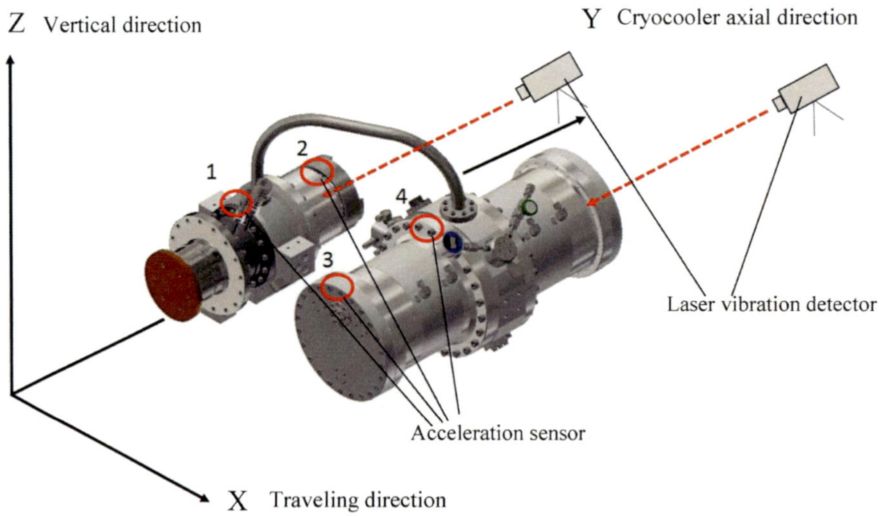

Figure 5. Three dimensional diagram of cryocooler and definition of the vibration directions and position of acceleration sensors.

10 to 1000 Hz which is exerted on the cryocooler. This test was conducted without operating the cryocooler. Figures 6, 7 and 8 show the measurement results of the response magnification factors in the X-axis (traveling direction), Y-axis (cryocooler axial direction) and the Z-axis (vertical direction), respectively.

Second, the natural frequency and the response magnifications of the moving cylinders in the compressor and the displacer in the expander and vibration absorber were measured. A sinusoidal vibration of 10 to 200 Hz and 0.5grms acceleration in the axial direction was applied to the cryocooler. The vibration of the cryocooler body and the moving parts was measured simultaneously using the two laser vibration detectors. The test was also conducted without operating the cryocooler. Figure 9 shows the response magnification and the phase with respect to the input of the compressor moving cylinder. A natural frequency of 33.1 Hz was observed and the maximum response magnification was about 10.8. Figure 10 shows the response magnification and the phase of the expansion displacer. As shown in Figure 10, the response magnification has a broader peak in comparison to the compressor. A natural frequency of 48.2 Hz was observed and the maximum response magnification was about 2.7. Figure 11 shows the measurement results of the response

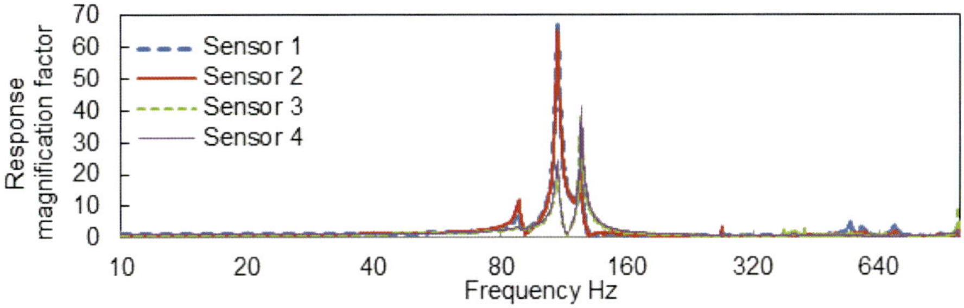

Figure 6. The response magnification factor of X-axis (traveling direction).

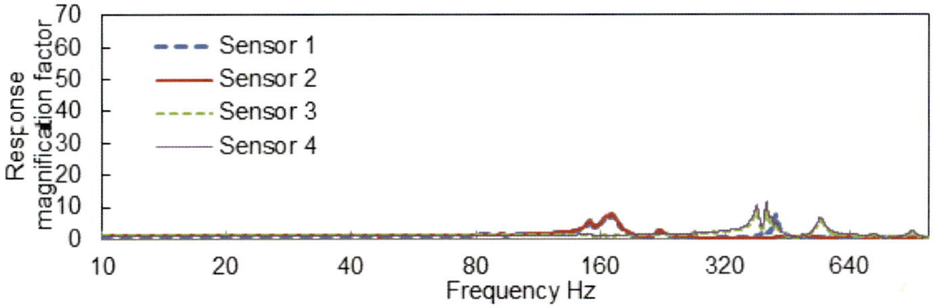

Figure 7. The response magnification factor of Y-axis (cryocooler axial direction).

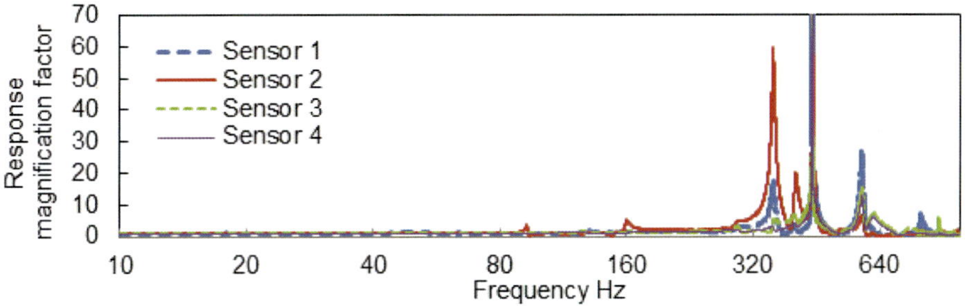

Figure 8. The response magnification factor of Z-axis (vertical direction).

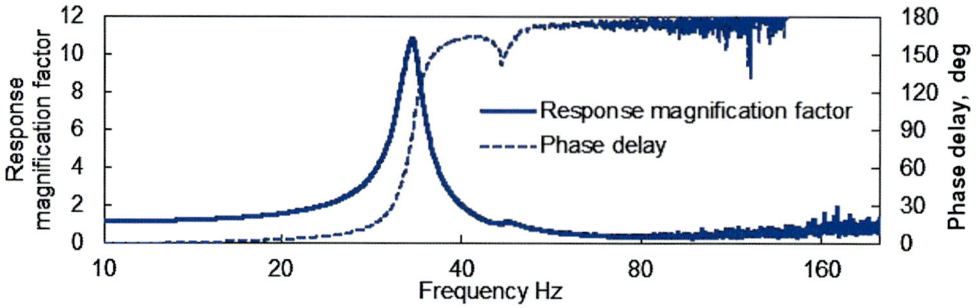

Figure 9. The response magnification and the phase delay of the compressor moving cylinder.

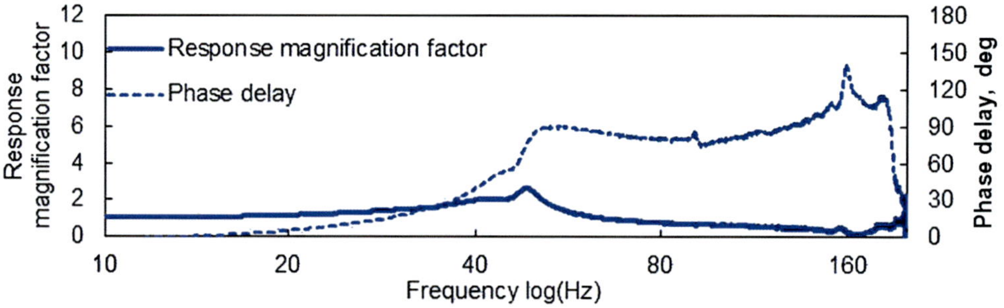

Figure 10. The response magnification and the phase delay of the displacer in expander.

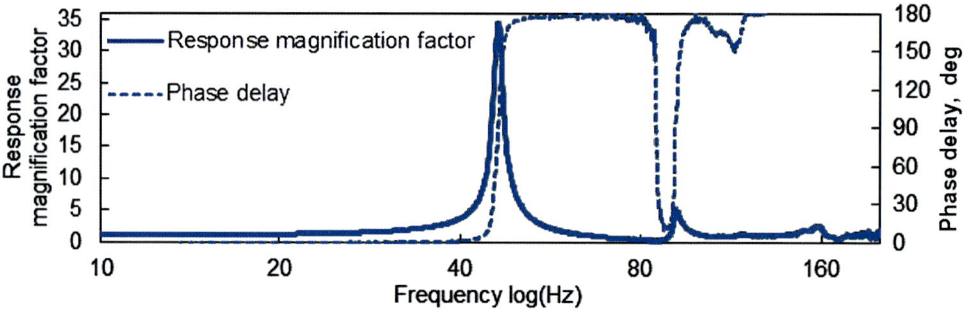

Figure 11. The response magnification and the phase delay of the vibration absorber.

magnification and the phase of the vibration absorber. As shown in Figure 11, the response magnification has an obvious peak at a natural frequency of 46.2 Hz and the maximum response magnification was about 34.6. The comparison of the test results and the natural frequency analysis which was calculated in advance is shown in Table 1. Differences between analysis and experimental results were considered to be due to imperfections in the analysis model, such as the viscosity of the gas and the constraint conditions. The measurement results will be utilized as data for next vibration analysis in our future work and will be feedback to the next design step.

Finally, the cooling capacity reduction rate due to vibration during cryocooler operation, was measured. The reduction rate was measured while a sinusoidal wave acceleration of 1 g in the range of 10 to 100 Hz was applied to the cryocooler. However, in the axial direction of the cryocooler, the displacer knocked the cylinder wall around 33 Hz when 1 g vibration was applied to. Therefore, for the axial direction, 0.3 g vibration test with full output and 1 g vibration test with 50% output of cryocooler, were conducted. Figure 12 shows the cooling capacity reduction rate at each frequency compared to that under no vibration. As a result, impact on cooling capacity due

Table 1. The comparison of the test results and the natural frequency analysis.

Item	Experimental results	FEM model analysis
Moving parts of compressor	33.1 Hz	20.2 Hz
Moving parts of expander	48.2 Hz	58.3 Hz
Vibration absorber	46.2 Hz	40.1 Hz

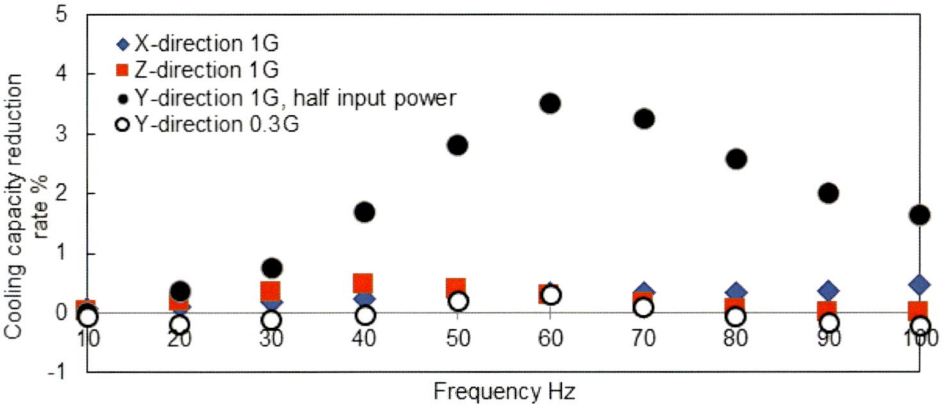

Figure 12. The cooling capacity reduction rate at each frequency.

to vibration was found to be small in the radial direction of the cryocooler (X-axis and Z-axis). In the cryocooler axis direction (Y-axis), cooling capacity reduction rate reaches its maximum around a vibration frequency of 60 Hz. The reason for the performance degradation around 60 Hz is still under investigation. Causes of knocking between the displacer and the cylinder wall around 33 Hz may affect the displacement and phase of the displacer and the cylinder wall. Figure 13 shows the state of the transmitted vibration to the displacer and the cylinder from vibration exciter. Table 2 shows the response magnification and the phase delay of the cryocooler in each frequency. With this information, the vibration of the displacer and the cylinder wall at each frequency are shown in Figure 14. In the low frequency range, for example around 10 Hz, the displacer will not knock the cylinder wall since there is no phase difference between the center position of the displacer and the cryocooler wall despite the large displacement. On the other hand, the displacer will also not knock with the cylinder wall since the vibration amplitude around 100 Hz is quite small. In the intermediate region, there is a risk of the displacer knocking. Since the natural frequency of the cryocooler's compressor cylinder is 33.1 Hz, the stroke of the compressor cylinder is increased in this frequency region. Hence it is believed that the stroke of the displacer is increased due to an increase in the compressor net input power.

Table 2. Response magnification and phase delay of the cryocooler in each frequency.

	G_1	G_2	G_3	G_4	θ_1 [deg]	θ_2 [deg]	θ_3 [deg]	θ_4 [deg]
10 Hz	1	1	1	1	0	0	0	0
33 Hz	1	1	4.97	1.58	0	0	90	23
100 Hz	1	1	0.33	0.66	0	0	172	79

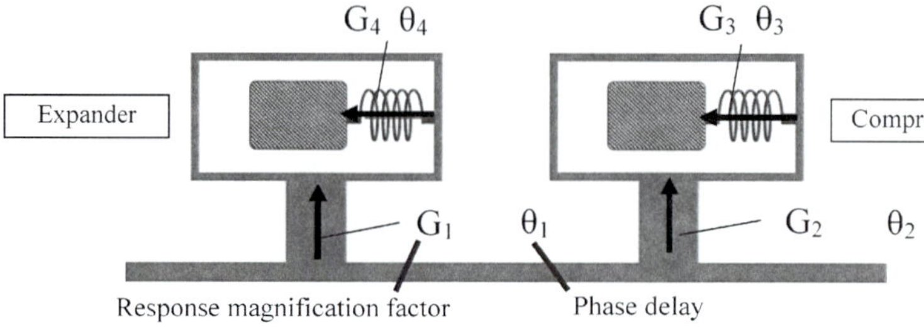

Figure 13. The state of the transmitted vibration.

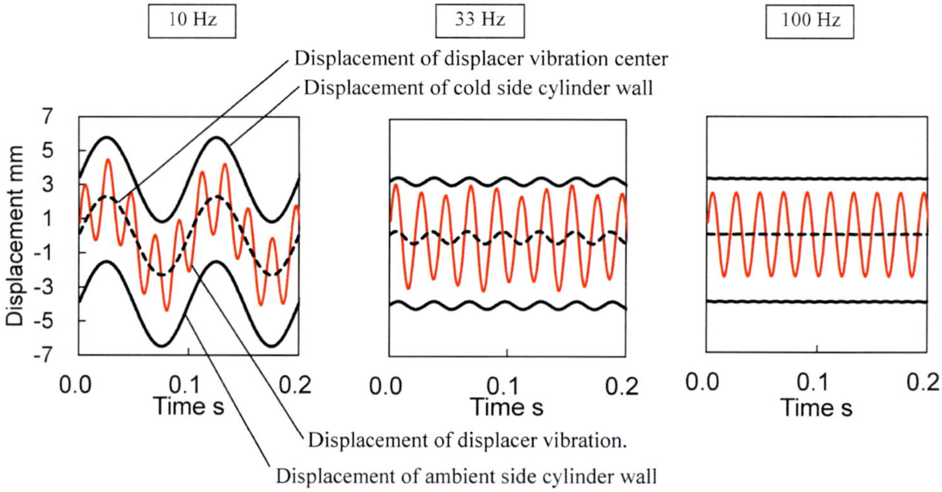

Figure 14. Vibrations of the displacer and the cryocooler wall.

SUMMARY AND FUTURE WORKS

This paper reports the reliability evaluation test results of a cryocooler for cooling a superconducting motor system. The cooling system is confirmed to be operated safely. The decrease in the cooling capacity was only a few percent after the system was continuously operated over 5,000 hours. Vibration tests were conducted, and the basic performance of the vibration resistance was obtained. In the future, a contamination measurement and anti-vibration design is scheduled to be performed.

ACKNOWLEDGMENT

This work was supported by Strategic Innovation Program for Energy Conservation Technologies Project of the New Energy and Industrial Technology Development Organization (NEDO) of Japan and a joint research with Sumitomo Electric Industries, Ltd.

REFERENCES

1. Y. Hiratsuka et al., "Development of 150 W at 70 K split Stirling cryocooler for high-temperature superconductors," *ISEC* 2014.
2. K. Yumoto et al., "Cooling system with a Stirling cryocooler for a high temperature superconducting motor," *IWC-HTS* (2015).

Proceedings Index

The International Cryocooler Conference has been publishing its works since its first Conference in 1980. The first seven conferences were typically sponsored by a government organization heavily involved in the field of cryogenics—typically a different organization for each conference. Most of these early proceedings were printed in limited quantities by the sponsoring government organization and are out of print at this time. However, scanned PDF versions of these proceedings are available on the ICC website's Past Proceedings tab (http://cryocooler.org/past-proceedings/).

Starting in 1994, the ICC began publishing its proceedings as hardcover books in the open literature via our New York publishers: Plenum, Kluwer and Springer. These are titled Cryocoolers 8 through Cryocoolers 13 and are available in print and eBook format from Amazon, and electronically as B&W PDFs from Springer's SpringerLink website. Links to SpringerLink for Cryocoolers 8 to 13 are also located on our ICC website's Past Proceedings tab.

Starting in 2006, the ICC began publishing its proceedings as hardcover books in the open literature via our own ICC Press; these books also include a CD of all the papers as color PDFs. Thus, for Cryocoolers 14 on, the author's papers are available in hardcover book form via Amazon, and as digital color PDFs via the book's CD. The papers in Cryocoolers 14 on are also available as PDFs from our ICC website's Past Proceedings tab.

For those attempting to locate references to earlier conference proceedings, the following is a listing of the eighteen previous proceedings of the International Cryocooler Conference.

1) *Refrigeration for Cryogenic Sensors and Electronic Systems*, Proceedings of a Conference held at the Nat'l Bureau of Standards, Boulder, CO, October 6-7, 1980, NBS Special Publication 607, Ed. by J.E. Zimmerman, D.B. Sullivan, and S.E. McCarthy, National Bureau of Standards, Boulder, CO, 1981.

2) *Refrigeration for Cryogenic Sensors*, Proceedings of the Second Biennial Conference on Refrigeration for Cryogenic Sensors and Electronics Systems held at NASA Goddard Space Flight Center, Greenbelt, MD, December 7-8, 1982, NASA Conference Publication 2287, Ed. by M. Gasser, NASA Goddard Space Flight Center, Greenbelt, MD, 1983.

3) *Proceedings of the Third Cryocooler Conference*, National Bureau of Standards, Boulder, CO, September 17-18, 1984, NBS Special Publication 698, Ed. by R. Radebaugh, B. Louie, and S. McCarthy, National Bureau of Standards, Boulder, CO, 1985.

4) *Proceedings of the Fourth International Cryocoolers Conference*, Easton, MD, September 25-26, 1986, Ed. by G. Green, G. Patton, and M. Knox, David Taylor Naval Ship Research and Development Center, Annapolis, MD, 1987.

5) *Proceedings of the International Cryocooler Conference*, Monterey, CA, August 18-19, 1988, Conference Chaired by P. Lindquist, AFWAL/FDSG, Wright-Patterson AFB, OH.

6) *Proceedings of the 6th International Cryocooler Conference*, Vols. 1-2, Plymouth, MA, October 25-26, 1990, David Taylor Research Center Report DTRC-91/001-002, Ed. by G. Green and M. Knox, Bethesda, MD, 1991.

7) *7th International Cryocooler Conference Proceedings*, Vols. 1-4, Santa Fe, NM, November 17-19, 1992, Air Force Phillips Laboratory Report PL-CP-93-1001, Kirtland Air Force Base, NM, 1993.

8) *Cryocoolers 8*, proceedings of the 8th ICC held in Vail, Colorado, June 28-30, 1994, Ed. by R.G. Ross, Jr., Plenum Press, New York, 1995.

9) *Cryocoolers 9*, proceedings of the 9th ICC held in Waterville Valley, New Hampshire, June 25-27, 1996, Ed. by R.G. Ross, Jr., Plenum Press, New York, 1997.

10) *Cryocoolers 10*, proceedings of the 10th ICC held in Monterey, California, May 26-28, 1998, Ed. by R.G. Ross, Jr., Kluwer Academic/Plenum Publishers, New York, 1999.

11) *Cryocoolers 11*, proceedings of the 11th ICC held in Keystone, Colorado, June 20-22, 2000, Ed. by R.G. Ross, Jr., Kluwer Academic/Plenum Publishers, New York, 2001.

12) *Cryocoolers 12*, proceedings of the 12th ICC held in Cambridge, Massachusetts, on June 18-20, 2002, Ed. by R.G. Ross, Jr., Kluwer Academic/Plenum Publishers, New York, 2003.

13) *Cryocoolers 13*, proceedings of the 13th ICC held in New Orleans, Louisiana, on March 29 - April 1, 2004, Ed. by R.G. Ross, Jr., Springer Science+Business Media, New York, 2005.

14) *Cryocoolers 14*, proceedings of the 14th ICC held in Annapolis, Maryland, on June 14-16, 2006, Ed. by S.D. Miller and R.G. Ross, Jr., ICC Press, Boulder CO, 2007.

15) *Cryocoolers 15*, proceedings of the 15th ICC held in Long Beach, California, on June 9-12, 2008, Ed. by S.D. Miller and R.G. Ross, Jr., ICC Press, Boulder CO, 2009.

16) *Cryocoolers 16*, proceedings of the 16th ICC held in Atlanta, Georgia, on May 17-20, 2010, Ed. by S.D. Miller and R.G. Ross, Jr., ICC Press, Boulder CO, 2011.

17) *Cryocoolers 17*, proceedings of the 17th ICC held in Los Angeles, California, on July 9-12, 2012, Ed. by S.D. Miller and R.G. Ross, Jr., ICC Press, Boulder CO, 2012.

18) *Cryocoolers 18*, proceedings of the 18th ICC held in Syracuse, New York, on June 9-12, 2014, Ed. by S.D. Miller and R.G. Ross, Jr., ICC Press, Boulder CO, 2014.

Author Index

Subject Index